Seashore
Ecology

John Morton

Seashore Ecology
of New Zealand
and the Pacific

John Morton

Scientific editor: Bruce W. Hayward

David Bateman

Sponsored by:
Department of Conservation
North Shore City Council
Auckland Regional Council
Northland Regional Council
Rodney District Council
Auckland Museum Institute, Conchology Section

Department of Conservation
Te Papa Atawhai

Rodney DISTRICT COUNCIL

NORTHLAND REGIONAL COUNCIL

Conchology Section
Auckland Museum Institute

Auckland **Regional** Council
TE RAUHITANGA TAIAO

NORTH SHORE CITY

Cover design by Shelley Watson/Sublime Design
Book design by Errol McLeary
Typesetting by Jazz Graphics, Auckland
Printed in China by Everbest Printing Company

To Pat, who has walked on many shores with me;
and for Matthew & Sam, and Tilde
who will discover them anew.

Charles Raven, Cambridge naturalist and theologian, was asked what feature drew out the quintessence of his homage and the secret of his real self. In a world with so much that is beautiful, his choice was the sea-board. *'Here in the marriage bed where land and water meet, life first had its origin; we hardly need the scientists to assure us of it; even mankind feels the sanctity of it and goes to the seaside as to his home, his nursery, his playground, where he can forget his dignity and be a child again, with a child's freedom to dream and to enjoy... Where sea meets earth under an ever-changing, ever-changeless sky, there if anywhere is the mystery that will reveal to us the parable of our being.'*

Contents

Foreword

Tena koutou katoa — greetings to you all.

The Department of Conservation is a major sponsor of this book and I am pleased to be associated with its publication. It brings up to date Morton & Miller's *The New Zealand Sea Shore*, the classic in the field of coastal marine biology of Aotearoa/New Zealand.

An update of this seminal work, first published in 1968, is well overdue as there has been much progress in the understanding of New Zealand's coastal ecosystems. I am delighted that, thirty years after the last revision, it is the original first author who again has managed to share his vast knowledge with us in his inimitable style. Not only has emeritus professor John Morton dominated marine biology in the last half century, he has also defined much of the work and the workers who have expanded our insights in the animals, plants and processes of Pacific coasts. Bruce Hayward, himself a scientist and science communicator of no mean repute, has been crucial in getting the revision to this stage.

This is a resource for a wide audience in Aotearoa and beyond: it makes available to and useable by us what is known about the fascinating zone that surrounds every island in our Pacific region. It also illustrates what we stand to preserve by the establishment of marine reserves in representative and unique areas along our coast lines.

Enjoy reading and using this taonga — treasure it.

H. E. the Hon. Sandra Lee-Vercoe
High Commissioner to Niue
Minister of Conservation in the Labour–Alliance Government 1999–2002

Niue, May 2003

Preface

This book aims to be the successor to Morton and Miller's *The New Zealand Sea Shore* (1968). It is far from being a revision, even if one could revise a book more than 30 years old. The better advice must be Hamlet's, to 'reform it altogether'.

Shore ecology in New Zealand today faces very different challenges from the 1960s. An entire new generation of marine biologists has inherited a large stake on the educational, economic and political fronts. The Resource Management Act has placed upon Regional Councils a mandatory concern for the shores. In 1968 Marine Reserves did not exist. At the time of writing, 18 of them have been achieved, with the hope of advocates — led by Dr Bill Ballantine — to see 10 percent of our whole coastline eventually become reserved.

Over these same years, marine biology has been enjoying a high priority in the universities, not least at Auckland. Scuba diving has led to a new understanding of the subtidal. In eastern Australia many marine biologists now at work have had Auckland beginnings. New Zealand has also shouldered some responsibility for coral reef studies in the South Pacific. A joint venture of Auckland with the University of the South Pacific owed much to the collaboration of Dr Uday Raj, a New Zealand graduate and until his untimely death a leader in marine studies in Fiji.

With such a background, what kind of book would seem most called for today?

Though still important, the primary need is no longer to identify species. Marine taxonomy has by now made progress in many, if by no means all, groups. Some excellent definitive guidebooks are now available, to mention only those on molluscs, fishes and marine algae. So a new work on shores can by now be in some sense synoptic, drawing patterns of species together, to reveal whole communities.

Each shore community has been broken into its principal life-forms. Photographs have been largely replaced by drawings, from which it is hoped that a large number of animals and plants can in a preliminary way be identified. As well as most rapid, a first viewing of illustrations can be satisfying and helpful, so long as it points to the need for confirmation in the proper taxonomic works.

For what kinds of readers could this book be destined? First it is hoped it may for its day be a useful reference work for shore ecology, conservation and management. It should give an opening to serious study at university level, while being widely readable at home, in schools, on coastal holidays, and in the Pacific abroad.

The outreach to the wider Pacific has brought a new focus on the unity of rocky and coral shores. Apart from some soft-bottom communities, such as mangroves visible above the surface, the buried communities of beaches and estuaries dealt with in the older book have been left aside. Sheer space apart, this could have two grounds. First, recent research has been growing chiefly though not wholly around hard shores. Second, sandy and muddy shores — with their high biological interest and productivity — have a species range almost wholly distinct, with very few overlaps with hard shores. Methods of investigation are in many ways different, as are often indeed the workers themselves. A new book upon soft shores must before long arrive, from naturalists specially skilled in their study.

The present work could be seen by some as old-fashioned, with something of the synoptic tradition of the 'New Naturalist' books that first introduced us to shores and other habitats half a century ago. Today's multiple specialism is part of a different biological world from the one I grew up in. *The New Zealand Sea Shore* was virtually written from four years in the field. Chapters on last year's discoveries were handed out as raw material for the next. In a small-staffed department, there were important biological groups still lacking specialists. But something we need not accept is that to teach any topic well, we must be engaged in first-hand research in it. Some of my best-remembered training came from Lucy Cranwell and Laurie Millener, who were both botanists.

Michael Miller brought to Auckland an experience of British marine studies in the forward-looking 1950s. Noted for his pioneering work on opisthobranch molluscs, he has been my earliest colleague since I arrived back in New Zealand in 1960. First to come was to be our linking of biology teaching with coastal field courses. It was Michael and Joan Miller that were the first to point to Whangarei Heads as a prime research and teaching centre. Thirty-five years later, week-long annual courses were still flourishing there, using the long out-of-print *New Zealand Sea Shore*, produced in 1968 and based on continuing field studies first developed around Auckland and Whangarei. That book owed so much to Michael's colour plates that I am glad to be able to present a selection of them in this book for the benefit of a further generation. It would be hard to think of a New Zealand shore book being complete without them.

John Gilpin Brown and Bill Ballantine were my friends and colleagues at Plymouth Laboratory in the fifties. Both came to New Zealand in 1961, to be the practical setters-up of Leigh Marine Station, where Ballantine in 1965 became the first director and supervisor of the laboratory's building and extension. He has been the staunch advocate in the founding of the growing chain of marine reserves. Through the same decade Derek Challis was my close assistant both in New Zealand and on the Royal Society Solomon Islands Expedition. In more recent years we have been indebted to Ron Cometti for his artistry of natural habitats, and I am grateful for those in this book, including shore geological forms.

Students and staff in those days came to know each other well. Field clubs under student impetus were alive and active, with time left available for them in proper vacations and 'after-degree' camps. From 1977 to 1983 a dozen or so Auckland students would pay their own way for an August fortnight on Fiji shores, aware that this formed no part of the syllabus and would offer no credit in final exams only three months' away.

Students in past years on field trips in New Zealand and Fiji will remember with appreciation John Walsby as a widely ranging natural historian as well as for his molluscan research and his special ability in shore teaching. For SCUBA and subtidal ecological studies, the fatherhood must belong to Wade Doak, soon followed by Tony Ayling and others active in the habits and taxonomy of fishes.

With these and many other shore workers, the term 'natural history' has taken on renewed importance, bringing together plants, animals and geology into the broad story. In marine biology this was the Stephensons' approach in the past century. It must not be allowed to disappear today.

A new feature of this book is thus Bruce Hayward's chapter dealing with the geology of shore formations. For perhaps the first time a work on biology has reached a detailed concern with the bedrock on which its story is securely laid. The naturalist's interest is no longer confined to taxonomy or even the living ecology. It has reached to the shorescapes, including ground formations and cliff sections cut back to reveal the distant past. Biologists are realising what Ferdinand Hochstetter, exploring geologist, first glimpsed in New Zealand a century and a half ago.

It seems appropriate — and not just an accident of domicile — that a Pacific-wide book should begin in New Zealand. Our mainland shoreline is as long and in some ways more concentratedly diverse than in the United States. With growing ecological travel and tourism, the study of sea shores is today becoming a global theme.

A few words may be said of the personal tradition from which this book came about. Half a century ago my happiest days were at England's Plymouth Marine Laboratory, an institution of scientific freedom, which the British government's economics of the eighties were virtually to destroy. My heroic figures in Britain were C.M. Yonge, Alan Stephenson and Alister Hardy. Later — leading the Royal Society Solomon Islands Expedition — was Cambridge's tropical botanist, John Corner.

Today's attempt at a comprehensive book must draw from such a past. This itself was to involve anxiety. As I continued to write during 10 years in retirement, enthusiasm was still there. But I was working more slowly, and I could spend less time in the field. Every year it seemed harder to keep alert to new species arrivals, changes of nomenclature and important lines of current research.

The time had come to seek help and advice; the friend I turned to in July 2000 was Dr Bruce Hayward. Primarily a palaeontologist, but with something like the late Charles Fleming's wide perspective, Bruce shared my whole Pacific outlook. He took on himself almost endless editorial labour, with submission of every section to its specialists, here and overseas, some — like Bruce and his wife Glenys — being my own former students. He has given wisdom, perspective and seemingly unlimited time. He has also contributed significantly to many parts, and most of all by his own chapter on shore geology.

For this whole book Bruce Hayward has done more than I can acknowledge or realise. I wish he had let me thank him as I wanted to, by drawing him in as co-author. But he will be rewarded in one reflection that can never be taken away. While any remaining errors or omissions must still be mine, without Bruce Hayward's final commitment this book — with whatever worth it may hold — could never have been adequately finished, or published at all.

John Morton
Auckland, March 2004

Acknowledgements

From colleagues and friends over 30 years I have drawn on their knowledge of the shore, as will be amply reflected in these pages. To many more, both in New Zealand and abroad, I am indebted for their generous checking, correcting and updating the scientific names used throughout the book. For this I am particularly grateful to:

Mike Barker (University of Otago, Dunedin); Chris Battershill (Australian Institute of Marine Science, Townsville); Fred Brook (Department of Conservation, Whangarei); John Buckeridge (Auckland University of Technology); Stephen Cairns (Smithsonian Institution, Washington, DC); Ewen Cameron (Auckland War Memorial Museum); Gene Coan (California Academy of Sciences); Steve Cook (University of Auckland); Peter Davie (Queensland Museum, Brisbane); Mike Eagle (Auckland War Memorial Museum); John Early (Auckland War Memorial Museum); Malcolm Francis (National Institute of Water and Atmospheric Research, Wellington); David Galloway (LandCare Research, Dunedin); Rhys Gardner (Auckland War Memorial Museum); Ray Gibson (Liverpool John Moores University, UK); Brian Gill (Auckland War Memorial Museum); Dennis Gordon (National Institute of Water and Atmospheric Research, Wellington); Michelle Kelly (National Institute of Water and Atmospheric Research, Auckland); Daphne Lee (University of Otago, Dunedin); Pat Mather (Queensland Museum, Brisbane); Don McKnight (National Institute of Water and Atmospheric Research, Wellington); Jim McLean (Los Angeles County Museum of Natural History); Wendy Nelson (Museum of New Zealand, Wellington); Dave Pawson (Smithsonian Institution, Washington, DC); Geoff Reed (National Institute of Water and Atmospheric Research, Wellington); David Reid (Natural History Museum, London); Dave Towns (Department of Conservation, Auckland); Richard Willan (Museum and Art Gallery of Northern Territory, Darwin); Keith Wise (Auckland War Memorial Museum). I owe much to the mollusc workers at Auckland Museum, notably Fiona Thompson, for valuable help with the text, Margaret Morley, Glenys Stace, Nancy Smith and technician Todd Landers.

Finally, it is a pleasure to thank my publishers for their encouragement and help and especially the chief editor, Tracey Borgfeldt, for her skilful advice and sympathy through all the production stages.

Students at work on shores at Aubrey Island, Reotahi Bay, Whangarei Heads.

PART ONE
NEW ZEALAND SHORES

CHAPTER ONE
The Pattern of Shores

Introduction

This is a book about the animal and plant communities set up where the margins of islands and continents slope into the sea. The primary story — to be told in greater detail — is concerned with the intertidal zone, the narrow belt over which the interface of land and sea regularly shifts back and forth, in most parts of the world twice daily.

This large realm of the intertidal is capable of division into two prime types of habitat. All the varieties of rocky foundation can be together referred to as *hard shores*. The *soft shores* comprise sand beaches and the soft flats of estuaries and harbours. These two prime types are almost totally different in their great assemblage of species. They have traditionally been assumed not to overlap and have been studied with different procedures and techniques, even by different people.

In *The New Zealand Sea Shore* (1968)[1] both hard and soft shores were presented in the fullest detail then possible. With the accrual of new facts, this present work has been designed with an alternative approach, virtually confined to hard shores, but in its purview carried beyond New Zealand over the total realm of the Pacific. First — as in the former book — New Zealand shores are systematically described in their regional detail, with the addition of material from 25 years' further investigation, particularly in the subtidal.

Then, with a broader brush, follows a comparative account of the equivalent hard shores, with their biogeography, around the entire Pacific Rim. The final section culminates with an account of the coral shores of the tropical Pacific, based first on the atoll of Aitutaki in the Cook Islands, easily reached from New Zealand and still — for the present at least — unspoiled. To do justice to the whole tropical Pacific in a nutshell — so to speak — has been difficult, but I believe worth trying. In the event, I needed a coconut shell.

The shore habitat

Rocky shores, in their upper reaches, can be aptly compared to mountain slopes. Both are stressful and demanding habitats, with the bedrock breaking through like knees and elbows left bare directly to the weather. Not only — like mountains — do rocky shores experience wide extremes of heat and cold, but in addition the shore undergoes the regular alternation of submersion and emersion: wet and dry. Moreover, as well as the tides, there is constant water movement, the strong flow of currents, and the constant battery of the waves. Above the water level at such time, the regular rise of splash and spray can raise the effective boundary of the strict 'intertidal' by several metres.

The stresses offered by the climate to life on the shore can be mitigated in several ways. The broken shore contour may provide standing pools and shade. Concealed under loose boulders there can live protected faunas richer by a whole order of magnitude than on the visible surface. There is finally the refuge offered by attached plants and animals themselves. In their shelter for a host of subsidiary species, they substitute for the primary rock-face a biotic substrate that is living or in virtually every part derived from living things.

The most numerous shore animals are small invertebrates; by far the majority are permanently fixed down. A few, such as mussels, are flexibly attached or able to move about. But the rest are *sessile*, literally 'squatters' sitting broad-based and immobile on the bedrock, or on the hard integuments of other species. Unless the rock itself — as sometimes happens — is penetrable, there are no substrates where animals can burrow or for intertidal plants to take root as on land. Those shore animals that have remained mobile, such as limpets and chitons and many of the snails, move so slowly that they can be regarded, just like sessile forms, as faithful to their particular zones and territories.

In a realm without soil all the metabolic needs of the sedentary species — food for animals, nutrients and dissolved gases for plants — must be supplied from the water column above. A majority of shore animals find a constant food supply that is easy to procure. This primary resource is plankton, from a system ultimately continuous with the ocean at large. Barnacles actively catch planktonic organisms by using their feathered limbs as casting scoops. Other sessile animals screen off or filter the plankton with a whole range of mechanisms employing mucus and cilia.

As a food source second only to plankton, food can be plentifully scraped off the surface, as from films of algae, wave-lodged diatoms or organic remains accumulating in crevices. A still more concentrated nutrition can be obtained from predation upon attached animals, by consuming them whole, shearing them with jaws or tooth-ribbon, or inserting the proboscis to digest the tissues *in situ* or suck out the body fluids.

Compared with chasing a diet on land, such food capture calls for little expenditure of energy. Indeed the most limited commodity on the shore must be not food but room to settle. Here we are faced with an 'ecology of space', with *Lebensraum* parcelled out frugally between competitors. The successful occupants live in a close mosaic, perhaps only a few centimetres from their possible limits of survival.

Shore communities are thus miniaturised not only in the *size* and space of individuals, but also in *time*. The building up of a complex community to its *climax* state may take as little as a couple of years. This amounts to an eon of time compared with the few days for protistan culture of bacteria or uni-cells, but it is ephemeral by the centuries-long span of a rainforest. Intertidal communities are in fact beautifully proportioned for short-term study. The seashore has provided the community ecologist with orders of time nicely attuned — it could be said — to a masterate or Ph.D. study!

Thus, in a variety of ways, the intertidal shore is an inviting laboratory. Its main species are sedentary enough to be numbered or labelled for ongoing study of growth or behaviour. Small and prolific in individuals, they can suffer repeated sampling without ill-effects. Students and amateurs can moreover find examples of nearly all the major invertebrate groups ready at hand. There is a largesse of animals like a rainforest's treasure trove of plants. But the seashore is a more compact frame than a forest, to show the salient properties of a community through space and time. These will entail competition and exclusion, population growth, ontogeny and the programme of ecological succession. There are also the trophic and energetic relations that the biology student early learns about, with the pyramids of the *ecosystem*.

To read a shore

The great pioneer of Australian marine biology, Professor W.J. Dakin,[2] has recalled for us Pope's lines:

Not chaos-like together crush'd and bruis'd
But as the world, harmoniously confus'd

The shore naturalist's first impulse must be to search for the order in this part of what Richard Hooker much longer ago called a *harmonious dissimilitude*. Faced with such complexity there can no longer be single-minded zoologists and botanists separately at work. The lives of plants and animals will invariably be found meshed and interwoven as, in T.A. (Alan) Stephenson's expression, we learn to *read a shore*.

The systematic procedure used in this book is one I was to teach over a good many years, as I first learned it from Alan Stephenson's artistry. For each shore is first constructed a measured zoning column. A rich store of information can be captured and noted for later recall. Much of its significance will only then be realised. This field method is an old and reliable one, used ever since geologists began to make their annotated stratigraphic sections.

For the shore naturalist, this method can be employed at any level of expertise. It is more widely availing than rapid camera-scanning, or winnowing out items of data to punch onto computer cards. Its virtue is in ensuring time to consciously observe. Eye and brain are engaged upon the spot, with analysis beginning where the complex of data is still at hand. Each observation becomes fertile in suggesting others. Even at the outset, the aim will be comprehensive, with the day's notebook a treasury to bring home.

The wider realm

The intertidal strip where marine biology was traditionally taught is only the top of a much longer continuity. This descends — as we realise today — to the foot of the subtidal cliff, at whatever depth the rock slope becomes cut off by sediments ultimately continuous with the continental shelf. The junction of the steep slope with horizontal sediments thus forms a major threshold. It occurs at different levels, from the diminished light at the lowest reach of scuba diving, up to where beach sand may cut off the rocky expanse somewhere above low tide mark.

Over the three decades past, the study of the hard shore has been carried far beyond the intertidal, with the probing of new and rich habitats by scuba diving. With students black-suited like penguins, marine ecology has long slipped away from its one-time concentration between the tides.

There is a lot of justification for a New Zealand-based book to take a wider Pacific view, even of places few readers may ever have the chance to see. For here is a branch of ecology that calls for continental comparisons. It is no bad way to travel the world, tracing the constancies and discovering the special character of hard shores. With all their local individuality the world's

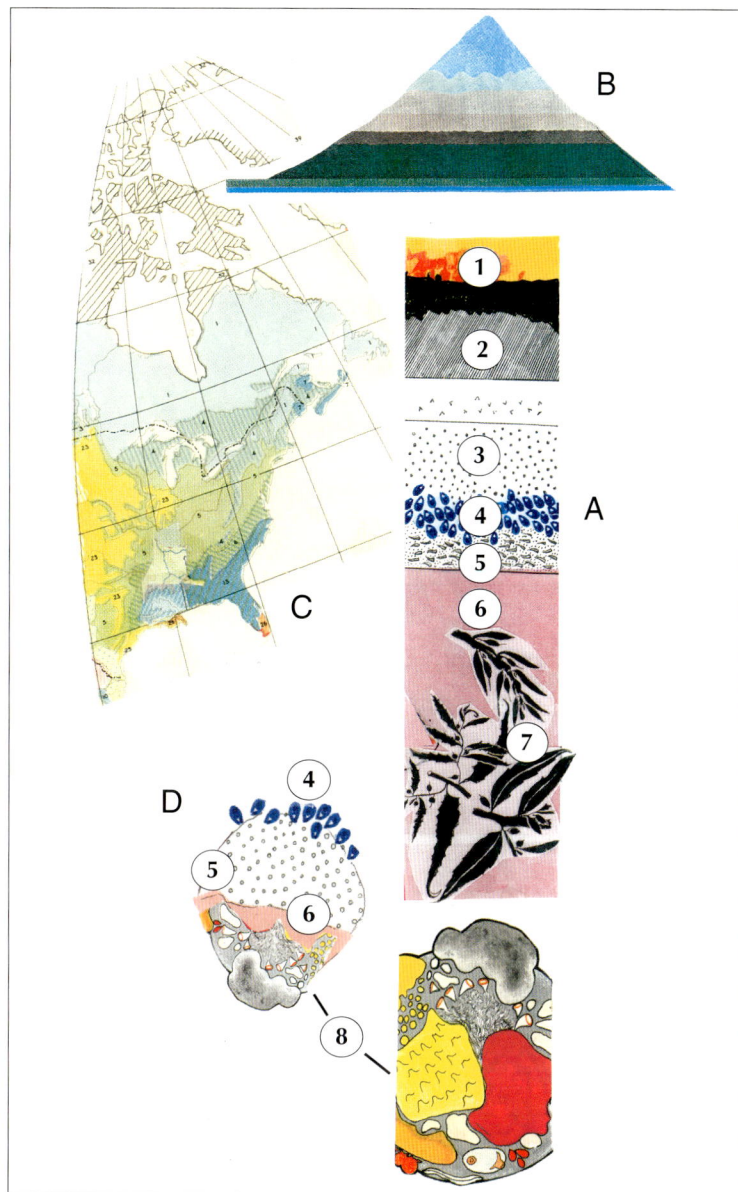

Fig. 1.1 The zoning sequence
The zoning of an intertidal shore (A) is related with the terrestrial vegetation zones: altitudinal on a mountain (B) and latitudinal on a continent (C). The extended shore zonation is related (**bottom**) to the zonules on the sides and undersurface of an intertidal boulder (D). 1 yellow lichens 2 black lichens 3 upper acorn barnacles 4 mussels 5 lower barnacles and tubeworms 6 crustose pink coralline algae 7 brown algae 8 sublittoral sponges and ascidians.

rocky shores present a finite range of themes. As with seeing a *Hamlet*, *Othello* or *Lear*, the high interest is in seeing how the major and minor roles have been filled, out of the plethora of species being locally cast.

The naturalist-traveller Alexander von Humboldt (1769–1859) was first to recognise the similar changes in the vegetation sequence with latitude and altitude. In the 8000 km from the tropics to the tundra, he observed vegetation zones comparable with the succession from tropical sea level up to a mountain top of 1500 m. Looking up an intertidal shore, through a span of only 2 m, he might have seen the shore as a mountain or continent in microcosm, with just such a set of changing life-forms of its own. Not only the total shore, but even — as we shall find — a separate boulder, from top to bottom, can present these same changes in ultimate miniature.

Beginning at the bottom of the shore, the gentlest microclimates are obviously found at around low tide. Extremes of heat and strong light hardly arise. Climatic stress is mitigated by long immersion, and by the shelter created by the mature biotic cover. Species diversity is high, with settlement opportunistic. The real constraints are biotic and competitive. Higher on the shore, on bare rock exposed directly to the sun, the converse reigns. Relatively few species live round high tide. Some of them are old relicts from quite primitive stocks, long adapted to confront the regular fluctuations of heat and cold, wet and dry. The constraints may be compared with those on the mountain tops and tundra. The main stresses to be coped with are no longer biotic but climatic.

Such a gradient of microclimate is a truism of ecology

with communities of all magnitudes between frigid and torrid, mountain peaks and lowland, high and low-tide marks, even in microcosm from the topside to the bottom of a boulder (Fig. 1.1).

Shore zonation

The shore between tide marks, with the land/sea boundary moving across it semi-diurnally, has the first property of being visibly *zoned*. Zonation does not cease beyond low water, even though scuba divers report it more difficult in the subtidal to pick out such well-defined zones, or even at first to believe in their existence.

Between tide marks, the narrow zones are by contrast concentrated and clear-cut. This is most notably true where the bedrock surfaces are smooth and intact, directly sloping, and free from the complications of a broken topography or the scatter of loose boulders.

The first intensive study of any New Zealand zoned shore was carried out in 1935 on the oceanic islands of the Poor Knights, with wave attack strong enough to sweep away all sediment and to carve surfaces where the bands are peculiarly sharp and distinct.

Cranwell and Moore's Poor Knights report began:

One of the most striking features of certain of the offshore islands of the North Auckland coast is the many-ranked and beautifully symmetrical zonation of intertidal

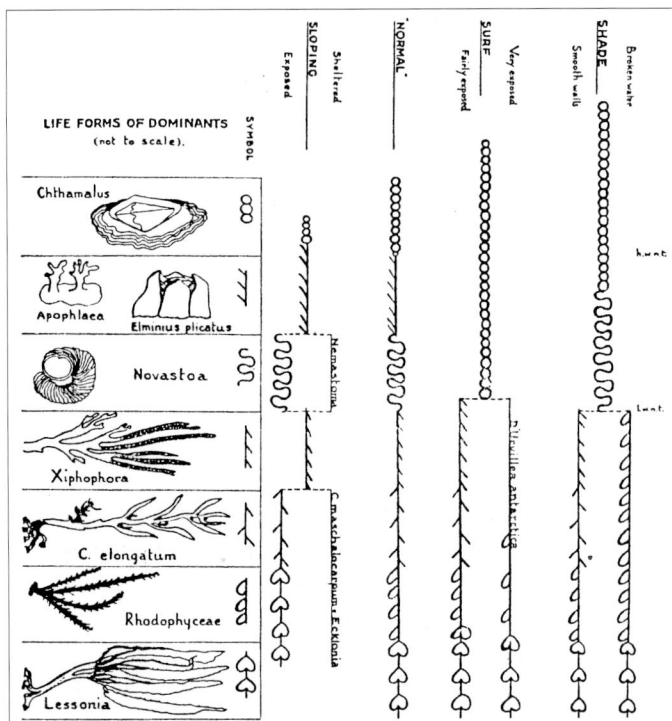

Fig. 1.2 The earliest account of zonation

The earliest schema of New Zealand intertidal zonation from the Poor Knights Islands, Northland.

(From Cranwell and Moore.)

communities. Bands of sessile shellfish and seaweeds run like white, red and brown ribbons around the shores, a striking local expression of the coincidence of certain major factors operating throughout the whole littoral region.[3]

The New Zealand-wide occurrence of comparable zones with a constant vertical order is unmistakable. Impressive above all — as we have today come to appreciate — is the presence of recognisably equivalent zones around the shores of a great part of the world. This does not imply anything like the identity of shores from place to place. Even within 100 m, in the same broad geographic location, the mix and proportions of the zones can become transformed by differences of wave exposure.

Upon such a constant basic theme, variations are continually being played. The vertical reach and relative extent of the zones, and often their component species as well, may alter spectacularly. Each locality, with its geology and physiography disparate from the next, will have its own individuality of zoning. Small-scale local effects can further be overlaid by regional variations, particularly in the impact of wind and waves. For the pattern of zoning is not often simply determined by the tides. Under major wave action, with the rise of surge, splash and spray, the zones will widely transgress the tidal limits. A point is reached where they cannot meaningfully be considered 'intertidal' at all. On coasts open to a long fetch from the sea, large waves, imposing their regime of mounting surge, breaking splash, and high-borne clouds of spray, can altogether supersede the control of the tides.

A universal system

The scheme of zoning of the intertidal shore, with the nomenclature to be used in this book, has with slight variation been well grounded in marine literature for some fifty years.

Our modern approach to shores between tides has stemmed from the lifetime studies of two British ecologists, T.A. (Alan) Stephenson and his wife Anne. Though they were not finally to visit every continent, they were able to look at far more shores, more comprehendingly than any student had done before. Their purview was confidently global. With its claim to universality their scheme of zonation stands as the grandest generalisation of descriptive ecology since the global travels of von Humboldt, Darwin and Wallace.

The Stephensons' shore studies were early to be initiated on coral reefs, with Alan's appointment to the British Museum's Expedition (1928–29) to the Great Barrier Reef of Queensland, where Anne accompanied him. For reasons that afterwards became clear, coral reefs formed a less than ideal introduction to the funda-

Fig. 1.3 T.A. (Alan) Stephenson

mental simplicities of tripartite zoning. Years were to elapse, with new insights needed, before the Great Barrier or any other coral reef could be faithfully assimilated into the universal system.

Straight after the Great Barrier Reef Expedition, Alan was appointed to the Chair of Zoology at Capetown. Here it was that a quarter century's reflection on the world's intertidal shores really began. As he was later to write:

A ten year sojourn in South Africa (1931–1940) provided an opportunity for the ecological survey of an almost unknown coastline of a particularly fascinating type. My fate as an ecologist was now sealed.

From such initial experience the Stephensons were to envisage their *Universal System of Hard Shore Zonation*, designed to have worldwide application. The South African studies resulted in a splendid series of pioneering papers (1936–47), developing a tripartite division of the shores, under the original terminology of **littorine** (periwinkle), **balanoid** (barnacle) and **laminarian** (brown seaweed) zones. Of major value was their comprehensive survey of the geographical distribution of plants and animals. From the cold-temperate south-west African coast, round the warm temperate Cape Province, to the subtropical shores of Natal, separate biotic elements were recognised as deriving from cold, mixed and tropical waters.

The papers from South Africa[4] provide us with perhaps the best pictorial representation of shore zonation that has ever been made. Their preparation takes account of an early interest in line-drawing and painting, of birds, anemones, flowers and seaweeds. Still a classic of illustration is Alan Stephenson's Ray Society volume *British Sea Anemones*, as well as a little book,

today a collector's piece, on *Sea Shore Design and Pattern.*

In 1949 a comprehensive paper on the universal features of shore zonation[5] was published setting the basis of the nomenclature used today, and grounded in an increasing understanding of shore zonation in Great Britain.

From 1950 on, with Alan Stephenson now at University College, Aberystwyth, papers were to flow from a tour of the Atlantic seaboard of North America, from Florida and South Carolina north to Nova Scotia and Prince Edward Island. Two reports on Vancouver Island saw the light in 1961, the year of Alan's death.

The rest of their joint American work, around Pacific Grove and La Jolla, was to be included in the volume left to Anne Stephenson to complete.[6] This was to collate and summarise their whole work, with some linking material from Australasia, where good literature now existed. It remained almost silent on Asia.

Thus was inaugurated with comprehension and clarity, but at no sacrifice of relevant detail, a still viable approach to hard intertidal shores. In an age of specialisation, with the sub-disciplines becoming ever more segregated, the authors were to acknowledge a debt to the past. Alan Stephenson wrote:

I am an admirer of the works of some of the leading naturalists of the late nineteenth and early twentieth centuries. These men would visit a country new to them, and would produce an account of its geology, climate, fauna and flora, often admirably illustrated, that showed a mastery in all these fields and was lively and readable without being unscientific. Perhaps it should be one of our aims to emulate such authors, making due allowance for the improvement in scientific methods and the increase in scientific knowledge that has taken place since their day.

Zones and terminology

Before embarking on our New Zealand shore survey, we must first explain the tripartite zonal terminology adopted for this book, with only slight changes made to the nomenclature used in *The New Zealand Sea Shore*. Terminology takes on real significance when it provides not just arbitrary labels for what may happen to grow at the three selected levels of the shore, but discloses evidently natural entities, strictly comparable over long geographical distances.

On virtually every hard shore, two thresholds at once stand out visibly. These provide us with the datum levels first to be looked for in embarking on any intertidal survey. Towards the top of the shore the **barnacle-line** can be identified as the highest level reached by acorn barnacles in quantity. Below this line the barnacles generally form an almost pure zone. Next, around the low water mark of an *average* low tide (midway, that is, between *spring* and *neap*) we shall be able to pick out the **brown algal-line**. This constitutes the upper limit of the continuous cover of large brown seaweeds that carries on unbroken into the subtidal. All authorities agree that the barnacle-line sets the upper limit of a **eulittoral zone**. Most would accept that the brown algal-line forms the lower limit of the same zone.

So defined, it is the eulittoral zone that makes up the great 'in-between' of the middle shore. This zone is sometimes simply referred to as the **littoral**. In *The New Zealand Sea Shore* it was styled the **mid-littoral zone**. Adjoining this zone, above and below, there are narrower bands, each referred to as a *fringe*. At the top there is the **littoral fringe**, which becomes submerged not daily, but only at higher than average tides. Conversely, the **sublittoral fringe** at the bottom is left uncovered only at lower than average tides. In contrast with such fortnightly intermittence, the eulittoral zone is covered and uncovered by the tides twice every 24 hours. It is thus sometimes described as a **semi-diurnal zone**.

Eulittoral zone

This middle reach of the shore lends itself to further subdivision. With biological criteria broadly constant from place to place, it is generally possible to speak of an *upper*, *middle* and *lower eulittoral*. Their characteristic species differ somewhat — in New Zealand — from north to south, and between exposure and shelter. With a few exceptions we can regard the **upper eulittoral** as being clad with barnacles alone.

The **middle eulittoral**, though barnacles are still present, is chiefly a bivalve preserve, with rock oysters in Auckland and Northland, but mussels in the cooler south. Still within the middle eulittoral, these bivalves are typically followed by a band of the tubeworm *Spirobranchus*, sometimes succeeded in the north by the sand tubeworm, *Neosabellaria*.

The **lower eulittoral** comprises that whole stretch closely turfed by the calcareous red alga *Corallina* or alternatively clad with non-calcareous red algae. An important associate of corallines is generally the Venus' necklace, *Hormosira banksii*.

Littoral fringe

Abutting the barnacle-line, this band (designated the *supralittoral fringe* by the Stephensons) is on most days moistened only by splash. Typically it is left as bare rock, but its upper reach is often darkened with the sooty black lichen *Verrucaria*. There are no sessile animals, but the periwinkles, *Austrolittorina*, are almost

invariably present as marker species. High level limpets and — in northern New Zealand — the black snail *Nerita* may also reach the littoral fringe.

Supralittoral zone

Above the littoral fringe, ideally in localities with a clean, smoke-free atmosphere, hard rock surfaces carry grey and pale green lichens, both encrusting and foliose. A bright band of orange or yellow lichens is generally interpolated below these, just above the black *Verrucaria* of the littoral fringe. Designated the *maritime zone* in *The New Zealand Sea Shore*, this lichen stretch will be referred to here by the Stephensons' original name of *supralittoral zone*. As well as lichens, it can have flowering herbs — usually succulent — growing in cracks where a little humus has collected.

The terrain above the supralittoral zone frequently carries a few species of coast-loving trees and shrubs, notably — in the North Island — pohutukawa. Essentially terrestrial, this stretch can conveniently be referred to as the **adlittoral zone**.

Sublittoral fringe

The boundaries and indeed the very need for separate recognition of a *sublittoral fringe* (called by the Stephensons the *infralittoral fringe*) have been called into question by some authors. J.R. Lewis and H.B.S. Womersley have pointed to the unbroken continuity of its large brown algae (in the New Zealand north, mainly *Carpophyllum* species) with the extended algal forest stretching beyond low water. They argued the difficulty of clearly demarcating the so-called *fringe* exposed briefly at low spring tides from this long subtidal sublittoral zone. Moreover, with scuba access beyond low water, many have today begun to regard low-tide mark (difficult to establish with any precision as it falls from day to day) as something of a biological nullity.

New Zealand students, in continuing to recognise a sublittoral fringe, can point for justification to brown algae (*Carpophyllum angustifolium* in the north and some of the *Cystophora* species to the south) that terminate rather sharply around low water spring. There are some that would regard the briefly emersed margin of brown algae alternatively as a *fringe*, or as part of a larger sublittoral *zone*, as the character of a particular shore seems best to favour.

(A further question of nomenclature involves the demarcation of the *sublittoral fringe*. Some would believe this term — if it is to be used — might better be applied, as in the European continental usage, to the division we have here designated the *lower eulittoral*. As a band characterised by coralline algae this could

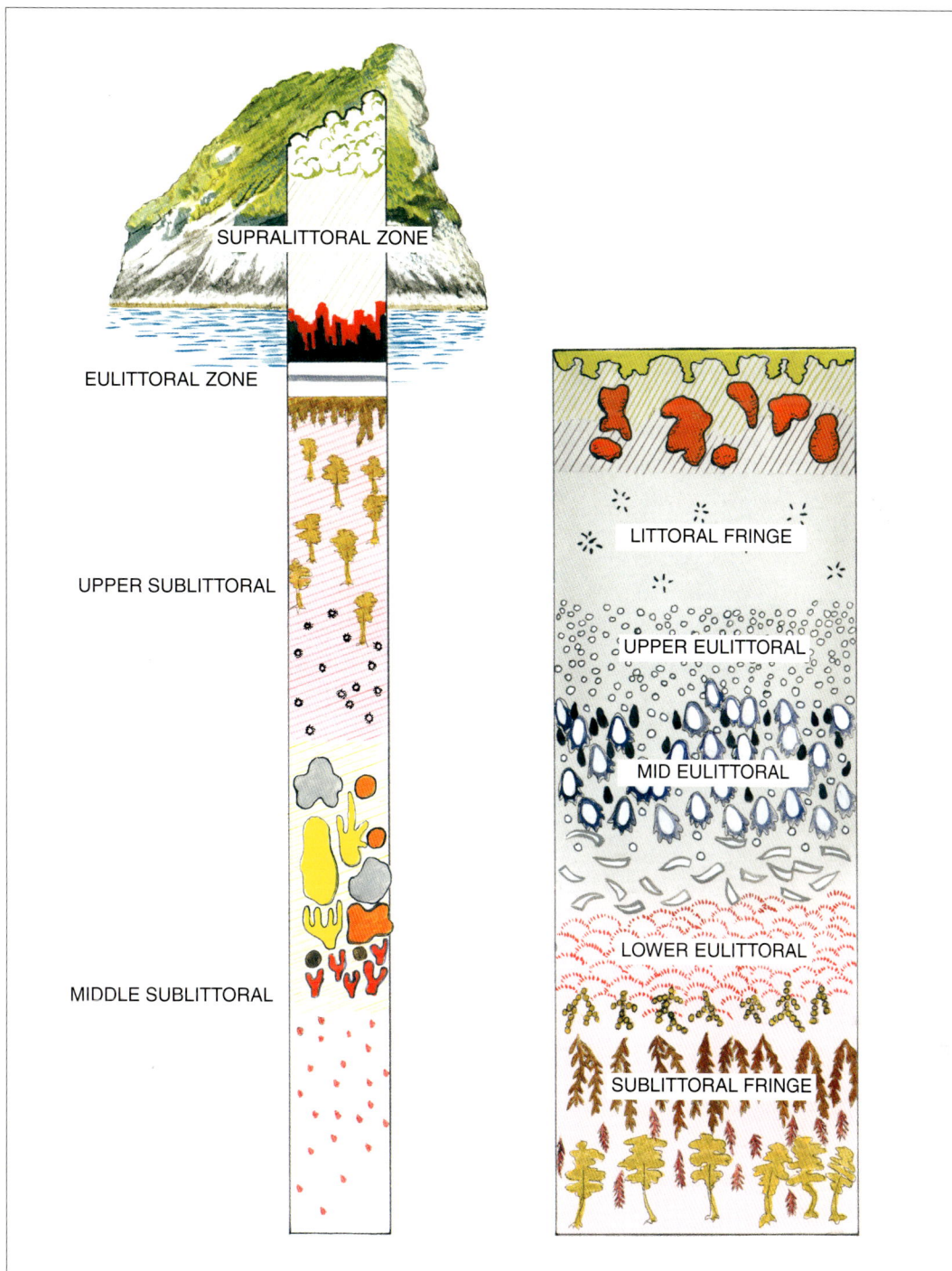

Fig. 1.4 The column of hard shore zonation
Supralittoral, eulittoral and *sublittoral*, with (**right**) detail from *supralittoral zone* to *sublittoral fringe*.

conveniently be demarcated at the top by a third threshold known as the *pink-line*. This marks the upper limit of the thin veneer of coralline algae applied like a pink paint to the rock surface. On shores reaching up to moderate exposure, it may carry a coralline turf or the veneer alone may reach a little above the turf.)

The *pink-line*, as a third reference level, is hardly less constant than the *barnacle-line* and *brown algal-line*. Under high wave attack a pink veneer without significant turf can extend appreciably above the brown algae.

In greater shelter, this veneer can be overshadowed by coralline turf or lost among the holdfasts of the uppermost *Carpophyllum*.

Into the subtidal, the same pink veneer continues a long way down, as deep as photosynthesis is possible. Its cessation marks the lower limit of an **upper sublittoral zone**. Clearly the whole extent of pink veneer could be seen as a continuous entity, upon which large brown algae may in varying richness and density be superimposed.

To recognise the sublittoral fringe as reaching up to the pink-line could have some logical appeal. Against this I have felt a reluctance to break from a usage well established in works in the English language, including *The New Zealand Sea Shore*.

The zonation column

From the foregoing description, a column of zonation can be picked out — often visible from some distance off — with its distinct colours. Below the deep green *adlittoral zone*, the *supralittoral zone* with its lichens stands out grey-green, then orange at the bottom. In the next stretch, the *littoral fringe*, bare bedrock shows up, below an initial band of black lichen. The *eulittoral* begins with the tawny grey of barnacles, with its middle reach white with oysters and tubeworms or in the south black with mussels. Pink-veneered bedrock begins with the *lower eulittoral* carrying corallines, and continuing with the dark brown algae of the *sublittoral fringe*.

The subtidal cliffs

The **upper sublittoral zone**, wherever we may decide to place its highest limit, descends beyond low water for distances varying with the water's turbidity and hence the reach of light. It thus reaches deepest on exposed outer coasts, especially of small offshore islands. Down to the limit of photosynthesis, a pink zone thus continues, diversely overlaid with large brown algae. At Auckland there is *Ecklonia* forest, with wide pink meadows kept grazed by kina or sea urchin, *Evechinus*.

Beyond the upper sublittoral zone, stretch the **middle sublittoral**, then the **lower sublittoral zones**. Both vary in vertical extent according to water clarity and illumination. One or both zones may be curtailed or drop out altogether where the steep cliff is cut off by the sediments of the continental shelf.

With the *middle sublittoral zone*, beyond the pink and kelp, the full dominance of sponges begins, with the bright colours that scuba diving has made familiar: scarlet, orange, tangerine, mauve and puce, backgrounded with browns or jet black. Plentiful also are filter feeding tubeworms, ascidians and bryozoans. Rich sponge gardens may extend out horizontally, growing up through the thin sediments of the shelf, until the bedrock is too deeply buried for any attached life.

The *lower sublittoral zone* is only complete where depth allows, mostly around offshore islands. Though sponges still abound, especially the white calcareous ones, this is chiefly a zone of cnidarians. Flashlight will bring to view yellow zoanthids, beige or puce sea fans, sometimes with swaying thickets of snow-white antipatharians, badly mis-named 'black corals' from

Fig. 1.5 W. R. B. Oliver
(Courtesy of Alexander Turnbull Library.)

their dead skeletons. High points of scarlet stand out, from tubeworm crowns, brachiopod shells and the discs of cup corals.

Our history of shore study

Credit must for all time belong to W. R. B. Oliver, Director of the Dominion Museum from 1928 to 1947, for the initial paper on 'Marine Littoral Plant and Animal Communities in New Zealand' that appeared in 1923.[7] Its author's scientific interests were many-sided, running from botany and biogeography to malacology and ornithology. In 1930 he published the long-used standard work on New Zealand birds. In his broad command of the natural sciences Oliver stands with our great pioneers, in a tradition carried down to more recent times by Charles Fleming.

With little of Stephenson's stylistic or pictorial skills, Oliver provided what was first needed. In an age before computers, he built up a voluminous retrieval system, with the many years of gleanings from often short visits to a range of shores. Out of this came a good conspectus of New Zealand's hard and soft shores, with the zones named from their key plants and animals, with adaptations to habitat noted where possible. While 'ecology' was still an adolescent science, with the early terminology of 'associations', 'consociations' and 'fasciations' borrowed from the North Americans, it was Oliver's comprehensive paper that first brought the New Zealand intertidal into the ambit of scientific study.

Fifteen years were to elapse before a second paper on shores appeared,[8] at once remarkable for its fresh viewpoint. Its authors, Lucy Cranwell and Lucy Moore, had moved far towards our modern comprehension of shore zoning. At their chosen site of the Poor Knights Islands, they were first able to recognise the importance of wave exposure. In later years the Poor Knights were again to become a 'first' for pioneering underwater studies; in 1981 the islands were to be gazetted as New Zealand's first island Marine Reserve.

The two Lucies, as they were to be known through the years, were botanists by first training. Lucy Cranwell (who later married and moved to America as Lucy

Fig. 1.6 The 'two Lucies': (*left*) **Lucy M. Cranwell, c. 1940)** (Courtesy of Alexander Turnbull Library.)**; Lucy B. Moore (1936)** (Courtesy of Landcare Research, Lincoln.)

Watson Smith) had a lasting influence on younger naturalists in the 1930s. As botanist at the Auckland Museum, she brought plant science out of dry herbaria into the bush and wetlands and importantly on to the shore. Lucy Moore became the leader of marine algology at DSIR Botany Division, but she never lost her zoological interests, particularly in barnacles, where she had named and described from the Poor Knights the new shore barnacle *Chamaesipho brunnea*.

Shore ecology came alive at Auckland with the arrival from Cambridge in 1946 of Valentine Chapman, as the university's first Professor of Botany. Algologist and salt-marsh ecologist, Chapman was soon to join forces with Victor Lindauer, then head-teacher at Russell School and builder of the university's algal herbarium that today carries his name. By the close of the 1940s, Chapman and his students had produced a set of Auckland shore studies, headed by the contributions of Vivienne Dellow, first on zonation at Narrow Neck Reef, and later on the marine algal ecology of the Hauraki Gulf.[9]

In 1953, another 15 years on from the Poor Knights study, the first South Island shore paper appeared, with George Knox's study of the intertidal ecology of Taylor's Mistake, on Banks Peninsula.[10] Three years later, Betty Batham was to give a detailed survey of the shore communities at Portobello, in Otago Harbour,[11] followed with an account of an exposed shore at Little Papanui.[12]

With the 1960s the pace was to quicken. At the four main New Zealand universities marine biology was already being taught. Many years before, Otago had been the first to establish a marine station, in the old Fisheries Laboratory at Portobello. Opened as a fish-hatchery in 1902, this had for many years survived as a public aquarium. In the 1950s Betty Batham, on her return from graduate study at Cambridge and at Plymouth Marine Laboratory, was appointed sole charge of Portobello, at first under the Department of Physiology. For the rest of her days she was single-mindedly devoted to creating a modern marine station, setting up its library, laboratory equipment and reference collections, personally carrying out its first research, and obtaining a trawler for studies offshore.

Elizabeth (Betty) Batham's name is to be written large, not least for her driving effort in circumstances she often sensed as discouraging. Her life was to be lost tragically early, in 1970s, in unaccompanied shore work on rough coasts near Wellington.

After Otago, Canterbury and Auckland were the next universities to set up marine stations. In 1959, marine biologist George Knox was appointed to the Zoology chair at Christchurch, and in 1960 John Morton returned from London and Plymouth to be the first Professor of Zoology at Auckland. In an era of low costs, pioneer enthusiasm was not to be too much held back by official delays. By the end of 1960 marine laboratories had been approved for both Kaikoura and Leigh. W. (Bill) J. Ballantine, who had been Morton's first research student in London, arrived in 1962, to become resident biologist and afterwards Director of the station at Goat Island Bay, Leigh. The early development of Leigh, today a well-staffed sub-campus of the university, gained impetus from Ballantine and from John Gilpin-Brown (trained at Plymouth, England) at Auckland from 1961 to 1967. Of key importance at Leigh was the first gazettal of a marine reserve, in 1975, from Cape Rodney around Goat Island to Okakari Point.

At Wellington, Victoria University had developing strengths in ichthyology and marine systematics, especially of crustacea and plankton, from the initial inspiration of Laurie Richardson (1945–60). With full-scale marine programmes developed by the DSIR from the New Zealand Oceanographic Institute at Evans Bay, the university's small station at Island Bay remained primarily as a teaching facility.

The New Zealand Sea Shore by John Morton and Michael Miller, first published in 1968, in a sense marked the end of this beginning. With information gathered through the 1960s — helped in major part by

Fig. 1.7 Elizabeth Batham

their own students on field courses — the authors aimed at a comprehensive habitat-based picture of all New Zealand's shore communities. The book included also beaches, tidal flats and estuaries, habitats virtually neglected until a pioneer tidal study by Don Woods, in 1962.

In the decades since the publication of Morton and Miller, the taxonomy of many groups of New Zealand's marine animals and plants has been brought to maturity, with the completion of definitive monographs. All build on major pioneering by early taxonomists, whose work must not be overlooked. The earliest and broadest advances were, as expected, with the Mollusca. Henry Suter's monumental *Manual of the New Zealand Mollusca* (1913) was only finally superseded in 1979 with the modern reference book from the hand of Baden Powell.[13] This owed much to Powell's own half a century of research and to the generation of malacologists he had brought forward, including the late Sir Charles Fleming, and R.K. (Dick) Dell.

Echinoderm systematics and ecology were studied through four decades by Barraclough Fell, and then by David Pawson and Alan Baker. For 30 years Professor George Knox of Christchurch laboured almost alone at the taxonomy of the New Zealand polychaete worms.

The traditionally 'difficult' sponges have been intensively studied over many decades from Auckland University by Professor Dame Patricia Bergquist and her students.[14]

Also from Auckland, Michael Miller and his students have advanced our whole understanding of the nudibranchs.

Our colonial hydroids were documented by Patricia Ralph in the 1940s and founding studies of our simple and compound ascidians were undertaken by Beryl Brewin in the 1940s and 1950s.

In the 1970s and 1980s the late Brian Foster completed thorough taxonomic and also physiological studies of the barnacles. Initial work on the Amphipoda and Isopoda was undertaken by Des Hurley in the 1950s and 1960s. The littoral and offshore crabs have drawn good attention, initially from Charles Chilton, E.W. Bennett and Laurie Richardson.

Since the 1960s the primacy in marine systematic study has been carried by the long series of memoirs of the New Zealand Oceanographic Institute, merged since the early 1990s into the National Institute of Water and Atmospheric Research (NIWA).

New Zealand's marine algal taxonomy is founded in the results of the French and British Antarctic expeditions' visits to our shores in the 1930s and 1940s. Our modern understanding of New Zealand's seaweeds can be largely attributed to the studies of Val Chapman and most recently Nancy Adams.[15]

New Zealand marine reserves

The idea of having areas of the marine environment set aside for complete protection is now well accepted in New Zealand. There are now (in 2004) 18 formally gazetted (Fig. 1.8) and other proposals are being investigated. The climate was very different back in 1965 when Professor Val Chapman of Auckland University first suggested that an area of coast adjacent to the newly established Leigh Marine Laboratory should be set aside as a marine reserve, so that research on the natural ecosystems could be undertaken free from interference by human harvesting and fishing. The government Marine Department's response to the idea was extremely negative, but this served to harden the resolve of university staff, who set in motion a campaign to collect information in favour of such reserves and to win over public support for them.

Six years of lobbying led to the Marine Reserves Act 1971. This provided a legislative base for the establishment of special purpose scientific reserves, like the one initially lobbied for at Leigh. The Act did not envisage their future popularity for education, recreation or fisheries stock replenishment. It provides for the setting up and management of reserves in their natural state as the habitat of marine life for scientific study. This Act is under review and it seems likely that the purposes of marine reserves will in the future be considerably broadened.

Under the 1971 Act, the University of Auckland Leigh Marine Laboratory applied for the creation of a Marine Reserve along the coast. In 1973 the application was finally advertised and was greeted with many objections. In hindsight it seems that most objectors just did not want any constraints on what they could do below high-water mark. They wanted to retain the rights to do as they pleased in and on the sea. Nothing was heard of the application and objections for two long years, until November 1975 when suddenly the Cape Rodney to Okakari Point Marine Reserve was gazetted as one of the Rowling Government's last actions. A year later a management committee was established and erected the first public notice of its existence. Today 'Goat Island Marine Reserve', as it is commonly known, is the showcase marine reserve — possibly the largest tourist attraction in the region with nearly a hundred thousand visitors each year coming to view the profusion of colourful fish and underwater life.

The obvious increase in fish and crayfish numbers in New Zealand's first marine reserve has brought support through the creation of many more such reserves around the New Zealand coast, particularly in the 1990s. The second marine reserve was set up at New Zealand's premier dive spot around the Poor Knights Islands in 1981, but in an effort to placate objectors fishing was still permitted with some restrictions. This fishing concession was finally removed in the late 1990s after drawn-out legal battles by the Tutukaka fishing fraternity.

The third marine reserve, around the Kermadec Islands, was not created until 1990 and at 748,000 ha it was the largest marine reserve in the world. The other 16 reserves around our two main islands cover a total of 15,000 ha, or less than 0.1 percent of our mainland territorial sea. This compares with the approximately 30 percent of New Zealand's land area that is protected in some form of reserve.

Frustrations over the delay and defects of the Marine Reserves Act 1971 led several groups in the late 1970s to seek alternative means to protect coastal marine areas. These led to the creation of similar marine protected areas using the Harbours Act or a special act of parliament, resulting in the Tawharanui and Mimiwhangata Marine Parks on the east coast of Northland in 1981 and 1984, and the Sugar Loafs Marine Park at New Plymouth in 1986.

A Marine Reserve is an area protected from the impacts of commercial and recreational fishing, trawling or dredging, as well as from any adverse effects of coastal marine developments. It is hard, indeed often impossible, to protect these reserves from organisms newly introduced from overseas by shipping, from pollution, and from the effects of increased fresh-water and sediment runoff from nearby land.

Clearly the main benefits from marine reserves are in the protection and natural restoration of the marine ecosystems. These can then be studied for education or research, and may attract low impact recreation and nature tourism. Reserves provide for the establishment of mature breeding populations, many of which release vast numbers of their dispersal phases (planktonic eggs, larvae, spores and the like) to be carried away by the currents. Thus settlement or metamorphosis occurs a long way from the parents and the reserve is effectively reseeding and benefiting all the surrounding marine

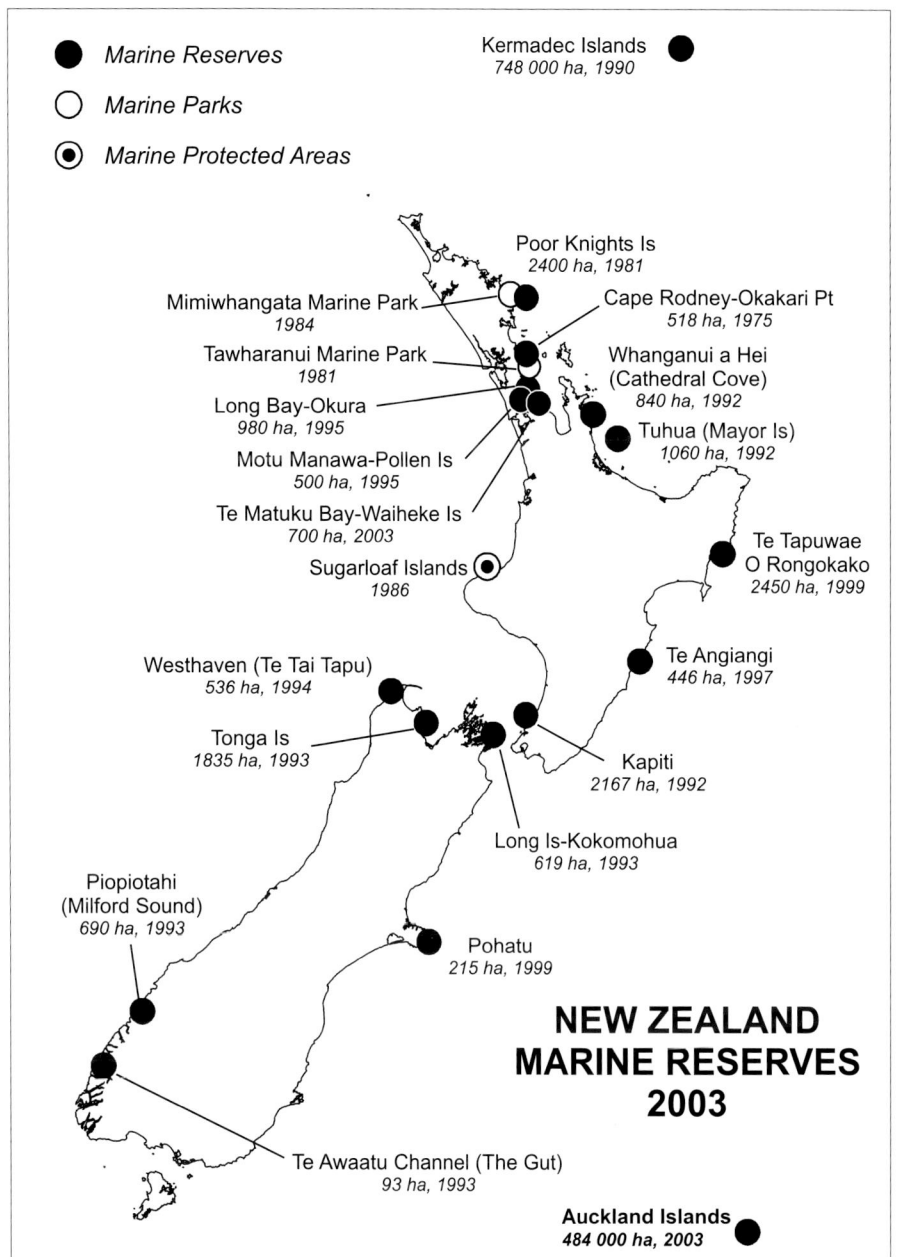

● Marine Reserves
○ Marine Parks
◉ Marine Protected Areas

Kermadec Islands
748 000 ha, 1990

Poor Knights Is
2400 ha, 1981

Mimiwhangata Marine Park
1984

Cape Rodney-Okakari Pt
518 ha, 1975

Tawharanui Marine Park
1981

Whanganui a Hei
(Cathedral Cove)
840 ha, 1992

Long Bay-Okura
980 ha, 1995

Tuhua (Mayor Is)
1060 ha, 1992

Motu Manawa-Pollen Is
500 ha, 1995

Te Matuku Bay-Waiheke Is
700 ha, 2003

Sugarloaf Islands
1986

Te Tapuwae
O Rongokako
2450 ha, 1999

Westhaven (Te Tai Tapu)
536 ha, 1994

Te Angiangi
446 ha, 1997

Tonga Is
1835 ha, 1993

Kapiti
2167 ha, 1992

Long Is-Kokomohua
619 ha, 1993

Piopiotahi
(Milford Sound)
690 ha, 1993

Pohatu
215 ha, 1999

NEW ZEALAND
MARINE RESERVES
2003

Te Awaatu Channel (The Gut)
93 ha, 1993

Auckland Islands
484 000 ha, 2003

Fig. 1.8 New Zealand marine reserves in 2003

environment. By natural design, very few of the dispersal phases will still be around to settle near their parents in the marine reserve — indeed most colonists in a marine reserve must arrive from somewhere else. Thus the need for not just one marine reserve, but a network so that they can provide the seed for colonising other nearby reserves.

The goal for Marine Reserves in the 21st century is to take up the challenge of Bill Ballantine and protect 10 percent of our coastal ecosystems in a network of reserves that encompass a fully representative range of New Zealand's marine environments and biodiversity.

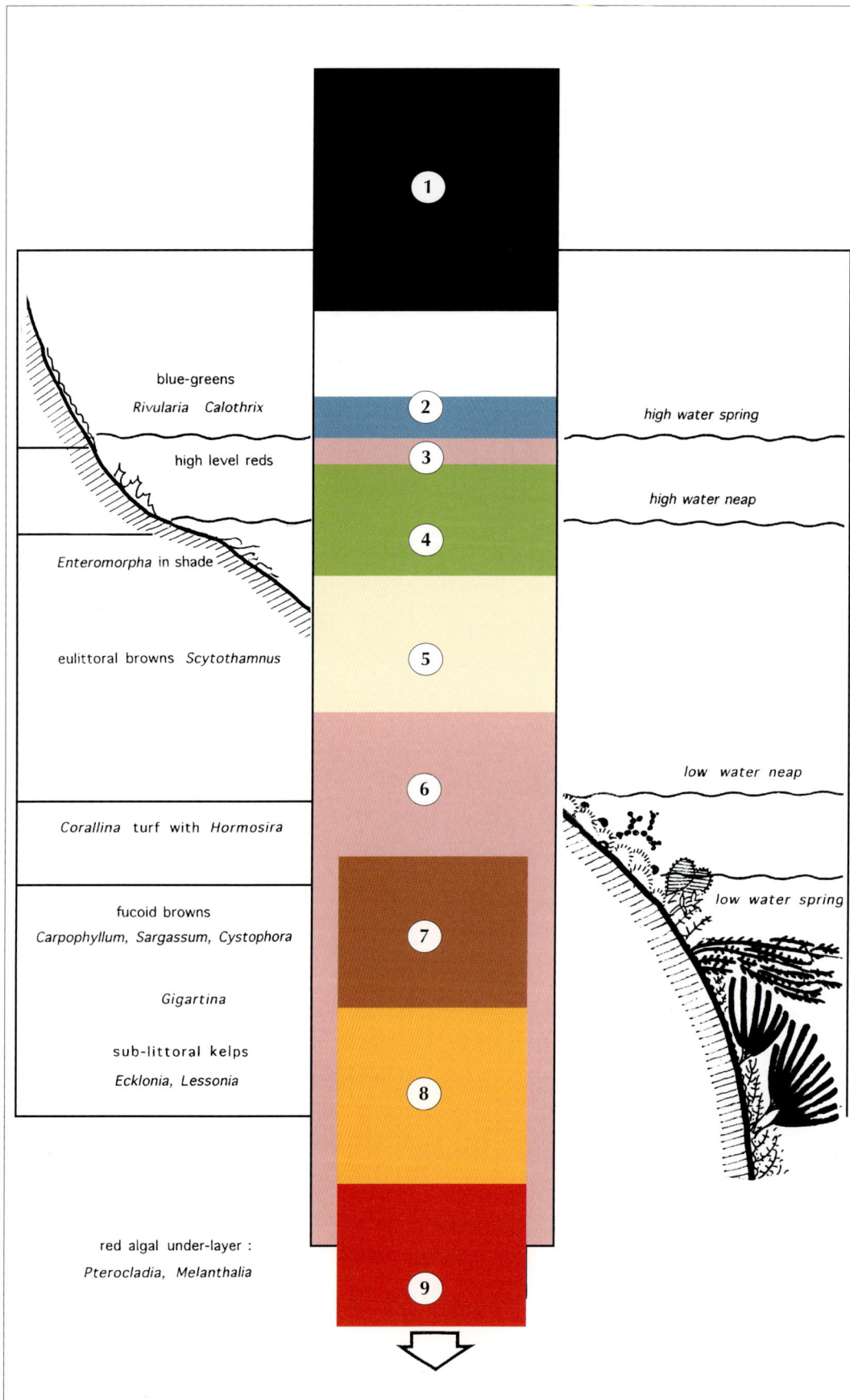

Fig. 2.1 The succession of plant colours in zones of the shore and subtidal
1 black lichen, *Verrucaria maura* 2 blue-green cyanobacteria 3 red *Bangia* and *Porphyra* 4 high level greens 5 eulittoral small browns 6 red algal veneer 7 fucoid browns 8 sublittoral kelps 9 red algal underlayer.

CHAPTER TWO

Life-forms of the Zoned Shore

The pasture of the algae: primary producers

By far the greatest algal biomass is to be accounted for in the unicells of the phytoplankton. From relatively big diatoms and dinoflagellates down to nannoplankton only a micron in length, it is on such elementary plants that the whole wealth of the zooplankton ultimately depends.* In the dark sub-photic layers where productivity of living algae has entirely ceased, dead plankton cells fall in a rain of detritus that is directly or indirectly the support of all life at the great depths.

In earlier times, living unicellular algae must have also fallen to the inshore bottoms, at depths where they could still photosynthesise. So the shallow seas could have seen one of life's huge innovations: the first evolution of macrophytes or large multicellular algae on well-illuminated rocky shores.

The shores between tides have a graded microclimate. Furthest up, with the sun's greater drying and warming power and higher illumination, the rock-attached algae (lithophytes) are especially vulnerable. They suffer here the effects of high evaporation, accelerated with 'wind-burning'; according to its quantity and spectral quality, light itself can be damaging. Far below the intertidal, the 100-m line represents the approximate depth to which, in clear water, plants can receive enough light for photosynthesis. Well before this, however, the more light-demanding algae have lost their predominance. Below wave base at around 50 m, water movement ceases to have real significance for the algae. Surge and currents can, however, have important effects for algae around the line of wave-break, but the representation of algae between the tides, and right down to their limits, is governed foremost by the nature and quantity of the light.

The intertidal shore profile (Fig. 2.1) could be likened to a graph expressing the regular gradations in emersion, illumination and exposure to desiccation. Most significant for our present purpose are the changes of illumination according to depth. It is with these that the broad lines of algal distribution are found best to accord.

Cyanobacteria (Fig. 2.2) is the modern name for those most primitive benthic shore plants that used to be called Myxophyceae or Cyanophyta. Their common name of blue-green 'algae' may remain, though we need to remember how remote as *akaryotes* they are from the *karyotes* that comprise not only the green, red and brown algae, but all the higher plant groups. The blue-greens are generally the highest 'algae' on the shore, growing in the *littoral fringe* under the conditions considered to be most stressful and limiting. These are the most elementary in structure of all multicellular algae

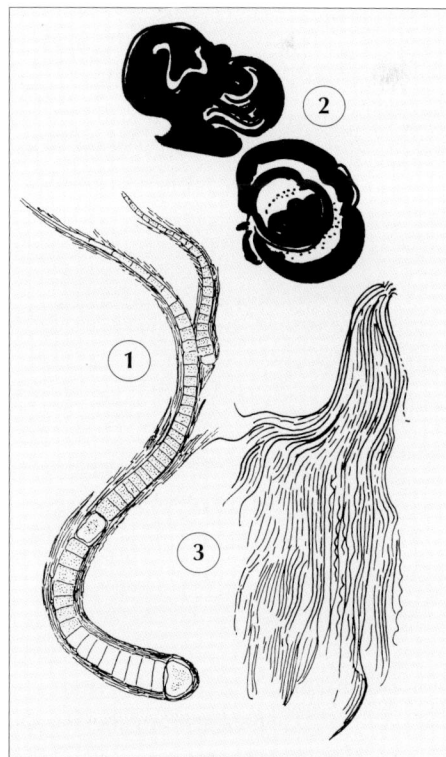

Fig. 2.2 Cyanobacteria (blue-greens)
1 *Calothrix parietina*, a single trichome 2 *Anabaena*, fleshy vesicles 3 *Lyngbya majuscula*, tuft of threads.

---

*Though the standing crop of phytoplankton fluctuates widely, some typical spring/summer figures for New Zealand inshore waters, in cells per ml of water, are cited by Bullivant: diatoms 50–500, dinoflagellates 20–50, coccolithophores (calcareous armoured flagellates) 20–100, naked flagellates c. 100–5000.

and with little doubt the most ancient. Many of their species are cosmopolitan.

With their traces going back to the oldest fossil-bearing strata, the blue-green algae stand apart as *akaryotes*, being without nuclei, and having the ribosomes, bearing their nucleic acids, distributed diffusely throughout the cell. They begin with the simplest forms (Chroococcales) that remain unicellular. Other Cyanobacteria have the cells arranged in simple or branched threads, known as trichomes. A more or less firm mucus sheath may bind these threads together into filaments. In the most advanced forms, simple thalli, moderately firm or fleshy, may be formed by the further binding of filaments with mucus. Reproduction is by vegetative multiplication of cells, including spore formation. No sexual processes are known to occur.

The green algae (**Chlorophyta**) in general occur next below the blue-greens, and higher up than the main onset of the browns. At this level most of the greens are of simple construction. There are the branched filaments of *Cladophora*, the large-celled and unbranched filaments of *Chaetomorpha* and the thin, tangled threads of *Rhizoclonium*. High on the shore, especially in the presence of freshwater seepage, come the *Enteromorpha* species, tubular or membranous in build. Towards mid-shore, notably in pools, grow the larger foliose thalli of the sea lettuce *Ulva lactuca*.

Much more elaborate in construction are the small green algae of low water. The *Codium* species — both finger-branching and adherent — belong to the sublittoral fringe. Here too, as also subtidally, grow the delicate, variously fronded *Caulerpa* species.

In their highest diversity, most of the red algae or **Rhodophyta** around low tide form an underworld to the canopy of large browns. Prominent wherever the brown algae are drawn aside is the pink basal crust of the calcified coralline species. From this simple veneer can be proliferated the jointed branchlets forming the coralline turf of the lower eulittoral. In the warm temperate and subtropics, this turf, with its wealth of associate species, increasingly takes over the territory held in higher latitudes by the fucoid browns.

Towards the very top of the shore two simple and primitive red algae, *Porphyra* and *Bangia*, must be mentioned, exceptions to the general rule that the reds grow low down under reduced illumination.

Around and beyond low water, the pink basal veneer is richly overgrown by brown algae (**Ochrophyta**) that grow upon it and largely conceal it from above. The calcareous reds continue far deeper than the browns down to the dimmest light possible for photosynthesis. Here, in the still waters below wave base, the soft red algae have a new manifestation. Their thin, often translucent thalli take on exquisite forms, with delicate colouring,

from carnation to pale pink. Not all the Rhodophyta look red, however. In the intertidal there are many kinds that turn to olive, different shades of green, maroon or purplish brown. *Gelidium* and *Stictosiphonia* are manifested as brownish swards right to the top of the eulittoral. The real profusion of the intertidal reds only begins, however, with the algal gardens of the lower eulittoral and sublittoral fringe. With clear water and lively wave action, this is the chief domain of *Gigartina*, with species both membranous and pinnate-branched. Their dull greens and purples serve as screens against a wider light spectrum than the reds can normally use.

Pigments and light

The changes of algal colours down the shore can make sense only with reference to incident light, just as their thalli have shaped themselves in response to waves and desiccation. Like all plants living by photosynthesis, algae possess **chlorophyll** as their basic pigment. Further, the marine algae carry other additional pigments in special cell organelles called *chromatoplasts*. These may be likened to the sensitisers added to photographic plates to make them react more quickly to a particular light.

Not only does the quantity of light diminish with passage through the water, its spectral composition alters too. The pigments of the algae reflect these changes, enough at least to support the sometimes over-simplified theory of *complementary chromatic adaptation*. Where land plants possess both *chlorophyll a* and *b* and *carotene*, the brown algae have **fucoxanthin** instead of *chlorophyll b*, while the red algae have also the two pigments **phycoerythrin** and **phycocyanin**. Light absorption requires pigments complementary in colour to the incident light to be taken up. Below the sea surface, red and blue light are extinguished rather rapidly, depending upon solar conditions, waves and turbidity. The green light that predominantly remains is utilised by the red pigment phycoerythrin, which also has a photosynthetic activity of its own.

It is thus the Rhodophyta that can most efficiently utilise the deepest-reaching green light. At intermediate depths brown algae use their fucoxanthin to absorb the yellow/orange light. This is complementary to chlorophyll which is essentially a red and blue absorber. Highest up, the Chlorophyta make the most use of the red and blue wavelengths and least of the green.

Algal pigments are also destroyed by light at rates depending on strength and spectral quality. Chlorophyll is the most resistant to light damage, phycoerythrin the least. Thus, when subtidal red algae are thrown high upshore in storms, they become brown, yellow or green. Lower down, some phycoerythrin remains to mask the

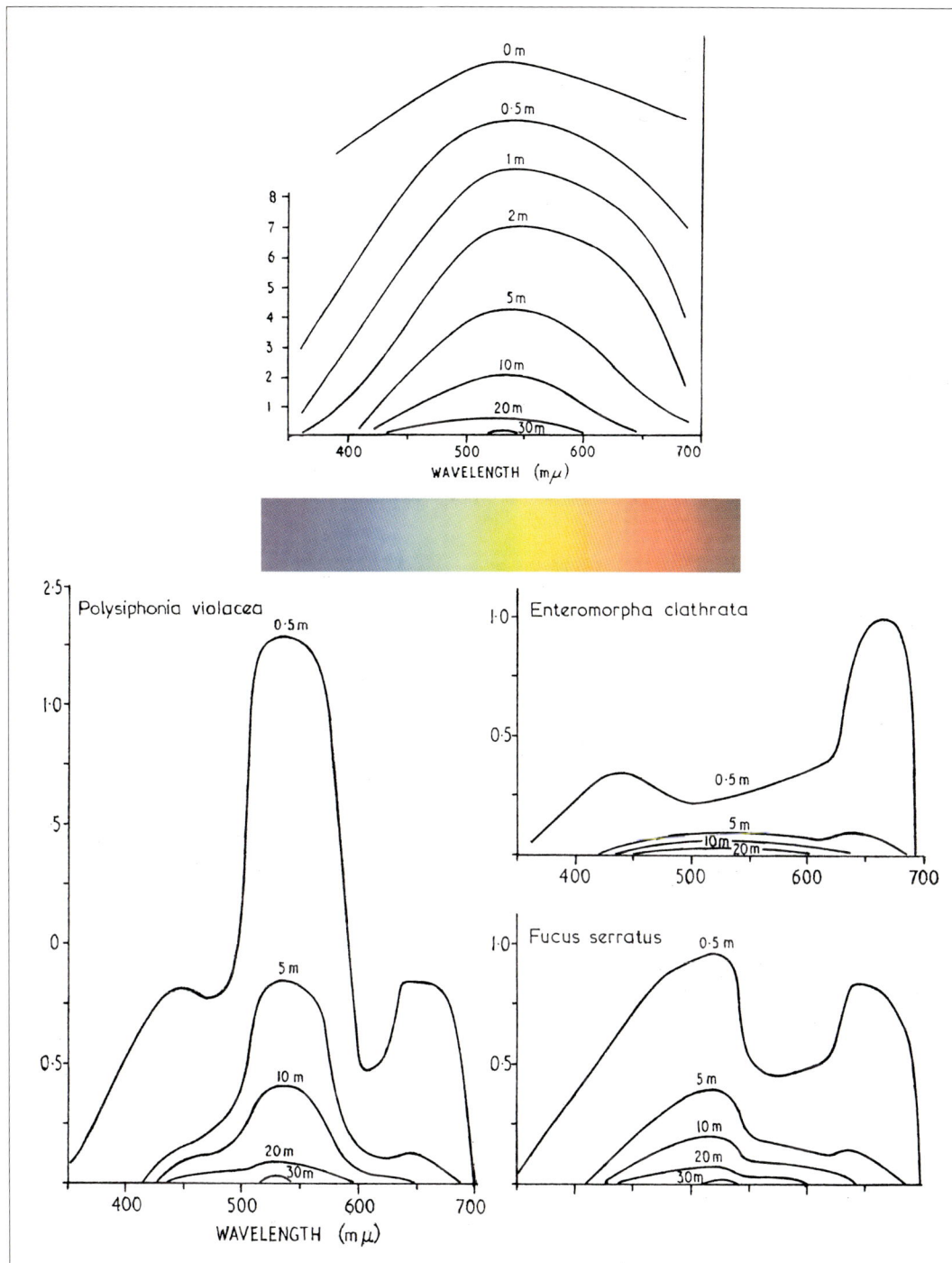

Fig. 2.3 Algae and illumination
Spectral assimilation curves of green (*Enteromorpha*), brown (*Fucus*) and red (*Polysiphonia*) algae at different depths. Spectral distribution of radiant energy at different depths (metres). (Based on data from Levring.)

chlorophyll. It can thus be seen why brown and red pigments would be useless on land. Phycoerythrin is also destroyed in fresh water, so laminarians become green, or red algae with little chlorophyll turn yellowish.

Ochrophyta

The **Ochrophyta** or brown algae have been kept to the last, not indeed for want of importance but because — as the largest and most visible algae of the shore — they call for a longer introduction. The wealth of the red

algae, as well as the greens, are discussed in sections where they become locally important.

The giants among all the seaweeds are the kelps or **Laminariales**. We shall discover later (Chapter 18) the cooler shores of the west coasts of the continents where the kelps hold a spectacular dominance over the sublittoral fringe. In New Zealand the laminarians scarcely enter the intertidal, but are confined to three native kelps (and one recent arrival) dominating the upper sublittoral zone. Our largest laminarian, but confined to

Fig. 2.4 The common brown algae in a transect of a northern shore

a–b *Chamaesipho columna*; b–c *Crassostrea glomerata* with *Scytothamnus australis*; c–d *Pomatoceros caeruleus*; d–e *Corallina* turf with *Leathesia* and *Ralfsia*; e–f *Hormosira banksii*; f–g *Carpophyllum plumosum* with a pelmet of *Xiphophora*

1 *Hormosira banksii*; 2 *Xiphophora chondrophylla*; 3 *Petrospongium rugosum*; 4 *Ralfsia verrucosa*; 5 *Hydroclathrus clathratus*; 6 *Leathesia difformis*; 7 *Colpomenia sinuosa*; 8 *Dictyota dichotoma*; 9, 10 *Carpophyllum plumosum*, two forms; 11 *Carpophyllum plumosum* var. *capillifolium*.

southern New Zealand, is the bladder-kelp *Macrocystis pyriformis*, forming dense forests offshore with its long intertwined axes ascending to the surface like slender pylons. Our smallest laminarian *Lessonia variegata* is entirely sublittoral, while the more familiar *Ecklonia radiata* — when found sporadically above low water —

is in danger of 'burning' on a warm day if a low tide coincides with strong winds.

The so-called bull-kelps *Durvillaea* are in fact not laminarians but fucoids. Like *Macrocystis* they are most at home in the colder south, but they live much less deep. On the wave-pounded coasts *Durvillaea antarctica* regularly moves up to the lower eulittoral. The second species, *D. willana*, where it accompanies *D. antarctica*, forms a tier below it.

Fucales

New Zealand's most widely evident brown algae belong to the Order **Fucales**. They reach from just above low-water mark, down through the upper sublittoral zone, where the kelps ultimately supersede them. *Durvillaea* excepted, they are smaller than laminarians, though still of good size, the largest forming streamers up to several metres long.

The Fucales have some salient differences from the Laminariales. The thallus springs from a small holdfast, and growth is apical, rather than from a meristem at the base of the blade, as in kelps. In the Fucales the visible plant is a sporophyte, whose spores can be said to function as gametes instead of producing a separate gametophytic generation. The reproductive organs are borne in pit-like conceptacles, and the egg is fertilised within the female conceptacle — to germinate without a free-swimming propagule.

In New Zealand most of our species of Fucales come in for only brief uncovering, during lower than average tides. In the colder North Atlantic, including Britain and east Canada, shores of moderate to low energy have a much extended representation of the fucoids. Different species, successively better adapted to resist desiccation, occupy the middle shore right to the top of the eulittoral.

The southern hemisphere Fucales, an element especially notable in Australasia, live predominantly in the sublittoral zone, in medium to high water movement. They differ notably from their North Atlantic counterparts by their generally leafy form, with their branching systems even recalling land plants. By contrast, Atlantic *Fucus* are simply flat, dichotomous fronds, mid-ribbed and — in some species — have bladders within the lamina.

The Family **Fucaceae** are themselves non-leafy. They are represented in New Zealand by the southern *Marginariella* species, forming dichotomous straps, with stalked bladders and spindle-shaped receptacles at the base; and by the simple dichotomous and bladderless *Xiphophora* (Fig. 2.4). In the Family **Cystoseiraceae** the thalli branch monopodially from a cylindrical (*terete*) axis, and bear narrow, compressed leaflets as well as bladders. New Zealand shares with Australia the large genus *Cystophora*, that comes to its maximum in cooler southern waters.

The great majority of the fucoids of northern New Zealand and the warm Pacific belong to the Family **Sargassaceae**. These have the most leafy and elaborate thalli of the whole order. Vesicles and reproductive branchlets are axillary to the leaves. Where in the north-ern hemisphere the Laminariales take command of the sublittoral fringe, in New Zealand's North Island this is the role of the important endemic genus *Carpophyllum*. In any degree of wave exposure, one or more of the four species will be found. Along with *Carpophyllum* in the north and *Cystophora* in the south grow several *Sargassum* species.

New Zealand's most thoroughly intertidal fucoid is the widespread and familiar *Hormosira banksii*. This is a specialised, amphibious plant with water-retaining bladders, able to respire and assimilate both in the wet and dry (Fig. 2.4).

Lesser Ochrophyta

The Order **Ectocarpales** shows us the smallest, simplest and by good inference the most primitive of the brown algae. Most are annuals with an easily visible diploid phase lasting only a few months, and a minute filamentous gametophyte persisting over the rest of the year. *Ectocarpus* and *Pilayella* are the fuzzy filamentous brown algae, so evident in upper tide pools. Some are endophytic in or upon large brown algae.

The *Ectocarpus* life cycle is described as *isomorphic*, with similar diploid and haploid phases, the first producing zoospores and the second the male and female gametes. The young stages of many other browns suggest a derivation from a filament like *Ectocarpus*, being *heterotrichous* and bearing colourless multicellular hairs. From such filaments have evidently evolved both crustose and cushion structures. In *Myrionema* erect branches spring from a prostrate base to form little circular discs. In *Ralfsia* threads coalesce to form dark brown leathery crusts, growing concentrically.

The **Leathesiaceae** have evolved to form hollow gelatinous sacs, built of a mucilaginous medulla with a thin cortex. *Leathesia difformis*, golden brown and brain-like, is common and cosmopolitan. Closely related to it is the small *Petrospongium rugosum*, forming circular pads with the medullary threads meshed by hyphae. Some of the Ectocarpales grow into branched cylindrical cords. In the bootlace-like *Myriogloea* a multiaxial medulla is encased in a cortex of branched assimilatory hairs. Near this group belongs the tubular *Splachnidium rugosum* filled with viscid mucus. The bladder-like *Colpomenia sinuosa*, superficially like *Leathesia*, has been placed in the separate Order **Scytosiphonales**.

The **Dictyotales** form a compact order characterised by their dichotomous foliose thalli. These are structurally far more simple than the Fucales, and their life cycle is isomorphic, with plants of the diploid and haploid generations superficially alike. *Dictyota*, *Glossophora* and *Zonaria* are familiar in the New Zealand sublittoral fringe.

Primary consumers: herbivores

In the broadest sense, herbivore is the term for all those animals cropping any part of the plant production of the sea. Two sorts of marine pasture exist. First, growing on the rocks, there are the multicellular algae sometimes known as **lithophytes**, and ranging from simple filaments up to the massive kelps. These are the food of the numerous mobile grazers, chiefly molluscs but including importantly echinoderms and fishes.

Second, and in aggregate biomass far more important, there is the plant food deriving from the **phytoplankton**. This constitutes the fodder of the sessile invertebrates that make up the zoned communities of the shore. It is these consumers that we shall introduce first, using filterers as the comprehensive term.

Filterers

Zoning animals that strain off a diet of plankton are arranged in predictable patterns, both in their order of settlement and in their level on the shore. **Operculate barnacles** are generally the earliest to settle and largely monopolise the upper eulittoral. In the middle eulittoral, dominance tends to pass to **bivalve molluscs**, typically mussels in our cooler waters and rock oysters in the warm north.

Directly below the bivalves, polychaete **tubeworms** appear: first serpulids, then, in the north, the sandy-tubed *Neosabellaria*. On our eastern offshore islands — as on tropical shores — the tubeworm habit has been adopted by the sessile, unwound gastropods called **vermetids**.

The lowest zoned of the filter feeders are usually the **ascidians** or sea squirts, beginning at the level of coralline turf. On open surfaces these are relatively unimportant in New Zealand, but come into their own chiefly under boulders.

Barnacles are the highest sessile animals on the shore, and generally the smallest-sized. Along with the bivalves they are the most effectively sealed against evaporation. At their upper levels their feeding time is limited to the brief arrival of waves or swash. Like all crustaceans, the barnacles lack cilia, and instead sweep for plankton with casting nets of paired limbs called *cirri*. Worked by muscular effort, these are far superior to cilia in yield per unit time.

Next down the shore, **mussels** and **oysters** grow larger than barnacles, with the smaller flea mussel (*Xenostrobus pulex*) living towards the top of the range. The shells can be tightly closed against evaporation by sustained contraction of their adductor muscles. Tubeworms and vermetids — exposed to the air for shorter time — are sealed less tightly, by a stopper-like operculum. The sea squirts have no shelly covering but instead a tough coat of *tunicin*. Left only for short intervals,

they enjoy a prolonged feeding time, and grow fast to large individual size.

Barnacles: Cirripedia

Hard-shelled and fastened to rock, the barnacles were in former times classed with molluscs. In 1830 the naval surgeon and naturalist J. Vaughan Thompson, using an early plankton net in Cork Harbour, discovered their six-limbed *nauplius* larvae, revealing their affinity with other crustaceans. At metamorphosis the final stage *cypris* larva fixes itself by its head and with special cement glands becomes attached. Shell plates are now secreted, in acorn barnacles taking the form of a conical tent, opening at the top by movable opercular plates. In their basic plan, the stalked barnacles are not essentially different from the operculates. In both types, dissection at once reveals a crustacean, highly modified for sessile life.

As occupants of the highest sites on the shore, the barnacles have a long history and today they are the most individually numerous eulittoral animals. Charles Darwin suggested the present might be called the Age of Cirripedes, just as past eras have been ages of Graptolites, Trilobites or Brachiopods.

The terminology of the operculate barnacle shell is explained in Fig. 2.5. The eight original plates of the *column* become fused to differing extents. The operculum is made up of two pairs of sutured plates, the *terga* and *scuta*, diverging along the mid-line to give exit for the limbs. These six pairs of jointed *cirri* fringed with bristles or *setae* form the sweeping net for feeding. The small body is slung beneath the scuta inside a wide mantle cavity, into which the cirri curl up when retracted. The three posterior pairs of cirri are longer and their setae strain the plankton as they sweep through the water. Brushed off by shorter anterior cirri, this food is passed to the mouth.

Barnacles are hermaphroditic, and are unusual among sessile animals in continuing to practise copulation. In operculate barnacles the slender penis is long enough when extruded to reach neighbours some distance away.

The largest order of the barnacles, **Thoracica**, has two suborders, the **Lepadomorpha** or stalked barnacles and **Balanomorpha**, sessile acorn barnacles. The **stalked barnacles** are the earlier group, having an ovate *capitulum* with shell plates, carried on a flexible *peduncle*. At first, there were no distinct capitular plates, only a thick investing mantle, as still to be seen in the small New Zealand *Heteralepas*. Almost as archaic is *Ibla idiotica*, a small stalked barnacle living in crevices, having chitinous terga and scuta and the peduncle invested with hairy bristles.

Evidently quite early, the stalked barnacles fixed

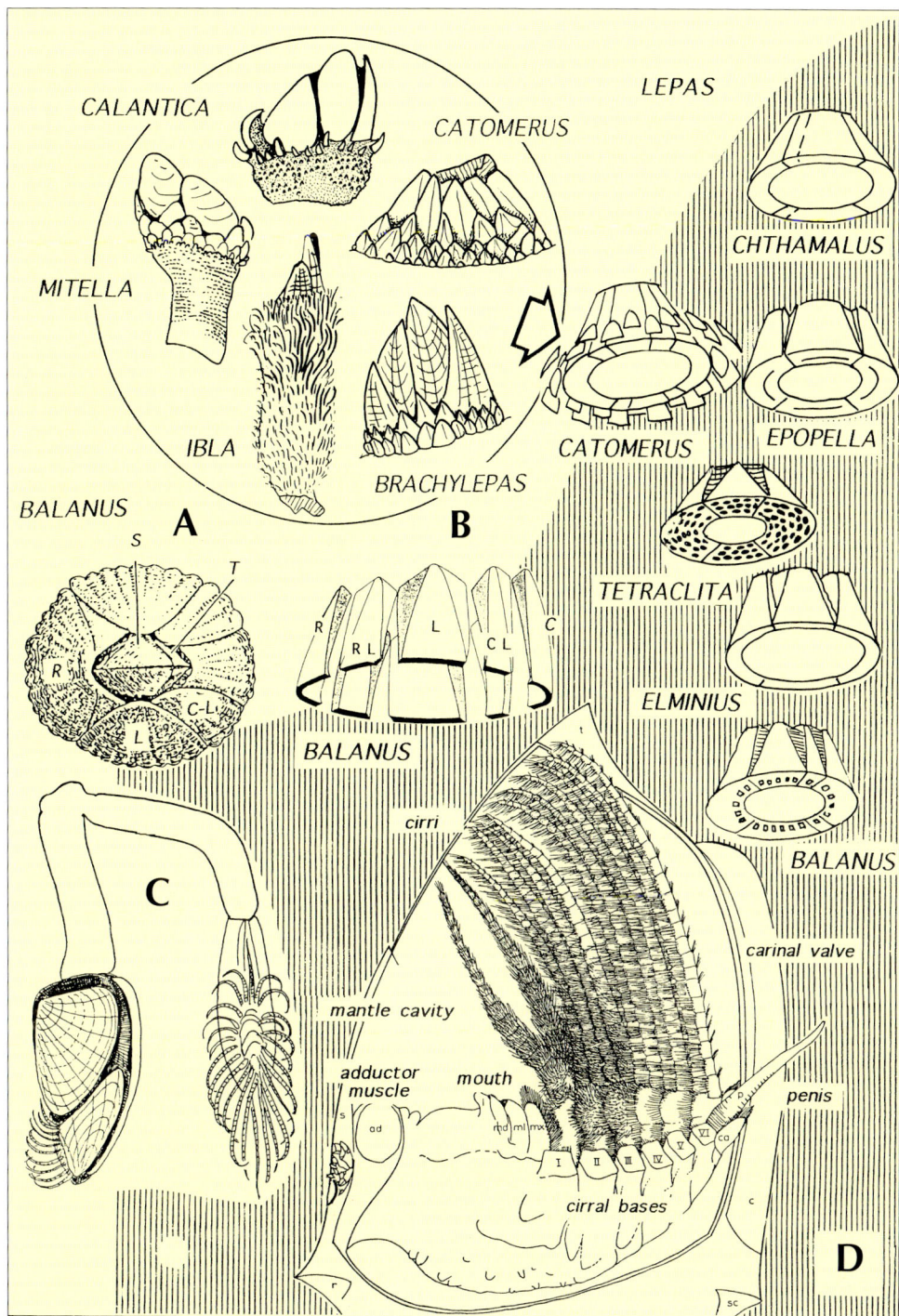

Fig. 2.5 Morphology of barnacles (Cirripedia)
(Top) Some primitive pedunculates with the derivation of operculate barnacles shown with detail of plates and in relation to shore level. **(Centre)** A *Balanus*, shell plates viewed from above B the eight column plates of an operculate barnacle before fusion (C carinal, CL carino-lateral, L lateral, R rostral, RL rostro-lateral, S scutum, T tergum) **(Bottom left)** C pedunculate barnacle, *Lepas anatifera*. **(Bottom right)** D internal anatomy based on *Calantica*. (After Brian Foster.)

upon the five-plated capitulum, that is today the standard pattern of the pelagic goose barnacles, **Lepadidae**. The shell plates have become reduced and functionless in the oceanic lepadomorph *Conchoderma*, fixed by its peduncle to the large balanomorph *Coronula*, that itself lives attached to whales.

Some of the most primitive stalked barnacles today are the *Calantica* species, secluded survivors in caves and crevices (Fig. 14.11). Their capitulum still remains many-plated. As well as the differentiated terga and scuta, median rostrum and carina, there are also tiers of small platelets round the capitular base.

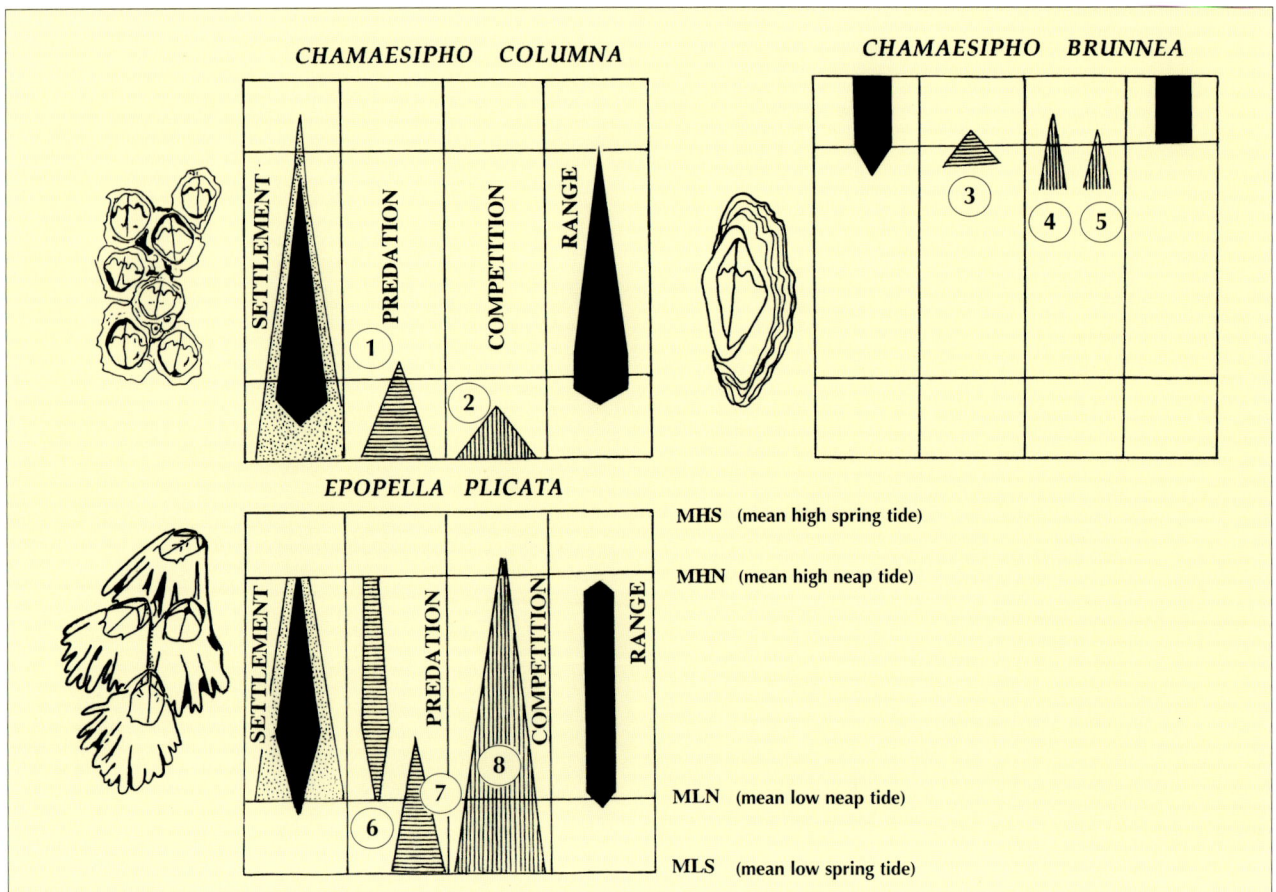

Fig. 2.6 Ecological relations of three zoning barnacles

For each species, *Chamaesipho brunnea*, *Chamaesipho columna*, *Epopella plicata*, is shown **(left)** the extent of normal settlement and **(shaded extension)** settlement on cleared experimental surfaces; and **(right)** final range, with the relative effects of predation and competition. 1 predation by *Lepsiella scobina*; 2 competition from *Epopella plicata*; 3 predation by *Lepsiella scobina*; 4, 5 competition from *Chamaesipho columna* and *Epopella plicata* 6, 7 predation by *Lepsiella scobina* and *Dicathais orbita* 8 competition from *Chamaesipho columna*.

(Based on findings by Penelope Luckens.)

It must have been surf-hardy lepadomorphs like these that shortened their peduncle and sat broadly to the rock, to give rise to the **acorn barnacles**. With the opercular plates kept unchanged, the column plates were reduced, first to a basic eight (antero-median *carina*, postero-median *rostrum*, and paired *carino-laterals*, *laterals* and *rostro-laterals*). Only a few of such early forms survive. *Octomeris* keeps its eight plates separate, while the remarkable east Australian *Catomerus polymerus* (Fig. 17.4) has eight column pieces with the base encircled by scaly accessory plates.

A very old balanomorph family, the **Chthamalidae**, has held to the hard terrain of the upper eulittoral. *Chthamalus*, with six column pieces, has no New Zealand species; in our close-related *Chamaesipho* these plates are fused into a single ring. In *Chamaesipho brunnea*, as the shell increases its diameter, the aperture is enlarged by erosion from the wetting and drying of the rim of the column.

The Family **Tetraclitidae** are par excellence the conical barnacles of the middle eulittoral. Calcite is laid down inside the four-piece column, to seal up the sutures which are still visible externally. In New Zealand they are represented by the large surge barnacle *Epopella plicata*, and by the wafer-like *Tetraclitella depressa* under boulders and in caves.

The **Balanidae** are the pink or mauve acorn barnacles of the sublittoral fringe and the subtidal. In *Balanus* species, the six column pieces never fuse but interlock by flanges or *radii*. On our open shores these include one of the largest living barnacles, the cosmopolitan *Megabalanus tintinnabulum linzei* (Fig. 5.3). To the same family belongs the small, highly successful opportunist *Austrominius modestus*, with the column-plates reduced to four. In the tropics *Pyrgoma* and *Creusia* are balanids embedding in corals, with the plates fused to a ring.

Physiology

The high level barnacles live in one of the most inhospitable habitats. Brian Foster wrote:[1]

Perhaps nowhere else in the world is the substratum so

subject to wide fluctuations of temperature and salinity and drying and wetting by fresh or salt water, with diurnal and tidal regularity and climatic irregularity. For a marine animal, the time spent feeding and reproducing must be small at high tidal levels: and in supralittoral regions (above tidal predictions) is dependent on the effectiveness of waves.

Chamaesipho brunnea manages to exist 4 m above predicted high water and endure several weeks unwetted. As well as having strong plates fused or interlocked, high tidal barnacles must also be *eurythermal* (tolerant of wide temperature fluctuations). During emersion the mantle cavity serves as a lung, expelling surplus water and taking in air by a small breathing hole between the opercular plates. Receptors in the thick integument at the edges of the operculum keep contact with the outside milieu; with light or tactile stimuli the valves reflexly close.

Long-term desiccation is avoided by tight closing. But the operculum must periodically open to respire, and when first emersed most barnacles expel a drop of water. Then they draw the opercular plates close together, periodically opening the small pore. Evaporative cooling and water conservation are thus conflicting interests. At the top of the shore, barnacles become more heat-tolerant, with higher upper lethal temperatures.

Feeding types

With different levels on the shore, barnacles show a hierarchy of feeding styles. Brian Foster distinguished the following types of **cirral activity:**

1 *testing*, slightly opening the opercular aperture
2 *pumping*, with cirri not extended but enough unrolled to make pumping strokes
3 *normal beat*, with the cirri rolling in fully after each stroke, opening out to
4 *fast beat*, with the cirri no longer fully rolled up between strokes
5 *extension*, with the cirri simply held outstretched and motionless.

All our intertidal barnacles test and pump. Low and subtidal balanids in addition practise normal and fast beat, with *Austrominius modestus* in quiet water excelling at the latter. In rough water, with food brought by currents and turbulence, active cirral beat becomes unnecessary. On high energy shores *Chamaesipho brunnea*, *C. columna* and *Epopella plicata* all use passive *extension* as the feeding mode. In turbulent habitats the body swivels to place the cirral net across the water current. This is also the habit of stalked *Calantica*.

Site adaptation

The effect of competition and predation on the species limits of barnacles was experimentally studied at Auckland by Penelope Luckens (1976).[2] The high level *Chamaesipho brunnea* is confined to surf coasts, and could not be induced to settle on cleaned rock surfaces below its normal adult range, even where the competitor *Chamaesipho columna* had been removed. In the Chatham Islands, where *C. columna* is missing, *C. brunnea*, however, ranges right down to the brown algae. This species is highly temperature-resistant, with its upper limit set by total dryness. The lower limit is normally imposed by competition from *C. columna*. *C. brunnea* suffers some predation from the thaid whelk *Dicathais orbita*, which can marginally overlap it from below.

The smaller *C. columna* settles on cleared rocks in place of (or contiguous with) *C. brunnea* and ranges further into shelter. Desiccation kills it at high levels, while competition from brown algae limits it below. *C. brunnea* or *Epopella plicata* offers little space-competition, since *C. columna* is able settle upon either. While *Epopella plicata* was found to have a better control of water loss than *C. columna*, the latter is able to reach higher, being small enough to nestle in humid microhabitats.

On level, poorly drained surfaces, algae and sheets of the flea mussel (*Xenostrobus pulex*) compete strongly with *C. columna*. Though vulnerable to over-treading and summer desiccation, the mussel competes favourably with *C. columna* by tolerating periodic sand burial.

Epopella plicata, the largest barnacle of the three, settles and grows best at lower levels. On open coasts it succeeds *C. brunnea*, and largely ousts *C. columna* down to mid-tide. Higher up it has limited survival, owing to its much larger size than *Chamaesipho*. At lower levels *Epopella* is predated by the whelks *Lepsiella scobina* and *Dicathais orbita*.

Ciliary feeders

Ciliary feeding can be extraordinarily efficient for small expenditure of energy and — though its rate varies with temperature — it is a continuous process for as long as the animal is submerged. Filtering mechanisms are highly diverse in detail, but their principle is basically the same. The first essential is a filtering screen with powerful *lateral* cilia lining its meshes and driving water through the screen from an inhalant space to the exhalant side. On the inhalant side the screen has also *frontal* cilia that carry away the particles strained off. This food is taken along grooves to the mouth, while excess or over-large particles are eliminated by rejecting cilia. In the bivalves, comb-like rows of inert *latero-frontal* cilia enhance the retention of particles on the screen. In

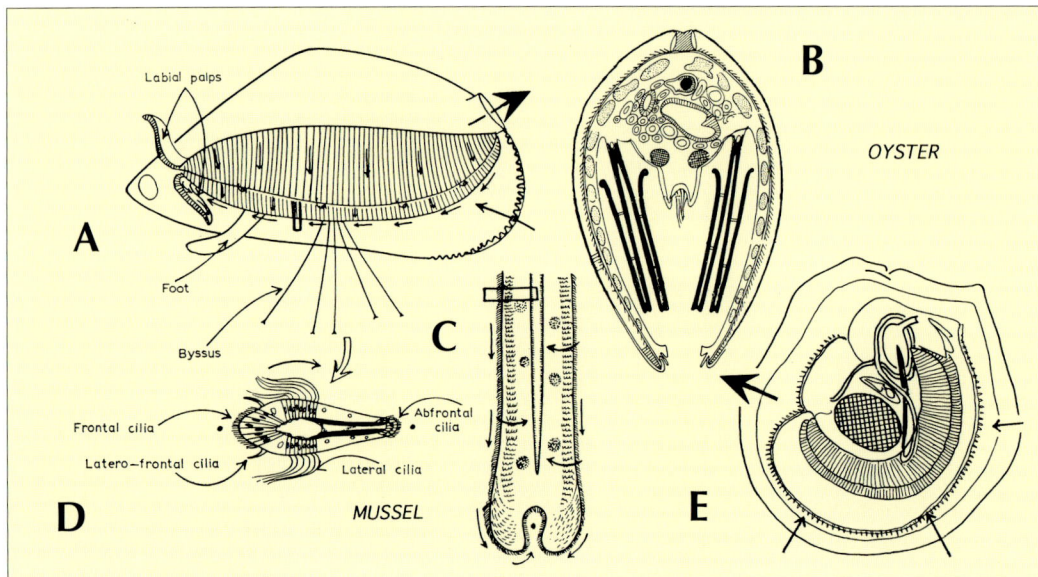

Fig. 2.7 Bivalve gill and mantle cavity
A. Mussel (*Perna*) with the gill in the mantle cavity, and its ciliary currents.
B. Transverse section of the mussel, showing the gill, with reflected inner and outer demibranchs.
C. Part of margin of inner demibranch, with a side view of two limbs formed by a double-bent filament, and the notch forming the marginal food groove.
D. Single gill filament in cross-section, showing ciliary tracts, and internal blood space.
E. Oyster (*Ostrea*), arrangement of the pallial organs, with location of inhalant and exhalant currents.

nearly all ciliary feeders, viscid mucus works with the cilia as a transport medium, a binding for waste, or — in ascidians — to retain particles upon the screen.

takes on a circular symmetry. Inhalant currents now enter nearly all round the mantle edge, with the exhalant flow concentrated at one point.

Bivalves

Some of the earliest bivalves must have lived attached to rock by the horny threads of the *byssus*, secreted at the base of the foot. Witness the antiquity of the mussel family (**Mytilidae**) where the shell is retracted against the rock by pedal muscles attached to the base of the byssus. From the byssus gland a stream of proteinoid secretion flows along a groove to the tip of the foot. Each implanted thread is hardened by chemical tanning. Threads can also be broken off by the foot's leverage against the rock, so the mussel — especially the juvenile — can readily change sites. Pulled down to the rock by the byssus, the mussel's anterior end is constricted and narrowed. Growth is in contrast accentuated at the rounded posterior end, where the inhalant and exhalant apertures are both located.

The oysters (**Ostreidae**) have evolved further. The settling larva, with a rounded shell, like a mussel, comes to rest with the right valve cemented down by a secretion from the pedal gland. The foot is then lost, and the shell begins to grow radial increments conforming to the shape of the substrate. The anterior adductor (already reduced in mussels) is in the oysters soon lost, and the posterior adductor shifts to the centre, as the animal

Polychaete worms

Two families, the **Serpulidae** and the **Sabellariidae**, contribute tubeworms to the zoned shore. Both have become highly modified, with the paired limbs (*parapodia*) reduced to struts holding the animal in place inside its tube. The head has become remodelled, with special appendages to entrap food particles from currents brought in by cilia. At the least disturbance fast muscles instantly retract the worm deep into its tube.

New Zealand's commonest shore serpulid is *Pomatoceros caeruleus*, with a white calcareous tube sharply spined above the mouth (Fig. 2.8). The head is crowned with two shallow funnels, each forming a wheel of radial gill filaments (*pinnae*). The expanded crowns of a whole colony form an indigo carpet, very sensitive to the touch. Each of the pinnae is fringed with pinnules, carrying two sorts of cilia. Long *lateral* cilia drive up currents between the spokes and towards the centre of the wheel. At the same time food particles impinge upon *frontal* cilia, beating down the pinnule, these travel in mucus strings to the base of each pinna and finally to the mouth. On the left, one of the pinnae is fashioned into an operculum to plug the tube when the worm is retracted.

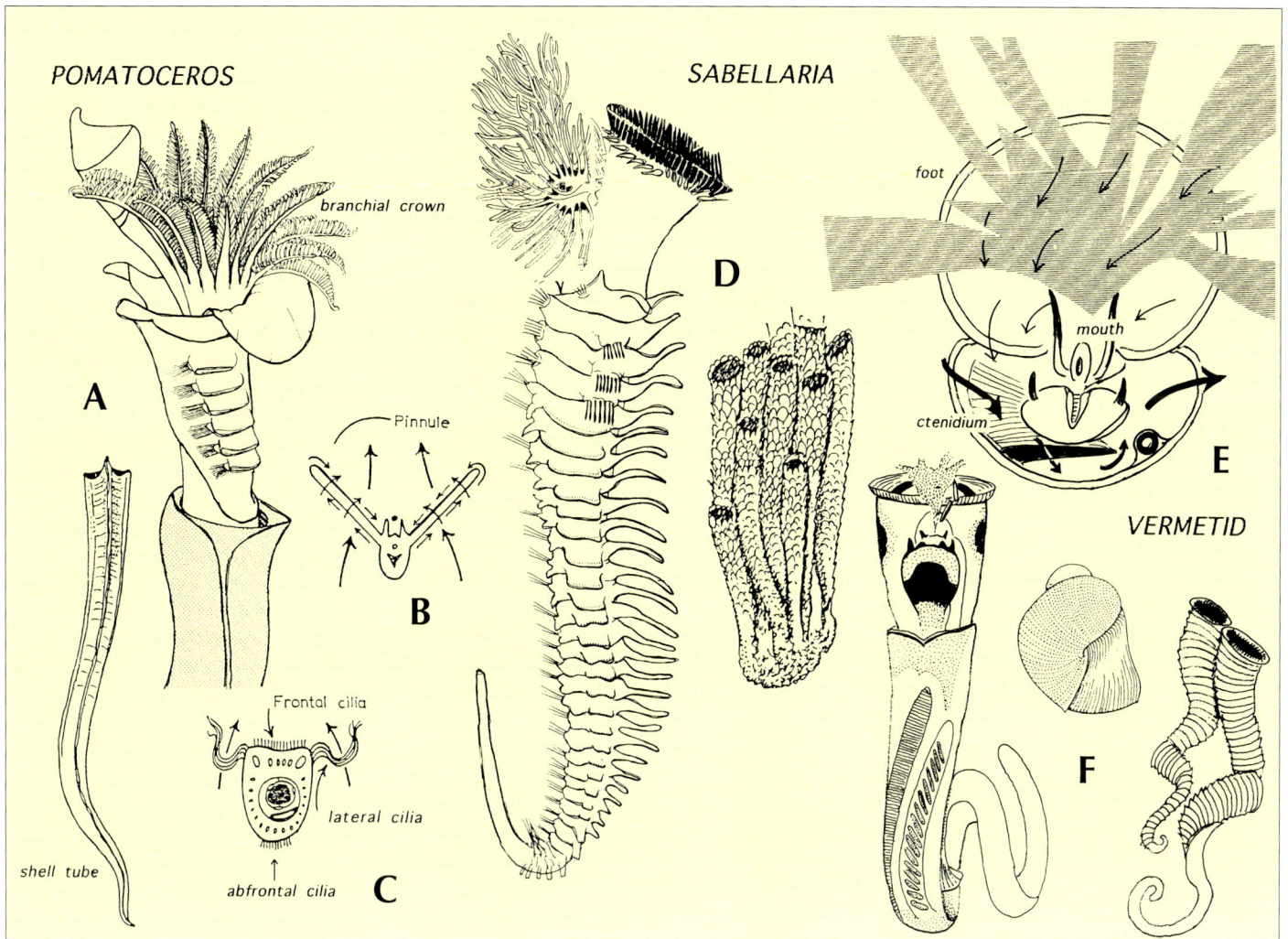

Fig. 2.8 Two tubeworms and vermiform gastropods

A. *Pomatoceros*, showing shell tube and animal with branchial crown; B. section of a filament with pinnules, C. a pinnule in section, with cilia.

D. *Neosabellaria kaiparaensis*, with head tentacles and operculum, and sand tubes.

E. Vermetid gastropod *Serpulorbis zelandicus*, showing head, foot and mantle cavity, with mucus feeding trap.

F. Vermetid gastropod *Dendropoma lamellosa* with animal, shell tubes and embryonic shell.

The sand-mason worm *Neosabellaria kaiparaensis* is found in the north straight below *Pomatoceros*. Sabellariids are recognised from their thick crusts of sandy tubes, with living worms only in the terminal reaches. Instead of a pinnate crown, the *peristome* (the first segment behind the head) is split into left and right lobes that retract over the mouth to form an operculum. Each carries on its outward face three crescents of chitinous bristles. From the inside springs a row of slender tentacles, entrapping small particles to be carried by cilia to the mouth.

The irrigating current bringing food and oxygen is driven by bands of fast-beating cilia on gills concealed within the tube as leaf-like outgrowths from the parapodia. As well as food, the peristomial tentacles intercept sand grains, to be cemented individually around the rim of the growing tube. The abdomen forms a narrow tail turned forward under the thorax to bring the anus to the opening of the tube.

Vermetids

These are gastropod molluscs with spiral shells loosely coiled, superficially apt to be confused with serpulid worms. They can, however, be distinguished easily by their unkeeled, perfectly circular tubes, with granular or lamellar sculpture, and are stamped as gastropods by their fine-sculpted embryonic shell (*protoconch*), visible with a good lens on tubes removed from the rock.

New Zealand's single zone-forming vermetid, *Dendropoma lamellosa*, is almost confined to our eastern offshore islands. The foot is surmounted with a saucer-shaped operculum. From a large pedal mucus gland, opening below the mouth, a mesh of threads is put out

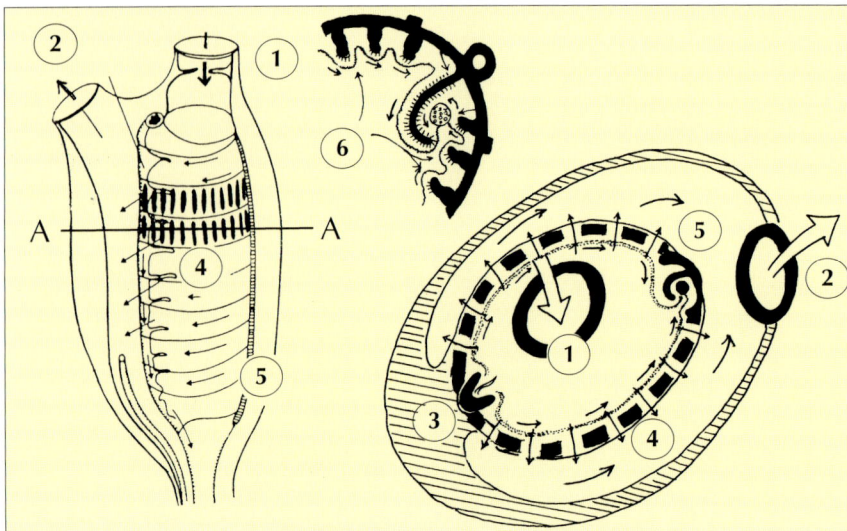

Fig. 2.9 Ascidian structure
(left) A simplified zoid seen transparently, based on *Clavelina*, with **(right)** cross-section A-A. **(centre)** Detail of hyperbranchial groove, with formation of food string. 1 inhalant siphon 2 exhalant siphon 3 endostyle 4 pharynx wall with stigmata 5 hyperbranchial groove 6 mucus sheets forming dorsal food string in groove.

to entrap plankton from currents drawn in by the gill. Such mucus traps largely supersede the role of cilia. Two slender pedal tentacles, held erect from the foot, help deploy the mucus strings, and pull them towards the mouth, where they are grappled by the small radula.

Ascidians

The sea-squirts or ascidians form the lowest tier of filter feeders between the tides. Nowhere in New Zealand do they dominate a shore zone, like the cunjevoi of east Australia (Chapter 17). But the sea-squirt *Pyura rugata* lives scattered in coralline turf and *Microcosmus squamiger* contributes plentifully to the sublittoral fringe, especially on wharves. In southern New Zealand, spectacular thickets of sea-tulips, *Pyura pachydermatina*, rise on metre-long stalks among the bladderkelp, *Macrocystis*. The greatest wealth of ascidians is, however, to be looked for under boulders and beyond low water.

The sea-squirt is enclosed in a tough jacket or *test*, made of a carbohydrate-bonded material called *tunicin* (hence the alternative name *tunicate*). Of the two siphons, the first or inhalant brings water into the greatly hypertrophied *pharynx* that serves as the feeding filter. The pharynx hangs in a surrounding cavity, the *atrium*, and its thin wall is close-meshed, like an airtex vest, with microscopic slits (*stigmata*) lined with current-driving cilia. Water from the mouth passes out through the stigmata into the surrounding atrium, and is jetted out from the exhalant siphon, by the contraction of muscles lining the test. The same expulsion shoots faeces and genital products clear of the nearby inhalant siphon.

Inside the pharynx a mucus sheet is continually secreted from a ventral groove, the *endostyle*. Spread over the perforate lining wall of the pharynx, this greatly assists particle retention. Dorsally beating cilia, and the raking movement of the internal longitudinal

bars of the pharynx, pass the mucus with food particles dorsally, to be rolled into a string by the cilia of the *hyperbranchial groove*. In this groove the cord is passed back to the opening of the oesophagus.

Grazing molluscs

The familiar herbivores of the rocky intertidal include the more primitive gastropod molluscs and also the chitons. Though not themselves to be called zoning species, they are all slow-moving, with some returning so regularly to a 'home' site as to serve as faithful markers of a visible zone of the shore.

The topshells (Trochidae), and turbans (Turbinidae), are not quite the earliest of the prosobranch gastropods; but they preserve what must have been the original feeding habit, in rasping off living algae or raking up organic detritus with the *radula*. This is a basic molluscan organ carried within the buccal mass upon a tongue (*odontophore*), as a chitinous strip bearing rows of microscopic teeth. The whole tongue is rolled forward as the mouth opens, and its two sides draw apart to expose the radula. The teeth engage on the substrate during the recovery stroke, to bring into the buccal mass their load of scrapings.

All the grazing molluscs — chitons and gastropods — feed in this basic way. Their differences are in the size, strength and number of the teeth and the types of food to which they thus have access.

Soft browsers: Archaeogastropoda

This is the first and oldest order of the prosobranch gastropods. The gills (*ctenidia*) are primitive and *bipectinate*, having a row of filaments along both sides of the axis. Eggs and sperm are shed directly and fertilisation is external. The larvae have a short or extended life in the plankton.

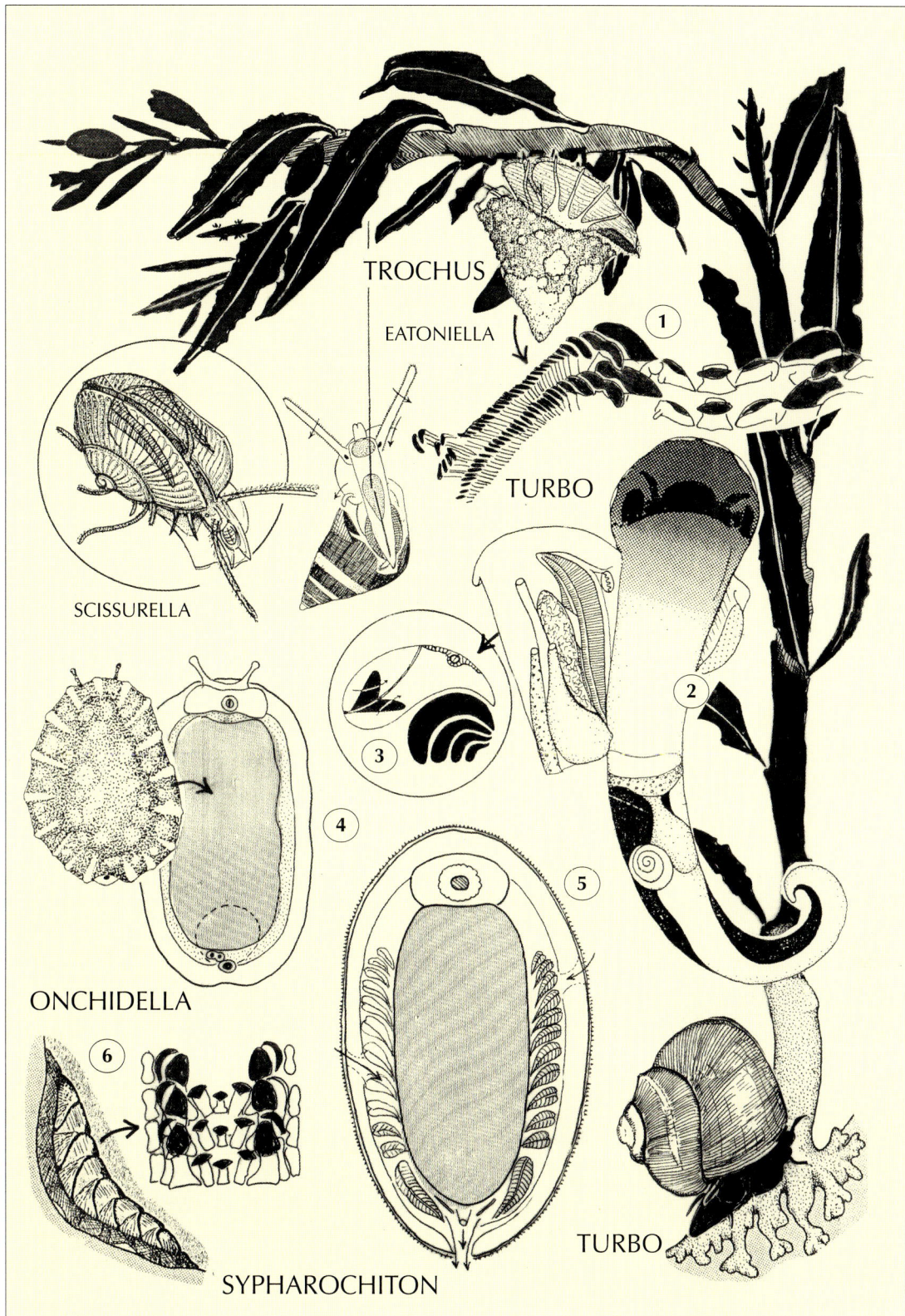

Fig. 2.10 Browsing and grazing molluscs

The algal browsers, not to the same scale, are represented on the fronds of *Carpophyllum flexuosum*.
1 *Turbo smaragdus*, two rows of radular teeth 2 *Turbo*, with mantle cavity opened to show gill and
other pallial organs (**left**) 3 *Turbo*, transverse section of mantle cavity with bipectinate gill
4 *Onchidella*, ventral view with posterior pneumatostome and site of 'lung' (**broken line**)
5 *Sypharochiton*, in schematic ventral view (gill number reduced) 6 radula teeth of *Sypharochiton*.
(With acknowledgements to John Walsby.)

The earliest gastropod radulae are called *rhipidoglossan*, from the way each tooth row is widened like a fan. Numerous slender marginal teeth browse the tissues of algae as the radula sweep and graze over broad surfaces without need of strong abrading.

The large Family **Trochidae** includes the common topshell, *Melagraphia aethiops*, familiar throughout New Zealand. Like all trochids it is to be recognised by the circular operculum which is thin and horny. *Melagraphia* is a snail of the upper and middle eulittoral, and browses widely over the open surface.

Our most widespread member of the **Turbinidae** is the cat's-eye *Turbo smaragdus*, living on coralline turf in the lower eulittoral, in algal-fringed pools and on large brown algae right to low water. It is distinguished at a glance by the convex shelly operculum, marked with emerald green. The newest *Turbo* settle as post-larvae on to *Corallina*, feeding on its fine epiphytic algae. Such juveniles have three nodulose ridges, later to be lost in the smooth adult. Becoming too large to shelter in the turf, shells are washed off by waves to settle on the open rock. The largest *Turbo,* forming breeding populations four years old, are found upon *Ecklonia* and *Carpophyllum* at the bottom of the sublittoral fringe. In February, males and females simultaneously release gametes, triggered by the stimulus of onshore wind and storms. The larvae swim for a couple of days, allowing dispersal by rough water before settlement onshore.

The warm-water Family **Neritidae** has one New Zealand species, *Nerita atramentosa*. The nearly hemispheric shell is ebony-black with a porcelain-white base, and the semicircular aperture has an orange, granulated operculum. This species reaches higher upshore than the trochids, and ranges through east Australia into the warm Pacific, and reaches appreciable numbers in New Zealand only in the north of the North Island. Like the periwinkles, *Nerita* is a gleaner of blue-green algae and lodged organic particles.

The Neritidae are unique among Archaeogastropoda in having complex genital ducts to transmit and receive sperm, and to enclose the eggs in tough capsules with calcified lids, to be found freely scattered underneath high level boulders.

Forebears

These spirally coiled archaeogastropods have some limpet-like forerunners still represented on the shore. The primitive slit and keyhole limpets (**Fissurellidae**) differ from the true limpets that carry a pair of equal bipectinate ctenidia. Only when the trochids and turbinids changed their bilateral symmetry for a spiral coil was the right gill of the primitive pair eliminated for lack of space. The Fissurellidae are hardly to be seen on our zoned surfaces, but in Pacific America (chapter 19) they are large and prominent. The small fissurellid *Montfortula rugosa* is found here in high level limestone pools notably at Kaikoura (Fig. 7.29). *Tugali* and *Scutus* (Fig. 13.8) live under boulders.

The most minute of the fissurellids, *Incisura rosea* and *I. lyttletonensis* (Fig. 8.7), though seldom noticed, are in fact easy to collect by careful washing of *Carpophyllum plumosum* or red algal fronds. Viewed by transparency with a binocular microscope, they reveal all the workings of the primitive, two-gilled mantle cavity. Both carry through life the spiral beginning that larger fissurellids lose, being in fact like tiny ormers, of the close-related **Haliotidae**. In these last the fissurellid exhalant slit has been converted to a line of holes.

Hard grazers: littorines

Virtually all over the world, the periwinkles (**Littorinidae***)* are the regular gastropods of the highest shore. The littoral fringe was in fact first designated by the Stephensons as the 'Littorine Zone'. All through New Zealand the common periwinkle at a high level in the fringe is *Austrolittorina antipoda*. Further south it is joined by the larger *A. cincta*.

The littorine radula is of the narrow *taenioglossan* type, standard in the Order Mesogastropoda, with seven teeth to a row, a central flanked by laterals, then two marginals (Fig. 2.11). Their tips are reinforced with chitin to rasp the hard surface, rather then sweep or scour. Just as with limpets, worn teeth must be rapidly replaced from a reserve length of radula in a long, spirally coiled caecum.

Only briefly in contact with water, the littorines have reduced the gill, so the mantle cavity is in effect a lung periodically opened to the spray-damp atmosphere. But full terrestrial evolution of the littorines has been precluded by the needs of their larvae.

Part of an ancient stock of mesogastropods, the littorines have like the nerites evolved internal fertilisation. Though the *Risellopsis varia* lays spawn jellies, our New Zealand *Austrolittorina* species shed fertilised eggs straight into the sea where the larvae are free-swimming.

Limpets

Low-pitched and conical, the limpets are the gastropods most at home under high wave exposure. The limpet shape has been resorted to many times. The true limpets (**Patellacea**) stem from spiral archaeogastropods, and begin life with a transiently coiled larval shell, before becoming cap-shaped. The foot takes a firm grip of the rock, drawing the shell down by its strong retractor muscles. In moving about, the shell margin is raised, revealing the mantle skirt with its fringing tentacles in

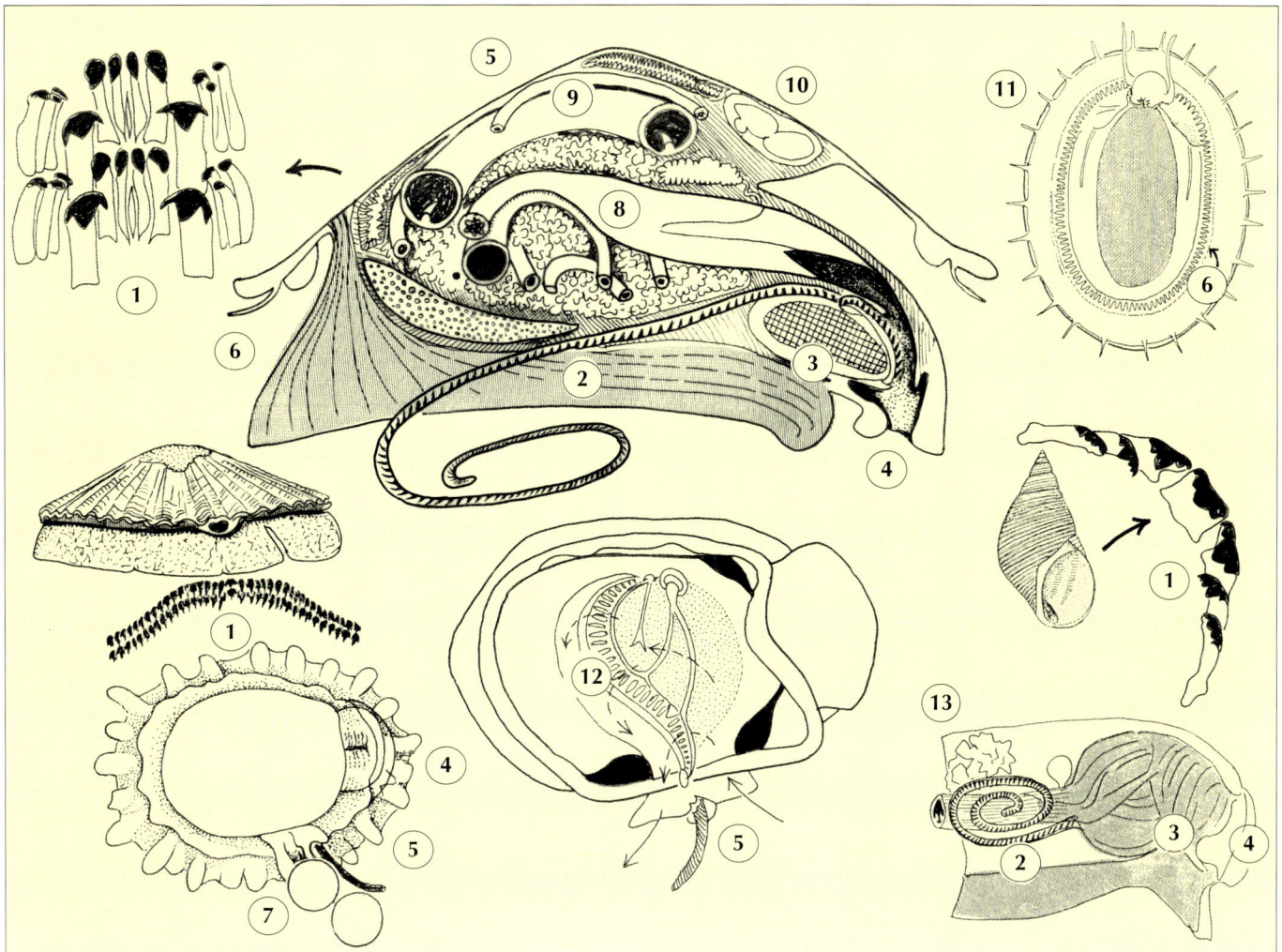

Fig. 2.11 Limpet and periwinkle structure
(**top**) *Cellana* in longitudinal section (**centre**), with radula detail (**left**) and undersurface (**right**).
(**bottom left**) *Siphonaria* from right side and undersurface, with radula detail below.
(**bottom centre**) Pallial organs as seen by transparency from above gill heart pneumatostome.
(**bottom right**) *Austrolittorina*, shell of *A. cincta*, with radula detail and section of head showing buccal mass and radula sac. 1 radula teeth 2 radula sac 3 buccal mass 4 mouth 5 anus 6 circumpallial gills 7 pneumostome (emitting air bubbles) 8 crop 9 stomach 10 heart 11 osphradium 12 gill seen through pallial roof 13 salivary gland.

all-round contact with the ground. The eye at the base of the head tentacle is no more than a simple pit.

The true limpets are divided between two families, **Patellidae** and **Acmaeidae**. Those of the former are the largest-sized and belong in New Zealand to the genus *Cellana*. Both families have certain specialisations built in to a basically primitive ground plan. The Patellidae have entirely lost the true gills (*ctenidia*), replacing them with a cordon of simple flaps just inside the mantle edge. Round the whole circumference inhalant currents are drawn by cilia between these lamellae, with a single exhalant point on the right side. The smaller limpets of the Family **Acmaeidae** lack the cordon of adaptive gills and retain a single bipectinate ctenidium.

On rocky shores right round New Zealand, the most widespread zoning limpets are *Cellana ornata*, reaching

high in the barnacle zone, and *C. radians*, ranging from the middle eulittoral down into the coralline zone. Other *Cellana* species are more geographically restricted.

A virtual limpet habit has also been evolved in the higher gastropod Family **Siphonariidae**, marine pulmonates where the mantle cavity not only serves as a lung but has re-acquired a secondary gill. Several features at once set off the siphonariids from true limpets. First, the foot can hardly be enclosed by the shell, and with its weaker suction comes off easily from the rock. On the left side, the shell has a triangular lip, beneath which is the narrow opening of the mantle cavity.

More than true limpets, siphonariids tend to line up in damp crevices. As in all pulmonates, the broad radula has small uniform teeth, adapted not for abrasion but for browsing succulent algae. Being pulmonates

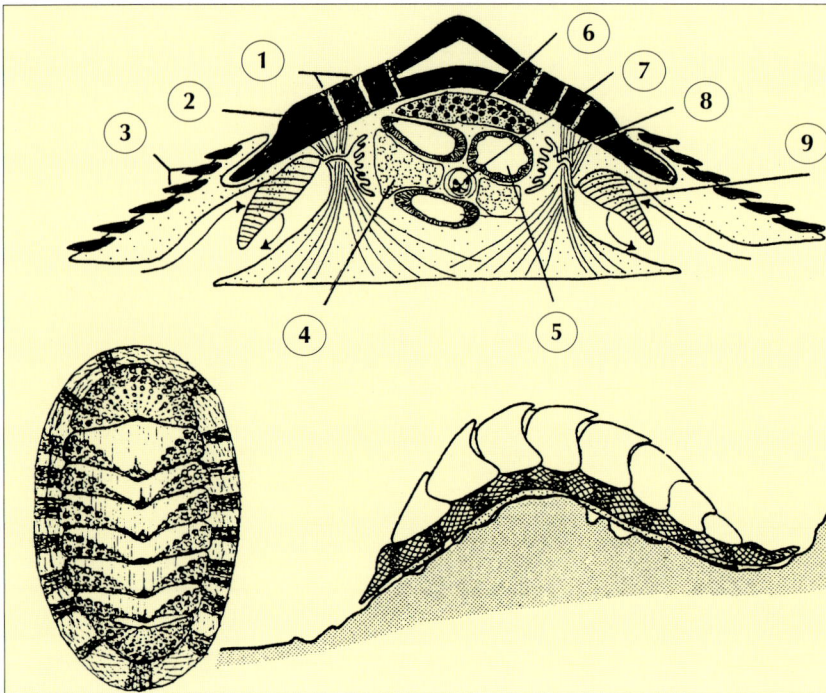

Fig. 2.12 Chiton structure

Sypharochiton pelliserpentis in dorsal and lateral view; and in transverse section (**top**).
1 aesthete sense organs 2 shell plate 3 girdle scales 4 digestive gland 5 stomach 6 gonad 7 radula sac 8 renal organ 9 gill.

(After John Walsby.)

Siphonaria limpets are hermaphrodite and reproductively advanced. The internally fertilised eggs are laid in a jelly crescent, attached to moist rock.

A more aberrant pulmonate, the oval slug *Onchidella* may be likened to a naked limpet, having instead of a shell a thick rubbery integument (*notum*), covered with small papillae. The common species *Onchidella nigricans* is black to olive green, often marbled with brown and grey. This slug crawls freely over shaded rock, gleaning diatoms, minute algal remains and organic detritus with the delicate radula. The mouth lies in front of the foot-sole, flanked by small head tentacles. At the posterior end, the dorsal skirt lifts up to reveal the rhythmically opening lung pore, just above the anus. The skin resists evaporation, and its small glands produce a milky secretion repellent to attackers.

Chitons

Chitons (**Polyplacophora**) are far more archaic molluscs than even the earliest gastropods, but in habit and diet they have much in common with limpets. They all possess eight articulated shell plates, enabling the flat body to adapt flexibly to irregular surfaces or, if detached, to roll protectively into a ball. A marginal girdle — in some chitons studded with small scales — holds closely to the rock, but can be lifted at any point to create a temporary inhalant path. The exhalant channel, carrying out faeces and renal and genital products, leaves at the posterior end. The foot is narrowly oval with the mouth in front and the anus behind. A deep groove at either side contains a row of gills, up to 70 pairs, each one equivalent to a single ctenidium.

The commonest chiton on open rocks is the snake-skin *Sypharochiton pelliserpentis*, ranging from high to low water neap and from extreme wave exposure to harbour backwaters. In the barnacle zone, large and deeply eroded specimens rest in scars like limpets. Younger individuals are radially ribbed, dull olive green, with the second valve often black. Low tidal *Sypharochiton* are of a distinctive brownish black, light-streaked, with the girdle banded in light blue and black. Traditionally separated as *Sypharochiton sinclairi*, these populations live in shade, under brown algae or boulders or on coralline paint, and are today considered a form of the single, highly polymorphic species *S. pelliserpentis*.

In the large, scar-forming individuals at high level, erosion of the valves has removed the microscopic shell receptors, sensitive in most chitons to light and heat. This loss permits the older chitons to frequent the open surface, homing like limpets. Further downshore, under stones, among oysters and in pools, and on coralline paint, lives a mixed population of medium-sized uneroded form *sinclairi*, along with many juveniles. These were found not to 'home', but merely to return by a negative light reaction to dark places. That response is weakened during submersion; when splashed by waves the chitons wander about freely. Unlike those of the upper shore, they have little resistance to desiccation.

Behaviour

Fodder for grazing molluscs comes in various kinds and sizes. Different herbivores are adapted to exploit all these efficiently, while each maintaining its optimal level on the shore. In high wave-exposure, limpets are almost the sole herbivores, though littorines can always obtain

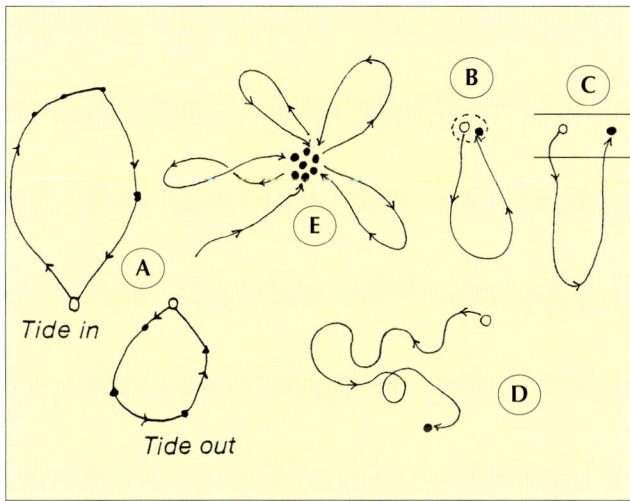

Fig. 2.13 Homing behaviour of gastropods and a chiton

A. Accurate return to home site or scar (most limpets, *Siphonaria* and *Sypharochiton pelliserpentis*).

B. Return to original area without accurate home locating.

C. Return to a given shore level after feeding forays (*Austrolittorina antipoda, Risselopsis varia, Diloma zelandica, Turbo smaragdus*, thaids, *Notoacmea pileopsis*).

D. Haphazard movements (*Melagraphia aethiops*).

E. Return to large aggregations on a home site. As well as a desiccation protection, this is an efficient way of sharing out the grazing with whole populations moving nomadically to a different site (*Nerita atromentosa, Onchidella nigricans*, as well as *Lepsiella scobina*, grazing on barnacles).

(From field drawings by Terry Beckett.)

micro-shelter in crevices or depressions. The size of a herbivore also governs its access to food. On rough surfaces, large limpets or chitons can scrape directly across the top of ridges. But a flexible chiton is also able to reach into concavities, where a large limpet is confined to level surfaces. A small snail can moreover graze between the ridges on finer food that a *Cellana* could not crop, while the smallest acmaeid limpets spend all their time scouring crevices. A niche closed to one species can thus offer a food cache to another.

At low tide, especially in the daytime, grazers must protect themselves from drying out. Many species stop feeding and stay motionless, holding a reserve of water under the shell. Cat's-eyes (*Turbo smaragdus*) line up in crevices. *Nerita atramentosa* cluster in large formations, neighbouring shells retaining water between their shells as well as underneath. The *Austrolittorina* species range too high to risk the water loss entailed in staying attached by the foot. Instead they retract, close the operculum tight, and either fall off the rock or attach just by

dried mucus on the shell lip. Thus protected, *Austrolittorina antipoda* can resist desiccation for several weeks at a spell. After a month kept dry in a matchbox, specimens re-activated and crawled away in half a minute at the first return of splash.

Terry Beckett in 1967 made a valuable study of the activity patterns of grazing molluscs. Most of these species disclose some homing behaviour, of greater or less precision. After feeding, *Nerita* and *Onchidella* were found to return to the same protective aggregations. One of the few species entirely haphazard in movements is *Melagraphia aethiops*, the only trochid species permanently at home on emersed surfaces.

Nearly all grazers were found to return to a given vertical level on the shore after feeding excursions. These include *Austrolittorina antipoda, Risellopsis varia*, the high level limpet *Notoacmea pileopsis, Turbo smaragdus*, and the trochid *Diloma zelandica*. Return to a standard level, though not to a specific 'home' is also shown by the carnivorous thaid snails *Lepsiella scobina* and *Haustrum haustorium* (Fig. 2.13).

Limpet homing

Many sorts of limpet show more precise degrees of homing behaviour. *Cellana radians* will return to its local area without actual homing to a scar. *Siphonaria australis* can also home accurately, as has been observed also for a number of Onchidiidae overseas.

The most accurately homing New Zealand limpet, *Cellana ornata*, was found by Terry Beckett to perform oriented movements whenever submerged. Every time the tide receded, individuals were back on their own scars, except sometimes under humid conditions at night. Wetting by the incoming tides at once initiated feeding movements. Individuals high on the shore showed a faster response than those lower down, moving to a level above their home, following the tide upwards and retreating with it at the ebb. The return might follow a different path from the outward foray, terminating as the ebbing tide passed below the level of the 'home'. At night *Cellana ornata* can move over damp surfaces with the tide out, often going below their home to take advantage of the new grazing and reduced desiccation downshore. If daybreak came before the next high tide, all the limpets were found re-sited by the next sunrise. Limpets covered by the tide just before daybreak occasionally remained off their homes. Return to site by *Cellana ornata* was always accurate, with limpets often crossing the home scars of other individuals on the way.

The sensory basis of limpet homing remains still one of the mysteries of animal ethology. The British limpet *Patella vulgata* was claimed by the ethologist W.H. Thorpe to be completely oriented throughout, appearing

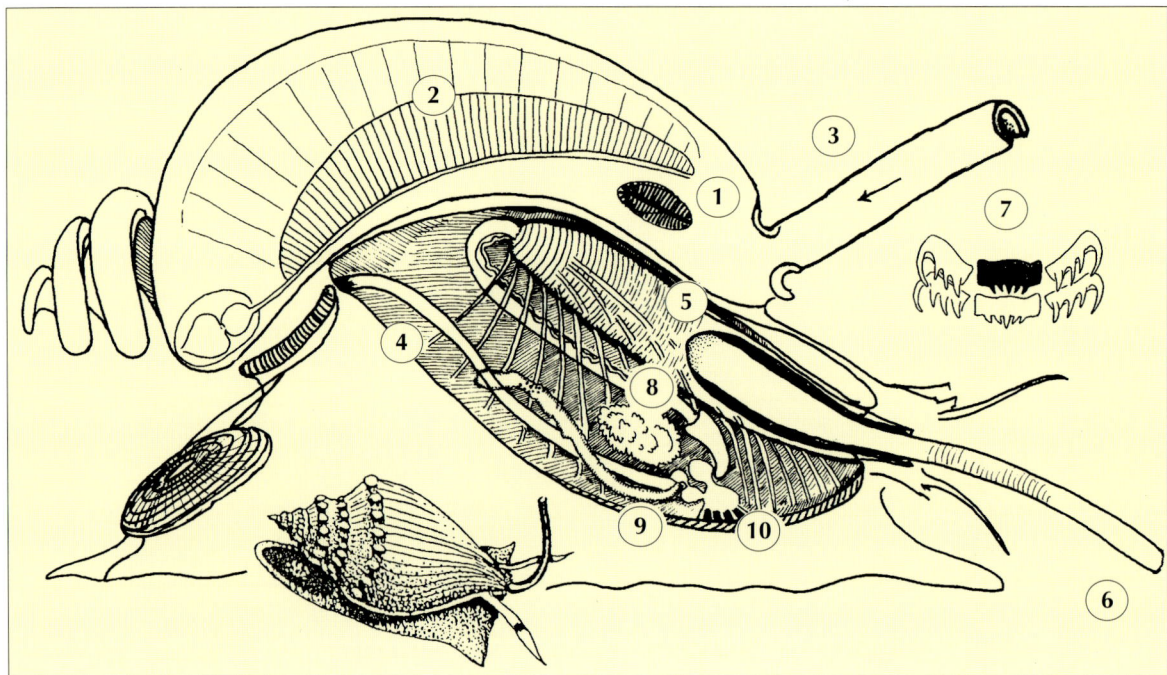

Fig. 2.14 The structure of a whelk
Based on *Cominella glandiformis* (shell, **lower**) dissected to show mantle cavity and anterior digestive system.
1 osphradium 2 ctenidium 3 inhalant siphon 4 oesophagus 5 proboscis sheath 6 proboscis everted 7 tooth row of radula 8 salivary gland 9 accessory salivary gland 10 circum-oesophageal nerve ring.

to show full appreciation of topography and environment, rather than depend on a single guiding stimulus. That limpet was often seen to shuffle and turn, as if trying to fit its shell into a precise scar. *Cellana ornata* made none of these tentative trials, but resumed its scar promptly, often coming to rest 180 degrees out of alignment. Even when the shell margin was chipped away, the return was unerring, negating the idea that the shell must exactly fit the scar. Nor did enlarging or altering the scar prevent a successful return. Removal of a scar with a chisel while the limpet was away resulted in a return to the general home area, with searching for several minutes before settling in the near neighbourhood.

Gastropod predators

The major predators of the sessile zoning animals and herbivorous molluscs are the carnivorous snails of the Order **Neogastropoda**; on almost every temperate and tropical hard shore, these are best represented in the **Muricidae**, notably in its Subfamily **Thaisinae**. The thaids are typically slow-movers, taking barnacles, mussels, oysters or other gastropods as their shelled prey. Like the grazing gastropods, they tend to stay faithful to their particular levels on the shore. A second family of predators, almost as familiar between tides as the thaids, is the Buccinidae or true whelks. Faster-moving than the

thaids, they feed chiefly on freshly dead carrion and moribund animal matter of many kinds.

The carnivorous gastropods present some notable differences from the grazers. First, the shell is longer and spindle-shaped (fusiform), generally with a tapered spire. Second, the mantle skirt is drawn out as an incomplete tube, forming the long inhalant siphon extended through the anterior canal of the shell. This serves as a movable nostril, ranging from side to side as the whelk advances. The incoming current, drawn by the gill, first flows across a pallial chemoreceptor, the osphradium, which lies in the mantle cavity just behind the siphon base, like a small pseudo-gill.

Third, the carnivorous gastropods have a long proboscis able to be rolled out or retracted like the eversible finger of a glove. Fully extended it has the mouth at its extremity, leading to the buccal mass with its radula. This tooth ribbon is of *rachiglossan* type, with three sharp-cusped teeth in each row, a central and two marginals. As well as for macerating soft tissues the thaids — and all muricids — employ the teeth to bore the shells of their prey.

Thaids are typically intertidal, with thick, blunt-pointed shells, engage in slow predation on shelled prey close at hand, generally using the radula to drill an opening to insert the proboscis. Though time-consuming, this brings ample rewards. The drilled hole is tightly

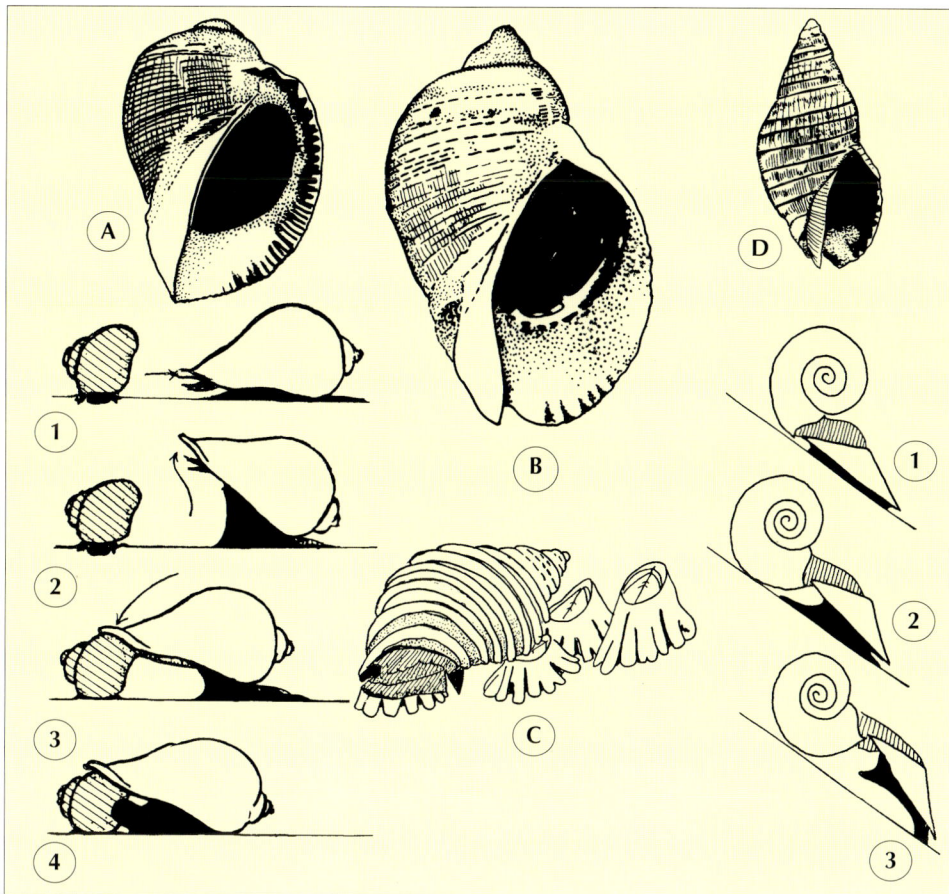

Fig. 2.15 Three predatory gastropods

(A, **left**) *Haustrum haustorium* in four stages of attacking *Turbo smaragdus*.

(B & C, **centre**) *Dicathais orbita* with *Epopella plicata*.

(D, **right**) *Cominella virgata*, in three stages of raising a limpet *Cellana*.

(By courtesy of John Walsby.)

filled by the proboscis so no body fluids leak to attract scavengers. Food being eaten is thus kept alive, with less essential organs such as the digestive gland consumed first. A special gland, the organ of Leiblein, secretes an acidic, shell-dissolving saliva, while the sharp radula teeth finish off the neatly bevelled hole.

The feeding habits of the three common New Zealand thaids have been studied by John Walsby. The smallest and most numerous is *Lepsiella scobina*, feeding when young mainly on the barnacles *Chamaesipho columna* and *Austrominius modestus*. *C. brunnea* and *Epopella plicata* escape predation because their wave-exposed sites are generally too precarious for *Lepsiella* to hold on. Older and larger *Lepsiella*, up to 15 mm long and more apt to be washed off by waves, are carried down from barnacle level to where oyster clusters give more shelter. Walsby found that at 10 mm, a *Lepsiella* will eat four *C. columna* a day. A large 25 mm whelk may take two days to drill through the shell of a mature oyster. *Lepsiella* also feeds on the tubeworm *Pomatoceros caeruleus*, the worm-like gastropod *Stephopoma roseum* and the flea mussel, *Xenostrobus pulex*.

Haustrum haustorium is a larger and shorter-spired thaid, with a rounded shell and wide aperture. Juveniles up to 15 mm live low on the shore in quite open areas where the small limpets *Patelloida* and *Siphonaria* abound. *Haustrum* feeds by flipping these over, sometimes first drilling the shell to kill or relax the limpet. Adolescent and adult *Haustrum* move up to the middle shore where they attack grazing gastropods: mainly *Turbo* and *Nerita*, but also *Melagraphia*. The *Cellana* limpets, *Sypharochiton* and *Lepsiella scobina*, can be taken too.

Haustrum is essentially an opportunist, waiting in shelter for the prey to come within range of its chemo-detection. As the concentration of stimulus increases, it judges when to strike, lifting the shell to position its wide lip over the approaching prey. The rounded aperture neatly fits onto a *Turbo*, *Nerita* or *Melagraphia*. With the shell over the prey, the thaid moves forward to grasp it with the front of the foot, then lets go of the ground and violently twists the prey to loosen its foothold (fig. 2.15). Its shell is then turned aperture up, so the operculum can be dislodged. Quite rapid with horny opercula, this takes longer with the shelly operculum of *Turbo* which is

45

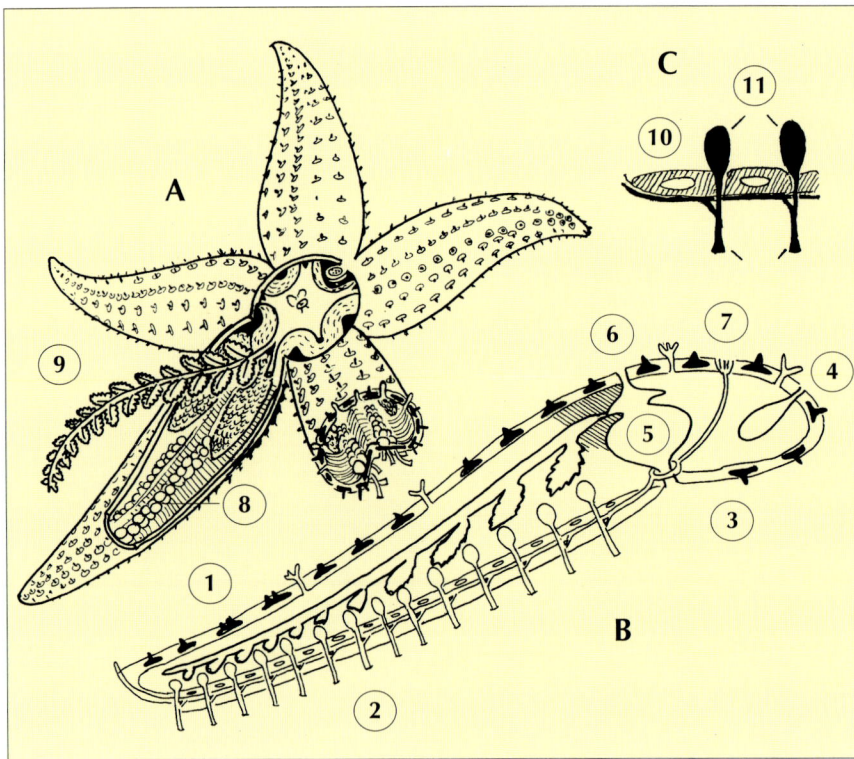

Fig. 2.16 Seastar structure
A. An asteriid seastar dissected to show a digestive diverticulum and gonads.
B. A radial section showing the structure of an arm and an inter-arm.
C. Detail of tube-feet.
1 papula 2 tube-feet 3 dermal skeletal plate 4 genital pore 5 stomach 6 anus 7 madreporite 8 gonad 9 digestive diverticulum 10 skeletal plate 11 ampullae.

sometimes bypassed by boring at the edge. Alternatively, a *Turbo* shell may be repositioned so it can be bored in its thinnest part at the inner lip. The proboscis is then withdrawn and the drill-hole covered by the foot to prevent release of chemo-stimuli to attract rival predators. Feeding then proceeds through the aperture.

The thaid lowest on the shore is *Dicathais orbita*. The juveniles feed at coralline level, flipping limpets over or dislodging the opercula of topshells or *Turbo* more expertly than *Haustrum*. Over 35 mm they no longer take gastropods, but move upshore to exploit surf barnacles, *Epopella plicata*, which are smothered by the foot, then drilled between the opercular valves. Empty shells can be found purple-stained with a secretion of the thaid that may have operated as a relaxant. *Dicathais orbita* can also be found on surf coasts, where other whelks cannot survive, feeding on the green mussel *Perna canaliculus*.

The true whelks, **Buccinidae**, include several species of *Cominella*. All these are highly mobile scavengers, also taking live prey as opportunity offers. Unlike thaids, they rove widely, swinging the inhalant siphon as a nostril and orienting to chemo-stimuli. Approach to food is rapid, and more mobile scavengers such as hermit crabs are often bulldozed away. Our several intertidal Cominellidae have overlapping ranges, from moderate exposure through to shelter. *Cominella virgata* is typical of cleaner shores, detaching limpets by hooking the shell lip under their margin, with the foot attached to both the prey and the ground. Rolling about its own axis, the whelk then levers off the limpet by a steady stretch on its foot, until it lets go of the ground.

Cominella maculosa and *C. adspersa* often occur together, in more sheltered sites. The first may extend on to silty sandflats, finally to be superseded on truly muddy shores by the smaller estuarine species, *Cominella glandiformis*.

Stars and urchins

The two sorts of echinoderm visible on our zoned shores — asteroids and echinoids — feed quite differently. But this whole Phylum Echinodermata is still so unified in basic plan that the two can be presented together. The asteroids and echinoids (as also brittle stars, sea cucumbers and crinoids) all share the spiny calcareous skin plates that give this phylum its name. In sea urchins (Class Echinoidea) long spines are the outstanding feature. In the starfishes (Asteroidea) smaller spinelets project through the skin.

Adult echinoderms are fundamentally radial, with a five-sided (pentamerous) symmetry. On such a ground plan, however, they ring nearly every possible change, as with spheres, discs, cups, cylinders long and short, as well as stiff and flexibly armed stars.

With a morphology so strangely unique, echinoderms could surely never have been predicted, had they not existed. First, their most unifying feature is the hydrostatic system of water vascular canals. Lacking conventional blood vessels, they have established a water circulation instead, filled up from outside through a sieve-plate, the madreporite, close to the anus on the aboral surface. From a ring-canal round the oesophagus, water is distributed along five radial canals (ambulacra).

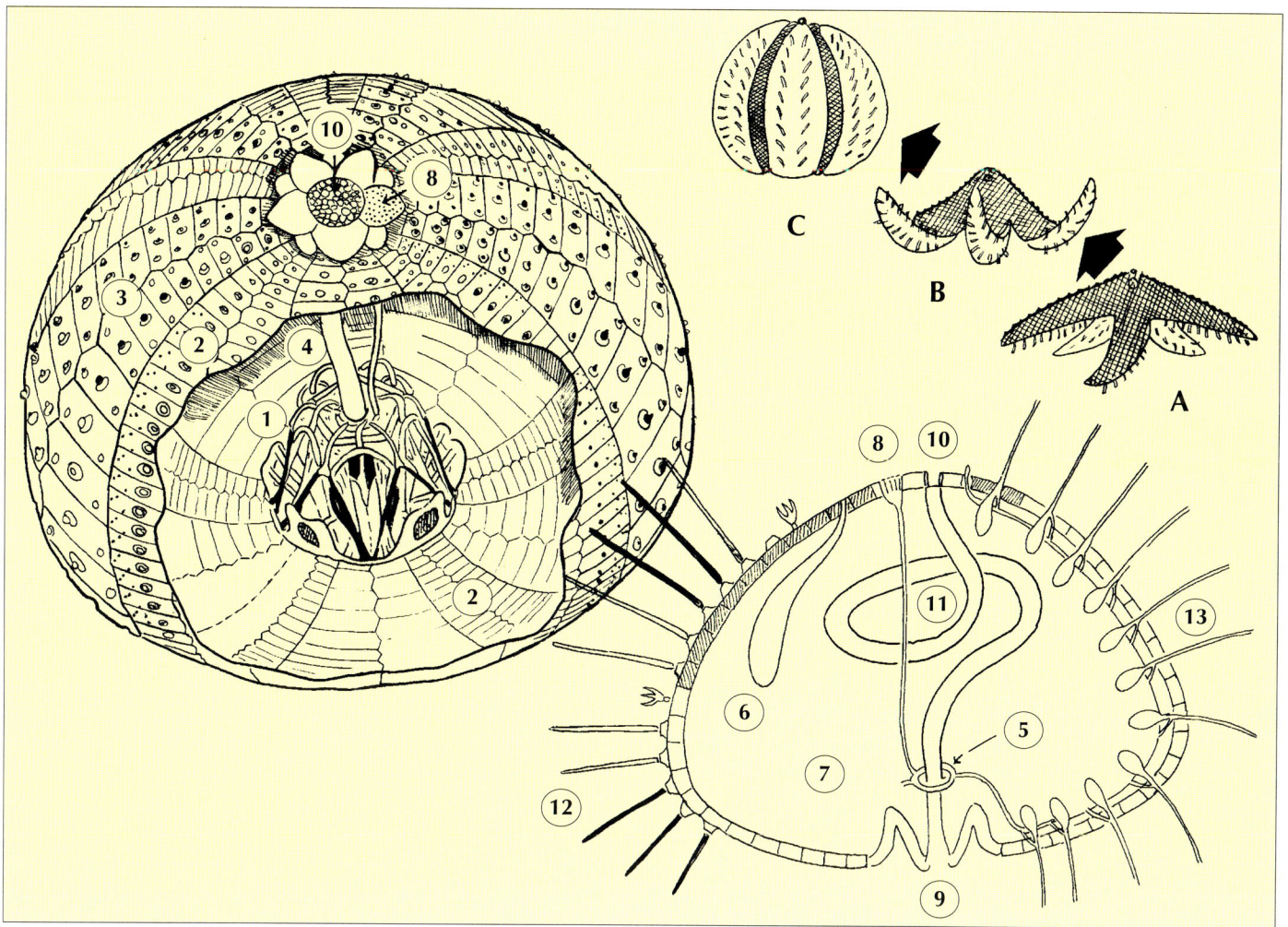

Fig. 2.17 Sea urchin structure
(top right) Suggested relation of an asteroid seastar to an echinoid sea urchin.
(left) Corona opened to show internal structure; **(lower right)** diagrammatic vertical section.
1 Aristotle's lantern 2 ambulacral plates 3 inter-ambulacral plates 4 oesophagus 5 ambulacral tube 6 gonad 7 coelom
8 madreporite 9 mouth 10 anus 11 intestine 12 spines 13 tube-feet.

From these open the numerous tube-feet (podia), each normally with a terminal sucker and a small water reservoir (ampulla). By the entry or withdrawal of water, their alternate extension and retraction of the podia is brought about, with fine intrinsic muscles controlling their direction and posture.

The **Asteroidea** or starfishes are basically flat with the radial arms (there may be more than five) all of one piece with the central disc. At the top of this disc lies the anus, and beside it the madreporite, together with five genital pores. Underneath are the central mouth and the radial ambulacral grooves, with their rows of tube-feet running beneath each arm. The tip of the arm has a light-sensitive pigment spot.

The commonest shore starfish is the cushion star *Patiriella regularis*, grazing over coralline turf, in pools or under stones. There are five short arms hardly emergent from the pentagonal disc. Rough with spinelets, the aboral surface varies from purplish brown or orange to blue grey, with the underside pale. *Patiriella*, unlike most starfish, is not a carnivore, but grazes over algal films, extruding the stomach to digest the food externally, then withdrawing the whole bag through the mouth.

On wave-beaten shores, particularly west coasts with the green mussel, *Perna canaliculus*, may be found the large ochre or pale mauve star *Stichaster australis*. The 10 to 12 arms are stiffly flexible and triangular in section, with their aboral surface beset with pin-head granules. Virtually unbreakable in its hold, *Stichaster* pulls down to the rock at each wave-break. Like any asteroid, it moves by the harmonious pointing of the tube-feet in a direction set by the currently leading arm. Locomotion is at these times under central nervous control. But on being pounded by a wave, all such total co-ordination breaks down, with the starfish becoming a 'republic of reflexes'. Each tube-foot is now left locally autonomous, shortening and tightening solely for hanging on. Tube-feet will now tear out of the animal rather than relax their grip.

Fig. 2.18 Sea urchin and asteroids
1 kina, *Evechinus chloroticus* 2 *Coscinasterias muricata* 3 *Stegnaster inflatus*
4 *Patiriella regularis* 5 *Stichaster australis*.
(By courtesy of Michael Miller.)

Stichaster is a predator of mussels. The tube-feet apply their combined grip as the star arches over a shell. With the opposing forces tugging each valve the adductor muscle sufficiently relaxes for the star to insert its everted stomach between the shell valves.

The **Echinoidea** are represented on our rocky shores only by the spherical or regular urchins. The urchin's morphology can be derived from a starfish by imagining the five triangular arms to have been drawn up and sewn together. The resulting sphere has exposed the oral surfaces of the star, with a small circle left at the upper pole, for the anus, madreporite and genital pores. The five meridians of the sphere are constituted by the ambulacral plates, carrying the pores for the tube-feet. These plates alternate with wider inter-ambulacral rows. The plates of both series are firmly sutured and bear small bosses where spines are attached, moved by their own extrinsic muscles. Both plates and spines — it must be remembered — constitute an 'endoskeleton', covered all over by the epidermis with a simple nervous network at its basement.

The kina, *Evechinus chloroticus*, is a secretively low tidal urchin. It can, however, make wide sorties into eulittoral coralline pools, as centimetre-wide juveniles up to 10 cm adults. Kina nestle in concavities, hollowed out by abrasion from the spines. Intermingled with the dull green spines are the long red tube-feet tipped with suckers. As well as attaching like guy-ropes, they also hold on to bits of pebble, shell and algae used as camouflage. Locomotion is primarily by the tube-feet, aided by the poling action of the spines.

Sea urchins graze on coralline or more delicate algae, as well as kelp fronds. Their abrading organ, known as Aristotle's lantern, is a structure as unique in its way as the radula in molluscs. Five jaw-pieces converge to act as a nest of chisels just inside the mouth. At the top of the lantern, these are linked by pieces called rotulae, overlaid by five radial compasses. Jaw retractor muscles can pivot the lantern in any direction as the mouth opens. Other muscles co-ordinate the opening and closing of the jaws, and also up-and-down movements of the lantern to circulate respiratory fluid in the surrounding coelome.

The internal morphology of a regular urchin is shown (Fig. 2.17) for comparison with that of a *Stichaster* starfish. Grazed food is carried from the lantern by peristalsis through the muscular oesophagus. Water passes in too, but short-circuits the stomach and arrives directly in the intestine by way of a narrow bypass, the siphon.

CHAPTER THREE
Some Basic Zoning

Shores in moderate exposure

Presented here are five New Zealand shores chosen to encompass the geographical changes in zonation from north to south. Each lies in a major embayment, so that — while waves are always significant — they are of less effect than on a fully open coast. These are all, in effect, shores of *medium energy*.

In such 'average' situations — so to speak — the biological zones can still be well enough aligned with particular levels of the tides. The *eulittoral zone* may be reckoned as delimited by *average high* and *average low* water, about half-way between neap and spring. Above it, the *littoral fringe* will be thus inundated only on days with higher than *average* tides. Conversely, the *sublittoral fringe* below will be exposed to the atmosphere only at tides lower than the month's average. This broad relation of zoning with the tides can be a useful rule of thumb for the intermediate shores we are about to see. A visit to a truly 'exposed' shore will show how inadequate it then becomes.

Auckland: Long Bay

East Auckland's shores of moderate shelter lie mostly on platforms consisting of moderately soft layers of mudstone alternating with harder sandstone, of the Waitemata sandstone formation of Miocene age.[1]

At **Long Bay** — gazetted a Marine Reserve in 1996 — a low-inclined slope reaches to seaward from the cliff-base. Towards the middle of the shore platform, there are pedestals standing as eroded remnants of a former high level capped by a harder sandstone bed. The bench drops at its seaward edge by a couple of metres to the subtidal. At this level its soft strata are being constantly bio-eroded.

Several landmarks of the zonation at once stand out. The tops of the pedestals are white with the rock oyster *Crassostrea glomerata*, serving to fix the *middle* band of the *eulittoral*. Seaward of these stretches the expanse of coralline turf that comprises the *lower eulittoral*. Furthest out, at the seaward drop can be seen the dark, briefly emergent brown algae of the *sublittoral fringe*.

Mainly *Carpophyllum* species, these are never left fully dry even at the time of low water. At the lowest of spring tides, the tops of small kelp *Ecklonia radiata* emerge for a short time above the water surface.

High upshore, where the mudstone dries out and crumbles on hot days, we shall look in vain for a well-developed *littoral fringe*. Such soft rock is a poor settling surface for lichens and algae, and the periwinkle *Austrolittorina antipoda* is to be found only in crevices and sheltered dips.

Such poverty is not total. Soft mudstone carries the cyanobacteria *Calothrix scopulorum*, greenish black and slippery with mucilage, or pallid with salt when it dries out. The cyanobacteria *Isactis plana*, *Entophysalis deusta*, with species of *Oscillatoria* and *Lyngbya*, may be scraped off from sandstone. Freshwater seepage is indicated by winter and spring swards of green *Enteromorpha compressa* or yellow-brown tubes of *Scytosiphon lomentaria*. In August or September two primitive red algae will appear, filaments of *Bangia* and membranes of a dwarf *Porphyra*; both are short-lived and will soon be left sun-dried against the smooth rock face.

Upper and middle eulittoral

The **eulittoral** begins with the appearance in quantity of the barnacle *Chamaesipho columna*. Often fragmentary on soft rock, its zone is best developed at Long Bay on outcrops of harder sandstone, splash-exposed and courting sunlight rather than shade.

The **middle eulittoral**, around mean sea level, is a zone of attached bivalves. First, above the rock oysters, and abutting into the barnacles, grows the flea mussel *Xenostrobus pulex*. Their small, shiny black shells are bound into carpets by byssus threads that can accumulate sand and shell fragments without too much harm. They can thus tolerate periodic sand burial, but suffer most from over-treading, and prolonged heat during summer.

The rock oyster *Crassostrea glomerata* is virtually confined to the Auckland province. More than the flea

Fig. 3.1 Zoned shore at Long Bay, Hauraki Gulf

Zoning species

a–b filamentous algae, including *Calothrix*; b–c *Austrolittorina antipoda, Bangia*; c–d *Chamaesipho columna*; d–e *Crassostrea, Epopella, Xenostrobus* and *Pomatoceros* ; e–f corallines, *Jania, Hormosira* and *Laurencia*; f–g *Carpophyllum, Cystophora* and *Sargassum*; g–h *Ecklonia radiata*

1 *Nerita atramentosa* 2 *Austrolittorina antipoda* 3 *Xenostrobus pulex* cluster, with *Chamaesipho brunnea, Austrolittorina, Risellopsis varia* and *Notoacmea parviconoidea* 4 *Crassostrea glomerata* 5 *Melagraphia aethiops* 6 *Lepsiella scobina* 7 *Haustrum haustorium* 8 *Turbo smaragdus* 9 *Trochus viridis* 10 *Cookia sulcata* 11 *Carpophyllum maschalocarpum* 12 *Sargassum sinclairii* 13 *Cystophora torulosa* 14 *Ecklonia radiata*.

mussel it chooses high spots or vertical faces. Untroubled by turbidity or low salinity, oysters require only to be raised above heavy sediment. Single specimens attach flat, but clustered shells can grow upright, only lightly inter-attached, and oriented for efficient filtering. Juveniles, the size of a 10-cent coin, have dark channelled spines along the growth lines. With even moderate wave action, *Crassostrea* thins out, and at Long Bay it is near its exposed extreme. Density increases in harbours, right up to the limits of hard substrate, including wood or concrete.

Just above the oysters, or replacing them in higher wave action, grows the barnacle *Epopella plicata*. Steep-sided and ridged, it is much larger than *Chamaesipho* (that may often settle upon it). *Epopella* continues out to wave-exposed shores, where it becomes low-conical. At Long Bay, it forms a tawny honeycomb on the sandstone tops, where currents and swash are strongest. Never reaching above EHWN (extreme high water of neap tides), *Epopella* appears — unlike *Chamaesipho columna* — to need full submersion each day.

The highest sited of the eulittoral algae is usually *Capreolia implexa*; brown-hued but in fact a red, in a low sward of pinnate branchlets, reaching in shade to the top of the barnacles.

Chamaesipho columna continues downshore among the oysters, finally to cut out near the coralline turf of the lower eulittoral. But ahead of the turf the middle eulittoral generally ends with a band of tubeworms. The calcareous tubes of the serpulid *Pomatoceros caeruleus* (Fig. 2.8) come higher, building up to white crusts on harder outcrops. Where sand collects in depressions, the sand-mason worm *Neosabellaria kaiparaensis* directly follows, with upright tubes massing into thicker crusts and ridges.

Lower eulittoral

The reef platform here levels out to the *Corallina* turf that continues as the main cover as far as average low water. Mauve-grey when clean and pale at their growing tips, *Corallina* plants gather much sediment at the base, giving to the northern shore beyond the oysters its familiar dullness of hue. Their jointed, pinnate branches can best be examined with a lens from a well grown tide-pool specimen. Plants in the turf tend to be trodden over or stunted by grazing. Basal to the turf or in patches grazed bare is the pink veneer by which *Corallina* first colonises. Jointed branches soon spring up, except in the highest exposure where crustose forms entirely substitute for turf. A thin coralline 'paint' may spread up over tubes of *Pomatoceros*. On a clean shore, its extreme upper edge forms the *pink-line*, one of the constant intertidal zone-markers.

Several sessile animal species vie with *Corallina* for space. The barnacle *Austrominius modestus*, that at the lower level supplants *Chamaesipho columna,* may for a spell take over bare-grazed sites. In the longer term, the pink algal veneer will again overgrow it.

At a lower level, the orange ascidian *Cnemidocarpa bicornuta* may appear in the turf. Several sponges are also common. Over the bedrock, the dull yellow *Hymeniacidon perleve* can spread widely, revealed by its upright villi with exhalant oscula. Other sponges of the turf may include yellow *Pseudosuberites sulcatus* and orange *Polymastia granulosa*.

Several smaller brown algae come into their own upon the moist coralline turf. The straggling bushes of *Scytothamnus australis* can reach up to the oyster clumps, drying out black and brittle during a midday low tide. Tan-brown crusts of *Ralfsia verrucosa* often peel away or disintegrate at the centre. There may be also the chocolate-brown fleshy disc of *Petrospongium rugosum* and grey, suede-like cushions of *Hapalospongidion saxigeneum*.

The nearly universal associate of *Corallina* is the Venus' necklace *Hormosira banksii*, lending to the outer turf a richer hue of olive brown. Highly plastic in growth and physiologically widely adaptive, its dichotomous strings of bladders can vary in different habitats from 5 cm to as much as a metre long. Over its whole range, in New Zealand and south-east Australia, this specialised alga is the only fucoid left emersed for a lengthy time in the eulittoral.

On sheltered coralline flats, *Hormosira* bladders are goose-fleshy and fluid-filled, giving protection from water loss, and growing larger with increased emersion upshore. Where they lie in layers, the bladders beneath stay turgid during low tide, with those on top small and contracted. *Hormosira* grows best away from waves, where its carpets spread down to the *Carpophyllum* line. At the limit of wave tolerance, it forms beaded strings with small solid bladders down to 5 mm diameter. In pools *Hormosira* can reach further upshore, even to the upper eulittoral.

The coralline turf takes on changing hues from its transient seasonal algae. In spring to early summer come the dark bootlaces of *Myriogloea intestinalis*. Two saccate brown algae, each cosmopolitan, appear about the same time, first as epiphytes on *Corallina*. The fleshy golden brown *Leathesia difformis* begins solid and spherical, soon to become convoluted like a small brain. Bladders of *Colpomenia sinuosa* are thinner and smooth, first appearing on the turf as pin-head swarmers. In winter and spring appear the brown tubes of *Scytosiphon lomentaria*, and the thin ribbons of *Petalonia fascia*, both also cosmopolitan. Towards the seaward edge, the turf can become pink with *Jania micrarthrodia* and the slightly cartilaginous rosettes of *Laurencia distichophylla*.

Green algae add some brighter effects. In cooler months the turf carries little tubular *Blidingia minima*, and there is sometimes the harsher green stubble of *Cladophoropsis herpestica*. Much softer are the cushions of *Boodlea mutabile*, formed of interlaced networks of microscopic branchlets. On the lower shore *Codium convolutum* is highly visible, beginning in the lower reaches of the corallines and entering the sublittoral fringe. Bottle green, with a soft, velvety surface, this plant forms thick-folded fleshy pads, up to 10 cm across. It regresses in winter, to reappear as circular discs in spring, building up by autumn into a clear-cut zone, sometimes a metre wide.

Sublittoral fringe

The final threshold of the zoned shore is at the *brown-line*, where large brown algae take over from the coralline turf, to continue well beyond low water. All around Auckland, these are chiefly of *Carpophyllum*, a genus peculiar to New Zealand, with three of the four species dominating in the middle range of exposure. Each has a small branching holdfast and a flattened, leaf-bearing axis with bladders and fertile branches in the leaf axils.

The widest spread species in the north is *Carpophyllum maschalocarpum* (sometimes called flap-jack) which is found everywhere save at the very extremes either of exposure or shelter. The narrow leaflets are blackish brown, and the oval bladders pointed at the tip. Though reaching deeper, this species belongs essentially to the sublittoral fringe, and at low spring tide lies laxly in branched streamers a metre long.

Carpophyllum flexuosum is longer and far looser in growth, with the yellowish brown leaves larger and faintly mid-ribbed. Usually staying submerged at low tide, it forms dense thickets under close shelter. The reproductive branchlets (*receptacles*), which are axillary in *C. maschalocarpum*, are in *C. flexuosum* marginally arranged upon special, narrower leaflets.

The third species, *Carpophyllum plumosum*, grows to only about half the length of *C. maschalocarpum*, with smaller and spherical bladders and wide variation of the leaflets. On higher energy shores, the fronds are cut pinnately into alternate segments. In sheltered waters they are more finely dissected, with thin, wiry segments (var. *capillifolium*). In the Hauraki Gulf, *C. plumosum* is one of the brown algae choking the wave replenished pools in the sublittoral fringe.

Best looked for in pools are two *Sargassum* species, common around Auckland. These can be distinguished from *Carpophyllum* by their cylindrical axis, never flattened, with repeated side-branching and pointed mid-ribbed leaves. Their axillary bladders are small and spherical and — as in *Carpophyllum* — have a leafy tip. *Sargassum scabridum* is the common species at Long Bay, with wavy leaflets and the main axis covered with fine prickles. *S. sinclairii* has a smooth axis, and first produces ovate basal leaves, then long streamers with smaller, strongly toothed foliage.

Mostly confined to surge channels or low tidal pools are the two north-reaching species of *Cystophora*. *C. torulosa* is golden brown with thick, cylindrical branchlets (ramuli) and *C. retroflexa* has narrower, attenuate ramuli of greyish brown. Both grow well at Long Bay, but flourish best on more open coasts, as in the clearer waters north of Whangaparaoa Peninsula.

Two smaller brown algae (of the Order Dictyotales) form part of a rich under-layer, in pools and on the open sublittoral fringe. *Glossophora kunthii* grows in thin dichotomous ribbons, developing at maturity the tongue-like villi carrying the reproductive structures. Under deeper shade grows *Zonaria turneriana*, in short fans with spatulate tips, edged with a 'zone' carrying the reproductive organs.

The same secluded reaches carry the underworld of red algae, richest on shores sub-maximally exposed. At Long Bay there are *Pterocladia lucida*, with wide, bluntly pinnate fronds, and *Pterocladiella capillacea*, with linear fronds branching almost at right angles. Tougher than either of these is *Melanthalia abscissa*, dark red to almost black with stiff fans branched in a single plane, and (as the specific name denotes) scissored off at the edge.

The only kelp of the Hauraki Gulf, *Ecklonia radiata*, is a plant of the upper sublittoral *zone* rather than the *fringe*. The flexible stipe holds up a smooth blade fringed with wide ribbons. Inshore plants, conspicuous among the thickets of *Carpophyllum*, have a stipe half a metre long, secured by a wide, branching holdfast, strong enough for sites not maximally exposed. Beyond low water, down to 25 m, *Ecklonia* grows larger, with a stipe up to 2 m long. This kelp is largely excluded from the intertidal by the excessive light, along with its sensitivity to dehydration where a low spring tide coincides with a warm, calm day. An hour's emersion will dry off the frond tips until they begin to erode, with the rest of the blade soon to follow.

Gastropoda

The common grazing snails each have their distinctive zoning levels. Black *Nerita atramentosa* is found scattered over the upper eulittoral and littoral fringe, especially on harder sandstone and grit. *Melagraphia aethiops* ranges essentially through the whole barnacle

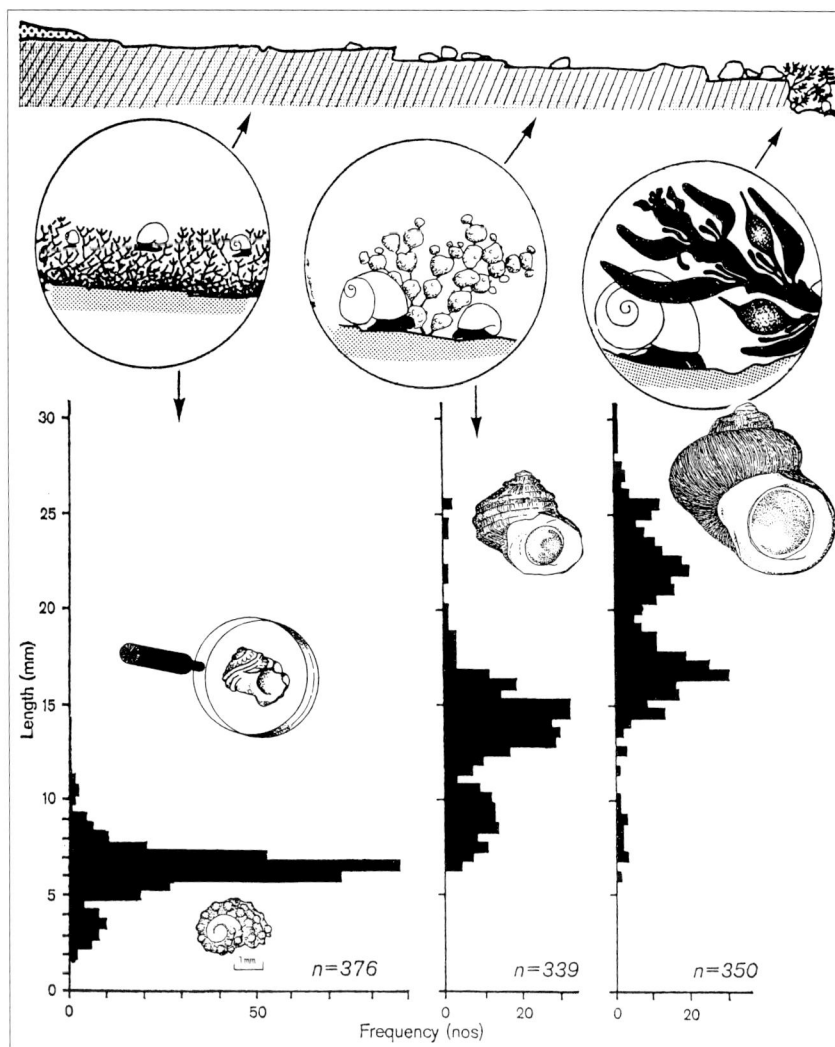

Fig. 3.2 Life history of the cat's eye _Turbo smaragdus_ (By courtesy of John Walsby.)

and oyster zone, on the upper half of the shore. Lower down, on coralline turf, the chief grazer is the cat's-eye, _Turbo smaragdus_, often crusted with the brown alga _Ralfsia_. Narrow-arched slipper limpets, _Maoricrypta monoxyla_, frequently attach to cat's-eyes. Corallines or their small epiphytes are grazed by the long-spired cerithioid snail _Zeacumantus subcarinatus_, common in pool-fringes and attaching viscid spawn strings among the coralline tufts.

Of all the mobile gastropods, this is the most faithful in its presence — at particular age and size — across zones of the middle eulittoral and sublittoral fringe of moderate energy shores. John Walsby has traced its detailed life history.[2] Breeding begins at three to four years with spawn release in February-March, triggered by sudden onset of a storm. The larvae spend only a day or two in the plankton, and the nodulose juveniles live for up to 18 months in _Corallina_ turf. An intermediate-sized population then finds refuge among _Hormosira banksii_ or other small macroalgae. Attaining a critical size for wave disturbance, adult _Turbo_ are finally carried downshore into the relative shelter of the sublittoral fringe with _Carpophyllum_ and _Ecklonia_. Here — at three to four years old and 25–30 mm high — they are

ultimately dislodged by strong waves and carried down sublittorally, beyond the limits of survival.

On wave-exposed shores, though it disappears from the open rock, _Turbo_ remains common in deep pools, with the young in the _Corallina_ of the pool-fringe and the intermediates and adults on the sides and bottom of the pool.

At the deeply sheltered extreme, _Turbo_ is able to survive in total seclusion from waves, in mangrove forests, grazing the trunk-bases and pneumatophores. With low recruitment from rare occasion of a spat-fall, _Turbo_ may live for 10–15 years, with smoothly rounded shells as big as large plums.

Cellana ornata is the common limpet from the top of the barnacles down to oyster level. Smaller across than _C. radians_ but higher-built, the shell has about 11 radial ribs, alternating with black riblets barred with white. The dark interior is rayed in black and silver. Older shells become steeply conical with their sculpture eroded.

Much more prevalent below oyster level is the wide, low-pitched _Cellana radians_, common all through New Zealand and more variable than others of the genus. The ground colour is greyish brown, darker along the 20–25 ribs; with the inside buff-centred, sometimes silver-

Fig. 3.3 *Lepsiella scobina*
This thaid is shown with three of its prey-species on the shore: *Chamaesipho columna* (**top**), *Crassostrea glomerata* (**centre**) and *Austrominius modestus* (**bottom**). Size distribution is independent of height on shore, but dependent on the micro-shelter offered by the prey species.

(By courtesy of John Walsby.)

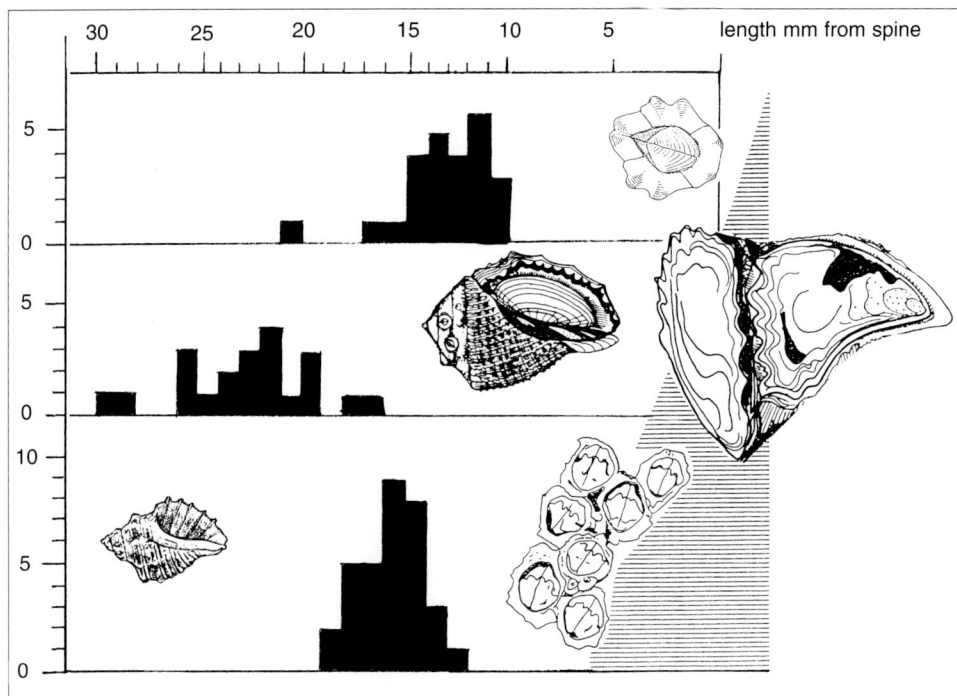

lustred. Young shells may be clear yellow with dark radial streaks.

Some smaller gastropods call for mention. Pressed in among flea mussels and oysters there is *Risellopsis varia*, a tiny, low-spiralled littorine relative grazing wave-lodged diatoms, fine algae and detritus. Unlike the *Austrolittorina* species it lodges oval spawn jellies in moist places. Pressing between the flea mussels is a still smaller shell, the high-pitched miniature form of the limpet *Notoacmea parviconoidea*. In oyster clusters and empty shells lives the small, north-confined trochid *Fossarina rimata*. The wide-mouthed shell is grey-green, prettily zigzagged with cream.

Several trochid snails browse on brown algae at or beyond low tide. *Trochus viridis* has a perfectly conical spire pink-encrusted and a pale green base pressed against the algal fronds to which the foot attaches by viscid mucus. The smallest, most attractive snails on kelp (today scarce in the inner Gulf, though 50 years ago abounding on *Ecklonia*) are two *Micrelenchus* species, no more than 8 mm long. *M. dilatatus* is pink to bronze and *M. sanguineus* is rich red or polished grey, with paler flecks. Both have a blue iridescent mouth.

Immature specimens of the large turbinid *Cookia sulcata* attach to kelp, distinguished by their low conical shape, spiral ribs themselves with cross-lamellae, and the ear-shaped shelly operculum.

The three thaid predators, *Lepsiella scobina*, *Haustrum haustorium* and *Dicathais orbita*, show their accustomed range and habits at Long Bay (Fig. 3.1). A smaller muricid, frequent but never numerous around low water, is the cream *Paratrophon quoyi*, with strong axial ribbing crossed by spirals, and a purple mouth (Fig. 13.8).

Two buccinid whelks are common from oyster level down to the sublittoral fringe. *Cominella maculosa* is plumper, green-grey and brown-tessellated, with a dull orange lip. The more slender *C. virgata* is ashen grey with spiral lines, and bright orange at the mouth. A small but very common whelk, the pink-encrusted *Taron dubius* (Fasciolariidae), often passes unnoticed on corallines. When lifted up, its scarlet foot at once identifies it. *Taron* has been found feeding on *Pomatoceros*, plunging the proboscis into the tube-mouth to relax and kill the worm.

The littoral whelks once considered a commonplace of rocky shores in Auckland and Northland have since the 1980s suffered ill effects from TBT [tributyl tin], a chemical poison released from bio-repellent anti-fouling paints at boat moorings. Claimed to be the most toxic substance ever deliberately introduced into natural waters, TBT will even at low concentrations induce 'imposex' [the development of male sex organs in females] in marine neogastropods. The occlusion of the oviduct impedes the deposition of eggs, with the final substitution of a wholly male condition. Species reported to be adversely affected and reduced in number are *Lepsiella scobina* [most impacted], *Dicathais orbita*, *Haustrum haustorium*, *Xymene ambiguus*, *Taron dubius*, *Cominella virgata*, and the sand-dwelling *Amalda australis*.[3]

Bedrock and habitat

Differences of bedrock can explain the presence or absence of many sessile and mobile species. Upon the 'papa' shore platform, as for example south of **Campbell's Bay**, Auckland east coast (Fig. 3.4), slower-

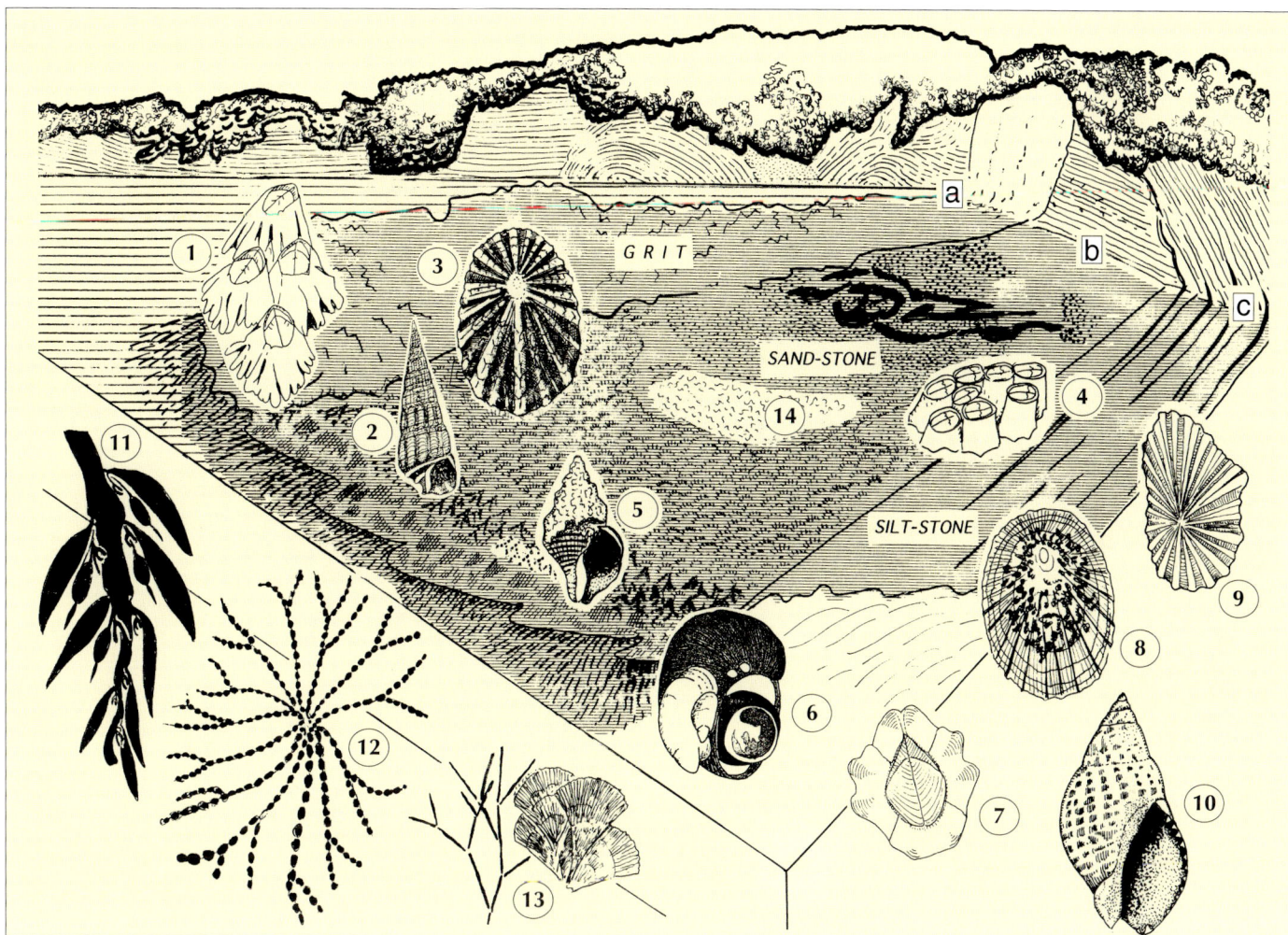

Fig 3.4 Campbell's Bay, Auckland: shore platform of grit (a), sandstone (b) and siltstone (c)

1 *Epopella plicata* 2 *Zeacumantus subcarinatus* 3 *Cellana ornata* 4 *Chamaesipho columna* 5 *Taron dubius* 6 *Turbo smaragdus* with attached *Maoricrypta monoxyla* 7 *Austrominius modestus* 8 *Cellana radians* 9 *Siphonaria australis* 10 *Cominella maculosa* 11 *Carpophyllum maschalocarpum* 12 *Hormosira banksii* 13 *Jania micrarthrodia* 14 *Pomatoceros caeruleus*.

eroding sandstone alternates with soft, friable mudstone. The cliffs may in addition have thick beds of harder, non-stratified Parnell Grit (Fig. 9.21) sending eroded salients across the reef.

Sandstone

The uniform expanse of sandstone is most attractive to settlement, with *Chamaesipho columna* and *Xenostrobus pulex* abundant, and to seaward of them broad sheets of *Pomatoceros caeruleus*, especially in shallow pans of standing water. Coralline turf continues towards low tide, overlaid with *Hormosira banksii*, or at the outer edge tinged pink with *Jania micrarthrodia* and *Laurencia distichophylla*.

Hard grit

Higher ridges of Parnell Grit left by erosion may remain standing above the adjacent papa. On this steeper, more rugged terrain, *Austrolittorina antipoda* and *Nerita atramentosa*, scarce on sandstone and siltstone, become abundant in the littoral fringe. *Chamaesipho columna* intensifies here, with rock oysters *Crassostrea glomerata* settled on high outcrops. The hard, irregular contour supports the barnacle *Epopella plicata* and the limpet *Cellana ornata*. In the upper eulittoral, among *Chamaesipho columna*, are large, scar-forming *Sypharochiton pelliserpentis*.

Mudstone

In contrast with sandstone or grit, mudstone carries little barnacle or mussel cover. Smooth eulittoral surfaces may be brown with filaments of *Ectocarpus* (often scoured with tooth-marks of grazing parore) or green-tinged by *Enteromorpha* in winter. Coralline turf appears only towards the seaward edge.

The pulmonate *Siphonaria australis* lines up in moist crevices, coming out to browse on small, more succulent green algae. The common barnacle on soft mudstone in the lower eulittoral is the fast-growing *Austrominius modestus*.

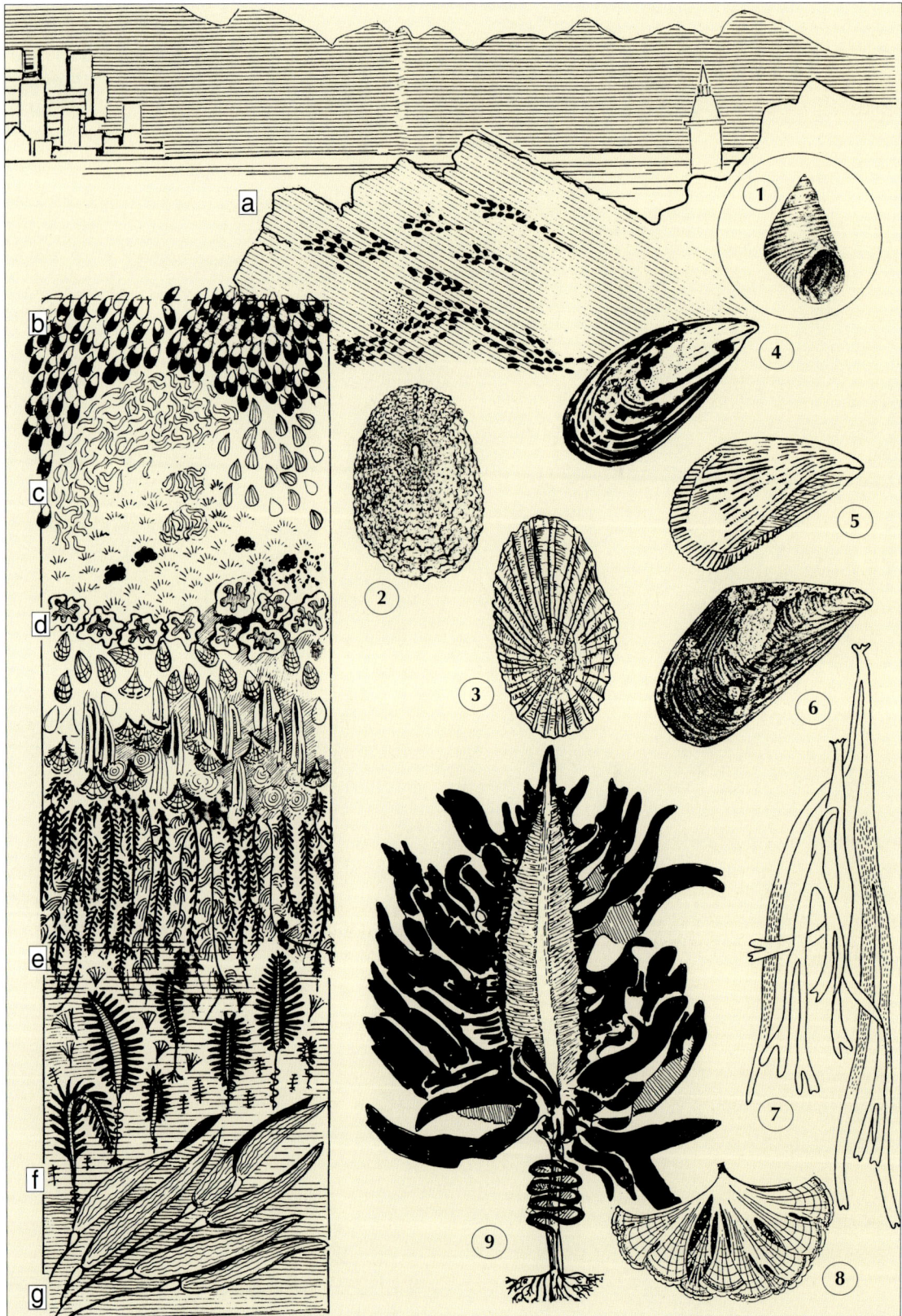

Fig 3.5 Zoned shore at Oriental Bay, Wellington Harbour

Zoning species

a–b *Chamaesipho columna*; b–c *Mytilus, Aulacomya* and *Pomatoceros*; c–d corallines, *Leathesia, Hormosira* and *Codium*; d–e *Perna, Glossophora, Zonaria, Ralfsia* and *Carpophyllum maschalocarpum;* e–f *Carpophyllum* (continued), *Cystophora, Melanthalia* and *Undaria;* f–g *Macrocystis pyrifera.*

1 *Austrolittorina cincta* 2 *Cellana denticulata* 3 *Benhamina obliquata* 4 *Mytilus edulis aoteanus* 5 *Aulacomya maoriana* 6 *Perna canaliculus* 7 *Glossophora kunthii* 8 *Zonaria turneriana* 9 *Undaria pinnatifida.*

Wellington: Port Nicholson

To an observer from the north, Wellington shores seem to belong to a different realm. Moving from Auckland to Cook Strait, we have in fact crossed a major biogeographic boundary, marking off the warm temperate from the cool temperate region of the south Pacific.

Wellington has also its more locally based differences from the north. First, the full extent of spring tides is no more than 1.5 m in Wellington Harbour, only half of the range at Auckland. Wellington's rocky shores are all built of hard greywacke, not readily erodible to form wide platforms as at Auckland and down most of the North Island east coast. The shore contours are thus steep, with the zonation on such complex, irregular slopes sometimes broken or disarranged.

Around Wellington the shores of moderate shelter all lie within the large embayment of Port Nicholson. The inner-city shoreline fronting **Oriental Bay** (Fig. 3.5) shows the lack of wide shelving, partly because of the short time since the 1855 earthquake uplifted the existing shore platform on which the coast road is built. The pervasive coralline turf of the north is here reduced to patches and tide-pool fringes. Instead of white rock oysters, a black mussel band dominates the middle eulittoral.

Along with the pale-banded *Austrolittorina antipoda* seen in the north, the **littoral fringe** has a second, larger periwinkle, the grey-brown and spirally striated *A. cincta*. At Wellington the former species still has the higher reach; but by Otago *A. cincta* will predominate, both in numbers and in vertical extent. On Auckland's cool temperate west coast *A. cincta* is common and enters the Manukau Harbour as far as Huia. In east Northland, on exposed coasts alone, two or three *A. cincta* might be found in a day's search. Very exceptionally, we have found it once inside the Hauraki Gulf, north of Castor Bay.

At Oriental Bay the barnacle zone of the **upper eulittoral** is made up of *Chamaesipho columna*. The **middle eulittoral** is conspicuous with its dark clusters of the navy mussel, *Mytilus edulis aoteanus*. Found on rocky shores of moderate shelter everywhere in the New Zealand cool temperate, this mussel is smaller and more triangular than the green-lipped *Perna canaliculus*. The periostracum is deep blue to black, with the beaks often pale-eroded, while the inside is bluish white with black outside the pallial line. Virtually unknown in the Hauraki Gulf, *M. edulis aoteanus* turns up in some odd pockets in east Northland, probably originating from ship bottoms, as around Russell in the Bay of Islands.

The flea mussel, *Xenostrobus pulex*, is scarce at Oriental Bay, whether by shortage of its preferred flat terrain, or perhaps from competition with the navy mussel. It is, however, zone-forming near the top of wharf piles, as at York Bay.

Intermingling with the margin of the navy mussel around low water neap lives a species found nowhere in the north, the ribbed mussel *Aulacomya maoriana*. Lower down it forms its own clusters or lives attached to holdfasts and under stones. The shell is dull purple to yellowish brown, with a triangular cross-section and squared-off lower side. Close-set ribs radiate from the beaks, some bifurcating. Towards low water, the green-lipped mussel, *Perna canaliculus*, reaches into Port Nicholson, being darker shelled than brown-rayed populations of open coasts.

Below the navy mussels or mingled with the ribbed, the tubeworm *Pomatoceros caeruleus* forms continuous crusts. The sand tubeworm, *Neosabellaria kaiparaensis*, has cut out at about Taranaki and Hawke's Bay.

Limpets

Along with *Cellana ornata* and *C. radians* in their part-overlapping, mainland-wide ranges, Wellington has a third patellid, *Cellana denticulata*, virtually confined to the east coast, mainly from Cook Strait to Kaikoura. The shell is strong and elevated, with 20–30 granulose, distinctly scaly ribs. Inside it is bluish-white, with dark streaks under the ribs and a buff centre. Most like *Cellana ornata* in site and habit, *Cellana denticulata* is tolerant of wave action, being commoner at Lyall Bay and Island Bay than in the Harbour.

As well as the small pulmonate limpet, *Siphonaria australis*, Wellington has a larger siphonariid, *Benhamina obliquata*, up to 4–5 cm long. Radially ribbed when young, the shell is oval to oblong with the apex two thirds of the way back, and mottled orange-brown, inside dark-blotched at the margin. The foot is yellow beneath and fits more completely within the shell than with *Siphonaria*. Living towards the top of the *Chamaesipho* zone, or clusters together in shaded sites, *Benhamina* produces a coiled spawn ribbon somewhat like that of a nudibranch.

Algae

Corallina officinalis is plentiful in moist pans and around pools, but has nothing of its northern profusion. In Cook Strait other species, including succulent reds, vie with it for space. The first competitor, beginning shortly above MLWN (mean low water of neap tides), is the widespread Venus' necklace, *Hormosira banksii*. From a lower level the fleshy, convoluted *Codium convolutum* breaks into the coralline zone, or in summer forms a continuous band just below it.

With the beginning of the **sublittoral fringe** two middle-sized brown algae, both members of the simple-structured Dictyotales, regularly appear. The fan-shaped tufts of *Zonaria turneriana* are far more prominent on the open surface than near Auckland. The thin, dichotomous ribbons of *Glossophora kunthii* are recognised when mature by their small, tongue-shaped reproductive papillae.

The *sublittoral fringe* and the succeeding *zone* have a quite different mix of brown algae from the north. *Carpophyllum maschalocarpum*, still abundant at and beyond low water, is joined far more regularly by *Cystophora retroflexa*, which now becomes the leading brown alga of the fringe and beyond. The fringe has also plenty of the chunky-branched *Cystophora torulosa*. *Carpophyllum plumosum*, not quite lacking in Cook Strait, grows only in small stands, reaching here the southern end of its range. *Carpophyllum flexuosum* is found only in quieter waters, established generally beyond low tide.

The cool temperate is better endowed with laminarians than the north. In Wellington Harbour, around low water, we meet with all three of New Zealand's traditional kelps — as well as one highly interesting newcomer. The shining blade and ribbons of *Ecklonia radiata* can be glimpsed at the lowest spring tides, but this warmer-water kelp is losing the wide sublittoral dominance it shows in the north. At many sites within Port Nicholson the fan-like heads of *Lessonia variegata* grow a metre or two beyond low water. In the north this kelp, smaller than offshore *Ecklonia*, occurs only on high energy coasts, but at Wellington, as for example on Eastbourne Wharf, it grows in sheltered sites right up to low-water mark. Well within Port Nicholson we can find also *Gigartina circumcincta* and the starfish *Stichaster australis*, both confined in the north to open coasts.

Wellington Harbour brings our first sight of the bladder-kelp, *Macrocystis pyrifera*. In southern harbour mouths, with strong tidal currents but subdued wave attack, we shall find this long-attenuate kelp regularly dominant. At Wellington, it grows in only small stands, though even here — at 2 or 3 m long — it is a relative giant among algae. The slender stipes are deep-attached by a complex holdfast, and produce long closely corrugated leaves, each with a pear-shaped bladder at the base. At low tide, a fringe of these leaves may float on the surface offshore, but at Wellington they remain mostly subtidal.

In 1987 Wellington Harbour became the first point of entry for a fourth kelp, *Undaria pinnatifida*, until around this time known only from cool temperate Japan and China. *Undaria* can be recognised at once by its flat blade with wide mid-rib and lateral pinnae, and by the undulant wings on the stipe bearing the sori. This large sporophyte reaches a metre long, to degenerate in summer after shedding the spores that give rise to minute gametophytes. After fusion of gametes, the resultant zygospore develops into the next season's sporophyte. In Japan, *Undaria pinnatifida* is the highly edible seaweed 'wakame', grown commercially in temperatures from 26°C in August to 2°C in February. In New Zealand, with temperatures of 10–16°C, it could be able to overwinter.

First recognised at Oriental Bay by Penelope Luckens,[4] *Undaria pinnatifida* is clearly an immigrant that should be carefully monitored. At Wellington, from low water down, it has already become an effective competitor for space and light with the traditional brown algae. Evidently arriving in Wellington on Japanese squid boats, it has now (1996) been reported right down the east coast of the South Island to Bluff and Stewart Island. It has also appeared in Tasmania (1988), having already (1985) been reported in France: it appears to be another of those Asian species that have travelled widely, like the Pacific oyster.

Nelson: Tasman Bay

Tasman Bay, like most of the shoreline of east Nelson, differs markedly from Cook Strait's northern seaboard. Where the Wellington shores are clean and wave-swept from the south-west, the Bay's great triangle is protected by the dissected landmass of Marlborough, and the great hook of Farewell Spit. From the wide land catchment, sediment is continuously brought down to be deposited on the shallow shelf. The Durville Current misses Tasman Bay as it sweeps south through Cook Strait, and its shores are immune from any strong flow that would sweep the shelf clean. The intertidal shore is kept partly sediment-free by wave action. But below wave-base, the subtidal takes its character from the overlay of the fine sediment resting lightly upon every surface, including algal fronds. Visibilty is reduced as these particles are constantly stirred up by gentle currents.

Aside from plankton, the prime food resource, far ahead of living tissue, is the available bulk of detritus, eventuating from algal decay. Wellington's pink-enamelled surfaces swept clean under high wave action are here exchanged beyond low water for a deposit-rich terrain. Bladder-kelp (*Macrocystis*) with its need for strong currents, the clear-water *Lessonia*, and (until the west shore of Tasman Bay) *Ecklonia radiata*, are wanting. *Carpophyllum flexuosum* dominates the **upper sublittoral** with its dense thickets.

The intertidal zonation, as shown for Cable Bay (Fig. 3.6), is relatively simple, and notable for its virtual lack of large algae. The topmost barnacle, *Chamaesipho brunnea*, forms only a narrow upper fringe, followed down the **eulittoral** by *C. columna*. The chief high level

Fig. 3.6 Zoned shore in Cable Bay, Nelson

a–b *Chamaesipho columna, Stictosiphonia, Porphryra;* b–c *Pomatoceros, Mytilus, Aulacomya;* c–d *Perna;*
d–e *Carpophyllum flexuosum;* e–f *Carpophyllum, Tethya, Evechinus, Stichopus*

1 spotted shag, *Stictocarbo punctatus* 2 *Patiriella regularis* 3 *Stegnaster inflatus* 4 *Cantharidus purpureus* 5 *Calliostoma punctulatum* 6 *Notolabrus celidotus* 7 *Phlyctenactis tuberculosa* 8 *Coscinasterias muricata* 9 *Stichaster australis* 10 *Stichopus mollis.*

algae are a thin felt of *Capreolia implexa*, small ulvoids and tufting *Ectocarpus*. The steep eulittoral slope has a good spread of *Pomatoceros caeruleus*, even upon the shells of mussels that intermix with the tubeworm patches. All four zoning mussels are present: *Xenostrobus pulex* highest up, then navy blue *Mytilus edulis aoteanus* (widely farmed in Tasman Bay), ribbed *Aulacomya maoriana*, and finally green-lipped *Perna canaliculus*, continuous beyond low water.

Often covered with the anemone *Actinothoe albocincta*, *Perna* ultimately give place to the brown algal zone of the upper sublittoral, reaching down some 2–3 m. The main species — in a dense entangling thicket beyond low water — is *Carpophyllum flexuosum*, distinguished by its lax growth form, tawny colour and large mid-ribbed leaves. A second important brown alga is *Cystophora torulosa*, like *Carpophyllum* thinly filmed with sediment. Common within the algal tangle is the long-armed decorating crab *Notomithrax peronii*. Subtidal patches of coralline turf carry silt-covered *Colpomenia sinuosa* and *Codium convolutum*.

Lightly attaching to algae may be found the wandering sea anemone, *Phlyctenactis tuberculosa*, our largest New Zealand actinian. This species is free-moving and only fastens temporarily by its base. Buoyed up and extended, it can lengthen to the dimensions of a jam-jar. The column is flaccid and covered with soft tubercles, rust brown, orange or grey. The attaching base is orange, and the tentacles and oral disc pale yellow.

Several browsing gastropods favour the live fronds of *Carpophyllum*, including *Turbo smaragdus*, *Trochus viridis*, and the dull-pink *Cantharidus purpureus*. The trochid *Calliostoma pellucidum* subsists as a deposit feeder, suspected even to have turned sponge-grazer. Feeding on subtidal algae at Pepin Island are the paua species, *Haliotis iris* in the open, and *H. australis* deeper in crevices, with the black shield limpet *Scutus antipodes* in the darkest sites.

Dense clusters of the urchin *Evechinus chloroticus* among accumulated algal debris are only the first of a notable range of echinoderms. Just beyond low water lives a common sea cucumber, *Stichopus mollis*. The soft, warty upper surface is mottled greyish brown to almost black, studded with pointed tubercles. The paler underside is flat, serving as a permanent sole with three ambulacral rows (Fig. 3.6) as well as diffuse tube-feet by which the animal progresses or fastens down. The two upper rows of tube-feet have been lost. The ring of 20 oral tentacles constantly pick up nutritive deposits from the ground film, the laden podia being passed one by one into the mouth and then withdrawn with their particles licked off. *Stichopus* is often first detected by its cylindrical faecal castings of fine grey deposits.

Several species of starfishes are regularly to be found in the sublittoral, most commonly the cushion star, *Patiriella regularis*, that feeds by browsing the surface. *Stegnaster inflatus* is a much larger star, forming a straight-sided pentagon some 10–12 cm across. The arms are united in a stiff web, smooth above and below, incised with the narrow ambulacral grooves. The range of colours is very attractive, some specimens are tangerine above and lighter below, others sage-green on top with cream underneath, usually blotched with chocolate or grey. The disc arches to a tent as the star applies itself to prey, usually bivalves, thaid whelks or crabs.

The common low tidal starfish, *Coscinasterias muricata* is a predator on gastropods and bivalves. Large specimens reach 15 cm across, with up to 11 arms varying in size and number with loss and regeneration. The upper surface is rust-red to grey and heavily spined, like the edges of the ambulacral grooves. The tube-feet are cream. A more stiffly built 11-armed star is the ochre-coloured *Stichaster australis*, that feeds on green-lipped mussels (Fig. 3.6).

The cloudy water favours sponges, dominant wherever illumination drops too low for algae. At shallow depths, loose rocks are welded with a dull orange crumb-o'-bread sponge, *Halichondria moorei* and red *Crella incrustans*, with hummocks that conceal the nesting mussel *Modiolarca impacta* (Fig. 13.14). Orange golf-ball sponges *Tethya aurantium* are common.

Beyond the sponges, the deposits thicken, to merge into a stretch of silty sand. Numerous sand-dollars or cake urchins (*Fellaster zelandiae*) lie just beneath the surface, with the common gastropods *Struthiolaria papulosa* and *Alcithoe arabica*, and the scallop *Pecten novaezelandiae*. The paddle crab (*Ovalipes catharus*) is abundant, along with the sole, *Peltorhamphus novaezeelandiae*. Shoals of goatfish (*Upeneichthys lineatus*) graze the bottom, with the spotty or paketi, *Notolabrus celidotus*, common in the weed, and sometimes the leatherjacket (*Parika scaber*).

Banks Peninsula: Akaroa

Akaroa and Lyttelton Harbours are Miocene volcanic craters, breached by giant landslides on the volcanoes' slopes and recently drowned by the rising sea level, after the last post-glacial ice-melt. Their basaltic slopes are still being eroded back, and these deep-indented shores on Banks Peninsula all lack broad intertidal platforms, possessing only a limited extent of soft flats. The rocky shores are kept clean of sediment by waves and strong tidal currents. With a good tidal rip far up-harbour and wind-funnelling to generate local waves, both these narrow harbours have medium energy shores like those of the Hauraki Gulf and Port Nicholson.

On Banks Peninsula we find bull-kelp reaching up to

Fig. 3.7 Zoned shore in Akaroa Harbour, Banks Peninsula

a–b *Porphyra, Scytosiphon*; b–c *Stictosiphonia, Chamaesipho columna*; c–d *Aulacomya, Pomatoceros*; d–e *Corallina, Hormosira, Ulva, Adenocystis, Perna*; e–f *Cystophora scalaris*; f–g *Macrocystis*

1 *Porphyra columbina* 2 *Adenocystis utricularis* 3 *Cystophora torulosa* 4 *Cystophora scalaris* 5 *Cystophora distenta* 6 *Cricophorus nutrix* 7 *Cystophora retroflexa* 8 *Cystophora platylobium* 9 *Hippocampus abdominalis* 10 *Pyura pachydermatina* 11 *Macrocystis pyrifera*.

sheltered limits where we would never look for it in the north. In Akaroa Harbour both *Durvillaea antarctica* and *D. willana* co-exist with *Macrocystis* up the western side as far as Oputereinga, while on the opposite shore *D. antarctica* reaches to German Point.

The site depicted at **Akaroa** (Fig. 3.7) is a secluded reach, protected by a high basalt stack. Away from the shore, swept by the tidal current but in relative calm, there are beds of *Durvillaea willana*. Much more extensive wherever the tides run strongly are the thickets of the bladder-kelp *Macrocystis pyrifera*, with leaves floating at the surface or at slack low water caught up like pointed pennants in a squall.

The **littoral fringe** carries the wavy-edged ribbons of the red alga *Porphyra columbina* (Fig. 3.7), typical for a medium energy southern shore and showing up dark at a distance. *Porphyra* comes to its maximum in July-August but regresses by November. The tissue-thin thalli dry out over the rock during low tides, leaving the top layer tight-stretched. The layers underneath remain mucilaginous and hence slippery underfoot. In winter, the brown tubular *Scytosiphon lomentaria* flourishes briefly, constricted into pods, and forming here a large variety as thick as quills. Alongside it grow festoons of green *Enteromorpha ramulosa*.

The **upper eulittoral** carries a thin scatter of the barnacle *Chamaesipho columna*. At the same level begin patches of algal cover, far thicker than would be expected in the north. Dull, reddish brown *Stictosiphonia arbuscula* forms a soft sward above and around the barnacles, increasing to a continuous band on steep or shaded faces. In sunnier spots, the rock is blackened with the stubble of *Lichina confinis*, the lowest ranging of the shore lichens. Both these plants harbour abundant amphipods and also the minute, pink or white bivalve *Lasaea hinemoa*. In the bushes of the *Stictosiphonia* nestle the gastropods *Zeacumantus subcarinatus* and *Risellopsis varia*.

All four of the zoning mussels regularly occur in Banks Peninsula harbours. First comes the flea mussel, *Xenostrobus pulex*, near the top of the barnacle band and just below *Stictosiphonia*. Clumps of the navy mussel, *Mytilus edulis aoteanus*, show up darkly against the barnacles, or the white filigree of *Pomatoceros caeruleus*. Below this tubeworm, forming clusters on coralline turf amid *Hormosira*, is the ribbed mussel, *Aulacomya maoriana*. Furthest down, where the turf thickens, green-lipped mussels, *Perna canaliculus*, are scattered, but this is a species far more at home on surf coasts than in harbours.

In steep, shaded places on Banks Peninsula, the tubeworm *Pomatoceros caeruleus* flourishes, taking command of the **middle eulittoral**, to the virtual suppression of mussels. At Akaroa, above the *Hormosira* zone,

Pomatoceros crusts occupy the vertical faces as well as the shallow pans left with water between tides. Each colony transforms to an indigo-blue carpet as the tentacle crowns open out when submerged.

At Taylor's Mistake, facing the open sea between Sumner and Lyttelton, there are shaded sites with *Pomatoceros* building a continuous layer, 25 cm thick, of chalky white tubes. The outermost tubes with living worms present their apertures, with the spine reduced, over a whole smooth surface.[5]

Below *Pomatoceros*, in the **lower eulittoral** fringe *Corallina* turf is always present, though other algae can seasonally overshadow it, as with the vivid green of *Ulva lactuca*, or the gold of *Leathesia* and *Colpomenia*.

A peculiar brown alga of cooler waters, unknown in the north, is the bladdered *Adenocystis utricularis*, growing best in runnels and elevated pools. Reaching marginally north to Wellington, but most familiar from Kaikoura south, *Adenocystis* forms bunches of thick pyriform vesicles, up to 5 cm long, short-stalked and greenish brown. Solid when young, these hollow out into tensely filled bladders. They have a tomentum of microscopic filaments, each with a basal sporangium. With no obviously close relatives among the brown algae, *Adenocystis utricularis* has a wide range in cool and subantarctic waters, at Cape Horn, southern New Zealand and Tasmania.

Sublittoral fringe

As is typical in quiet southern waters, the brown algal fringe is *Cystophora*-dominated. *Carpophyllum maschalocarpum*, still found sparsely south from Kaikoura, has at Banks Peninsula reached its southern extreme. *C. flexuosum*, growing further offshore, is no more than occasional, and *Ecklonia radiata* exists only in rather reduced beds in the sublittoral.

Cystophora is a genus of some 20 species, confined to Australia and New Zealand, chiefly on cool temperate shores. The main axis is round in section (terete), whereas in *Carpophyllum* it is generally flat. *Cystophora* carries its tufts of cylindrical branchlets (ramuli) on alternate zigzagged branches. On Banks Peninsula, tawny *Cystophora retroflexa* and golden *C. torulosa* are as common as at Wellington, the first with slender ramuli, over 3 cm long, the second with the ramuli thick and club-shaped.

The third common species, nowhere seen in the north, is *Cystophora scalaris*, growing in dull brown to grey clumps. This plant is distinguished by its stepped main axis, giving the specific name 'stair-cased'. The zigzag branchlets have their ramuli often twisted or bent, giving the plant a somewhat harsh texture. A less frequent species, *Cystophora distenta*, appears from

Cook Strait south, distinguished by its pod-like ramuli with small constrictions.

A final species, *Cystophora platylobium*, stands out from the rest of the genus by its flattened ramuli pointed at the tips. This species is seldom intertidal but may be locally common in South Island drift.

Every southern harbour with appreciable tidal currents will have white and magenta sea-tulips *Pyura pachydermatina*, bobbing at the surface through the thicket of bladder-kelp, *Macrocystis*. With a flexible stalk up to a metre long, this is the largest of our ascidians, though shorter specimens can enter the *Cystophora* fringe close inshore. Mounted on its tough cable, the strong-ridged test reaches up to 12 cm long. The inhalant siphon, conical when closed, lies to one side, with the exhalant siphon behind it, recurved as a short funnel.

The brown algae around low water shelter a range of fishes and Crustacea, along with some anemones that deserve special mention here. In the deepest pools, mostly among *Cystophora*, lives our largest New Zealand actinian, the wandering sea anemone, *Phlyctenactis tuberculosa*, as already described at Cable Bay (Fig. 3.6). The small brooding anemone *Cricophorus nutrix* sits in the axils of *Cystophora*, and sometimes on *Carpophyllum*. The short column is yellow, orange or deep brown, closing down to a small button. The oral disc and tentacles are translucent green or brown, often phosphorescent. Long white filaments (*acontia*) from the internal mesenteries may stream from the mouth. The young are brooded externally in a ring around the base of the column.

The attractive pink and white anemone *Epiactis thompsoni* lives in sheltered pools on southern rocky coasts, sometimes fastened in coarse gravel. The smooth column is striped in marzipan-pink and the oral disc dark red, with 60 or more tentacles mauve-tipped. This anemone is viviparous, with young ones escaping from the coelenteron of a large specimen.

The camouflaged spider crab *Notomithrax ursus* is common in algal pools. The carapace grows to 60 mm long, and is clad thickly with springy hairs serving to attach small slips of growing algae.

Living in nearly all kelp-beds, but perhaps commonest in the south among *Macrocystis* or *Cystophora*, is the charming but bizarrely shaped seahorse, *Hippocampus abdominalis*. Encased in a firm bony frame overstretched by shiny skin, a seahorse attaches with its tail-tip to the algae, holding the body upright and still. In swimming, rapid undulation of the dorsal fin is accompanied by vibrations of the pectorals. The whole posture is like riding up an escalator. Seahorses feed by picking out individual items from the zooplankton, swaying and bobbing, with the mouth tube used as a pipette. The female is distinguished by the small anal fin. In the same spot, the male has a distended pouch into which the female transfers the eggs. Their entire care is performed by the male, until the moment of birth when the young swim away as tiny replicas of the adult.

Otago Harbour: Portobello

Enclosed by the long Otago Peninsula, the Harbour offers a narrow fetch, through which strong winds can funnel north-east or south-west. Waves of appreciable size are whipped up during intermittent squalls, after a few hours subsiding to a flat calm with the drop of the wind.

Portobello Point, which is the site of the Otago University's Marine Biological Station, is attached to the Peninsula as part of a drowned divide including Quarantine and Goat Islands. The divide is in the eroded centre of the Miocene Dunedin Volcano and is formed from a mix of eroded lava flows and breccia-filled volcanic necks. The reef below the Laboratory is steep and irregular, with its slopes fanning out to broken promontories swept by tidal currents.

With the absence of strong waves, the bladder-kelp *Macrocystis pyrifera* dominates the low-water-line from the harbour mouth up to the Portobello divide. From Cook Strait south, *Macrocystis* is as constant in harbours with strong current flow as is *Durvillaea* on the open, wave-pounded coasts. *Macrocystis* far exceeds the bull-kelp in length, reaching up to 35 m, and anchoring at more than 20 m deep. The buoyant leaves align with the flowing tide, the topmost ones just breaking the surface as the breeze scuds across the water.

The bladder-kelp has none of the tough resistance of *Durvillaea*, being adapted not to wave-pounding or wrenching, but rather to fast current flow. In the deeper harbours *Macrocystis* forms subtidal beds of many hectares, although avoiding the most open coasts. North of Otago, its largest beds lie in the harbour mouths of Port Nicholson, the Marlborough Sounds and Banks Peninsula. In small stands, *Macrocystis* reaches north to Castlepoint and Kapiti Island.

The zoned shore

In the **littoral fringe** around high water of spring tides the yellowish brown bedrock is darkened by a film of the lichen *Verrucaria*. The single important periwinkle species at Portobello is *Austrolittorina cincta*. *A. antipoda* occurs in Otago chiefly outside harbours, and in smaller numbers than to the north.

At the same level lives the splash limpet *Notoacmea pileopsis sturnus*. This cold-water subspecies is distinct from the northern *N. p. pileopsis* in the extreme anterior position of the apex. Crevices contain the littorine ally *Risellopsis varia*, notably larger at 8 mm diameter than in the north and reaching well above barnacle level. The small black spire-snail, *Zeacumantus subcarinatus*, clusters densely in the soft swards of *Stictosiphonia arbuscula*. Above the barnacle-line, this plant is widespread and conspicuous at Portobello as on most cool southern shores. Its clumps teem with the minute bivalve *Lasaea hinemoa*.

Several cyanobacteria (blue-green 'algae') are important on high level rocks. Bright green vesicles of *Rivularia australis* grow saccate and fleshy from January to March, while autumn brings filamentous *Oscillatoria corallinae*, and felt-like patches of *Lyngbya semiplena*.

Down to the middle eulittoral the basalt bedrock carries thin and inconstant patches of the barnacle *Chamaesipho columna*. Below this lives *Epopella plicata*, thickening to a total cover in places where currents are unusually strong. The **eulittoral** carries some notable seasonal algae, including *Porphyra columbina* higher up, and towards the middle shore, *Scytosiphon lomentaria*, *Leathesia difformis* and the two southern species *Adenocystis utricularis* and *Scytothamnus fasciculatus*.

The grazing herbivores are mostly species familiar in the north: *Melagraphia aethiops* wandering at large without homing, *Onchidella nigricans*, venturing out in damp weather, and *Sypharochiton pelliserpentis*, regularly returning to its permanent scar. The two mainland-wide limpets, *Cellana ornata* and *C. radians*, have their usual ranges, respectively in the upper and middle eulittoral. Otago's most abundant patellid is, however, the southern limpet *Cellana strigilis* in its mainland (and Stewart Island) subspecies *redimiculum*. With a shore range similar to *C. ornata*, this limpet is recognised by its 20 strong radial ribs with weaker ribs in the interspaces; the interior is light yellow with darker rays showing through from outside. An important limpet in crevices and shade is the large siphonariid *Benhamina obliquata*, often first detected by its pale egg spirals.

An exposed stack is shown in Fig. 3.8, crowned by *Epopella plicata*, followed by the several bivalves that dominate the lower half of the eulittoral. The principal mytilids at Portobello are the navy mussel, *Mytilus edulis aoteanus*, and the ribbed mussel, *Aulacomya maoriana*, that intermingles with it below, or further down forms clusters of its own. The other intertidal mytilids, *Xenostrobus pulex* and *Perna canaliculus*, belong in quantity to exposed shores. The fringed mussel, *Modiolus areolatus*, can be found scattered just beyond low water at Portobello; its salient feature is the shining epidermis of chestnut brown with stiff, hair-like tufts over the posterior end.

Below the clustered navy mussels spreads a wide band of the commonly called Dunedin rock oyster, *Ostrea lutaria*. Previously styled *O. heffordi*, this had been regarded as a species peculiar to Otago, with its relation to the other Ostreidae (never an easy family to classify) not clear. It is distinct from the northern *Crassostrea glomerata* in being smaller and circular, with the left valve deep-cupped and the right one a flat lid. The shell is greyish white outside and olive within, with the edge not crenate but entire.

Pomatoceros caeruleus, found sparsely in Otago underneath or at the sides of boulders, does not form a conspicuous zone.

Continuous algal cover begins a little below MLWN, with the Venus' necklace, *Hormosira banksii*, mingled with large, bright green *Ulva lactuca*. On these colder coasts, *Hormosira* appears to have become mainly a plant of the sublittoral fringe, completely covered at neap tides, in contrast with its eulittoral position further north. Its chief space competitor would thus appear to be not *Corallina* but low tidal *Ulva*. *Gigartina decipiens* (Fig. 6.7) appears in the sublittoral fringe, along with smaller red algae much more diverse than at comparable places in the north.

The colour and variety of life are lavishly increased at low water of spring tides. Forests of sea tulips gird the reef inside its fringe of bladder-kelp, springing in clusters from any available rock and swaying on their long cables with the movements of the water.

Carpophyllum is scarce in Otago, represented occasionally by the one species *C. flexuosum*. *Ecklonia radiata* forms small beds in the sublittoral zone, dwarfed as compared with its tall groves at Auckland. *Desmarestia ligulata* is abundant through spring and summer. Under extreme shelter the bladder-kelp is replaced by the *Cystophora* species, three of them common in Otago: the New Zealand-wide *C. retroflexa* and *C. torulosa*, and the southern *C. scalaris*, with its prominent 'stepped' axes and zigzagged side branches.

From intimate knowledge of her home shore, Betty Batham has emphasised in her pioneer paper[6] the brightness and variety of its small algae and the profusion of free and sessile invertebrates.

In the **lower eulittoral**, coralline paint lends its pink hue to the rock and produces patches of turf, with *Corallina* yielding place to the more delicate *Jania micrarthrodia* and related species. Here too grow the encrusting sponges, *Hymeniacidon* and *Haliclona*, along with white networks of *Clathrina*, especially in crevices. Clumps of the leathery ascidian *Pyura suteri* attach to the rock.

A prolific crevice and boulder fauna inhabits the **sublittoral fringe**. Common in Otago are the compound ascidians *Aplidium adamsi*, *A. benhami* and *Hypsistozoa fasmeriana*, the scarlet-splashed tubeworm *Galeolaria hystrix* and the mall urchin *Pseudechinus*

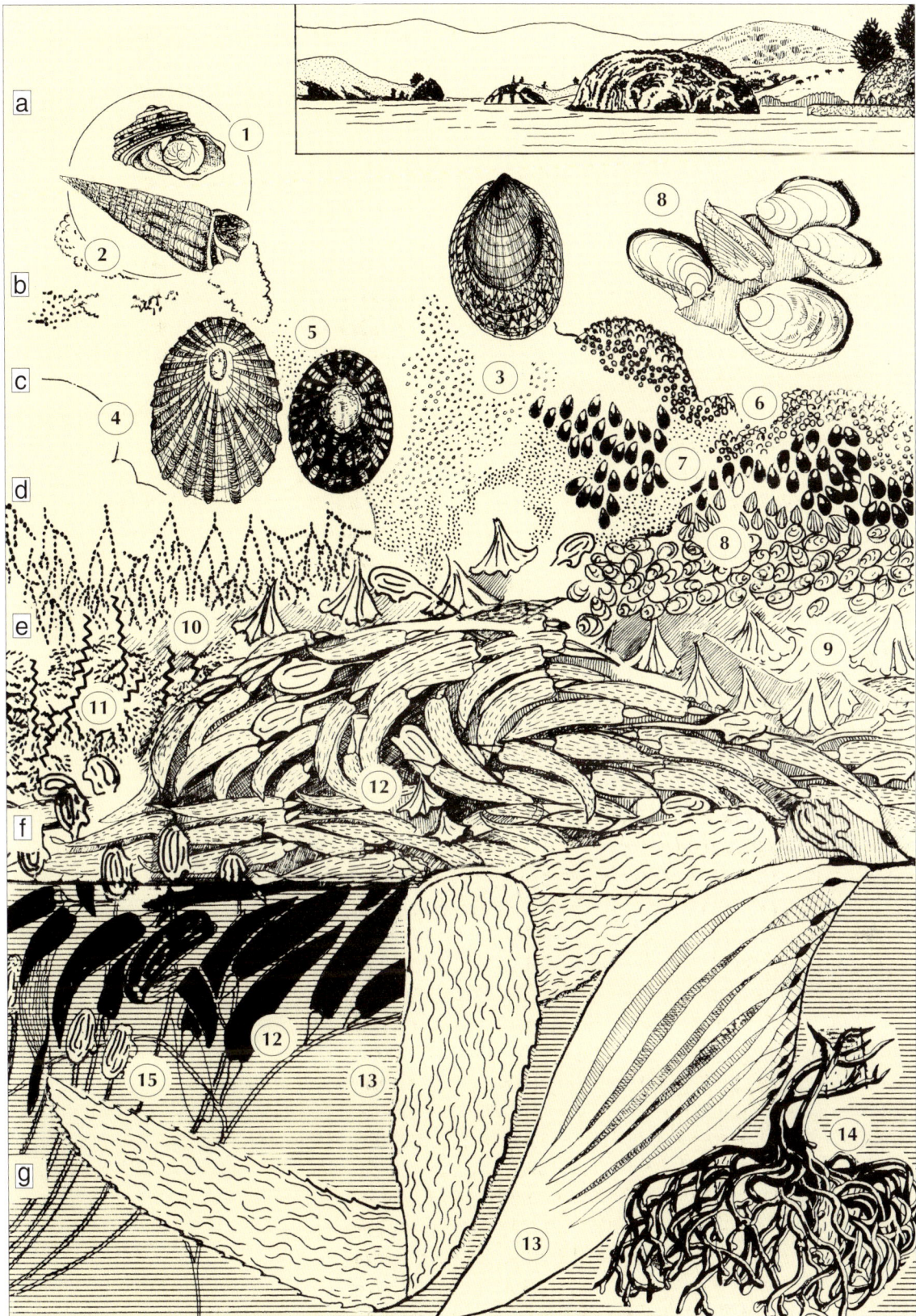

Fig. 3.8 Zoned shore at Portobello, Otago Harbour

a–b *Stictosiphonia;* b–c *Chamaesipho columna;* c–d *Mytilus, Aulacomya, Ostrea;* d–e *Hormosira, Corallina, Ulva;*
e–f *Cystophora scalaris;* f–g *Macrocystis, Pyura*
1 *Risellopsis varia* 2 *Zeacumantus subcarinatus* 3 *Notoacmea pileopsis sturnus* 4 *Cellana strigilis redimiculum* 5 *C. s. redimiculum* (immature) 6 *Epopella plicata* 7 *Mytilus edulis aoteanus* 8 *Ostrea lutaria* 9 *Ulva lactuca* 10 *Hormosira banksii* 11 *Cystophora scalaris* 12 *Macrocystis pyrifera* 13 *M. pyrifera,* new frond division 14 *M. pyrifera,* holdfast 15 *Pyura pachydermatina.*

novaezelandiae. Anemones in shade include the fine *Anthopleura rosea*, as well as the striped *Diadumene neozelanica*. On patches of silty gravel lives the attractive scarlet biscuit star, *Pentagonaster pulchellus*. The small pink synaptid cucumber *Taenigyrus dunedinensis* and the brittle star *Amphiura amokurae* are found burrowing in soft patches.

The bases of *Macrocystis* hold a rich encrusting and nestling fauna. Among the branched fingers sponges and ascidians live, especially *Corella eumyota* and *Aplidium benhami*, the tufted ectoproct bryozoan *Caberea zelandica* and terebellid tubeworms. Entwining or slowly on the move are nereid and syllid worms, with the buff or mauve brittle star *Ophiomyxa brevirima*, distinctive among ophiuroids by its smooth slippery skin.

Most colourful of the holdfast species is the soft and attenuate hemichordate *Balanoglossus australiensis*, twining through the holdfast with a combination of ciliary and weak muscular movement. The slender proboscis is flame-red, followed with a cylindrical collar, enclosing the mouth at the proboscis base. The long hind-body is lighter red, darkened by the brown of the intestine. *Balanoglossus* feeds on fine detritus, picked up by the ciliary tracts of the proboscis or grazed directly by the mouth.

Macrocystis leaves carry an abundance of hydroids and bryozoa. *Obelia geniculata*, *Silicularia rosea*, *Orthopyxis crenata*, *Plumularia setacea*, and a *Clytia* species represent the thecate hydroids, while the principal bryozoans are *Celleporella delta* and sheets of *Membranipora membranacea*. The serpulid worms *Spirorbis* and *Hydroides* may at times be very common.

Upon the sea tulips (*Pyura pachydermatina*) there are brown tufts of the ctenostome bryozoan *Elzerina binderi* and golden beards of the hydroid *Amphisbetia bispinosa*. Here too grow some delicate red algae, *Myriogramme* sp., *Schizoseris griffithsia*, *Phycodrys quercifolia* and *Porphyra subtumens* being especially characteristic.

Of the wealth of small subordinate animals and algae of the sublittoral fringe, (here given modern species names) Betty Batham wrote:

Among the swirling fronds of *Macrocystis* and the swaying beds of *Pyura* with their dense algal load one looks down on rock faces carrying an amazingly rich and colourful range of organisms. Algae and filter-feeders abound. *Enteromorpha*, *Ulva* and their allies flourish, ranging up to great sheets of *Ulva lactuca*, two metres long or more. Broad, dark red sheets of *Gigartina* sp. contrast in texture with the delicate foliose reds on *Pyura* stalks. Pink, bilobed fronds of *Rhodymenia obtusa* may clothe a rocky face. Abundant filamentous reds range from finest *Anotrichium crinitum* to coarser *Ceramium* and *Anotrichium*, *Lophurella* and *Polysiphonia* species. Encrusting coralline algae look like dabs of pink paint on many stones.[7]

Contrasting with the colours of algae are the bright oranges and purples of sponges and ascidians. The large mauve branching sponge, *Callyspongia latituba*, thrives in sheltered hollows. A grey siliceous species is abundant round Tip Rock. Other siliceous sponges are brick red, biscuit-coloured and vivid vermilion. Pink, pointed mounds of *Darwinella* and an encrusting, allied yellow species abound, as do the pink and yellowish spheres of *Tethya*. Calcareous sponges, solitary and branching, range from cream to dirty grey.

Ascidians are as colourful and varied as the sponges, a score of species having been recorded from Aquarium Point by Beryl Brewin.[8]

Abundant among these were clusters of purple, pendent columns of *Hypsistozoa fasmeriana* and neat, orange colonies of *Aplidium benhami*. Apricot clumps of *Aplidium adamsi*, with their common cloacal apertures on raised mounds, are larger but less frequent. White encrustations of *Didemnum candidum* abound. Everywhere one looks down on the delicately vermilion-tinged siphons of *Corella eumyota*, and frequently on to maroon-striped siphons of *Asterocarpa humilis*. Tall, translucent *Ascidia aspersa* is more local.

Where the current is swifter the delicate algae give place to athecate hydroids, with wide sheets of *Hydractinia parvispina*, *Turritopsis nutricula* and branching tufts of *Bougainvillea muscus*, while the thecate *Amphisbetia fasciculata* may form bushes more than three feet high. Twenty species of hydroids in all have been recorded from Portobello. There is also the octocoral *Clavularia* sp., with its stolons forming thick, violet brown patches four or five inches across, adhering firmly to shells and stones. Each is no more than a millimetre thick, giving off small polyps up to 5 mm long, each provided with eight pinnate tentacles.

CHAPTER FOUR
The Tides and Waves

The tides are produced by the gravitational force of the moon and sun, pulling the water on the earth's surface first one way, then another. Semi-diurnal rise and fall is the usual result, with a high and a low tide recurring twice in a little over 24 hours.

The tide-raising force of the moon is about twice that of the sun. At new and full moon, when the attractive force of the two bodies acts in a line, we have the larger movements known as **spring tides**. At first and fourth quarters, when these forces act at right angles, the tides are of smaller extent and are called **neap tides**. The tidal wave in an ocean may be compared with water swinging back and forwards in a dish. Because of the earth's rotatory movement, oscillation is not simply from end to end across a line. The tidal wave also acquires a rotatory course, moving clockwise in the southern hemisphere, around a node called the *amphidromic point*.

New Zealand, unlike some Pacific shores, has no irregular or multiple tides. The tidal regime is uncomplicated, showing two tidal maxima and two minima in the lunar day, with a lag of a little under an hour between the corresponding tides on successive days. The vertical ranges of springs and neaps can vary greatly on different parts of the coast. Wellington Harbour has a spring range of less than 1.5 metres. Auckland Harbour enjoys double this.

Of no small importance to the plants and animals of the sublittoral fringe must be the time of day when — at the lowest spring tides — they are laid open to the maximum stress of the atmosphere. A low spring tide at dawn obviously presents a smaller hazard than one at noon.

The *New Zealand Government Annual Tide Tables* show the times and heights of high and low water at the main ports for each day of the year. Corrections are then given for more than 100 places round the coast. From further tables can be calculated the height of the tide at a specified time of day, or the time at which the tide will reach a specified height.

The height of a given tide is stated by reference to *Chart Datum*. For New Zealand shores this may be taken as the low water tidal plane of the largest scale Admiralty Chart of the area in question. Chart Datum

approximates then to the zero prediction of tidal levels for each place. Tides will occasionally fall below this, and such an exceptionally low tide is indicated by a minus sign. Computer software is now available to find tidal information for any location for years in advance.

The predicted tidal level may on some days fail to be realised. Strong onshore winds may have the effect of keeping the water level abnormally high, while offshore winds can lower it. Barometric pressure of the atmosphere regularly affects the water level; a rise equivalent to 25 mm of mercury will depress it by 0.3 m.

Shore levels

The landward boundary of the intertidal zone may be set at the **extreme high water of spring tides**, abbreviated to EHWS. This is the average height of the two highest tides at the period of the year when the range of the tides is greatest. For the same period we may derive the **extreme low water of spring tides** (ELWS), a level beyond which the water-line never retreats. **Mean high** and **mean low water of spring tides** (MHWS, MLWS) are calculated from the averages of each pair of successive high waters or low waters during that period of about 24 hours in each semi-lunation when the range of the tide is greatest. By a similar procedure **extreme high water** and **extreme low water of neap tides** (EHWN, ELWN) are obtained by reference to the times of the year of the smallest tidal range; MHWN and MLWN refer to the period in each semi-lunation when the tidal range is smallest.

A mean neap or spring tide is then a twice monthly event, an extreme a yearly one. We should remember that EHWN (sometimes written E(L)HWN) is the lowest of all high tides; and ELWN (or E(H)LWN) the highest of all low tides. **Mean sea level** (MSL) is the average level of the sea from all states of the tide. **Mean tide level** (MTL) is a similar reckoning made for a particular locality.

The genesis of zonation

The tide levels are obviously the shore's simplest vari-

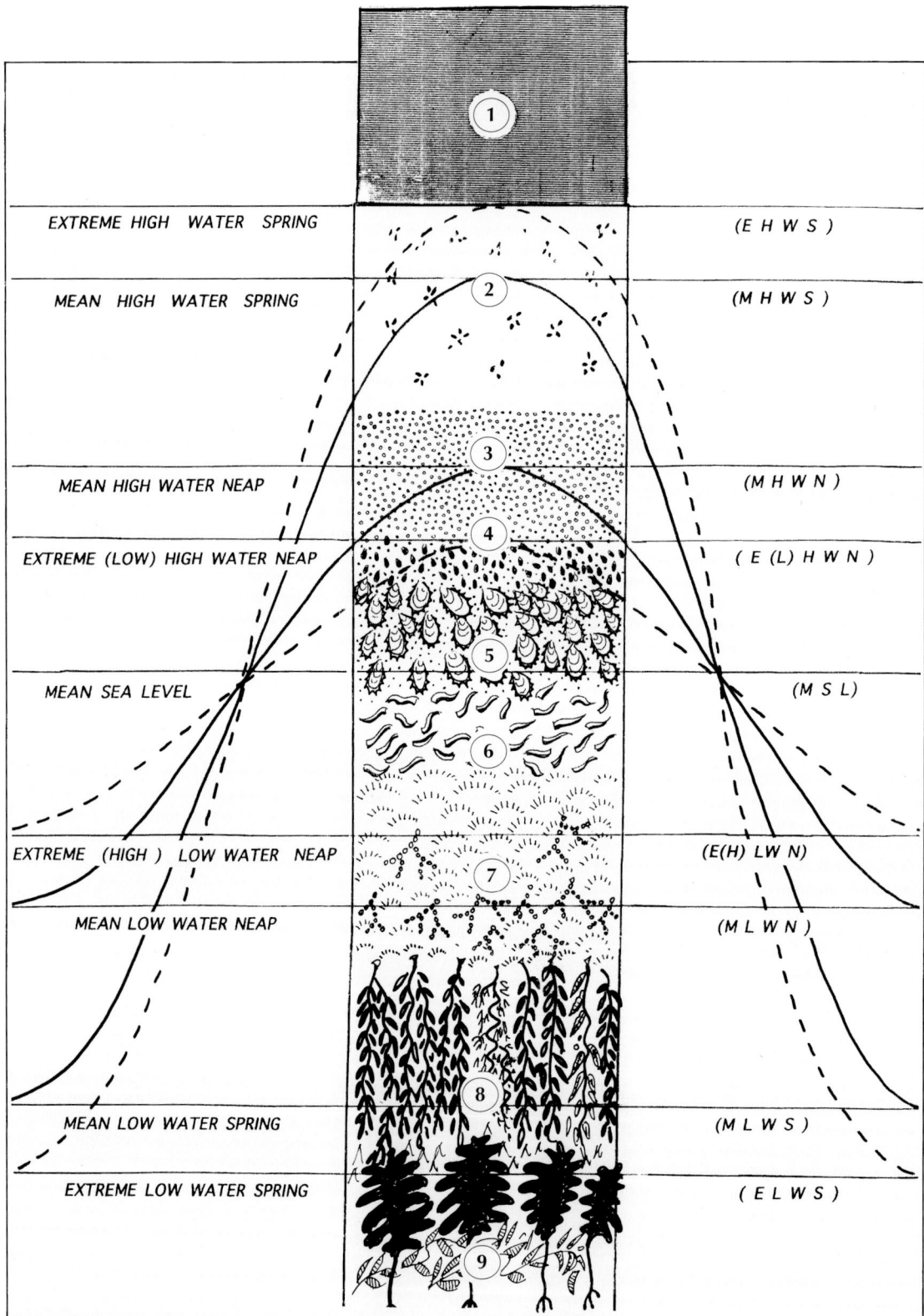

Fig. 4.1 Tidal levels and zonation

The column represents a northern shore of moderate (or 'average') exposure.

1. **Supralittoral zone** with *Verrucaria*.
2. **Littoral fringe** with *Austrolittorina*.
3. **Upper eulittoral zone** with *Chamaesipho columna*.
4.–6. **Middle eulittoral zone** with *Xenostrobus pulex* (4), *Crassostrea glomerata* (5) and *Pomatoceros caeruleus* (6).
7. **Lower eulittoral zone** with *Corallina* and *Hormosira*.
8. **Sublittoral Fringe** with *Carpophyllum maschalocarpum*, *Sargassum* and *Cystophora*.
9. **Sublittoral Zone** with *Ecklonia radiata* and *Carpophyllum flexuosum*.

ables to measure. In a first approach to a zoning pattern, the student will not unnaturally try to relate each of these living bands to a particular tidal level. The islet off Taurikura Bay, at Whangarei Heads, locally called Little Hat, became for three decades of students the site for their first day's study of zonation.

It seemed at the outset useful to assign particular tidal levels to each zone. But after the first attempt to use the plants or animals as a natural tide gauge, it needed only a walk round a promontory to where waves bear in more directly to find that the biological zones are not inflexibly tied to the same tidal levels.

Only 100 m away from our starting point, with increased wave action on the islet's open side, the top of each zone — lichen, periwinkle, barnacle and *Apophlaea* — was elevated by up to half a metre in relation to the tide. Moreover, rock oysters were now scarce, with *Hormosira* becoming small-bladdered. Alternatively, a move towards increased shelter, with the water becoming clouded, we could have found *Carpophyllum maschalocarpum* replaced by *C. flexuosum*, along with *Ecklonia radiata*.

How far then can we usefully speak of 'tidal' zonation at all? For despite the breakdown of correspondence with particular tidal levels, the bands have a regular vertical *order* that seldom fails, being reversed only rarely and locally for some special and usually quite apparent cause.

What are the causative factors of such a regular sequence of zones across the shore? It would still seem reasonable that the principal initial ones are tidal, and that the first determinant of the order and composition of the zones must be the moving interface of air and water.

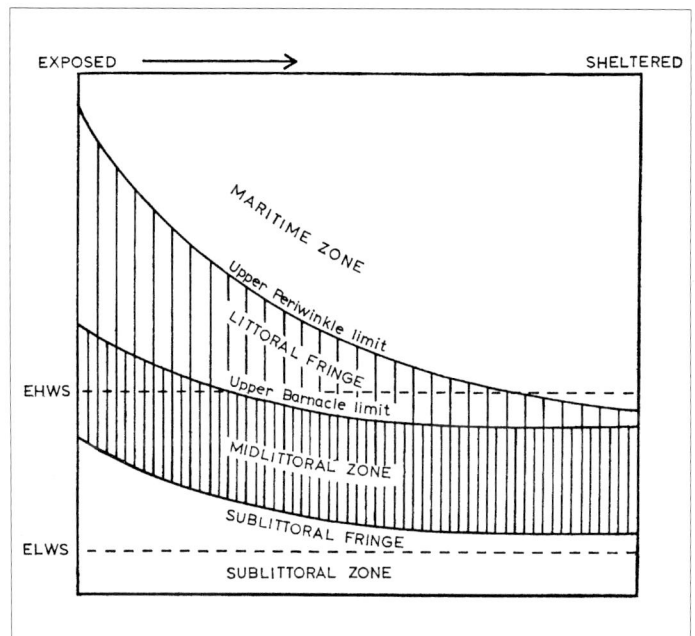

Fig. 4.2 The nomenclature of shore zones
(Adapted from Lewis's modification of the Stephensons' scheme.)

The straightness of zoning

The zones have not only their special *order*, but rather clear-cut **boundaries**. It is these that give a tidal shore its most notable character, with the reference levels *'barnacle-line', 'red-line'* and *'brown algal-line'* plainly to be picked out on almost any strip of coast.

While there are secondary species that can spill from one zone into the next, the sharp visible replacement of horizontal zones is still impressive, being most prominent on high-exposed shores and islands. But the semi-diurnal tidal curve is normally graded and *continuous*.

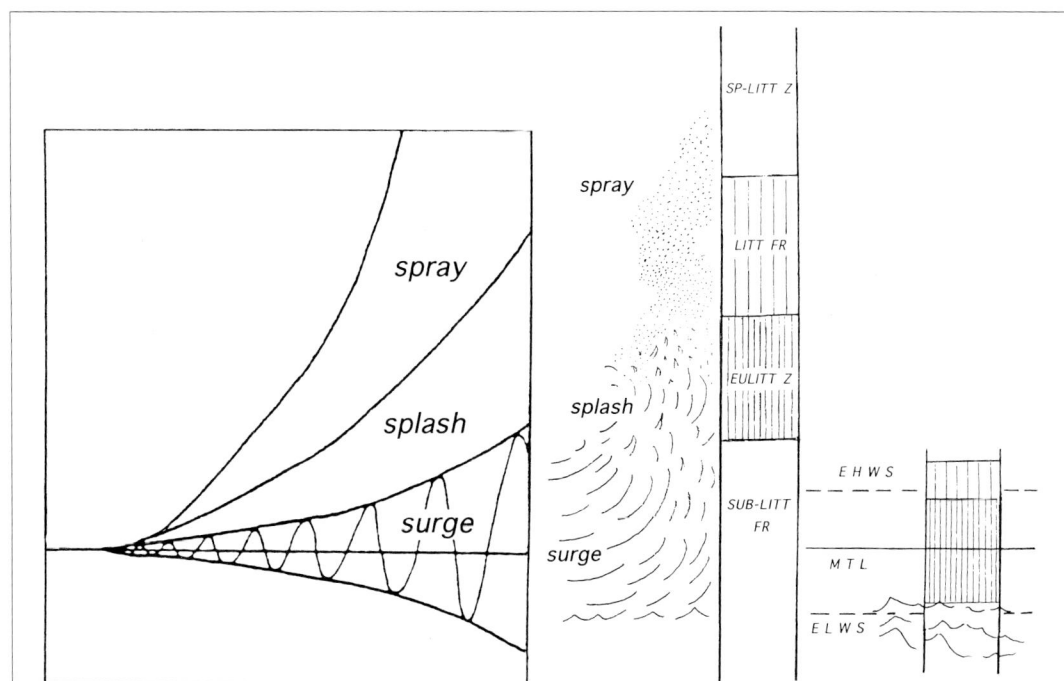

Fig. 4.3 Waves supersede the tides (1)
The want of regular correspondence of zones with tides is shown in two locations with low and high extremes of wave action.

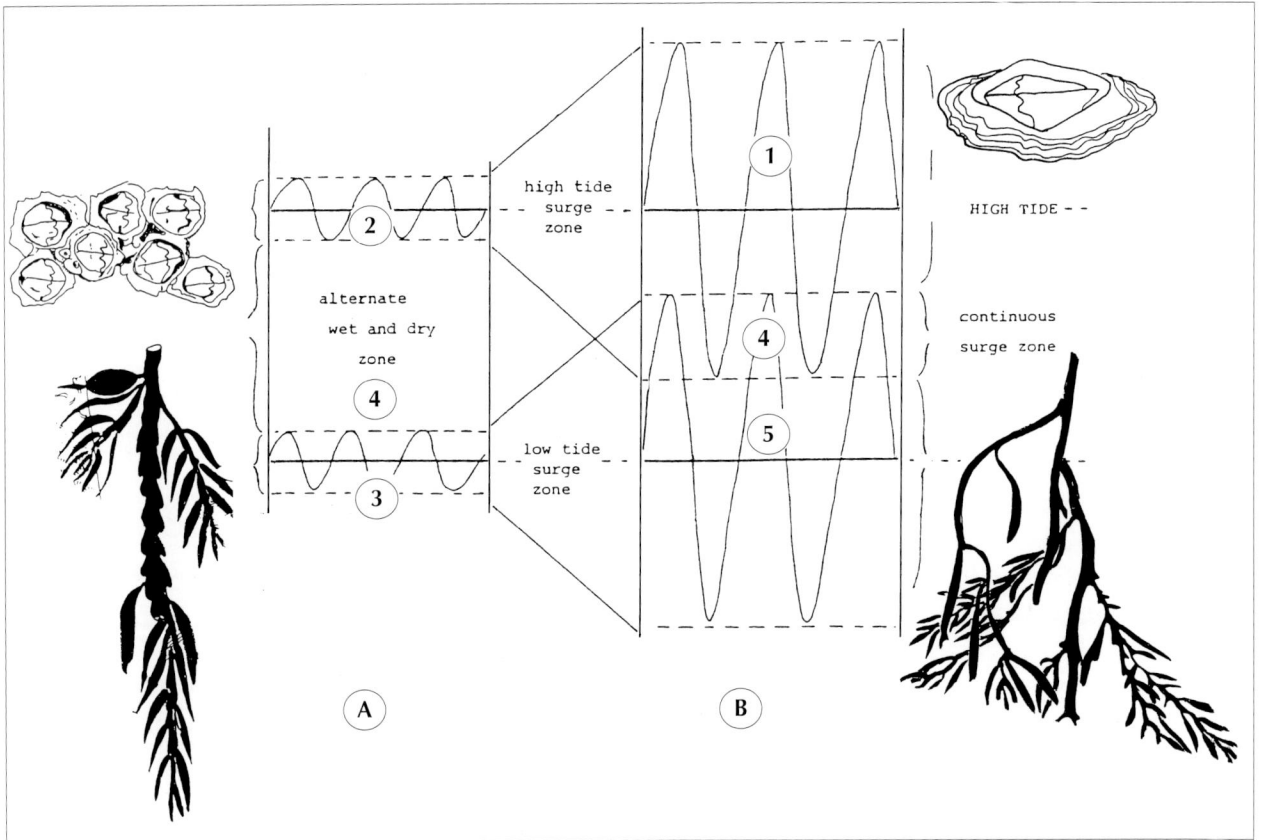

Fig. 4.4 Waves supersede the tides (2)
Diagram showing the relation of wave amplitude with tidal range. A. Waves much less than tidal range.
B. Wave amplitude greater than tidal range. Biological zones: 1 *Chamaesipho brunnea* 2 *Chamaesipho columna*
3 *Carpophyllum maschalocarpum* 4 tufting and coralline algae 5 *Carpophyllum angustifolium.*
(By courtesy of Bill Ballantine.)

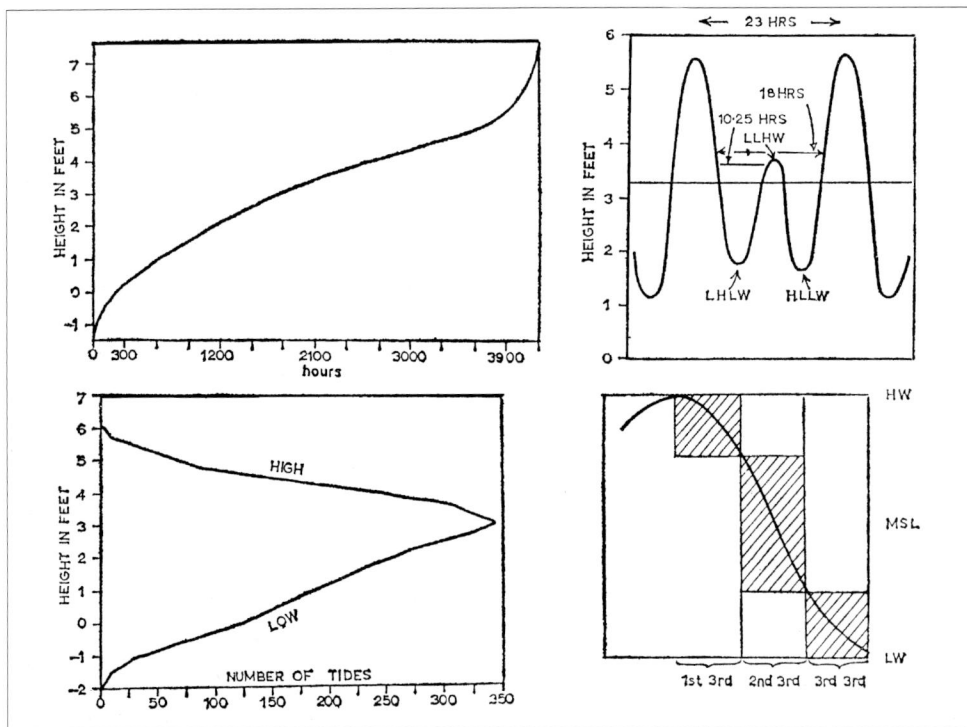

Fig. 4.5 The effect of dodge tides
Curve of an irregular or 'mixed' tide at San Francisco. With the short tide intervening between two normal high tides,
adjacent levels on the shore have widely different spells of emersion.

What are its properties that could help explain the step-wise *discontinuities* of the biotic zones?

'Critical levels'

The flow of a single tide is fastest during the middle two hours of the six. It slackens to about half this speed in the first and sixth hours. Given such a curve, some have suggested that certain levels of the shore present transitions between rather different lengths of immersion time, and that the presence of such a threshold could establish a biologically **critical level**.

Plots could be made of the *accumulated* hours of emersion over six months experienced at each level of the shore from high to low water. Alternatively, we could determine the *frequency* of immersion and emersion by the number of high and low tides effective at each level, with increasing distance above and below mean tide level. Neither of these curves seems to offer any discontinuity that would support a theory of *critical levels*.

It is easy, however, to identify shore levels where a more-than-average number of species meet their high or low cut-off. Thus, for each shore level we could record the number of species with an upper or lower terminus, and by subtracting the non-terminating species obtain for that level a ratio of *criticality*.

On certain shores, as in parts of western North America, the two successive tides of the day reach very disparate levels. This phenomenon of **dodge tides** was invoked in the 1940s by the marine ecologist Maxwell Doty to explain some postulated critical levels in California. In some places — as he was able to show — a couple of points only a short distance apart vertically could experience two or three hours' difference in total daily emersion. Here, it was contended, a critical cut-off might be established, with significant differences of microclimate above and below.

Theories of such critical levels were at that time much canvassed, even though dodge tides are far from general, and hardly occur in New Zealand at all. Some supporters of the tidal hypothesis were *un*critical enough to overlook the strong modifying effects of wave action on the notional predicted level of the tide. Wave action must always be in some degree significant, but it takes over the commanding influence on open shores where the zones are straightest and most sharp.

Critical levels, even corrected for the true extent of wetting, have less appeal as a hypothesis today. With growing understanding of the physiology and behaviour of the zoned species, and of the *biotic* influence of other animals and plants, few naturalists would seek to explain zonation by a strait-jacket of *climatic* forces. Once more we realise the truism: 'a species lives not where it *can* but where it *may*'.

Maintaining the zones

We have already found — as in barnacles — a zonal limit is imposed by a predator, or by space-competition from neighbouring species. Yet even the sum of external biotic agencies, competitive, predatory or grazing, may not be decisive in determining a zone. We must be alert as well to a species' own selective behaviour in the maintenance of its territorial limits. Most significant could be the site preferences shown at the time of recruitment.

First, some mobile gastropods can maintain their shore level by orienting behaviour. In the periwinkle, *Austrolittorina antipoda*, studied by Brian Foster, larvae were found to range lower than adults. They will apparently settle at any level, later to migrate upshore. Marked specimens transferred to lower levels travelled up to 1.5 m to regain their former shore level after 24 hours. With a marked dislike of submersion, they retreated before the advancing tide up to 0.6 m above the contemporary high tide, relying on surge for moisture. In rough weather their top level could be extended by more than 4 m. Above this, random grazing might take some individuals higher, so long as they were kept moist by splash and spray. *Austrolittorina* protects itself from desiccation by a tight-fitting operculum, sealed by dried mucus. With a stimulus of moisture, periwinkles kept in a dry matchbox for six weeks responded to the softening of their mucus seal by prompt emergence.

Settlement and recruiting

To perpetuate a sessile zoning, enough larvae must regularly settle at their parents' level. For all we yet know, site-finding by algae may be largely unguided. Hazards may, however, for some algae be reduced by the timing of spore release to avoid disadvantageous tides or currents. Some sporelings obviously settle and begin to grow where ultimate survival would be impossible. In October *Durvillaea* plants 6–10 cm long may be found well up in the barnacle zone, where summer desiccation will bring early doom. Seasonal brown algae proliferate in spring at levels untenable in summer. There follows a time of wholesale failure and destruction of tender eulittoral algae, offering as they die back important fodder for herbivores.

The larvae of sessile animals used to be counted lost, like the seed in the Parable, if they settled on unfavourable ground. We are finding now that nature is not always so wasteful, and that most larval mortality happens from predation in the plankton, rather than from mistakes of settlement. A wide tally of species, including bryozoans, oysters, mussels, *Spirorbis* and barnacles, can employ a several-days probationary period, during which they can sample a substratum by crawling over it, then take off to swim once more.

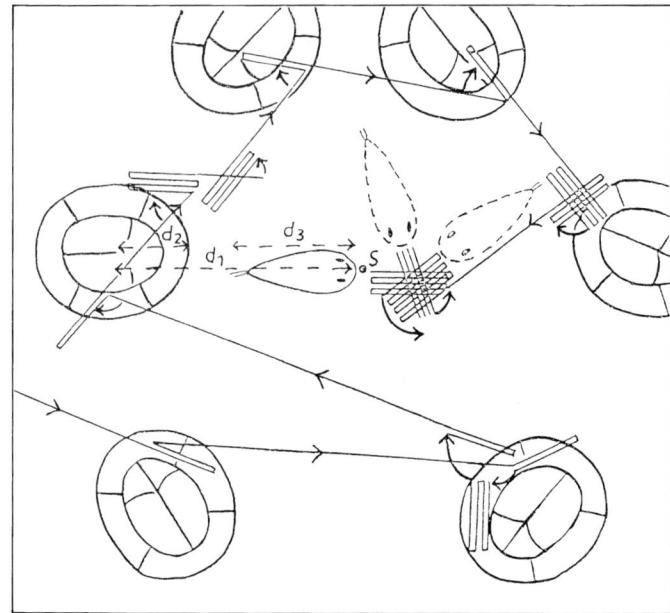

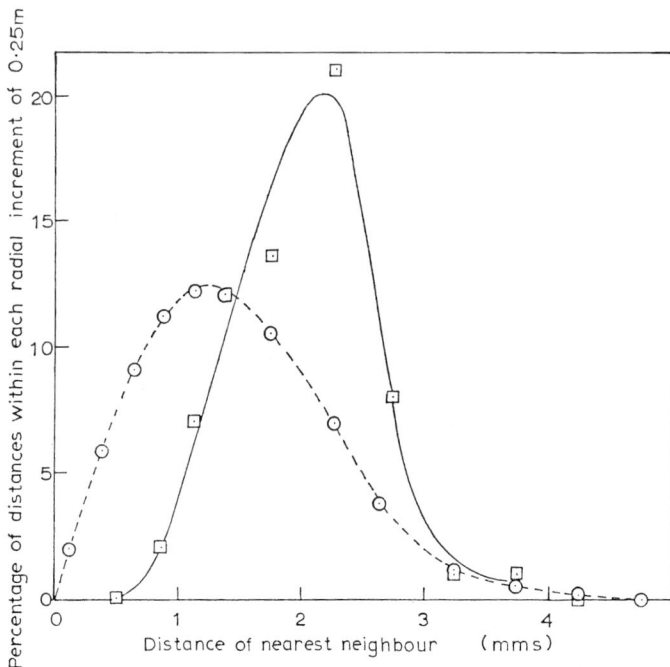

Fig. 4.6 Settlement of barnacle larvae

(top) Larval movements of *Balanus balanoides* are shown in relation to 6 attached spat. Straight lines denote swimming or walking of a larva entering from the left. To-and-fro movements are shown by close parallels, and swivelling movements by curved arrows. The final settling point is at 'S'. (d 1), territorial separation of attaching larvae from adjacent spat (d 2) radius of existing spat (d 3) length of larva.

(bottom) The spacing behaviour of larvae in relation to the nearest neighbour is expressed as the percentage of the total within each radial increment of 0.25 mm. The dashed curve shows the expected distribution in a random settlement, the continuous line the actual distribution found. (From D.J. Crisp.)

The planktonic stage is thus primarily exploratory. Bryozoan and ascidian larvae are poorly equipped to take food in the plankton but have site-finding sense organs and structures for attachment. Larval *Phoronis* have been shown to choose limestone rocks, and *Alcyonidium*, *Gregarinidra* and *Spirorbis* to settle on a narrow range of algal sites. Larvae of pholad bivalves will selectively bore into the corks of moored bottles. Young mussels after the first settlement may be carried about by strong currents, being finally attracted to the byssus threads of shells already attached.

Choice of site is, moreover, a gregarious process. Larvae will settle where their own adults are already established, helping to rectify boundaries and reduce straggling. In some still classic experiments in the 1940s, Crisp and Knight-Jones put out glass settlement plates for the cyprid larvae of barnacles. Plates that carried adult shells or even the tanned proteins left from previous shell bases were always preferred by larvae of the same species.

Similar results for sessile molluscs, worms and other invertebrates suggest that gregarious settling is a widespread ability. Individuals of a species thus assist each other in the maintenance of a pure zone.

Discontinuity

The sharpness of boundaries, such a clear fact of shore zonation, has something in common with the major biological discontinuity of the species boundary. Since nature — as we were taught — abhors jumps, we might have predicted both zones and species to exhibit a graded continuity. The reverse seems true. Between species we find not just steepened intergrades, but sharp divides. Every species, it would seem, by its structure, behaviour or genetics, employs isolating mechanisms that will identify and sharpen its boundaries. Thus is avoided all the wastage of hybrids doing neither one thing nor the other as efficiently as the parent. Niche adaptation demands highly effective specialists as, compared with generalist 'Jacks-of-all-trades' with a wide niche, specialists can flourish only where there is little or no competitive overlap. Where living space is tightly filled — as on the shore — by a plethora of competing species, there will be a high premium on specialist efficiency. This will be favoured by a narrow, homogeneous zone. Over too wide a space, any special advantage would be diminished.

On the shore, then, efficiency would be lost if zones were open to a wide entry of species cropping a similar resource. A uni-species regime, with individuals close-assembled, will allow the optimum for each strategy of food-collecting or space-utilisation. Sessile species thus tend to settle together in tight, exclusive bands. More open spacing, permitting the entry of other species, would reduce the autonomy of the first colonists, exposing them to smothering or over-growth, reducing their feeding efficiency, or — with barnacles — lessening an individual's chance of cross-fertilisation. All these optima are promoted by straight, homogeneous zones.

Spacing

Gregarious settlement by itself, however, could be rough and ready, leading to congestion, with too little room for the crowded young to grow. Oysters indeed often suffer in this way, becoming stunted and misshapen, and often prematurely lost. Crisp and Knight-Jones[1] were the first to show that settling larvae are able to space out optimally before final attachment, giving them more chance to attain maturity. The acorn barnacle *Balanus balanoides* was found to settle in greater numbers where squares had been cleaned of existing barnacles, and in lower concentration where there was already a dense occupation. So long as there was clear space, the larvae would avoid high densities; only under crowded conditions would they resort to close settlement.

This spacing ability is clearly intraspecific. *Balanus balanoides* (Fig. 4.6) will thus primarily avoid near contact with others of its own kind. Only 15% were found to settle within 0.5 mm of others, with the median distance apart being 2.5 mm. But where they settled among

Austrominius modestus as many as 76% of spat remained in actual contact with the other species, preferring especially the shell's deep radial grooves. An even higher proportion settled in contact with *Chthamalus stellatus*. This species, being a member of a different family (the Chthamalidae), is apparently not recognised for avoidance at all, while *Austrominius*, being one of the same Family Balanidae, gets a partial taxonomic recognition.

Waves and tides

On high energy shores, as we have found, the notional levels predicted for the tides can have small biological relevance. Above the level of full wave attack, surge can sweep so continuously that a slope is never dry between runoff and wave return. **Surge** rushes furthest up channels and small gullies, prolonging through these places the reach of barnacles or coralline paint. Above the ultimate action of surge, the surface remains wet with the fall of **splash**. Still higher, for many metres above the splash, the air is permanently humid and the surface damp with salt **spray**.

Upon the tidal wave, with its semi-diurnal period, are thus imposed the shorter frequencies of wind-generated waves. In the more sheltered waters, these are generally of such small amplitude that the tidal wave retains full command. But on high energy shores, the distance

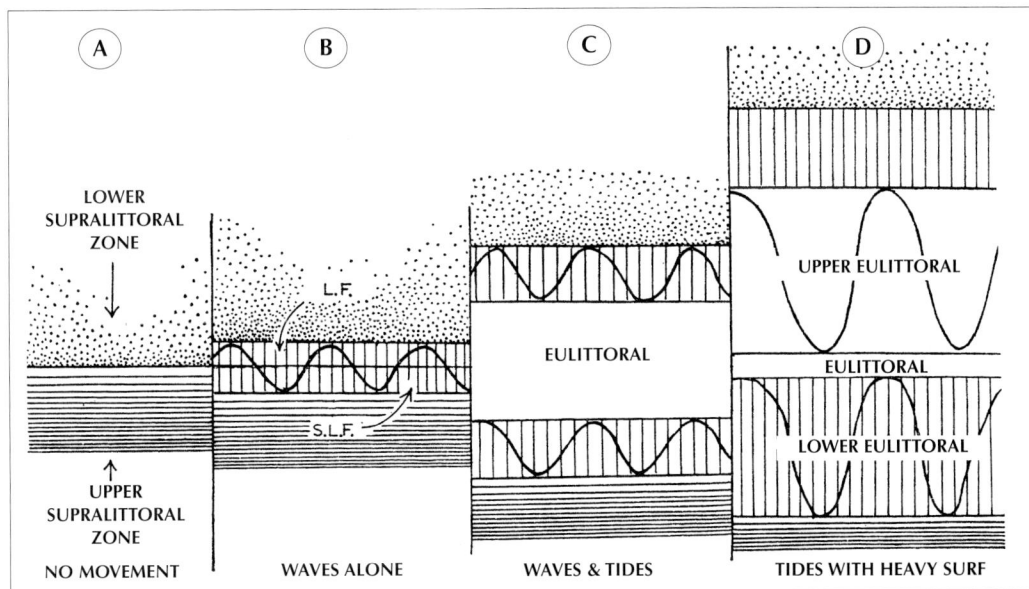

Fig. 4.7 The role of waves and tides in shore zonation
(A) shows a hypothetical situation with no water movement at all, but with *supralittoral* and *sublittoral zones* produced at either side of the air/water interface. In (B) with action of waves alone, a *littoral fringe* (LF) and a *sublittoral fringe* (SLF) are introduced. With the addition of tides in (C), a distinct *eulittoral zone* now appears. On a surf-pounded shore, as in (D), the high waves assume the dominant role; the *sublittoral fringe* is now raised clad by wave action to take over the lower extent of the eulittoral zone, while the barnacle-clad upper eulittoral extends far above notional high water under the influence of splash.

between wave trough and crest can exceed the whole daily tide curve. With a regimen thus wave-controlled, the term 'tidal zonation' is no longer sufficient.

At the opposite extreme, not only may wind-waves be lacking, but there are also parts of the globe where the tides have effectively disappeared (Fig. 4.8). Even without tides or waves an elementary zonation can be set up by the existence of an interface between air and water. Above the water-line, there is a graded change in the influence of spray and salt. Below the water there are gradations in the amount and spectral quality of light, and in the effects of freshwater runoff, with dissolved or suspended solids from the land.

Such a stationary interface can produce the two primary divisions: a *supralittoral zone* above, and a *sublittoral zone* below the water-line. The introduction of waves, but still without tides, will bring regular short-term alternations of immersion and emersion. There is thus intercalated a narrow strip between the primary zones, corresponding pretty well to the combined *littoral* and *sublittoral fringes*. It remains for the further addition of tidal action to interpose a normal *eulittoral zone*, regularly (and usually twice daily) immersed and emersed.

On a high energy shore, with waves predominant over tides, the *eulittoral zone*, that in moderate shelter has the three divisions, *upper*, *middle* and *lower*, tends to be simplified. The brown algal *sublittoral fringe* has become wave-elevated, virtually to suppress the lower half of the eulittoral. The upper half of the eulittoral now consists essentially of a barnacle zone, surge-elevated to displace upwards in turn much of the littoral fringe. Splash and spray will in their own order carry the successive belts of lichens still higher (Fig. 4.7).

Towards a 'tideless' sea

There are a few parts of the world where this extreme could be reasonably put to the test. The Mediterranean Sea is not — as often supposed — quite 'tideless', but the diurnal range is by normal comparison much reduced. It was in fact in Italy, remote from the Pacific scene, that this chapter came first to be written.

Fig. 4.8 is from the four shores we visited. At **Naples**, the total spring tide range is only 0.5 m. On the high energy shore at **Sorrento**, wave effects so predominate over tides as to produce a shore with the two fringes, *littoral* and *sublittoral,* but with the *eulittoral* failing to appear. The littoral fringe is clad with the alga *Porphyra umbilicata* and blue-greens, while the sublittoral fringe is pink-veneered with corallines, showing up vividly as the waves recede. Close up, this is found to carry the barnacle *Balanus perforatus* and the gastropod *Vermetus triqueter*, in the site tubeworms occupy on our New Zealand shores.

The island of **Ischia**, off Naples, with the same abbreviated tides as Naples, is far less wave-exposed. The *supralittoral zone* has the black lichen *Verrucaria* and the high-level periwinkle *Littorina neritoides*. The *littoral fringe* supports both *Porphyra umbilicata*, and the high-level limpet *Patella lusitanica*. The pink *sublittoral fringe* carries abundant algae, with a turf of *Corallina mediterranea* and tufts of the cosmopolitan *Pterocladiella capillacea*. The two commonest molluscs are the limpet *Patella caerulea* and the topshell *Monodonta turbinata*. The *sublittoral fringe* stands out by its rich brown algal cover of *Cystoseira* and *Sargassum*. Beyond these algae the shelving *sublittoral zone* is dominated by the sea-grass *Posidonia oceanica*.

Ischia has a few situations such as the concrete jetty below the Castello, where the barnacle *Chthamalus stellatus*, in a strip only 5 cm wide, is intercalated between the supralittoral and sublittoral fringes. Here is the most modest rudiment of the *eulittoral zone*, to be expected from the imposition of a minimal tide.

In the Adriatic Sea, the tidal range is extended to about a metre. At **Ancona**, a fishing port since Phoenecian times, there is a high-energy shore largely under wave control. A black *supralittoral zone* of *Verrucaria*, elevated 2 m by splash, is followed by a *littoral fringe* with *Patella caerulea* while the *sublittoral fringe* is rich with *Cystoseira*, *Petalonia*, *Dictyota*, and a diversity of rhodophytes. With the enhanced tidal effect, a *eulittoral zone* has been added with the black mussel *Mytilus galloprovincialis*, mingled with bright green *Ulva*.

Our Adriatic study was carried right up the tidal canals of **Venice**. On the steps in front of St Mark's Piazza the *eulittoral* becomes more complex — with the tidal range approximately 1 m, but under low wave energy. In the upper half is the barnacle *Balanus amphitrite* and in the lower *Mytilus galloprovincialis*, with the limpet *Patella caerulea*. The black *littoral fringe* has *Littorina rudis* with blue-green algae and *Porphyra*. The *sublittoral fringe* carries *Scytosiphon*, *Gracilaria* and *Ulva*. The canals, with a tidal range of about 1 metre, are sites of high eutrophication. On each wall is a bright green *supralittoral fringe* of *Enteromorpha intestinalis* and *Cladophora glomerata*, with a lower band of large ulvoids and *Lyngbya*.

Wave exposure

Marine waves are set in motion by wind acting on the surface of the sea. They may thus travel long distances either as 'forced waves' actively pushed by the wind or 'free waves' or swell continuing under their own momentum. The movement of water particles at a given moment incorporated in a wave is distinct from the forward movement of the visible wave-form. They move

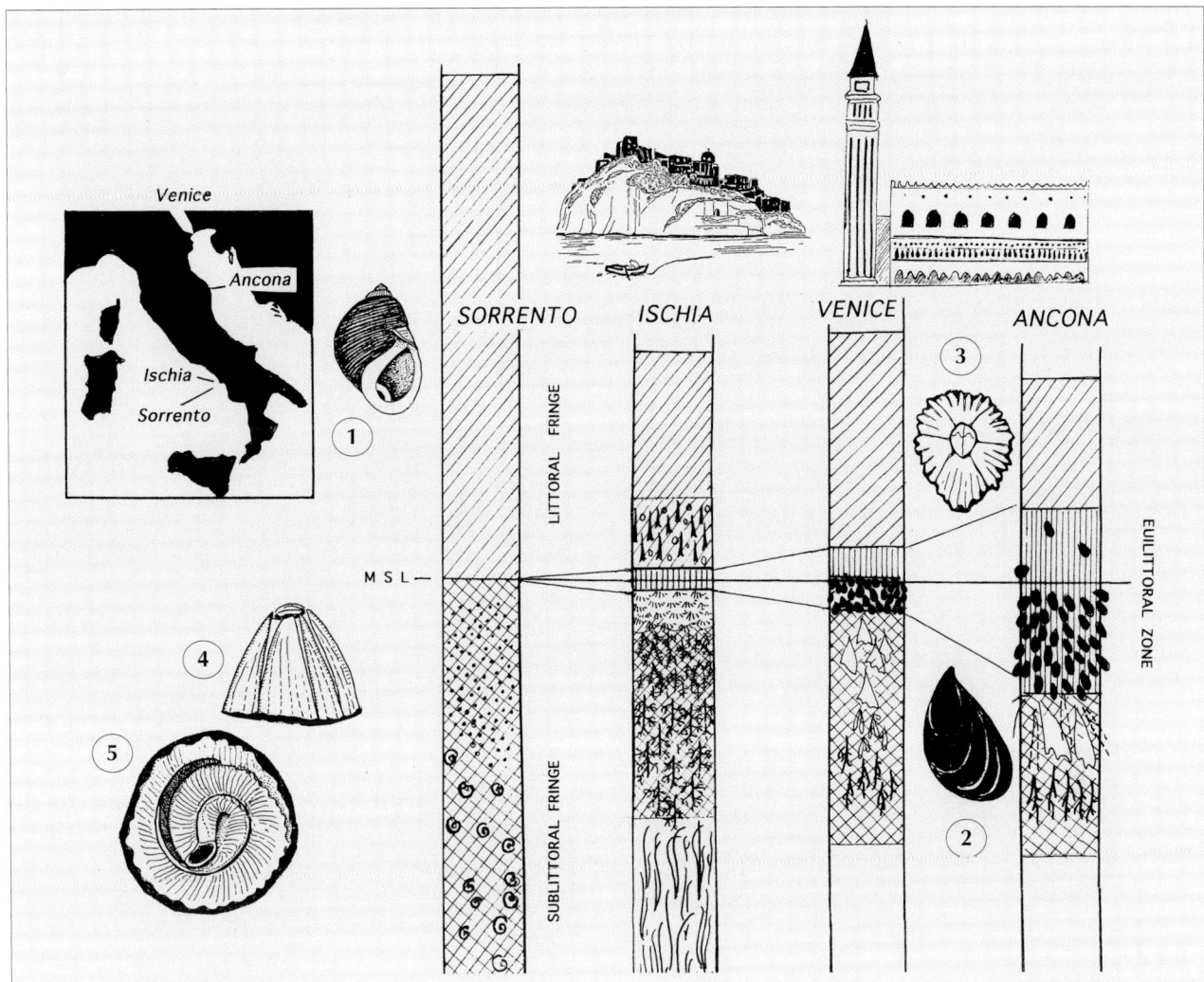

Fig. 4.8 With diminished tides
The zonation columns are shown for Ancona and Venice (Adriatic Sea) and Ischia and Sorrento (Tyrrhenian Sea). Notice the effect of minimal tides in intercalating a narrow eulittoral zone at Ischia and its expansion with tides of up to a metre in the Adriatic. (For zoning details, see text).
1 *Littorina neritoides* 2 *Mytilus galloprovincialis* 3 *Chthamalus stellatus* 4 *Balanus perforatus* 5 *Vermetus triqueter.*

clockwise in circular orbits, each particle being on the side of its orbit nearest to the approaching wave crest (Fig. 4.9). The crests are thus heaped up while the troughs between them lose water. The size of the orbits diminishes with depth until — at a depth of half the wave length — orbital movement ceases. The height of a wave is equal to the diameter of an orbit at the surface. In the low swell of the open sea, this may be some 2 m, rising to more than 10 m in heavy seas. The wave length (i.e. the distance between one crest and the next) may be 70 to 200 m with forced waves, and much greater with swell. The wave period is the time elapsing between the passage of one crest and the arrival of the next. The height of an ocean wave is dependent for a particular wind-speed on the 'fetch' or length of sea over which it is generated, varying from a few kilometres across harbours to thousands of kilometres across an open ocean. The graph in Fig. 4.9 relates wave height and

wave period to length of fetch. An important ratio is the *steepness* of a wave, determined by the height over the length. On its steepness rather than its amount of energy depends the constructive or destructive effect of a wave on the shore.

Waves on the shore

The term *exposure*, as applied above to wave effects on high-energy shores, obviously covers a whole complex of physical variables. First, breaking waves increase the frequency of wetting. Swash flows up from the break-point to a level determined by the properties of the wave itself (height, wave length, velocity), by the shore profile and by the force and direction of the wind. Exposure involves increase in both wave force and the upreach of swash, splash and spray.

Inundation by periodic waves or swash is far from

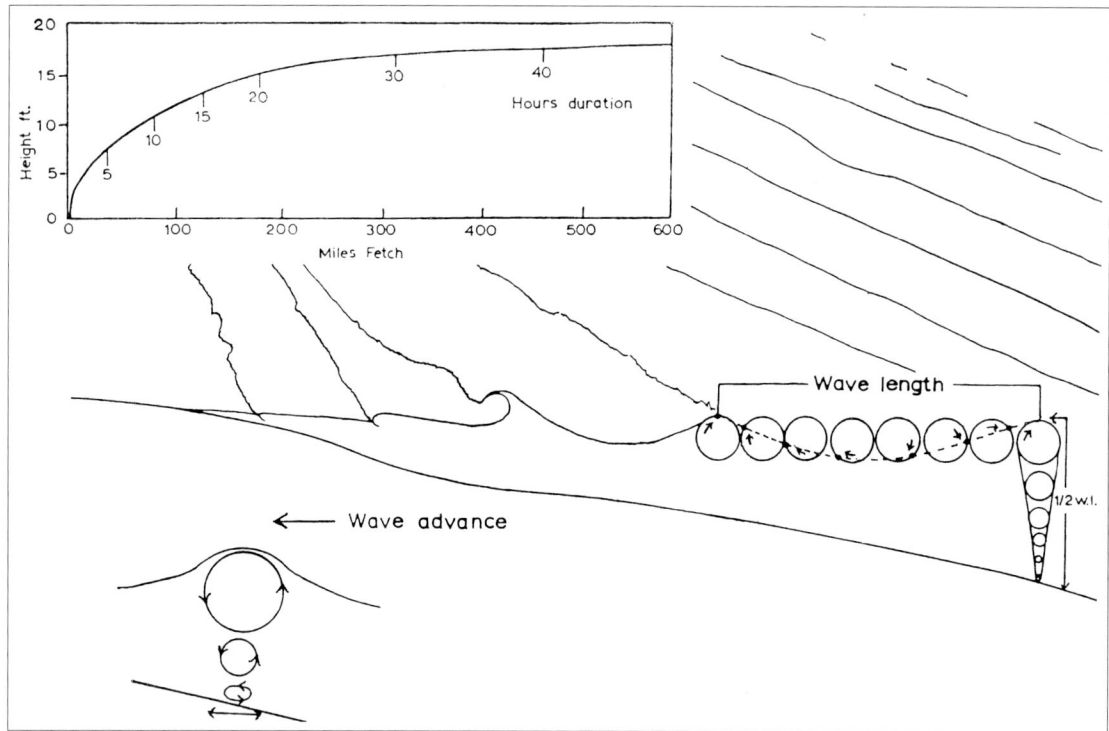

Fig. 4.9 Marine waves
Orbits shown at intervals in the propagation of a wave, with the simultaneous position of water particles and the profile of the water surface. **(bottom left)** Motion of water particles on reaching shallow water. **(inset, top left)** Relation between wave height and length of fetch, at maximum height of waves raised by a 50 km/hr wind. (From Sverdrup, Johnson, Fleming.)

equating in its effects with sustained tidal immersion. The quantity and spectral quality of light, including UV radiation, are different on wave-swept and fully submerged surfaces. The algae of the swash-inundated eulittoral thus form a community entirely distinct from that of the permanent sublittoral zone.

The damaging effects of waves include abrasion, pressure and drag. Abrasion will be greatest where organisms are scoured by suspended sand and shell fragments. Algae will be damaged where their flexibility — itself necessary under constant wave impact — allows them to be whipped by the waves against the surrounding rock.

Moving water exerts *hydrostatic pressure*. This will be augmented as the air in the hollow wave-face of the breaker is compressed. Not acting directionally, the pressure will be exerted uniformly on the plants and animals, which — being fluid-filled and incompressible — may suffer little damage.

Drag is the frictional force of water over the surface of the organism, with the dynamic pressure exerted against it in the direction of flow. In the absence of abrasion, drag must present the greatest wave hazard for shore organisms. Though acting at all stages of life, drag will increase as an organism grows and presents more resistance.

Measuring wave effects

There have been various approaches to measuring the total factor-complex of 'exposure'. Something can first be learnt from map study. As a provisional estimate of *wave exposure* we could calculate the number of days per hundred that the wind blows into the *exposure aperture* of the locality in question, this being taken as the seaward aperture measured at a distance of a kilometre. Thus, a straight shore, presenting an aperture of 180 degrees, with wind direction uniform round the compass, will have an *exposure factor* of 50.

We must consider not only the wind's direction and frequency but also its strength, and the length of fetch over which it has acted to generate waves. Fig. 4.9 shows the theoretic wave-raising power of winds over varying fetches. If we assume that a full oceanic wave can be got up in 320 km, and multiply by 180 for a full exposure aperture, then division by 57,600 will reduce the co-efficient value for maximum exposure to a convenient unity. Shorter fetches or smaller open sea apertures will then produce fractional values.

By watching from the shore we can usefully reckon the number of tiers of simultaneously breaking waves, or — against a rock face — the number of waves reaching a specified height. Sustained rough weather may have more effect than intermittent bursts of heavy seas.

Reliable shore observations can be made by rule of thumb: as of the height above Chart Datum reached by splash and surge on calm and rough days. The upreach of swash above predicted high tides will generally show some correspondence with the position of the *barnacle-line*; and the level below which swash does not recede on a calm day appears critical for some low tidal algae on exposed shores.

A biological exposure scale

The effects of wave exposure are of transcending importance, despite all the complexities of their final analysis. Bill Ballantine wrote:

Wave action effects tend to be so pervasive that the whole pattern of community structure will change with any general increase or reduction in wave movement.

Any physical assessment of wave exposure would assume it is known which components are the most important. A *biological scale* of shore exposure avoids this problem by starting from the opposite end. A given shore, despite its individuality, will not have a random collection of organisms growing on it. We have found predictable patterns on shores where high waves break, and on those that are quiet and land-locked. The existence of a biological spectrum from high to low energy sites was one of our earliest realisations about the shore. It did not need measurement to be first obvious. The Stephensons in the 1930s had drawn the first diagrams of coasts from exposure to shelter, showing that organisms towards the exposed side reached higher than the same species on the sheltered side.

Over the years since, it was to become clear that the left-right axis, conventionally from high to low energy, is as important for the distribution of animals and plants as the vertical axis from high to low water. With such a justification, Ballantine in 1961, from a study of shores at Dale Fort, south-west Wales, under a wide exposure/shelter range, proposed a *biological exposure scale*.[2] Using the occurrence, shore levels and density of selected animal and plant species, it was moreover assumed that between the two extremes of *very exposed* and *very sheltered* there would exist an orderly progression.

The *biological scale* was first designed around the limiting factor of *wave action*, chosen for its universality of effect. It thus leaves out of prime account other factors such as *illumination*, fall of *salinity*, and local effects of *sand-scour*. Beginning with wave action on a uniform hard slope and using the ranges and abundances of chosen species, shores can be ranked in an order of exposure, so as to uncover and organise new information. Being itself a tool for further discovery, the system has high *heuristic* value.

The assigned *exposure units* 0 to 7 cannot be conformed to a strict arithmetic scale; though they are placed at convenient intervals, their intercepts are not necessarily equal. We cannot deduce from the scale that one shore is twice as exposed as another. Still, attempts to correlate this scale with estimated physical data have shown good agreement.

Each of a series of neighbouring shore transects is first quantified from the abundance of chosen indicator species down a vertical line. The transects are then arranged in order, first by the abundance of each species separately. Such orders can be checked for consistency by using rank correlation methods. The final ordering will come from the averaging of the single species orders, and this is divided up into convenient segments to provide a horizontal numerical scale.

Suitable organisms must be chosen for the particular geographic region. For first use of the biological scale in Wales, Ballantine employed the following forms in order: the brown algae *Laminaria* and *Alaria* and fucoid species; other algae; the barnacles *Balanus balanoides* and *Chthamalus stellatus*; limpets, littorines, topshells, other animals.

From Norway to Spain

The biological scale was found to work well in Britain up to approximately 150 km from Dale, where it was first derived. Anomalies developed with distances further afield, revealing latitudinal or biogeographic trends that had to be brought into account if the scale was to be given wider use. Adjustments to the original scale were now provided through a span of 3000 km, from localities Ballantine studied from the Bergen-Hammerfest coast of Norway to Santander in North Spain (see Fig. 4.10).

The latitudinal trend so emerging was in some ways like the local shelter/exposure trend at Dale. Species showed a progressive change in their exposure-tolerance from north to south, so that if the Dale exposure scale had been applied unmodified, one would have been led to believe that there are no exposed shores in Norway and no sheltered shores in Spain.

Latitudinal change was found to be first concerned with the *fucoid: limpet: barnacle* balance. Limpets graze sporelings of fucoid algae. Fucoids tend to smother barnacles, and barnacles compete with limpets for space and so reduce their feeding efficiency. Competition between the three may exert a control just as effective as wave action on the distribution and abundance of any one. The relative efficiency of all the species must change with latitude, but at different rates. Ballantine's early

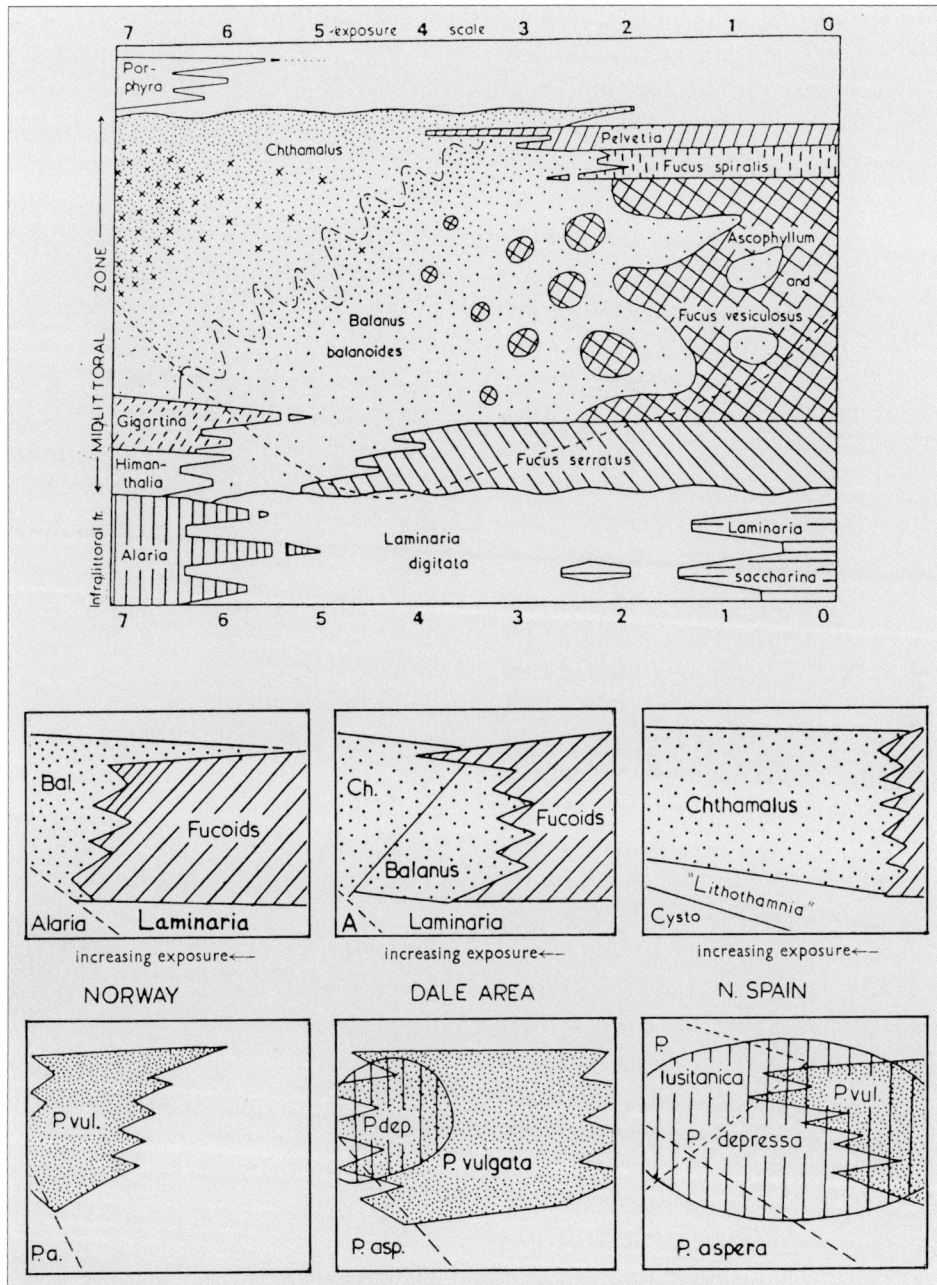

Fig. 4.10 From Norway to Spain
Ballantine's first presentation of a biological scale for west European shores, showing **(top)** alteration in the barnacle-fucoid balance and **(bottom)** associated changes in the range of *Patella* limpets.

diagrams (Fig. 4.10) show the changes in barnacle, limpet and fucoid balance from north to south.

North-south differences produce changes not only in proportions and relative balance of species, but also in their presence or absence. Well-documented changes from cold to warm temperate had been shown in the Stephensons' studies from Nova Scotia and Prince Edward Island to Florida Keys (1950s) and from cold South West African coast to the warm and subtropical coastlines of Natal (1930s).

East Northland

The biological exposure scale developed by Ballantine in 1959 has been taken much further by his researches in New Zealand. Today we have for these shores some of the best zoning analysis anywhere achieved. Not all of Ballantine's work has yet been definitively published. The chief source is still in a report for east Northland (1973),[3] based primarily on shores at Whangarei Heads and Mimiwhangata. Thirteen principal organisms are listed, and 15 more taken into account, with their exposure characteristics listed. It had early been realised by the Stephensons that zonation limits and tidal levels do not exactly correspond. Differences shown in local situations — as on the two sides of Brandon Island, near Vancouver — seemed primarily due to shade and illumination. Ballantine's great advance was to see changes of the level and composition of zones as the most fundamental feature of the intertidal. Of all the world's habitats, the intertidal must be the most extended and also the narrowest. It is under the tight constraint of overriding primary factors, being stereotyped in a way few other great habitats show. The factors ultimately concerned are physical and hence capable of quantitative measurement in setting them out. The ultimate determining factors lie in the effects of

wave action that transcend the action of tides, or of small, essentially local variations in shade, salinity and turbidity.

Thus in the Biological Exposure Scale is shown the vertical distribution of zone-forming animals and plants as dependent on the exposure of their shores to wave action. Thus a continuous range of habitats runs from the quietest water with little wave action, as in the extreme shelter of Whangarei Harbour, to the most exposed outstanding islets at the Poor Knights.

Waves tend to reach higher as the shore slope has increased — in the first place by wave action — to a high angle. Conversely, protected or sheltered areas, with reduced or negligible wave action, are left at a low angle of slope or almost flat. Ballantine's diagram (Fig. 4.11)

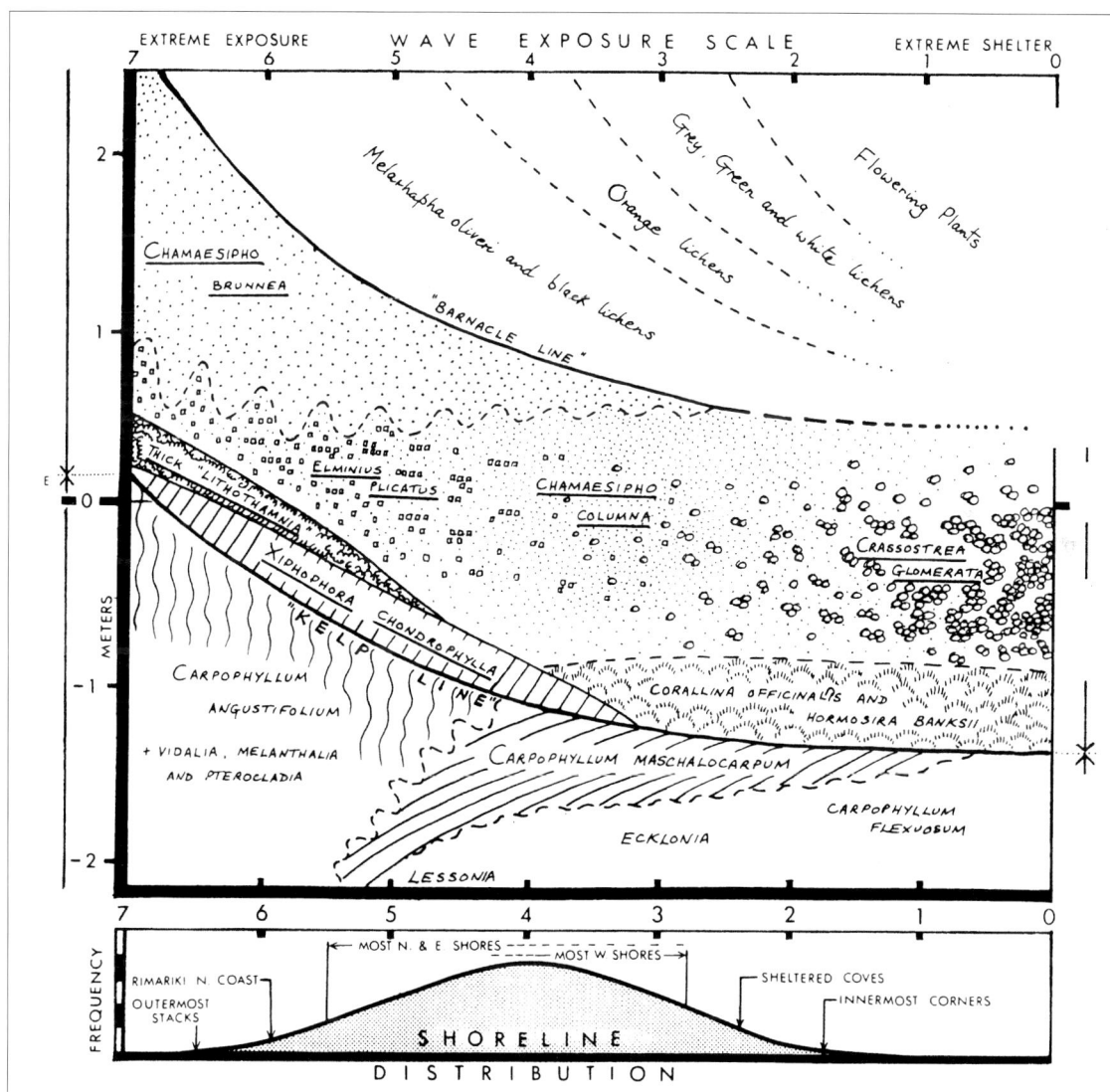

Fig. 4.11 Ballantine's biological exposure scale for east Northland
A continuous range of habitats is defined for organisms from the quietest water with little wave action (0–1 on the scale), as in the inner Whangarei Harbour, to the strongest wave action (6–7 on the scale), around rocky islands as at the Poor Knights. The frequencies plotted below for the shores of Mimiwhangata fall mainly between 5 and 3 on the scale, with a smaller number of shores towards either extreme. (From Ballantine 1973.)

shows the relation of shore slope to wave exposure and further that the common distribution of shorelines falls mainly between 5 and 3.

Wave impact is then a function of shore geography, with all the variant shapes and contours of the coastline. Not only do we find west-to-east differences, according to onshore or offshore prevailing winds, but local importance will attach to differences from place to place in the *open sea angle*. There are coasts entirely unprotected, as of offshore islands or headlands; coasts that are straight or indented; and coasts that are semi-enclosed or finally entirely land-locked.

The Biological Exposure Scale thus expresses quantitatively what no previous shore scale had attempted. Each of the eight sections of the scale is characterised by the key levels reached on the vertical axis, first established by tides but effectively taken over by waves. In tables setting out the general characteristics of the scale for use on rocky shores in New Zealand, Ballantine has presented a wealth of information derived from widespread observation and shown to agree from place to place for comparable shore types. For each shore type we are given its frequency of occurrence, and also its accessibility, and problems in working, as well as the amount and variety of epifauna and flora, and finally the sciaphilic fauna under stones.

A second table turns to the Key Biological Characteristics for each group from 7 to 0, beginning with 10 zoning species, followed by 13 unattached or mobile species.

Each species will have numerical differences over a convenient *abundance scale*, **A C F O R**, related to a species' 'normal' numbers, near the centre of its geographical and ecological range. This establishes the top of the **common** category. **Abundant** is more than this 'normal' value, and **frequent** is less, followed by **occasional** and **rare**, in a geometric scale down to the lowest quantity that one day's observation could be expected to yield.

Thus, for zone-forming browns, such as *Carpophyllum*, *Ecklonia*, *Xiphophora*, *Hormosira*:

A more than a third of the rock covered, more than half by the large browns at some level

C up to a third of the rock covered, forming a clear zone

F less than 5% cover, zone confused or interrupted

O scattered plants in no apparent zone

R a few plants found in 30 minutes' searching

Nerita atramentosa:

A more than 50/sq m, spreading throughout the barnacles

C 10–50/sq m, confined to upper half of shore

F 1–10/ sq m, very patchy

O less than 1/sq m, confined to boulders or crevices

R only a few found in 30 minutes' searching

The Biological Exposure Scale also serves as a frame to plot the relative abundance of a taxonomic group or trophic class. Thus, for Mimiwhangata, Fig. 4.11 shows the location and density of the major grazing molluscs.

The degrees of exposure

From high level to low we have thus been finding every shore *vertically* stratified into zones with their order broadly under the control of *tides*. But as distinct from the tidal levels, we may recognise *horizontal* differences between shores some geographic distance apart or — it may be — just separated around an islet or large rock-stack. These are determined primarily by their exposure to *waves*. Here we shall be using the term *exposure*, not for uncovering to the atmosphere when the tide goes out, but for the energy and impact of waves.

The living cover of the shore — in the way a basic tidal theme is locally rendered — is then under wave control. As we have already found, to remark that a shore possesses *Carpophyllum angustifolium* or rock oysters is in itself to make a clear statement about its wave regime.

Every biological story is then based on a distinctive topography. As well as their direct effect on the organisms — the waves have had a primary effect in the shape of the shore. They have determined the course of erosion with bedrock of differing hardness, and thus the shore's overall slope, with its accidents of contour, and the accumulation of mobile particles, from silt and sand grains up to cobbles and boulders.

Types of shore

Studies of shore topography and microclimate have given us a Classification of Hard Shores based on the eight index points of the Biological Exposure Scale. These have been classified as 7 to 0, based on increasing wave attack, and running from shores designated by Ballantine *Extremely Exposed* to *Extremely Sheltered*. The eight types of shore will be broadly characterised as follows:

7 Extremely exposed [A]

Confined to hard rocks on headlands or islands and receiving continuous large swell averaging 2 m. Open sea angle of more than 120 degrees and overall slope more than 40 degrees. Always high-cliffed and surface totally devoid of loose cover.

6 Very exposed [B]

Confined to mainland headlands, with continuous swell averaging 1.3 m. Open sea angle of 60–120 degrees. Overall slope greater than 30 degrees. Access to shore

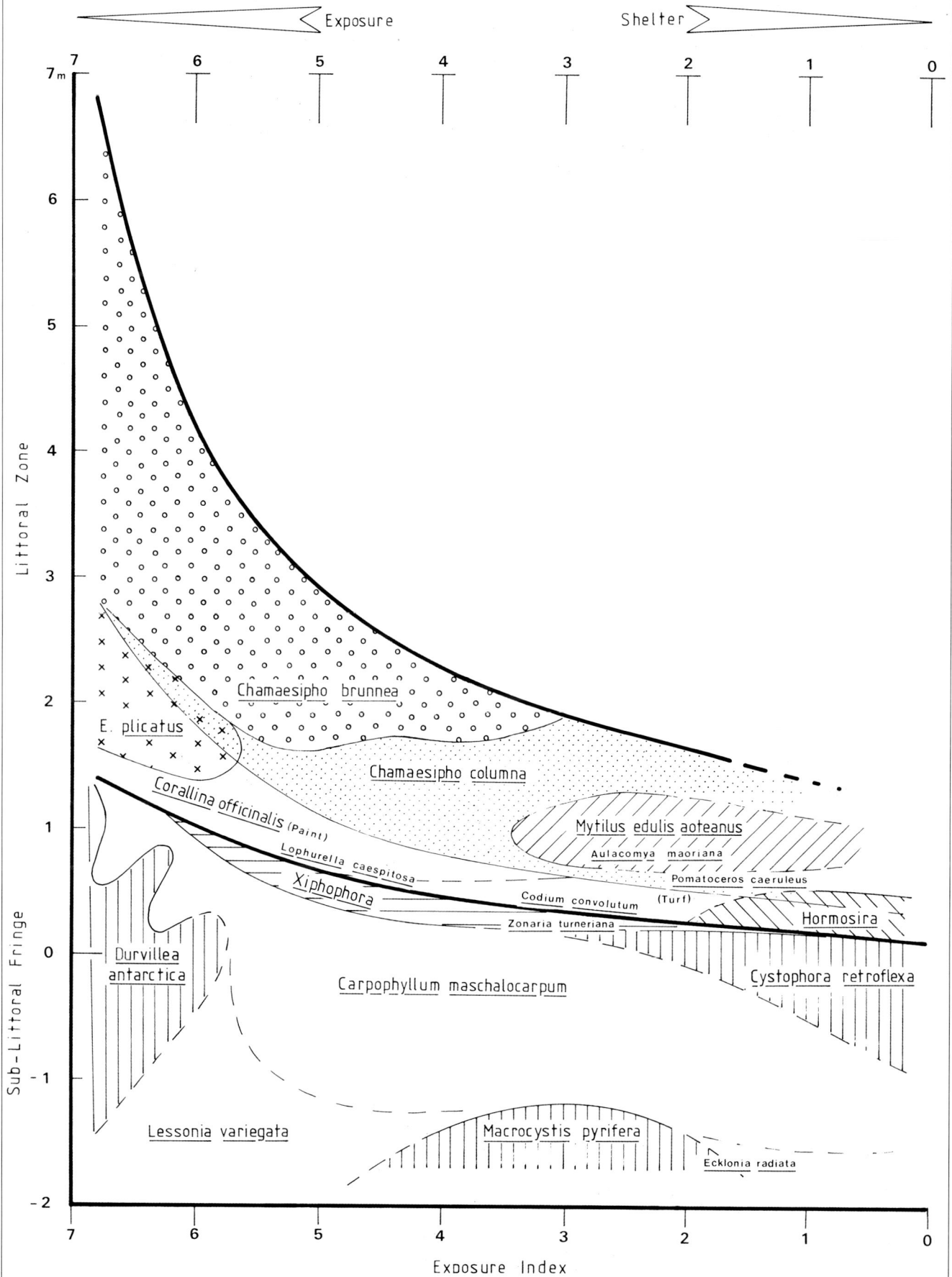

Fig. 4.12 Biological exposure scale of Wellington shores (By courtesy of Blair Dickie.)

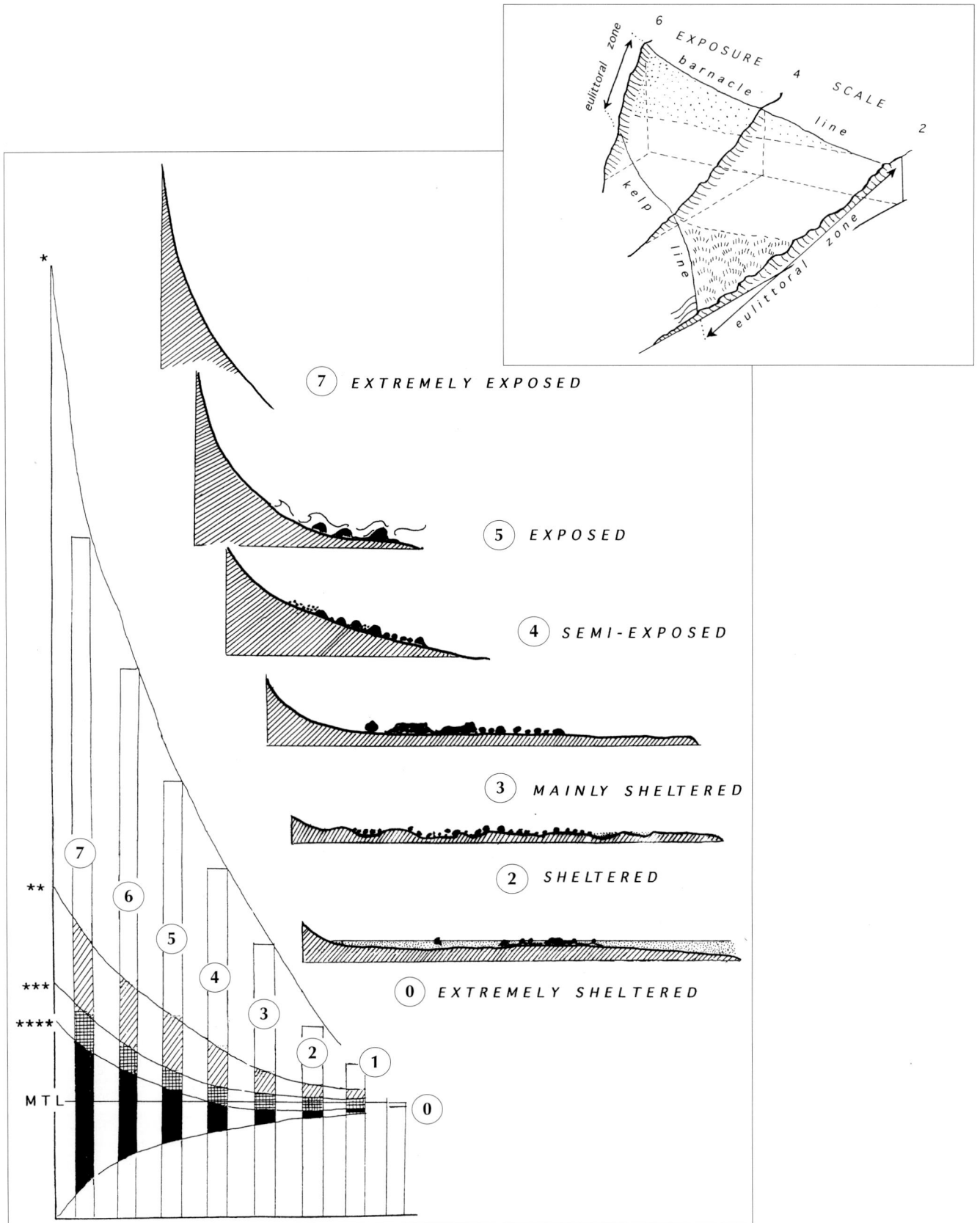

Fig. 4.13 The vertically zoned shore

The vertical extent of the primary zones is shown for the divisions of Ballantine's biological exposure scale. The upper limit of the *littoral fringe* *, the *barnacle-line* **, the *pink-line* *** and the *brown algal-line* **** are shown as standard boundaries. MTL is the mean tide level. (By courtesy of Bill Ballantine.)

Inset: The sloping shore in exposure and shelter

The progressive lowering of slope is shown, along with barnacle line and kelp line, from Exposure Scale 6 to 2.

(After Ballantine.)

types 7 and 6 is not easy and with 7 often impossible. There is little epifauna and no movable stones. Working difficulties are often great.

5 Exposed [C]

Located on straight, open coasts with large swell-waves common, averaging 0.7 m. Overall slope averaging 30–60 degrees, and boulders occur only as large, hardly movable blocks. Occasional sand-scour.

4 Semi-exposed [D]

On indented but still largely open coasts with swell waves averaging 0.4 m, generally for less than a third of the time. Open sea angle less than 5 degrees and overall slope between 20 and 60 degrees. Movable boulders present, with sand scour becoming common but without silt.

3 Mainly sheltered [E]

Semi-protected and with no open sea angle, and swell waves across enclosed waters only occasional, and averaging 0.3 m. Overall slope 10–40 degrees with boulders and big blocks accumulating. Sand-scour common, but silt no more than local or occasional.

2 Sheltered [F]

Protected shores only rarely with low wave swell. Overall slope widely variable, from 5–30 degrees. Boulder fields with stones of any size, and scour common, and silt-fall now frequent.

(In 3 and 2, with stones long stable, there are good collecting conditions with both epifauna and hypofauna increasing with reduced wave disturbance. In 1 and 0 diversity begins to fall as silt and mud become increasingly widespread.)

1 Very sheltered [G]

Shores to a high extent enclosed by the land. No swell, wave extreme is 0.3 m. Long low-sloped expanses locally varying but with overall angle of 5–20 degrees. Boulder fields of any size, with muddy sand and silt accumulating.

0 Extremely sheltered [H]

Shores totally land-locked with wave extreme no more than 0.15 m. Overall slope less than 10 degrees, with much of available surface covered with sand or mud.

The shores of 'moderate energy', already surveyed in Chapter 3, belong to groups D, E and F. These may be regarded as 'average' for most geographical regions. They are generally the richest and most convenient shores for a first study; they also occur more frequently as shown by Ballantine for the variegated coastline of east Northland.

As distinct from these shores of 'average' exposure, there are those towards the higher and lower extremes called *exposed* and *sheltered*, to be described in the following two chapters. Then (in Chapter 8) we shall devote space to some *semi-exposed* shores, favoured by their high diversity of algae, most notably with the reds. They owe this largely to their varied topography, freedom from sediment and lively flow of currents. Most of these rich algal shores would fall on our scale at D, with some reaching C.

A valuable shore exposure scale was presented in 1982 by Blair Dickie, in a thesis on the ecology of three *Cellana* species at Wellington (see Fig. 4.12).

Fig. 5.1 Trees and lichens of the adlittoral and supralittoral zones

In the vertical column (A) pohutukawa, *Metrosideros excelsa* (B) grey-green lichen zone (C) yellow lichen band (D) black lichen crust (E) rock-face with *Lichina*.

Plant details: 1 pohutukawa, *Metrosideros excelsa* 2 taupata, *Coprosma repens* 3 *Ramalina celastri* 4 *Heterodermia obscurata* 5 *Rimelia reticulata* 6 *Xanthoria parietina* 7 *Verrucaria maura* 8 *Lichina confinis*.

CHAPTER FIVE

High Energy (Exposed) Shores

Under the highest wave attack, it becomes hard to speak strictly of an 'intertidal' zone at all. From trough to crest, the short-period wave can have a greater amplitude than a whole semi-diurnal tide. We shall be looking now at the expanse of shore lying between the furthest draw-back of the receding wave, and the highest upreach of wave return. All exact relation to notional tidal levels has broken down. The successive zones widen with increased elevation as we pass up the shore, now primarily determined by the reach of surge, splash and spray.

Maximal wave impact is received on outlying coasts, usually steep-cliffed, where the 20 m line is in some places less than a kilometre offshore. Fronting an ocean fetch of perhaps hundreds of kilometres, these seaboards will experience a continuous onshore swell with an average wave height of 2 m, but sometimes much more. Loose boulders and fine sediment are non-existent, though there may still be scouring effects from wave-borne sand. Submerged reefs of bedrock offshore may break or moderate the wave attack further in.

On a vertical rock-face, an attacking wave can make a slap-up impact with little foothold for erosion. On an inclined face a wave will send a head of swash far upshore. The water level is constantly changing by the minute-to-minute rise and fall of surge; after a moment's access to lower down, the observer must scramble upshore to avoid the next mounting wave. Seconds later the wave draws back, to leave rivulets of well-oxygenated water pouring through runnels and narrow defiles.

No such open shores will be found in embayments even as large as the Hauraki Gulf. In the eastern North Island only the outer bastions and islands, from the far north to East Cape, experience the full force of the Pacific. The mainland east coast, with its prevailing south-westerlies, enjoys a majority of calm days. The strongest, most erosive wave attack is confined to the minority of easterly and north-easterly blows.

Zoning patterns

East Northland is famed for its high scenic coast. The elevated shore zones run straight and horizontal for long distances. First is the tree profile, traditionally of pohutukawa, next the pastel colours of lichens, then — below an interval of bare rock — a tawny barnacle zone, followed in turn by pink corallines and closed by a dark curtain of *Carpophyllum*. Each band stands out clean-cut, like the elemental zoning first described for the Poor Knights (Fig. 1.2).

On the high offshore islets, as at Bream Islands off Whangarei Harbour Heads, such straight horizontality is lost. Differential wave exposure results in marked elevation on the windward side. Beneath a bush cover, displaced high up by wind and spray, there can be seen from afar the steep lichen-clad fall of the *supralittoral zone*. Two prime requisites have been achieved for good lichen growth, a hard bedrock and clean unpolluted atmosphere, remote from urban smoke.

The lichen symbiosis

A lichen is a composite of two once independent entities, a fungus and a photobiont (an alga or a cyanobacterium). Their symbiosis is one of the most intimate biological partnerships ever evolved. In some lichens the photobiont is a photosynthetic bacterium formerly called a blue-green alga; and in other lichens the photobiont is a green alga. These cells are concentrated in a definite layer in the upper part of the thallus. It is then the algae that determine the lichen's form, with fungal hyphae enmeshed within the mucilaginous algal sheath. There are other lichens with their green algal cells concentrated in a definite layer in the upper part, while the thallus shape is determined by the fungus. Lichen growth forms can be crustose, foliose, squamulose or fruticose. Shore lichens especially favour the two first.

The photosynthetic algae or cyanobacteria produce a surplus of carbohydrate, which is used by the fungal mycelia, with a return of carbon dioxide to the photobiont. The hyphal structure of the fungi also provides the photobiont cells with shelter. Though we know few fossil lichens, the fungus-photobiont relation must date from the earliest colonisation of the land. The photobiont unicells had not yet advanced to forming filaments — the fungal hyphae had to supply these. In turn the algae would have taken the place of the photosynthetic cells the fungal ancestor had early lost. The photobiont cells would thus have ordered the hyphae into multifilamentous shapes, constituting — as it were — 'the seaweeds of the land'.

The fungus produces its own reproductive spores in surface bodies called *apothecia* or in deeper buried *perithecia*. As spores are blown away and germinate, the new established hyphae must capture wild photobionts, so that some sorts of lichen have to be resynthesised with each generation. There are other lichens that liberate a powder containing both hyphal fragments and photobiont cells.

In the extended **supralittoral zone**, sun-warmed but constantly damp with spray, the lower reach is picked out by orange and yellow lichens abutting on to the black littoral fringe just beneath. Of lighter yellow-green are sheets of foliose *Xanthoparmelia* species with circular plaques of the world-ranging *Xanthoria parietina*, which are yellow-orange, and the deeper orange-red rosettes of *Xanthoria ligulata*.

The yellow lichens give place higher up to others that are dull green, pale grey, or almost white. These may be crustose, foliose or a few of them profuse and bushy. Some common grey and white lichen crusts of the north include *Buellia*, *Lecidia*, *Lecanora* and *Pertusaria* species and the distinctive *Ochrolechia parella*. The important foliose species are dull grey *Parmotrema crinitum*, *Heterodermia obscurata* and *H. speciosa*. Always prominent are the shaggy ribbons of *Ramalina celastri*.

Below the supralittoral zone, lichens may come to occupy parts of the **littoral fringe**. Moistened by splash at every high tide, this fringe is no longer — as on protected shores — restricted to a narrow band at either side of EHWS; it now extends upwards over clean rock, sun-warmed and drenched with spray, to constitute the wide domain of the cosmopolitan black lichens.

The widest spreading of these is the dark *Verrucaria maura*, a thin, sooty crust while dry, but thicker and more gelatinous when moistened. Below *Verrucaria,* on the otherwise bare rock where littorines aggregate, a black stubble of *Lichina confinis* may spring up. Small, blunt branchlets, no more than 5 mm tall, are massed in brittle crusts when dry. Lower down, in the barnacle zone, they retain water, and shelter a minute bivalve *Lasaea hinemoa*, mites and collembolans. *Lichina* apart, our only true intertidal lichen appears to be *Pyrenocollema sublitoralis*, a tarry crust sometimes blackening limpets and barnacles.

East Northland

The high energy shores of the north can be typified on the hard greywacke, below the cliff-faces north of

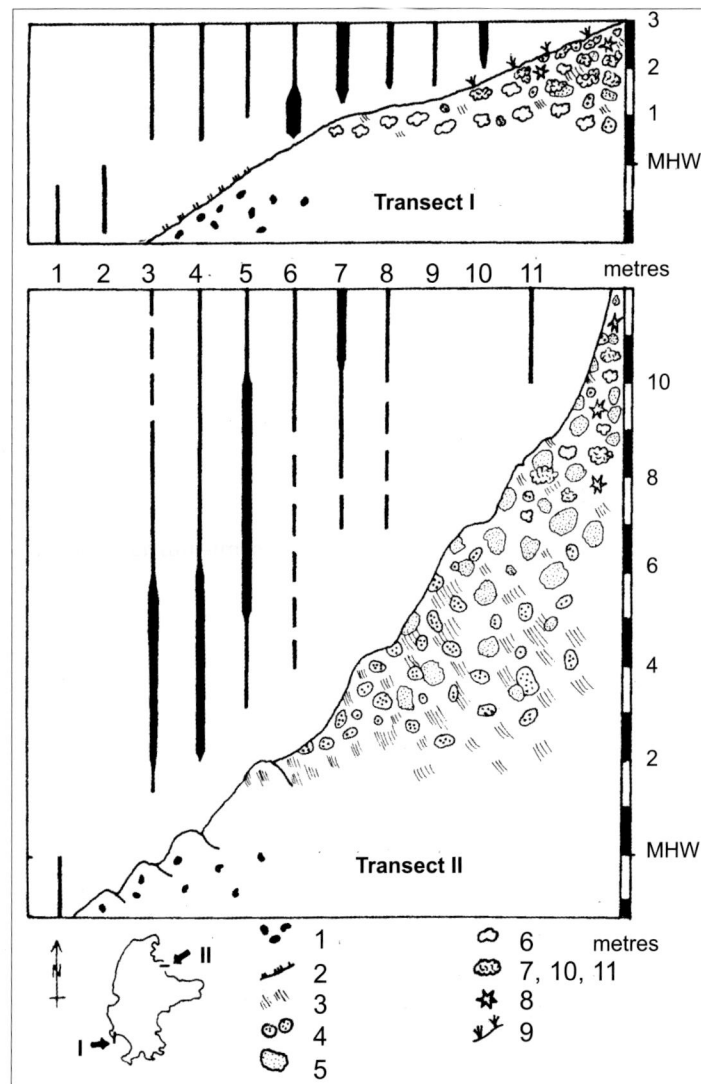

Fig. 5.2 Supralittoral lichen zonation, Slipper Island, east Coromandel Peninsula

Lichen transects through the upper eulittoral and supralittoral zones on the andesite rocky foreshore on **(top)** the sheltered south-west side of Slipper Island, and **(bottom)** its exposed north-west coast.

1 black *Verrucaria maura* 2 black stubble *Lichina confinis* 3 black crustose *Verrucaria* spp. 4 white and cream crustose *Buellia* spp. 5 cream crustose *Pertusaria* spp. 6 yellow rosettes of *Xanthoria parietina* 7 yellow-green foliose *Xanthoparmelia australasica* 8 grey foliose *Heterodermia* spp. 9 grey-green clumps of *Ramalina celasteri* 10 grey-green foliose *Rimelia reticulata* 11 blue-grey *Pannaria elixii*. (From Hayward and Hayward 1974.)

Whangarei Heads (Fig. 5.3). The high cliff-edge is fringed with pohutukawa (*Metrosideros excelsa*) familiar in its December shocks of crimson all through the north, and on Coromandel Peninsula and down to East Cape. Below this tree-line the taupata or mirror-leaf (*Coprosma repens*) stands out deep green against the pallor of the lichens. In its cliff-form this small tree becomes wind-reduced to a rigid mesh of branches, prostrate against the rock.

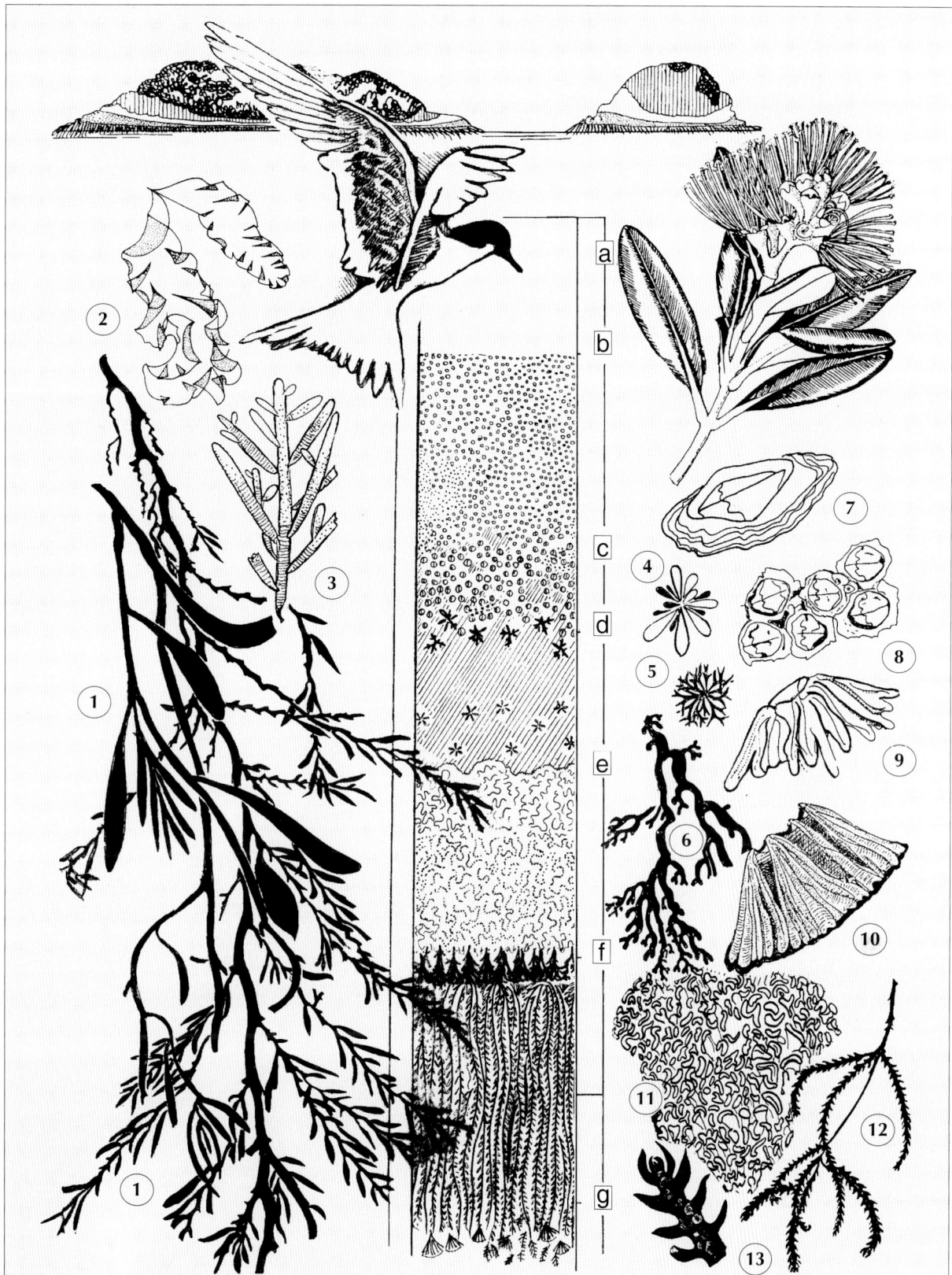

Fig. 5.3 A high-exposed shore, east Northland

The steep zoning strip is based on the open shores north of Whangarei Heads, with profile (**top**) of the inward facing sides of the offshore Bream Islands. The zoning shown is related to the shore column extending above and below it. (**Top left**: the white-fronted tern, tara, *Sterna striata*. **Top right**: pohutukawa, *Metrosideros excelsa*) a–b bare rock with littorines; b–c *Chamaesipho brunnea* and (lower) *C. columna*; c–d *Epopella plicata*, *Chamaesipho columna* with *Splachnidium rugosum*; d–e coralline veneer with *Catenellopsis* and *Nothogenia*; e–f *Lithophyllum arcuatum*; f–g *Xiphophora* above *Carpophyllum angustifolium* 1 *Carpophyllum angustifolium* 2 *Porphyra columbina* 3 *Splachnidium rugosum* 4 *Catenellopsis oligarthra* 5 *Nothogenia fastigiata* 6 *Xiphophora chondrophylla* 7 *Chamaesipho brunnea* 8 *C. columna* 9 *Epopella plicata*, exposed growth form 10 *Megabalanus tintinnabulum linzei* 11 *Lithophyllum arcuatum* 12 *Osmudaria colensoi* 13 *Osmundaria*, reproductive branchlet.

At the cliff-base the **supralittoral zone** with taupata and coloured lichens is followed by a **littoral fringe**, first with black *Verrucaria maura*, then by the bedrock for a space left bare or studded with *Lichina* tufts.

North of Whangarei Heads, in the bare fringe below the lichens, a morning's search may turn up two or three of the southern periwinkle *Austrolittorina cincta*. The smaller *A. antipoda* is common in rock cracks and — given slight shelter — the black snail, *Nerita atramentosa*, will also abound.

The barnacle-line is sharp and horizontal, with the whole **upper eulittoral** dominated — in maximal exposure — by the surf-loving *Chamaesipho brunnea*. Up to 10 mm across, this barnacle is flat-topped, with a wavy suture between terga and scuta. The column plates, being fused in a ring, become worn away like ivory, the only way the aperture can now enlarge with increase of shell diameter. Settling individuals orient fore-to-aft along the direction of surge, for optimum employment of the cirri in the passage of swash and backwash. Smaller *Chamaesipho columna* intermingle with the lower *C. brunnea* zone, below which they may form a pure band of their own. Their relative proportions vary with the degree of wave action.

The third eulittoral barnacle, *Epopella plicata*, occurs at **middle eulittoral** level. Compared with its high columnar form on protected shores (Fig. 3.4) it looks a different barnacle, now depressed and streamlined in contour. *C. columna* mingles with it throughout its range and may settle upon the larger *Epopella* shell.

On the highest energy shores, the barnacles are directly followed with a pink belt of all-encrusting corallines. Few branching coralline species can remain, but some, with their joints rigidly seized up, lie prostrate and fused with the rock. The prevailing corallines grow in heavy crusts, up to 4–5 cm thick. There is usually a complex of several species including the form assigned to *Lithophyllum arcuatum*, with a small crescentic relief.

Algal gardens, with their lavish *Gigartina* (Fig. 7.28) and other low tidal reds, disappear under higher wave attack. The large robust *Pachymenia lusoria* is often the last to persist. Except under maximal wave action, *Xiphophora chondrophylla* forms its familiar pelmet immediately above the *Carpophyllum* curtain.

In spring and early summer the lower barnacle zone becomes spangled with small succulent algae. Tubes of *Splachnidium rugosum*, filled with thick mucus, are first golden brown, to shrivel by late summer to dark remnants. Fleshy pads of brown *Petrospongium rugosum* reach 2.5 cm across. The two most miniature algae, though yellowish hued, are in fact reds. *Catenellopsis oligarthra* forms rosettes of drop-shaped bladders up to 10 mm long. *Nothogenia pulvinata* is about the same size, with stiff dichotomous branchlets drying brittle in the sun.

In east Northland the high-exposed **sublittoral fringe** is dominated by the brown alga, *Carpophyllum angustifolium*, ranging south to East Cape. This is the toughest, most sparsely leaved of the four Carpophylla. Its thongs are not flattened but cylindrical, tan to almost black and as strong as leather. They fall in a curtain from up to 2 m above slack low water, lifting and swaying with each rising wave. The leaves are thick and elliptical, with the basal ones smaller and narrow. Bladders are very scarce, being unneeded for buoyancy in strong surge.

Among *Carpophyllum* holdfasts in the coralline crust grows our largest operculate barnacle, the pink to mauve *Megabalanus tintinnabulum*, in its New Zealand subspecies *linzei*. The column can reach 7 cm across the base, and has strong varicose ribs. On exposed places, it forms a depressed cone (Fig. 5.3) but grows taller and bell-shaped in local shelter. In New Zealand, *M. tintinnabulum linzei* is confined to our open north-eastern coasts as at Cape Brett, Karikari, Whangarei Heads, Cape Rodney and the offshore islets.

Under less exposure, *Carpophyllum angustifolium* is supplanted by *C. maschalocarpum*; or — where clean sand has built up — by *C. plumosum*, and in the shallows by *Glossophora kunthii*. Beneath the brown curtain grows an understorey of wave-resistant reds: *Melanthalia abscissa*, *Pterocladia lucida* and the stiff, saw-toothed blades of *Osmundaria colensoi*. The last is a high-energy species confined to the north, including west Auckland.

Limpets and chitons

Except for littorines and (in the north) *Nerita*, the grazing force on high-energy shores is composed of limpets and chitons. The two mainland-wide patellids flourish. *Cellana ornata* reaches to the top of the barnacles and far into exposure. *Cellana radians* is lower but overlapping, being more typically a limpet of shelter. The star limpet (*Cellana stellifera*) is confined to the coralline paint level and alone in its genus reaches the subtidal. The shell is low-pitched, oval to almost straight-sided; when not pink-encrusted it is reddish brown, with granular ribs, taking its name from the pale star at the apex. The interior is silver, with the centre chestnut-brown.

The smaller acmaeid limpets come into their own on open coasts. The highest occurring is *Notoacmea pileopsis*. Its shade form is the smooth, high-arched *N. p. pileopsis*, in clusters, well above the barnacles, especially on Auckland's west coast (Fig. 5.6). A flatter, more cramped form with radiate ribs, *N. p. cellanoides,* lives lower down, bedding tightly among *Chamaesipho brunnea*. Our smallest limpet, the radially streaked

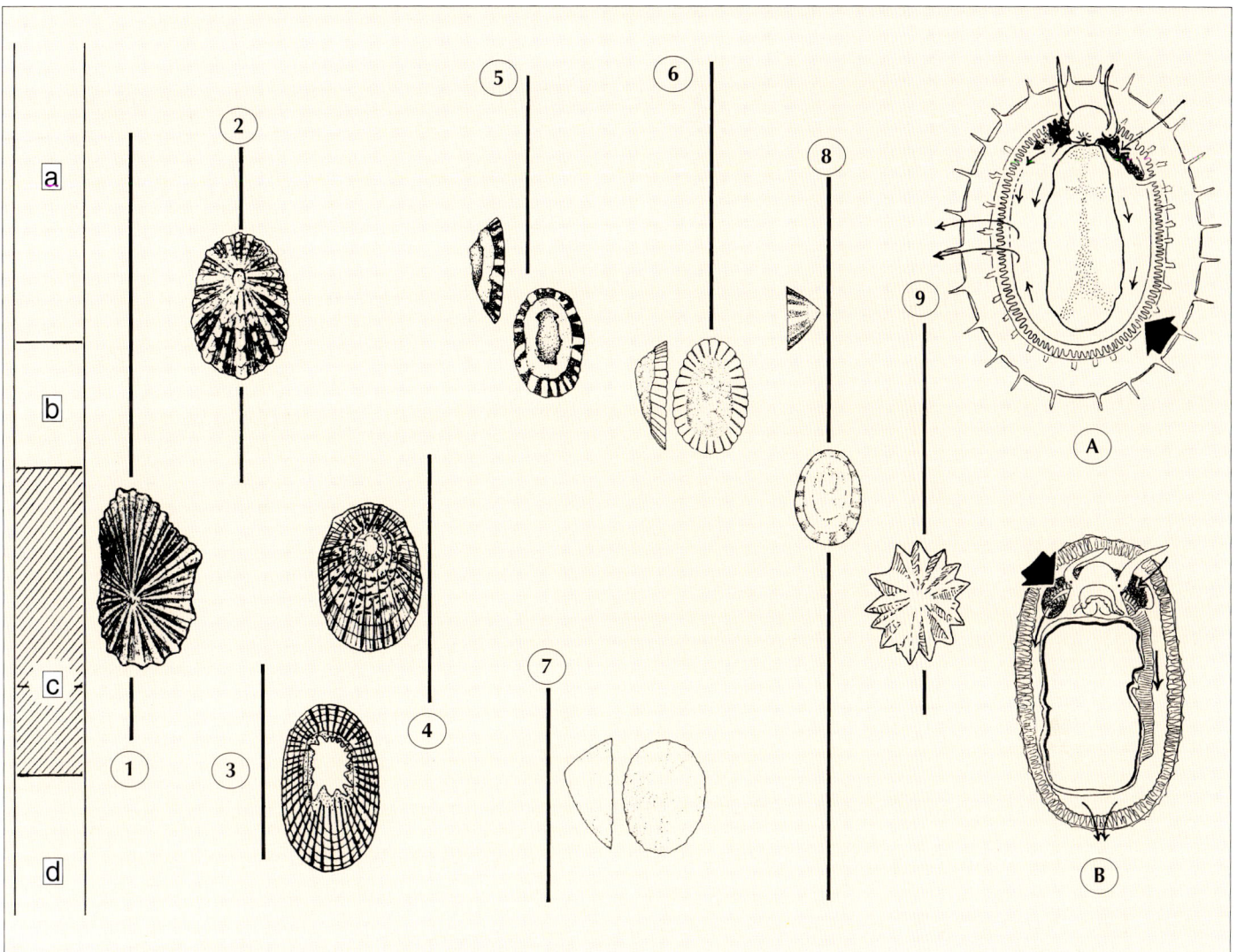

Fig. 5.4 Ranges of limpets on an exposed northern shore
(a) upper eulittoral zone (b) middle eulittoral zone (c) lower eulittoral zone (d) sublittoral fringe.
1 *Siphonaria australis* 2 *Cellana ornata* 3 *Cellana stellifera* 4 *Cellana radians* 5 *Notoacmea pileopsis* (*cellanoides* form)
6 *Notoacmea scopulina* 7 *Radiacmea inconspicua* 8 *Notoacmea parviconoidea* (small, high conical form)
9 *Patelloida corticata* (large-ribbed *corallina* form). **Right**: A, a patellid and B, an acmeid limpet, viewed from beneath to
show circum-pallial and true gills. (With acknowledgements to John Walsby.)

Notoacmea parviconoidea, is found inserted among barnacles throughout the *Chamaesipho* zone. In miniature, it is tall-conical, attaching to *Xenostrobus pulex* mussels (Fig. 5.6), while the larger form *nigrostella*, with an apical star, lives among *Perna* (Fig. 5.7).

Coralline paint is the special preserve of pink-encrusted *Patelloida corticata*. Crenate or star-shaped, with seven radial ribs, this species makes permanent home-scars. Higher up, a depressed, unevenly eroded variety lives among *Chamaesipho columna*; in coralline pools there is a higher-built pink form, with smaller ribs and crenate margin. The steeply conical *Radiacmea inconspicua* lives lower down on coralline surface. Polished pink inside, it has radial ribs obscured by algal crust, and is common in the subtidal, often on shells of *Haliotis* or *Cookia*.

The most limited in range of the acmaeids is the pale, strongly ribbed *Notoacmea scopulina*, living deep among *Chamaesipho brunnea* on steep to vertical faces. A sure recognition is offered by the deep yellow sole.

The pulmonate limpet *Siphonaria australis* will always be found in high level sites on open coasts, often betrayed by its lemon yellow spawn crescent.

The snakeskin chiton, *Sypharochiton pelliserpentis*, has a wide range from close shelter to extreme exposure. Large, eroded specimens live in the open among *Chamaesipho columna* and after grazing return to home scars kept clear of barnacles.

Two common open-shore chitons belong to the Family **Mopaliidae**, with scaleless, leathery girdles carrying tufts of bristles. The smaller *Plaxiphora caelata* is found on the coralline veneer and — on the west coast —

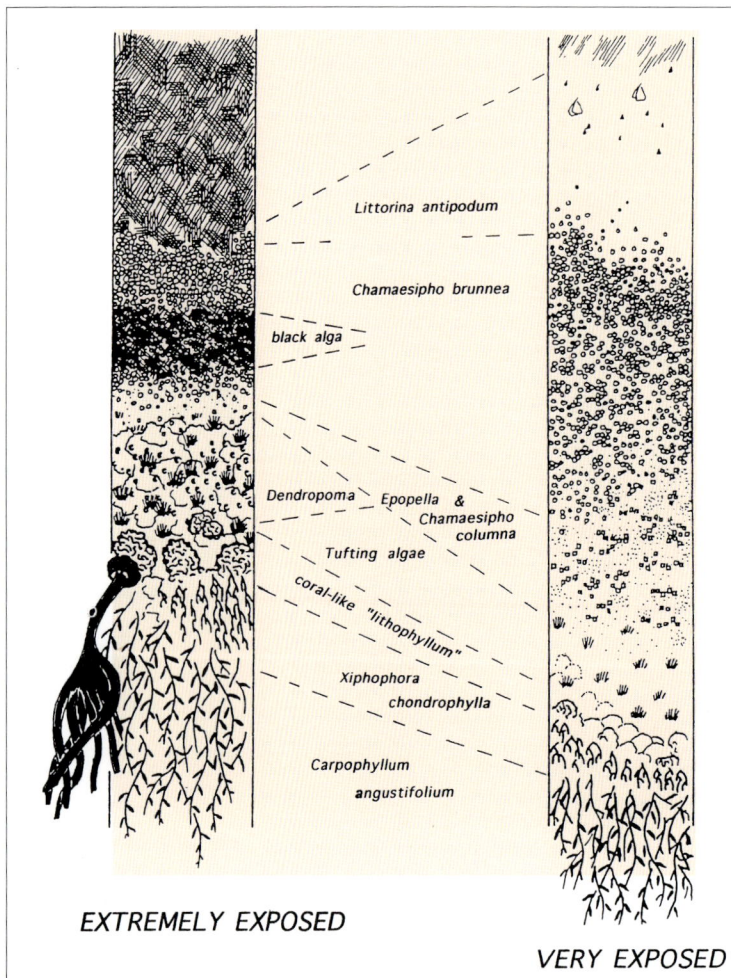

Fig. 5.5 A shore under extreme exposure at Mokohinau Islands
The typical zonation pattern of an 'extremely exposed' shore of an offshore island is compared with that of a 'very exposed' shore from the mainland in north-eastern New Zealand.
(By courtesy of Bill Ballantine.)

extends down to the *Durvillaea* bases. At lower level, around *Carpophyllum*, lives the larger *Plaxiphora obtecta*, with the girdle black-stubbled, and dark green valves, pale-streaked in the mid-line. At the same level can be found our largest chiton, *Eudoxochiton nobilis*, up to 10 cm long, with the leathery girdle sparsely bristled. The valves are wider than in *Plaxiphora obtecta*, and generally coralline-crusted.

Extreme exposure: offshore islands

For the highest extremes of shore exposure we must look with few exceptions to offshore islands near the edge of the shelf. They are difficult to reach, and few naturalists have worked them. The best account has been given by Bill Ballantine and Bob Creese for the Mokohinau Islands, outside the Hauraki Gulf. Shores with the same characteristics were first recognised at Poor Knights and at Needles Point, the most exposed extremity of Great Barrier.

Ballantine has compared the zonation of an 'extremely exposed' shore at the Mokohinaus with that of a 'very exposed' mainland shore. The Mokohinaus picture has been generalised by addition of *Durvillaea antarctica*, an extreme exposure species from the Poor

Knights and Three Kings, though not yet observed at the Mokohinaus.

As environments, the isolated offshore islands are remote from any river runoff, giving them higher and more constant salinity, greater light penetration, and less sediment and land-supplied nutrients. Local species populations will have a lower recruitment from outside, and less human pollution or exploitation. Being closer to deep water their drop-away is steep, often sheer, both on shores and subtidally.

In their biology, extremely exposed shores have features at first unexpected. Though the heights reached by lower zones (*Carpophyllum angustifolium*, *Xiphophora*, and encrusting lithophylla) show predictable elevation as surge, splash and spray increase, that of *Chamaesipho brunnea* does not. This heavily dominant barnacle is cut off sharply at the top, by a dense mixture of small algae including *Bangia fuscopurpurea*, blue-greens and in cooler seasons *Porphyra columbina* abutting above directly upon the black lichen *Verrucaria maura*. No bare rock remains, as on other high-exposed shores, to form a littoral fringe above the barnacle-line. This algal luxuriance seems due to the smaller numbers of *Austrolittorina antipoda*, here relatively uncommon. *Notoacmea pileopsis* is also scarcer than on the mainland. Low grazing may also explain the presence in places of a black alga encrusting *C. brunnea* to cover all but the opercula.

Chamaesipho brunnea, where it begins, is so densely packed as to leave little if any space even for small limpets; through the upper eulittoral, *Cellana ornata* as well as *Sypharochiton pelliserpentis* and thaids are absent though a few *Siphonaria australis* are to be found. The limpet *Notoacmea scopulina*, with white, fine-ribbed shell and bright yellow foot, is by contrast abundant in extreme exposure.

The barnacles *Chamaesipho columna* and below it *Epopella plicata* are much reduced as compared with mainland exposed shores. Below them, the **eulittoral** carries a three-banded complex. First is a novelty in the coralline-encrusted or buried vermetid gastropod *Dendropoma lamellosa*. This effectively forms a covering **middle eulittoral** zone, present on north-eastern islands and reaching to the Chathams, but found on the mainland only at Lottin Point, in most of its features 'almost an island'.

Below *Dendropoma* comes a large array of small algae, including tufts of finely branched corallines, slippery turf-patches of *Lophurella caespitosa* and soft tufts of *Ceramium*. There are also small tufting *Gigartina* (like a miniature *G. alveata*), a few patches of larger *Pachymenia lusoria* and groups of tiny *Nemastoma feredayi*. The **lowest eulittoral** is notable for a band of encrusting corallines much more developed than on any exposed mainland shore. These include a massive form like a brain-coral and forming clumps and ridges up to 10 cm thick.

The **sublittoral fringe** is dominated by *Carpophyllum angustifolium* underlaid by coralline crust and the red underworld of *Melanthalia*, *Osmundaria* and *Pteroclaldia lucida*. The brown pelmet of *Xiphophora chondrophylla* is present at the Mokohinaus, though generally with its maximum at less than extreme exposure. The ultimate exposed feature, not always to be seen, is *Durvillaea antarctica*, appearing in small northern stands on a few extremely exposed shores.

The increase of the sublittoral fringe brought about by surge will extend downwards as well as up. With considerable wave action, increases in light, aeration and water movement occur well below tide levels. The diagram (Fig. 5.5) shows how with surge effective well below as well as above low water, the 'central' or eulittoral zone can virtually disappear.

The *Durvillaea* coasts: West Auckland

In the cold south, we shall find the bull-kelp *Durvillaea antarctica* on shores of only moderate exposure. Its North Island distribution is far more limited. Farthest north it appears at high-exposed Cape Maria van Diemen and Cape Reinga, and on the west side of Spirits' Bay. To the east *Durvillaea* grows at the most exposed point of the Poor Knights, at The Needles on the Great Barrier and at the extreme tip of Coromandel Peninsula. There are also records from Cape Runaway and Lottin Point (Fig. 10.8) but it is at Table Cape on the Mahia Peninsula that *Durvillaea antarctica* begins its southward continuity along high-exposed shores.

At **west Auckland**, *Durvillaea* coasts can be reached by a 40 minute drive from the city to Te Henga (Bethells), Anawhata, Piha or Karekare. Part of a colder biogeographical realm, 600 sea kilometres remote from east Auckland's warm temperate, these shores receive from the Tasman Sea the direct force of the prevailing south-westerly winds.

The sea temperature of the west coast averages 2–3°C below east Auckland's, with the cold Westland Current (Fig. 10.3) washing this coast as far as Kaipara Heads. Southern fur seals may be sighted in the west, or come into the Manukau Harbour as far as Cornwallis. Cold upwelling off the Tasman coast raises nutrients to the surface, and a rich phytoplankton is brought in by onshore winds. The sandy stretches of the west coast north from Muriwai form the traditional toheroa beaches, with the once flourishing bivalve *Paphies ventricosa*, now depleted and closed from harvesting.

Zonation

Upreach of splash and spray bring about the usual elevation of the successive zones. At Taitomo Island (Nun Rock), Piha, the mounting surge funnelling through The Gap is pushed to extreme heights. Spray-moistened *Austrolittorina antipoda* reach some 12–14 m above low water. *A. cincta* is as common at Piha as in the south, cutting out at the top before *A. antipoda*, but ranging further down, at least to the bottom of *Chamaesipho brunnea*. As on exposed shores in the east, the high level red alga *Porphyra columbina* flourishes at the top of the *Chamaesipho* band in early spring but regresses by summer.

The pink veneer from around *Durvillaea* abuts at the top with the barnacle *Epopella plicata* intermixed with *Chamaesipho columna*, which in turn reaches higher, until it gives place to a solid pavement of *C. brunnea*. Amid the barnacles is a jet black mosaic of the small flea mussel *Xenostrobus pulex*, but at its lowest limit this bivalve is spattered with a crust of pink coralline. Easily tolerant of turbidity and sand-scour, it is also found flourishing here under the highest wave exposure.

On steep, shaded faces, away from strong waves, the two bands of tubeworms become prominent. *Pomatoceros caeruleus* spreads in a white scatter just below

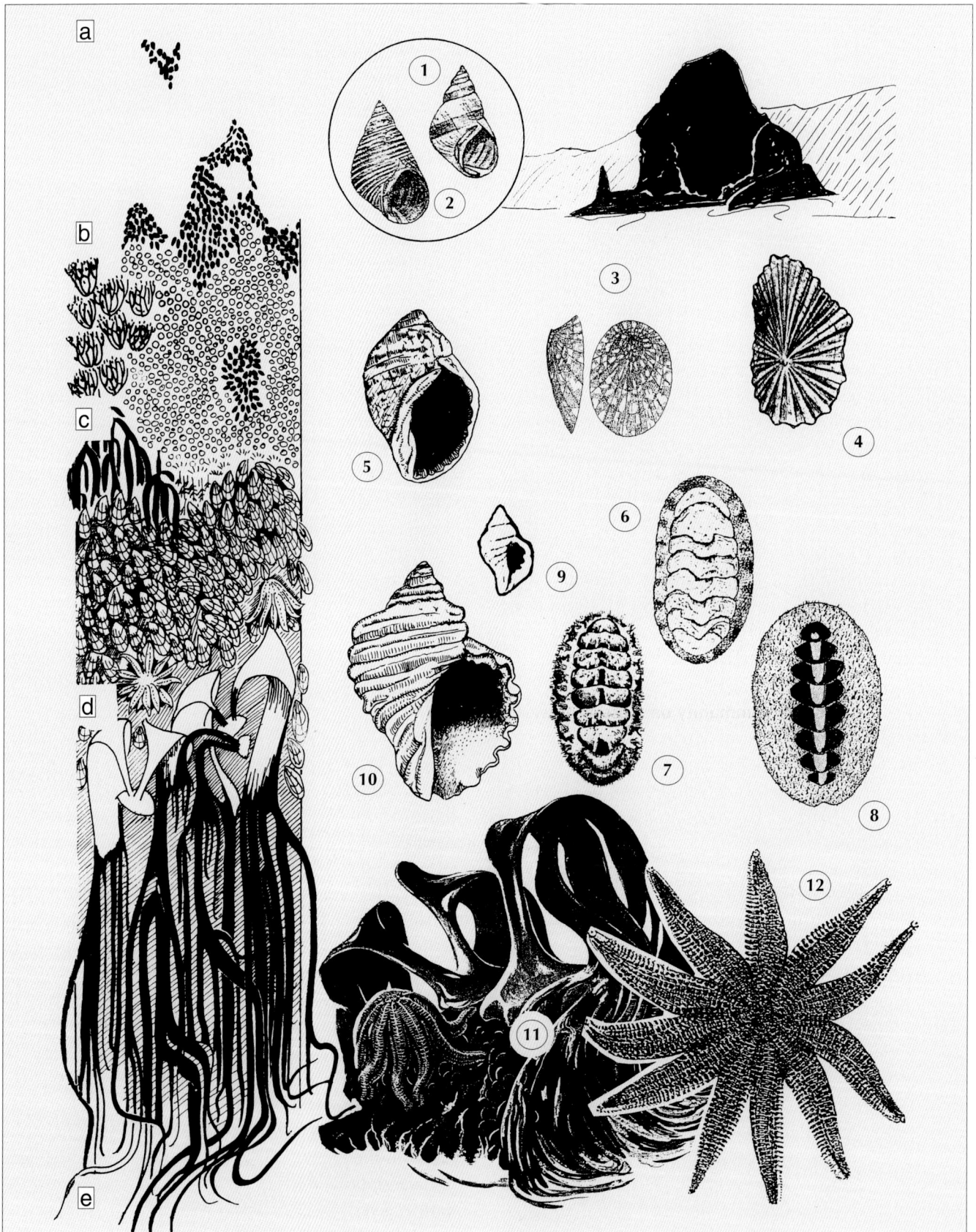

Fig. 5.6 High-exposed shore, Piha, Auckland west coast

a–b *Chamaesipho brunnea, C. columna* and (lower) *Xenostrobus pulex;* b–c with *Epopella plicata, Xenostrobus pulex* and *Gigartina alveata;* c–d with *Perna canaliculus* and *Pachymenia lusoria;* d–e *Durvillaea antarctica*

1 *Austrolittorina cincta* 2 *A. antipoda* 3 *Notoacmea pileopsis,* in typical *pileopsis* form 4 *Siphonaria australis* 5 *Lepsiella scobina* (open coast form) 6 *Sypharochiton pelliserpentis* 7 *Plaxiphora caelata* 8 *Plaxiphora obtecta* 9 *Paratrophon cheesemani* 10 *Dicathais orbita* (open coast form) 11 *Durvillaea antarctica* 12 *Stichaster australis.*

(With acknowledgements to John Walsby.)

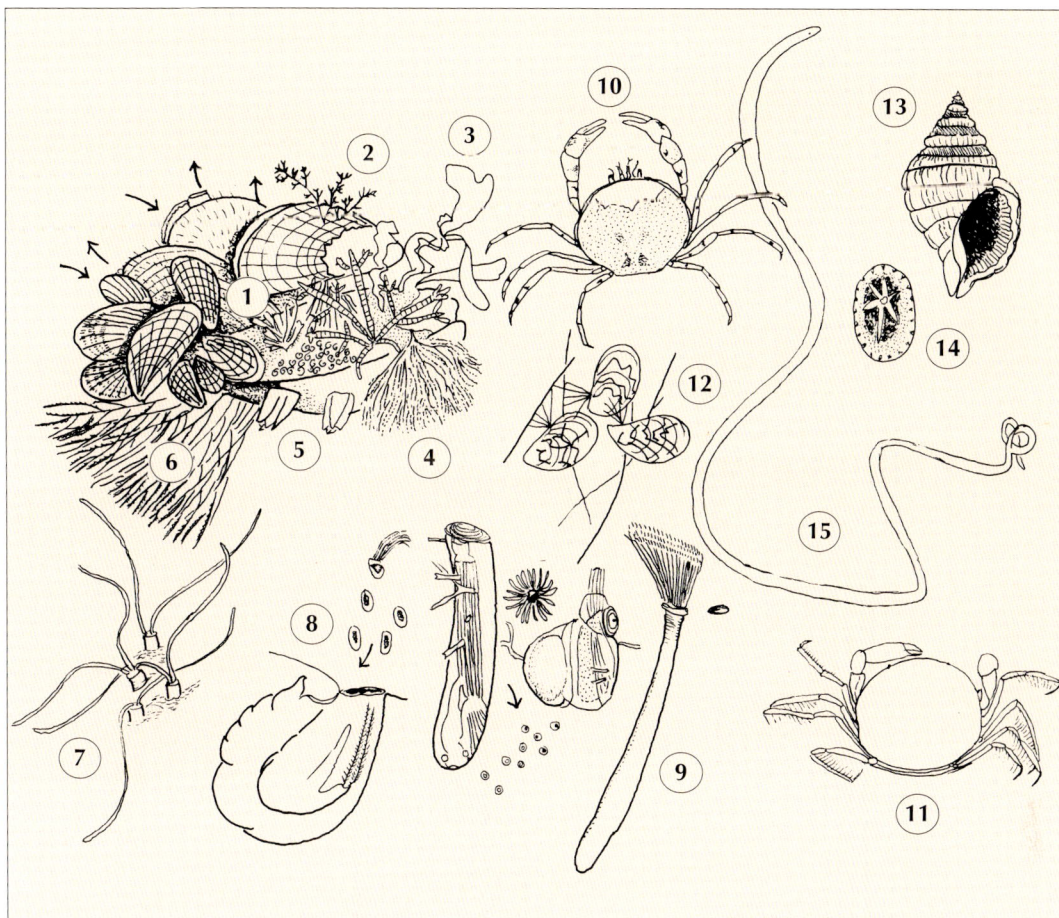

Fig. 5.7 The community on mussels at west Auckland
1 *Perna canaliculus* 2 *Chondria macrocarpa* 3 *Ulva spathulata* 4 *Polysiphonia* sp. 5 *Balanus trigonus* 6 *Amphisbetia bispinosa* 7 *Boccardia* sp. 8 *Australophialus melampygos* 9 *Phoronis ovalis* 10 *Halicarcinus innominatus* 11 *Pinnotheres novaezelandiae* 12 *Perna canaliculus*, juvenile 13 *Paratrophon cheesemani* 14 *Notoacmea parviconoidea* form *nigrostella* 15 orange nemertean.

Chamaesipho columna and *Neosabellaria kaiparaensis* follows it directly, with its iron-grey crusts most evident near supplies of building sand. Just below *Neosabellaria*, often part-buried in sand, may be found the large anemone *Isocradactis magna*, with its column studded with shell fragments, and the oral disc pink, lilac or sage-green.

Durvillaea

The bull-kelp is a giant among the fucoids and indeed the heaviest, most massive of all the algae. Much more than the true kelps, it can ascend to virtually intertidal positions, reaching sites a metre or more above the sublittoral. Juvenile *Durvillaea* may briefly establish in winter within the barnacle zone, to succumb to rising temperatures by early summer.

All the *Durvillaea* along Auckland's west coast were killed off by unusually high sea temperatures in the late summer of 1998. They took two years to recolonise and regain their former size.

Durvillaea is structurally simple as compared with the smaller Fucales. It attaches singly or in clusters by a hemispheric holdfast only removable with a hatchet or strong knife, and the cylindrical stipe expands to a leathery blade that in turn divides into long straps. These taper to flat or near-cylindrical thongs 10 or more metres in maximum length.

Both the blade and thongs of *Durvillaea antarctica* are honeycombed with air-spaces between their leathery skins. Parted with a knife the blade's two faces were used by the Maori as water pouches or muttonbird bags. At high tide, the buoyant blade stays aloft, with its thongs curling in the white eddies of reflectant foam. Light is thus captured as its thongs spread at the surface. The simple thallus thus has three elemental roles: flotation, flexible response to waves, and unbreakable prehension.

Close-set ranks of *Durvillaea* emerge at a low tide, with the stipes erect and the shining, arched blades breaking into their serpentine thongs. Under their sweep there is little else but the coralline paint. *Pachymenia lusoria* appears at Piha in a band just above *Durvillaea*. Further down and in gullies slightly more sheltered than kelp are the strong, coriaceous red algae *Osmundaria colensoi*, *Melanthalia abscissa* and *Pterocladia lucida*, all tolerant of heavy wave action.

Pachymenia thongs and the smaller red algae carry the tiny bivalves *Gaimardia finlayi*, a northern outlier of a distinctive subantarctic family (Fig. 7.16). The radial-ribbed *Philobrya munita*, classed near the Arcidae, attaches to holdfasts and to the fronds of *Osmundaria colensoi*.

The laminarian kelps, and all the fucoids *Xiphophora*, *Carpophyllum*, *Sargassum* and *Cystophora* are rare or entirely missing from large parts of the exposed west Auckland. But between north Te Henga (Bethells) and Muriwai small *Lessonia variegata* are to be found in a few sheltered reaches, and *Carpophyllum maschalocarpum* occur in pools or protected gullies. Elsewhere, the bull-kelp stands alone.

The mussel community

As on all New Zealand's cold shores, clusters of the green-lipped mussel, *Perna canaliculus*, accompany *Durvillaea*. Open coast mussels are small compared with dredged specimens from the Gulf, and a brighter blue-green, with reddish-brown radial streaks. Their byssus threads become loaded with the newly settled juveniles, pearly white and brown zigzagged, only a couple of millimetres long. Hydroids abound on the mussels, including feathery *Plumularia*, silvery fringes of *Orthopyxis*, and the long golden 'mussel beards' of *Amphisbetia bispinosa*.

The important mussel predators (humans apart) are the starfish *Stichaster australis* and the whelk *Dicathais orbita*. Congregating in low tidal clefts, the thaids have their typical open coast form, being shorter, more reddish and with strong spiral cords, as compared with those from shelter. As well as mussels, *Dicathais* attacks the surf barnacle *Epopella plicata*. Among *Perna* shells, search should also be made for the small muricid *Paratrophon cheesemani* (Fig. 5.7).

The open coast starfish *Stichaster australis* is a constant predator of mussels along both Islands' open west coasts. Its arm action alternates between firm prehension to the rock or the mussel-bed, and slow, secure mobility. The tube-feet apply their combined grip as the star arches over a shell. As opposing forces tug each valve, the adductor muscle eventually relaxes, until the star can insert its everted stomach between the valves.

Eroded mussel beaks become riddled with small penetrant organisms. A boring barnacle *Australophialus melampygos* makes oval pits 1 mm across, the female alone putting out cirri and carrying dwarfed males in her mantle cavity. In old barnacle galleries, a miniature sabellid worm may be found sitting in a jelly investment. The bryozoan *Penetrantia irregularis* also bores mussels, detectable by its tiny circular apertures, with the zooids connected by a network of stolons within the calcite of the mussel shell.

Penetrating deeper into older shells is *Phoronis ovalis*, a filter feeder with a lophophore not unlike that of a bryozoan, but folded in a double horse-shoe. A little smaller than *Phoronis* is the spionid worm, *Boccardia* sp., a shell-boring annelid, with its double-bent tube eroding into the shell and having two chimneys just projecting from the surface. From the inhalant opening, two rhythmically waving feeding tentacles are put out.

The Durvillaea coasts: South Island west coast

Around all its open rocky shores, the South Island is girded by the bull-kelp, *Durvillaea antarctica*, no more confined to the highest exposure as in the north, but entering some way into harbours, as we may see at Akaroa and Lyttelton and in the estuary at the Catlins. Beyond New Zealand *Durvillaea antarctica* reaches to the subantarctic islands, and to Chile, Patagonia, the Falklands and New Georgia.

A second bull-kelp, *Durvillaea willana*, is endemic to the New Zealand mainland and Stewart Island. Very restricted in the North Island, it reaches sporadically as far as Castlepoint. Somewhat smaller than *D. antarctica*, it seldom exceeds 5 m long. On open coasts in the south the two confront the surge together, *D. willana* being found slightly further down, as well as some way offshore. The stipe, as thick as an arm, widens into a broad blade, deeply split to the base, where that of *D. antarctica* is at first undivided. *D. willana* is best recognised by the side-blades projecting from the stipe as oval paddles. The stipe typically arches above the water at low tide with its paddles depending from it. *D. willana* lacks air-filled buoyancy tissue, and its holdfast is flatter and discoid.

Cape Foulwind, out beyond Westport, presents one of the South Island's wildest, most spectacular shores. Built of hard Karamea granite (Figs. 5.8, 9.12), the 'cape' is itself a cluster of pointed stacks, girding the coast like shark's teeth. Under gale-force onshore winds, these stacks are constantly under wave-break, and with the seas normally in tumult, close-up access is possible only on a rare calm day.

The Foulwind shore shows from a distance its high-elevated tripartite pattern. A tawny barnacle zone under continual splash and spray is followed by a pink zone

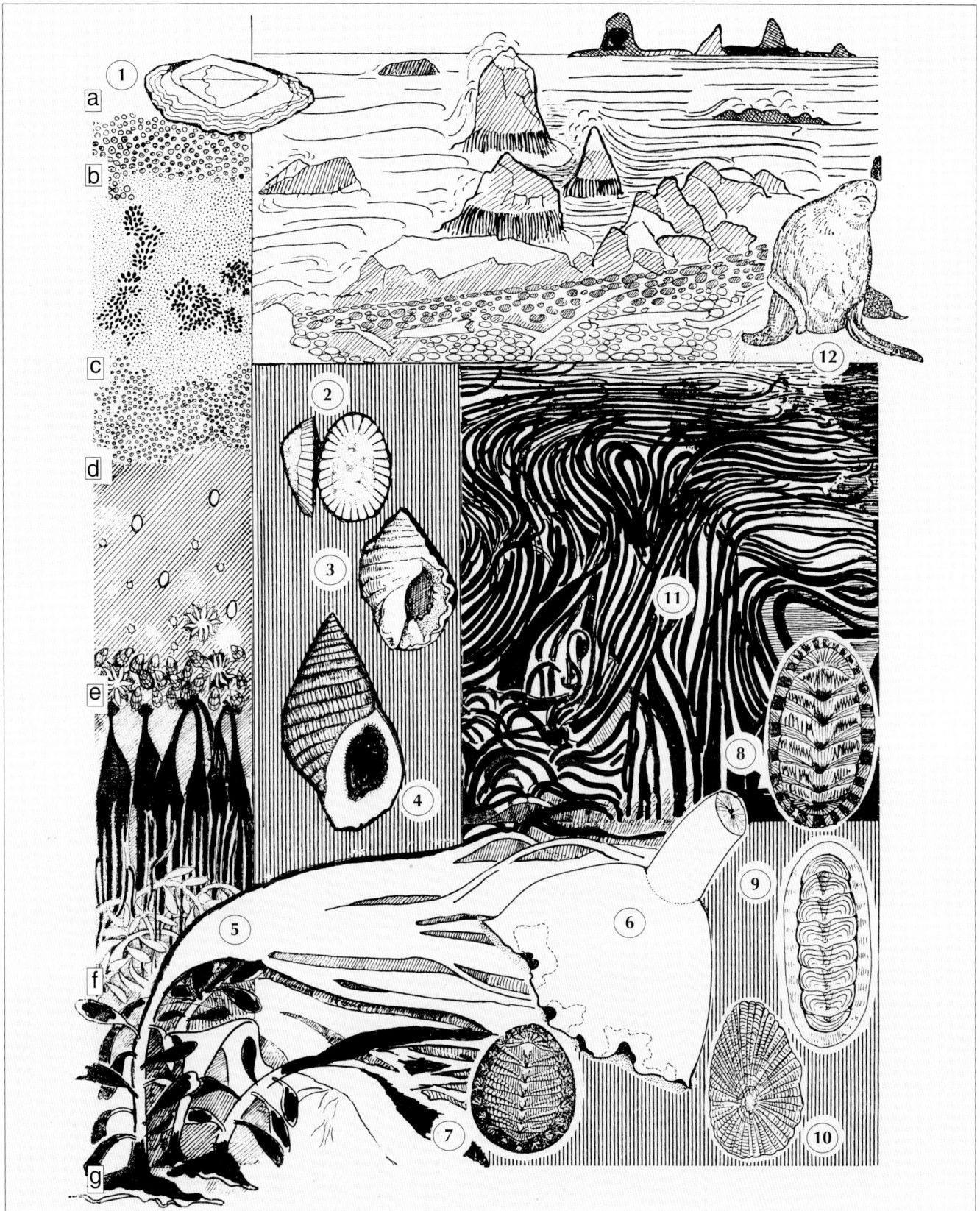

Fig. 5.8 High-exposed shore at Cape Foulwind, Westland

a–b with *Chamaesipho brunnea, C. columna* and (lower) *Xenostrobus pulex;* b–c *Chamaesipho columna;* c–d *Fpopella plicata* and *Xenostrobus pulex;* d–e with coralline veneer and *Perna canaliculus;* e–f *Durvillaea antarctica;* f–g *D. willana*
1 *Chamaesipho brunnea* 2 *Notoacmea scopulina* 3 *Paratrophon patens* 4 *Lepsithais lacunosus* 5 *Durvillaea willana*
6 *D. antarctica* holdfast, with the following species: 7 *Plaxiphora egregia* 8 *Sypharochiton pelliserpentis* 9 *Onithochiton neglectus* f. *opiniosus* 10 *Siphonaria australis* 11 *Durvillaea antarctica* 12 New Zealand fur seal, *Arctocephalus forsteri.*

swept with surge, and a long, straight-topped *Durvillaea* zone, briefly emersed at wave retreat, and curtaining a high vertical face.

The northern barnacle, *Chamaesipho brunnea*, is thinning out to its final loss south of Cape Foulwind, and the widest barnacle stretch is of *C. columna*. Above the barnacles, *Porphyra columbina* favours the littoral fringe. The middle shore is pink-veneered, first with broad swathes of the flea mussel *Xenostrobus pulex*, and then lower down the surf barnacle *Epopella plicata*. Below these, in the metre above the bull-kelp, the welded corallines form a gnarled pavement. Smaller red algae are sparse. Along the level baseline of *Durvillaea antarctica* are green-lipped mussels, *Perna canaliculus*, with their predator star, *Stichaster australis*. *Durvillaea willana*, a darker brown than *D. antarctica*, stands just offshore.

Gastropods

Among the mussels, the northern *Dicathais orbita* is supplanted through the whole south by the smaller *Lepsithais lacunosus*, with flat spiral ribs, and deep grooves. Of the small *Paratrophon* whelks, the west Auckland subspecies *Paratrophon cheesemani cheesemani* is replaced by *P. cheesemani exsculptus* from Taranaki to Wanganui, and on the South Island west coast by the larger *P. patens*. *Paratrophon quoyi* (Fig. 13.8), a quiet water species, ranges over the whole North Island east coast, including the Hauraki Gulf.

The exposed coast limpet *Notoacmea scopulina*, found from Auckland to Charleston, is especially abundant on surf-beaten rocks near Westport.

The holdfast fauna of *Durvillaea antarctica* includes several species found nowhere else. The firm tissues are scarred out by molluscs or tunnelled by nereid and euniciid worms and the isopods *Phycolimnoria insegnis*, *Isocladus* and other sphaeromids, until the eroded base is finally wrenched away. The chitons *Onithochiton neglectus* (in its holdfast form *opiniosus*) and *Sypharochiton pelliserpentis* (in its uneroded *sinclairi* form) come out from their narrow fastness to open rock. The more retired *Plaxiphora egregia* is permanently confined. Almost as wide as it is long, this chiton excavates a scar, but can be pulled readily from the rock; its adhesive power is too weak for life outside. The pulmonate slug *Onchidella nigricans* presses into irregular spaces, as does *Siphonaria australis*, in the weaker ribbed form. The pallid white *Gadinia conica* (Fig. 11.14) is here evidently confined to holdfasts. Abundant at Stewart Island in holdfasts, and wandering out over the rock, is one of the cold-water trochids, *Margarella antipodea rosea*, greenish white with blood-red spirals and splashes.

CHAPTER SIX

Close Shelter

In the ultimate heads of harbours and inlets, the shores are virtually immune from wave attack. In the larger embayments transient waves seldom more than half a metre in height can still get up over short stretches by intermittent squalls. In the ultimate arms, the shore receives ongoing deposits from the land, with the shelving intertidal covered with sediment beyond the power of the small waves to remove. In tide-scoured channels the coarser particles, including dead shells, that can accumulate, are effectively stripped of the finer grades. In the harbours and estuaries that we have now to describe, only a few sites on current-swept outposts stay clean enough from fine sediment to develop a full zonation.

Even here we shall lose some of the classic distinction between *hard* and *soft* shores. Shelving platforms may have depressions and secluded reaches filled in with fine sediment. In the Northland and Auckland inlets, still among rocky outcrops, mangroves take root; and in all our harbours a mosaic of soft patches supports the bivalves *Austrovenus* and *Nucula*, that are not properly speaking part of the hard shore biota at all.

As well as depositing fine sediment, runoff from the land has other biological effects, in reducing salinity, and supplying nitrate, phosphate and silicate, all nutrients favouring phytoplankton growth. Conversely, productivity is limited by reduced transparency from the stirring of particles into suspension.

The dissolved substances of sewage effluents must be counted towards productivity, at least until regressive effects ensue from eutrophication. Death and decay of macroalgae will then deplete the oxygen supply, and sharply lower the pH. Chemical wastes from industry bring ill-effects from the outset in fouling backwaters, deprived of oxygen and deserted by their normal biota. Harbour works and shore construction cause permanent depletion of substrate, and other necessities for life. Road embankments cut off pockets where deep mud permanently accumulates, beyond the reach of currents or tides.

Zoning of hard shores

Rocky shores of harbours have thus a curtailed species list, marked, however, with some forms that flourish in such sediment-laden waters, virtually 'ecological slums'. Harbours with overseas' shipping may be constantly open to a stream of new arrivals. On the debit side, large brown algae are disadvantaged, by the reduced depth and transparency offshore, and also by the film of silt that falls on their fronds. There are few delicate red algae and virtually no lithophylla. The most flourishing green algae are those such as *Enteromorpha*, *Ulva* and *Cladophora*, *Monostroma* and *Rhizoclonium*, favoured by high eutrophication.

In northern harbours, *Corallina* turf for a while suffers no setback, being untroubled even by quite heavy silt-fall. With it can flourish the seasonal brown algae *Leathesia*, *Colpomenia*, *Splachnidium* and *Scytosiphon*. *Hormosira banksii*, wide-ranging and ecologically plastic, takes on a new vigour in land-locked waters, and may remain to dominate the eulittoral, when *Corallina* is left only at the margins of pools. Thin swards of *Capreolia implexa* continue abundant in shade.

Northland

Secluded harbour shores, as at **Whangarei** or the east side of the **Kaipara** (Fig. 6.5), have some notable differences from those of moderate energy. There is first a marked contraction of the **littoral fringe** and **supralittoral zone**. With wave action subdued, adlittoral pohutukawa may reach out to overhang the shore, carrying on their lowest boughs the barnacle *Austrominius modestus*, now the marker species of the top of the **eulittoral zone**. Only on open sites near harbour mouths can *Chamaesipho columna* enter the zonation. *Epopella plicata* is even scarcer, but forms spasmodic patches.

Through the **middle eulittoral**, still with a scatter of *Austrominius modestus*, runs a broad band of rock oysters across mean sea level and up to a metre in vertical range. Until the 1970s these would have consisted of pure *Crassostrea glomerata* that until recently has been assumed to dominate the upper half of the band. Over 20 years, a second species has moved in, with the introduction of the 'Pacific oyster', *Crassostrea gigas*.

On low-sloping benches, as of muddy limestone in the Kaipara, *Crassostrea* is succeeded by loose clumps of

Fig. 6.1 Zonation of a sheltered northern shore

a–b *Crassostrea glomerata, Hormosira banksii* and *Austrominius modestus*; b–c chiefly *Crassostrea gigas* and *Pomatoceros caeruleus*; c–d *Corallina* turf (top) and *Codium convolutum* (bottom); d–e *Ostrea lutaria* and *Microcosmus squamiger*; e–f *Ecklonia radiata* and *Carpophyllum flexuosum*

1 flax, *Phormium tenax* 2 pied shag, *Phalacrocorax varius* 3 *Austrominius modestus* 4 *Hormosira banksii* 5 *Crassostrea gigas* 6 *Cominella adspersa* 7 *Microcosmus squamiger* 8 *Carpophyllum flexuosum* 9 *Cantharidus purpureus* 10 *Cabestana spengleri* 11 *Ranella australasia*.

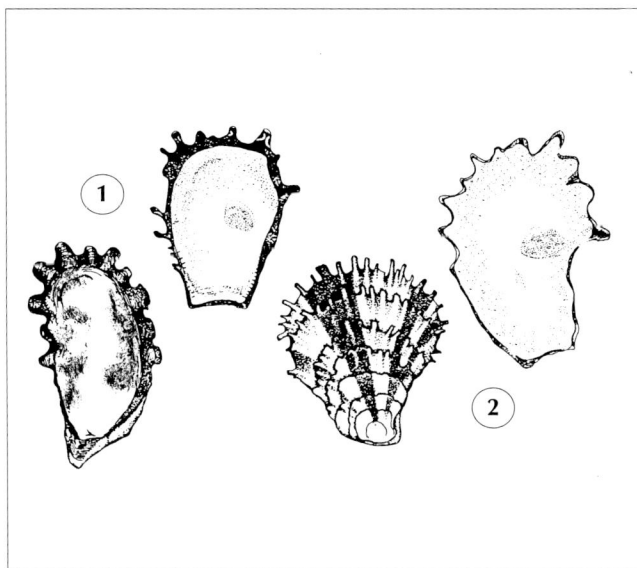

Fig. 6.2 The two rock oysters
The two species *Crassostrea glomerata* (1) and *C. gigas* (2) are shown in external and internal views. The traditional gap between these two species is no longer so clear-cut, with the current alignment in habitat.

Pomatoceros caeruleus. The long tubes have a flaring, luxuriant growth and rise clear of the ground, each cluster beginning with two or three individuals, usually growing from a single oyster shell or pebble. Between them, the permeable limestone bedrock carries a stubble of tubes of the spionid polychaete worm *Boccardia* (Fig. 8.16). Low crusts of *Pomatoceros caeruleus* can also spread horizontally, in a regular zone, where a platform has shallow standing pools.

In the **lower eulittoral**, *Hormosira banksii* finally dominates over *Corallina*, a trend increasing with shade and reduced salinity. On hard shores, a coralline turf can still generally be distinguished, often with yellow oscula of the sponge *Hymeniacidon perleve* visible above its sediments. Beyond *Corallina*, *Codium convolutum* patches may in some places fuse into a continuous band, filmed with silt in its deep interstices.

Around low water of spring tide, two species abound, the flat oyster *Ostrea lutaria* and the sea-squirt *Microcosmus squamiger*, sometimes clustering together, but with the oyster usually the higher. The flat *Ostrea* species are the oysters of the cool temperate (like the British 'native' *Ostrea edulis*), generally found sublittorally and scarcely ever above extreme low water.

The cup oysters (*Crassostrea*) in contrast belong typically to the eulittoral of warm temperate as well as tropical shores. The New Zealand *Ostrea lutaria* can be first recognised by its near-circular shape. Lacking the purple corrugations of *C. glomerata*, it is edged with small shingles of scaly periostracum, with the shell beneath thin and often knife-sharp. The interior is 'oyster'-coloured rather than clean white or purple-tinged. *Ostrea lutaria* attach to each other in shelving piles. Seldom more than 6 cm across, they are evidently a dwarfed variant of the commercial oyster, dredged subtidally in Marlborough Sounds and at Foveaux Strait.

The sea-squirt *Microcosmus squamiger* forms clusters towards the silt-line below *Ostrea lutaria*. Distinguished from other pyurids (Fig. 12.4) by its plumply rounded form, it has the siphons lined with iridescent reddish blue.

The low tidal brown algae form a distinct fringe only in good tidal flow. On open coasts dwarfed specimens of *Ecklonia radiata* may occasionally appear in pools above low tide. Within harbours this kelp can regularly occupy the **sublittoral fringe** where turbidity cuts illumination to the low level *Ecklonia* prefers. The stipe is now feeble, no more than 4–5 cm long, and bears a loose cabbage-head of thin, crinkled fronds (Fig. 6.3) often filmed with silt. These can carry the thecate hydroid *Obelia longissima*.

Found along with *Ecklonia*, the commonest fucoid of harbours is *Carpophyllum flexuosum*, long and loose-growing and usually remaining submerged at low tide. Its freer branching, and larger, light brown leaves, slightly mid-ribbed, clearly set it off from *C. maschalocarpum*. Far into some northern harbours, *Sargassum scabridum* remains a common plant at and beyond low water.

Hormosira banksii

This specialised eulittoral fucoid is common in harbours, where it develops notably larger bladders than on the open coast. Physiological studies have revealed the adaptations of *Hormosira* for life between tides. Capacities for desiccation-resistance and atmospheric photosynthesis are each efficiently adjusted to the growing level on the shore. In withstanding desiccation, *Hormosira* is far superior to *Ecklonia*, as revealed by comparing the water loss curves for the kelp with those for *Hormosira*, of variant bladder sizes, in the lower, middle and upper eulittoral. *Hormosira* can respire and photosynthesise efficiently both in water and in the atmosphere. Aerial respiration reaches its maximum at a dehydration level corresponding to 20% loss of weight. In subtidal *Ecklonia* by contrast respiration falls off rapidly as drying begins.

Taking three successively higher levels within the eulittoral, the hourly rate of water loss in *Hormosira* was found to decrease as the period of tidal exposure lengthened. Thus, whatever the duration of air exposure, a maximum of 60% dehydration was found to have been reached at the return of the tide.

As well as respiration, photosynthesis is also affected

by dehydration. In the atmospheric phase, it is found to reach a peak corresponding to the 20% dehydration point, which was — for any level of the shore — attained in one quarter of the emersion time. The *compensation point*, where the respiratory output of CO_2 equals its photosynthetic uptake, is reached — at whatever the level on the shore — at 40% dehydration.

Gastropoda

The periwinkle *Austrolittorina antipoda* penetrates into small harbours only where high level rocks remain relatively clean. The black snail *Nerita atramentosa* barely enters the Waitemata Harbour and in the Manukau has been occasionally picked up at Huia, Cornwallis and Mill Bay on harder rock substrates. But in east Northland harbours as at Whangarei, this snail — with its warmer temperature preference — ranges right through the eulittoral, even under boulders; eroded specimens are found on silted shores under extreme shelter.

The topshell *Melagraphia aethiops* can live on the clean upper shore or among oysters. Lower down, with more silt, it is replaced by its smaller estuarine counterpart, *Diloma subrostrata*, more depressed and spirally ridged, usually silty or green-filmed. The pale yellow lip is blotched with black, as Baden Powell used to say, 'like a dirty thumb-nail'. Typically found inside empty *Crassostrea* shells is the small trochid *Fossarina rimata*, restricted to Auckland and east Northland.

Below the mud topshell, *Diloma subrostrata*, there is a smaller trochid *Micrelenchus huttoni*, globose or near-conical, with leaden grey spiral threads and blue iridescent mouth. Belonging to the Cantharidus group of topshells, this species is most at home on sea-grass flats. Around low water, on *Carpophyllum flexuosum*, a common browsing topshell is the dull pink *Cantharidus purpureus*.

Our only limpet persisting far into harbours is the small, high-built *Notoacmea helmsi*. An old cockle or pipi shell gives it a 'mini-reef' where the constantly renewing algal film seems enough for lifelong grazing.

Wherever there is *Corallina*, the cat's-eye *Turbo smaragdus* is likely to be found. Indeed it is in harbours, even among mangroves, that *Turbo* reach greatest size, with very old specimens as large as plums, 50 mm across. The black spire-snail, *Zeacumantus subcarinatus*, browses among *Corallina* or on sediments. Upshore and in high shelter it gives place to the larger *Z. lutulentus* that has a preference for silty sand or mud.

The *Cominella* species take their food dead or moribund rather than living. *Cominella maculosa* is not infrequent among oysters, but much commoner is the plumper *C. adspersa*, with the mustard yellow lip heavily thickened in old age. Furthest out on silty flats lives the common scavenging snail *Cominella glandiformis*. Brown or mauve, often dark-banded, its whorls are shouldered, with nodules and low axial ribs. The aperture is chocolate brown and the lip purplish. These whelks converge in scores on dead animal remains, or insert the proboscis through the siphonal gape of cockles. Their number in well-frequented harbours has been recently depleted by toxic tributyl tin (TBT).

Common under *Zostera* or with oysters and *Pomatoceros* is the small muricid *Xymene plebeius*, only a centimetre long and sharp-lipped, with a fine-netted sculpture. Though able to drill young oysters, its principal food seems to be small cockles and the nut-shell, *Nucula hartvigiana*. Subtidally the small, axially ribbed *Cominella quoyana* is found in silty shell gravel.

The largest predatory gastropods around low water in our northern harbours are the trumpet shells, **Ranellidae**, that have, in their essentials, evolved like the whelks. The heavy *Cabestana spengleri*, spiral-corded with fawn epidermis and rounded varices, is frequent in some classic northern localities in the Manukau Harbour, and at Whangarei Heads. Among rocks on muddy sand or beneath *Ecklonia*, *Cabestana* feeds on the ascidian *Microcosmus squamiger*, grasping the test with the foot and cutting a hole with the radula to insert the proboscis.

In similar places can be found the shaggy, spirally ribbed *Cymatium parthenopeum*, attacking the ascidian *Cnemidocarpa bicornuta*. It feeds too on the burrowing bivalve *Ruditapes largillierti*, thrusting its proboscis down the siphons between the part-open valves.

Auckland Harbour: Black Reef

At **Westmere**, beyond the Auckland Harbour Bridge, a narrow ribbon of basalt lava flow strikes across the low tidal flats of mudstone. Reaching more than a kilometre across the Harbour, it rises some 1.5 m above the flats. Up to mid-tide, its zoned surface is filmed with silt; above this line the basalt stands out black and largely bare.

At the lowest spring tides, the tip of the Black Reef (Te Tokaroa, also known currently as Meola Reef) can be reached by a walk across the flat, and offers a wonderful vista of seclusion in the midst of built-up urban surrounds. The quiet may be broken by the robust whistling of a flight of oystercatchers. A white-faced heron may stand watchfully in the shallows. Kingfishers from nearby cliff-edges scan the whole tidal flat. A caspian tern is generally to be seen wheeling or descending, while at the reef crest a pair of pied shags may be seen drying outstretched wings.

Fig. 6.3 Zonation of Black Reef (Te Tokaroa), Waitemata Harbour

(left) 1970 a–b *Austrominius modestus, Xenostrobus pulex* and *Capreolia implexa;* b–c *Crassostrea glomerata* and *Austrominius modestus;* c–d *Codium convolutum* and *Leathesia difformis;* d–e *Ostrea lutaria, Microcosmus squamiger* and sponges

(centre) 1995; b–d *Crassostrea gigas,* pools with *Hormosira* and mud with *Onchidella nigricans*

(right) e–f stretch with *Crassostrea-Microcosmus* clusters, with *Halichondria moorei* at their sides, and *Tethya aurantium Aaptos confertus; Ecklonia* and *Sargassum*

1 *Hormosira banksii,* compared with 2 open coast form 3 *Capreolia implexa* 4 *Gelidium caulacantheum*
5 *Sargassum scabridum* 6 *Ecklonia radiata* (high shelter form) 7 *Aaptos confertus* 8 *Tethya aurantium* 9 *Ostrea lutaria*
10 white-faced heron, *Ardea novaehollandiae.*

The **littoral** and **sublittoral fringes** are curtailed: the first by the whole reef's low contour and the lack of wave action, the second by encroachment of silt. At the highest points, a few *Austrolittorina antipoda* gather in crevices. The sparse zoning barnacle is *Austrominius modestus.* Shade patches of *Gelidium caulacantheum* and *Capreolia implexa* share space with the flea mussel, *Xenostrobus pulex.* Further down among rock oysters the straggly brown alga *Scytothamnus australis* darkens and dries out in the sun.

The Pacific rock oyster, *Crassostrea gigas*, has notably out-competed *C. glomerata* since the 1980s. But since the 1920s the zoning dominance at Black Reef has changed even more. The tubeworm *Pomatoceros caeruleus* was in those times well to the fore. In 1923 Oliver published a photograph of Black Reef showing its pale convex mounds like sheep. All these stretches are now in the full possession of abundant *Crassostrea gigas*. In past years *Pomatoceros* flourished around Auckland in places with high organic runoff. In the mid 1940s its 'sheep' grew near the sewer outfalls at St Leonards and Black Rock, Takapuna, and at Bastion Reef, off Auckland's Tamaki Drive. Today it is essentially absent from the Waitemata Harbour.

Hormosira banksii is everywhere large at Black Reef, but most so in pools where its chains grow to half a metre, with bladders up to 3 cm across. Further up the Harbour, *Hormosira* can lose its holdfasts and lie free. In late winter and spring, coralline turf is encroached on by fleshy green *Codium convolutum*. Below this prominent band comes dull pink coralline paint, much filmed by silt from below, and in spring, becoming golden yellow with the vesicles of *Leathesia difformis*.

A **sublittoral fringe** with brown algae exists only at the outermost edge of Black Reef, or in gaps with tidal currents. *Ecklonia radiata* here displays its sheltered ecotype, with a small, vestigial stipe and loose cabbage heads lying directly on the bottom. The common fucoids are *Sargassum scabridum*, with thin, undulant leaves and finely prickled stems, and *Carpophyllum maschalocarpum*.

The narrow sublittoral fringe at Black Reef is pre-eminently a domain of sponges, the small, low-tidal rock oyster (*Ostrea lutaria*) and ascidians. *Microcosmus squamiger* fuses into tuberous masses, with oysters packed in tiers at the sides and beneath. Nearly concealed with silt, the ascidians betray themselves by thin, desultory jets of water when disturbed during low tide.

In these low-lit, turbid waters, sponges become the new and long-stable dominants, moving into their own prominence as the brown algae fail. The lightly built sponge *Halichondria moorei* lends its dull orange to shaded underhangs, crusting the rock like waterlogged bread. To a general prospect of grey, the sponges lend their reds, chromes and ochres. As well as under boulders, *Ophlitaspongia*, *Microciona* and *Plocamia* at the Black Reef here contribute their more vivid reds to the open surface. Cherry-coloured knobs of *Crella incrustans* grow luxuriantly. Golf-ball sponges, *Tethya aurantium*, and chocolate *Suberites perfectus* like tennis balls appear on every hand. Dull purple hemispheres of *Aaptos confertus* live just beneath overhangs, and *Suberites axinelloides* spreads over empty oyster shells.

New migrants

Through the long past, New Zealand shores have been enriched by the arrival and colonisation of the dispersal stages of many marine organisms from east Australia and the warm Pacific. These were often the larval stages carried in by the warm Tasman Front and East Auckland Current. In the last 200 or so years there have been many other marine species arriving as assisted migrants from not only the traditional source countries, but also from the northern hemisphere. These new migrants have mostly come in as fouling species on the hulls of ships and less frequently in ballast water discharged on arrival. Without shipping, warm temperate species from the northern hemisphere would never have been able to pass through the tropics and reach New Zealand.

Around 150 species of new migrants are now recognised[1] as living around New Zealand's coast, having been introduced by shipping or occasionally on purpose. New Zealand's largest port at Auckland receives the most shipping and has many receptive habitats within the surrounding Waitemata Harbour for the establishment of newly arrived migrants. Indeed the Waitemata Harbour has the largest number of adventive species with over 60 identified new arrivals living within it.[2]

Nineteen of the new migrants are marine algae,[3] several of which have very local distribution patterns around nineteenth century whaling stations, such as Port Underwood in Marlborough or Paremata, near Wellington. These would appear to have been some of the earliest arrivals of fouling species arriving attached to the hulls of wooden whaling boats.

Some of the more notable new migrants that have arrived in New Zealand within the last 50 years are documented below.

The ectoproct bryozoan *Watersipora arcuata* was the first of the novelties, noted in New Zealand in 1957. It soon had a rapid bonanza in sheltered northern waters. In a good season this flourishing arriviste could take over any lower eulittoral space where turbidity had cut down the light. As often happens, initial success has followed in recent years by relative decline. Under boulders in the lower eulittoral and sublittoral fringe, *Watersipora* colonies typically form a dark periphery just inside the shaded edge. Their scrolls are brittle to the touch, but its black pigmentation gives *Watersipora* an ability to spread out into light. It can even begin to spread a fragile canopy over coralline turf, advancing so fast as to seal off *Pomatoceros* or *Sypharochiton pelliserpentis* (in its form *sinclairi*) alive.

By 1968 the ascidian *Ciona intestinalis* was already at Auckland, after recording by Brewin from Lyttelton Harbour in 1940.

The file shell *Limaria orientalis*, today a significant food for snapper in the Auckland Harbour, had turned

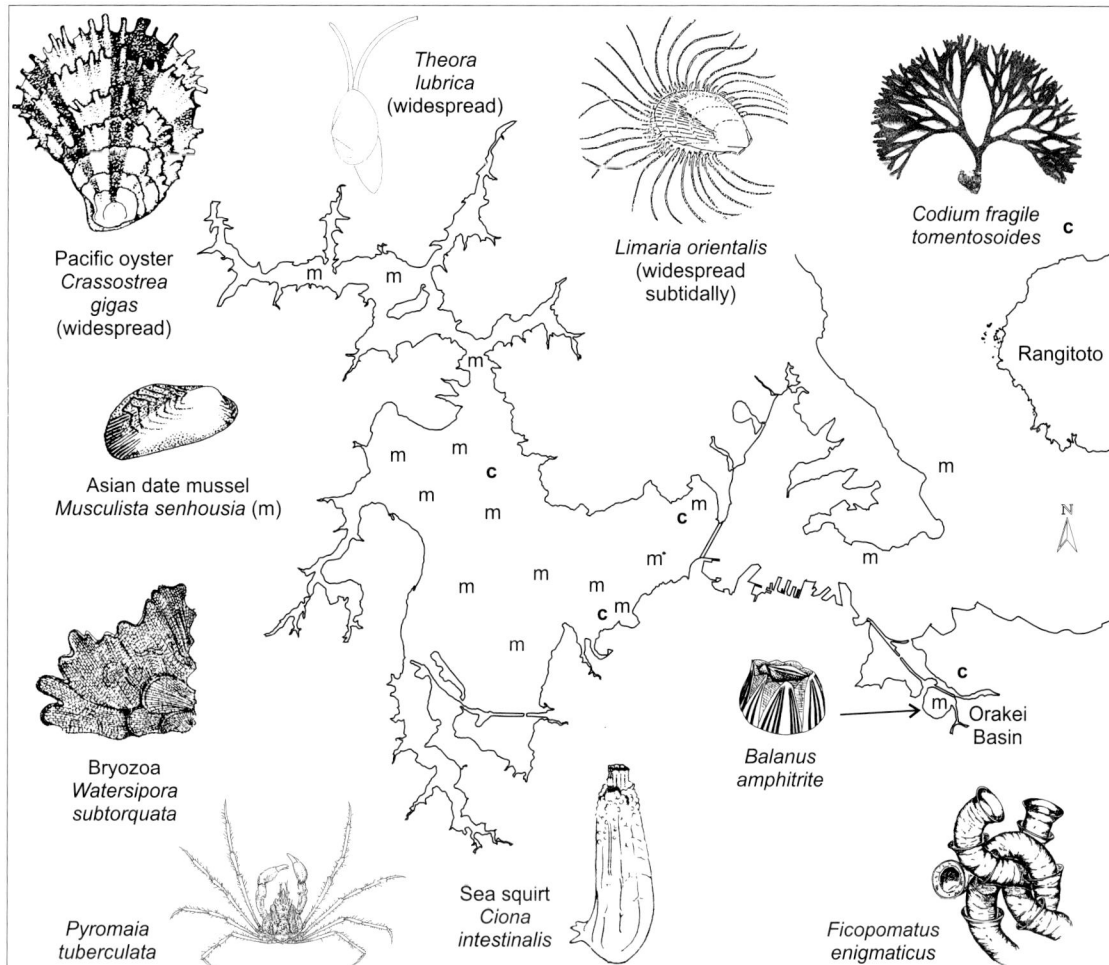

Fig. 6.4 Recent immigrants, Waitemata Harbour

Some of the variety of marine organisms that have arrived recently as assisted immigrants with shipping, and are now established in Auckland's Waitemata Harbour. The 1990s distribution of *Musculista senhousia* (m) and *Codium fragile tomentosoides* (c) within the harbour is shown. (From Hayward 1997.)

up from Japan or Australia by 1972, being present today in the sublittoral around the Waitemata Harbour and in Northland. Far from a first arriver, it had been known from fossils in the Miocene and Pliocene. *Limaria* builds a nest of byssus fibres and shell debris and also has its family ability to swim effectively by expelling water, either beside the hinge or from the ventral gape. The foot and the long pallial tentacles and velum are scarlet.

Some time after *Limaria*, about 1978, arrived the small Japanese spider crab *Pyromaia tuberculata*, now established in the inner Hauraki Gulf. Around 1972, the semelid bivalve *Theora lubrica* arrived, probably also from Japan. Today it abounds as a dominant in subtidal and low tidal mud of the Waitemata and other northern harbours. Small and thin-shelled, *Theora* is an attractive object for live study under a stereo-microscope, showing by its transparency all the workings of the gills, palps and mantle cavity.

The anemone *Haliplanella lineata*, common in the warm south-west Pacific in silty and even brackish waters, arrived in the Waitemata Harbour in the late 1970s, evidently in company with the Pacific oyster. The column, contracting to a small dome, is dark bottle green, and orange-striped, with the tentacles lighter green.

The green alga *Codium fragile tomentosoides*, originally from Japan, introduced itself around the same time. With microscopic cell differences from the long-established New Zealand *C. fragile*, this subspecies is now seasonally common at Auckland in the lower eulittoral and on wharves and buoys. The tropical Pacific brown alga, *Hydroclathrus clathratus*, arrived in the late 1970s, being noted first on coralline flats at Whangarei Heads. Close akin to *Colpomenia sinuata*, it can be distinguished by the perforate lattice of the mature thallus. The related *Colpomenia durvillaeae*, of Japanese origin, was recorded from Leigh in 1982.

The small black mussel *Musculista senhousia* was first recognised at Black Reef in 1981, and is now widespread in the north, between the Bay of Islands and Coromandel. The thin greenish shell is prettily flecked and zigzagged in red brown. More slender and fragile than *Xenostrobus pulex*, it avoids bare rock but is prolific in silt-lined pans, bound into mats with byssus threads. Originating in south-east Asia, this mussel has today an extended Pacific range.

Ficopomatus enigmaticus, a serpulid tubeworm, with a potentially major fouling impact in brackish waters, first appeared in the tidal basin at Whangarei in 1967, and has also invaded the water intake at Otara Power Station in Auckland.

In 1997, large quantities of the 10–20 cm long parchment tube of the unusual polychaete worm *Chaetopterus* began washing up on the beaches such as Long, Okoromai and Omaha bays in the middle Hauraki Gulf. In subsequent years it has overrun large areas of shallow seafloor from the eastern Coromandel to Leigh Harbour. It forms extensive intertwining thickets of tough tubes loosely bedded into the surface sediment at depths from low tide to at least 30 m. The worm is also becoming a major inhabitant of low tidal sea-grass beds in sheltered harbours and can be found in considerable numbers attached to low tidal rocks living in close association with coralline turf. Recent taxonomic studies are indicating that it is probably a recent shipping-assisted migrant from the northern hemisphere.

A resurvey in the 1990s[4] of the soft-bottom communities of the Waitemata Harbour, first documented in Baden Powell's classic study of the 1930s,[5] showed considerable changes over 60 years. Twelve mollusc species are found to have disappeared, with the elimination of two communities from the outer harbour and the reduction of the turret shell *Maoricolpus roseus* community in the harbour's central channel. At least nine New Zealand species have newly colonised the inner harbour, including beds of the horse mussel *Atrina zelandica*. Three of the dominant molluscs in various parts of the harbour's seafloor are recently arrived migrants — *Limaria*, *Theora* and *Musculista*.

Some of these changes are attributed by the authors of the resurvey to channel dredging, increased sediment and freshwater sediment runoff, ballast water discharge, pollution and TBT poisoning. On the ameliorative side has been the closure of long-standing sewerage outfalls and the resulting increase in water quality.

Rock oyster shores

Since the mid-1960s the new industry of oyster farming has transformed the vista of many Northland shores. The **Kaipara Harbour** is intricately embayed, carved into limestone benches and covered up to mean tide level with sediment. These spacious shores — it has been remarked — could accommodate in one corner all the French oyster leases of the traditional Gironde.

After the Marine Farming Act in the 1960s, the old small-scale government harvesting was superseded by private oyster cultivation in foreshore leases. Wooden racks at half-tide support thousands of horizontal battens. Special racks are reserved for spat-collecting, with the larvae settling on bunches of sticks laid out in midsummer. The Kaipara is the highest-esteemed spatting harbour; as well as being on-grown locally, sticks with centimetre-sized young are regularly taken to other growing areas.

After six weeks the spat bundles are broken up and the sticks moved to growing racks further downshore. Mature oysters lightly attached can easily be knocked off the oldest stakes with harvesting from April through to Christmas. Half are live air-freighted overseas, most of the rest being sold on the half-shell or intact and frozen.

The industry began hesitantly in the sixties with culling of natural *C. glomerata* laid loose on wire fattening-trays. Growth to market size took four to five years. In the early 1970s the prospect was transformed with the chance arrival of the Pacific oyster, *Crassostrea gigas*. Virtually excluding *C. glomerata* from its lower range, the new oyster now commands the market. It can feed longer at each tide and matures in 12–18 months.

Early records from Whangarei (1958) and the Kaipara (1965) could suggest *C. gigas* had been liberated, as in east Australia, to boost an industry. But its New Zealand arrival seems to have been as fortuitous as it was well-timed. By 1978 its spat-fall in the Mahurangi Harbour exceeded that of *C. glomerata*, and the new oyster was competing well with the older, often over-smothering it. Its best growth is made lower down but still contiguous with the native oyster.

The New Zealand success of *C. gigas* has been only part of its quarter century of expansion. Once regarded as a fouling pest to native oysters, *C. gigas* has spread through both the Pacific and Atlantic, becoming the cultivated species at Hong Kong (Fig. 18.8) and commercially overtaking the renowned Whitstable 'native', *Ostrea edulis*, in England.

Both *C. glomerata* and *C. gigas* are confusingly variable, and they are thought to hybridise. Shells over 7 cm long with sharp or frilly growth ridges can be reliably called *C. gigas*, but in the eroded state the two are externally similar. Empty *C. glomerata* shells reveal cross-ridges near the hinge, that give a grating sound when scratched. *C. glomerata* shells are less frilled, with a corrugated purple edge, and the inside pure white, or purple-marked, while the interior of *C. gigas* is more 'oyster' coloured (yellowish brown to cream). By 2000 it had become difficult to find convincing specimens of the New Zealand *C. glomerata*. Among a single clump of oysters are specimens clearly identifiable as *C. gigas* and

Fig. 6.5 Northern oyster shore on the Kaipara Harbour
(bottom) Muddy limestone shore platform at Whakapirau, buried with silty sand up to mean tide level (MTL).
(top right) an oyster farm.
a–b scattered slabs with *Diloma* and *Zeacumantus;* b–c *Crassostrea gigas* and *Pomatoceros caeruleus;* c–d muddy lower
shore, with *Pomatoceros* clusters
1 *Crassostrea gigas* 2 *Pomatoceros caeruleus*, free-standing cluster 3 *Ostrea lutaria* 4 *Zeacumantus lutulentus*
5 *Diloma subrostrata* 6 *Notoacmea helmsi* 7 *Cominella adspersa* 8 *Xymene plebeius* 9 *Cominella glandiformis*
10 *Cominella quoyana* 11 white-faced heron, *Ardea novaehollandiae*.

others with cross-ridges near the hinge that appear to be intermediate hybrids between *C. gigas* and *C. glomerata*.

The invasive power of *C. gigas* was dramatically shown in the Manukau Harbour's north-eastern reach, where it was a beneficiary of nutrients from the nearby oxidation ponds. In the 1980s it took possession of the middle and lower eulittoral at French Bay. Unlike rounded *C. glomerata*, young *C. gigas* shells are narrow and rectangular, slightly expanded at the ends. They raise themselves as a fragile cover clear of sediment, until — with the increasing surface presented — small waves can break off their flimsy pig-a-back clumps. As a result a beach debris of thousands of empty shells has built up, often with the hinge still intact. Round much of the Manukau shoreline, old cockle shell beaches have been replaced with thick piles of razor-sharp oyster shells; and intertidal platforms are disappearing beneath

thick growths of Pacific oysters and the drifts of mud that accumulate around them.

By 2000, Pacific oysters had spread around the North Island coast and were reaching the Marlborough Sounds. They continue to extend in the Waitemata Harbour, growing to giant size in quiet subtidal estuarine channels and in Orakei Basin.

Mangrove shores
In Northland, the Hauraki Gulf and the Bay of Plenty, the heads of the harbours are held in possession by an intertidal scrub, that at its zenith becomes a forest. The single mangrove species, *Avicennia marina* subsp. *australasica*, forms the seaward extreme of a formation running back through rush-beds and salt meadow to plants entirely terrestrial.

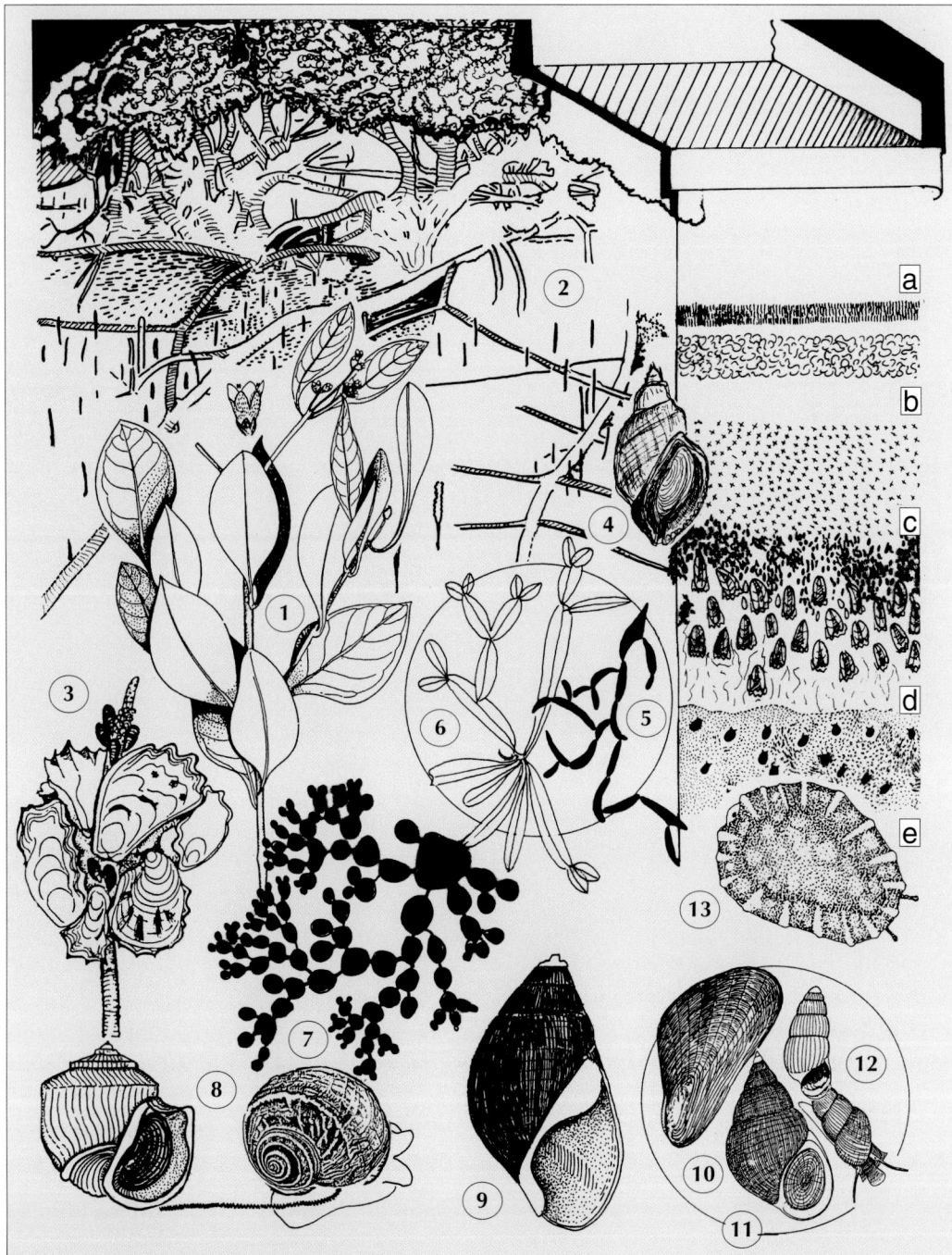

Fig. 6.6 Estuarine zonation in high shelter, Whangarei Harbour

a–b blue-green cyanobacteria; b–c *Austrominius;* c–d *Catanella* and *Crassostrea gigas;* d–e low tidal silt with *Onchidella*

1 mangrove, *Avicennia marina* 2 mangrove root system 3 mangrove pneumatophore with *Crassostrea gigas* 4 *Lepsiella scobina*, extreme sheltered form 5 *Catenella nipae* 6 *Caloglossa leprieurii* 7 *Hormosira banksii* 8 *Amphibola crenata* 9 *Melanopsis trifasciata* 10 *Xenostrobus securis* 11 *Potamopyrgus estuarinus* 12 *Potamopyrgus pupoides.*

The bronze-green fringe of *Avicennia* — or what remains — is today a nostalgic element of Auckland's tidal inlets. Mangroves knit together the sprawling suburbs in a way more like a tropical estuary. Our childhood memories go back to exploring their channels by canoe, when mangroves were still mysterious, even darkly sinister.

At Whangarei and further north, *Avicennia* is a considerable tree, reaching 10 m tall along the main tidal courses. Away from these margins, the wide flats carry a low 'mangrove park', of very uniform height, with bushes of less than half a metre. Far older than juvenile, their growth has been evidently retarded by the sluggish drainage at a distance from the well-oxygenated tidal channel. Strong radial roots form a stable horizontal base just below the surface, and put up the pencil-like breathing roots called *pneumatophores.*

Mangroves do not reach south of the winter frostline. Beyond the Kaipara, fringed by some of its best stretches in the whole north, *Avicennia* appears on the west only in small stands as far south as Raglan. In east Northland and the warm Hauraki Gulf, mangroves command most of the natural backwaters. Over the extensive shallows of the Tauranga Harbour, they have dwindled to bushes, to reach their southern outpost with numerous small plants in Ohiwa Harbour, between Whakatane and Opotiki.

Up-harbour zonation

Mangrove trunks and roots themselves offer a settling ground, in less than a metre's depth at low tide and protected from strong currents. These apart, the hard surfaces for colonisation are mostly concrete breast-works and bridges.

Near Huia, in the **Manukau Harbour**, the high-level band of algae, discoloured with silt, consists of inter-mixed small greens, a filamentose *Rhizoclonium* and a tubular *Enteromorpha*. Below these come several of the blue-greens, typified by crusts of *Oscillatoria nigroviridis*, built up of microscopic filaments in stiff mucilage.

Around high water neap, large and well-grown barnacles, *Austrominius modestus*, initiate a **eulittoral zone**. An oyster band cuts through these, with *Crassostrea gigas* in recent years predominant. Among and above the oysters the small eulittoral rhodophyte *Catenella fusiformis* typically hangs in chains of purplish brown segments. The similar-sized *Caloglossa leprieurii* is paler, and distinguished by mid-ribbed leaves. Both are cosmopolitan in warm seas, typically and most often found as a small community upon mangrove boughs.

The oyster borer, *Lepsiella scobina*, follows *Crassostrea* far into enclosed waters, assuming a taller, eroded spire distinct from its open coast form. In northern harbours (though not at Auckland) small eroded *Nerita atramentosa* become common in harbours. Further down-shore *Melagraphia aethiops* turns up where the rock is moderately clean. Littorines and limpets are usually wanting.

The oyster band is cut off below by greasy mud, in some places filmed with a bright cyanobacteria, like green paint applied with a fine brush. In the shade the pulmonate slug *Onchidella nigricans*, creeps about actively to graze the fine deposits. The same mud can be studded with burrows of snapping shrimps *Alpheus*; or from a hard foundation beneath, the anemone *Anthopleura aureoradiata* (Fig. 13.22) will often emerge, to open its oral disc and tentacles.

Older mangrove pneumatophores carry their own outposts of mini-zonation, being loaded with *Austrominius modestus*, and the flea mussel *Xenostrobus pulex*. Rock oysters may crowd upon mangrove trunks to almost half a metre above ground. Pneumatophores have also their typical flora of *Catenella* and *Caloglossa*, with wefts of *Rhizoclonium implexum* and sparser *Enteromorpha* and *Chaetomorpha capillaris*.

On rocky outcrops near low water, *Austrominius modestus* may, near Auckland, be accompanied by the larger pink *Balanus amphitrite*, a cosmopolitan ship-fouling barnacle. Found at the same level, the small estuarine mussel *Xenostrobus securis* continues upstream as far as the tides, wherever rock or half-exposed tree roots offer a hard base. Longer and lighter than the common *X. pulex*, it is more posteriorly flared, differing too in its varnished brown periostracum and the pink flush of the interior.

Two sorts of brackish water gastropods live at stream mouths. The hydrobiid *Potamopyrgus estuarinus*, grazing over organic films, is teemingly abundant on hard outcrops. The smaller, cylindrical *P. pupoides* can in some places be found under stones. The much larger *Melanopsis trifasciata*, a fast-stream snail, comes down to brackish estuaries to graze under stones. The shell grows to 2.5 cm long, with a broken ('decollated') tip, black epidermis and china blue interior.

Where fluid mud surrounds the pneumatophores, the mangrove backwaters are the haunt of the mud snail (*Amphibola crenata*), the tikito of the Maori. This is a primitive pulmonate, with the pallial cavity converted to a lung, the larva free-swimming and an operculum retained in the adult. *Amphibola* grazes on the finest surface deposits, triturated for nutrient extraction by the muscular gizzard. The faecal string, often trailed for half a metre behind the progressing snail, is rich in bacteria, with more nutrient production already set in train. Like every pulmonate, *Amphibola* is hermaphroditic, with small eggs (some 7000 to 10,000) deposited in a circular 'nidus' made of mud moulded between the shell rim and foot.

Sheltered southern shores
Catlins

South of Dunedin, an interesting Otago inlet is the estuarine lake of the **Catlins River**, formed by drowning as the sea level rose at the end of the Last Ice Age. Its waters are essentially sheltered, though periodic storms may blow across it from the north side.

On the south shore, tilted greywacke sandstone just inside the estuary from Jack's Bay shows a southern species-mix: the higher level red algae *Stictosiphonia arbuscula* and *Apophlaea lyallii* are succeeded by the barnacle *Epopella plicata*, followed in turn by a band of

Fig. 6.7 Shore community in deep shelter in Catlins Estuary, Otago
1 sea-grass, *Zostera capricorni* 2 *Durvillaea antarctica*, juvenile plant newly settled 3 *Durvillaea antarctica*, unbranched estuarine form 4 *Xiphophora gladiata* 5 *Ulva* sp. 6 *Scytosiphon lomentaria* 7 *Myriogloea intestinalis* 8 *Gigartina decipiens* 9 *Notoacmea helmsi* form *scapha* 10 *Micrelenchus huttoni*.

the navy mussel, *Mytilus edulis aoteanus*, with a scatter of ribbed mussels, *Aulacomya maoriana*.

Low water is dominated, in a way never to be seen in a northern harbour, by the bull-kelp *Durvillaea antarctica*. At the Catlins these plants are thoroughly adapted to quiet water, and markedly different from those on the

open coast in form. The shortened stipe expands to a heavy, almost rectangular apron with its undulant margin carrying only short straps. Further offshore, secluded from waves, a cordon of *Macrocystis pyrifera* shows up by its floating leaves. In the shallows behind it, large *Durvillaea*, initially fastened to small boulders,

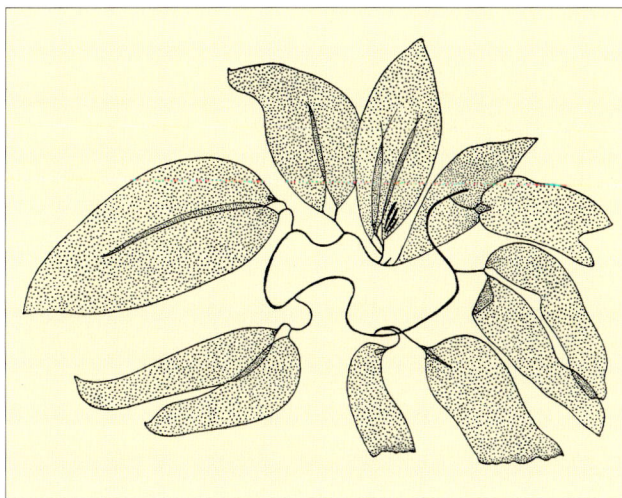

Fig. 6.8 *Macrocystis pyrifera*
The loose-lying bladderless form of *Macrocystis pyrifera* from harbours in the far south.

remain stationary under their own weight. Young *Durvillaea* just settled are heart-shaped or circular, altogether unlike the juveniles of open coasts.

The common attribute of all the quiet water algae at Catlins is their lax, attenuated growth, and usually large size (Fig. 6.7). The elliptical sheets of an *Ulva* species reach a metre or more long, are perforate and narrowly attached. Tubular *Enteromorpha* and *Scytosiphon lomentaria* grow exceptionally long, as do also the attenuated bladders of *Adenocystis utricularis*. Attached to rock under low tidal sand are the slippery, sinuous cords of *Myriogloea intestinalis*. Further out, among free-lying *Durvillaea*, is an estuarine form of *Xiphophora gladiata* with its long branches perfectly straight. The common *Gigartina* of southern harbours is *G. decipiens*, with dichotomous straps bearing slender pinnules.

In the shallows inside the shelter of bull-kelp, long, streaming leaves of the sea-grass *Zostera* (normally a strange companion for *Durvillaea*!) break the surface at low tide. On *Zostera* leaves the estuarine limpet *Notoacmea helmsi* grows narrow and parallel-sided in the form *scapha*, to fit exactly along a leaf. Typical also of *Zostera* flats is the leaden grey trochid *Micrelenchus huttoni* (Fig. 6.7), while two minute rissoid snails, *Pisinna zosterophila* and *Eatoniella limbata*, are found in teeming numbers, browsing on the small *Zostera* epiphytes.

In the most sheltered reaches of the far south, such as Port Pegasus and Paterson Inlet (part a marine reserve), Stewart Island, *Macrocystis pyrifera* (Fig. 6.8) takes a peculiar loose-lying form, with large, thin leaves lacking bladders. The holdfast is also missing, the plants resting immobile on sheltered sand, and shifted only slightly by gentle up and down movements of the tides.

Land-locked basins

The most sheltered of all marine shores are those of impounded basins altogether secluded from waves, and sometimes cut off from regular tides. Where the tides rise and fall with no strong horizontal movement, the mud or gravel at low water neap (LWN) may be occupied by a peculiar free-lying form of *Hormosira banksii*. From its lack of an attaching holdfast, Lucy Moore named it 'ecad *libera*', in first describing it from Picton Harbour. In the heavy clumps up to 30 cm across, almost every bladder proliferates, initially with up to a dozen separate buds. A branch 20 cm long may thus have several thousand growing tips. The colour may be different on top and bottom, and the underparts often rot away with the rest left healthy. This free-lying form appears to be sexually sterile, with smooth bladders lacking the usual goose-flesh of conceptacles. Vegetative proliferation occurs as clumps break asunder. With so little water movement, plants lie in one place, merely rising and subsiding with the tides, and in the north they may be confined by a palisade of mangrove pneumatophores.

Algae grow large in these undisturbed basins. Above the *Hormosira*, *Enteromorpha ramulosa* at Picton reaches a metre long in spring, and sheets of an *Ulva* species in summer attain 2 m across.

Where the tides are excluded by sluice-gates or impounding, *Hormosira* disappears and free-living green algae may achieve a bonanza of rapid growth promoted by eutrophication. Cast adrift in summer, these will rapidly break up and rot. Offensive smells of sulphuretted hydrogen attend their decay, which results in the build-up of black anaerobic mud. In Orakei Basin, Waitemata Harbour, the cycle of growth was found to begin in February and March with *Gracilaria secundata* for the first six weeks red with an epiphytic *Polysiphonia*, and afterwards pale green with attached diatoms. It broke away to float loose, with a maximum bulk in July. *Enteromorpha intestinalis* succeeded it as gas-buoyant mats in late spring, followed after decay by an algal complex matted with green filaments of *Rhizoclonium implexum*. These had regressed by mid-summer. In the 1990s an introduced red alga of the Family Solieriaceae was also found flourishing in Orakei Basin.

Opisthobranchs

Regressing green algae or *Zostera* are of special attraction to shelled opisthobranchs, such as the bullomorphs (Order **Cephalaspidea**) as well as one of the sea-hares, *Bursatella leachii*. These periodically come inshore to copulate and spawn, building up temporarily high numbers.

The bullomorphs are the first forerunners of the naked sea-slugs. They are primitive in retaining the thin,

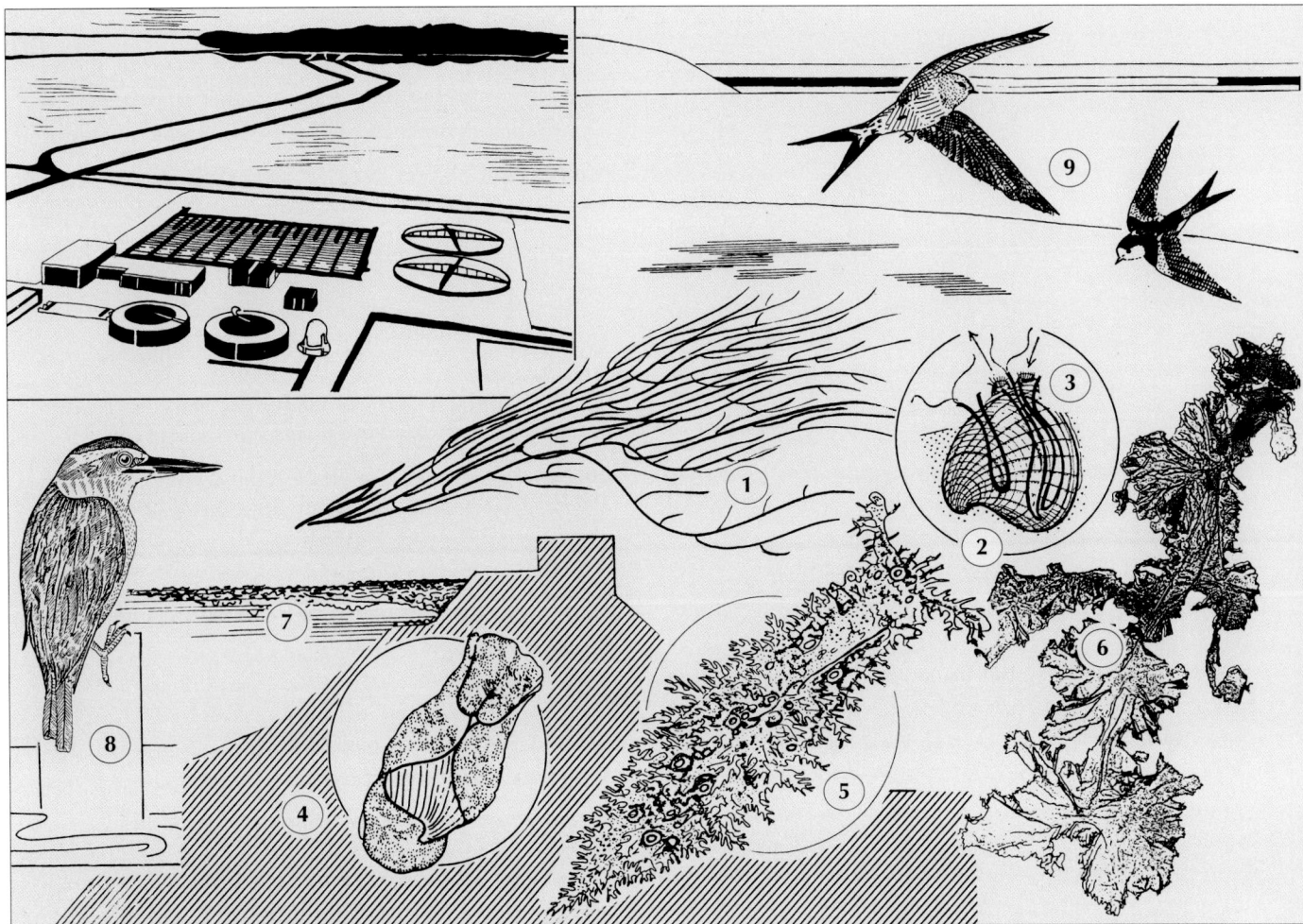

Fig. 6.9 Harbour flat in deep shelter adjacent to outfall of Mangere sewage treatment oxidation ponds, Auckland
1 *Gracilaria secundata* 2 *Austrovenus stutchburyi* with spionid worm *Boccardia* (3) 4 *Haminoea zelandica*
5 *Bursatella leachii* 6 *Ulva* sp. 7 decaying algae accumulated in enclosed basin 8 kingfisher, *Halcyon sancta*
9 welcome swallow, *Hirundo tahitica*.

spiral shell, mostly with an insunken spire from which they take the name bubble shells. *Haminoea zelandica* is slug-like and depressed, greyish black and flecked with brown. The globular shell is fragile and transparent, normally covered from view by parapodia. The animal has in front a fleshy head-shield, hard inside from the presence of a box-like gizzard lined with shell plates. The thin drawn-out spawn string spirals through a clear sausage-shaped jelly. A second bullomorph, more associated with *Zostera* than decaying algae, is *Bulla quoyii* with a thicker, ovoid shell marbled in purplish brown or grey, with darker spiral bands. The parapodia and head shield are speckled in yellowish brown, and plainer yellow beneath.

The sea-hare *Bursatella leachii* stands at an intermediate level between the bubble shells and the shell-less nudibranchs. The soft, flask-shaped body is greyish green, with dark blotches and emerald specks, and covered with short-branching dendrites. Often silt-covered, *Bursatella* can wholly submerge in the surface mud, taking in respiratory water by a siphon formed from the narrow opening through the fused parapodia. Its forerunners are the aplysioid sea-hares, that retain a vestigial internal shell and mobile parapodia, grazing succulent living algae on rocky shores. *Bursatella* comes ashore on *Zostera* or in large masses of dead ulvoid greens. Like all opisthobranchs these sea-hares are hermaphroditic, copulating in chains of up to half a dozen overlapping so the right sides are thus in contact, each individual acting as 'male' to the 'female' in front. The spawn forms a tangled yarn of jelly strings entwined upon the surface.

CHAPTER SEVEN
Shores With Enriched Algae

The most flourishing communities of algae between tides are on reefs where the bedrock offers a contour of gullies, overhangs and channels. Equally important to diversity is the lively action of surge and splash. The major wave force is often broken further to seawards and thus comes in with reduced attack. Broken topography also offers protection from insolation (heating by the sun), from evaporation with the dangers of wind-burn and from the harmful effects of light. Clean water will not carry enough sediment to reduce illumination, but may stir up sand or shell fragments to lacerate delicate fronds. Several algae are, however, positively tolerant to sand; for just a few species a sand layer over the rock is a condition for normal growth.

Hard greywacke or igneous rocks, irregular in topography, tend to be more hospitable to algae than soft or friable sandstone or siltstone. Large, stable boulders shaded at the sides offer protected microclimates, and the channels around them promote faster flow. Algal diversity may increase in higher latitudes; compare our northern species lists with those for the cooler waters of Kaikoura, Banks Peninsula, Otago and Stewart Island.

The *pink-line* — marking the top of a continuous algal veneer — is an important biological threshold. Below it the rock surface never becomes fully dry between tides. In this book the uppermost part of the ensuing pink stretch has been equated with the lowest reach of the eulittoral. Some have preferred to treat this band as the upper part of the sublittoral fringe.

Up to the top of the coralline veneer, the lower eulittoral (like the sublittoral fringe) is dominated by algae. Broken topography and fast water movement combine to favour their growth until — most notably in the cold south — there is a total dominion of the algae. Above the pink-line, in contrast, we have already found the zoned shore — with only a few exceptions — pre-empted by sessile animals.

With the reef-edge direct wave impact, the more delicate of the red algae are less favoured. Only the tough *Carpophyllum* and *Xiphophora* and one or two coriaceous or calcified reds can withstand the constant wrenching and battery. Likewise, the platforms of our northern coasts, clad with corallines, tend to be too directly sun-warmed for really high algal diversity. Still, golden *Hormosira banksii* can crowd their moist runnels; and their sun-lit pools may develop a veritable wealth of algae, often with their own depth stratification.

Rhodophyta

Before starting our geographical survey, the large and diverse Phylum **Rhodophyta** calls for an overview. Small in biomass compared with the browns, the red algae are on nearly every temperate shore the richest in species. Many rhodophytes are simple and filamentous, and the largest are of only middling size. They are often exquisite in form and colour. Far more than the greens or the browns, the red algae are looked on as the denizens of the lowest intertidal and the subtidal. The isolated **Bangiophyceae** apart, almost all the other rhodophytes, constituting the far larger Class **Florideophyceae**, grow below the pink-line.

The pink coralline base continues down far below the tides or the effect of waves, and indeed beyond any significant water movement. Whether carrying a *Corallina* or *Jania* turf, or simply as a painted veneer, this base runs without a break through the upper sublittoral zone, down to the ultimate depth — some 20 m for an open coast — at which photosynthesis must cease for want of light.

Away from strong wave-break rhodophyte meadows can reach well up between the tides. It is here, under lively surge and splash, that the **Gigartinaceae** come into their own, increasing in diversity until on cleaner shores or further south they come to supplant the corallines. The pigments of intertidal *Gigartina* are moreover adapted for an illumination of different quantity and quality from that of the subtidal where most rhodophytes operate best.

All the red algae possess two pigments, red *phycoerythrin* and blue *phycocyanin* that visually mask out their chlorophyll. The red and blue have an accessory role to the primary photosynthetic pigment, transmitting energy at red wave lengths to chlorophyll while they absorb other light at wave-lengths of their own. Phycoerythrin is easily destroyed by light, so the *Gigartina* species in the emersed sublittoral fringe have an increased proportion of phycocyanin. Chlorophyll may also show up until we have — in effect — 'green rhodophytes'.

At wave-break line, beneath the brown canopy, the red algae are tough and resilient. Three common northern species, *Melanthalia abscissa*, *Osmundaria colensoi* and *Pterocladia lucida*, become almost cartilaginous in texture. Phycoerythrin is destroyed less than at the higher level, so they remain purple or wine-red with their green component masked. Many rhodophytes — as we shall see — have joints reinforced with calcium carbonate. In some the segments become fused into rigid crusts and lamellae. Over this hard internal skeleton spreads the living skin over which the water flows restlessly.

In subdued light, in the quiet waters below wave disturbance, all this changes. Delicate and clear pink rhodophytes grow fern-like with elaborately feathered pinnae, or as translucent finely veined leaves. They are often carmine or carnation pink, with high phycoerythrin values, absorbing the green wavelengths that reach in greatest proportion to these low-illumined levels.

Bangiophyceae

These small upper shore reds are evidently of ancient origin. They agree with the higher **Florideophyceae** in the presence of phycoerythrin and phycocyanin and the manufacture of the carbohydrate floridean starch. But their reproductive organs are primitive, with the rhodophytan carpogonium and trichogyne only in the simplest form. In both *Bangia* and *Porphyra* the carpospores give rise to creeping filaments that penetrate shells and have been identified with the once supposed separate genus '*Conchocelis*'. From this early stage, *Porphyra* is widely cultivated in Japan.

The simplest Bangiophyceae are uniseriate filaments, as in the epiphytic *Erythrocladia*, a bright scarlet epiphyte on brown algae. *Bangia* too has unbranched filaments, though their cells often divide lengthwise. A dark red or violet *B. angia* (traditionally called *atropurpurea*) appears on rocks or concrete in the spray zone in winter or spring. A blackish estuarine or freshwater form has been recorded as *B. fuscopurpurea*. Both can resist desiccation and tolerate large salt fluctuations.

The *Porphyra* species appear not as threads, but as wide, frilled membranes still only one cell thick. The cells divide not apically, as in the higher reds, but diffusely. Like that of *Bangia*, their taxonomy in New Zealand is complex and still incomplete. *P. columbina*, with a much expanded membrane or tufted rosette, lives

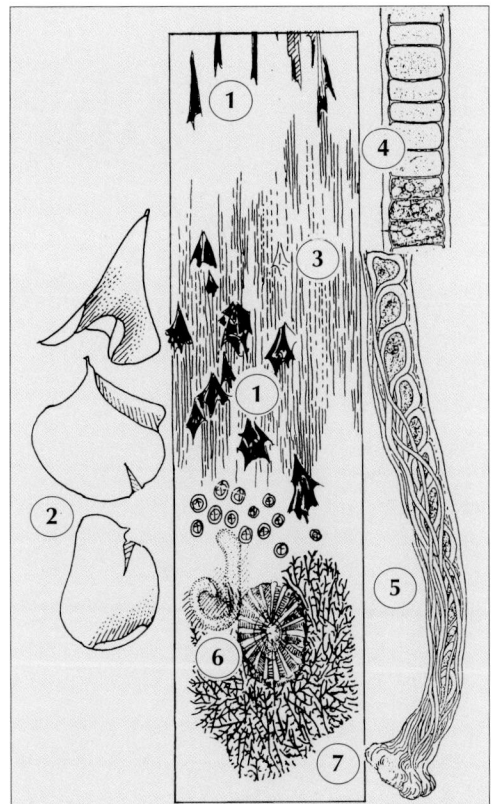

Fig. 7.1 Small high tidal algae in August, Castor Bay, Auckland
1 a small short-lived *Porphyra* 2 the same enlarged 3 *Bangia atropurpurea* 4 *Bangia* young filament 5 *Bangia*, base of thread with rhizoid formation 6 *Siphonaria australis* with spawn coil 7 *Capreolia implexa*.

high in the supralittoral fringe. The longer, wavy-edged ribbons of *P. subtumens* are to be found near low tide as an epiphyte on *Durvillaea*.

Florideophyceae

These higher red algae must also have begun their evolution from branching filaments. A few, such as *Acrochaetium*, have remained as simple as this, and constitute the small first order of the Florideophyceae, the **Acrochaetiales**. Certain *Acrochaetium*, where only asexual reproduction is known, are suspected to be the alternate generation of other reds, such as *Liagora*. Even the advanced and diverse **Ceramiales** have some members with uniaxial filamentous structure, such as *Anotrichium* and *Antithamnion*.

Most of the filamentous reds have, however, become *heterotrichous*, with both upright and prostrate systems of threads. In many encrusting reds, it is mainly the prostrate system that survives, with the upright threads used only for reproduction. But the crustose thalli of *Peyssonnelia* and *Melobesia* have been built up with the compaction of dense upright threads.

More generally, the prostrate system is suppressed

and the upright threads take part in more advanced tissue construction. In the **Florideophyceae**, we can basically distinguish *uniaxial* and *multiaxial* growth, according to the number of strands in the central axis. Thus, while the apical cells divide only at right angles to the plane of elongation, other axial cells can split tangentially, to form *pericentral* cells. In *Polysiphonia*, these can still be recognised unchanged, but in many reds they have given rise to side-branches of limited growth; or — instead of lengthening out — have remained short, investing the axis or *medulla* with a *cortex* of small photosynthetic cells. Fig. 7.2 shows *Wrangelia* with downgrowing hyphae from inner cortical cells.

This has been the principal mode of tissue-building in the more advanced Rhodophyta. Richly branching lateral filaments have pressed together to build a *pseudoparenchyma*, which is an alternative to the true parenchyma of higher plant construction. There are minute pores between adjacent cells of a filament, but none between filaments lying side to side. This whole structure condensed upon the axis becomes invested with mucus, sometimes as firm as a tough plastic.

Nemaliales

The condition just described is found in this early order of the Florideophyceae. In the worm-like cords of *Nemalion* the multiple axis can be freed up by pressing the growing tip with a coverslip to show the cortex, with its branched lateral threads. The pale dichotomous branches of *Liagora* (Fig. 7.3) are more compacted, and reinforced as well with lime grains. *Galaxaura* is an important tropical genus, represented at the Kermedecs, having species — like *Liagora* — lightly calcified but flexible. All *Galaxaura* have dichotomous branches, some cylindrical and jointed, others flattened. Tufts of crimson hairs may cover much of the plant. To this order belong the small, rigidly tufted *Nothogenia fastigiata* (Fig. 7.3) found on exposed coasts throughout New Zealand, and *N. pseudosaccata* on east South Island coasts. Finally, there are the *Scinaia* species, with long, dichotomous branches with a slippery cortex. *S. firma* (Fig. 7.3) and *S. berggrenii* are both summer-seasonal.

Corallinales

In this important order, calcification has reached its high point. In the familiar turf-forming **Corallinaceae**, the erect branches rising from the basal crust are built of long segments (*intergenicula*) with uncalcified joints (*genicula*), with the whole micro-structure multiaxial. Fig. 7.4 shows the cell pattern with elongate parallel filaments revealed after decalcification. The intergenicula carry reproductive conceptacles opening by apical or

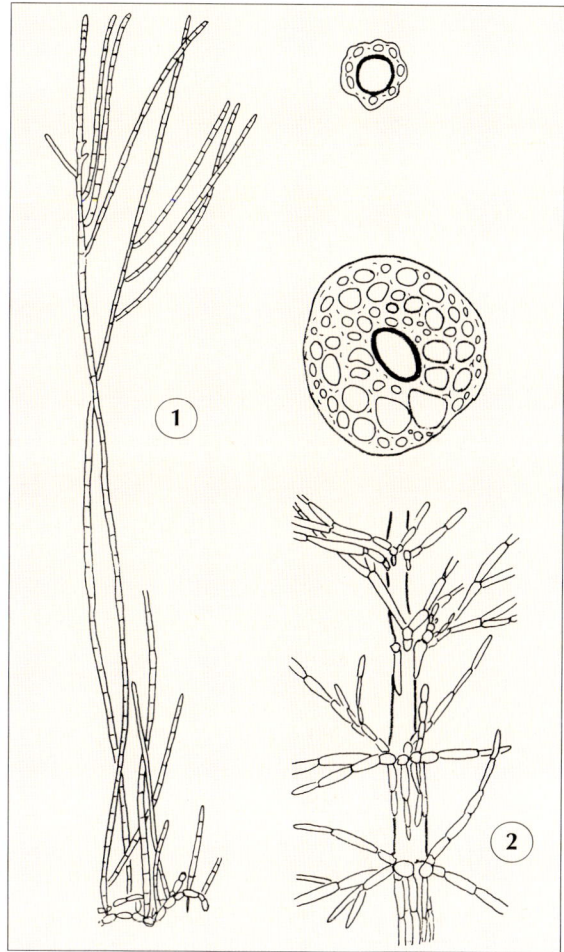

Fig. 7.2
1 *Audouinella*, a monosiphonous red alga 2 *Wrangelia*. Filaments invest the main axis to build a cortex (see cross-sections).

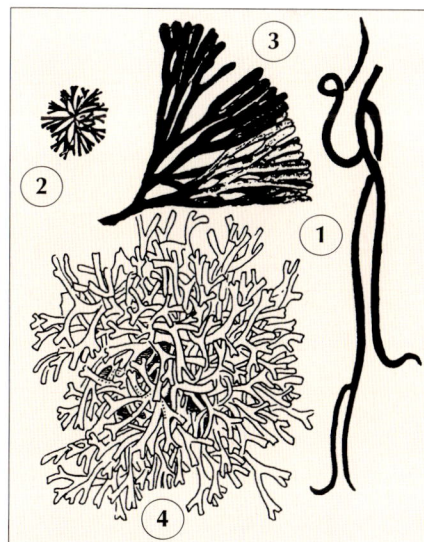

Fig. 7.3 Nemaliales
1 *Nemalion helminthoides* 2 *Nothogenia fastigiata*
3 *Scinaia firma* 4 *Liagora harveyana*.

marginal ostioles. The New Zealand shore species involve at least six genera:

Corallina branches pinnately, with its conceptacles

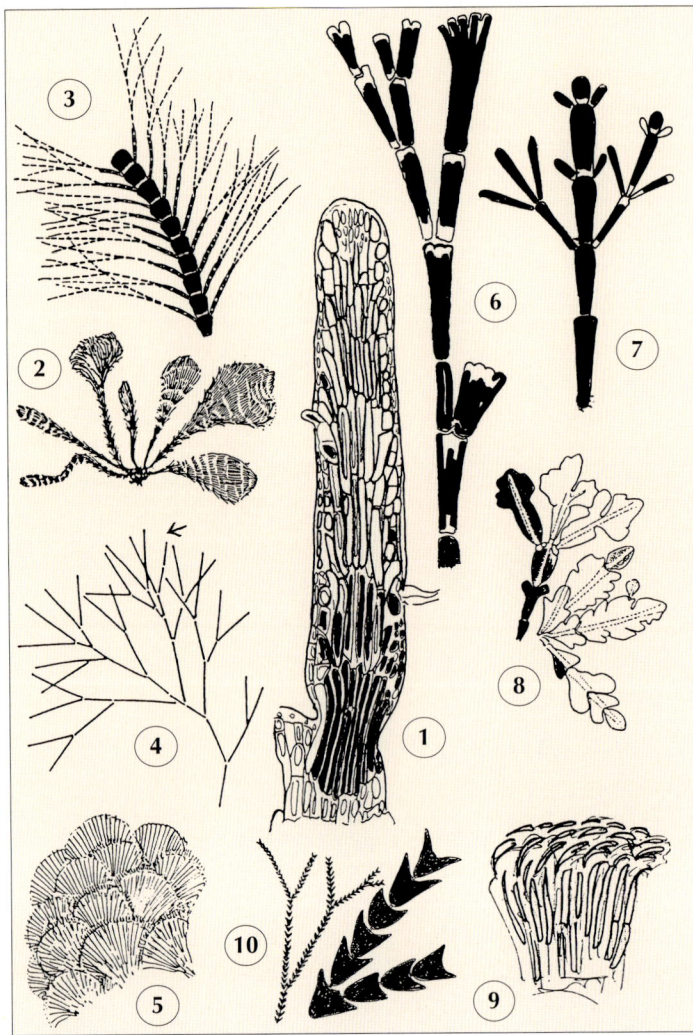

Fig. 7.4 Corallinales

1 *Corallina*, longitudinal secretion of terminal segment and joint 2, 3 *Haliptilon roseum* 4, 5 *Jania micrarthrodia* 6 *Amphiroa anceps* 7 *Corallina officinalis* 8 *Arthrocardia corymbosa* 9 the same, growth habit 10 *Cheilosporum sagittatum*.

located in the swollen tips of the side-branches. *Jania* — in contrast — is dichotomous, with the enlarged end cells of the cortex growing out into hairs (*antennae*). New segments arise from the upper corners of the female conceptacles which thus become lodged in forks of the thalli. The male conceptacles are at the tips of special branches in the forks. *Haliptilon* has branching pinnate at the base but dichotomous above.

Arthrocardia corymbosa shows the sides of the segments expanded into lamellae. *Cheilosporum* grows side-horns that harbour the intercalary conceptacles. The fronds of *C. sagittatum* have segments like strings of tiny arrowheads.

Amphiroa, has long, dichotomous branches, as typified in *A. anceps*, the heaviest of our New Zealand jointed corallines.

In contrast with all the above are the non-articulate coralline crust of light or massive build. Thus *Melobesia* forms thin patches upon other algae or *Zostera*. Each

erect thread carries at the top a flat '*cover cell*'. The lamellate, coral-like *Lithophyllum carpophylli* (Fig. 7.5) is like a small coral, and envelops the basal stem of brown algae.

In the massively calcified thalli of *Lithophyllum* and *Lithothamnion* two layers have been detected. A basal *hypothallium* of prostrate threads gives rise to the vertical threads of a *perithallium*. The top cell of each thread has chromatophores; the cells just beneath carry starch grains, while the deepest are in large part dead. The mauve nodules of *Lithophyllum polymorphum*, studded with clubbed branches, are often washed up from subtidal channels. A heavy veneer of *Lithophyllum arcuatum*, with its small crescentic sculpture, is common on northern exposed shores.

Species identification of the encrusting forms is specialist's work, calling for sectioning or at least recourse to reliable herbarium material. Many current names are dubious, with genera not yet in a condition to be dealt with in Adams' (1994) authoritative book.

Gelidiales

This is a small order with three main genera, pinnately branched. Our most familiar species are the tufting *Gelidium caulacantheum* and the low, patch-forming *Capreolia implexa*. *Pterocladia* species live at or beyond low water, *P. lucida* being long pinnate and lightly cartilaginous, and *Pterocladiella capillacea* finer branched, soft and bushy.

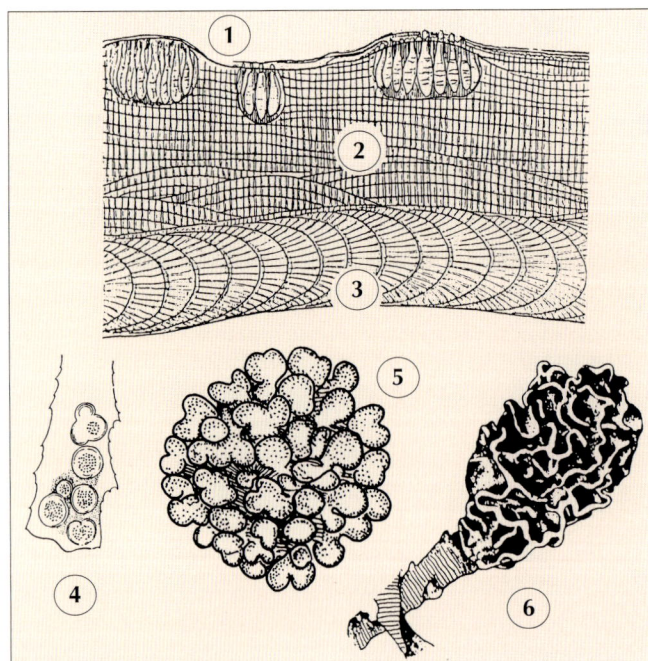

Fig. 7.5 Corallinales

Lithothamnion, vertical section 1 tetrasporangium 2 perithallium 3 hypothallium 4 *Melobesia* 5 *Lithophyllum polymorphum* 6 *Lithophyllum carpophylli*.

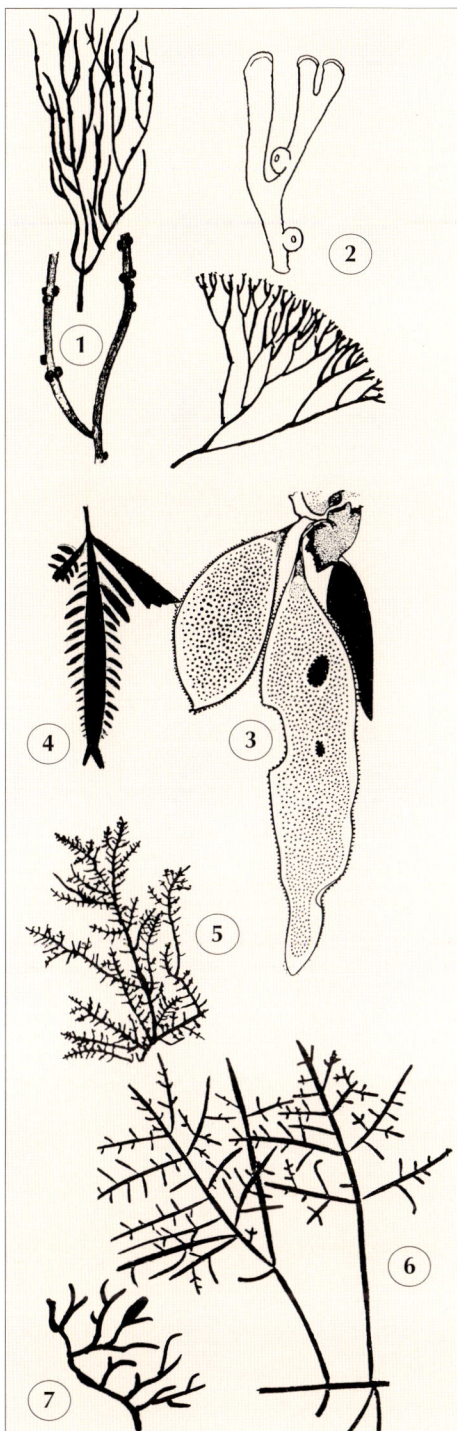

Fig. 7.6
Gracilariales
1 *Gracilaria secundata* 2 *Melanthalia abscissa*, both with cystocarp detail
Gigartinales
3 *Gigartina circumcincta* 4 *G. marginifera*
Gelidiales
5 *Gelidium caulacantheum* 6 *Pterocladiella capillacea*
7 *Capreolia implexa*.

Hildenbrandiales

Two genera classed here have heavy encrusting cartilaginous thalli. *Hildenbrandia dawsonii*, growing at the tops and sides of boulders, is reddish to black, smooth and glossy when wet. The *Apophlaea* species (*sinclairii* in the north and *lyallii* in the south) are like crusts of dried, congealed blood, putting up antler-like branches.

Gracilariales

These are robust plants of firm, cartilaginous texture, with a '*pseudoparenchyma*' built of juxtaposed vertical filaments. The *Gracilaria* species, with their slender whip-like branching, are typical of sheltered harbours and include the agar-rich *Gracilaria chilensis*. The erect and robust *Curdiea* species found at low tide belong to this order, as do also the tough coriaceous thalli of the low or subtidal *Melanthalia abscissa*.

Gigartinales

This order is rich and diverse in New Zealand, with species mainly endemic. Some extend New Zealand-wide, but there are also northern and rather more southern-limited species, as well as several confined to the subantarctic islands. The *Gigartina* species are collectively some of the most important red algae of the clean water sublittoral fringe; towards the south they increasingly enter sheltered harbours. They take two prime forms, with either pinnate/dichotomous branching, or expanded blades (Fig. 7.28). The female plants have cystocarps growing from side branches or scattered like warts over the whole surface.

With new insights into structure, the *Gigartina* species are being reclassified into a number of genera. In the traditional *Gigartina*, the pinnate or dichotomous species, branched in a single plane, include (North and South) pinnate, flattened *G. livida* (S), *G. marginifera* (N) dichotomous, *G. decipiens* (NS), *G. ancistroclada* (S), *G. tuberculosa* (Antipodean), *G. alveata* (N), *G. clavifera* (S), *G. macrocarpa* (N). The chief leafy bladed species are *G. circumcincta* (NS), *G. atropurpurea* (N) and *G. lanceata* (S).

In addition, the southern South Island and Stewart Island have two long-bladed species of the southern, cold-water genus *Iridaea*.

In the wider **Gigartinales** are now included the **Caulacanthaceae**, represented in intertidal pools by the short-tufting *Caulacanthus ustulatus*, purplish black and drying out brittle. Of the small *Catenella* species, *C. nipae* is epiphytic on mangroves and *C. fusiformis* lives on *Apophlaea lyallii*.

Halymeniales

This order has a diverse range of algal forms. In the Family **Halymeniales** are the *Grateloupia* species, varying from tubular fronds to flattened blades, simple or much branched, pinnately or dichotomous. All are slippery, some being filled with clear mucilage, as in the hollow *G. intestinalis* (Fig. 7.17). Very different from these thin sheets are the tough, cartilaginous *Pachymenia* species and *Aeodes nitidissima*, with dark red, almost sessile lobes, like thick steaks.

The **Kallymeniaceae** are typified by pink sublittoral forms, *Kallymenia berggrenii*, erect and foliose with irregular side blades, and the fan-shaped *Callophyllis* species, dichotomously branched into narrow segments.

In the **Phyllophoraceae** the principal New Zealand species are of *Gymnogongrus*, forming wiry, dichotomous tufts, strongly corticated and rising from a cartilaginous holdfast. All are found on open coasts, generally on rocks periodically covered by sand.

The **Plocamiales** — all subtidal — have a well-defined character, being delicate pink and finely divided, with two to five secondary or tertiary branchlets set alternately at the sides of the primary series (Fig. 7.7).

To a separate family belong the small, plate-like *Peyssonnelia* species, mostly tropical, where a thin thallus, pinkish brown on top, loosely attaches by rhizoids underneath. The basal layer is lightly calcified. Mostly tropical *P. 'rubra'* is the name at present used for the species found in northern New Zealand, including the Three Kings and the Kermadecs.

Rhodymeniales

This order has multiaxial families rather distinct in habit. In the **Champiaceae** the tissues break down at the centre to leave the frond hollow, filled with mucus and in *Champia* divided by septa. *Lomentaria* has no cross-septa, but is constricted into pods attached by rhizoids to the ground. In **Rhodymeniaceae**, the membranous fronds are flattened, with palmate or dichotomous branching, and short proliferations from the margin. Most species live subtidally in harbours with sediment, but *Rhodymenia leptophylla* is found in low tidal pools on open coasts.

Bonnemaisonniales

All these algae are subtidal, having delicate pinnate fronds, arising from creeping stolons. *Asparagopsis* has tropical affinities with one species at the Kermadecs and one New Zealand-wide. The *Delisea* has four species, one at the Kermadecs, and others taking in New Zealand or southern islands species; also northern and Kermadecian, *D. plumosa* ranges to Stewart Island. The third genus, *Ptilonia*, has one northern and one southern species. In lieu of a holdfast curled modified spines serve for attachment.

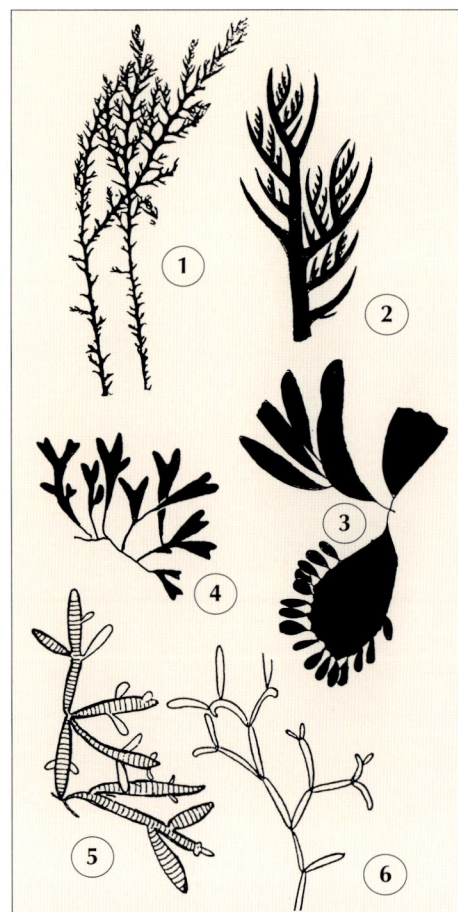

Fig. 7.7 Plocamiales and Rhodymeniales
1 *Plocamia costatum* 2 branching detail
3 *Rhodymenia* sp. 4 *Rhodymenia leptophylla* 5 *Champia laingii* 6 *Lomentaria umbellata*.

Ceramiales

Here we have reached one of the largest orders of reds, generally placed high among the **Florideophyceae**, and itself containing a wealth of seemingly unlike forms. All begin life with a single bipolar filament that remains prostrate until the growing plant becomes erect, with each family then taking on its distinctive character.

First, the **Ceramiaceae** remain filamentous and uniaxial. *Anotrichium* and *Antithamnion* form richly branched tufts, 2–5 cm tall, with a habit as simple as the filamentous greens and browns. The delicate *Euptilota* and *Plumaria* produce lateral branching. The branched filaments of *Ceramium* can be recognised with a lens by their thickened annuli of cortical cells and small terminal pincers. *Ballia* is delicately pinnate, heavily corticated by a felt of rhizoids.

In contrast, the Family **Delesseriaceae** is notable for its delicate foliose growth, with mid-rib and veins recalling the leaves of higher plants. Sublittoral and often perennial, its species are abundant in the southern hemisphere, as in cooler New Zealand waters. The thallus is

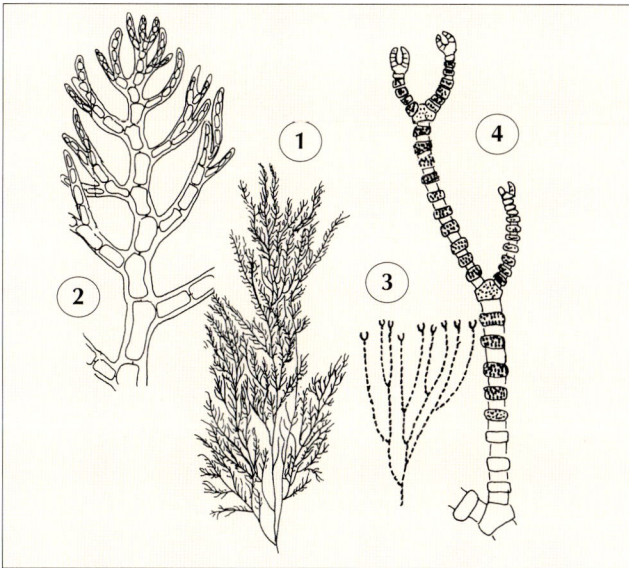

Fig. 7.8 Ceramiales: Ceramiaceae
1 *Anotrichium* with (2) detail 3 *Ceramium* 4 *Ceramium* with terminal detail.

formed by the fusion in one plane of numerous branches from the axial filament, and — except for the mid-rib and veins — remains one-layered. *Delesseria* has membranous pink fronds, with branches rising out of the mid-rib and the main axis can grow to a metre long. In *Phycodrys* the fronds are lobed like oak-leaves. The small-lobed *Myriogramme* is epiphytic on other algae, while the tiny mid-ribbed leaflets of *Caloglossa* grow on mangrove roots in warm and tropical seas. The tropical *Martensia* has a complex leaf-spread, formed by a network of radiating but interlinked lamellae.

The Family **Rhodomelaceae** are basically polysiphonous, and possess trichoblasts as well as multicellular branches. The species of *Polysiphonia* have the cells tiered, forming a standard number of primary (and

sometimes secondary) pericentral *siphons*. A scalpel section across a main axis, or a count of the siphons in surface view, can easily be made in a fresh filament. Polysiphonous axes — or parts of them — creep over the substrate, and can take on a dorsi-ventral form. *Stictosiphonia* thus has part of the thallus prostrate and the upgrowing apices curled in towards the substrate.

The pericentral siphons may divide horizontally, as in *Streblocladia* and *Lophurella*. In *Chondria* and *Laurencia* the thallus branches are fleshy and cartilaginous, as the pericentrals progressively branch into a parenchymatous cortex. A cell or papilla lodged in an apical depression still, however, indicates the early uniaxial structure.

A bilateral and dorsi-ventral structure has finally taken over in *Pterosiphonopia* and *Herposiphonia*, with the adults prostrate and attached by rhizoids. Branched segments arise alternately; where their bases fuse with the parent axis, leading to the fine, almost fern-like *Pterosiphonia*. Some of the creeping species live as small epiphytes as in the exquisite *Dasyclonium incisum* and *D. adiantiformis*.

Dorsi-ventrality appears also in some forms with wings growing out in one branching-plane. In *Lenormandia* (Fig. 7.30) the wings are keeled and one-layered, with branches arising from the ribs or main axis. *Osmundaria* is stiffly cartilaginous, with a several-layered cortex, and wings forming the saw-toothed margin.

Algal patterns on shore
Northern coralline flats
The prolonged *Corallina* flats of the north are generally poorer in species than the higher energy reef margins.

Fig. 7.9 Ceramiales: Delesseriaceae

1 *Laingia hookeri*
2 *Delesseria crassinervia*
3 *Phycodrys quercifolia*
4 (the same) section of mid-rib.

117

Fig. 7.10 Ceramiales: Rhodomelaceae (1)

1 *Cladhymenia oblongifolia* 2 *Lophurella caespitosa*
3 *Echinothamnion* 4 *Laurencia* with (5) detail
6 *Dasyclonium* sp. 7 *Chondria* (section with cortification
and growing apex).

But even under low wave energy and in a warm atmosphere, their water-retaining turf can maintain its own range of small algae. Outside the harbours, as water movement and clarity increase, species diversity is enhanced.

The leading brown alga is always the pervasive *Hormosira banksii*. There are other common Ochrophyta too. Where the surface has been grazed bare, patches of *Ralfsia verrucosa* are the first to re-appear, sometimes with the smaller, fleshy pads of *Petrospongium rugosum*. Through spring and early summer the platforms show the golden tint of fleshy *Leathesia difformis* and the thinner, saccate *Colpomenia sinuosa*. Close-allied to *Colpomenia*, the fleshy lattice of *Hydroclathrus clathratus*, from the tropical Pacific, has been turning up in Auckland and Northland since 1974.

Green tints are imparted to *Corallina* by small epiphytic *Enteromorpha* and the tufted meshwork of *Boodlea mutabile* (Fig. 7.23). Since the late 1960s, a novel green alga *Codium fragile tomentosoides*, has appeared in northern New Zealand (Fig. 6.4).

Towards the platform edge, more constant surge brings up the pink pastel hues of *Jania micrarthrodia* and *Laurencia distichophylla* and the olive green in *Laurencia thyrsifera*. There may be also the harsh green

stubble of *Cladophoropsis herpestica*, and at the seaward drop bottle-green *Codium convolutum*, sometimes interspersed with dark brown tassels of *Halopteris paniculata*.

In the north, the predominant fucoid species are those of *Carpophyllum* and *Sargassum*. The widest ranging — from submaximal exposure to relative shelter — is *Carpophyllum maschalocarpum*. Far more form-variable is *C. plumosum*, intermediate in exposure between *C. maschalocarpum* and *C. flexuosum*. *C. plumosum* in its 'pinnatifid' variety has fronds cut into membranous lobes. In more sheltered sites grows the 'capillifolium', variety, with the lobes narrowed to twisted branching filaments. The form *quercifolium* (Fig. 10.11) has broad lobes like tiny oak leaves. On rich algal shores, the short, palm-like kelp, *Lessonia variegata*, tends to replace *Ecklonia radiata* in the fringe, the more so as we go further south.

On sublittoral sand and pink-cobbled Lithothamnion, as off Long Beach, Russell, there occurs the unusual kelp *Ecklonia brevipes*, up to a metre high.

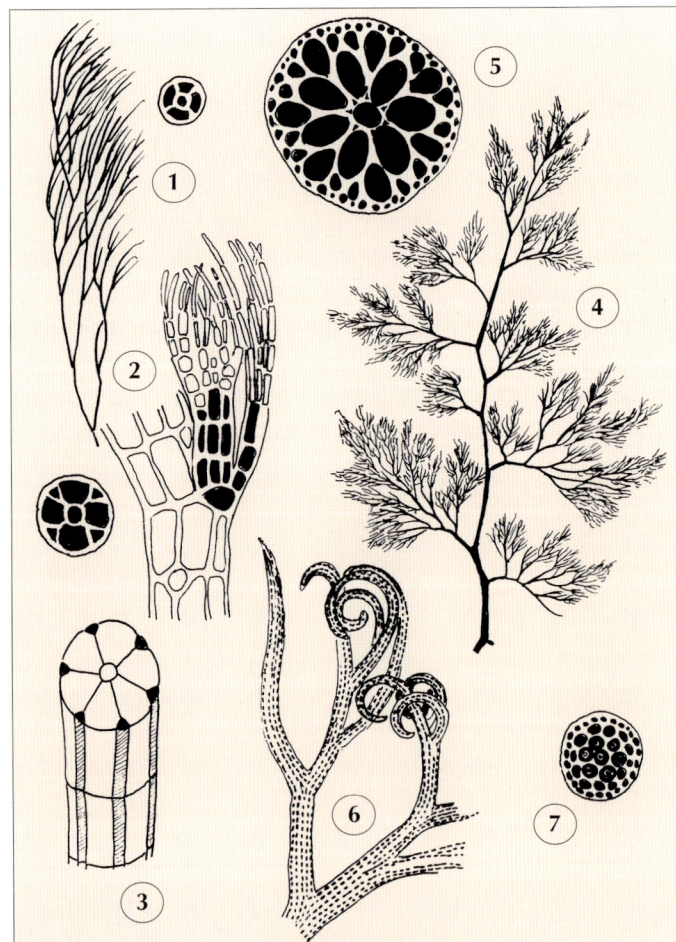

Fig. 7.11 Ceramiales: Rhodomelaceae (2)

1 *Polysiphonia strictissima* 2 *Polysiphonia*, detail with cross-sections 3 *Polysiphonia*, pericentral and corticating cells 4 *Streblocladia glomerulata*, with (5) section
6 *Stictosiphonia arbuscula*, with (7) section.

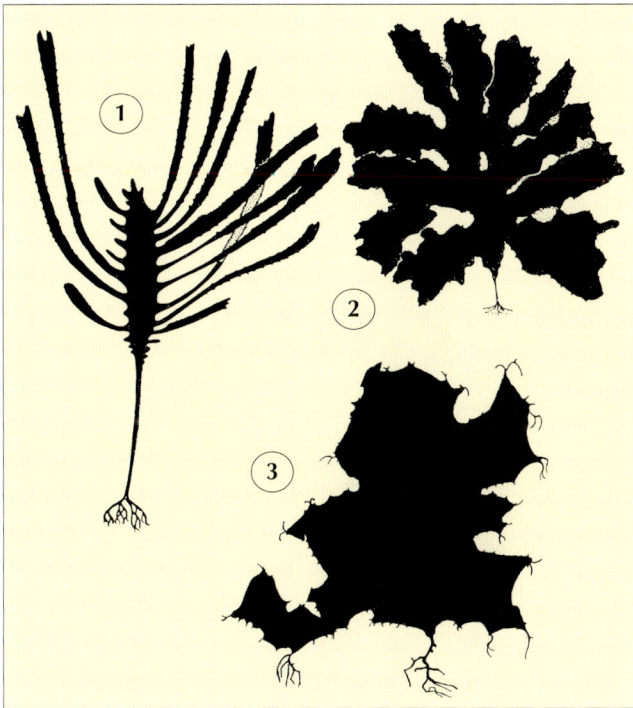

Fig. 7.12 Ecological form range of *Ecklonia*

1 *Ecklonia radiata* with narrow blade and lateral lobes from pools on open coast 2 *Ecklonia radiata*, small-stiped form in turbid harbour waters 3 *Ecklonia brevipes*, sheltered, muddy substratum, Bay of Islands.

With a delicate, easily torn blade, the thallus has lost its stipe, and springs direct from a widened holdfast. The margins of the pinnae can produce small adventitious holdfasts, and in vegetative reproduction these take their own foothold and detach.

Bay of Islands shores are good habitats for the green algae *Caulerpa*, established just beyond low water, but only sparsely entering the sublittoral fringe. *Caulerpa flexilis* is clearly visible around Russell, just beyond low water; it has lateral branching with close, pointed leaflets. From the prostrate rhizomes, erect green shoots mass into soft cushions. Also found subtidally is the very differently constructed *Caulerpa geminata*, with its bunches of grape-like appendages. Forming cushions or broad carpets, this alga can be gathered and eaten fresh as a salad.

Algal gardens of the north

To discover the northern red algae in their full diversity, we must leave the big harbours and go beyond the Hauraki Gulf. Good enriched shores can be reached in two directions from Auckland city. A 30 km drive brings us into the cool temperate, with the oceanic shoreline from south Muriwai to Karekare. It is some 900 ocean kilometres from this cold shore to Auckland and Northland's open east coast that forms part of the Pacific warm temperate. Its algal-rich shores closest to the city are those at Goat Island (outside Cape Rodney), on Great Barrier Island, and on the rocky outcrops in Bream Bay.

The choicest algal sites are in places of higher than normal energy, but with less than maximal wave action. Topographic diversity, especially on volcanic rocks or greywacke, offers the needed range of shade, surge and current flow.

My own memorable introduction to an algal garden was in company with Professor Val Chapman at Lion Rock, Piha, in 1946. That site was for many years to be a place of distinction for student teaching, duly commemorated in *The New Zealand Sea Shore* (Fig. 100). Today, its richness has been all but destroyed by repeated mussel-gathering and increased algal-stripping.

As typified at **Lang's Beach** (Fig. 7.14) or on the **Piha** coast, the assemblage beyond the *pink-line*, becomes increasingly diverse. From mid-tide level down, the coralline turf is replaced by two larger red algae that in the Auckland Province stand as indicators of enriched shores. First, *Gigartina alveata* forms a swathe through the middle of the eulittoral, even above the pink-line, directly following or intermingled with *Epopella plicata* or *Xenostrobus pulex* and itself often succeeded by a second red alga, *Pachymenia lusoria*. Their relations are shown at Whale Bay, south of **Raglan** (Fig. 7.13), where the set of the coast gives some protection from the straight-on impact of south-westerlies. *Pachymenia lusoria* is here followed downshore by small, solid-bladdered *Hormosira banksii* and *Carpophyllum maschalocarpum*.

Fig. 7.13 Semi-exposed shore at Whale Bay, south of Raglan

1 *Gigartina alveata* 2 *Pachymenia lusoria* 3 *Hormosira banksii* 4 *Carpophyllum maschalocarpum* (**inset**) 5 *Paratrophon cheesemani exsculptus.*

Fig. 7.14 Zoned shore at Lang's Beach, Bream Bay, east Northland
a–b *Xenostrobus pulex* and *Epopella plicata;* b–c *Gigartina alveata;* c–d *Pachymenia lusoria;* d–e *Corallina* and *Jania;*
e–f *Gigartina* garden; f–g *Carpophyllum plumosum;* g–h *Lessonia variegata*
1 *Gigartina alveata* 2 *Pachymenia lusoria* 3 *Gigartina marginifera* 4 *Gigartina atropurpurea* 5 *Pterocladiella capillacea*
6 *Pterocladia lucida* 7 *Champia novae-zealandiae* 8 *Gigartina circumcincta* 9 *Lessonia variegata* 10 *Carpophyllum
plumosum* 11 *Melanthalia abscissa* 12 (northern) variable oystercatcher, *Haematopus unicolor.*

Gigartina alveata reaches the highest intertidal level of any of its genus, and is confined to the Auckland Province, hardly reaching south of Whale Bay or East Cape. Bottle green thalli attach fast to the rock in springy, fist-sized clumps, drying out between tides to near black. The narrow, linear branches fork dichotomously, finally dividing and recurving at the tips. Down

each branch runs an incised channel, a water-retaining adaptation in this species for an elevated site.

Immediately below *G. alveata* comes *Pachymenia lusoria,* the largest and most robust of our intertidal rhodophytes. The flat thongs are smooth and shiny, forking several times and with tattered side-processes. Leathery in texture, they resist wave-wrenching, and are

Fig 7.15 Gradation of algae in surge exposure, Te Henga, west Auckland
1 *Durvillaea antarctica* (winter-settled young plant) 2 *Pachymenia lusoria* 3 *Gigartina alveata*
4 *Scytothamnus australis* 5 *Hormosira banksii* (exposed form).

nearly impossible to pull from the rock. *Pachymenia* is seasonal; by the end of summer it is often sheared off by wind-burn, and its regressing foliage turns from wine-red to olive-green.

These two species have a constant relation to wave impact. As at north **Te Henga** (Fig. 7.15) where surge sweeps over a low saddle, *Gigartina* occupies the middle distance with *Pachymenia* and young *Durvillaea* on the wave-break side, *Scytothamnus australis*, *Hormosira banksii* and the mussel *Xenostrobus pulex* in succession to leeward.

Formerly given the species name *Pachymenia himantophora*, the northern forms of *Pachymenia* — though different in detail — are now held to belong to the mainland-wide *P. lusoria*. The genus has two more species, both restricted to maximally exposed coasts: *P. laciniata*, a wedge-shaped cartilaginous disc, and the far northern *P. crassa*, thick and prostrate, covered with blunt knobs.

The fucoid brown alga *Xiphophora chondrophylla* is typical though not always constant on enriched shores, growing as a pelmet above the *Carpophyllum* curtain. We are shown here the ultimate fucoid simplicity with neither leaves nor bladders. The narrow fronds are

strong but flexible, almost cartilaginous, and branch with a sort of 'rococo' asymmetry. *Xiphophora* is wave-resistant and hard to detach. It drops out only under the highest wave action.

Below the shining, dark-red *Pachymenia*, is presented first an enriched turf with several jointed corallines. *Corallina officinalis*, dominant on the horizontal flats and in pools, becomes supplanted by other species on cleaner shores. The most common species is often the pink or grey *Jania micrarthrodia*, with brush-like clumps of long narrow segments. *Jania* species differ from pinnate *Corallina* in their dichotomous side-branching, where *Corallina* is pinnate.

Arthrocardia corymbosa, found only in patches, is clearly recognised by the broad side-wings packed together like small shingles to build up a hard turf a couple of centimetres thick. *Cheilosporum sagittatum* can be at once identified by its linear chains of segments like rows of tiny arrowheads (Fig. 7.4).

Our largest, most coarsely built coralline, *Amphiroa anceps*, lives on Northland's roughest coasts. The thick segments are dichotomous, uneven in length, and distally flattened.

Beyond this band of jointed corallines, the basal pink

veneer continues, and on this foundation the shore enriches. Its dapple of rhodophytes varies through wine-red, golden brown, olive and bottle green. This is the domain of low tidal *Gigartina*, some species distinctive to the north, others New Zealand-wide. As well as high level *G. alveata*, two more narrow-branching species are regularly found in Northland. The dark red *Gigartina marginifera* forms narrow, dichotomous straps, fringed with long pinnules that carry the reproductive organs. In the grass-green *Gigartina macrocarpa*, the linear branches fork several times, and are fringed with stalked cystocarpic knobs.

Of our foliose *Gigartina* species, the commonest in the north — and right through New Zealand — is *G. circumcincta*, standing out as elliptical sheets of pale olive or maroon. Fastened by short petioles they are often tattered or perforate. The cystocarpic thalli are beaded over both faces and round the margins. *G. atropurpurea* is more membranous, with wine-red fronds deeply divided and with marginal lobes. The petiole is channelled, as distinct from *G. circumcincta*.

Beneath these recumbent *Gigartina*, the compressed, tubular branches of *Champia novae-zealandiae* may be found to scramble over the rock, here and there taking point-attachment. The *Champia* tubes, khaki brown towards the tips and deep crimson at the base, are slightly translucent, easily recognised by holding up to the light to see their internal cross-septa. The smaller *C. laingii* — confined to pools — can at once be spotted under water by its brilliant peacock iridescence.

Below the olive or dull purple *Gigartina*, the brown algal band is generally made up of the tresses of *Carpophyllum plumosum*. Thin ribbons of *Glossophora kunthii* intermingle with it towards shelter. *Carpophyllum angustifolium* succeeds *C. plumosum* only in near-maximal exposure. Just beyond low water may be glimpsed the golden brown fans of the small kelp *Lessonia variegata*.

Under the *Carpophyllum* curtain, the prevailing hue deepens to wine-red. This is the beginning of the rhodophyte underworld shaded and overtopped by the large browns. Here the reds are narrow-branched, as typified by the common *Pterocladia lucida*, lightly cartilaginous with flat, pinnate fronds. Pale pink to dull-red, it is a prolific source of agar. The close-related *Pterocladiella capillacea* is bushy and finer in texture, with several orders of pinnate branching. Both species extend beyond low water and are well adapted to wave action.

Two accompanying species are still more wave-robust. The stiffly flexible *Melanthalia abscissa* has fans of repeated dichotomous branching, almost black if dried out. The terminal segments are expanded at the tips, and flattened into a fan as if trimmed off by scissors. The deep crimson *Osmundaria colensoi* is common

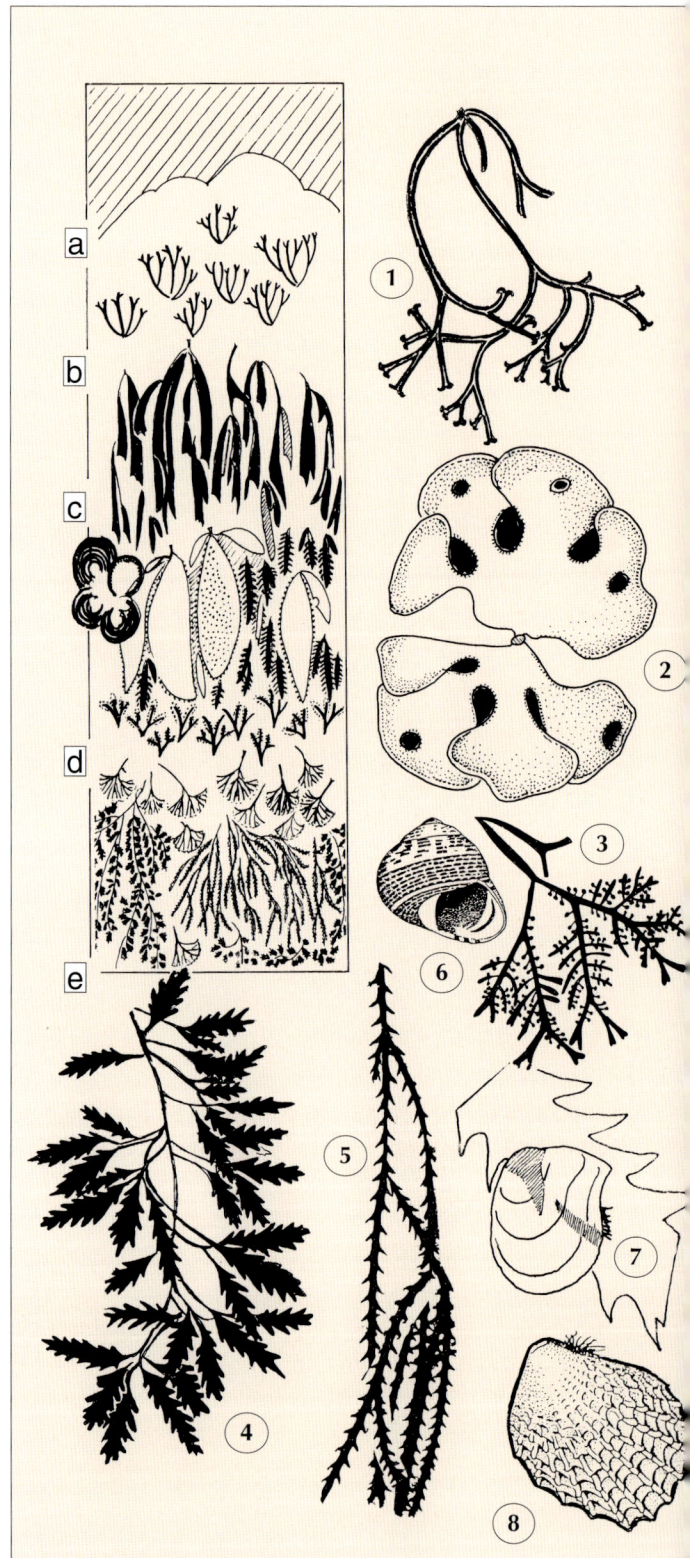

Fig 7.16 Algae of a deep shaded grotto, Te Henga, west Auckland

a–b *Gigartina alveata;* b–c *Pachymenia lusoria;* c–d *Gigartina circumcincta, G. marginifera, G. macrocarpa, Aeodes nitisissima;* d–e *Melanthalia abscissa, Landsburgia quercifolia* and *Osmundaria colensoi*
1 *Gigartina alveata* 2 *Aeodes nitidissima* 3 *Gigartina macrocarpa* 4 *Landsburgia quercifolia* 5 *Osmundaria colensoi* 6 *Cantharidella tesselata* 7 *Gaimardia finlayi* on *Osmundaria* 8 *Philobrya munita* from algal holdfasts.

on northern exposed shores. The narrow and linear stems each branch from the mid-rib of an older one. They are springy and lightly cartilaginous, like saw-blades toothed at either side.

In cave-mouths and grottoes under very reduced illumination, *Carpophyllum* disappears from the scene. At some such sites as at north **Te Henga** the single large brown alga is the oak-leaf weed, *Landsburgia quercifolia* (Fig. 7.16), normally subtidal and only entering the sublittoral fringe in deep shade. Its structure cannot be mistaken, with its rounded stems, and rich brown, flattened leaves, with a weak mid-rib. Dense clumps of saw-toothed *Osmundaria colensoi* stand out within the sublittoral fringe. Likewise exposed at low tide are the fleshy lobes of *Aeodes nitidissima*, in colour and texture like thin beef-steaks.

Sand effects

Where sand collects in depressions or builds up around low water to truncate the rocky slope, the algal garden is abbreviated. A few less luxuriant species form an alternative community. One of the simplest red algae, the filamentous *Audouinella*, comes up from a rock-face thinly buried in sand.

The *Gymnogongrus* species could be called the counterparts in sand of the Gigartinaceae. Never leafy, they form wiry clumps with linear dichotomous branching. *G. furcatus* — up to 5 cm tall — has bead-like cystocarps on the smooth, wiry stems, while *G. torulosus* is taller and sharply side-branched.

A small handful of larger algae are very tolerant if not actually requiring of sand. The branched cords of dull red *Nemalion helminthoides* appear in summer coated in slippery mucus and periodically sand-buried. Its brown counterpart, attached to rock beneath sand, is *Myriogloea intestinalis*, like slippery bootlaces covered when young by velvety hairs. In east Northland, the slippery tubes of *Grateloupia intestinalis* spring up around sand-buried rocks. In similar places may be found *Gracilaria secundata*. Two *Gigartina* species, *G. decipiens* and *G. macrocarpa*, especially favour sand-strewn rocks.

Where abrasion or burial by sand has effaced the algae, young *Xenostrobus pulex* rapidly settle, sometimes mingled together with mature patches of the same mussel. As shown on the Auckland west coast (Fig. 7.17), the wiry *Gymnogongrus*, and pink filamentous *Audouinella* may be expected around the sand-line. The small *Gigartina laingii* typically appears through the sand in pink foliose tufts.

Seasonality

The winter intertidal can look very different from that of full summer, with several eulittoral algae not destined to survive the spring. August shows a high *Porphyra*

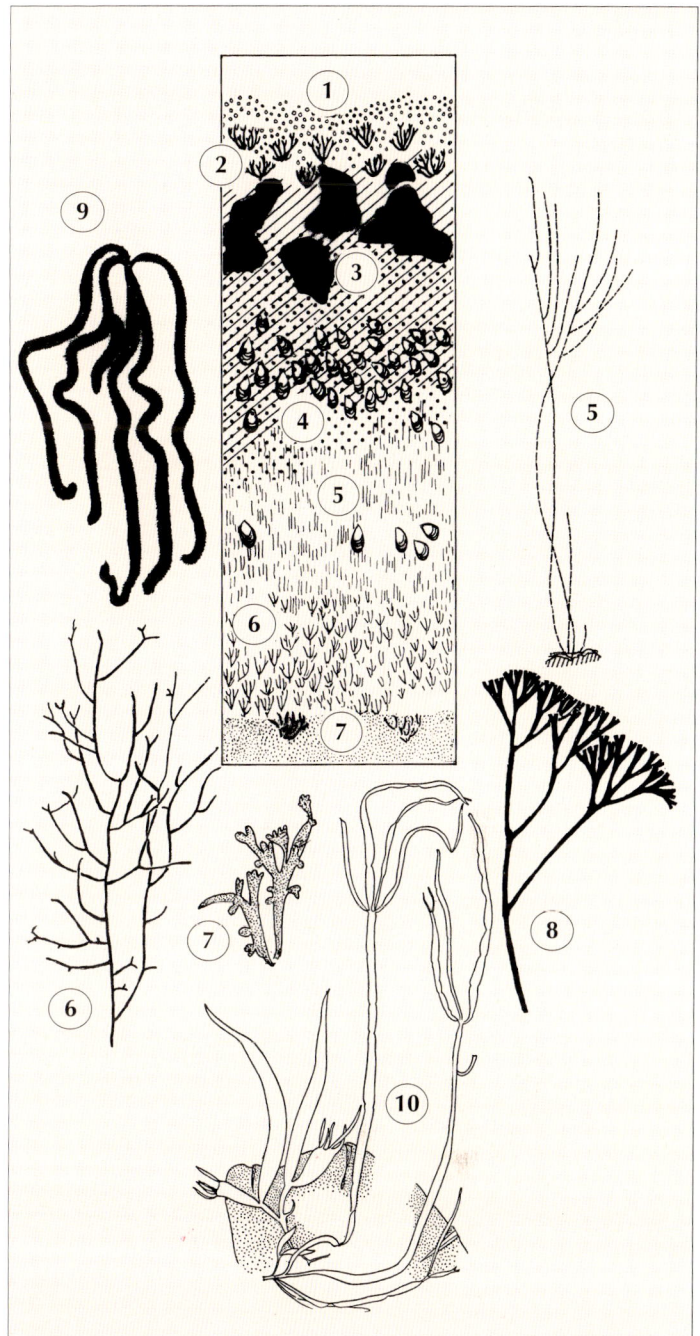

Fig. 7.17 Effects of sand burial, Auckland west coast
1 *Chamaesipho columna* with *Gigartina alveata* 2 clumps of *Xenostrobus pulex* 3 *Xenostrobus* spat recolonising cleared area 4 *Perna canaliculus* 5 *Audouinella* sp. 6 *Gymnogongrus furcatus* 7 *Gigartina laingii* 8 *Gymnogongrus humilis* 9 *Myriogloea intestinalis* 10 *Grateloupia intestinalis*.

columbina band destined to regress by summer. Juveniles of *Durvillaea antarctica*, 25 cm long, have attached above the pink zone amid *Epopella plicata*. Two small brown algae appear through the whole eulittoral during winter, both cosmopolitan with a tropical to cool temperate range. *Petalonia fascia* is known by its narrow slightly undulant leaves, smooth and of soft texture. *Scytosiphon lomentaria* forms long straw-coloured tubes typically constricted into pods, but sometimes

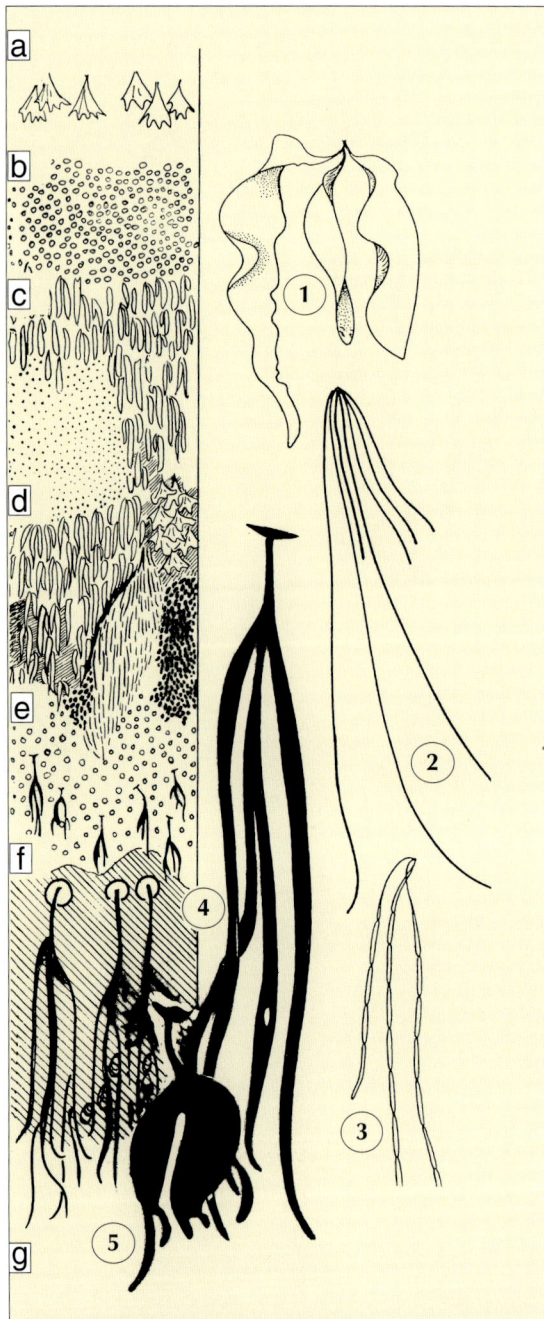

Fig. 7.18 Seasonal patterns of algae, late winter to spring, Te Henga, west Auckland

a–b *Porphyra columbina;* b–c *Chamaesipho brunnea;* c–d *Chamaesipho columna* with *Petalonia fascia;* d–e *Xenostrobus pulex* with *Scytosiphon lomentaria;* e–f *Epopella plicata* with high-level juvenile *Durvillaea;* f–g larger *Durvillaea antarctica*
1 *Petalonia fascia* 2 *Scytosiphon lomentaria,* unconstricted form 3 the same, constricted into pods 4, 5 young, winter-settled *Durvillaea antarctica.*

unsegmented. Like some *Enteromorpha*, this species is favoured by freshwater seepage.

The regular seasonality of some of the smaller browns is brought out by Betty Batham's record (Fig. 15.6) of the annual occurrence at Portobello of the four species *Adenocystis utricularis, Leathesia difformis,*

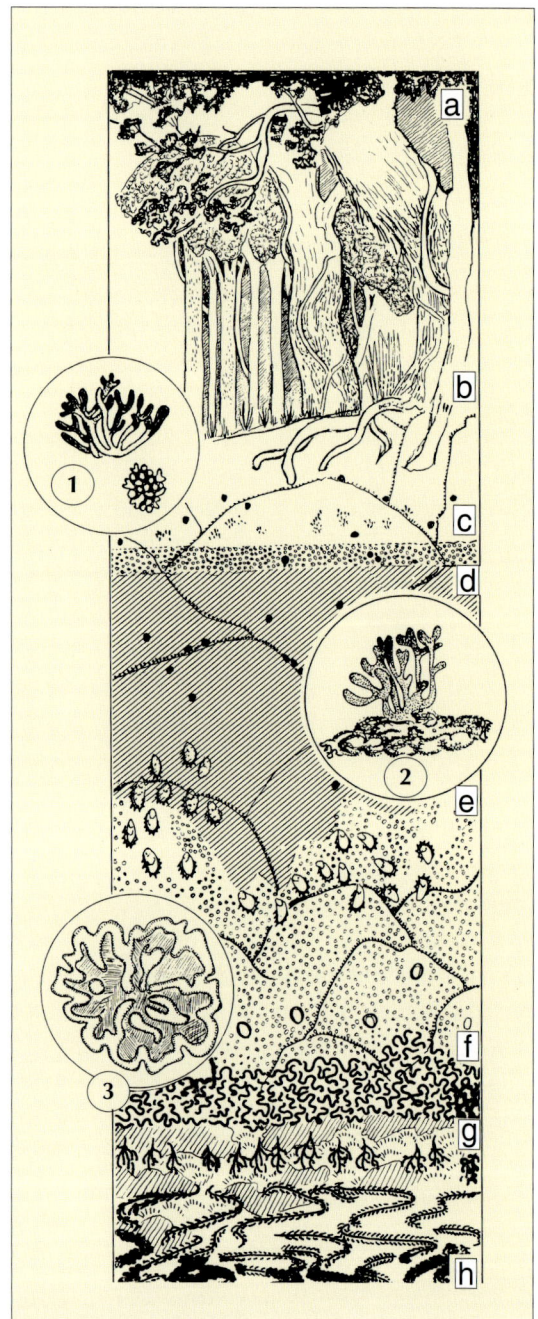

Fig. 7.19 Zonation on steep, deeply shaded shore at High Island, Whangarei Heads

a–b pohutukawa and karaka; b–c *Lichina confinis* and *Nerita atramentosa;* c–d strip of *Chamaesipho columna;* d–e *Apophlaea* and *Nerita;* e–f *Crassostrea glomerata* on *Chamaesipho columna;* f–g *Codium convolutum* and *Corallina;* g–h *Xiphophora chondrophylla* and *Carpophyllum maschalocarpum*
1 *Lichina confinis* 2 *Apophlaea sinclairii* 3 *Codium convolutum.*

Scytothamnus fasciculatus, Scytosiphon lomentaria, with the addition of the red *Porphyra columbina.*

Shade and aspectation
Moving from the outer coasts to deeply shaded shores of only moderate exposure, we have shown a northern

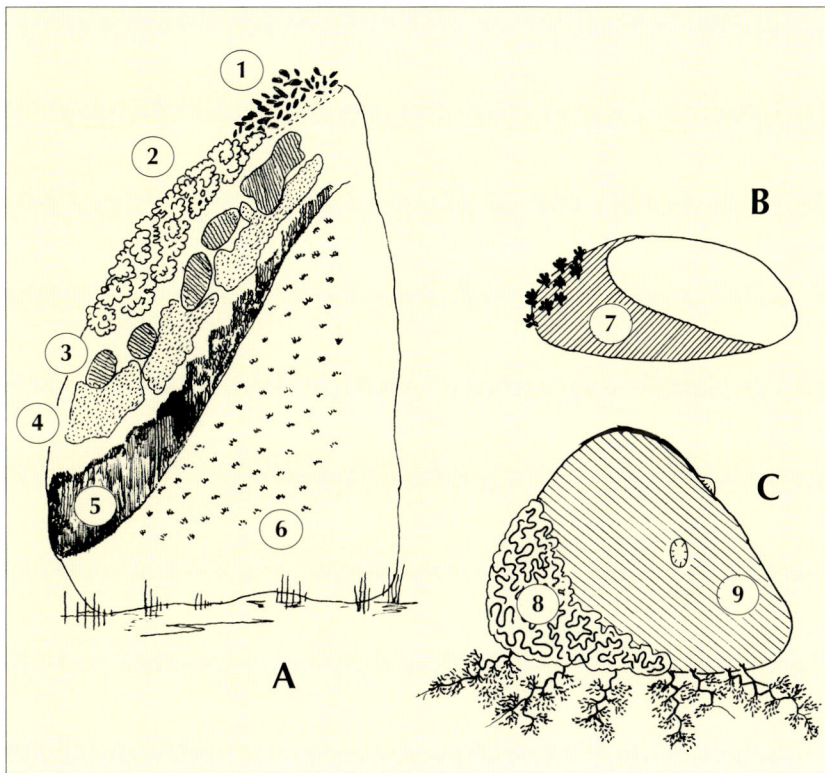

Fig. 7.20 Aspectation effects
A. Lichen zonation, in light (left) and shade (right) on an andesite rock stack, 2.5 m high, crowned with 1 the fern *Pyrrosia serpens*. Lichen zonation: 2 *Parmotrema, Rimelia, Heterodermia* (grey-green foliose) 3 *Buellia, Pertusaria* (crustose) 4 *Xanthoria parietina, X. ligulata* (orange-yellow) 5 *Verrucaria maura* (black crustose) 6 *Lichina confinis* (black tufted).
B. *Apophlaea sinclairii* 7 on the shaded (upshore) face of a beach boulder (Coromandel Peninsula).
C. low tidal boulder (Opunake, Taranaki) showing upshore green *Codium convolutum* 8 and seaward coralline veneer 9.

example at High Island, in the Whangarei Harbour (Fig. 7.19). The cool, south-facing slope displays two encrusting algae, the hard overlay of red-brown *Apophlaea sinclairii* in the middle eulittoral and the softer, fleshy *Codium convolutum* in the sublittoral fringe.

The andesitic rock-face is shaded from coastal bush by karaka descending right to the supralittoral (Fig. 7.19). The barnacle band of *Chamaesipho columna*, about 15 cm wide, is followed by pure *Apophlaea sinclairii*. Both have the upper cut-off perfectly straight. At its lower boundary *Apophlaea* is sparsely indented by rock oysters, *Crassostrea glomerata*, with the spread of *Chamaesipho columna* resumed among them. Black *Nerita atramentosa* numerously dot the *Apophlaea* and *Chamaesipho* belt. The limpet *Cellana radians* is common among the lower barnacles.

The bottle-green *Codium convolutum* has its straight upper margin at the foot of the lower eulittoral. Spreading seaward at a more gentle slope, it mingles below with coralline paint and turf. A short pelmet of *Xiphophora chondrophylla* follows, then comes *Carpophyllum maschalocarpum*. On these shaded and wave-sheltered shores, old *Turbo smaragdus* at coralline level can reach the size of small plums. Secured from wave dislodgment the largest *Turbo* at High Island can remain in the upper eulittoral, no longer moving downshore with the tides.

Apophlaea sinclairii is evidently averse to soft sedimentary rocks. On the hard boulder beaches of the Coromandel Peninsula (Fig. 7.20) it covers the shaded aspect of each boulder, so that when viewed from upshore, the whole stretch looks reddish brown, while from the seaward side, only the bare bedrock is visible. *Codium convolutum* has developed the same shaded aspectation at Opunake boulder beach (Fig. 7.20).

Fig. 7.20 shows also the aspectation, according to light, of the lichens of the *supralittoral zone* and the *littoral fringe*. At High Island the common form on shaded faces is the black stubble of *Lichina confinis*. On a sun-warmed side, we find in upward succession black *Verrucaria maura*, yellow *Xanthoria parietina* and the grey-green foliose lichens *Parmotrema* and *Rimelia*, topped by the oval-leafed climbing fern *Pyrrosia serpens*.

Cook Strait

The clean, current-swept shores of Cook Strait outside Wellington have notably rich algal communities. At Lyall Bay and Island Bay, looking out direct to the cold southern ocean, a few minutes' drive from Wellington city brings us to shores more prolific in algae than any others so nearly urban. A full checklist of the seaweeds has been given by Nancy Adams.[1]

Wellington's rocky shores are all constructed of hard eroded greywacke, with strata much tilted or upended. Currents are strong, but wave impact is broken by the reef's seaward complex of rocky outposts. On the most exposed of these promontories the bull-kelp, *Durvillaea antarctica*, confronts the highest seas with its dark thongs buoyant in the wave-break. Closer in, a confusion of stacks and gullies has produced every combination of steep inclines, vertical drops and overhangs, as well as pools either quietly secluded or being constantly replenished by waves.

Fig. 7.21 Zonation of shore with enriched algae, Island Bay, Wellington

a–b **(top)** narrow band of *Chamaesipho brunnea*, then *C. columna*, with *Porphyra columbina* and **(bottom)** *Calantica spinosa*; b–c **(top)** brown *Hildenbrandia*, then pink coralline veneer, with *Ulva lactuca*; c–d coralline turf and *Codium convolutum*; d–e *Xiphophora chondrophylla, Gigartina, Champia, Glossophora, Cystophora*; e–f *Carpophyllum maschalocarpum, Cystophora retroflexa, C. torulosa, Marginariella urvilliana*; f–g *Desmarestia willii, Macrocystis pyrifera* and *Lessonia variegata*.

1 *Porphyra columbina* 2 *Ulva spathulata* 3 *Ulva laetevirens* 4 *Gigartina atropurpurea* 5 *G. decipiens* 6 *G. livida* 7 *Marginariella urvilliana* 8 *Landsburgia quercifolia* 9 *Desmarestia willii* 10 *Cladhymenia oblongifolia* 11 *Lessonia variegata* 12 yellow poppy, *Glaucium flavum*.

The compressed tidal range at Wellington, hardly more than 1.5 m at springs, has left the sequence of zoning unchanged. Moreover, with an hour's gain of the tide raising the water-line less than 25 cm, the lowest zones remain — on a calm day — accessible for much longer.

At **Island Bay** the barnacle-line is regular and horizontal. The upper 20 cm forms a sharp band of the high energy barnacle *Chamaesipho brunnea*, giving place abruptly below to the prevailing *C. columna*. These barnacles abut below with a 10 cm strip of chocolate *Hildenbrandia*. This is in turn succeeded at the bottom by a coralline veneer that continues downshore like a pink rough-cast.

From spring to early summer, the **eulittoral zone** has its small transient seasonal algae. Miniature clumps of stiff, coriaceous *Nothogenia fastigiata* and the bladder-rosettes of *Catenellopsis oligarthra* (despite their golden brown colour these are both 'reds') grow amid the long branched tubes of *Splachnidium rugosum*. Pale gold in spring, filled with clear mucus, these last dry out by mid-summer to reddish brown. Shaded faces may carry a stubble of *Capreolia implexa*. During early spring at the top of the barnacles are found membranes of pinkish grey *Porphyra columbina*, stretched tautly over the rock as they dry in the sun.

Apophlaea is scarce at Wellington, though *A. sinclairii* occurs at Owhiro and Cape Palliser, Wellington in vain. *A. sinclairii* appears to come no further than Castlepoint and Titahi Bay, while *A. lyallii* (Fig. 7.30) belongs to the South Island and the Chathams.

Low tidal surge gullies carry their own profusion of small red and green algae. *Corallina* turf has each of its common associate species as in the north. Bright green *Ulva lactuca* is seasonally much more evident at Wellington, and *Ulva laetevirens* gracefully festoons the edges of low and mid-tidal pools. *Champia novaezealandiae* forms a scrambling ground-cover, with blue iridescent patches of the smaller *C. laingii* in pools.

The foliose Gigartinaceae at Wellington are the much-lobed sheets of *Gigartina atropurpurea* and the oval blades of *G. circumcincta*. The latter species increasingly moves into shelter towards the south. In Port Nicholson it reaches York Bay, while in the far southern harbours it can produce thalli a metre in length. The commonest branched species at Wellington is *G. decipiens* with flat, dichotomous fronds, fringed with reproductive sori. The narrow-branched, pinnate *G. livida* is also frequent. Below the *Gigartina* level grow *Cladhymenia oblongifolia*, *Melanthalia abscissa*, *Pterocladia lucida* and *Pterocladiella capillacea*.

The green algae of Cook Strait are rich and diverse compared with those of the Hauraki Gulf. The softest, most delicate filamentous forms are the branched *Cladophora*, of which *C. feredayi*, up to 40 cm long, is

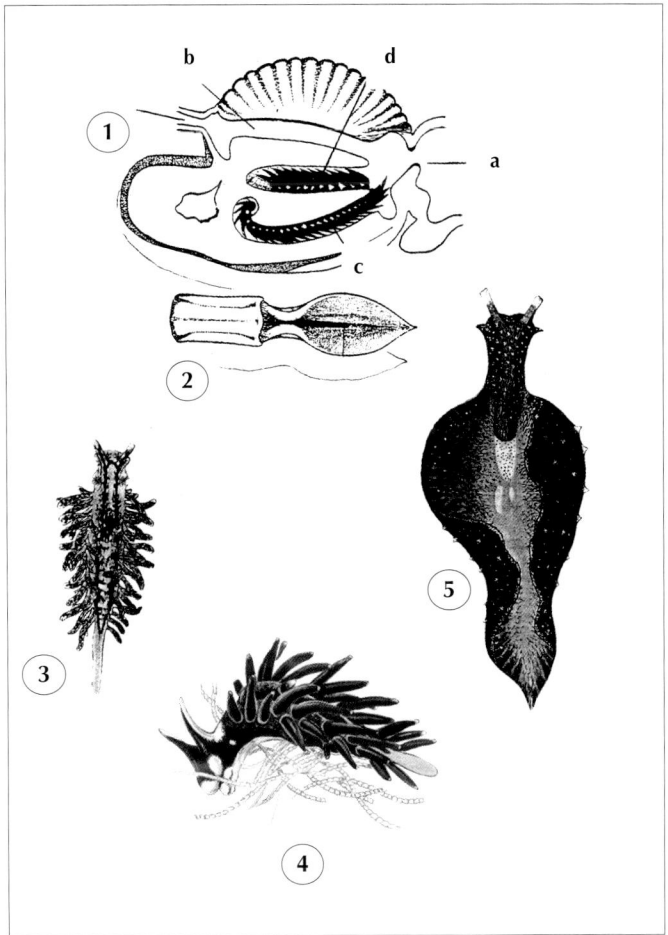

Fig. 7.22 Sacoglossan slugs from green algae
1 *Pharynx in sagittal section* a *mouth* b *oesophagus* c *radula sac* d *ascus sac* 2 *a single tooth* 3 *Placida dendritica* 4 *Stiliger felinus* 5 *Elysia maoria*.

common on the northern shore of Cook Strait. To the same family belong the *Chaetomorpha* species, with single unbranched filaments made up of bead-like or barrel-shaped cells. *C. aerea* produces tufts of slender filaments, while *C. coliformis* growing near the wave-front has straight solitary filaments with cells up to 1–3 mm in diameter. *C. linum* forms long unattached tangles in pools. The thick cell walls of all *Chaetomorpha* can be recognised at once by their brittleness between the teeth.

The smallest, most exquisite greens in pools are the *Bryopsis* species, *B. plumosa* and *B. vestita*, erect and with pinnate fronds. The coarsest green is the stiff stubble-like *Cladophoropsis herpestica*, at or beyond low water in wave-exposed spots.

The *Codium* species (Fig. 7.25) have developed far more tissue solidity than the other greens by the weaving of tubular filaments into a firm parenchyma. These include both prostrate and upright branching forms. The fleshy and irregularly lobed *C. convolutum* is as common at Wellington as through the rest of New Zealand in moderately sheltered sites. The smaller, pale green cushions of *C. cranwelliae* are confined to coasts north of East Cape. Three species of dichotomous finger *Codium* live within and below the sublittoral fringe. The New Zealand subspecies of the cosmopolitan *Codium fragile*, growing on open coasts, is thicker-branched, dark green and velvety. It was joined recently at Auckland by a widely distributed subspecies *C. fragile tomentosoides*, probably arriving from Japan. The grey *C. gracile* is found on open coasts, often subtidally, while the slender, dark green *C. dichotomum* is a plant chiefly of bays and harbours.

Four species of *Caulerpa* are to be found at Cook Strait (Fig. 7.24). Their prostrate rhizomes send up erect shoots, clearly distinct for each species. *C. brownii* ('sea rimu') has a close-set foliage of pointed leaflets. In *C. flexilis* these are wider-spaced, on fronds with pinnate side-branching. *C. geminata* has rows of bladders like tiny grapes. The scarcer and subtidal *C. articulata* has a stem with bead-like divisions, carrying opposite rows of long, cylindrical vesicles.

Sacoglossa (Fig. 7.22): Both *Codium* and *Caulerpa* can be searched with profit for these small sea slugs, which are specialised herbivores, sucking the cell contents of green algae. Their diversity of body form hides their essential unity as a group, as denoted in the specialised buccalmass and radula. All sacoglossans feed by lancing one by one the cell walls in a green filament with the single row of radula teeth like scalpel-blades. Favouring *Chaetomorpha* in pools is the exquisite jet black *Stiliger felinus*, only half a centimetre long. Daintily mobile despite the two rows of heavy club-like

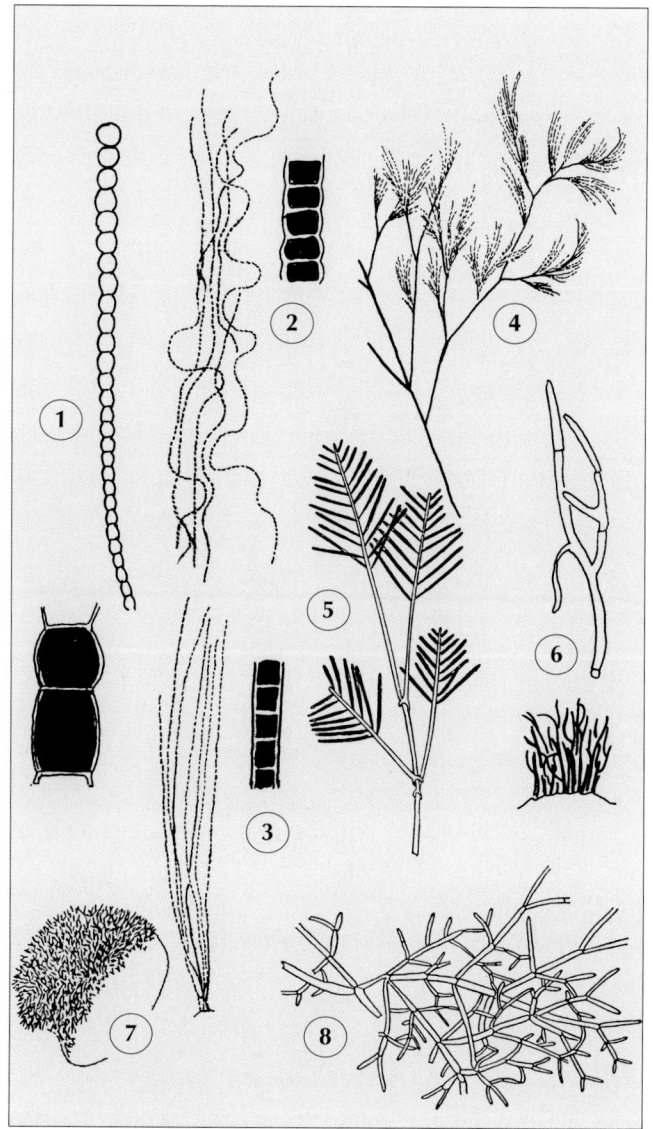

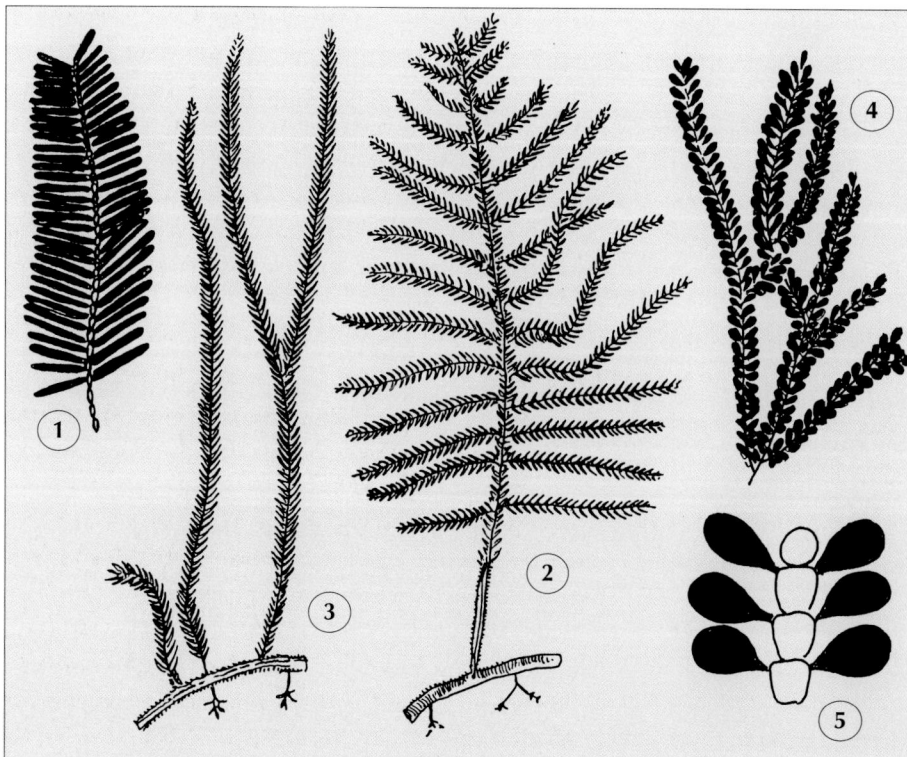

Above: Fig. 7.23 Green algae: Cladophorales

1 *Chaetomorpha coliformis*
2 *C. linum* 3 *C. aerea*
4 *Cladophora feredayi*
5 *Bryopsis plumosa*
6 *Cladophoropsis herpestica*
7 *Boodlea mutabile*
8 the same, detailed structure.

Left: Fig. 7.24 Green algae: Caulerpales

1 *Caulerpa articulata*
2 *Caulerpa flexilis* 3 *Caulerpa brownii* 4 *Caulerpa geminata*
5 the same, detail.

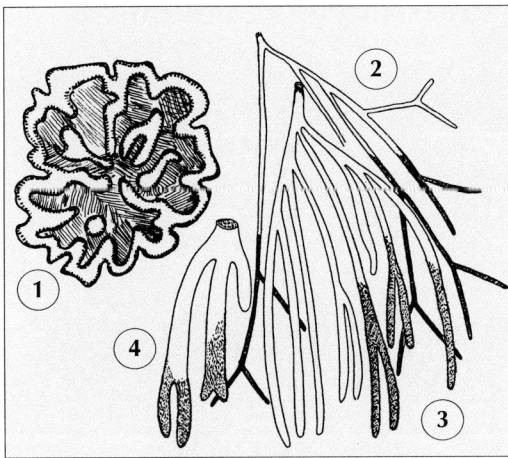

Fig. 7.25 Green algae: Codiales
1 *Codium convolutum* 2 *Codium dichotomum* 3 *Codium gracile* 4 *Codium fragile*.

appendages, this slug creeps or twines over the alga, with the lips passing a single filament over the mouth.

In the deep interstices of *Codium convolutum* the larger, bottle-green sacoglossan *Elysia maoria* can often be discovered up to 2 cm long. Thin and leaf-like, *Elysia* becomes heart-shaped when its side-extensions are spread flat. The near-perfect colour-match with *Codium* is not shared by the pallid spawn coils, easily visible on the plant. Attaching to the branches of *Codium fragile* lives the slender green *Placida dendritica*, with dorsal outgrowths smaller and more numerous than in *Stiliger*. Finally, *Caulerpa geminata*, in Northland as well as at Wellington, has yielded specimens of the primitively shelled sacoglossan, *Oxynoe viridis*.

Of the lesser brown algae around low tide, *Xiphophora gladiata* subsp. *novae-zelandiae* replaces *X. chondrophylla* from Cook Strait south. Like the northern species, it regularly grows as a pelmet overhanging *Carpophyllum* and *Cystophora* in more-than-average water movement. In the shade just below *Xiphophora* may be found also the thin ribbons of *Glossophora kunthii*, generally with the related *Dictyota dichotoma*. The clustered fans of *Zonaria turneriana* are much more prominent than in the north.

In Cook Strait — and increasingly on exposed shores to the south — the brown *Halopteris* species form an important element around low tide. Belonging to the smaller Order **Sphacelariales**, they take the form of dark tufts or tails, with hairy branchlets dividing and redividing, with sporangia single or clustered within the axils. The New Zealand-wide *Halopteris virgata* is close-tufted and bushy, often forming well-defined bands. *H. paniculata* is an Auckland and northern species, laxly branched and less dense. Common in Cook Strait are the densely haired *H. funicularis* with short stiff branchlets, the bushy *H. congesta*, often forming a continuous low

turf on exposed shores, and the paler, delicately pinnate-branched *H. novae-zelandiae*.

Most of all it will be the brown algal drift coming in after a blow that brings home salient algal differences of Cook Strait from the north. The single important *Carpophyllum* species is *C. maschalocarpum*. *C. plumosum* reaches Cook Strait only sparsely, and *C. flexuosum*, though continuing much further south, is virtually excluded from northern Cook Strait by strength of water movement. The lone *Sargassum* species is *S. sinclairii*.

Important low tidal Fucales in Cook Strait are the two northern *Cystophora* species, *C. retroflexa* and *C. torulosa*, both abounding under submaximal exposure. *C. distenta* and *C. congesta* have also been recorded from the shores of Cook Strait, while *C. scalaris* has turned up as far north as Castlepoint, a terminus for several southern organisms. Deep-water *C. platylobium* is thrown up after storms.

Immediately beyond low water we find an important fucacean, the strap-weed, *Marginariella boryana*, with its northern limits at Castlepoint and Kapiti Island. The flat stipe divides unequally into long, dichotomous straps, tough and widely saw-toothed at the margins. Clusters of these golden-brown streamers reach 2 m long, among *Lessonia* and right up to the sublittoral fringe. One edge carries ovoid bladders and there is a basal fringe of pod-like receptacles. A second species, *M. urvilliana*, has spherical vesicles and broader, coarsely toothed straps, lying spirally curled at the plane of the surface.

A fucoid at once distinctive, the oak-leaf weed, *Landsburgia quercifolia*, found sublittorally from North Cape to Banks Peninsula, is common in Cook Strait. Thin cylindrical stems spring from a conical holdfast, and carry deciduous leaflets coming away to leave scars on the axis. Bladders are wanting.

In the sublittoral zone the kelp *Ecklonia radiata* is seldom lacking but has lost its northern dominance to the smaller, short-stiped *Lessonia variegata*. The open coast alternative to *Ecklonia* in the north, *Lessonia* ousts it at Wellington even in sheltered waters, as in Port Nicholson right up to York Bay. *L. variegata* is thus a familiar Wellington kelp from moderate up to all but the most extreme exposure, confined to the sublittoral zone, beyond the maximum of either *Carpophyllum* or *Cystophora*. Only emergent at spring tides, *Lessonia* can be distinguished from *Ecklonia* by its short, woody stipe, buttressed holdfast, and thicker, smooth blades. New blades are formed by dichotomous splitting at the base. A good-sized plant reaches little more than a metre long.

The genus *Desmarestia* (with its separate order in the Ochrophyta), belongs to cold southern waters. One of our species, *Desmarestia willii*, is confined to the subantarctic islands, but the other, *D. ligulata*, reaches north to Cook Strait, from the deeper subtidal up to channels

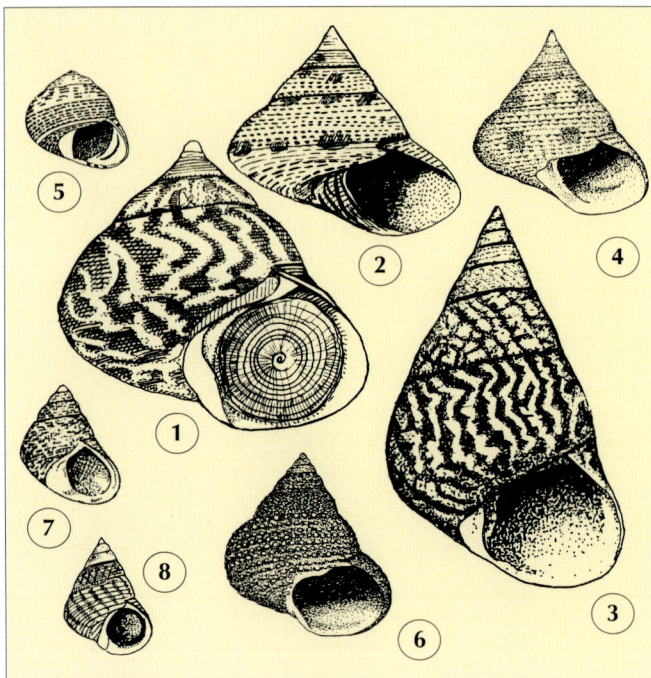

Fig. 7.26 Algal-grazing trochid gastropods
1 *Calliostoma tigris* 2 *C. selectum* 3 *Cantharidus opalus* 4 *Calliostoma pellucidum* 5 *Cantharidella tesselata* 6 *Calliostoma punctulatum* 7 *Micrelenchus dilatatus* 8 *Micrelenchus sanguineus.*

just beyond low water. The flat axis is golden brown, pinnately branched, with the main and sometimes the secondary branches toothed. *Desmarestia* is an annual, first to be seen in early spring, when its margins will be fringed with fine hairs. Half a metre long by January, it trebles in size by autumn, with the axis now 2 cm wide. Broken loose and cast up on the beach, *Desmarestia* turns verdigris green, to soften and release free sulphuric acid, that will decompose any algae collected with it.

Topshells: The low tide and sublittoral algal grazing gastropods are notably abundant in Cook Strait. Most of the Cantharidus group of the Trochidae are found on brown fronds, like the smaller *Micrelenchus sanguineus* and *M. dilatatus* already mentioned. The still smaller dome-shaped *Cantharidella tesselata* is found intertidally in pools or browsing on succulent reds. Much taller and sharp-spired is the dull pink *Cantharidus purpureus*, on kelp and *Carpophyllum* in relative shelter. The largest and most beautifully patterned of all our kelp trochoids is *Cantharidus opalus*, frequently found sublittorally on rather exposed coasts. The shell may reach 35 mm tall, being pink to mauve, with zigzagged axial stripes in purple. The aperture is brilliantly iridescent.

To the nearby genus *Calliostoma* belong the larger brown or beige topshells, thinner-shelled, pointed spired, and sculptured with small close-set granules. The largest and most robust is the handsome *Calliostoma tigris*, axially banded with russet on a cream ground.

The wide, low-pitched *C. selectum* is fawn or pink, finely brown-speckled. Both are found under ledges or boulder bases, and are thought to have become sponge-grazers, as may also two smaller and commoner species, the pinkish buff *C. pellucidum* and the brown *C. punctulatum* with beaded sculpture and round whorls.

Kaikoura and Banks Peninsula

Rocky shore studies perforce become centred on the two hard salients that break the long continuity of sand or gravel on the Canterbury coast. Since the opening of the Edward Percival Marine Laboratory in 1961, much Canterbury University research has been focused on the complex of shores at Kaikoura and the immediate south.

Kaikoura Peninsula is an oceanographic and biological nodal point, lying in the convergence zone of the subtropical and subantarctic water masses. There is a mix of species, northern and southern, and a richness of fauna and flora unrivalled in New Zealand. Every one of New Zealand's intertidal patellid limpets, mussels and barnacles, and virtually all the large brown algae, are to be found associated within the narrow confines of the peninsula shore.[2]

Above the intertidal, flat tops support breeding populations of red-billed gull and white-fronted tern as well as a colony of the New Zealand fur seal (*Arctocephalus forsteri*), the only one in the southern hemisphere to breed north of 48 degrees south, and with Kaikoura as one of its mainland winter sites.

At Kaikoura, the landmass drops steeply to the sea, with the 200 m contour only 3 km offshore. Rich in plankton and the fishes that depend on it directly, the coastal waters attract the dusky and Hector's dolphin, both of which school in South Bay. Throughout the year there are bachelor congregations of young sperm whale, while the killer-whale *Orca* regularly comes in to the coast. The now rare humpback whale is from time to time to be seen.

Between the tides — as we have elsewhere described it[3] — the Kaikoura terrain offers much in little. It has developed level platforms, pools and deep channels, stacks, bluffs and folds with overhangs, as in their variant ways the Amuri limestone and Miocene siltstone have been folded and eroded. As well as in its geological foundations, the Peninsula shows also the differences between sheltered and wave-exposed faces.

A wave-exposed seaward face is depicted in Fig. 7.27 at **Seal Rock.** The four intertidal mussels (*Xenostrobus pulex, Mytilus edulis aoteanus, Aulacomya maoriana* and *Perna canaliculus*) are present in relatively distinct bands. *Durvillaea antarctica* is dominant along the wave front, with *Carpophyllum maschalocarpum* the only member of its genus in relative shelter, though *C.*

Fig. 7.27 Seal Rock, Kaikoura: exposed algal zonation

1 *Xenostrobus pulex* 2 *Mytilus edulis aoteanus* 3 *Aulacomya maoriana* 4 *Durvillaea antarctica* 5 *Carpophyllum maschalocarpum* 6 *Lophurella caespitosa* (also detail) 7 *Halopteris funicularis* (also detail, with sporangia) 8 *Streblocladia glomerulata* (also detail). **(Top)** southern fur-seal, *Arctocephalus forsteri*.

flexuosum can occur sparsely in high shelter. Small algae immediately below the bull-kelp cover the slope in wide expanses. These have narrow, much-divided foliage, robust under wave assault and fast runoff. The largest species is one of the Ceramiales, the red *Streblocladia glomerulata*, with lightly cartilaginous streamers up to 0.5 m long. A dwarf by comparison is the bottle-green *Lophurella caespitosa*, forming slippery surf-drenched swards. The same slopes are also prime sites for the short brown *Halopteris* of which *H. funicularis* is here the leading species.

South Bay

In contrast with Seal Rock, South Bay (Fig. 7.28) at the Peninsula's south-west extremity, is one of the few Kaikoura sites that can be called sheltered. Here the sharp folded and tilted bedrock of Amuri limestone slopes steeply seawards.

In the **upper eulittoral**, through winter and early spring, *Porphyra columbina* is drawn tightly over the rock when dry. Probably one of several species within the currently styled *P. columbina*, it differs from the northern form in its long, undulant ribbons. During low tide the piled thalli are shiny greenish grey on top, with a slippery, mucilaginous underlayer. By November this alga has reproduced and begun to regress. In sites with

fresh seepage large *Scytosiphon lomentaria* grow as quill-like tubes, festooned in between with green *Enteromorpha ramulosa*. In November (when Fig. 7.28 was drawn) the **middle eulittoral** was still piled with a green overlay of *Ulva lactuca*, with its piles of thalli tattered and disintegrating.

The highest of the prominent all-year-round algae, *Hormosira banksii* forms a leaden grey band in exposed sites but takes on its more familiar golden brown where sheltered by limestone knolls, as also on wide platforms and in pools. At Kaikoura, as widely through New Zealand, *Hormosira* may carry *Notheia anomala*, a partial parasite from its own family. Stringy and monopodially branched, this plant penetrates the host bladders by a root-like organ in place of a holdfast.

The highest on shore of the southern *Gigartina* is the wide-branched dichotomous *G. decipiens*, growing in a dull green to mauve zone beyond *Hormosira*. Next below it begins a rich algal swathe dominated by *Cystophora torulosa*, with the big chunky pinnules typical of its southern form. This is the most intertidal species of its genus, and its clumps are interspersed with pale pink *Champia novae-zealandiae*, vivid green *Ulva*, and the olive *Gigartina livida*, with pinnately set marginal lobes. Dark brown *Echinothamnion lyallii* grows epiphytically upon the yellow *Cystophora*.

131

Fig. 7.28 Zoned shore at South Bay, Kaikoura Peninsula
a–b sparse *Chamaesipho columna*; b–c *Ulva lactuca*; c–d *Hormosira banksii* and *Gigartina*; d–e *Cystophora torulosa*, *Gigartina*, and *Echinothamnion*; e–f *Cystophora scalaris*; f–g *Carpophyllum maschalocarpum*; g–h *Marginariella urvilliana*
(right) The southern Gigartina: 1 *Gigartina lanceata* 2 *G.* sp. 3 *G. circumcincta* 4 *G. ancistroclada,* with (5) channelled petiole 6 *G. clavifera* 7 *G. livida* 8 *G. decipiens* 9 *G. dilatata.*

The **sublittoral fringe** is marked by a changeover to nearly pure ashen-grey *Cystophora scalaris*. Short leafy straps of brown *Carpophyllum maschalocarpum* follow in turn, but in only a token reminder of its dominance around northern shores. Its final southern limit is at Banks Peninsula. *Carpophyllum flexuosum* is apparently lacking at Kaikoura, though along with *Ecklonia radiata* it turns up sublittorally as far south as Otago and even to Stewart Island. At South Bay, the upper sublittoral is dominated by a rich brown thicket of *Marginariella boryana*, with its narrow streamers and pointed bladders. This species is increasingly accompanied in the south, as in surge channels around Kaikoura, by *Marginariella urvilliana*, distinguished by its wider fronds and spherical bladders (Fig. 7.31). The pinnate fronds of *Desmarestia ligulata* seasonally appear offshore at Kaikoura. With faster water movement but out of direct wave action, leaves of the bladder kelp *Macrocystis pyrifera* break the surface at low tide, accompanied increasingly towards the south by sea tulips, *Pyura pachydermatina*. The two form a subtidal thicket far stronger than in Cook Strait.

Gigartinaceae

The *Gigartina* family becomes important in cool waters, presenting an increasing diversity as we progress south. This large genus has two sections with very distinct thallus forms. First, there are the branching forms. *G. decipiens*, *G. dilatata* and *G. clavifera* have dichotomous fronds markedly flattened, while in *G. livida* and *G. ancistroclada*, they are narrow and forked, with the side-branches pinnate.

Of the foliose Gigartinaceae, the thin and lobate *G. atropurpurea* is present at Westport and Kaikoura. The thicker, elliptical *G. circumcincta* is common right through New Zealand, entering harbours and growing larger in the south. Two other southern species have very thick blades. The strap-like *G. lanceata* is fleshy and pliable, fringed and covered over the surface with hooked or branched proliferations. In the subantarctic islands there is the species sometimes listed as *G. circumcincta*, with an orbicular, cartilaginous blade, with mature specimens often concave like a dish.

Limpets

Special note must be made of the Kaikoura limpets. Each of the mainland patellids — northern and southern — finds a foothold here. The New Zealand-wide *Cellana ornata* and *C. radians* have their normal range. The latter shows its southern *perana* form, with numerous radial ribs all small. The exterior is olive green and the inside silvery grey, dark-edged and sometimes white-callused. Kaikoura is the northern limit of the cold-water limpet *Cellana strigilis redimiculum*, and the southern cut-off for *C. denticulata* on siltstone; the golden limpet, *C. flava*, keeps faithful to east coast calcareous bedrock from East Cape south. Also from the north, the star limpet, *C. stellifera*, still turns up subtidally at Kaikoura. The slit limpet, *Montfortula rugosa* (Family Fissurellidae), is unusually common at Kaikoura, in middle and upper limestone pools.

South of Kaikoura

The shore stretching south from Kaikoura past Oaro to Haumuri Bluff continues to be well accessible and just as prolific in algae. At Oaro the Amuri limestone platform is some 400 m across, with plentiful pools. As well as *Hormosira*, these are rich in *Codium dichotomum*,

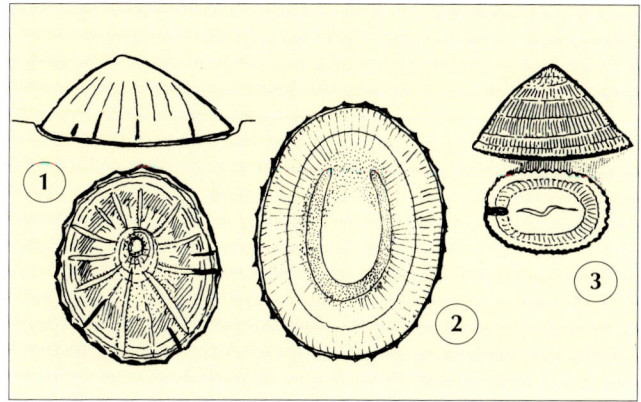

Fig. 7.29 Three limpets from Kaikoura
1 *Cellana flava* 2 *Cellana radians* form *perana*
3 *Montfortula rugosa*.

Gigartina atropurpurea, *Cystophora* species, and the finely filamentous *Centroceras clavulatum*.

The seaward edge of the limestone platform has an irregular relief with outcrops tilted towards the land, and dissected by channels and pools of every size. At the outward margin, the steep faces drop to the sublittoral, carrying — according to depth — either of the two bull-kelps *Durvillaea antarctica* or *D. willana* (Fig. 5.8). These — like the rock barrier itself — afford some protection to the inshore reef platform. On their surge-facing slopes, the outcrops carry *Gigartina decipiens* and *G. livida*. On calm days at spring tides their upper sublittoral slopes can be examined with wetsuit and snorkel. Here are to be found abundant populations of *Cladhymenia oblongifolia*, the foliose *Gigartina* species, *Streblocladia glomerulata*, *Cladhymenia oblongifolia*, *Laurencia thyrsifera*, and *Ballia hirsuta* with epiphytic *Chaetomorpha coliformis*, *Lophurella hookeriana* and *Grateloupia stipitata*.

Similar diversity is available just west of Spy Glass Point on **Haumuri Peninsula** at the foot of the limestone cliffs. Here the shore is scattered with large, perfectly smoothed boulders, mostly stable with their broad tops, pink-veneered at extreme low water of neap tides. Down their sides the algae include *Gigartina livida* and *G. decipiens*, with occasional patches of *Caulerpa brownii*, short *Carpophyllum maschalocarpum* and the *Cystophora* species. Browsing gastropods *Haliotis iris* and *H. australis* abound between and under the boulders, with large turban-shells, *Cookia sulcata*, and the shield limpet, *Scutus antipodes*.

The algae of the **sublittoral zone** at Kaikoura have been described by Fenwick (1974) for the Peninsula, Nine Pins and St Kilda rocks. On large areas of gently sloped limestone bedrock, in 8–15 m, with some influence of ocean water, there is a dense sublittoral forest of *Marginariella urvilliana*.

The delicate understorey of foliose reds includes *Hymenena* species, *Phycodrys quercifolia*, *Schizoseris*

Fig. 7.30 Zonation of moderately exposed shore at Jack's Bay, South Otago
(top) Adlittoral cliff-margin with *Lupinus arboreus* and yellow-eyed penguin (hoiho, *Megadyptes antipodes*).
a–b *Stictosiphonia arbuscula*; b–c *Chamaesipho columna, Apophlaea lyallii, Scytothamnus fasicuulatus* and *Epopella plicata*; c–d *Gigartina* and *Pachymenia lusoria*; d–e coralline veneer and tufting reds; e–f *Durvillaea antarctica, Codium* and *Lenormandia*; f–g subtidal reds
1 *Apophlaea lyallii* 2 *Adenocystis utricularis* 3 *Scytothamnus fasciculatus* 4 *Pachymenia lusoria* 5 *Durvillaea antarctica* 6 *Codium gracile* 7 *Lenormandia angustifolia* 8 *Modelia granosa* 9 *Gigartina clavifera*.

griffithsia and several *Plocamia* (chiefly *P. cartilagineum* and *P. leptophyllum*). Recorded at St Kilda were *Delisea elegans, Rhizopogonia asperata, Ptilonia willana*, and the brown algae *Spatoglossum chapmanii*. The last is one of the Dictyotales, fan-like when young, then split into wedge-shaped corrugated segments.

Otago

In contrast with Canterbury the Otago seaboard is almost entirely rock-bound. Above the zoned shore at the south end of Jack's Bay, in the Catlins district (Fig. 7.30), the coastal scrub is yellow in spring and summer with its spikes of flowering lupin (*Lupinus arboreus*). In places where it has not yet suffered destruction, the adlittoral bush serves as a breeding refuge for hoiho, the yellow-eyed penguin, the rarest of its whole family and today on New Zealand's list of endangered species with its very survival under threat.

Much of the Otago **littoral fringe** is clad with a thick pile of the brown-coloured rhodophyte, *Stictosiphonia arbuscula*. The individual thalli form miniature bushes a few centimetres high, with the fine branchlets incurved at the tips. Below *Stictosiphonia*, the two periwinkles, *Austrolittorina cincta* and *A. antipoda*, appear in the **fringe** and **upper eulittoral**. At this level there are two prominent zoning algae distinctive to Otago and Southland. *Apophlaea lyallii*, confined strictly to the south (and the Chathams), has a much freer growth than its Auckland cousin *A. sinclairii*; its ground crust puts up brittle antlers reaching 10 cm long. The brown *Scytothamnus fasciculatus* occurs patchily from Wellington south, but comes to its maximum on Otago shores in late spring. Finer in texture than the northern *S. australis*, it differs also in the hollowing out of the main axis as growth proceeds.

Through the top of the **middle eulittoral** a band of *Porphyra columbina* was found in December, though already at this date starting to regress and break up. Common at the same level are the firm, yellowish vesicles of the peculiar southern brown alga *Adenocystis utricularis* (Fig. 7.30). Many of the small pools between tides become filled with these bladders by December. In this same month *Gigartina dilatata*, with its flat branches frilled at the edges, is also plentiful in the middle eulittoral. The barnacle *Chamaesipho columna* and the mussels *Mytilus edulis aoteanus* and — less abundantly — *Aulacomya maoriana* — have their usual places in the upper and middle eulittoral.

In the **lower eulittoral** appears the red alga *Pachymenia lusoria*. Absent in central New Zealand, it is evidently conspecific with the northern forms traditionally known as *P. himantophora*. The southern form appears to be simpler in outline and to lack the ragged side branches, being dichotomously divided like inverted Vs.

The 'pink-line' on these open shores is indented and irregular, but the coralline veneer is generally continuous from the bottom of *Pachymenia* down. From this base spring the jointed thalli of *Corallina*, *Haliptilon* and *Jania*, with the bushy tufts of *H. roseum* generally most evident. On the pink surface, tiny rosette species appear in late spring. The green algae, *Ulva lactuca* and *U. laetevirens*, are also regularly present. Just below this level a full algal garden begins, with slender branching *Codium fragile* forming its uppermost tier.

An outstanding feature of the shores of Otago, Southland and Stewart Island is their wealth of small rhodophytes, both in quiet pools and in the shelter of large browns in the sublittoral fringe. The Rhodophyta at low tide considerably out-top those of the north, and increase in size and delicacy of form in the upper sublittoral.

The Ceramiales are rich in examples, notably *Lenormandia angustifolia*, universally common in Otago around low tide. The narrow secondary leaflets characteristically spring from the mid-ribs of the older. In *Marionella prolifera* lateral branches arise from the leaf edges. The *Hymenena* species form branched ribbons, often iridescent, with strong mid-ribs and a fine tracery of veins.

Of the Rhodymeniales, the pink forked straps of *Rhodymenia obtusa* bear the reproductive organs on small leaflets scattered over the surface. *Rhodymenia* species vary from the narrow forked leaflets of *R. leptophylla* to dichotomous fans, with frequent marginal leaves or pinnate branching. *Hymenocladia chondricola* has longer expansions, forming dark red streamers up to a metre long, often attached to sea tulips, *Pyura pachydermatina*.

Some of the most important tufted rhodophytes are the *Polysiphonia* species, ranging from the small patch-forming species such as *P. rudis* to the robust *P. decipiens*, up to 15 cm, epiphytic on *Hormosira*, *Cystophora* and *Xiphophora*, and *P. muelleriana*, with tresses up to 40 cm, in tide pools. The uniaxial Ceramiaceae include the species of *Ballia*, with soft hair-like branching. *Ballia scoparia* grows in long dark tails and *B. callitricha*, producing large feathered fronds, is often epiphytic on browns.

Delicate cut-leaved fronds are the mark of the lace-like *Plocamia* species, generally abundant subtidally and brought in with beach drift after every blow. The long stems of *Streblocladia glomerulata* are terminally fine-branched and *Heterosiphonia concinna* in the *Durvillaea* zone breaks into crimson fern-like fronds, terminally fine-branched. *Euptilota formosissima* is a lovely wine-red species, with fine regularly pinnate fronds.

Perhaps the most exquisite of all the small reds are the epiphytic *Dasyclonium* species, fringed with tiny leaflets at either side like some of the liverworts. The leaflets are ovate in the northern *D. ovalifolium* (growing on the coralline *Arthrocardia corymbosa*), and serrate-triangular like maiden-hair fern in *D. adiantiformis* (on *Xiphophora*, *Gymnogongrus* and *Marginariella*). In *D. incisum* (on *Pterocladia* in the north) they are reduced to combs of minute ribs.

Young *Durvillaea antarctica* establish over winter at Jack's Bay at a level too high for long survival. Lower down at slack water old plants lie at low water like heavy leather aprons over the smaller algae on the pink veneer (Fig. 7.30). Here *D. antarctica* shows little of the

Fig. 7.31 Zoned shore at Murphy's Nugget, Stewart Island

(above) exposed face a–b *Porphyra, Apophloea, Scytothamnus;* b–c *Adenocystis, Ulva,* corallines; c–d *Xiphophora* and *Durvillaea;* d–e *Marginariella*

(below) sheltered face f–g *Apophloea* and *Scytothamnus;* g–h *Hormosira* and *Adenocystis;* h–i *Cystophora scalaris;* i–j *Marginariella*

1 *Marginariella urvilliana* 2 *M. boryana* 3 *Sargassum sinclairii* 4 *Cystophora scalaris* 5 *Xiphophora gladiata* 6 *Sargassum verruculosum* including basal leaves 7 *Hormosira banksii* 8 *Cystophora torulosa* 9 *Pachymenia lusoria.*

serpentine configuration developed in high exposure. On the shore depicted, it forms smooth sheets a metre across, sometimes with fist-sized holes and with the edge scalloped or raggedly fringed. Massive to lift and movable only as waves return, the bull-kelp form a strong rampart at low tide, too slippery to allow a foothold.

Stewart Island (Rakiura)

With important settlement only at Oban, Stewart Island still keeps its bush and seaboard almost pristine. The tree fringe of the adlittoral zone typically comes down closely to overhang the shore. Shaded *Pilayella* and *Porphyra* grow on the lower branches of small southern rata (*Metrosideros umbellata*), and the green algae *Wittrockiella* and *Cladophoropsis*, brown *Sphacelaria* and blue-green *Rivularia* extend up over the peaty humus.

The zonation pattern is compactly shown on small, high islets (nuggets), set close in to the shore. **Lonneker's Nugget** sits mid-way along **Half Moon Bay**. **Murphy's Nugget** is a similar prominence (Fig. 7.31) at the end of **Ringa Ringa Bay**. Below the bush-line the rock is painted with a white lichen, with no trace of yellow. Black *Verrucaria* sharply cuts off the white below, followed by the honey-coloured granitic rock, carrying the two periwinkle species, *Austrolittorina antipoda* and *A. cincta*. Next below, from winter through spring, appears a dull mauve band of *Porphyra columbina*, in summer regressing to drab brown.

Abutting with the eulittoral barnacle *Chamaesipho columna* grows a stubble of black *Lichina confinis* and clusters of the brittle antlers of *Apophlaea lyallii*. Broken rock stacks carry a soft sward of *Stictosiphonia arbuscula* on their shaded faces. Mussels are scarce around Half Moon Bay with *Mytilus edulis aoteanus* the main or sole species.

The bedrock slope of the **eulittoral** displays a band of *Scytothamnus fasciculatus*, mingled in winter or spring with yellowish brown *Scytosiphon lomentaria*, and often with bright green *Enteromorpha* towards its lower margin. Slightly lower, and in relative shelter, there is abundant *Ulva lactuca* in winter and spring, regressing by the summer. Dull yellow bladders of *Adenocystis utricularis* spring up seasonally in pockets of shade, mingled with leaden grey to brown *Hormosira banksii* reaching high up the eulittoral. Even in low wave action isolated *Pachymenia lusoria* persist. With greater water movement, *Hormosira* gives place to a pink coralline veneer, sloping off to deeper water.

At the upper margin of the pink zone small *Corallina* and *Jania* clumps give some reminder of the extensive turfs further north. Below the pink-line dense thickets of brown algae girdle the shore, thickly piled and widely emergent at low water to give the visible character to the whole fringe. The brown algae carry a wealth of epiphytes, and a shaded underworld of small reds flourishes beneath their canopy.

Beneath the corallines is a fringe of the southern species *Xiphophora gladiata*, distinguished from the thicker northern *X. chondrophylla* by its attenuate, linear branches, well adapted to current flow.

Even in the quieter spots there is *Durvillaea antarctica*, in its southern sheltered form, that can be lifted to reveal the long dichotomous ribbons of *Marginariella*. Both the species are present on Stewart Island, *M. urvilliana* with wide straps, deeply serrate and carrying spherical bladders, and *M. boryana* with the streamers narrower and the vesicles elliptical.

Outside *Durvillaea antarctica* grows the second bull-kelp, *D. willana*. At slack low tide its branched stipes arch above the surface, weighed down by their side-paddles, or sometimes found broken off, looking in silhouette like the stumps of a drowned forest. Under shelter, *Marginariella* is preceded by two species of *Cystophora*; furthest inshore there is golden *C. torulosa* with the pinnules unusually large, and below it *C. scalaris*, ashen brown and with zigzagged stipe and side-branching.

Stewart Island has three sorts of *Sargassum* occurring patchily on the shore and in the shallow **sublittoral**. These include the New Zealand-wide *Sargassum sinclairii*, and a sporadic form of *S. undulatum* with finely lacerate leaf margins. The third is an evidently introduced species, *S. verruculosum*, patchily ranging from Kaikoura south. It is well distinguishable, with its earliest fronds trichotomous-pinnate and later ones forming long tufts of branching filaments.

Of the *Carpophyllum* species, Stewart Island has only *C. flexuosum*. The smaller kelps *Lessonia variegata* and *Ecklonia radiata* are far less common than to the north, tending to lie in shallow stretches under rather close shelter.

Outside all the other sublittoral browns grow the bladder-kelp *Macrocystis pyrifera*. Its loose-lying bladderless form (Fig. 6.8) occurs in the deep-enclosed Port Pegasus. Stewart Island is among our best studied phycological areas and has one of New Zealand's richest algal lists. The authors of the most recent account of the seaweeds[4] pay tribute to Mrs Eileen Willa whose collecting in the early 1940s laid the foundations of our modern knowledge.

The luxuriance of small algae in the understrata of pools must be specially noted. The greens include species of *Cladophora* (*C. verticillata* and *C. daviesii*), *Chaetomorpha* (*C. aerea*, *coliformis* and *linum*) and the plumose *Bryopsis vestita*. Small brown algae are represented by tufted *Halopteris congesta*, and the filamentous *Ectocarpus*, *Pilayella* and *Feldmannia* species. There is an abundance of filamentous and other small reds, as of *Polysiphonia* and *Ceramium* species, and *Echinothamnion lyallii*.

Fine sand covering the bottoms of pools is the habitat of many small browns — among them *Ptilopogon botryocladus*, *Cladostephus spongiosus*, *Microzonia velutina* and *Chordaria cladosiphon*. Here too can be found numerous rhodophytes: the slippery tubes of *Grateloupia intestinalis*, the stiff-branched *Gymnogongrus humilis* and *G. torulosus*, with the leafy *Lenormandia angustifolia* and *L. chauvinii*. Pink *Audouinella* filaments grow up as a fine pile from rocks strewn with sand.

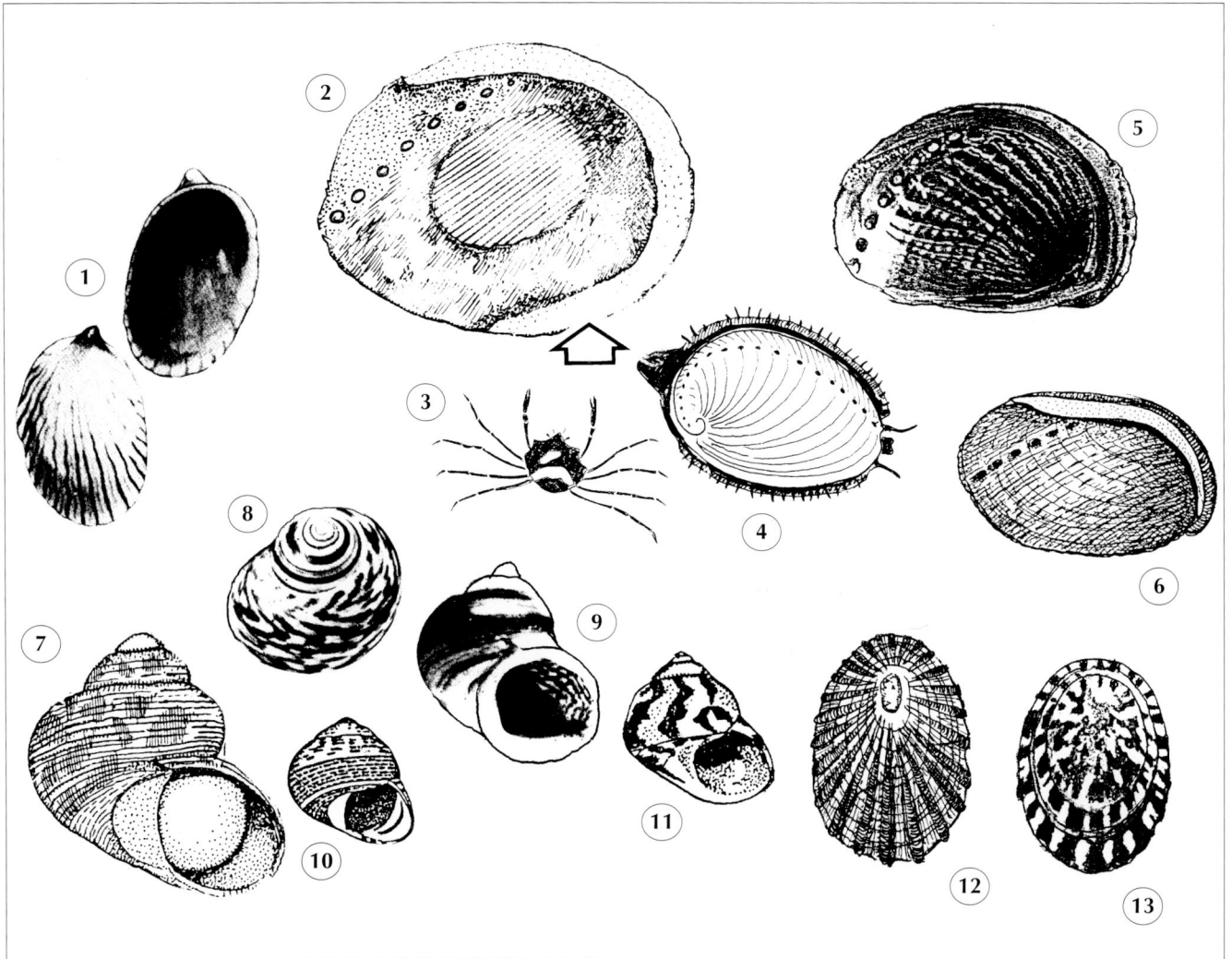

Fig. 7.32 Grazing and browsing gastropods of Otago and Stewart Island
1 *Kerguelenella stewartiana* 2 *Haliotis iris* with 3 *Elamena producta*, with commensal resort beneath the shell rim
4 *Haliotis iris* in progression with extension of pedal marginal tentacles 5 *H. australis* 6 *H. virginea* 7 *Modelia granosa*
8, 9 *Margarella antipoda rosea* 10 *Cantharidella tesselata* 11 *Margarella turneri* 12 *Cellana strigilis redimiculum* 13 *C. s. redimiculum*, young example showing ocellate pattern.

Many of the simplest filamentous browns are epiphytes. Profuse tufts of *Sphacelaria* species and *Elachista australis* grow upon *Xiphophora*, with *Sphacelaria* on *Cystophora*, as well as *Myrionema*, *Herponema* and *Hecatonema* species forming minute brown cushions. Among the epiphytic reds, there are both simple and complex thalli: filamentous *Acrochaetium* on *Zostera*, and — among the Ceramiales — a host of species, with some of them needing expert identification. Littoral and sublittoral epiphytes are represented by species of *Anotrichium*, *Antithamnion*, *Ballia*, *Platythamnion*, *Griffithsia*, *Microcladia* and *Ceramium*.

The leafy Delesseriaceae found in drift at Stewart Island include delicate *Schizoseris* and *Myriogramme*. The large but delicately textured Kallymeniaceae: *Callophyllis calliblepharoides* and *C. hombronia*, are both common, the latter on the stalks of *Pyura pachydermatina*.

CHAPTER EIGHT

Pools, Crevices and Borings

We have already seen how the patterns of plants and animals are influenced not only by the states of shelter and exposure but also from the shore's geology. Such features of the substrate, with the structures they form, may not often characterise a whole shore, but will give rise to what we must now examine as special habitat types including pools, crevices and rock-borings.

The most important and widest occurring of such special habitats are generally contributed by the rock pools and the algal turf characteristic of their walls and floors. Such locations, with their continuous submersion, are almost always well-lit. There are in addition other habitats wholly concealed from light, as well as being protected from intermitttent temperature rise or evaporation. These include crevices and borings that will be described towards the end of this chapter, leading often to the geological process of bioerosion.

Rock pools

First, however, our study must centre upon rock pools. The intertidal shore is seldom left fully dry at low water. Pools remain, giving a haven against the effects of the atmosphere; their plants and animals will thus reach a higher shore level than they could achieve on bare rock. There are plenty of intertidal species, however, that could not exist in pools. Far from being a safe replica of the larger sea, the pool environment has hazards of its own. Pools high on the shore run to far wider extremes than the open sea. They can create an even harsher regime than periodic exposure to the atmosphere.

At high water spring (HWS) pools can become tepid and highly saline, renewed by wave splash only at an exceptional high tide. Alternatively, high pools can become heavily diluted by freshwater seepage or rainfall. With high photosynthesis by algae in pools isolated during daylight, pH and oxygen concentration build up

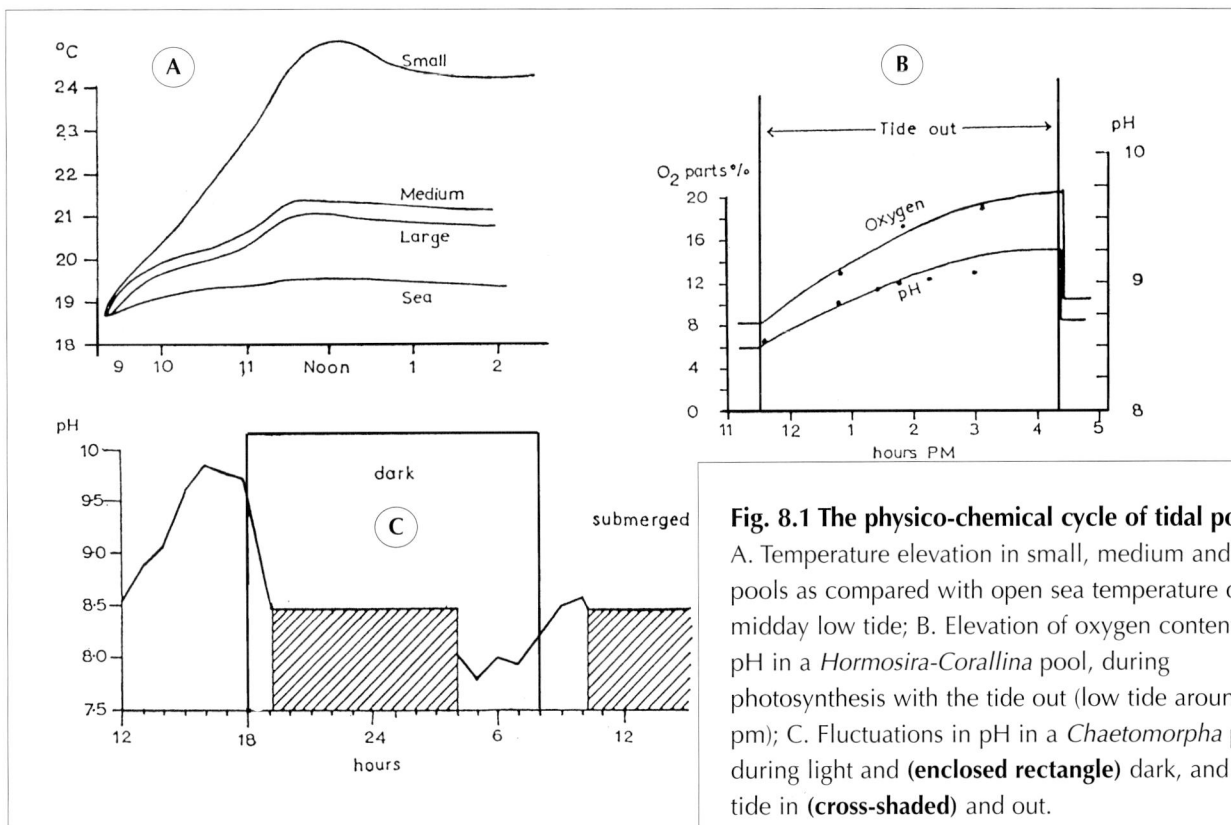

Fig. 8.1 The physico-chemical cycle of tidal pools (1) A. Temperature elevation in small, medium and large pools as compared with open sea temperature during a midday low tide; B. Elevation of oxygen content and pH in a *Hormosira-Corallina* pool, during photosynthesis with the tide out (low tide around 2 pm); C. Fluctuations in pH in a *Chaetomorpha* pool, during light and **(enclosed rectangle)** dark, and with tide in **(cross-shaded)** and out.

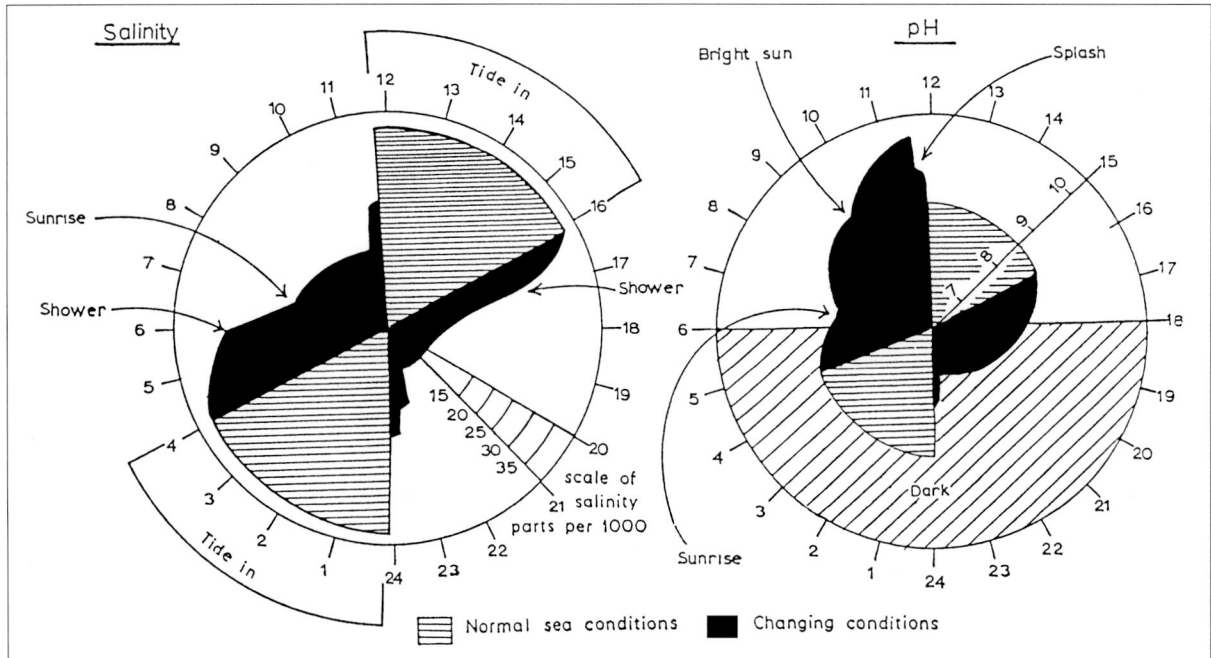

Fig. 8.2 The physico-chemical cycle of tidal pools (2)
Data from 24 hours' observation of a *Chaetomorpha* pool in the line of freshwater seepage in the upper eulittoral zone, Auckland's west coast. Diagrams show fluctuations in salinity **(left)** and in pH **(right)**, with the tide out **(black)**; and uniform regime with the tide in **(horizontal hatching)**.

steeply. Conversely, in stagnant pools with organic decay, oxygen may be totally depleted, with high hydrogen sulphide content and depressed pH.

We may first consider the day and night fluctuations in a shallow eulittoral pool fringed with *Corallina* and *Hormosira*. During an afternoon low tide we find a steep rise in pH and oxygen content. With the net removal of carbon dioxide in photosynthesis, pH can climb to even 10, still within the tolerance of some algal sporelings. Photosynthesis will result in a net surplus of oxygen above its consumption in respiration. At times a pool can thus become hyper-saturated, with delicate algal fronds buoyed up with bubbles of oxygen coming out of solution. Both oxygen and pH values are thus highest in algal-dense pools.

During a nocturnal low tide these trends are reversed, with oxygen depleted with cessation of photosynthesis, and unutilised carbon dioxide responsible for a depression of pH.

Salinity fluctuations can also be related to the time of day. With evaporation negligible at night, any dilution by fresh runoff is now apparent, most so in small pools. Temperatures may rise steeply during a daytime low tide, even on a dull day. Conversely, the inrush of a returning tide can bring a sudden temperature fall of several degrees in a few seconds. Temperature variation is also influenced by the volume of the pool in relation to surface area, and the extent of shading algae. Algal stratification within a pool owes much to temperature and lighting, producing a part-replica of the shore at large.

The cycles (Fig. 8.2) are based on a middle eulittoral pool with algae, over 24 hours, with a high and low tide both in light and dark. Both salinity and pH are highly sensitive to the incidents of the weather. Salinity is shown to fall steadily by fresh seepage after the ebbing tide before dawn, followed by a more rapid drop during a rain-shower. Evaporation after sunrise retarded this fall, until a sudden rise was brought by wave flooding, ahead of the flowing tide at noon. At a low tide in late afternoon, with evaporation reducing, seepage dilution was evident, being partly reversed by wave entry just before the return of the tide at midnight. After sunset, pH steadily fell, and the same trend was seen between the next withdrawal of the tide and sunrise. A sharp pH rise followed daylight, especially steep after a burst of sunshine.

Animals and plants in pools may be raised to a vertical level well above their open rock limits. Thus, in the upper eulittoral where barnacles, *Xenostrobus* and oysters possess the dry surface, the pools will be fringed with *Corallina* and *Hormosira*. In the littoral fringe with the rock bare of sessile animals, the pools may have small rock oysters and the tubeworms *Pomatoceros*, *Hydroides* and spirorbids. In the lower eulittoral, over the *Corallina-Hormosira* flat, the brown algae of the sublittoral fringe occupy the pools. Extreme examples of the elevation effect can be shown by splash pools within a high supralittoral zone. In one such pool of a surf-pounded northern coast, 8 m above low water, *Nerita atramentosa* was found round the water-line, with *Zeacumantus subcarinatus* feeding on the diatom film of

Fig. 8.3 Tidal pools with algae and animals at Army Bay, east Northland

A. Shallow pool in the upper eulittoral zone with *Chaetomorpha linum* and *Stiliger felinus*.

B. Deeper pool in the lower eulittoral zone with (a) *Carpophyllum maschalocarpum* (b) at deeper level *Ecklonia radiata* form *biruncinata* and (c) on higher ledge *Corallina officinalis*.

C. Deep, shaded pool in the lower eulittoral zone with light stratification of algae (a) *Hormosira banksii* (b) *Bryopsis plumosa* (c) *Pterocladiella capillacea* (d) *Halopteris* (e) *Carpophyllum plumosum* (f) *Zonaria turneriana*.

1 *Carpophyllum maschalocarpum* in pool B, with (2) enlarged terminal leaflets bearing filamentary epiphytic brown alga and (3) smaller leaflets at shade depths 4 *Hormosira banksii* from upper level in pool B 5 *Ecklonia radiata* of the narrow-leafed form *biruncinata* 6 *Zonaria turneriana* 7 *Pterocladiella capillacea* 8 *Stiliger felinus* 9 *Lissocampus filum* (short-snouted pipefish) 10 *Stigmatophora macropterygia* (long-snouted pipefish) 11 *Hippocampus abdominalis* (seahorse) 12 *Acanthoclinus fuscus* (rockfish).

Fig. 8.4 Surge-replenished tidal pool at Oaro, North Canterbury

1 *Porphyra columbina* 2 *Gigartina* sp.
3 *Cladhymenia oblongifolia*
4 *Gigartina* sp. 5 *Gigartina circumcincta* 6 *Gigartina* sp.
7 *Durvillaea antarctica* 8 *Ulva lactuca*
9 *Marginariella urvilliana*
10 *Glossophora kunthii*
11 *Streblocladia glomerulata*
12 *Carpophyllum maschalocarpum*.

the bottom. Scattered on the sides were small *Crassostrea glomerata*, transported as larvae by splash.

The high-saline pools of the **littoral fringe** reach their greatest extremes of temperature and salinity only during a few tides monthly. Their principal animals are dipteran larvae and pupae, as of the vexatious salt pool mosquito *Opifex fuscatus*, swarming here in spring and summer. Fringing green algae, *Enteromorpha* and *Cladophora*, become luxuriant in spring, bleaching and finally disintegrating by summer to a tepid organic broth, removed only when a spring tide reaches the pool.

On open coasts, the highest **eulittoral** pools are freely replenished by waves, while evaporation is compensated by spray, keeping salinity more constant. Up to a high level, shallow pans are prettily blotched with pink coralline crusts, and contain a number of grazers and browsers. Small radially streaked *Notoacmea parviconoidea*, larger than those on open rock, are the principal limpets. *Siphonaria australis* may cluster freely in saucers and runnels and here their yellow spawn shows up conspicuously. High level pans are nurseries for young *Austrolittorina antipoda*, and in the north for the pinhead-sized juveniles of *Nerita atramentosa*.

Pools in the **upper eulittoral** support heavy fringes of algae. On open coasts there are the brittle filaments of *Chaetomorpha aerea*, *C. linum* and tubular *Enteromorpha compressa*. Shallower pools may develop a soft pile of the filamentous brown algae *Ectocarpus* and *Pilayella*, increasing from small tufts in July to a total blanket from spring to autumn. Buoyant with oxygen bubbles, these algae may increase till they choke the pool. Green algae of upper eulittoral pools include the coarser stubble of *Cladophoropsis herpestica* and

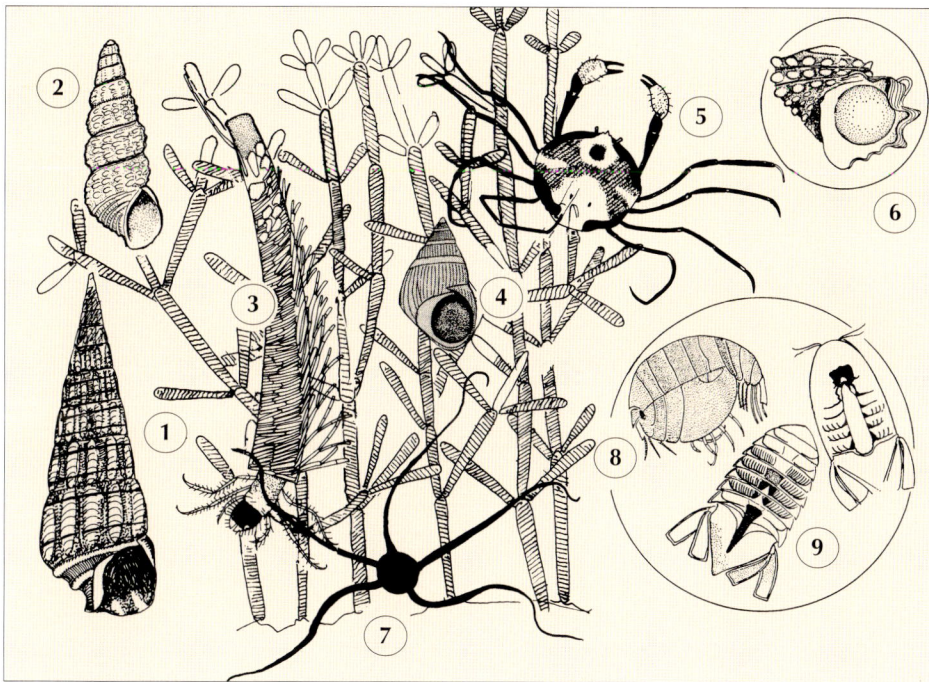

Fig. 8.5 The fauna of _Corallina_ in pools and turf
1 _Zeacumantus subcarinatus_
2 _Zebittium exile_ 3 _Philanisus plebeius_ 4 _Eatoniella olivacea_
5 _Halicarcinus innominatus_
6 _Turbo smaragdus_ (juvenile form) 7 _Amphipholis squamata_
8 _Tetradeion crassum_
9 _Isocladus_ sp.

almost microscopic forests of delicate _Bryopsis plumosa_ and — in the north — _B. derbesioides_. Fresh seepage will encourage the hyposaline community of _Enteromorpha intestinalis_, _E. compressa_ and a diminutive upper shore form of _Scytosiphon lomentaria_.

In the **lower eulittoral**, pools are fringed with _Corallina officinalis_. Except on the roughest coasts, this gives place near the water surface to _Hormosira banksii_, a species as versatile in pools as in the open. Most _Hormosira_ are said to extrude gametes only when emersed, yet pool populations have fertile conceptacles, and the osmotic effects of salinity change may be the trigger of gamete release. In mid-tidal pools, _Hormosira_ has 'normal' large bladders and short connectives. In high-level splash-pools with regular splash, the axes are long and attenuate, with reduced branching and small, thick-walled bladders. In shallow pools and runnels, _Corallina_ is excluded and _Hormosira_ appears alone, with large bladders for water conservation. Where the centre of a plant lies submerged, its bladders are reduced, with larger distal ones high and dry above the water.

Spring epiphytes of _Hormosira_ and _Corallina_ include the small, pin-head 'swarmers' of _Colpomenia sinuosa_. The species of _Myrionema_, _Herponema_ and _Hecatonema_ (brown **Chordariales**) are small obligate epiphytes, never growing on rock. The brown alga _Notheia anomala_ (Fig. 10.8), living on _Hormosira_, is generally held to be a true parasite.

Pools of the **lower eulittoral** and **sublittoral fringe** differ in their deep algae according to wave action. In the Hauraki Gulf, they may be filled with _Carpophyllum_ species, less often with _Cystophora_. Fig. 8.3 shows such a pool at Army Bay, Whangaparaoa Peninsula. Under a fucoid canopy, the deep-shaded species include _Glossophora kunthii_, fan-shaped tufted fans of _Zonaria_

turneriana and the soft, dense _Pterocladiella capillacea_.

From Oaro, south of Kaikoura, Fig. 8.4 shows a pool of southern character, in both the floral composition, and its abundant variety. Noteworthy are the dwarfed bull-kelp, _Durvillaea antarctica_, as a permanent feature within the pool.

Intimate fauna of algae

The algae of the pools and sublittoral fringe are favoured habitats for the lightest and daintiest animals. As a refuge, near at hand, algae have many advantages, as a launching site or place of quick return. Their fronds give temporary attachment for swimmers, and the holdfasts deeper shelter for nestlers or burrowers. Food is available from the plant's own tissues or the small epiphytes and animals that are attached. Light-bodied animals with prehensile limbs can while still securely fastened reach out to grapple with plankton in the water bathing the fronds.

The richest micro-fauna is that of _Corallina_. The calcified segments, themselves a poor food, carry a wealth of small epiphytes, for grazing, and the sediments round the base support numerous deposit feeders. _Corallina_ in pools forms a well-stratified habitat. The common browsing molluscs in its fronds are young cat's-eyes, _Turbo smaragdus_, and the long-spired snail _Zeacumantus subcarinatus_. Both ingest whole _Corallina_ segments, but are primarily concerned with the delicate epiphytes that grow thereon. Abundant in coralline turf in the north, often in silted places, is a common spire shell, _Zebittium exile_, only 5 mm long, classed near _Zeacumantus_, in the Family Cerithiidae.

The small, attractive pill-box crab _Halicarcinus innominatus_ moves with a stilted gait over bushy _Coral-_

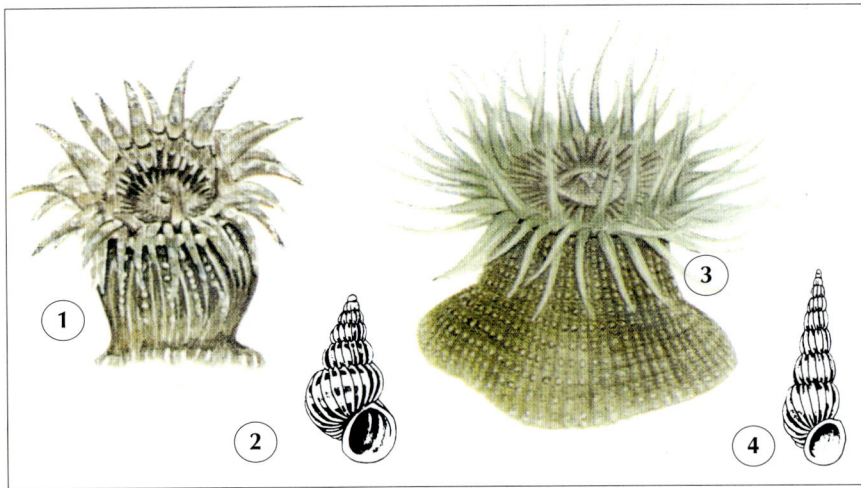

Fig. 8.6 Two anemones and their predators in coralline turf
1 *Anthopleura aureoradiata*, with (2) *Epitonium tenellum* 3 *Isactinia olivacea*, with (4) *Epitonium jukesianum*.
(after Michael Miller)

lina. Like all its family, it is flat-backed and somewhat polygonal, with frail, spindly legs. Older individuals may be bottle green to olive, with the back opaque white, or light-blotched. In others the carapace is mottled brown or jet black, in contrast to the pale, almost invisible legs. The smallest individuals are black and white pied, sometimes tinted with rust-red.

A marine insect found with *Corallina* in pools is the common marine caddis fly *Philanisus plebeius*, in its horn-shaped larval tubes open at either end. The tube membrane is studded with sand or shell, and with segments of *Corallina* or *Jania* set neatly side to side. Larger pieces give the tube a ragged camouflage. The larva clings by its thin legs, or scrambles among the fronds in search of diatoms, dinoflagellates and ciliate Protozoa. Larger ones, above 8 mm, were observed to ingest coralline fragments as well. The adult caddis hovers languidly over pools or walks about on the turf. John Leader has published a study of the larva's osmoregulation.

At the *Corallina* bases a plethora of small worms take shelter, including carnivores, ciliary feeders and deposit-swallowers. *Perinereis camiguinoides* twines freely among the branches, and a small, deep brown hesionid, *Ophiodromus angustifrons*, moves with rapid undulations. Sabellid tubeworms stand up freely, putting out plain or striped crowns. Terebellids extend tentacles from more flimsy parchment tubes. Burrowing in silt, with no tube at all, is a small cirratulid. The scale-worm, *Lepidonotus polychromus*, performs its slow undulant walk at the coralline bases. Small syllids may be discovered as well.

A slender orange nemertean twines round the bases, with the proboscis intermittently active. Where the silt content increases, the sipunculan worm *Themiste minor huttoni* becomes common. Feeding on diatoms and other small particles is the tiny brittle star, *Amphipholis squamata*.

From Wellington south, coralline tufts may be highlighted with the flame red of the proboscis worm *Saccoglossus otagoensis* (Fig. 13.10). Delicate and hard to remove whole, it twines freely between the fronds, and subsists on the fine organic deposits near their base.

Mollusca

Classically grouped together as 'rissoids' are the miniature gastropods gleaning particles both under stones and on fine algal fronds. The sculptured rissoids and their kin are commoner under stones. The commonest is relatively large *Eatoniella olivacea* (2 mm long), which is widely at home in both habitats. It swings from fronds by a mucus thread secreted from the foot. This species

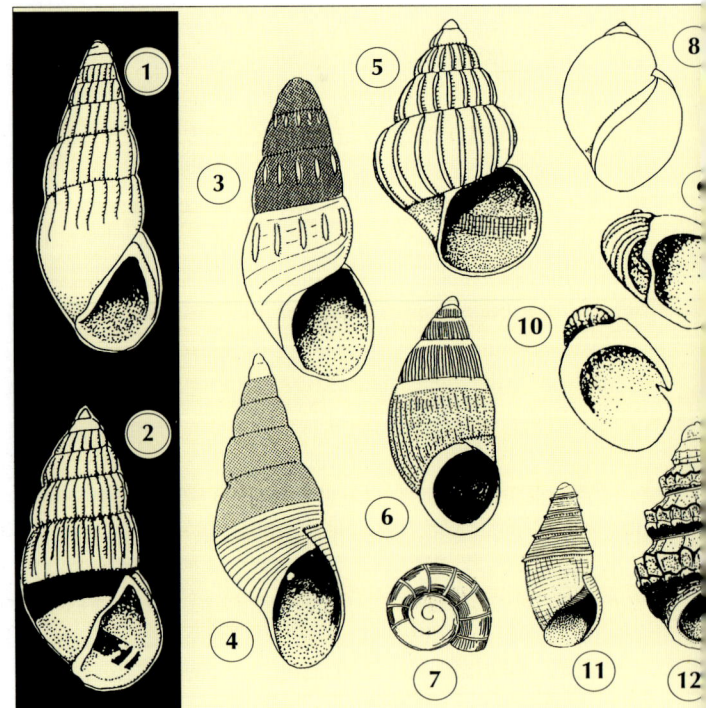

Fig. 8.7 Minute gastropods of algae and tide pools
1 *Rissoina chathamensis* 2 *R. anguina* 3 *Fictonoba rufolactea* 4 *F. carnosa* 5 *Pusillina hamiltoni* 6 *Pisinna zosterophila* 7 *Omalogyra fusca* 8 *Rissoella cystophora* 9 *Incisura rosea* 10 *I. lytteltonensis* 11 *Anabathron hedleyi* 12 *Merelina lyalliana*.

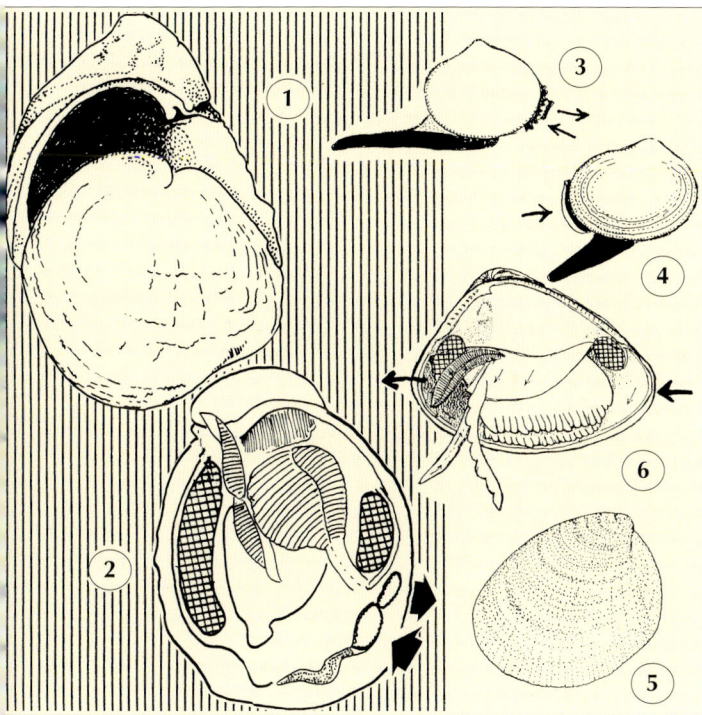

Fig. 8.8 Bivalves among coralline algal turf
1 *Cleidothaerus albidus*, with left valve displaced 2, the same, pallial organs 3 *Nucula hartvigiana* 4 the same, pallial organs 5 *Neolepton antipodum* 6 *Lasaea hinemoa*.

can be readily collected in washings from algae or with a brush from under small stones. The smoother and lighter species such as *Pusillina hamiltoni* can be gathered from algal washings. The largest, most elegant *Eatoniella* is the pink-marbled *E. flammulata*, common on *Carpophyllum plumosum*. Here too can be found the white zigzagged *E. limbata*, and the pretty pink *E. roseola*. There are also the low, wide-umbilicate shells of *Eatoniella pfeifferi*, the thin-shelled *Rissoella* species (*R. elongatospira* and *R. rissoaformis*) and the tiny planorbiform *Omalogyra fusca*.

The commonest small bivalve in coralline turf is perhaps *Neolepton antipodum*, with a rounded shell only 2 mm long, and red-blotched at the beaks. The long mobile foot has no byssus, being held enmeshed by its sticky mucus. At high level *Neolepton* is often replaced by the white or pink *Lasaea hinemoa*, more common in the tufts of *Stictosiphonia arbuscula*. With heavy silt, both give place to *Nucula hartvigiana*, a primitive deposit-feeding protobranch (Fig. 8.8). In the north, the smallest of all our bivalves, *Pachykellya minima*, less than a millimetre long, is found in low tidal *Corallina*.

Isopoda and Amphipoda

With its backdrop for cryptic and disruptive camouflage, *Corallina* is rich in small, well-concealed crustaceans. The mauve-grey algal segments have paler growing tips, and around the bases mottling is added by lodging of bluish chips from mussels, pink from scallops or *Balanus*, the white of other bivalves, and the dull green of broken urchin spines.

From such a dappled setting, numerous sorts of amphipods come out in washings: most samples include the black, comma-shaped *Tetradeion crassum*, pale, pupiform lysianassids with long side-skirts, light pink *Lembos* and the small buff *Paracalliope novizealandiae*.

The isopods in corallines include — in the north — a vari-hued *Cerceis*, handsomely marked in red, gold and brown, a *Cilicaea*, and the almost ubiquitous backspined *Isocladus* species (Fig. 8.9), with *I. armatus*, *I. dulciculus* and *I. calcareus* most widespread. The camouflage, based on a fawn ground pattern splashed with black, is in no two individuals alike.

New Zealand's amphipods and isopods are still incompletely assessed. New species, especially for the evidently fast-evolving Amphipoda, continue to turn up freely. On open coasts, the familiar amphipods in

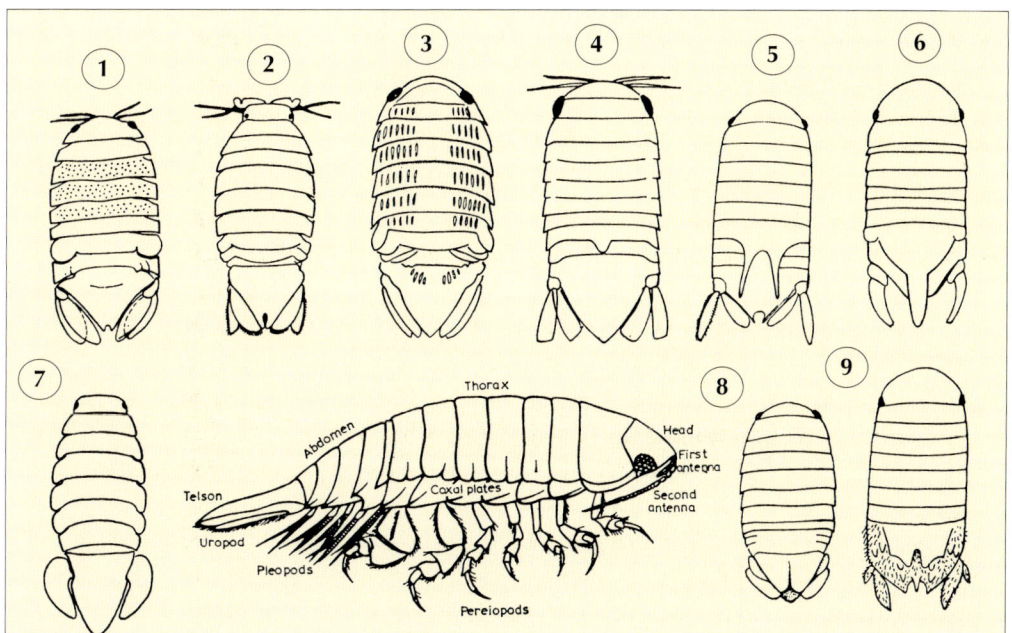

Fig. 8.9 Some isopods from algae
Structural features of an isopod. 1 *Dynamella huttoni* 2 *Dynamenoides* sp. 3 *Exosphaeroma obtusum* 4 *Isocladus armatus* (not yet adult) 5 *Cymodopsis* sp. 6 *Cilicaea* sp. 7 *Scutuloidea kutu* 8 *Cymodopsis* sp. 9 *Cymodoce hodgsoni*.

145

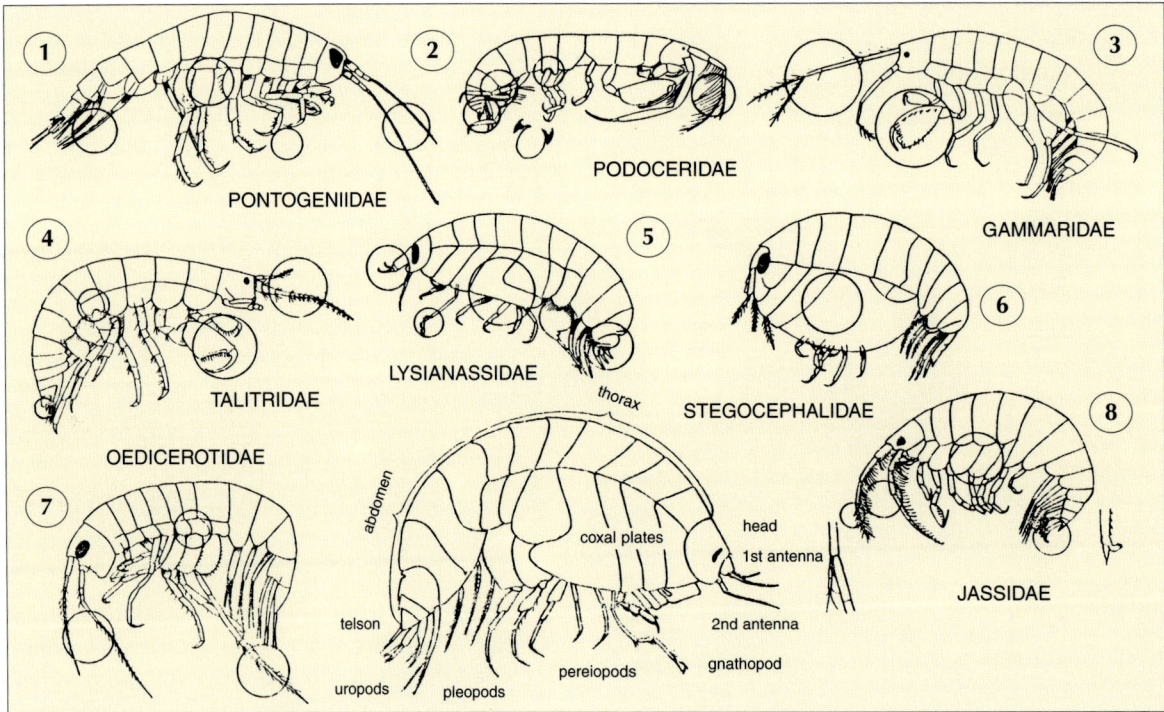

Fig. 8.10 Families of algal-dwelling Amphipoda

1 **Pontogeneiidae** Slender and compressed. Antennae long, almost equal. Pereiopods long, increasing from front to back. Pleopods prominent.

2 **Podoceridae** Antennae long and second larger. Posterior pereiopods turned sharply backward. Side-plate 3 bilobed. Uropod 3 reduced to small hook.

3 **Gammaridae** Slim-bodied, with slender antennae, first with accessory flagellum; gnathopod 1 much smaller than 2.

4 **Talitridae** Antenna 1 much shorter than 2. Male with large second gnathopod. Side-plate 5 bilobed. Uropod 3 uniramous.

5 **Lysianassidae** High, evenly vaulted. Antenna 1 with accessory flagellum, and short, thick peduncle. Gnathopods feeble. Side-plates deep, 4 the largest. Urosome short, thick and convex above.

6 **Stegocephalidae** High and compressed. Side-plates large and deep, especially 4.

7 **Oedicerotidae** Tumid with small side-plates. Antennae setose, first usually shorter. Pereiopod 5 turned back, long and styliform.

8 **Jassidae** Smooth with deep side-plates. Antennae long and densely setose, second stronger, first with tiny flagellum. Uropod 3 biramous and uncinate.

Gigartina alveata are the talitrids *Hyale rubra*. Their shining olive bodies glisten with each move as they half hop, half wriggle their way under fresh clumps of *Gigartina* or hide beneath *Letterstedtia* or *Ulva*. *Dynamella huttoni* is also a common isopod in *Gigartina*, when it changes its normal orange to match itself to the green surrounds. The rose-pink isopod

Scutuloidea maculata is to be found clambering among the fronds.

The low tidal algal garden with *Gigartina* (Fig. 7.28), *Pterocladia* and *Laurencia* and *Jania micrarthrodia* is also rich in amphipods. The families Pontogeneiidae, Stegocephalidae, Lysianassidae, Podoceridae, Ischyroceridae and Calliopiidae are all commonly represented. Their rich red and paler pink blend with the algal background. A few species are semi-transparent, with brown disruptive camouflage. *Tetradeion crassum* stands out in jet black.

Fine red algae, especially *Pterocladiella capillacea*, form a camouflage screen for those bizarre amphipods, the Caprellidae. Well named the skeleton shrimps, they progress like looper caterpillars, taking a forward grip by their big chelae, while the rear end relaxes its hold to

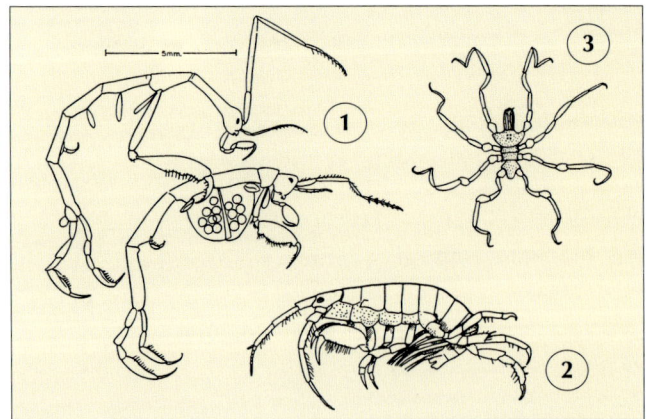

Fig. 8.11 Specialised Crustacea and pycnogonid from red algae

1 *Caprella* (male, **top**; female, **bottom**) from *Pterocladiella capillacea* 2 a podocerid amphipod 3 a pycnogonid or sea-spider.

draw up behind. Swimming is performed in this same way, like a mosquito larvae. The caprellid thorax has seven cylindrical segments, with the small head fused to the first. The limbless abdomen is a vestigial knob. Obvious behind the long antennae are two pairs of chelate gnathopods, each with strong chelae. The female caprellid carries a ventral brood pouch on segments three and four. The posterior pereiopods, generally three pairs, have subchelate claws, giving anchorage as the foreparts are swung forward to seize copepods or other swimming prey. Living nets of caprellids may thus hang poised in vantage places of moving water. Fig. 8.11 shows the wine-red *Caprella* from *Pterocladiella capillacea*.

The seaweed amphipods of the Family Podoceridae have curiously converged with the caprellids. Long and straight, they have the gnathopods strongly chelate, and the last pereiopods turned back and prehensile. The females are noticeably swollen towards the middle. The common species in New Zealand is the red-brown *Podoceris manawatu*.

The caprellids have their isopod counterparts in the strangely shaped 'skeleton lice', Astacillidae. Living among fine low water algae, they resemble L-shaped stick insects with raptorial antennae, and the thoracic legs in two sets, four in front and three behind, with an elongate fourth segment between. *Astacilla tuberculata*, our common species, holds out its antennae like caprellid chelae, flicking the fore-body at intervals by its hinge.

Alga rinsing from the sublittoral fringe brings to light a number of tight-clinging Isopoda, narrowly linear and camouflage. The **Idoteidae** or 'sea centipedes' are denizens of both brown algae and the largest reds. Stiff and parallel-sided they have short, prehensile limbs and gnawing mouthparts. The abdomen is fused into a single tail-piece pointed behind, and the pleopodal gills are protected by modified uropods, hinging forward to open like valves (hence the name of this Order Valvifera). *Batedotea elongata* take on the brown of *Carpophyllum* or the wine red of Rhodophyta, swimming by the oscillations of the covered pleopods, or by flicking the tail-piece like a shrimp. The olive *Paridotea ungulata* on *Carpophyllum* has a characteristic notch at the end of the tail-piece, and its ovigerous females are swollen at the middle. The body is held like a looper caterpillar and with a humping gait. The small idoteid *Cleantis tubicola*, only 7 mm long, lives in pools on *Pterocladia* and *Jania*.

The most highly adapted seaweed isopod is the sphaeromid *Amphiroidea media* (Order Flabellifera). Smooth and kelp-brown, it fastens by its short legs to *Ecklonia* blades so firmly as to resist all dislodgement, or may let go to swim slowly in graceful arcs before re-attaching. The antennal bases form broad plates, and

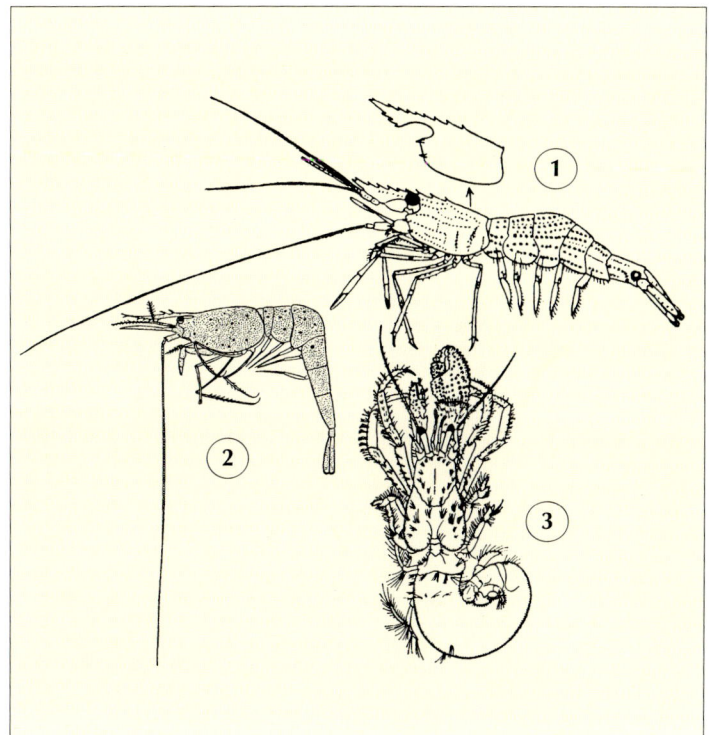

Fig. 8.12 Three decapod Crustacea of rock pools
1 *Palaemon affinis*, with carapace outline 2 *Hippolyte bifidirostris* 3 hermit crab *Pagurus novizelandiae*, removed from its adopted shell.

the head narrows like the neck of a vase. Sharp-edged coxal plates expand beyond the limbs, and the outer uropods are claw-like.

A much smaller flabelliferan, *Plakarthrium typicum* forms an oval disc flattened against the kelp, with marginal plates constructed of antennal scales, coxae and uropods.

Decapoda

The most familiar crustacean of rock pools is the light-bodied prawn *Palaemon affinis*. Ever on the watch to pick up animal food, it swims cautiously by the pleopods, half-walking on the tips of the thoracic limbs. It may be coaxed from cover by the offer of food, which is seized by the slender chelae, then dragged away as the prawn darts backwards with quick flexion of the abdomen.

Camouflaging algae at low tide are the refuge of the small aesop prawn, *Hippolyte bifidirostris*. Though more active than the isopods, they remain rather unadventurous, darting back into the fronds of low tidal greens, reds or browns. *Hippolyte* can be recognised by its permanently bent abdomen, sharply humped at the third segment. This species is olive to bright green, sometimes scattered with bright blue spots.

Clinging upon floating *Sargassum* fronds may be found the small oceanic grapsid *Planes minutus*, the

Fig. 8.13 Tide pool fishes of three families
Blenniidae: 1 crested blenny *Parablennius laticlavius*; Tripterygiidae, triple fins: 2 variable triplefin *Forsterygion varium*; Clinidae, weedfishes: 3 crested weedfish *Cristiceps aurantiacus*.

(By courtesy of Tony Ayling.)

crab first noted far out from land by Christopher Columbus. Pale honey coloured or brown blotched and only 2 cm in carapace length, it can swim effectively by the silken fringes of the legs.

Pycnogonida

These so-called 'sea spiders' form an isolated group without close relatives, but classed loosely alongside the Arachnida, with the true spiders and their kin. Pycnogonids are partly suctorial, partly seizers of solid prey, living typically among algae where hydroids grow epizoically. The body is slung down from four pairs of thin, attenuate legs with prehensile claws. The segmented thorax carries a short cylindrical abdomen, while the head terminates in a stout proboscis, with paired appendages differing among the genera. The New Zealand *Pallenopsis obliqua* and *Pallene novaezelandiae* have strong chelate mandibles and lack palps. *Achelia dohrni* has a pair of long, eight-jointed palps, with the mandibles rudimentary.

Pycnogonids seize and swallow small hydroid polyps, but feed also on anemones, tunicates and holothurians, sucking out the body fluids by the piercing proboscis. At the back of the proboscis are filtering setae that strain off all but fluid tissues. The capacious stomach, as in spiders, gives off digestive diverticula into the leg bases. Most pycnogonid females attach the eggs to two or more of the male's legs. In *Pallenopsis*, the six-legged pyriform larvae creep after hatching into the body cavity of a *Coryne* or *Hydractinia*, there to encyst and undergo further development.

Small fishes

New Zealand's most active fishes in low and subtidal algae, and sometimes in tide pools, belong to the Family **Tripterygiidae**, closely akin to the blennies, and called 'triplefins' from their three dorsal fins. Darting to cover with strokes of the pectoral fins, they have a near-perfect camouflage when at rest. There are several common shore species. The familiar 'cockabully', *Forsterygion varium*, is diversely marked, with a small tentacle above each eye as well as over the nostrils. The thripenny, *Gilloblennius tripennis*, differs from others in its uniform dark hue. The twister, *Bellapiscis medius*, so-called from the dark zigzag along the side, lacks the eye tentacle. The topknot, *Notoclinus fenestratus*, has oblique dark bars on a bright orange body; the first dorsal fin is highest in front and narrows behind.

To the related Family Clinidae or weed fishes belongs the orange *Ericentrus rubrus*, not uncommon among kelp in low tidal pools. An undivided dorsal fin runs the whole length of the back.

Our single true blenny, *Parablennius laticlavius*, common in low tidal pools, is gold with black and silver flanks, and long dorsal and anal fins. The pectorals carry two free barbels, and the snout and forehead are tufted.

Two of the pipefish family, **Syngnathidae**, live entwined in the algae of low tidal pools. Using the snout with its small terminal mouth as a pipette, they suck in minute copepods and crustacean larvae. *Lissocampus filum*, up to 10 cm long and squarish in section, is dark brown or black, twisting among algal fronds. The snout

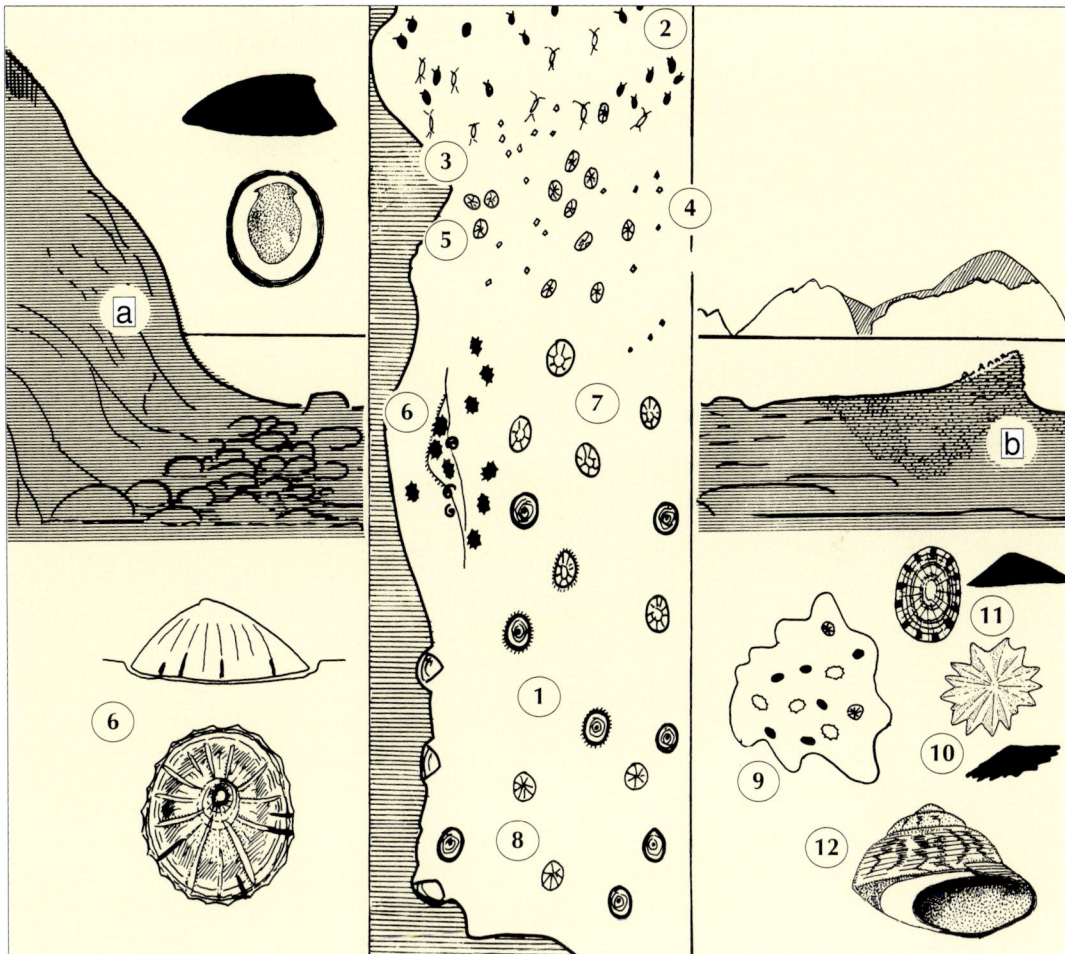

Fig. 8.14 Gastropods on soft mudstone, Tolaga Bay
a soft eroding cliff slope; b intertidal platform with beginning of barnacle cover
1 *Cellana flava* 2 *Onchidella nigricans* 3 *Ligia novaezelandiae* (in shade) 4 *Austrolittorina antipoda*
5 *Notoacmea pileopsis* 6 *Siphonaria australis* (in crevice) 7 *Cellana ornata* 8 *C. radians* 9 etched
pool with *Patelloida* and *Notoacmea* 10 *Patelloida corticata* 11 *Notoacmea parviconoidea* 12
Diloma coracina on sand-strewn rock surface.

is short and stubbed, and the dorsal fin very small. Pectorals, ventral and caudal fins are all minute, and the male has an abdominal furrow where the female implants the eggs. A second pipefish of low tidal pools, the long-snouted *Stigmatophora macropterygia*, differs from *Lissocampus* in its circular section and olive green colouring. Anal, ventral and caudal fins are all wanting, but the fish can swim upright or obliquely by undulating the rather long dorsal fin.

Benches and rock-faces

Broad intertidal platforms, cut out of soft Tertiary sandstone and siltstone, are the typical coast form from East Cape south, and in east Auckland and parts of Northland. Porous and weakly indurated, they are susceptible to regular wetting and drying with atmospheric weathering during low tides. Abrasion by wave-carried particles can also gradually reduce a broad surface. A long,

graded profile results, from cliff-base to low water, with its curve reflecting the level of the water table. The seaward edge drops by several metres, from both wave-cutting and bio-erosion (Fig. 9.30).

The Miocene cliff-profiles at Tolaga Bay have some notable differences from Auckland's. With strata of hard sandstone lacking, the soft blue-grey mudstone erodes quickly, with the receding profile not delaminating or forming crevices, but smoothly rounded. The friable surface is inhospitable to barnacles. Instead the algal film at the cliff-base offers a rich limpet fodder. The golden limpet, *Cellana flava*, is always pre-eminent, with its home-scars on every siltstone or muddy limestone shore, from Te Araroa to Kaikoura. Closest to *C. radians*, this is an east coast species, standing out by its apricot or honey colour, sometimes with dark radial streaks.

The golden limpet mingles below with *C. radians*. Further up, around high water neap, lives *Cellana ornata*, along with small *Siphonaria australis*, clustered

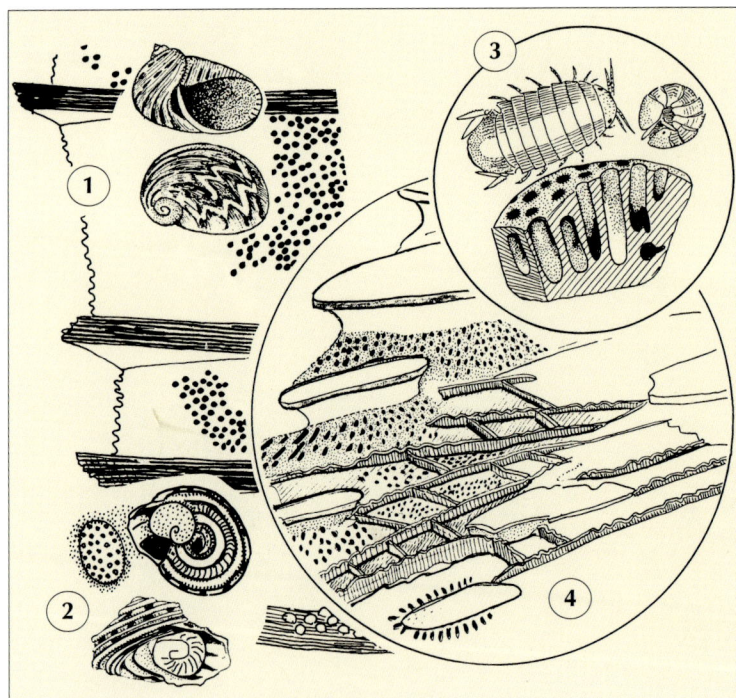

Fig. 8.15 Crevice fauna of Waitemata sandstone

1 *Fossarina rimata* 2 *Risellopsis varia*
3 *Sphaeroma quoyanum*, with its siltstone burrows 4 The soft rock base is shown pitted by *Sphaeroma* borings, with blocks and strata separated by dark, harder limonite.

in crevices with their spawn jellies. The topmost limpets, around high water spring, are high-convex *Notoacmea pileopsis*. Mixed with these are *Onchidella nigricans*, *Austrolittorina antipoda* and the constantly mobile isopod *Ligia novaezelandiae*.

Shallow etched pools on the cut platform contain *Patelloida corticata*, *Siphonaria australis* and hosts of small, dark *Notoacmea parviconoidea*. To complete the limpet tally, *Zostera* leaves in sand pockets have the narrow *Notoacmea helmsi scapha*.

The small, concave-based topshell *Diloma coracina* attaches to rocks in sand, into which it makes outward sorties half-buried.

South of East Cape, the stalked ascidian *Pyura spinosissima*, a smaller relative of the southern sea-tulip, lives in clusters at low tide, in runnels or attached to holdfasts. The stalks reach 12 cm long, and the dull red test is covered with pointed papillae.

Crevice faunas

Along jointing planes deep fissures open up between strata, both by biological penetration, and also by air pressure under wave action. The resulting crevices are an important habitat space, often developing its own internal zoning.

Furthest back, there is room only for flattened species such as the small beach centipede, *Tuoba xylophaga*, to press between the strata. The euniciid worm, *Marphysa depressa*, glides through its long, often forked burrows as smoothly as an earthworm. *Marphysa* is recognised by its red dorsal cirri ('blood gills') behind the head, and by the iridescent cuticle and median head tentacle, all euniciid features.

Crevice space in-filled with sediment has two more sorts of polychaetes. The **Terebellidae** are recognised by their bunches of long peristomial tentacles. The body is soft and contractile, with plump 'thorax' and longer, sometimes coiled 'abdomen'. Irrigating currents are produced by peristalsis. A ciliated groove runs along each feeding tentacle, that can at any point be flattened into a temporary attachment disc. Particles are wiped off at the mouth by drawing each tentacle between the lips.

The **Cirratulidae** live where black anoxic deposits have collected in crevices. *Cirratulus nuchalis* and *Timarete anchylochaeta* are both over-delicate worms to remove without damaging. Unlike terebellids they have no head tentacles, only long tentacular filaments down the sides, emerging to rove about at the surface. There are also red blood gills, though both sorts of filament have been considered respiratory, with food ingested directly by the lobes of the pharynx.

In the open crevice mouth, away from in-fill, lives a mobile fauna, with three elements. First, there are temporary entrants such as the slug *Onchidella nigricans* and the polychaete *Eulalia microphylla*; second, permanent crevice-dwellers of marine origin; and finally a number of air-breathing arthropods, refugees from the land.

Arthropoda

Small pockets of air remain locked in crevices as the tide comes in. These are used by marine spiders and other arthropods. High in the littorine zone, web-lined cavities are inhabited by the spider *Amaurobioides maritimus*, feeding on the fast-running isopod, *Ligia*. In mid-tidal crevices lurks a second spider, *Desis marina*, spinning a bag of silk with a door at one end. Found also beneath

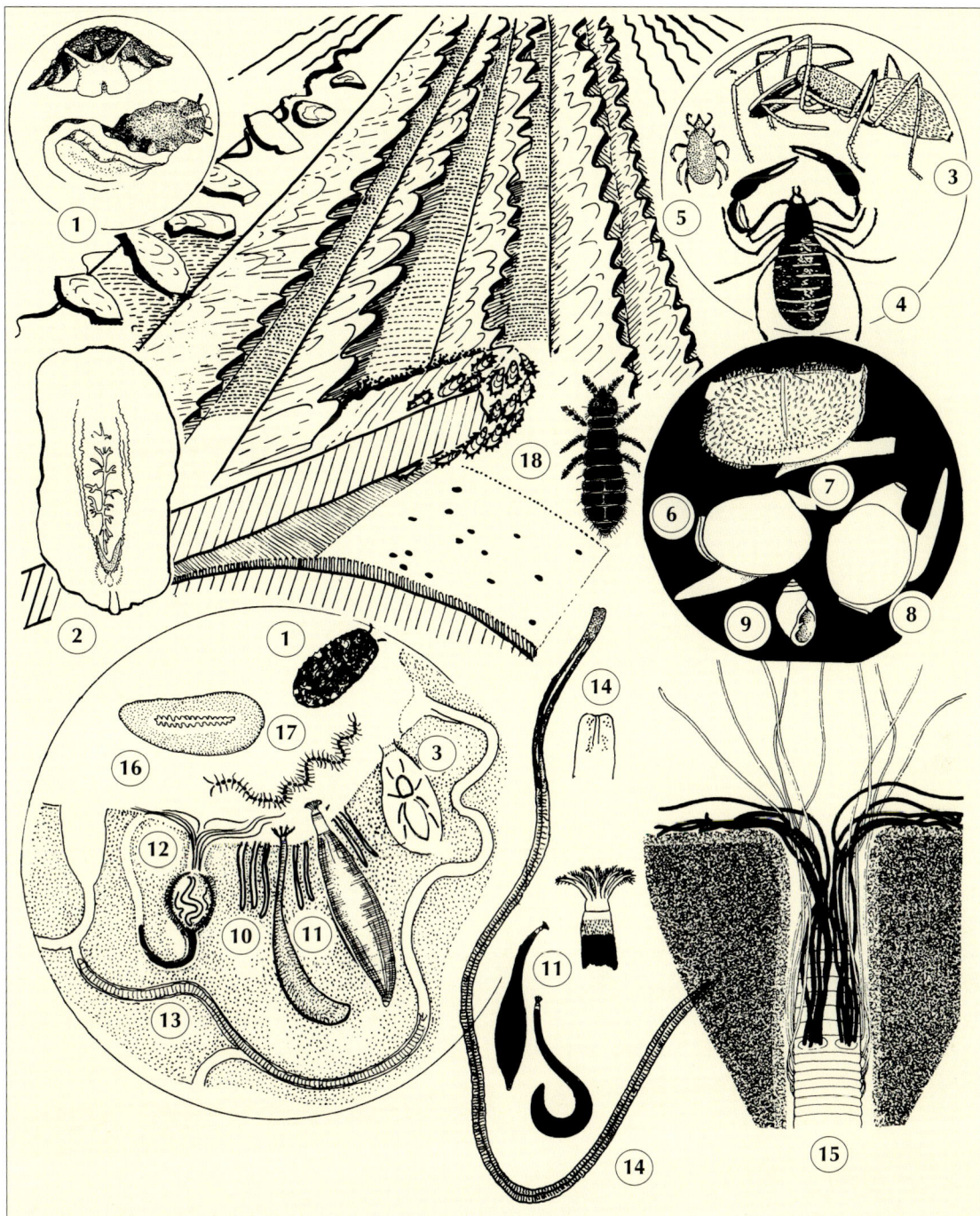

Fig. 8.16 Crevice fauna of Waitemata sandstone
1 *Onchidella nigricans* 2 flatworm, *Stylochus zanzibaricus* 3 *Desis marina* 4 a pseudoscorpion 5 mite *Hydrogamasus* sp. 6. *Lasaea hinemoa* 7 *Scintilla stevensoni* 8 *Kellia cycladiformis* 9 *Leuconopsis obsoleta* 10 *Boccardia* sp. 11 *Dendrostoma aeneum* 12 *Terebella* sp. 13 *Marphysa depressa*, in burrowed shaft 14 nemertean *Amphiporus*, with head detail 15 *Cirratulus cirratus* with feeding tentacles (black) and gills.

oyster clusters, *Desis* runs about freely when the tide is out, catching amphipods as its principal food.

Rock crevices have two arachnids more primitive than the spiders in the miniature pseudoscorpions *Maorichthonius mortenseni* and *Opsochernes carbophilus*. Both retain a segmented abdomen (*opisthsoma*) and brandish slender chelae or *pedipalpi* behind the flat, back-walking body.

Those minute — and also advanced — arachnids, the mites or **Acarini** are teemingly numerous on the rocky shore. In cooler places inside crevices the polished brown mites, *Hydrogamasus kensleri* and *Fortuynia elamellata*, congregate. Over the open, sun-warmed rock, among *Chamaesipho columna*, can be found the larger red mite *Tangaroellus porosus*.

All these crevice Arachnida live on newly killed or

moribund animals, either sucking their juices or liquefying their tissues by external digestion.

The small, yellowish brown beach centipede *Tuoba xylophaga*, 2–3 cm long, ventures out of the crevice's narrowest reaches to feed — among other foods — on dead barnacle remains.

The insects of marine crevices are generally pickers-up of decaying fragments. During high tides these can retain air on the body in a tomentum of unwettable hairs or in a silvery sheet, the *plastron*, under the abdomen. The commonest crevice insects belong among the primitive, wingless **Apterygota**. *Nesomachilis maoricus* — 15 mm long, excluding the long caudal filaments — resembles a silverfish or bristle-tail. The most gregarious is the blue-black springtail (Collembola), 3 mm long, with stumpy legs and short divergent antennae.

Mollusca

The smallest and by far the most numerous crevice bivalves are the pale or colourless *Lasaea maoria* and *L. hinemoa* living also in empty barnacle shells, beneath oysters or in tufts of *Stictosiphonia* or *Lichina*. A larger and rarer species from the same group Leptonacea is the glistening white *Scintilla stevensoni*, occasionally found by prising apart creviced strata. The thin, rectangular shell is invested by the finely papillose mantle. The valves normally remain open at 180 degrees, with the mantle edges fused into a false foot. From a small pedal gape, a true foot can be protruded for active locomotion.

The primitive pulmonate Family **Ellobiidae** has two minute colourless species in crevices, *Leuconopsis obsoleta* and *Microtralia occidentalis*. Almost restricted to oyster clusters on northern shores is the small, prettily marked trochid *Fossarina rimata*.

Bio-erosion

Sandstone and siltstone are intensively penetrated, first by animals that bore the soft matrix, and then by successors that nestle in the burrows opened up. Bivalves, crustaceans and polychaete worms riddle the soft rock with tunnels, weakening the matrix by lowering its density and increasing its porosity, thus opening it to salt-weathering and mechanical erosion.

A simple bio-erosion occurs on low-inclined faces, where homing limpets, *Cellana ornata* and — on the east coast — *C. flava*, together with *Siphonaria* and *Sypharochiton*, deepen their scars.

A widespread bio-eroder at high levels is the spherical pill-louse *Sphaeroma quoyanum*. A vertical face of soft siltstone between hard laminae of limonite, leached out along the joint planes (Fig. 8.15) becomes riddled by this isopod below mean high water, with the shelves of limonite deeply undercut. At the bottom of a pit, a pill-louse can be found rolled in a ball, with the saw-edged uropods projecting as spikes. The abraded soft rock is seen to be scratched with fine striae. Small branch pits contain juvenile *Sphaeroma*.

Secondary occupants of *Sphaeroma* pits may include the littorinid *Risellopsis varia*, the small flea mussel *Xenostrobus pulex*, *Sypharochiton pelliserpentis* and the pulmonate *Onchidella nigricans*. Enlarged borings may hold the anemones *Anthopleura aureoradiata* and *Isactinia olivacea*, along with the tubeworms *Pomatoceros* and *Hydroides*.

Around low water neap, the siltstone face is riddled with the borings of piddocks, bivalve molluscs of the Family **Pholadidae**. Wide enough to admit a fingertip, these run 6–8 cm nearly horizontally into the soft rock. A water jet from a siphon will locate a live piddock amid a honeycomb of old shafts. Of the three pholads common in siltstone, *Barnea similis* and *Pholadidea spathulata* begin at about low water neap and deeper; the smaller *P. tridens* lives from low springs to beyond the tides, often to be found in mudstone with the holdfasts of cast-up kelp.

Piddock boring is entirely mechanical. Throughout life in *Barnea*, or in *Pholadidea* until active boring ceases, the shell retains an open gape in front for the cylindrical foot. The forequarter of the valve has abrasive sculpture. As the foot takes grip by its disc-shaped end, the shell is rotated by alternate contractions of the pedal retractors to the left and right. A clean, circular drilling results. The shell is engaged with the rock by the divergence of the anterior ends of the valves with the contraction of the posterior adductors. The valves thus move in a side-to-side plane on a fulcrum point at the hinge. The anterior adductor has spread partly outside the shell, so its contractions serve to divaricate rather than close the valves. In *Barnea*, this externalised muscle is protected by an accessory shell plate. During shell movements, the mantle cavity is kept closed around the pedal gape to maintain a pressure-head of water through which the posterior adductor can transmit a thrust to the front of the shell. In *Pholadidea* boring happens only in the sub-mature stage. The adult has closed the gape by a thin extension of the shell.

Where boring together, the two piddocks can be told apart by the tip of the extended siphon, rimmed in *Pholadidea spathulata* but in *Barnea similis* round-tipped. Also, in the *Pholadidea* the siphon base is invested with a tube, horny in *P. spathulata*, but calcified and with a three-pronged fork in *P. tridens*.

The siphon base or fused mantled of *Barnea* should be searched for the tiny leptonacean bivalve, *Arthritica crassiformis*. With free access to water currents, these are protected by the shell margin from crushing as the pholad bores.

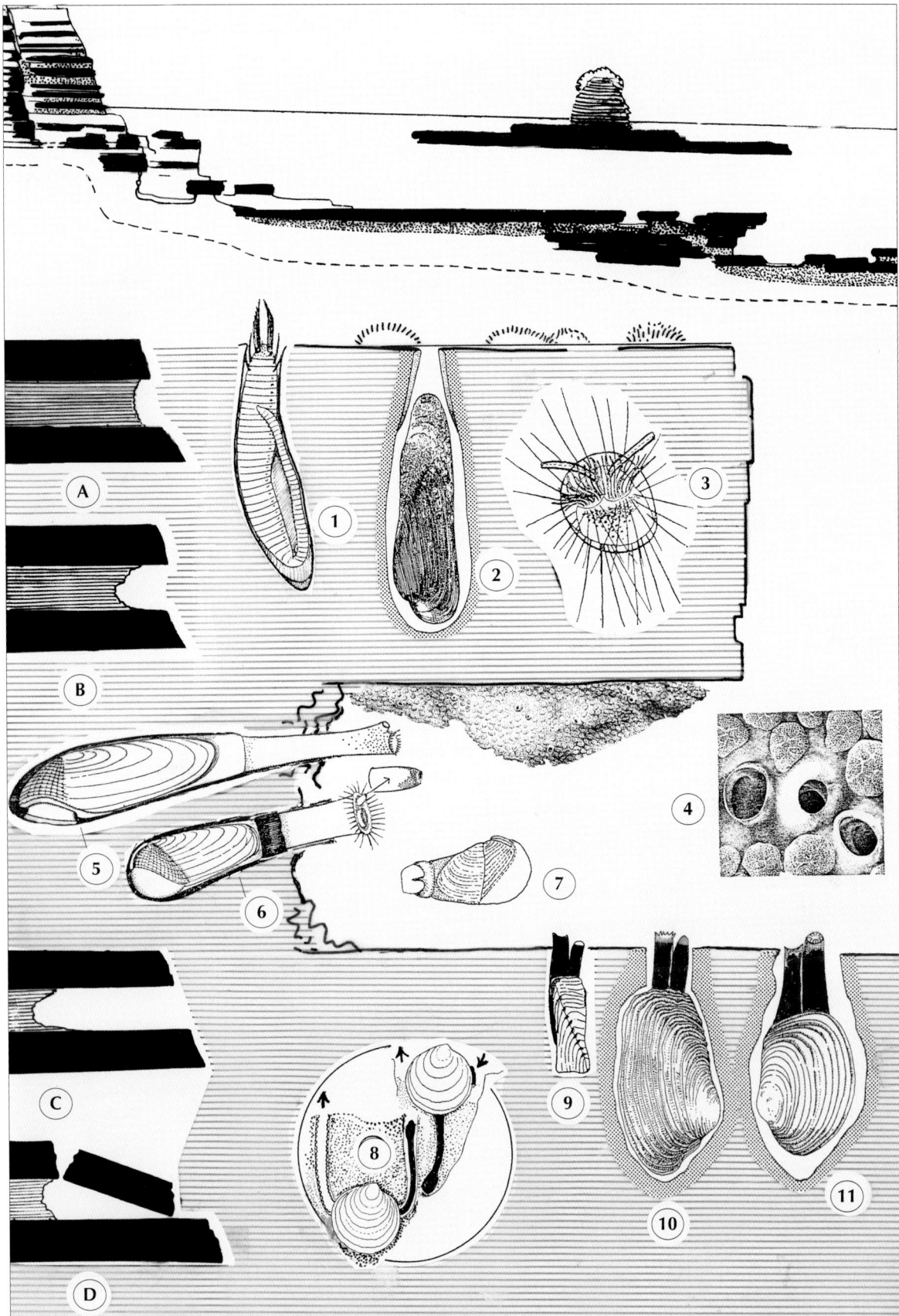

Fig. 8.17 Bioerosion of Waitemata sandstone/siltstone at Auckland

Vertically penetrant organisms are shown in more resistant sandstone, with piddocks boring horizontally into alternating layers of siltstone. A to D shows the break-off of sandstone slabs so loosened up.

1 *Pherusa parmatus* 2 *Zelithophaga truncata* 3 head detail of *Pherusa* 4 *Cliona celata* on roof of crevice and with pustule detail (right) 5 *Barnea similis* 6 *Pholadidea spathulata* 7 *Pholadidea tridens* 8 *Diplodonta striatula* 9 *Hiatella arctica* 10 *Irus elegans* 11 *Irus reflexus*.

The cut-back of alternating sandstone and siltstone is represented in Fig. 8.17. The harder sandstone is attacked by chemical erosion, and is also undercut as pholads riddle and loosen the siltstone. In the space opened, the yellow sponge *Cliona celata* erodes the sandstone from below. Pseudopodia from its mesenchyme dissolve calcite, passing fragments into their tissues to be ejected in the sponge's exhalant current.

Meanwhile other species have been boring sandstone from above. One is the date mussel, *Zelithophaga truncata*, narrow and cylindrical with its heavy brown epidermis. Byssus threads give it a breakable attachment to the burrow wall. The boring is heart-shaped in section, showing the shell does not rotate. Erosion is the work of acid phosphatase secreted from the mantle margin, possibly. Loose fragments are passed to the opening of the burrow, by ciliary tracts in the pallial cavity, to be cemented by secretion into a papier-mâché-like funnel over the posterior end of the shell (Fig. 8.17).

Fitting snugly into vertical borings in sandstone are polychaete worms *Pherusa parmatus*, detected by their emerging head bristles. Spread in a fan, these filter particles from the current brought in by the gill tentacles. Two grooved prostomial palps pick up food from the fan and carry it to the mouth. *Pherusa* is inert when taken out of the burrow, and has the abdomen turned forward to bring the anus to the front. The burrow is hard-lined and evidently bored by chemical means. Belonging to the Family **Chlorhaemidae**, this worm has the green blood pigment chlorocruorin. It also produces a fine green luminescence, from a patch under the thorax and a chevron below the mouth.

A much smaller penetrant worm is *Boccardia* sp., making innumerable hairpin-shaped borings, by abrasion — it is believed — from modified setae of the fifth segment, possibly helped by acidic secretion.

Several bivalves nestle in the holes already established in sandstone. The small venerids *Irus reflexus* and *I. elegans* use their strong sculpture for additional abrasion. In-filled with silt, cavities can be occupied by the thin-shelled bivalve *Diplodonta striatula*, one of the **Lucinacea**. This bivalve lacks normal siphons and uses the vermiform foot to fashion inhalant and exhalant shafts. Bivalves sitting in silted cavities include also the small semelid *Leptomya retiaria* and the fragile leptonacean *Kellia cycladiformis*.

In sedimentary rocks the small, oft misshapen bivalve *Hiatella arctica* has a limited abrading ability, with a ridge of spines running from the umbones. The fused siphons are orange. *Hiatella* lives also in wood, holdfasts and hard clay, where the young shell attaches by a byssus and proceeds to widen its cavity.

Extensive bio-erosion near low tide is performed by kina, *Evechinus chloroticus*. With their spines and strong chisel-teeth of Aristotle's lantern, they abrade hemispherical pits in underhangs, and pools, coalescing into the maze of channels by which the rock surface is progressively broken down.

CHAPTER NINE
Coastal Bedrock

by Bruce Hayward and John Morton

More than with any other biological realm, the habitat-space of the shore is grounded upon the original bedrock. Our study must thus involve some basic geology and coastal geomorphology.

The coast of New Zealand today is extremely young in geological terms — less than 7000 years old. Its outline and diversity of forms has been in a state of constant change for millions of years and is still changing even today.

As we will see in the pages that follow, the shape and character of today's shorelines have been determined by the interplay between a number of significant factors. New Zealand's location on a tectonically active plate boundary has resulted in numerous volcanic eruptions, and in uplift or subsidence of various regions, with consequent major impact on our shorelines. Throughout our geological history, sea level has never stayed still for very long and this has been accentuated during the regular climate cycles of the Ice Ages of the last 2.5 million years. The repetitive rise and fall of sea level, often in excess of 100 m over a 40,000 to 100,000 year cycle, has been the most significant factor in shaping our modern coasts.

On a more local scale, the physical properties of the coastal rocks themselves have played a major role in determining where there are rocky reefs, points, cliffs or caves and where there are bays, beaches or boulder spits. The character of individual sections of rocky shore or beach has been strongly influenced in just the last 7000 years (since sea level has been at its present height), by local variations in the processes of coastal erosion, accretion and weathering interacting with the different kinds of rock that are present.

The physical properties of the rocks strongly influence erosion and weathering patterns in the intertidal and shallow subtidal zones and determine the presence and character of micro-habitats that may allow for the successful colonisation by various species. In addition, some organisms (such as limpets and barnacles) require harder rock substrates for colonisation, whereas others (such as rock-boring bivalves and isopods) require relatively soft rock substrates.

The separation of New Zealand from Gondwana-land and our subsequent isolation from the rest of the world by large oceanic barriers played a major role in determining the composition of New Zealand's native terrestrial biota. To a lesser extent, this oceanic isolation has also influenced the composition of New Zealand's coastal marine biota, which is a mix of endemic, cosmopolitan and temperate or cool subtropical Southwest Pacific elements.

A few endemic species (such as ostrich-foot gastropod, *Struthiolaria*) are derived from Gondwanaland ancestral stock, but most endemic species have ancestors that successfully dispersed across wide oceanic barriers (as pelagic propagules or juveniles in currents or attached to flotsam) from the west or north-west since the formation of the Tasman Sea, c. 50 million years ago.

Taxa with more widespread distribution patterns have also been able to disperse successfully across the oceanic barriers but have been either slow to evolve or have arrived relatively recently (last million or so years).

Much of the warm subtropical component of our marine biota off the north-east of the North Island has arrived and established since the end of the Last Ice Age (last 10,000 years). An increasing number of northern hemisphere temperate taxa, previously with no natural means of dispersal through the tropics, are new migrants that have been introduced to New Zealand in the last 250 years as boat fouling or in ballast water.

Nature and origin of the rocks that form our coastline

The rocks and their contents tell us that the Earth was formed about 4500 million years ago, that New Zealand's history began at least 540 million years ago, and that a complicated series of events since then has combined to produce the country we know today. These events not only produced the rocks, landforms and shape of New Zealand, but also played an important role in the development of its unique plant and animal life on land and around our coasts.

Following its origin, the molten Earth has cooled on the outside to form a solid crust of rock. Much of this

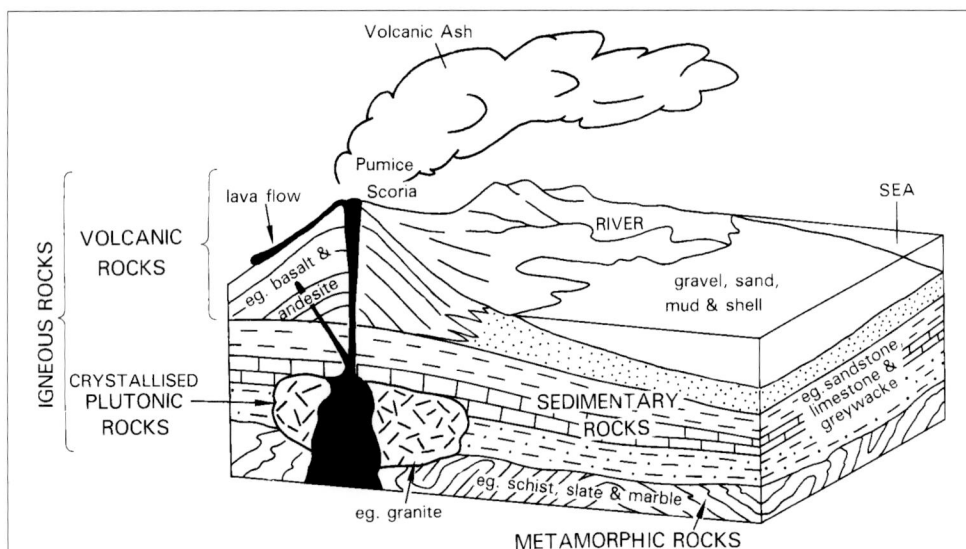

Fig. 9.1 Rock types
How the major groups of rocks are formed at and beneath the Earth's surface.

crust has been eroded by natural forces, the resulting fragments have been carried by rivers and redeposited, mainly in the sea as sedimentary rocks. Some of the rocks have remelted at depth, then solidified again as they intruded upwards to form plutonic rocks, or erupted at the surface as volcanoes. Many of the deeply buried crustal rocks have been greatly changed by the intense pressures and heat down there, forming metamorphic rocks.

The oldest, or *basement rocks*, that form the deep backbone of New Zealand, consist of hardened sedimentary rocks (greywacke and argillite), ancient plutonic rocks (granite and diorite), and metamorphic rocks (schist, slate and marble). On their eroded surfaces, younger rocks were progressively laid down, producing softer, sedimentary mudstone, sandstone, limestone and conglomerate, mixed with lava and ash from periodic volcanic eruptions. All these later rocks comprise the 'overmass', contributing by far the greater part of our contemporary coastline.

Our jigsaw of basement rocks (540–100 million years ago)

Large thicknesses of basement rock (mostly greywacke, schist and granite) underlie almost all of the New Zealand mini-continent. They can be seen at the surface where they have been pushed up high by fairly recent (last 5–10 million years) tectonic forces along the modern plate boundary. These uplifted rocks form the axial greywacke and schist ranges of the North and South Islands, as well as the deeper backbones of Northland and Coromandel Peninsulas, and uplifted hills in Waikato, Nelson, Westland, Otago and Southland.

New Zealand's basement rocks occur in 10 or more huge elongate slabs or slivers (terranes) of crust, separated from one another by major faults, or zones of bro-

ken rock (Fig. 9.2), rather like pieces in a giant jigsaw puzzle. The rocks in each separate terrane accumulated or erupted together in one region, and subsequently underwent the same history of deformation, intrusion, and possibly metamorphism. Each terrane is believed to have formed in a different place, sometimes hundreds to thousands of kilometres apart, mostly along the eastern margin of the supercontinent of Gondwanaland.

Throughout most of the Paleozoic and Mesozoic epochs (540–100 million years ago), the eastern margin of Gondwanaland was a convergent (collision) plate boundary between the Gondwanaland and Phoenix (ancient Pacific) plates. The various terranes were brought together by conveyor-belt like movement of the plates along the plate boundary. Westward subduction consumed the largely oceanic crust of the Phoenix Plate beneath the eastern edge of the lighter, continental crust of the Gondwanaland Plate. Some of the oblique compression between the two plates was translated into enormous sideways displacements of the thicker, crustal rocks (terranes) along large faults parallel to the plate boundary (similar to the modern Alpine Fault). Over hundreds of millions of years, these plate tectonic processes transported the various New Zealand basement rock terranes, like pieces of a jigsaw, into a region off the coast of ancestral Victoria, Tasmania and Antarctica (Fig. 9.2). Here they were crunched and welded together into the basic configuration we see them in today.

Subduction along the Gondwanaland-Phoenix plate boundary resulted in the periodic generation of magma, which rose to the surface and erupted as volcanic arcs. Magma that did not reach the surface slowly cooled at depth in the magma chambers, forming large plutons of coarsely crystalline granite or diorite.

The collision forces between the plates periodically pushed up the margin of Gondwanaland as coastal

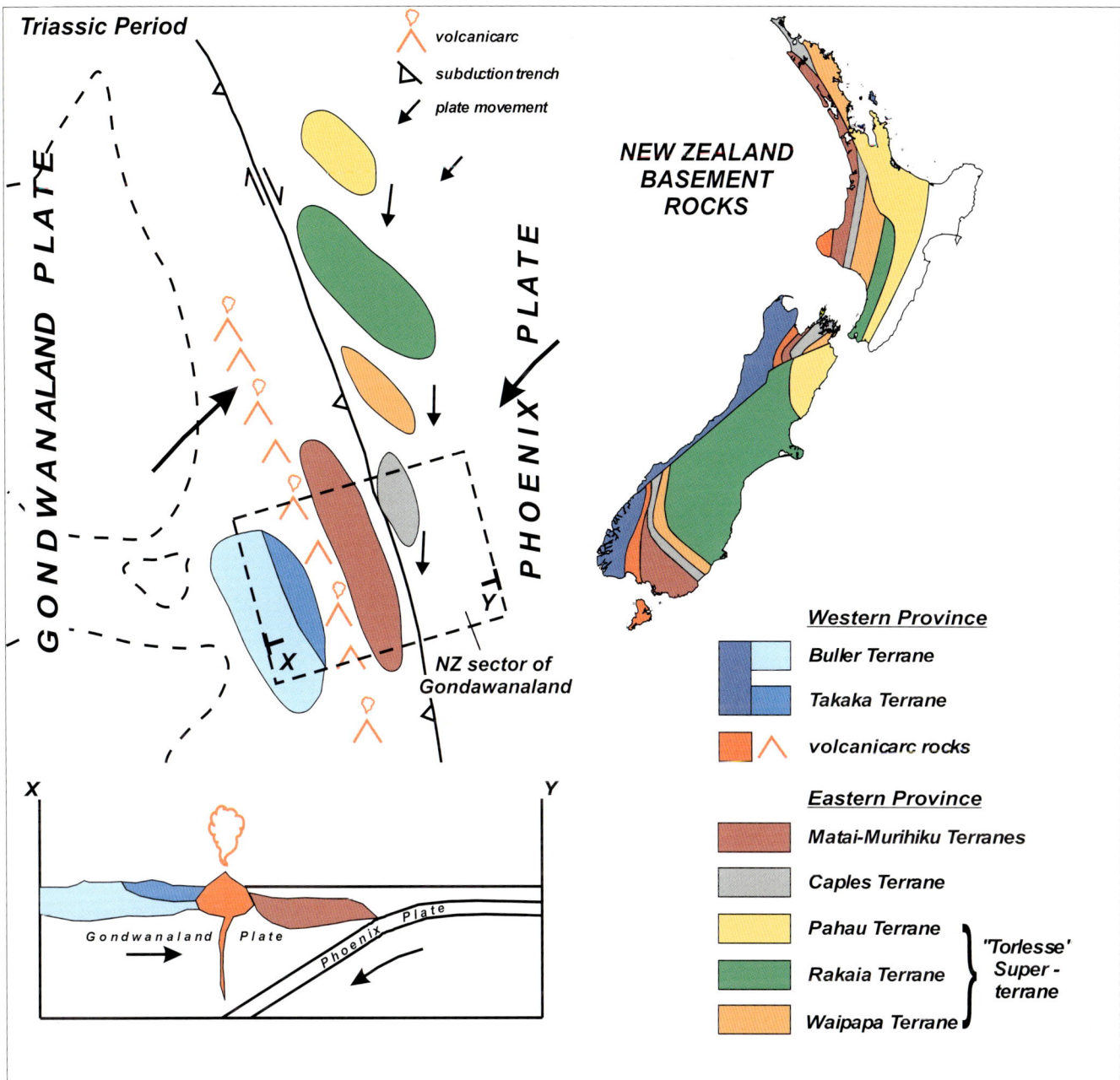

Fig. 9.2 New Zealand's basement rocks come together

(left) Schematic map and cross-section of the eastern margin of the supercontinent of Gondwanaland, 200 million years ago (Triassic Period). New Zealand's basement rocks are a group of giant crustal slabs (terranes), comprising sedimentary, plutonic, and volcanic rocks that accumulated at various places along the margin of Gondwanaland during most of the Paleozoic and Mesozoic (Cambrian to early Cretaceous, 540–100 million years ago). These terranes were gradually brought together, deformed, and sometimes metamorphosed, by plate tectonic movements along the collision boundary between the Gondwanaland and Phoenix Plates. The sector which later broke away from Gondwanaland to form the New Zealand mini-continent is outlined as a dashed rectangle.

(right) Present-day distribution of the different blocks (terranes) of basement rock that now form New Zealand. In many places they are buried beneath accumulations of younger (less than 100 million years old) volcanic and sedimentary rocks (not shown).

mountain ranges. Subduction along the boundary pulled down the seafloor offshore, forming a deep trench out east of the coastline. Erosion of Gondwanaland, particularly its uplifted mountains, and the plate boundary volcanoes, produced huge volumes of gravel, sand and mud that were carried into the sea and deposited in enormous thicknesses offshore along the coast, particularly in the trench. These were the major processes that initially formed New Zealand's basement rocks.

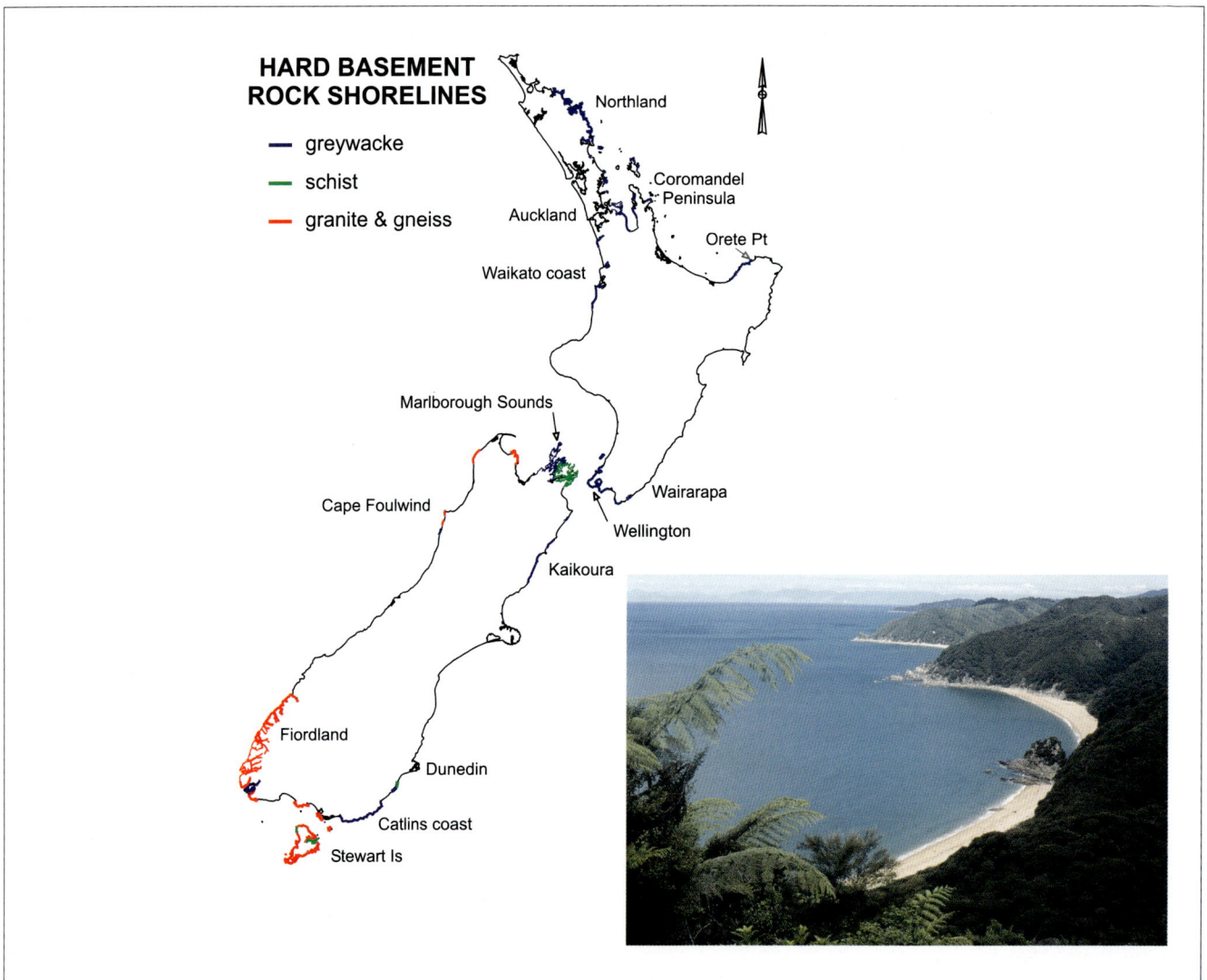

HARD BASEMENT ROCK SHORELINES

— greywacke
— schist
— granite & gneiss

Northland

Coromandel Peninsula

Auckland

Orete Pt

Waikato coast

Marlborough Sounds

Cape Foulwind

Wairarapa

Wellington

Kaikoura

Fiordland

Dunedin

Catlins coast

Stewart Is

Fig. 9.3 Basement rock shorelines
The distribution around the New Zealand coast of basement greywacke, schist, granite and gneiss that form most of our hardest rocky shorelines.
Photo: Rocky granite coastline, Abel Tasman National Park, north-west Nelson.

Western Province

New Zealand's basement rocks are divided into the older Western Province terranes separated, by several crunched and dismembered volcanic arc terranes, from the younger Eastern Province terranes (Fig. 9.2). The largely Paleozoic (540–300 million years old) Western Province consists of quartzose greywacke and slate that had been deposited as sand and mud in the ocean depths (Buller Terrane), and greywacke, marble and volcanic rocks that had accumulated in a shallow coastal setting (Takaka Terrane). Both of these terranes are intruded by large granite and diorite plutons.

These terranes were brought together and added to the eastern margin of Gondwanaland in the late Silurian, about 370 million years ago (formerly called the Tuhua Orogeny). Metamorphosed plutonic rocks (gneiss) of Fiordland are also part of the Western Province and form much of the coastline of our

southern fiords. Rocky granite coast-lines of the Western Province occur in Stewart Island, northern Westland and north-west Nelson (Abel Tasman National Park). Coastal erosion of the quartz-rich granite produces the renowned golden sands of Nelson's Golden Bay beaches (Fig. 9.3).

The oldest rocks and fossils in New Zealand are mid-Cambrian (c. 540 million years old) and outcrop in the Cobb Valley, north-west Nelson, as part of the Takaka Terrane. They cannot be seen down on the coast. Buller Terrane greywacke forms coastal rocks in several parts of Westland.

The volcanic arc rocks that separate the Western and Eastern Provinces are a mix of plutonic rock (that solidified beneath the volcanoes), lava, ash and sediments (largely derived by erosion of the volcanoes). These form our coastal rocks in parts of east Nelson, central Southland and northern Stewart Island.

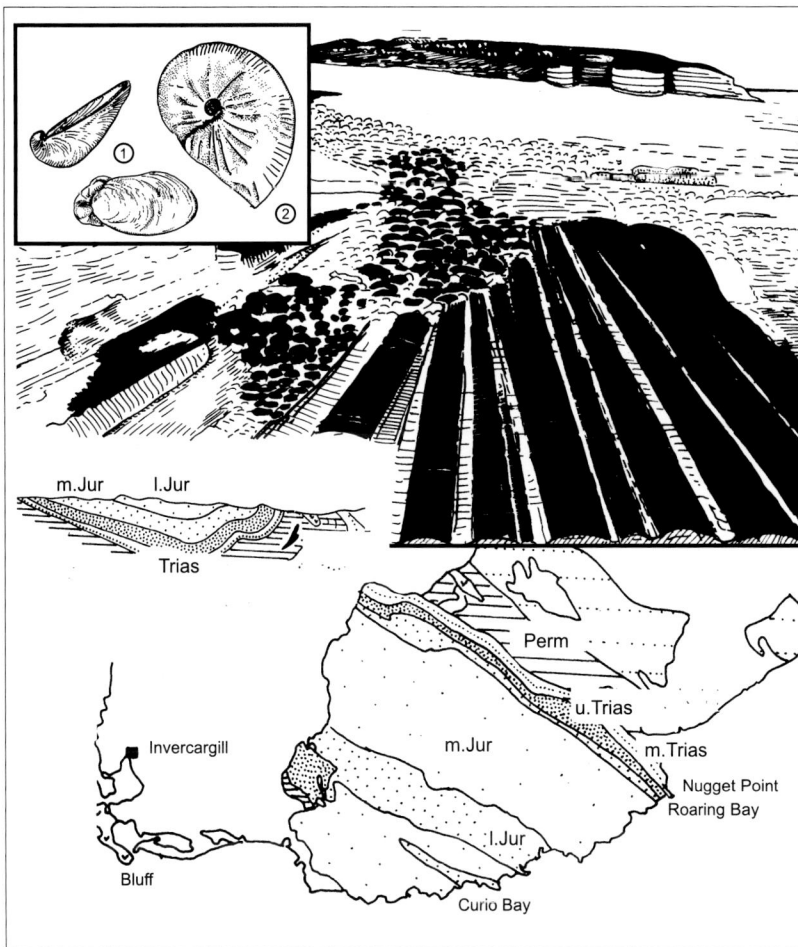

Fig. 9.4 Jacks Bay, Catlins Coast, east Southland

(top) Looking north towards Hayward Point, across a shore platform cut in tilted, but otherwise little deformed, beds of indurated Jurassic sandstone and mudstone (greywacke and argillite) of the Murihiku Terrane. These strata contain a wide variety of fossil molluscs (**inset: 1** bivalve *Buchia* **2** ammonite, *Uhligites*).

(bottom) Geological map and cross-section of South Otago and Southland. Rock ages: (Perm) Permian (m.Trias) middle Triassic (u.Trias) upper Triassic (l. Jur) lower Jurassic (m. Jur) middle Jurassic.

Eastern Province

The Permian to early Cretaceous (300–100 million-year-old) rocks of the Eastern Province occur in six to eight elongate terranes that were assembled together during the Jurassic, and particularly in the early Cretaceous, about 140–100 million years ago (formerly called the Rangitata Orogeny). These terranes can be lumped into two groups. One group is dominated by well-bedded, often fossil-bearing, greywacke and argillite, that accumulated as sand and mud in relatively shallow seas and coastal areas close to the volcanic arcs (such as the Murihiku Terrane). This group of terranes contains rocks that are little deformed and probably were not transported far by plate boundary forces. They form the picturesque rocky coast of the Catlins in south Otago (Fig. 9.4) and some of the harder rocky headlands down the North Island's west coast from Port Waikato to Kiritehere.

The second group of Eastern Province terranes (referred to by many as 'Torlesse') is characterised by vast thicknesses of often highly deformed, mostly non-fossiliferous, greywacke and argillite derived by erosion from non-volcanic mountains made of granite and older sedimentary and metamorphic rocks. These are believed to have accumulated in deeper water, often in the submarine trench or on the deep ocean floor, off the coast of Gondwanaland far to the north, perhaps off the sector that later was to become Queensland. Rocks of these 'Torlesse' terranes are intensely folded and faulted and cut by numerous small, white quartz and zeolite veinlets. Typical Torlesse greywacke coasts occur along the uplifted, eastern side of much of Northland and Auckland (Fig. 9.6), around parts of Coromandel Peninsula and the Marlborough Sounds, and more rugged sections of the Kaikoura coast. The coast of the entire Welling-

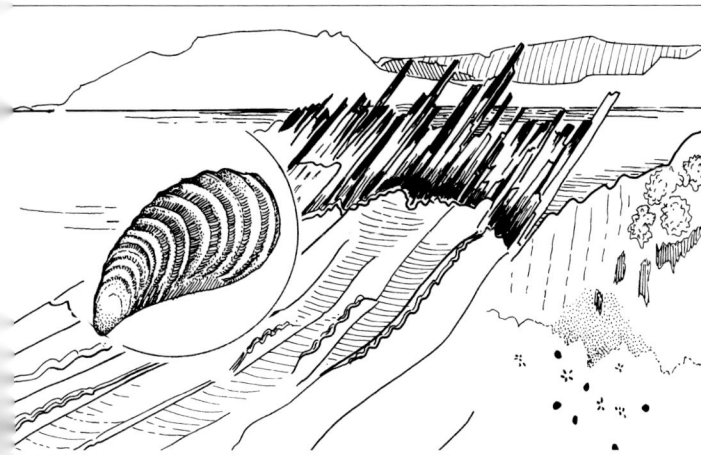

Fig. 9.5 Orete Point, eastern Bay of Plenty

Tilted, but otherwise undeformed, Cretaceous greywacke sandstone beds form a high tide platform on the west coast of Raukumara Peninsula. (**inset**) fossil bivalve, *Inoceramus*, which occurs in these strata.

Fig. 9.6 Eastern Province greywacke, Leigh, Auckland

The shore platform at Leigh Harbour, east Auckland, eroded out of fractured and tilted 'Torlesse' greywacke. Around low tide are the golden beads of *Hormosira banksii*, while clinging to the hard substrate are the mid-tidal limpets *Cellana radians* and *C. ornata*, and barnacles *Chamaesipho brunnea* and *C. columna*. At higher tide levels is the black snail *Nerita atramentosa* and black tufted lichen, *Lichina confinis*. This lichen only lives on hard rock substrates like greywacke and igneous rock.

(Painting by Ron Cometti.)

ton Peninsula from Paekakariki to the Wairarapa, including Wellington Harbour, is also formed of the hard and intensely fractured Torlesse greywacke.

Greywacke is a general term for hardened sandstone. These rocks are rich in quartz, feldspar and the ferro-magnesian minerals chlorite, hornblende and biotite. The fracturing in the rocks allows water and oxygen to enter and break down the feldspar (aluminium silicate) to clay (kaolin) that, with the quartz grains, eventually washes away. The residue of ferric minerals leaves laminae of rust-coloured limonite (a hydrated iron oxide) along the former joint planes.

In some places the vast thickness of sediment that accumulated within the Torlesse terranes resulted in the deep burial of the earlier deposited sandstone and mudstone. With the increased temperature and pressure at depth these rocks were metamorphosed to schist. In recent times some of these have been pushed up from great depths and with erosion are now exposed on land in Central Otago and form parts of the Southern Alps. Schist forms the coastal rocks in the eastern Marlborough Sounds (Fig. 9.7) and south of Dunedin.

New Zealand adrift and sinking (100–25 million years ago)

The great southern supercontinent of Gondwanaland (comprising South America, Africa, India, Australia, Antarctica and New Zealand sectors) began to break apart about 120 million years ago. The plate collision zone of mountain uplift, subduction and volcanoes, that had existed along its eastern margin for at least 400 million years, became inactive. In its place a new exten-

Fig. 9.7 Schist shoreline, Queen Charlotte Sound

In New Zealand the main area of schist coastline is in the eastern Marlborough Sounds, where the tilted schistosity planes strongly influence the angle of low cliffs and shore platforms. Schist is sandstone that has been metamorphosed by prolonged pressure and temperature of deep burial. The metamorphism produces new cleavages with flat crystals of silvery mica (muscovite), black mica (biotite) and green chlorite, aligned along the new planes.

(Painting by Ron Cometti.)

Fig. 9.8 New Zealand set adrift
(top left) The ancient basement rocks of New Zealand were formed along the eastern margin of the southern supercontinent of Gondwanaland between 540 and 100 million years ago.
(top right) Between 85 and 55 million years ago the New Zealand sector split away from Gondwanaland with the formation of the Tasman Sea and Southern Ocean. The New Zealand mini-continent had begun its independent existence 'adrift' in the Pacific Ocean.
(bottom left) Some 90% of the New Zealand mini-continent currently lies beneath the waves forming a submarine plateau surrounding our meagre islands.
(bottom right) Geological time scale and major events in New Zealand's geological history.

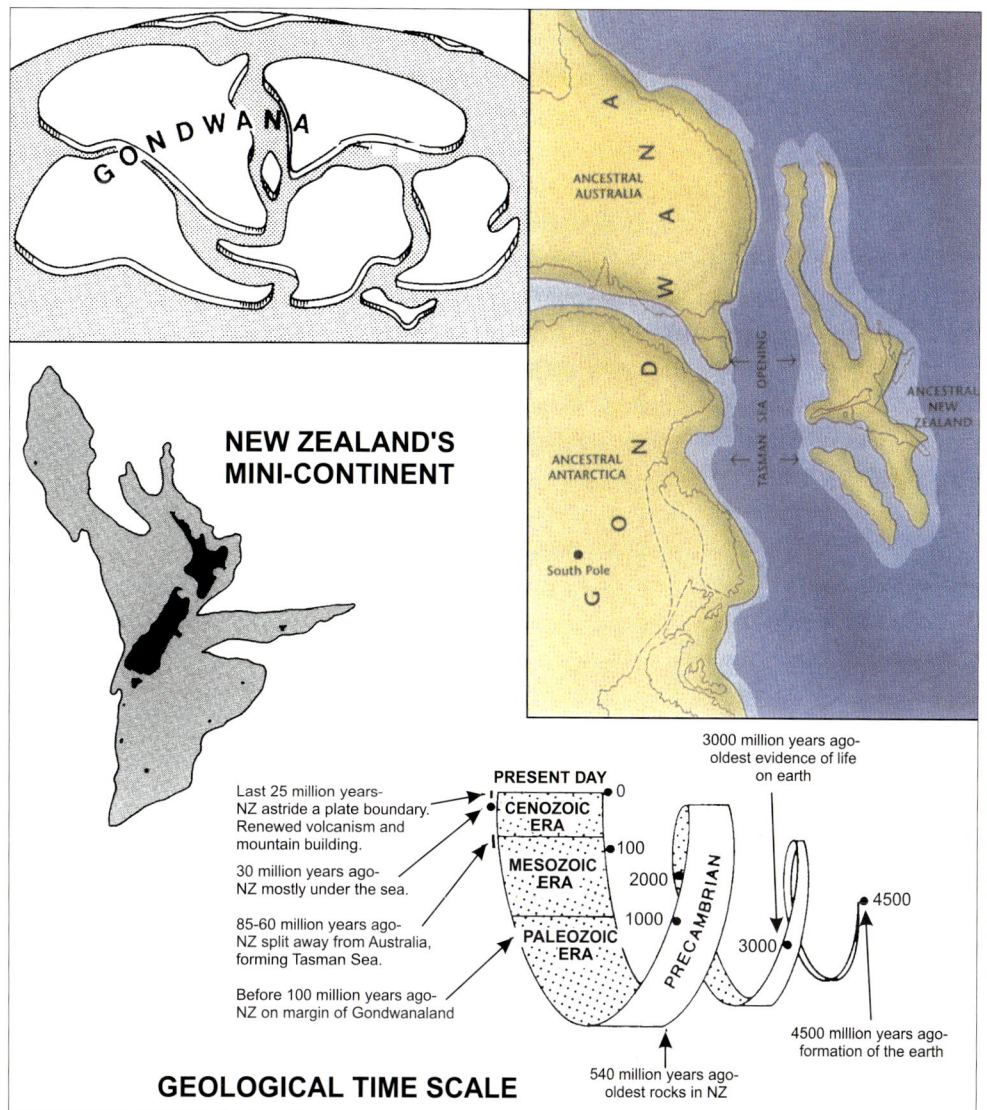

GONDWANA

ANCESTRAL AUSTRALIA

GONDWANALAND

ANCESTRAL ANTARCTICA

TASMAN SEA OPENING

ANCESTRAL NEW ZEALAND

South Pole

NEW ZEALAND'S MINI-CONTINENT

3000 million years ago-
oldest evidence of life
on earth

Last 25 million years-
NZ astride a plate boundary.
Renewed volcanism and
mountain building.

PRESENT DAY

CENOZOIC ERA — 0

30 million years ago-
NZ mostly under the sea.

MESOZOIC ERA — 100

2000

85-60 million years ago-
NZ split away from Australia,
forming Tasman Sea.

PALEOZOIC ERA — 1000

PRECAMBRIAN

4500

3000

Before 100 million years ago-
NZ on margin of Gondwanaland

4500 million years ago-
formation of the earth

540 million years ago-
oldest rocks in NZ

GEOLOGICAL TIME SCALE

sional boundary developed some 1000 km inland to the west. Plate tectonic forces deep in the Earth began opening this extensional boundary, moving the New Zealand sector north-eastwards away from the Antarctic and Australian sectors of Gondwanaland. Between 85 and 55 million years ago, the New Zealand mini-continent was set 'adrift' in the Pacific Ocean, as new seafloor was generated in the widening gap that became the Tasman Sea and Southern Ocean.

The New Zealand mini-continent, composed of basement rocks from the edge of Gondwanaland, covers about 3 million sq km (Fig. 9.8) or about 10 times our present land area. Today most of it is under water. The tensional stretching of the crust that preceded New Zealand's split from Gondwanaland resulted in the mini-continent being considerably thinner than normal continental crust. Following separation, this thinner crust slowly cooled and sank beneath the ocean. It reached its maximum submergence in the Oligocene, 30–25 million years ago. By this time just a few small islands existed and these only in places where thicker crust remained, such as beneath east Auckland and central Otago.

The last phases of collision along the eastern margin of Gondwanaland, in the early Cretaceous about 140–100 million years ago, had pushed up a chain of coastal mountains (Fig. 9.10). As the New Zealand sector was stretched from beneath and then split off, erosion was eating into these mountains at the surface, and over tens of millions of years helped wear them down to the small, low-lying islands that were left by the Oligocene.

Stretching of the crust in the mid-Cretaceous to Paleocene (100–55 million years ago), both prior to and during New Zealand's break away from Gondwanaland, resulted in the opening of large, elongate, tensional gashes (or rifts) in the surface of the mini-continent. Extensive valley systems, like the present-day Hauraki Plains, formed in these subsiding rifts. They filled with large thicknesses of river gravel, and flood plain sand and mud that was eroding from the nearby mountains (Fig. 9.9). Freshwater swamps in these valleys accumulated peat, that with deep burial became some of the coal that has been mined in the West Coast, Otago and Southland.

Elsewhere on the New Zealand mini-continent dur-

Fig. 9.9 Conglomerate coast, Wharariki Beach, north-west Nelson

Islands, arches, guts and points have been carved by marine erosion out of these fluvial conglomerate and sandstone beds. The sedimentary rocks were deposited in a rapidly subsiding rift valley as ancient New Zealand was ripped away from the edge of Gondwanaland in the late Cretaceous, about 70 million years ago.

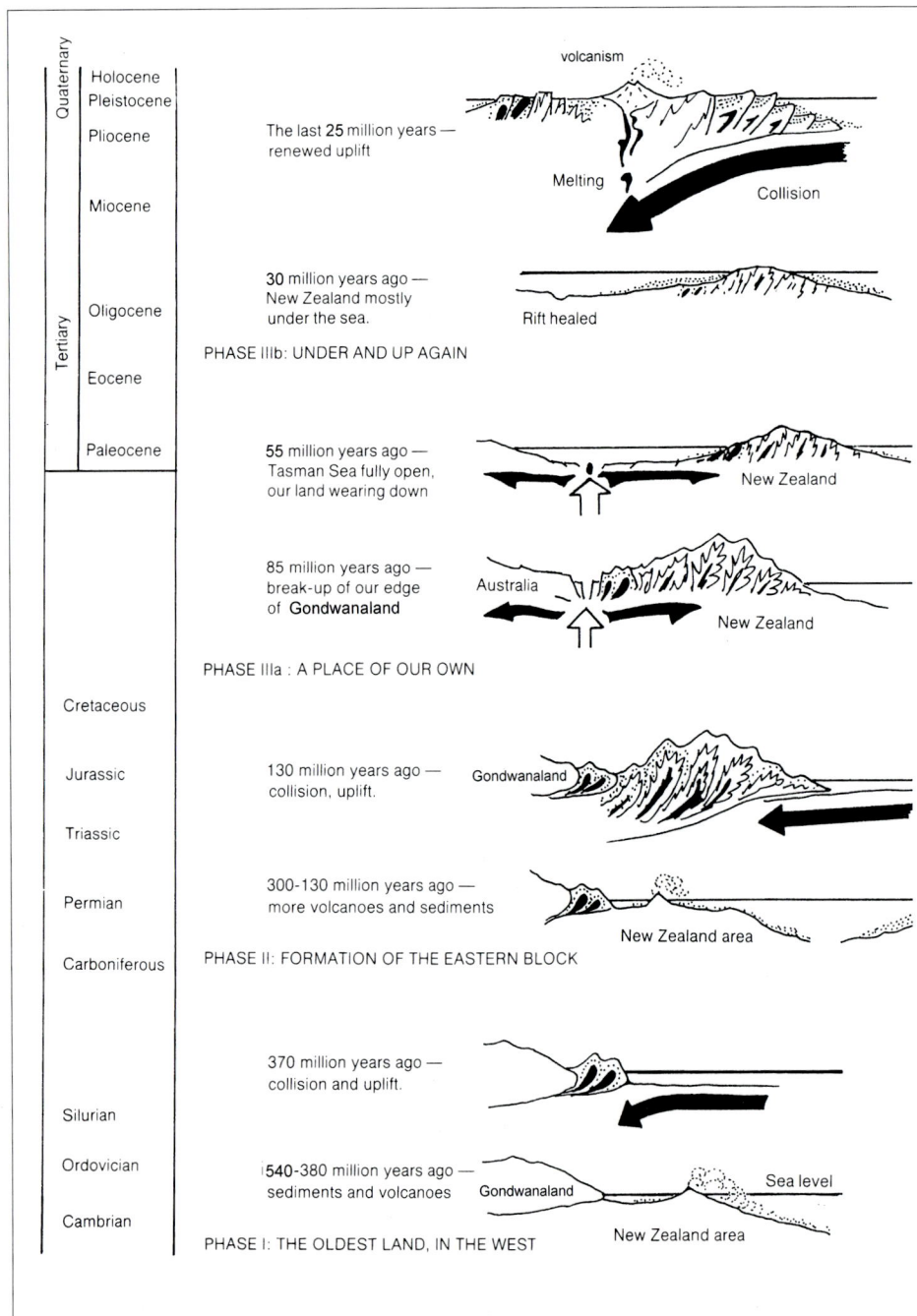

Fig. 9.10 A simplified geological history of New Zealand

Sketched west-east cross-section through the New Zealand region illustrating the major phases in its evolution over the last 600 million years.

(From *The Reed Field Guide to New Zealand Geology*, by Jocelyn Thornton, 1985.)

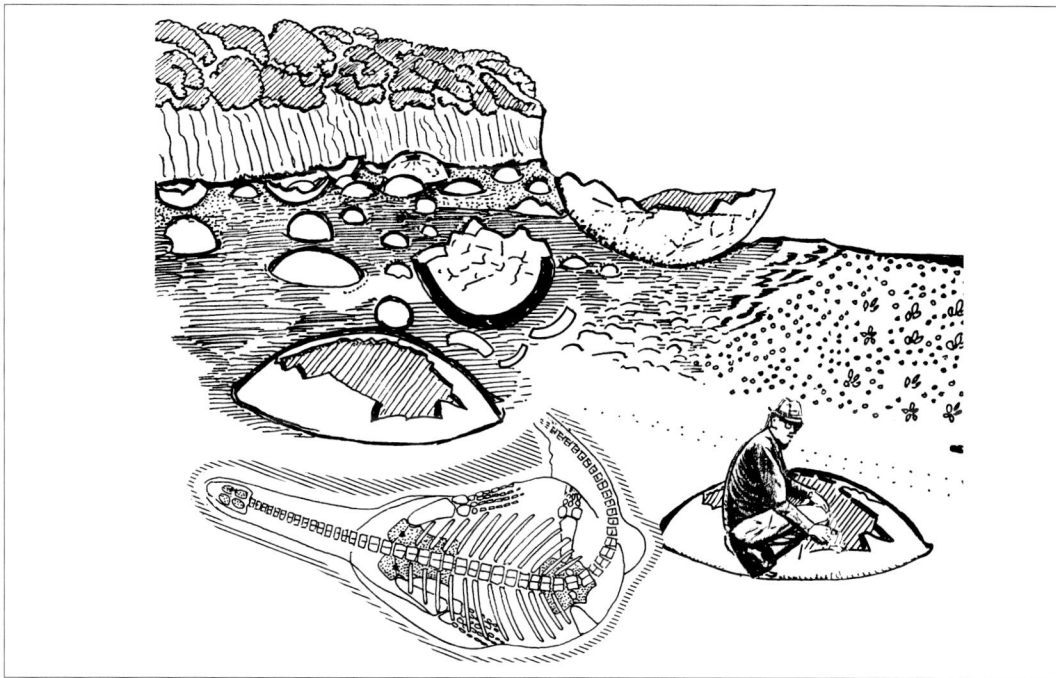

Fig. 9.11 Shag Point concretions, North Otago
The rounded tops of hard, erosion-resistant concretions stick out of the softer sandstone shore platform. The shards of several broken concretions litter the shoreline and may take many years to break up and be removed by the sea. The hard substrate of the concretions is colonised by a wide diversity of barnacles, limpets and other tidal organisms, whereas the soft sandstone is home to fewer species.

These concretions, like the slightly younger Moeraki Boulders to the north, were formed by crystal growth around a small pebble or fossil within the soft permeable rocks. Crystals of calcite and siderite were precipitated out of the carbonate-saturated ground water, cementing the surrounding grains of sand or mud together and gradually growing into larger and larger spheres. Some of these concretions at Shag Point have grown around skeletal remains of extinct marine reptiles that were buried by sand on the seafloor during the late Cretaceous (end of the Age of Dinosaurs). The near-complete skeleton of a small plesiosaur (shown) was excavated from one of these concretions in the 1980s.

ing this long period of slow submergence, sand and mud was deposited in thick layers on the seafloor that surrounded the land areas from which it was eroded (Figs. 9.11, 9.12). Further offshore on the more deeply submerged fringes of the mini-continent muddy ooze accumulated, largely composed of the calcareous skeletons of microscopic zooplankton (foraminifera) and phytoplankton (coccoliths). With burial these oozes hardened into chalk, and eventually became relatively hard, fine-grained limestone. Some of these limestones have subsequently been pushed up out of the sea and can be seen forming intertidal platforms and cliffs in widely scattered localities on the Kaikoura (Fig. 9.13), Wairarapa, and west Northland coasts.

By the Oligocene (30 million years ago), there was very little sand or mud eroding from the small areas of low-lying land that remained above sea level (Fig. 9.17). The extensive areas of shallow sea that surrounded the islands were home to numerous bottom-dwelling organisms, many of which produced hard calcareous shells or other skeletal parts. When they died, their hard parts accumulated as shell banks in the shallow water, because

there was little or no sand or mud to mix with them or bury them. Over several million years, drifts of shell, tens of metres thick, were deposited over large areas. They were later buried, and recrystallised into hard, erosion-resistant limestone beds. During recrystallisation, the calcium carbonate in the shell slowly dissolved in ground water and new calcite crystals grew in the pore spaces cementing the fragments together. Subsequently some of these limestones have been pushed up and uncovered by erosion to form much of the spectacular limestone country that we see around New Zealand today. These coarsely crystalline, shallow water limestones form inspiring rocky coasts in parts of north Otago, Westland, north-west Nelson, east Northland and west Waikato (Figs. 9.14, 9.15).

Riding a plate boundary (the last 25 million years)

A major turning point in the history of New Zealand came towards the end of the Oligocene, about 25 million years ago. The change transformed the country

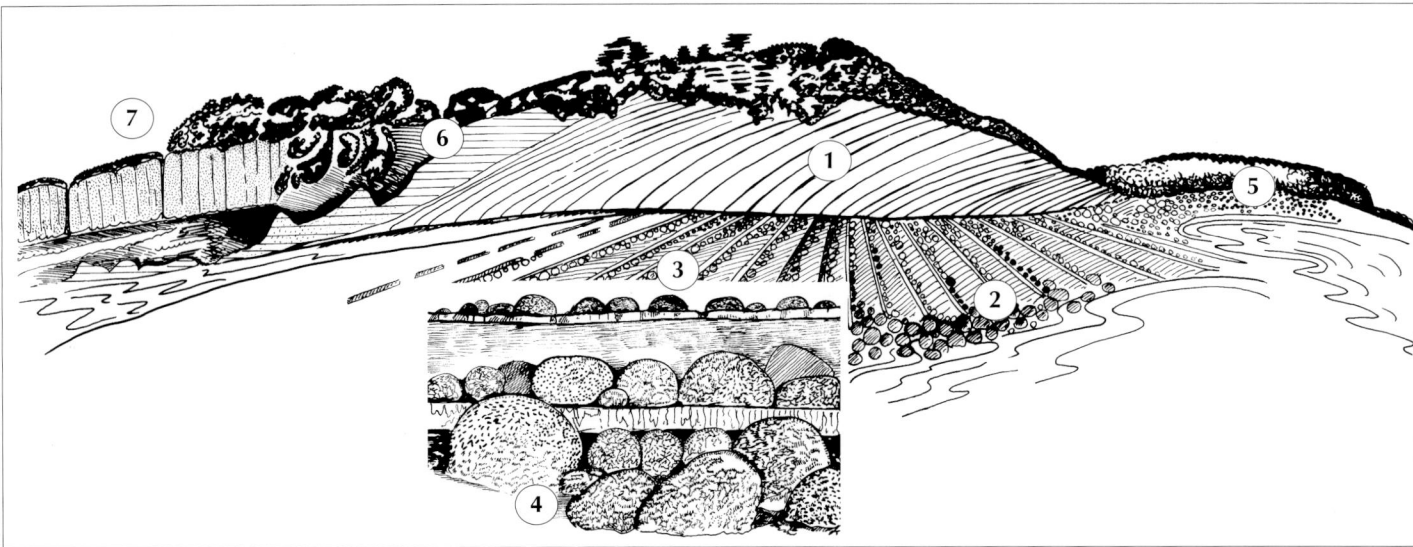

Fig. 9.12 Cape Foulwind, Westland

Many New Zealand shorelines contain a mix of rock types that combine together to provide a diversity of substrates and habitats. Here on the north side of Cape Foulwind, the cliffs and shore platform are eroded in well-bedded Eocene siltstone (1), overlain by a thin band of Oligocene limestone (6), and relatively soft Miocene mudstone (7). Boulders of older granite (4) have been broken from the cape to the south by the forces of the Tasman Sea; they have been tossed, rolled and rounded in the boulder beach (5), and thrown up along the edge of the shore platform (2). Some of the granite boulders have been trapped in the eroded grooves along the strike of the softer siltstone beds (3).

from a quiescent, mostly submerged and low-lying region to a shaky scenic land with erupting volcanoes, high mountains ranges and rapid erosion — the New Zealand we know today (Fig. 9.16).

This turning point was once again a result of changes in plate tectonic forces in this part of the globe. About 45 million years ago, Australia began to split away from Antarctica and move northwards at a rate much faster than New Zealand. This resulted in the creation of a new plate boundary through New Zealand, with the Australian Plate, including western New Zealand, moving north-east relative to the Pacific Plate, including eastern New Zealand.

Initially the effect of the new plate boundary was minor, with small extensional marine basins opening up in the south. By 25 million years ago, the motion had become more of a north-east–south-west shear resulting in the creation of the Alpine Fault. In northern New Zealand the boundary became more convergent, the edge of the Pacific Plate was subducted beneath the Australian Plate, and for 10 million years (25–15 million years ago) a wide volcanic arc erupted along the length of Northland Peninsula. Onset of convergence began buckling the crust, resulting in areas being uplifted out of the sea to form new hilly land, and other areas rapidly subsiding to become deep coastal marine basins, that filled with mud and sand that was washing off the newly created nearby hills.

For the last 25 million years, New Zealand has been

subjected to varying combinations of vertical and sideways displacements on fault-lines that criss-cross the country. These forces, together with massive volcanic eruptions in the northern half of the North Island, have largely been responsible for building the country we have today.

Although hilly areas were uplifted and eroded in many places from the early Miocene on (last 23 million years), the main displacement on the Alpine Fault and the convergent uplift and erosion of the Southern Alps has only taken place, with increasing tempo, in the last 5–7 million years. Indeed the main axial ranges of the North Island (Rimutaka, Tararua, Ruahine, Kaimanawa, Urewera, Raukumara Ranges) are also extremely young, having been pushed up out of the shallow sea in just the last 2–3 million years. Equally young has been the large subsidence of the Wanganui region (filled up with sediment) and the Hauraki Plains-Firth of Thames (partly filled with sediment).

The main mountainous or hilly spines that have been pushed up on both islands have had most of their covering of soft, younger rocks removed by erosion to expose their underlying core of hard basement greywacke or schist. Also recently pushed up out of the sea alongside the axial ranges in some places are extensive areas of soft, grey Miocene and Pliocene mudstone and sandstone (colloquially called 'papa'). Hills of these soft rocks readily slip and rapidly erode, and form the distinctive, often cliffed, coastlines of North Taranaki,

Fig. 9.13 Muddy limestone shore platform, Kaikoura Peninsula
The cliffs and wide shore platform forming the middle, and particularly the seaward end, of Kaikoura Peninsula are composed of relatively hard, sometimes intensely fractured and tightly folded, muddy limestone. This limestone is composed of billions and billions of microscopic calcareous shells of planktonic organisms, that sank to the floor of the ocean well away from land during the late Cretaceous, Paleocene, Eocene and Oligocene periods (70–23 million years ago). Common wildlife visitors are the red-billed gull and southern fur-seal. Trailing over the rocks are the bushy box thorn, *Lycium ferocissimum*, and pink-flowered horokaka or ice plant, *Disphyma australe.* (Painting by Ron Cometti.)

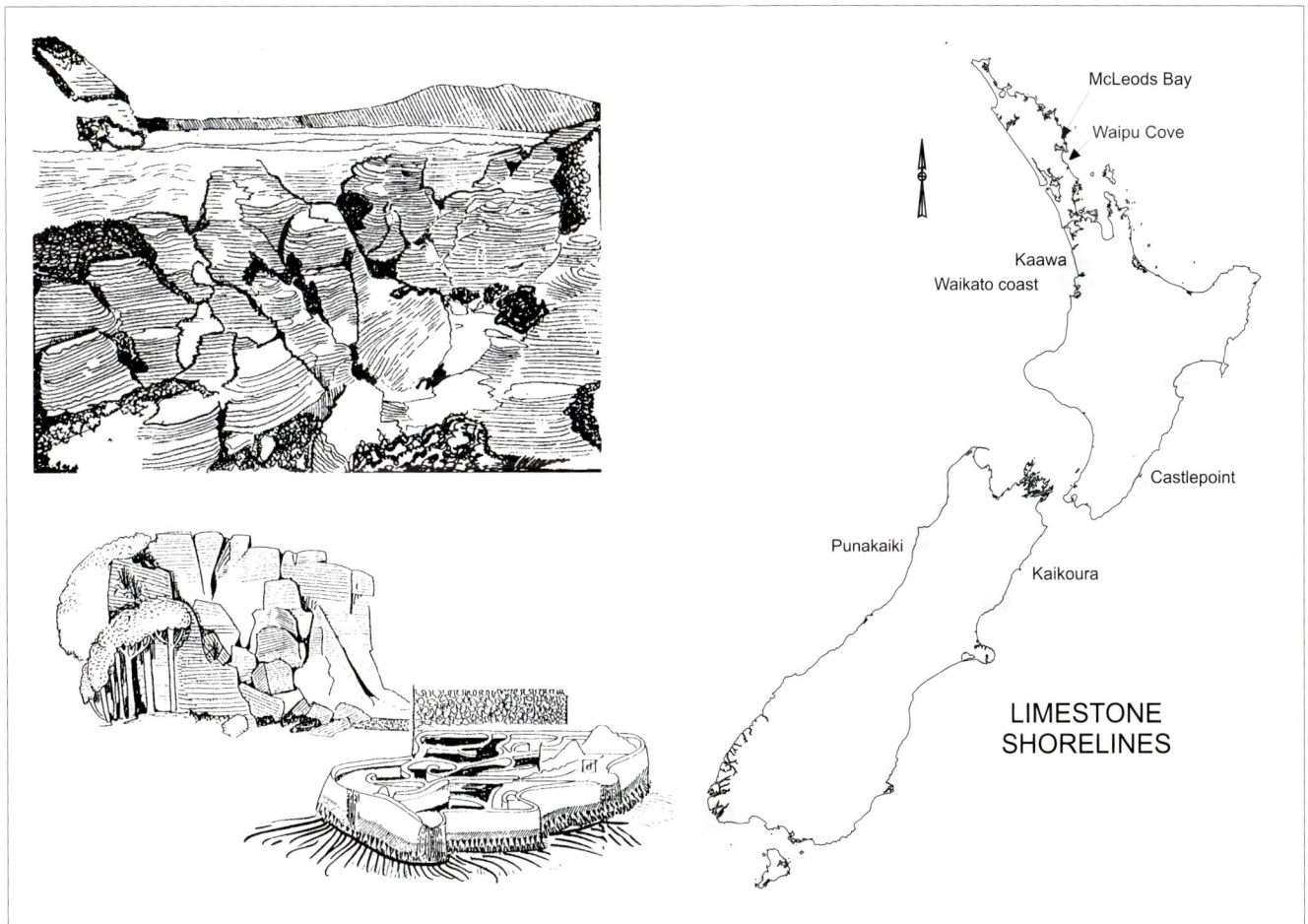

Fig. 9.14 Hard crystalline limestone coasts
Map showing the distribution around New Zealand of spectacular coastal karst shorelines, that have been eroded and dissolved out of crystalline limestone. The limestone formed from drifts of shell debris that accumulated on the floor of the shallow seas around small low-lying islands in the New Zealand region, in the Oligocene, 33–23 million years ago.
(top) Deeply incised, flaggy limestone shore at Kaawa, on the Waikato west coast.
(bottom) Flaggy limestone cliffs at Waipu Cove, Northland and deep-etched tidal-pool system.

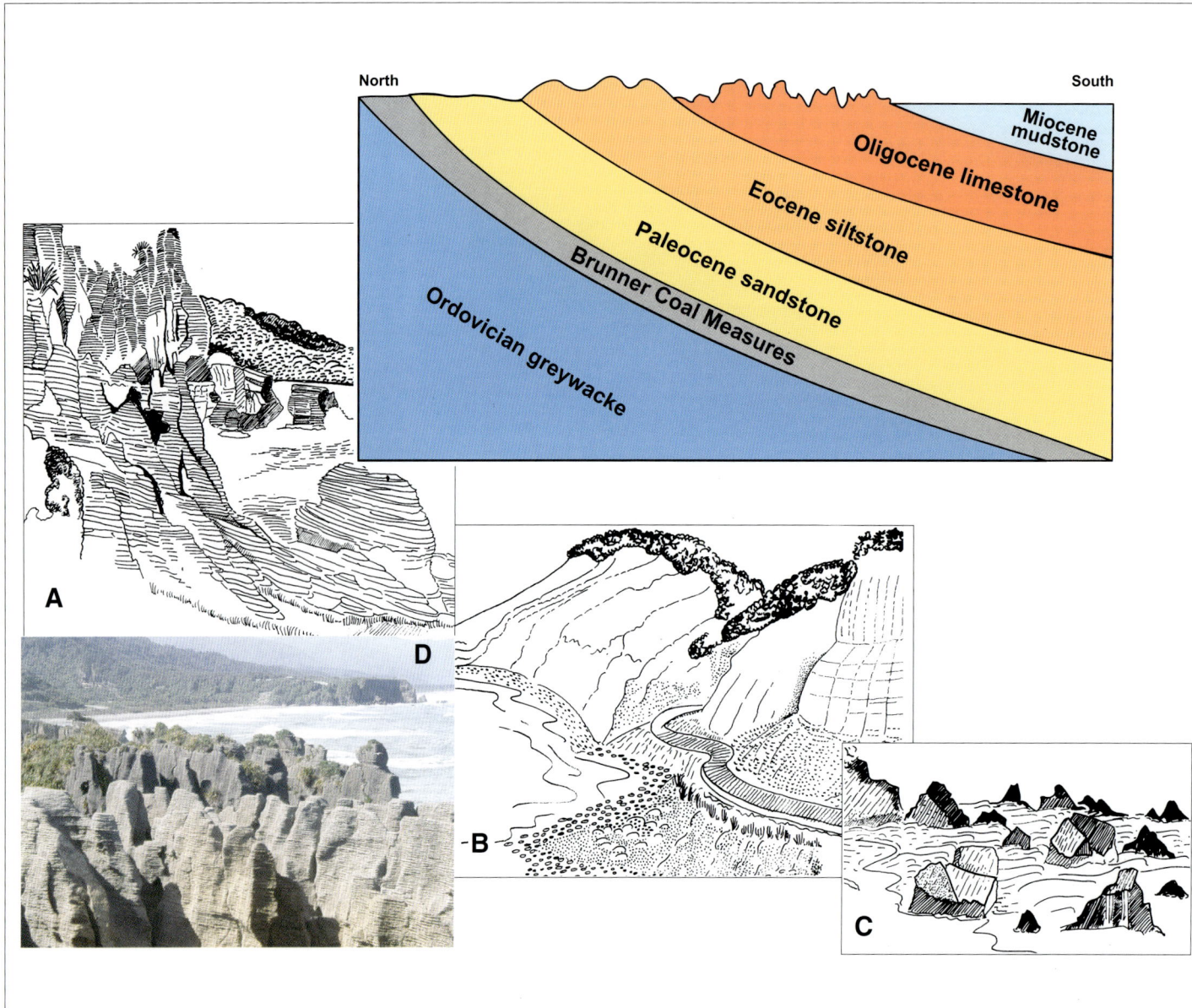

Fig. 9.15 Coastal rock sequence, Westport to Greymouth

(top) North-south cross-section along the coast between Barrytown and Greymouth, showing south-dipping sequence of Cretaceous to Miocene strata, unconformably overlying basement greywacke of the Western Province.

(bottom) Topographic sketches showing the character of the shoreline with different rock types:

A. Oligocene flaggy limestone near Punakaiki.

B. Eroding cliff-line of soft Eocene Kaiata Siltstone.

C. Offshore stacks of Ordovician basement greywacke near Fourteen Mile Bluff.

D. Pancake Rocks, Punakaiki. This coarsely crystalline limestone accumulated on the seafloor in the Oligocene as drifts of shell debris. The erosional shape of these coastal rocks is strongly influenced by their high lime content, their flagginess and the orientation and spacing of prominent sets of near-vertical cracks (or joints). The flagginess (or pancake layering) results from preferential weathering out of thin mudstone seams from between the more resistant limestone layers. The slightly irregular mudstone seams were formed within the rock during burial by a pressure-induced solution process. The blowholes, caverns and pinnacles were formed by the gradual dissolving of the limestone rock over the last few tens to hundreds of thousands of years. Most of this solution occurred when sea level was lower and the area covered in forest. Rain water, percolating through leaf litter, becomes slightly acidic and this dissolves the rock as it seeps down along vertical joints and through more porous limestone layers. Over a long period an underground system of caves, sometimes with collapsed roofs and sinkholes, was formed. In the time since sea level rose to its present height (7000 years ago), after the end of the Last Ice Age, the crashing waves of the Tasman Sea have carved into this rocky ridge and exposed the caves, which now act as blowholes.

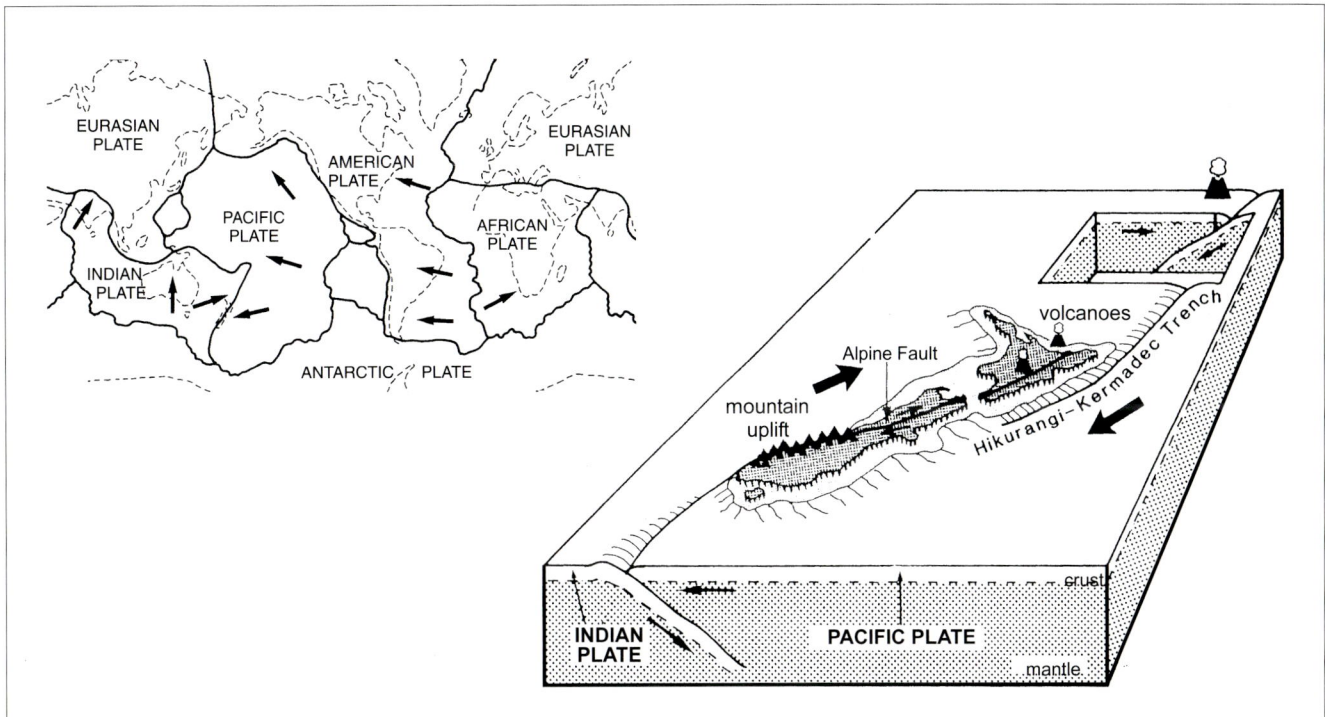

Fig. 9.16 New Zealand's modern plate boundary

(inset) The Earth's crust is composed of a number of large rigid plates moving in different directions.

New Zealand owes much of its present shape and character to the constructional processes of mountain uplift, faulting and volcanism that result from its position astride a collision boundary between the Australian and Pacific plates. For the past 25 million years tremendous forces have been concentrated within this broad collision zone. They have pushed up the high mountain chains, split the country in two and slid the eastern part hundreds of kilometres south (along the Alpine Fault). In the north, the leading edge of the Pacific Plate is being pushed beneath the Australian Plate. As it descends, molten magma forms at depth and rises to the surface to erupt as volcanoes in the North Island and the Kermadec Ridge. South of New Zealand, the reverse is happening with the leading edge of the Australian Plate being subducted beneath the Pacific Plate.

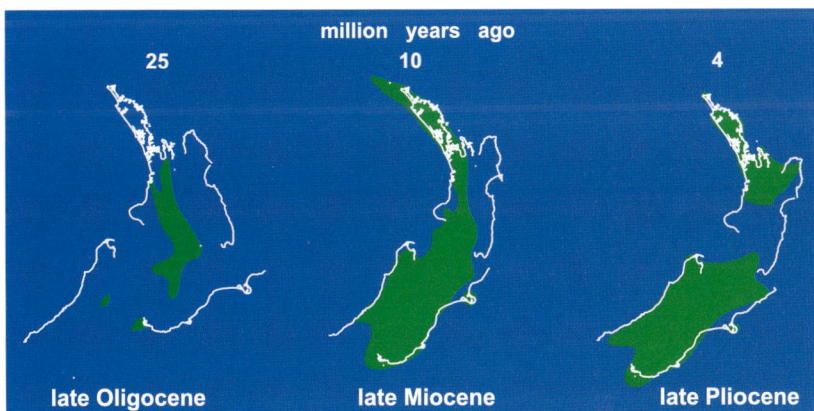

Fig. 9.17 New Zealand emerges from the sea

This set of three maps depicts the changing outline of the New Zealand land area as it was pushed up out of the sea by the collision forces between the Pacific and Australian plates over the last 25 million years. The present coastline of New Zealand is shown for reference purposes only.

(from left) In the late Oligocene (25 million years ago), New Zealand was several small, low-lying islands surrounded by extensive shallow seas, where our future limestones were accumulating as shell banks. Subduction beneath northern New Zealand would soon result in the eruption of the Northland volcanic arc (23–15 million years ago).

By the late Miocene (10 million years ago), large parts of future New Zealand had been pushed up to form hilly country. The Alpine Fault sliced through the South Island and volcanoes were active at Dunedin, Banks Peninsula, offshore north Taranaki and Coromandel Peninsula. Thick deposits of sandstone and mudstone were accumulating along the East Coast and in Taranaki.

In the Pliocene (4 million years ago), the northern North Island and the South Island were taking shape, but there was still a lot of horizontal displacement to take place on the Alpine Fault and the main axial ranges of both islands had not yet been pushed up. The southern North Island area was still under the sea and thick sequences of soft sediment were accumulating in Taranaki, Wanganui and the North Island east coast.

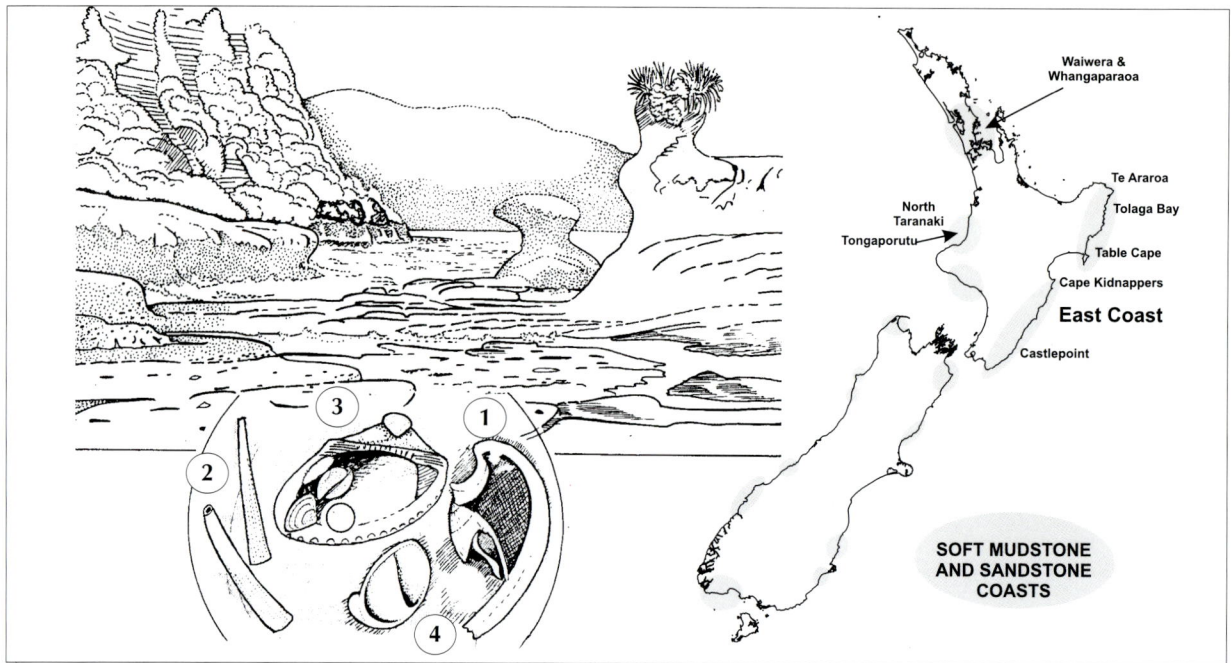

Fig. 9.18 Soft sandstone and mudstone shorelines

Map showing the distribution of the relatively soft Miocene and Pliocene sandstone and mudstone shorelines around New Zealand. They are particularly common along the recently uplifted east coast of the North Island.

(top left) Relatively smooth shore platforms and low stacks are characteristically developed in massive, unjointed, calcareous mudstone, such as here at Te Araroa, East Cape.

(inset) The overlying sandstone beds in the adjacent road cuts contain numerous late Miocene (10 million-year-old) fossils. 1 extinct, thick-shelled bivalve, *Cucullaea* 2 tusk shell, *Dentalium* 3 slipper limpet, *Maoricrypta*, and 4 olive shell, *Amalda*.

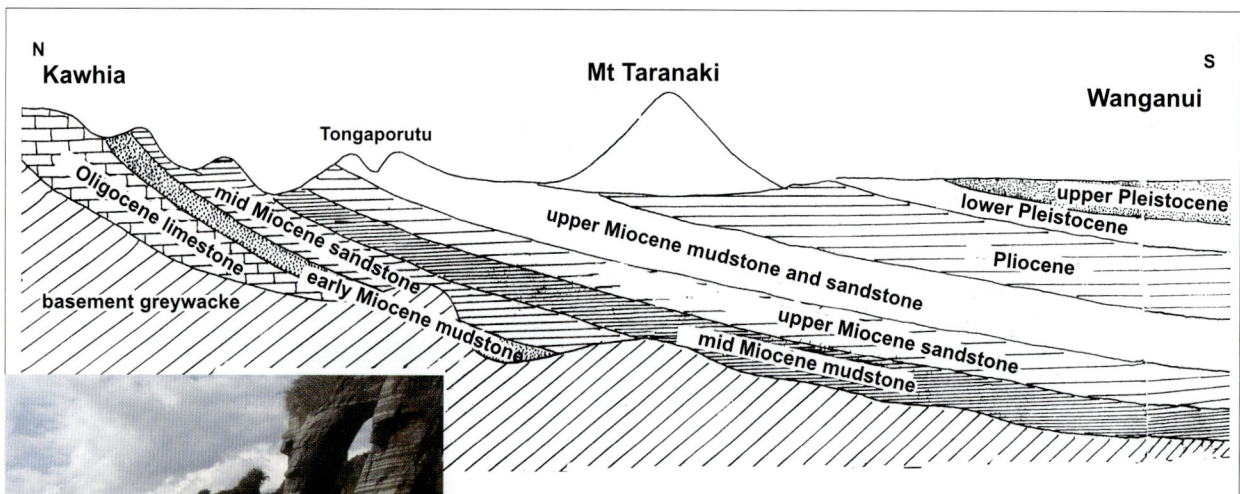

Fig. 9.19 Taranaki's coastline

Most of the north Taranaki and Wanganui coastline is fringed by cliffs carved into soft sedimentary rocks. This schematic cross-section (vertically exaggerated) illustrates the general southerly tilt on the strata with the oldest (Oligocene limestone) exposed as coastal rocks in the north (around Kawhia Harbour) and progressively younger (Miocene to Pleistocene) sandstone and mudstone forming the cliffs further and further to the south. These strata underlie Mt Taranaki, but the semicircular coast surrounding the volcano from Waitara to Hawera is composed of laharic volcanic breccias and bouldery beaches eroded from them.

(bottom left) The photogenic stacks at Tongaporutu, north Taranaki, are composed of relatively soft sandstone and thin mudstone beds (late Miocene). Erosion has preferentially carved its way into the cliffs along vertical joints and fault planes creating numerous sea caves and tunnels, which eventually collapse, leaving free-standing stacks.

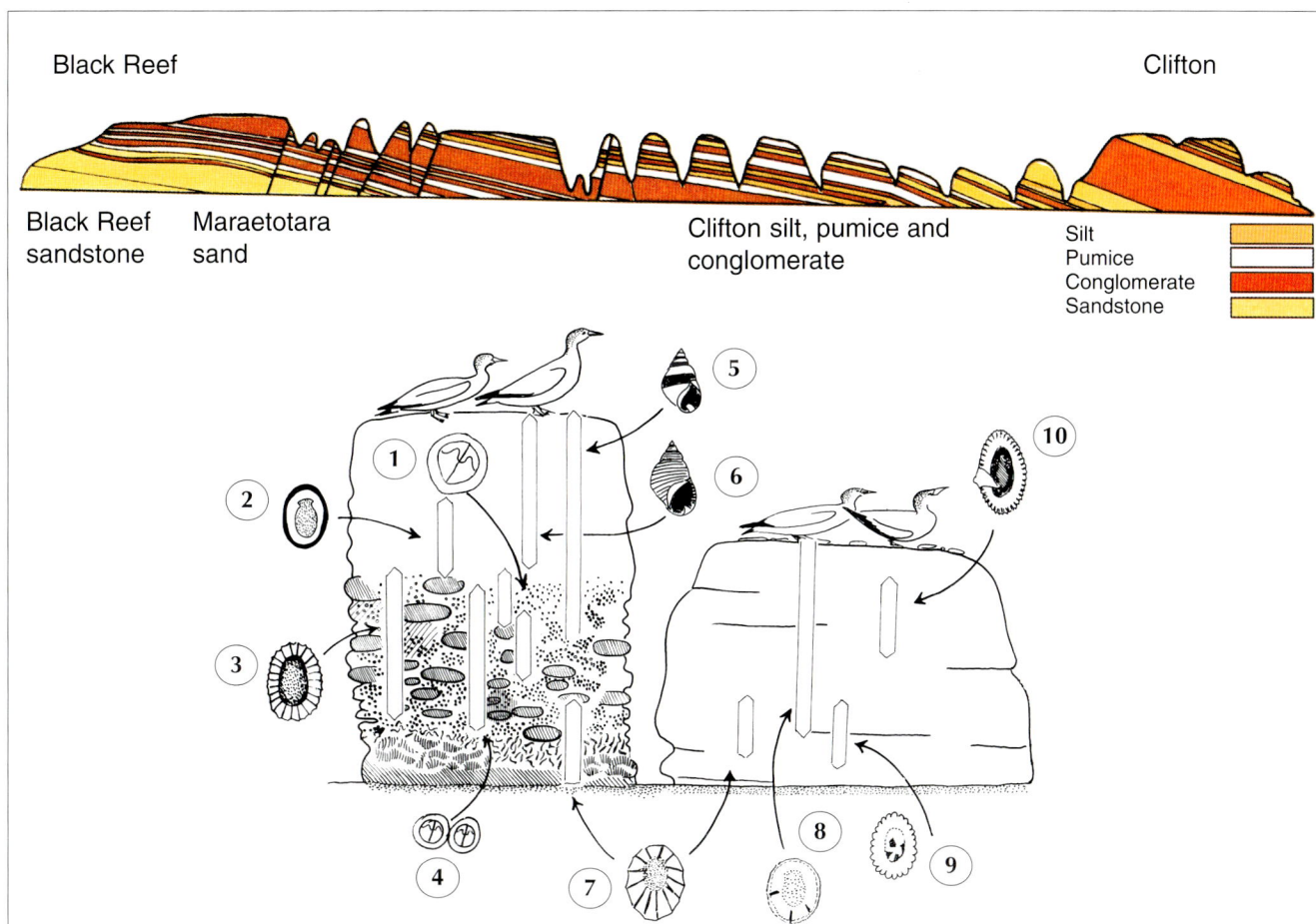

Fig. 9.20 Conglomerate and sandstone stacks, Cape Kidnappers
(top) The scenic trip by vehicle or foot to Cape Kidnappers gannet colony passes along the beach beneath spectacularly bedded, multi-coloured cliffs of Pleistocene conglomerate, siltstone and pumice. At Black Reef these softer strata overlie slightly harder beds of Pliocene (3 million year old) sandstone and conglomerate, which form the reef and a cluster of small stacks **(bottom)**. The flat tops of the stacks, now used by nesting gannets, were formerly part of an intertidal shore platform that was uplifted by a large earthquake 2300 years ago. Subsequent erosion is carving out a new platform at the new intertidal level.
The vertical ranges of barnacles and molluscs on the sides of the stacks are illustrated: 1 *Chamaesipho brunnea* 2 *Notoacmea pileopsis* 3 *Cellana ornata* 4 *Chamaesipho columna* 5 *Austrolittorina antipoda* 6 *A. cincta* 7 *Cellana radians* 8 *C. flava* 9 *Patelloida corticata* 10 *Benhamina obliquata*.

Wanganui and particularly the east coast of the North Island (Figs. 9.18–9.22).

Subduction-related andesitic and rhyolitic volcanism began in the Northland volcanic arc in the early Miocene, moved to the Coromandel volcanic arc during the middle and late Miocene (16–5 million years ago), and finally shifted, in the last 2 million years, to its present location running northwards from Ruapehu through Taupo, Rotorua and offshore into the Bay of Plenty and beyond to the Kermadec Islands.

Hot spot, mostly basaltic, volcanism produced two huge shield volcano complexes at Dunedin and Christchurch (Banks Peninsula) in the middle and late Miocene (13–6 million years ago). Smaller 'hot spot' basaltic fields have also been active in the last five million years at Timaru and in several places in northern

New Zealand (Kaikohe, Whangarei, Pukekohe, Auckland). The young Auckland field erupted 48 small volcanoes in the last 200,000 years and is still considered to be active.

The different magma compositions and different styles of eruption produced a wide diversity of volcanic rock types and resultant shorelines (Fig. 9.23). Rocks formed by solidification of molten lava in a flow or intrusion usually are hard and erosion-resistant, but shot through with contractional cooling cracks, which marine erosion often takes advantage of (Figs. 9.24, 9.25).

Andesitic stratovolcanoes are often surrounded by gently sloping ring-plains of rubbly breccia, that has been carried down the volcano slopes by lahars. With coastal erosion, the boulders and cobbles of andesite are released from the breccia matrix and commonly accu-

169

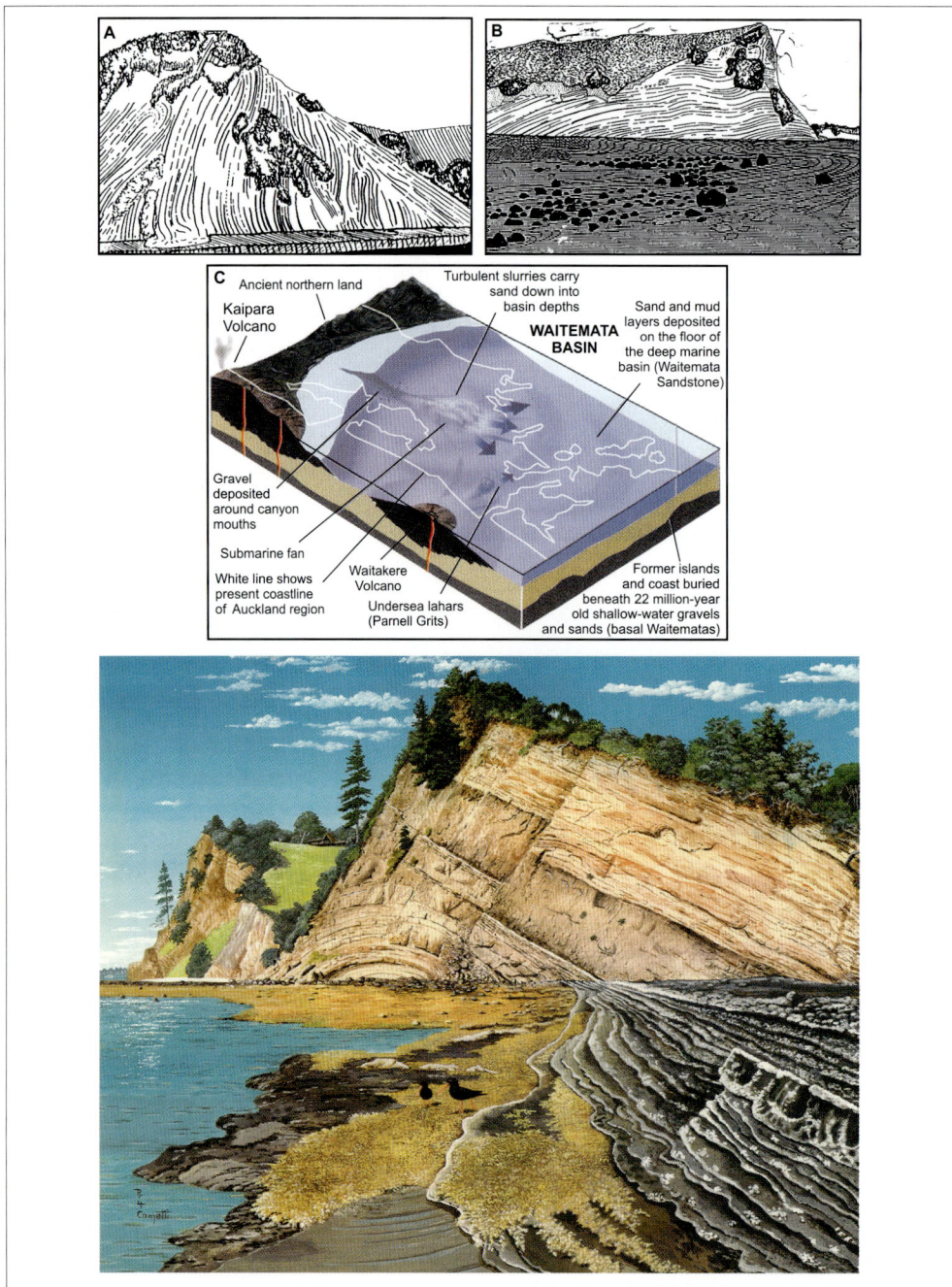

Fig. 9.21 Waitemata Sandstone and Parnell Grit shores, Auckland

The soft rock shore platforms and cliffs around Auckland are made of interbedded sandstone and mudstone layers (called Waitemata Sandstone) that accumulated on the floor of the 1–2 km deep Waitemata Basin, which covered the region (C) during the early Miocene (22–17 million years ago). Out west, the Waitakere Volcano was actively building a large submarine edifice capped by one or more rocky islands. Periodically slumps of volcanic debris slid down the slopes of the Waitakere Volcano and were deposited as a thick layer of volcanic gravel and grit (called Parnell Grit) on the basin floor.

A. Wide intertidal platforms are characteristic of Waitemata Sandstone shorelines, like this one cut into vertically tilted sandstone and mudstone strata, at Whangaparaoa Head, Shakespear Regional Park.

B. In many places the Waitemata Sandstone strata are intricately folded and faulted creating swirling patterns in the intertidal platforms, as seen here east of Army Bay, Whangaparaoa Peninsula. Dark grey basalt cobbles and boulders are eroding out of a Parnell Grit bed, which forms part of the shore platform.

C. Geography of the Auckland region during the early Miocene, 20 million years ago. (From *A Field Guide to Auckland*, Godwit.)

D. Tilted Waitemata Sandstone strata form high cliffs and a wide shore platform at Waiwera. The most prominent reefs and points around Auckland are composed of the more resistant beds of thick sandstone or Parnell Grit. Here the thick bed in the centre of the cliff and the cobbly upper part of the shore platform (right) are a Parnell Grit bed, derived from the volcanoes out west. Two variable oyster catchers search the *Hormosira* flat for food. (Painting by Ron Cometti.)

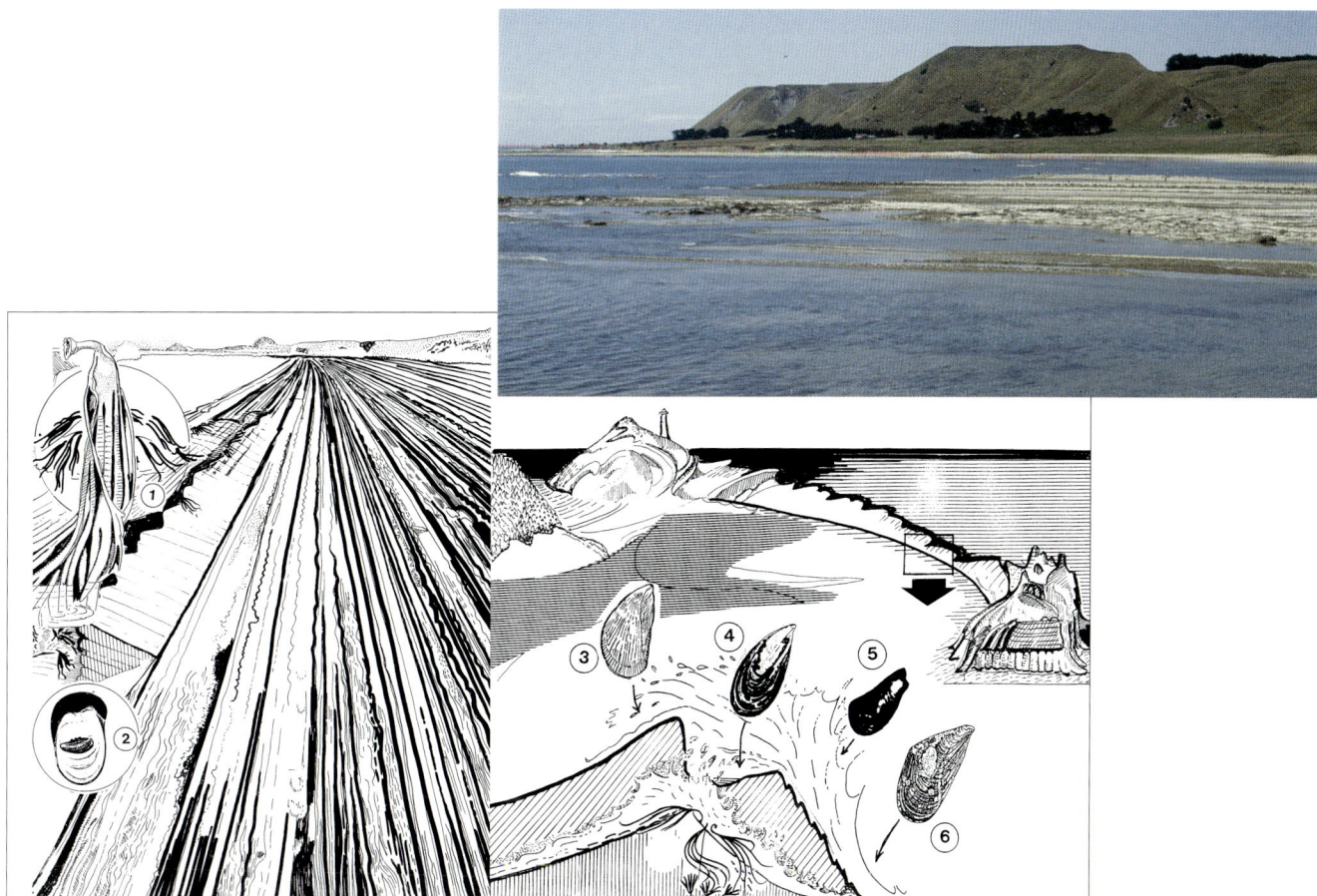

Fig. 9.22 Shore platforms in young sedimentary rocks
(top; and bottom left) Broad intertidal platform carved out of tilted Miocene mudstone, alternating with ridges of inclined sandstone, set in perfectly straight furrows parallel to the shore at Table Cape, north-eastern extremity of Mahia Peninsula, northern Hawke's Bay. **(bottom right)** A tilted spine of Pliocene limestone protects a sandy beach and terminates in the lighthouse promontory at Castlepoint on the Wairarapa coast. **(inset bottom left and right)** Vertical distribution of mussels that grow attached to the limestone spine in the wave-break zone. 1 Kelp, *Durvillaea antarctica* 2 *Nerita atramentosa* 3 *Aulacomya maoriana* 4 *Mytilus edulis aoteanus* 5 *Xenostrobus pulex* 6 *Perna canaliculus*.

mulate as bouldery shorelines and beaches, particularly around the extensive shores of young Mt Taranaki. Older, more cemented, examples of these volcanic breccias are erosion-resistant and form substantial cliffs and rocky coasts, as around Whangaroa and the Waitakere Ranges (Fig. 9.23).

Volcanic ash is usually soft and may be well-bedded, whereas rhyolitic ignimbrite is commonly massive, and moderately welded together by the heat from its eruption. Ash and ignimbrite form coastal platforms and cliffs (Fig. 9.23) often indistinguishable from those eroded from young mudstone and sandstone of similar hardness.

Ice Ages and the shape of the coast

The tectonic uplift, subsidence and volcanism of the last 25 million years created the general form of modern New Zealand. The intricate shape of its coastline and landforms, however, are the result of much younger erosion and deposition processes that occurred during the Ice Ages, and particularly since the peak of the Last Ice Age, just 18,000 years ago.

The world has experienced alternating periods of cold and warm climate during the Ice Ages of the last two and a half million years. These cyclical climate fluctuations have been driven by variations in the amount of solar energy reaching different parts of the Earth's surface, due to the changing position of the Earth's axis in its varying path around the sun.

There have been at least 40 of these cold-warm cycles in the last 2.5 million years. Each cycle lasted 40,000 or 100,000 years and included a warm period similar to the present day and a cold or glacial period when large ice caps formed on northern hemisphere continents. These ice caps froze large amounts of the world's water on land and resulted in major worldwide drops in sea level of 100–120 m during each Ice Age period. Sea level has only been up at around its present level during the peaks of the warmer periods, for about 15% of the time in the last 1 million years. It has probably never risen

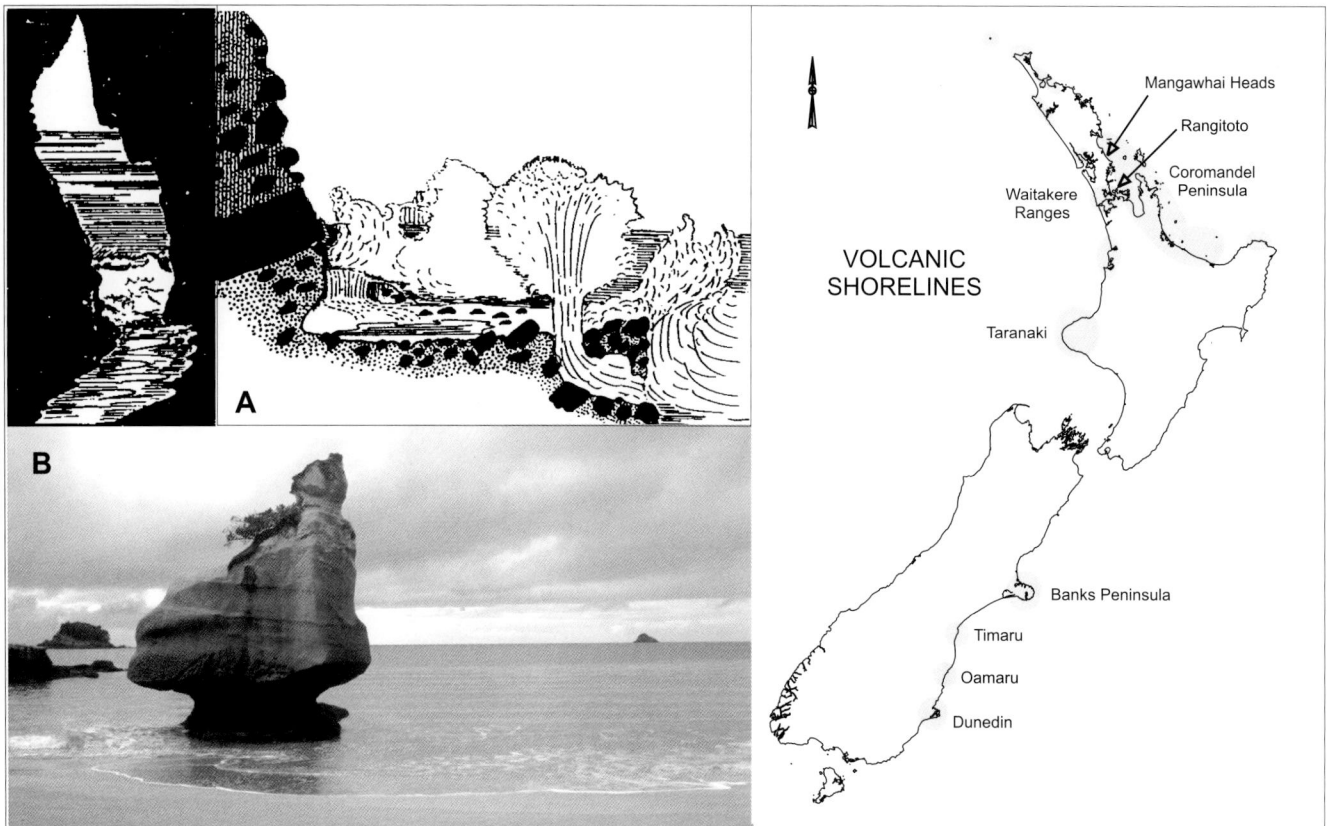

Fig. 9.23 Volcanic rock coasts

Map showing the concentration of volcanic rock along the coastline of northern New Zealand, with outliers at Banks Peninsula, Timaru, Oamaru and Dunedin.

A   Erosion-resistant, cemented volcanic breccia forms much of the coast of the Waitakere Ranges, west of Auckland. It accumulated on the submarine slopes of the large Waitakere Volcano (Fig. 9.21), during the early Miocene (20 million years ago). Here we see a high tide platform and small blowhole at The Gap, South Piha. The pounding action of waves and compressed air have blown a tunnel (**left**) along a columnar-jointed dike of andesite, right through nearby Taitomo Island.

B. Some of the cliffs along the east coast of the Coromandel Peninsula, like Cathedral Cove near Hahei, are eroded from massive, slightly welded deposits of ignimbrite. Ignimbrite is a mix of glassy volcanic ash and fragments of pumice that were erupted as hot, gas-propelled, pyroclastic flows from rhyolitic caldera volcanoes. Their near-featureless character results in smooth shore platforms and cliffs undercut around high tide level. Erosion along the occasional joint plane results in the formation of guts, caves or tunnels.

more than 50 m higher than what it is today in the last 2.5 million years.

During the coldest part of the Last Ice Age, just 18,000 years ago, sea level fell to about 120 m lower than present (Fig. 9.26). At that time, glaciers and snowfields covered the mountains throughout the South Island and southern North Island, while forest grew over most of the northern North Island and around the coast of the northern South Island. Many of today's harbours were forested valleys with streams flowing seaward across the broad coastal plain.

Following the peak of the Last Ice Age, the world's climate began to warm, the ice caps slowly melted and the world's sea level rose correspondingly. Sea level reached its present level about 7000 years ago and has remained close to that ever since. A slightly warmer period about 4000–5000 years ago resulted in sea level rising to nearly 1 m above what it is now, but it has since receded.

The cycles of wildly fluctuating sea levels had a major impact on the shape of the world's coasts, and particularly the New Zealand coast. During each of the Ice Age periods, erosion on land was greatly increased because of reduced forest cover. Sediment poured down the rivers and was spread along the coasts by long-shore drift. This was particularly true in the South Island where increased glacial erosion of the Southern Alps resulted in vast quantities of sediment building up to form the Canterbury Plains in the east. On the west coast of the South Island much of the increased sediment supply was transported northwards by long-shore currents and built a wide coastal plain, now drowned as the continental shelf south and west of Taranaki.

In northern New Zealand large volumes of volcanic ash and ignimbrite were erupted and eroded from the Taupo and Bay of Plenty region. This sediment was carried down the Waikato River to the west coast and at

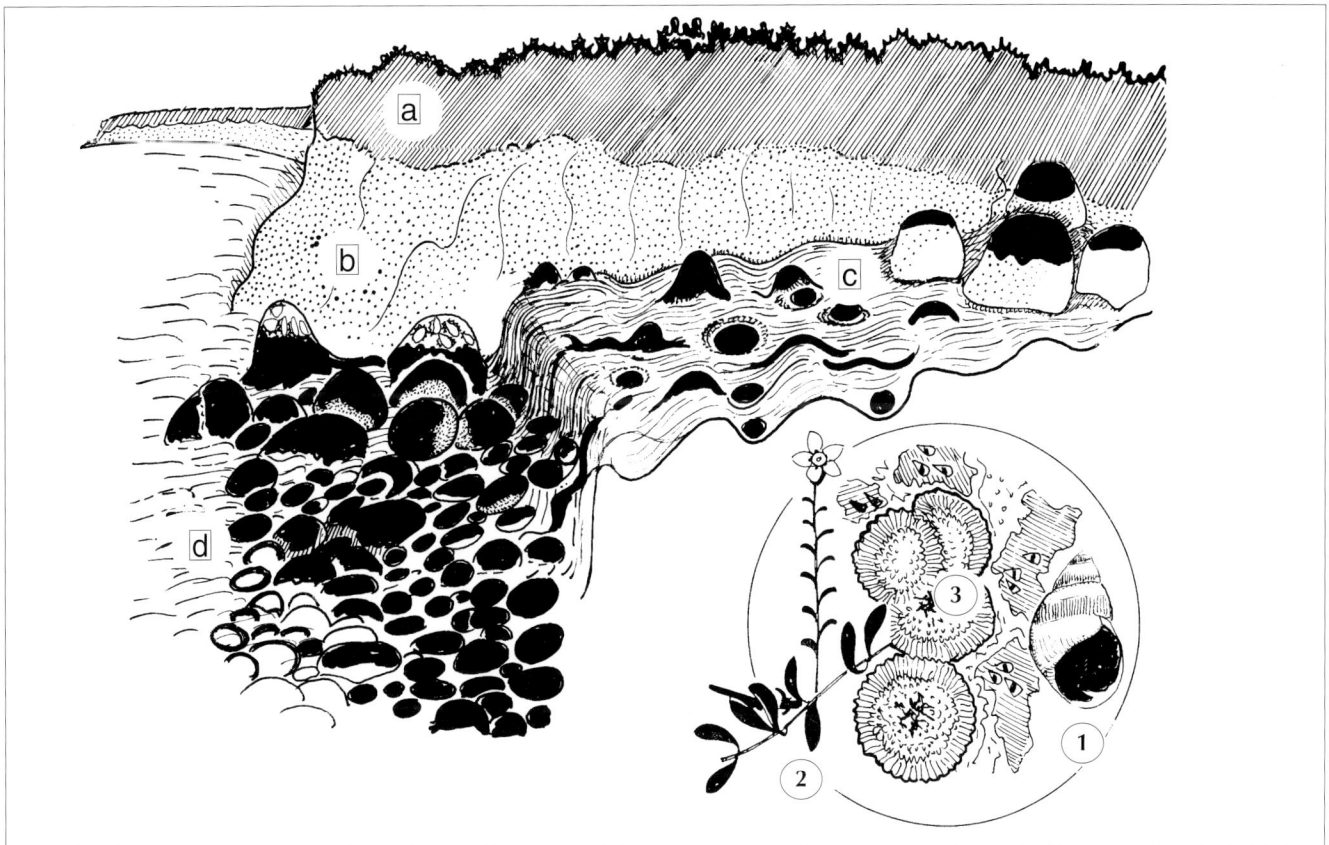

Fig. 9.24 Basalt shore, Thornes Bay, Takapuna
The black basalt coast between Takapuna and Milford beaches on Auckland's North Shore is composed of young lava flows erupted from Pupuke Volcano about 200,000 years ago. a bare littoral fringe of scoriaceous basalt; b eulittoral *Chamaesipho columna*; c undulating surface with mobile boulders in eroded potholes; d seaward zone of wave-rolled cobbles. **(inset)** Biota of scoriaceous littoral fringe: 1 *Austrolittorina antipoda* 2 maakoako, *Samolus repens* 3 yellow lichen, *Xanthoria parietina*.

other times into the Hauraki Gulf via the Waikato's alternative route through the Hauraki Plains. Over the last 2 million years, the copious supply of sand (including the black sand mineral titanomagnetite) from Mt Taranaki and the Waikato River has been thrown up as a series of sand dunes along the west coast of the North Island. These have straightened out the coast and formed barriers across major bays, creating the large Kaipara and Manukau Harbours.

At the end of the Last Ice Age the rising sea encroached over the land and the sand that had built up along the coast was swept shoreward. Shallow valleys that flowed out across the former coastal plain were drowned and rapidly filled with sediment. For several thousand years after the sea reached its present level, vast quantities of sand were thrown up against the land to form beaches, barriers and dunes. Where there was a plentiful supply, whole valleys were filled or large sand barriers created enclosing estuaries and shallow harbours (Fig. 9.27). Examples include Omaha, Wenderholm and Orewa sand spits near Auckland, Ohiwa Harbour, Farewell Spit, Heathcote estuary, Christchurch, and Bluff Harbour. In other places, gravel barriers and spits

quickly built up to enclose sheltered inlets. Examples include Ahuriri Inlet, Napier; Nelson boulder bank; Wairau Bar, Blenheim; and Lake Ellesmere, Canterbury.

Where there was less sand available, the incised river valleys were drowned to become our modern embayed coastline and harbours (Fig. 9.28). The meandering and branching shape of these former river valleys is still recognisable in the extensive headwaters of many northern harbours (such as Hokianga, Kaipara, Manukau, Waitemata and Kawhia) and the Marlborough Sounds, Lyttelton and Dunedin Harbours. The extensive inland waterways of Fiordland were created by the drowning of steep-sided, U-shaped valleys that had been carved out by glaciers when sea level was much lower during each cold Ice Age period.

Most of the cliffs around our modern coast are particularly young and have been eroded out of the sloping hillsides in just the last 7000 years. The Waitemata Sandstone cliffs around Auckland are eroding back at rates of 1–5 cm per year. The softer mudstone cliffs of the North Island's east coast are being cut back at an even faster rate. The intertidal reefs in front of these cliffs, extending up to 100 m out to sea, are an indication of the

173

Fig. 9.25 Columnar-jointed basalt, St Clair, Dunedin

Marine erosion of the black basalt lava flow is strongly influenced by the four-, five- and six-sided columns. These were produced by contraction cracks that formed during cooling and solidification of a lava flow erupted from the large Dunedin Volcano some 10–13 million years ago. Columns that have been broken from the reef now form mobile boulders being rounded by the breaking waves on the beach. Rising from the waves is a line of surging bull-kelp, *Durvillaea antarctica*. (Painting by Ron Cometti.)

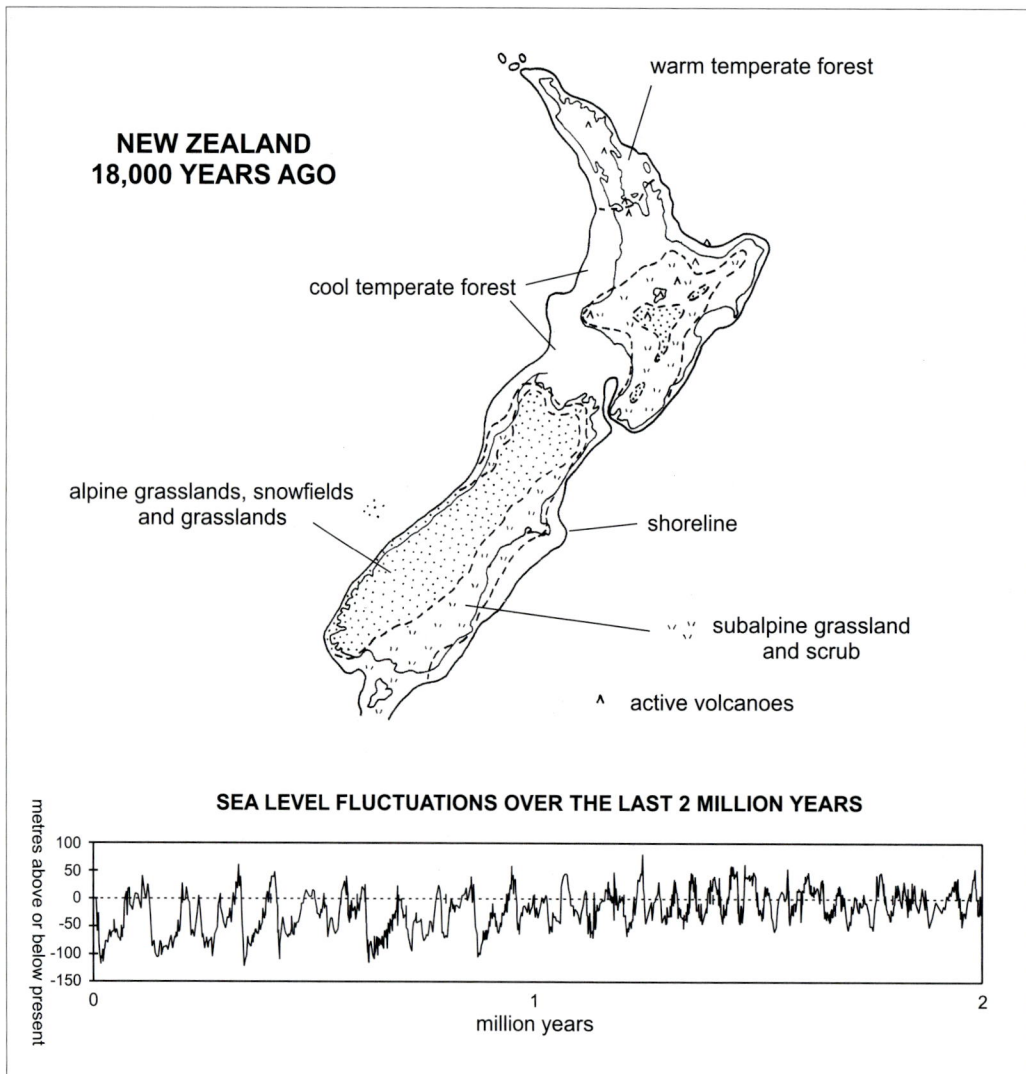

NEW ZEALAND 18,000 YEARS AGO

warm temperate forest

cool temperate forest

alpine grasslands, snowfields and grasslands

shoreline

subalpine grassland and scrub

^ active volcanoes

SEA LEVEL FLUCTUATIONS OVER THE LAST 2 MILLION YEARS

metres above or below present

million years

Fig. 9.26 Coastline changes since the Last Ice Age

Map showing the position of New Zealand's coastline just 18,000 years ago at the peak of the Last Ice Age, when sea level was c. 120 m below present level. The present-day coast has only been in existence for 7000 years.

(bottom) Graph of fluctuating sea level over the last 2 million years. The 40,000-year cycles of high and low sea levels (warm and cold periods) were replaced by 100,000-year cycles about 900,000 years ago. Note that sea level has been lower than the present for 85% of the last 1 million years.

Fig. 9.27
Tolaga Bay sand spit and tidal inlet
Coastal sand and gravel spits and their enclosed inlets and harbours are some of New Zealand's youngest landforms. Virtually all have been formed within the last 7000 years, since sea level reached its present level after the end of the Last Ice Age.

Fig. 9.28 Tamaki Estuary, Auckland
The many meandering arms of New Zealand's harbours were eroded by rivers and streams during the long periods of lower sea level during the Ice Ages. They have been drowned by the sea on many occasions during warm episodes within the last 2 million years and were most recently flooded about 7000 years ago by rising sea level after the end of the Last Ice Age.

Fig. 9.29 Wide shore platform, Mahurangi Island, Waiwera
Virtually all the coastal shore platforms around New Zealand have been cut out of sloping hillsides within the last 7000 years, since sea level stabilised at about its present level. Here the width of the intertidal shore platform cut into the Waitemata Sandstone is a good measure of the rate of marine erosion and cliff retreat over that period. The widest shore platforms are usually cut out of the softest rocks.

amount of cliff retreat since sea level rose (Fig. 9.29). Some of the higher cliffs in harder rocks, such as those along the west coast of the Waitakeres and Wellington, would have been carved back during a number of periods of higher sea level and became frittering inland bluffs during the intervening Ice Age intervals.

Today our youthful coast is still changing, in places eroding and elsewhere growing, as nature continues to respond to the post-Ice Age rise in sea level and to the variable patterns of winds, waves, long-shore currents and ongoing tectonic uplift and subsidence.

Shore platforms

Shore platforms are formed by a combination of marine erosion and subaerial weathering. Physical erosion by the waves is most pronounced along the seaward edge of shore platforms, whereas subaerial chemical and physical weathering mostly occurs in the cliffs behind. The shape of the platform and its speed and manner of formation are determined by the properties of the rock. The hardness of most igneous rocks (granite, gneiss, basalt, andesite) and indurated basement sedimentary strata (greywacke, schist) retards the rate of erosion.

Under strong wave attack, the amount and alignment of rock joints can be significant in determining the shape of the rocky shore. Crashing and surging waves produce repeated alternations of hydraulic pressure directed along joint planes in hard rocks. This loosens blocks which drop out and are washed away, eventually creating a gut or cave.

Water-saturation in the pore spaces of permeable softer rocks (sandstone, siltstone, volcanic ash, ignimbrite) also retards erosion, by reducing the amount of air that can enter the rock to destroy it by chemical decay. In these softer rocks, the fastest weathering and erosion usually occurs in and just above the littoral fringe, with repeated splash-wetting at high tide followed by drying out at low tide. The cliff then retreats by undercutting and collapse, leaving its profile steep. Where high-energy waves are refracted onto exposed headlands, such weathering occurs at higher levels to cut a bench just above high water.

New Zealand's shore platforms can thus be classified into narrow, high-tidal benches and broad intertidal platforms. The first, no more than 5–15 m across (Fig. 9.30B), are usually cut in harder rocks and fall off by a sea-cliff of 3–4 m. The best examples of these are in

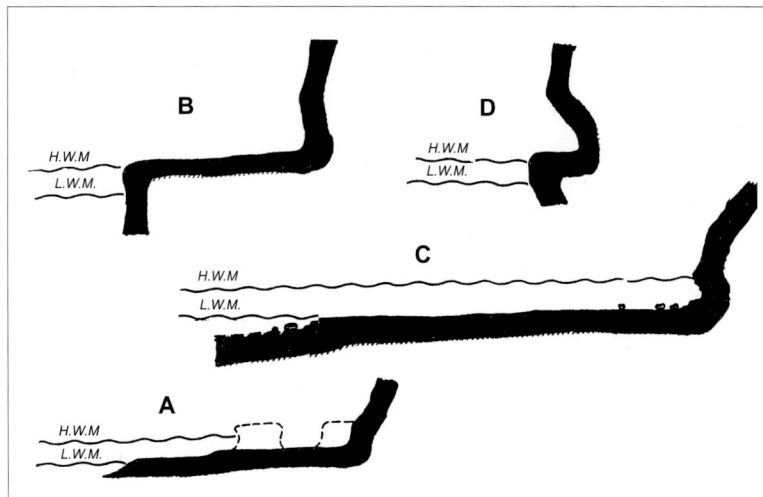

Fig. 9.30 Shore platform profiles
A. Intertidal horizontal platform, sometimes stepped and incised into an older platform.
B. High-water horizontal platform with low-tide cliff.
C. Broad low-angle intertidal platform, sometimes with slight cliff-foot notch and small low-tide cliff or step.
D. High-water notch. A variant of B, on some steeply cliffed coasts.

(After Healy.)

Fig. 9.31 Old Hat Island, Mangawhai Heads
This islet of weathered, columnar-jointed dacite (volcanic rock) has a characteristic old hat profile, with the remnant core of the island surrounded by a wide brim — a high tide bench more resistant to erosion and chemical weathering because of the permanent water saturation in the underlying rock.

Fig. 9.32 Ancient and modern greywacke shore platforms
At Leigh Marine Reserve, north-east of Auckland, the modern greywacke shore platform (3) was originally formed 20 million years ago in the early Miocene. Back then the ancient coast was subsiding and soon buried by beach gravel (2), sand (4) and silt (1), which with continued subsidence and further burial hardened into sedimentary rocks. In the last few million years the Auckland region has been pushed up again and today coastal erosion is exhuming the ancient shoreline by preferential erosion of the softer overlying sedimentary rocks.

greywacke, but they have been formed also in hard granite and cemented andesitic breccia (Fig. 9.23A).

By contrast, broad intertidal platforms (Fig. 9.30C), cut out of soft sandstone, siltstone or fine-grained limestone, are the typical coast form down the east coast of the North Island and around Auckland. Porous and weakly indurated, they are susceptible to regular wetting and drying with atmospheric weathering during low tides. Abrasion by wave-carried particles can also gradually reduce a broad surface. A long, graded profile results from cliff-base to low water, with its curve reflecting the level of the water table. The seaward edge often drops by several metres, a result of both wave-cutting and bio-erosion.

Reef platforms are not always perfectly planed, but may have large steps, with residuals left from part-eroded older platforms (Fig. 9.30A). Where the bedrock consists of alternating soft and hard layers, often mudstone and sandstone, the resulting shore platform may

be stepped or ridged depending on the angle of tilt on the beds. Joint and fault planes in these softer sedimentary rocks are commonly the locus of preferential erosion by the surging waves and hydraulic pressure changes, forming elongate guts through the shore platforms and caves in the base of cliffs.

On low-angle platforms of massive softer sedimentary rock, layer weathering is all important, as broad, shallow pans progressively merge together. Pool bottoms and surfaces within wave-reach stay permanently wet, while emergent rocks dry out, with salt crystals growing on their surface. Stress within the matrix causes disintegration, with the loosened material removed and fresh surfaces then exposed to weathering.

'Old-hat' profiles may develop around an island stack in a variety of rock types. A bench may develop right around the islet, with planation controlled by the level of permanently saturated pore space (Fig. 9.31).

Fig. 10.1 Bathymetry and islands in the New Zealand region

(Reproduced with permission of NIWA.)

Kermadec Is

Three Kings Is

Challenger Plateau

Chatham Rise

Chatham Is

Stewart I

The Snares

Bounty Is

Campbell Plateau

Antipodes Is

Auckland I.

Campbell I.

CHAPTER TEN

Islands and Biogeography

New Zealand, with its string of outlying islands from the Kermedecs to the cold south, extends over some 22 degrees of latitude. At the edge of the coral seas and in the near-subantarctic, these island extremes are remote in character from the mainland. But even around New Zealand's two main islands, we have already noted the important differences in the shore biota from north to south.

First, despite the broader continuity of theme, Auckland and Northland are in major ways distinguishable from all the rest of New Zealand. From Cape Reinga down the west coast to Kawhia, and on the east coast around the Bay of Plenty to East Cape, the winters are frost-free. The resulting biotic differences, best understood so far for the marine algae, the mollusca and the fishes, will from the previous pages already have become plain.

With a *warm* centre of distribution thus to be recognised for Auckland, Northland and the Bay of Plenty, and a *cool* southern one taking in the rest of the North and the whole of the South Island, together with Stewart Island and the Snares, New Zealand exhibits two primarily distinct sorts of coast. Their divide (even as revealed from the shoreline of east to west Auckland) is one of major import on a Pacific scale, separating the two primary biogeographical regions of the **South Pacific Warm** and **Cold Temperate**.

Current systems

The coastline of New Zealand, over its 22 degrees of latitude, experiences a wide range of sea temperatures from an annual mean of 21°C at the Kermadecs to 9°C at Campbell Island. The east Northland seaboard is reached by Subtropical Water, as evidenced by the high surface temperatures encountered on the open coast. In the Hauraki Gulf, with the added effect of landlocking, the highest summer isotherm reaches 24°C. By contrast, the subantarctic islands are bathed with Subantarctic Water. The Tasman Sea has an anticlockwise circulation of Subtropical Water, modified by its passage south and affecting middle New Zealand by sweeping eastward

with one branch flowing up the New Zealand west coast and the other flowing south and around the bottom of the South Island.

From this circulation pattern, our system of coastal currents is derived. Warm water from the **East Australian Current** sweeps eastward across the northern Tasman Sea as the Tasman Front, around North Cape and southwards to the Bay of Plenty (**East Auckland Current**), and from Cape Reinga down the west coast to near the Kaipara Heads and sometimes beyond (**West Auckland Current**). A warm **East Cape Current** continues southwards to the seaward of the cool **Canterbury Current** derived from subantarctic water that flows from Banks Peninsula to Gisborne. The **Southland Current**, a combination of warmer water from the south Tasman Sea and cold Subantarctic Water mixed together in the Subtropical Front, passes from Fiordland through Foveaux Strait or around Stewart Island and up the east coast of the South Island. Part of the near-shore Southland Current flows through the Mernoo Saddle over the Chatham Rise to feed the north-flowing Canterbury Current. A major part of the Southland Current turns eastward offshore at 43 degrees south and flows along the Chatham Rise in the vicinity of the Subtropical Front.

Along the west coast from Fiordland to approximately the Manukau Heads runs a **Westland Current** of south Tasman sea (modified subtropical) water; its northern limit fluctuates, marked by a *convergence* with the West Auckland Current off the Northland to Waikato coast. A **Durville Current** of Tasman Sea water flows from the west into the Wanganui Bight and thence south through Cook Strait.

South of mainland New Zealand, the Campbell Plateau and subantarctic islands are bathed in colder and less saline Subantarctic Water separated from the warmer Subtropical Water by the Subtropical Front. This front is usually located at about 45 degrees south, but bends southward around the South Island and Stewart Island before passing northwards along the continental slope of the eastern South Island. The Subtropical Front heads east again with its position apparently

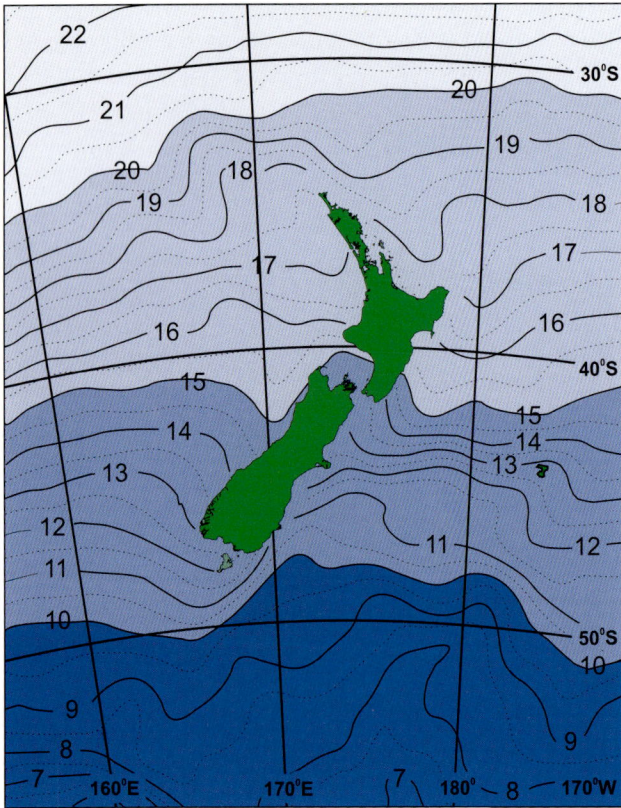

Fig. 10.2 New Zealand mean sea surface temperatures

locked by the west-east Chatham Rise. The Chatham Islands lie right in the path of the Subtropical Front, sometimes bathed in eddies of warm and sometimes cold water.[1]

Historical

The first attempt to divide the New Zealand coastline into formal biogeographic provinces was made by Harold Finlay, on the basis of the Mollusca, as early as 1927. His system was modified by Baden Powell in 1937, and its divisions were first correlated with currents and sea temperatures by Charles Fleming in 1944.

An **Aupourian Province** rich in warm-water species was proposed for the North Island's east coast as far south as East Cape, with its west coast boundary left with a good deal of elasticity somewhere between Powell's original choice at Ahipara and Manukau Heads. Pawson, from a later study of the echinoderms, preferred to set a southern boundary as far south as Cape Egmont, while Lucy Moore, from the evidence of algae, would have brought it to Albatross Point, off Kawhia.

Powell's second division, the **Cookian Province** was seen as an 'intermediate region of mixed waters', taking in the rest of the North Island, and the South Island with the exclusion of the cold shores of Otago and Southland. These last, along with the Snares Islands, were for some years held to constitute a third entity, the **Forsterian Province**. There now seems to be no justification for a separate Forsterian Province, if significant endemism is used as the accepted criterion for recognition of a marine province. Clearly forming a single southern centre of distribution, the full Cookian Province (southern North Island, South and Stewart Island) has a 33% endemism among its Mollusca.

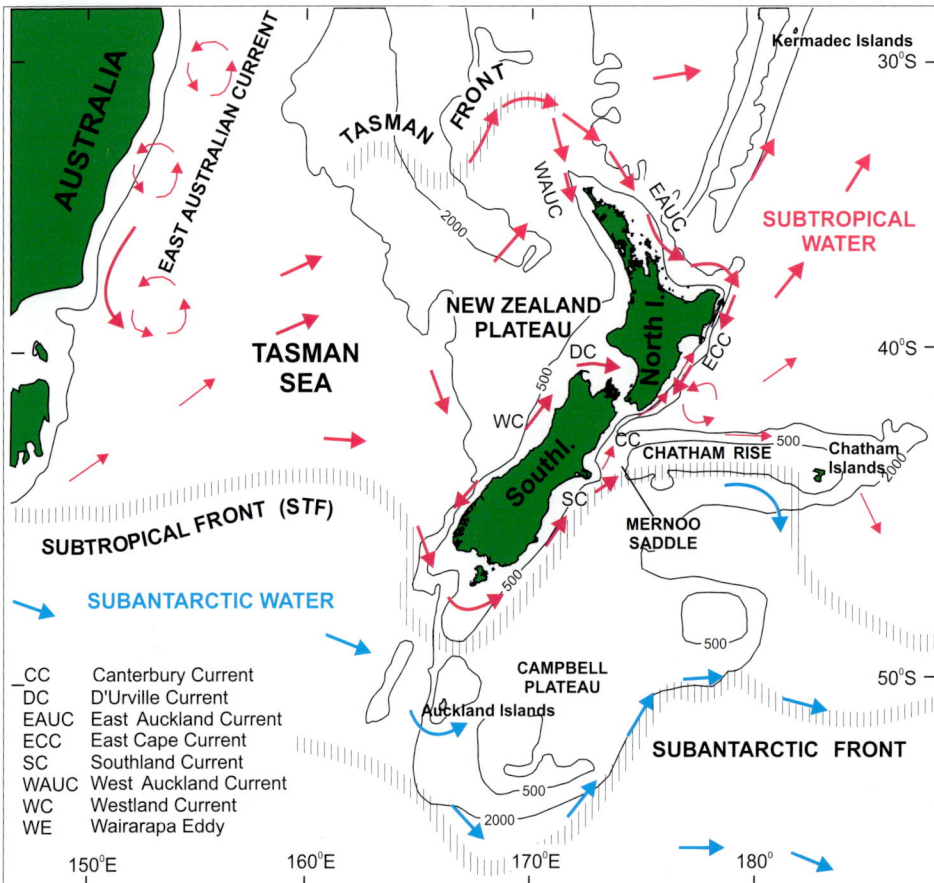

Fig. 10.3 The surface current systems of the New Zealand region

CC	Canterbury Current
DC	D'Urville Current
EAUC	East Auckland Current
ECC	East Cape Current
SC	Southland Current
WAUC	West Auckland Current
WC	Westland Current
WE	Wairarapa Eddy

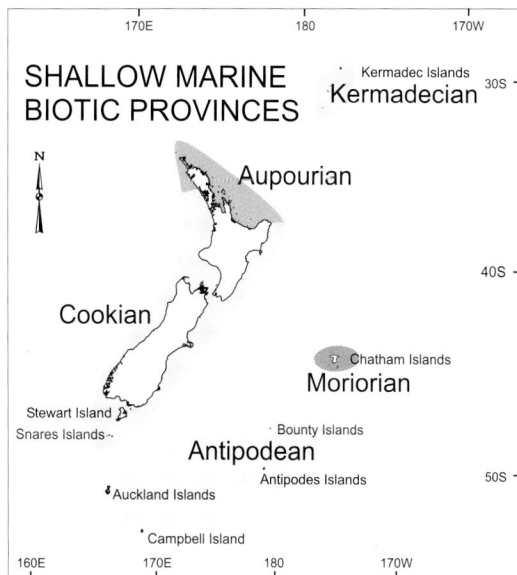

Fig. 10.4 New Zealand's marine biogeographical provinces

Powell recognised two offshore provinces for New Zealand: an Antipodean (originally Rossian) to include the Antipodes, Bounty, Campbell and Auckland Islands; and a Moriorian for the Chatham Islands alone.[2]

Northern New Zealand region

For the New Zealand territory falling within the South Pacific Warm Temperate Region, two marine biogeographic entities are now recognised, the Aupourian Province (called by some the Aucklandian), consisting of the mainland along with its close-offshore islands, including the Three Kings; and the Kermadecian Province, reserved for our northernmost, not-quite-tropical dependency.

Even from above the tidal shore, an Auckland seaboard is obviously distinctive enough from that of the south. Its adlittoral bush or forest is richer and more diverse, and though few of its coastal trees are exclusive to this province, pohutukawa (*Metrosideros excelsa*)

reaches its full magnificence here at the fully exposed coastal edge. In sheltered dips and bay-heads the trees include puriri, taraire and karaka, with kohekohe, titoki and whau; in Northland kauri can itself sometimes reach the coast. The northern estuaries are fringed with *Avicennia marina*, New Zealand's only species of mangrove, that requires an environment largely frost-free.

The littoral and shallow water mollusca of the Aupourian Province add up to a significantly longer list than for the south, with much endemism and a growing accession of newcomers from the warm Pacific. The broad features of the intertidal zoning pattern include the regular presence of *Nerita atramentosa* and (in shelter) *Crassostrea* spp., the virtual absence of *Mytilus edulis aoteanus* and total lack of *Aulacomya maoriana*. A considerable number of molluscs, earlier mentioned in these pages, remain confined to this Province.

Algae so near endemic as to be regionally characteristic of northern shores include *Carpophyllum angustifolium*, *Osmundaria colensoi*, and in *Gigartina* the species *alveata*, *macrocarpa*, *marginifera* and *laingii*. These and some other northern species drop out with the loss of the East Auckland Current at East Cape. For the Three Kings, *Perisporochnus regalis* and *Sargassum johnsonii* are unique.

Warm affinities

The current circulation explains the close relation of New Zealand's Aupourian Province with the warm temperate Peronian Province of South Eastern Australia (Fig. 17.3) The warm East Australian Current has brought to these shores a constant migration of Peronian organisms of almost every taxonomic group, many of them fluctuating over time, or reappearing after long breaks in their occurrence.

Prominent among these is the large component of subtropical fishes with their southern limits off east Northland and the Bay of Plenty, like snapper, tarakihi, red gurnard, red moki, Sandager's wrasse, leatherjacket,

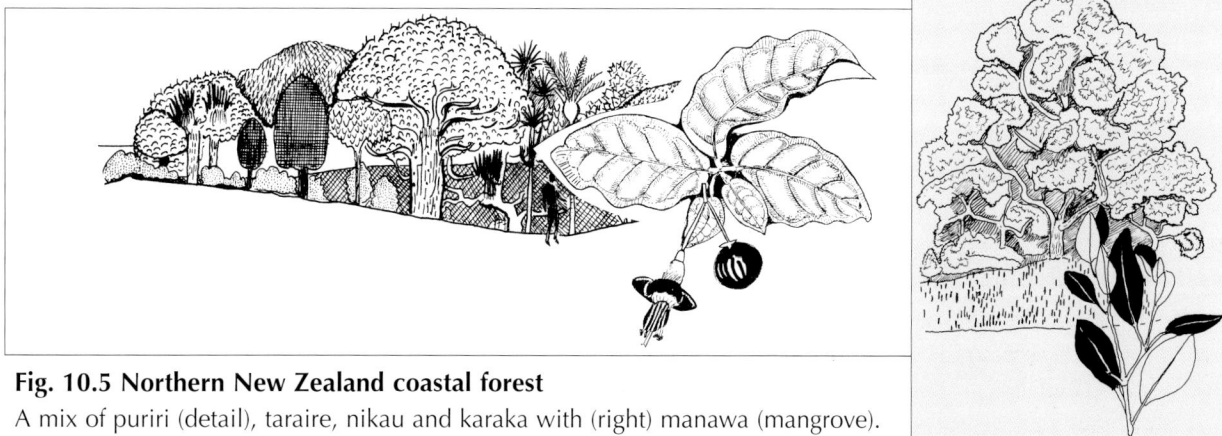

Fig. 10.5 Northern New Zealand coastal forest
A mix of puriri (detail), taraire, nikau and karaka with (right) manawa (mangrove).

181

kingfish, koheru and trevally. There is also a group of inshore subtropical fishes with a rather limited distribution around the offshore islands and exposed rocky headlands of east Northland and including the colourful reef species: crimson cleaner, elegant wrasse, orange wrasse, blue knifefish, black angel fish and gold-ribbon grouper.[3]

The Aupourian Province of New Zealand — as well as its primary and more important relation with south-east Australia — has also an affinity with the Pacific to the north-east. In fact, although tropical influence is strong in the Poor Knights enclave, the northern province has achieved nothing like a majority of eury-thermal tropical species, as from Tonga or Samoa. The largest contingent of our shore fish is the Tasman element we share with east Australia, including nearly all those restricted to the Bay of Plenty and Northland. Recent migration of echinoderms appears to have occurred from south-east Australia to northern New Zealand. Indeed a large proportion of all those Australian species capable of pelagic migration seem to have found suitable habitats in New Zealand.

Most shallow marine plant and animal groups have a number of widespread subtropical or north temperate species with their southern limits extending into the Aupourian Province. There is also a significant number of species that are essentially endemic to this Province, some of which we will meet in the pages that follow. Examples of Aupourian endemic intertidal fish are the bigeye, pink brotula and New Zealand flesh fish.

Aupourian Province: The northern islands

We may designate as the 'offshore islands of the north' the chain that takes in the Three Kings, furthest away to the north-west, the greywacke and volcanic islands from the Cavallis to the outer Hauraki Gulf, and continues east of Coromandel Peninsula to the Mercury and Cuvier Islands, and reaches to Mayor Island and White Island in the Bay of Plenty.

All these islands are washed to varying degrees by the warm **East Auckland Current**, flowing south-east along the edge of the continental shelf. This could bring from the near-tropics such a rich accession of larvae that a species enjoying such dispersal might maintain its numbers by constant replenishment rather than on-site reproduction. The status and abundance of a species population on an island can thus depend either on its fecundity or its ability to recruit new individuals.

The chanciness of such recruitment is responsible for many absences and anomalies. Thus, low crayfish numbers on the Poor Knights would seem attributable not to

overfishing but to poor recruitment. The absence of *Chamaesipho columna* at the Chathams, and of *Sypharochiton pelliserpentis* from several offshore islands, despite the obvious suitability of the terrain, may be ascribed to difficulty of larval transport. A small black chiton species, still unnamed, that Bob Creese has found common on islands but absent on the mainland, has avoided dispersal risks by cutting out the larval stage and brooding the eggs. The same would seem true of the island vermetid gastropod *Dendropoma (Novastoa) lamellosa*.

Northern labrid fishes commoner on, or confined to, offshore islands have clearly been brought by the East Auckland Current. Temperature fluctuations (+ or -2°C in an annual range of 6°C) control their transport and chances of survival. Thus Howard Choat found many subtropical fishes common in the 1970s were totally wanting in the 1980s.

Our northern offshore islands share some biological characteristics arising from their small landmass and remoteness from the mainland. Both these reduce the freshwater runoff, with consequent effects on salinity, water clarity and nutrient supply. Thus, with greater transparency offshore, the kelp *Ecklonia radiata* has increased its depth range from 1 m in the Waitemata Harbour to 20 m outside Goat Island, off Leigh, and to 50 m at the Poor Knights.

Small islands also experience greater wave action, focused by refraction from almost any direction of the wind, and giving an 'exposure' co-efficient higher than theoretically possible on a straight mainland coast.

Biotic differences from the mainland will be proportional not mainly to an island's distance offshore, but to its distance across the continental shelf, and hence the degree to which (regardless of distance from the mainland) it may become 'oceanic'. Thus, the Poor Knights, though only as far offshore as Little Barrier, are much nearer the edge of the shelf, closer to deep oceanic water. Hence they experience stronger currents, larger and more frequent upwellings, lower seasonal range in temperature and generally milder climate. Extended deep-water habitats on steeper subtidal cliffs will also be available. The same features occur where the 50 m isobath draws in against Lottin Point, East Cape, in effect an 'island', though at present land-attached.

It is not only from their scenic quality and sense of solitude that small islands get their allure, but from their biotic individuality, each with a peculiar species-mix and fortuitous local survival. Though distant enough from sustained human attack, their diverse communities are still vulnerable to raiding and disruption. Even more than for the mainland, it is the diversity of small island shorelines that cries out for protection under the Marine Reserves Act.

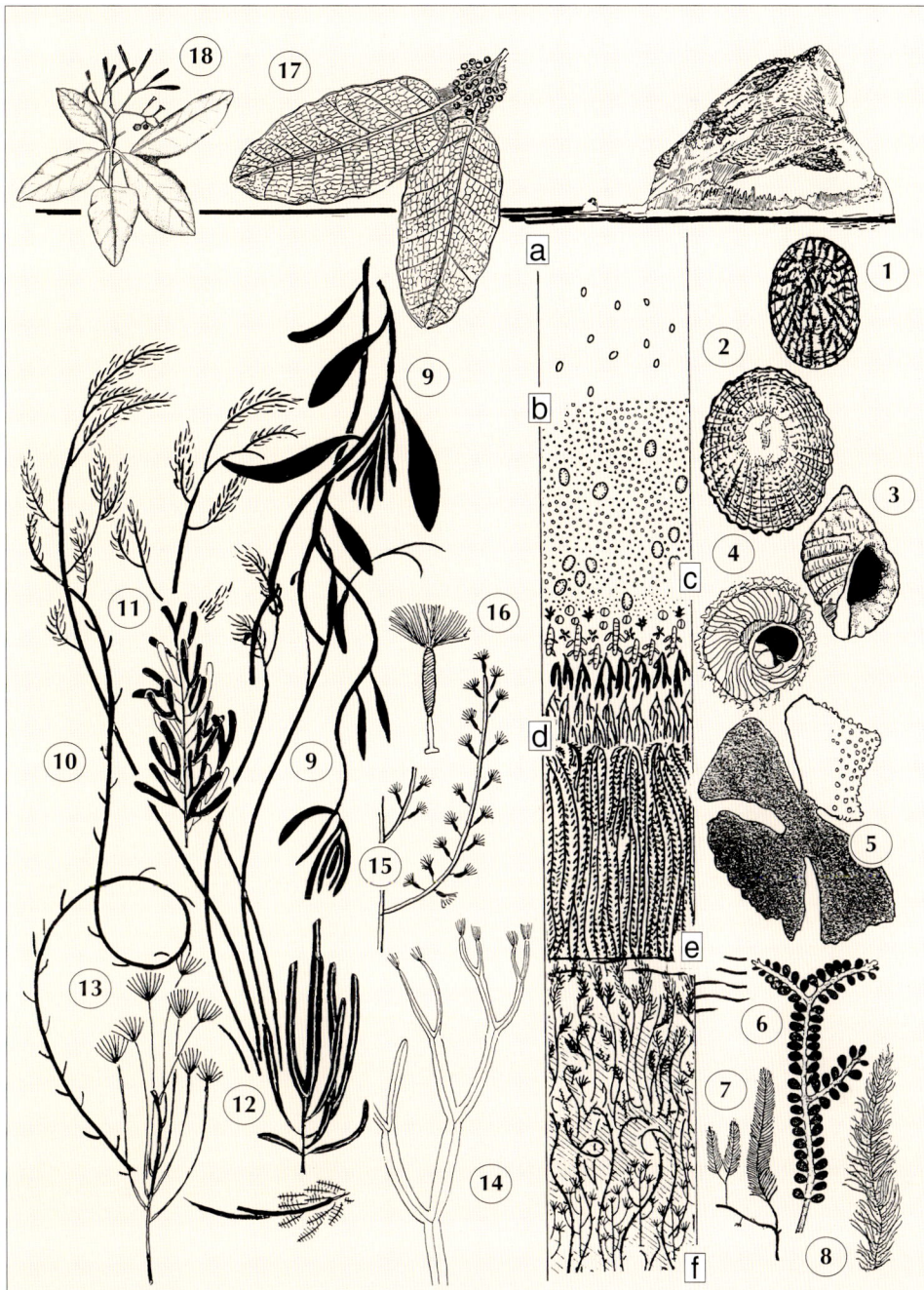

Fig. 10.6 Three Kings Islands: shore zonation at Great Island

a–b *Notoacmea pileopsis cellanoides;* b–c *Chamaesipho columna* and *Cellana denticulata;* c–d *Splachnidium,* *Catenellopsis* and *Epopella plicata;* d–e *Xiphophora chondrophylla* and *Carpophyllum angustifolium;* e–f *Sargassum* and *Perisporochnus*

1 *Notoacmea pileopsis cellanoides* 2 *Cellana denticulata* 3 *Neothais smithi* 4 *Dendropoma lamellosa* 5 *Pachymenia crassa* 6 *Caulerpa geminata* 7 *C. sertularioides* 8 *C. longifolia* 9 *Carpophyllum angustifolium* 10 *Sargassum johnsonii* 11, 12, the same, variant leaflets 13 *Perispoprochnus regalis* 14 *Carpomitra costata* 15 *Sporochnus stylosus* 16, same, detail of receptacle 17 puka, *Meryta sinclairii* 18 parapara, birdcatcher plant, *Pisonia umbellifera*. (Details kindly provided by Roger Grace and Richard Willan.)

The Three Kings

This cluster of small islands lies 72 km north west of Cape Reinga at 34 degrees south latitude, with a winter isotherm of 21°C. They are of outstanding interest for their high endemism and unique terrestrial vegetation and offer a classic example of how major ecological changes can be produced at a relatively small distance offshore.

Open over a wide arc to warm and cold currents, the Three Kings receive both cold upwelling and relatively warm water of subtropical origin. There is a resulting mix of 'cold' and 'warm' species, the latter including many subtropical fishes also found off warm east Northland.

The bull-kelp, *Durvillaea antarctica*, must owe its presence here both to continuous wave action and to cold upwelling with the resultant fog. The density of large seaweeds rivals that of the cold south. Though most of the mainland large brown algae extend to the Three Kings, there are also some 20 subtropical or east Australian algae not known from the mainland.[4] While the five species of *Caulerpa* are to be expected, the subtidal presence of *Desmarestia ligulata* has been attributed to cold upwelling within the East Australian Current flow.

Intertidal zonation is illustrated (Fig. 10.6) for the Landing, Tasman Bay, on the **Great King**. With a whole high-exposed shore, it is only the promontories under maximal wave attack that support *Durvillaea*. Most of the sublittoral fringe is curtained continuously with *Carpophyllum angustifolium*.

To be noticed first is the absence of many familiar mainland species, foremost the barnacles *Chamaesipho columna* and *Epopella plicata*. Chitons, trochids, turbinids, *Austrolittorina cincta* and *Lepsiella scobina* are likewise wanting. *Dicathais orbita* shows a remarkable elevation into secluded crevices from which it can feed on *Chamaesipho brunnea*. The distinctive far northern *Neothais smithi*, dull cream with broad spiral ribs, occurs at and beyond low water.

The **littoral fringe** has a strong black lichen cover of *Verrucaria maura*. Young *Austrolittorina antipoda* are immensely numerous in crevices, but adults are rare on open faces. *Notoacmea pileopsis*, in its subspecies *cellanoides* with strong radiate ribs, is common (100/sq m) under shaded ledges.

The **upper eulittoral** has only a patchy (50%) cover of *Chamaesipho brunnea*, interspersed with *Verrucaria* from above and invaded by *Apophlaea sinclairii* below. The dominant limpet of this zone at the Three Kings is rather remarkably *Cellana denticulata*, in the critical absence of the mainland *C. ornata*. We have noted already the spasmodic spread of this Cook Strait species up the east coast, with small, often non-breeding, populations from local spatfalls. From pre-European times, *C. denticulata* are prominent in middens in east Coromandel.

The large, fast-running crab *Leptograpsus variegatus* enjoys its normal high levels, taking refuge in rock fissures. In these shaded places is found the red anemone *Actinia tenebrosa* and the black snail *Nerita atramentosa*. Small bladders of the red alga *Catenellopsis oligarthra* and the little horny tufts of *Nothogenia pulvinata* reach the upper eulittoral.

Towards the bottom of this sparsely occupied zone appears *Cellana radians*, always fewer and lower sited

than *C. denticulata*. In the **middle eulittoral** *Catenellopsis* and *Nothogenia* thicken, and the reddish crustose *Apophlaea sinclairii* first appears. The **lower eulittoral** presents a pink band, up to 0.5 m wide. Coralline turf is scarce on the exposed coasts, supplanted by the jointed *Haliptilon roseum* and *Amphiroa anceps*. The gnarled crusts of lithophylla often have embedded in them the tubes of the vermetid *Dendropoma (Novastoa) lamellosa*, though this is nowhere as continuously zoned as at the Poor Knights (Fig. 10.7). There are also abundant pink-crusted *Patelloida corticata*.

The rich mix of red algae in the **sublittoral fringe** reaches to a metre above low water, in full continuity with the subtidal. In shade *Pterocladia lucida* dominates at highest level, displaced below by *Melanthalia abscissa*. Both are replaced in stronger light by long, parallel-branching *Xiphophora chondrophylla*. Below this short pelmet, a metre-wide unbroken curtain of *Carpophyllum angustifolium* descends from just above low water neap. As its thongs are parted in the waves, the pink paint is revealed, carrying wine-red *Pterocladiella capillacea*, *Melanthalia abscissa* and *Osmundaria colensoi*, brown *Glossophora kunthii* and green *Cladophoropsis herpestica*.

The brown algal forest of the **sublittoral zone** is highly distinctive in its composition at the Three Kings. *Ecklonia radiata* — where it occurs — is peculiarly short-stiped. The bushy *Sargassum johnsonii*, confined to these islands, is like none of its New Zealand kindred. The long, spiralling axes bear narrow, alternate-pinnatifid leaves, with leaf-like ramuli and minute elliptic vesicles on more apical branches. The deeper plants are larger-leaved.

Amid the dense groves of *Sargassum* lives a second endemic Three Kings brown alga, *Perisporochnus regalis*, belonging to its own small Order Sporochnales. The thick main axis repeatedly gives off side-branches that in turn bear whorls of branchlets of limited growth, with terminal plumes of hairs.

Other species of the order, found subtidally at Three Kings as well as on the mainland, are the more slender *Sporochnus stylosus* and *Carpomitra costata*, New Zealand's single member of that genus. The mid-ribbed dichotomous branches terminate in tufts of delicate hairs.

In maximal exposure, especially where the slopes are less steep, *Durvillaea antarctica* can replace *Carpophyllum angustifolium* and *Sargassum*. With a continuous *Durvillaea* curtain, the fringe above is generally not *Xiphophora* but *Pachymenia lusoria*.

On low-pitched boulder beaches, other brown algae attain prominence, including *Zonaria turneriana* with *Glossophora kunthii*, *Carpophyllum plumosum* (here found only subtidally) and *Landsburgia quercifolia*.

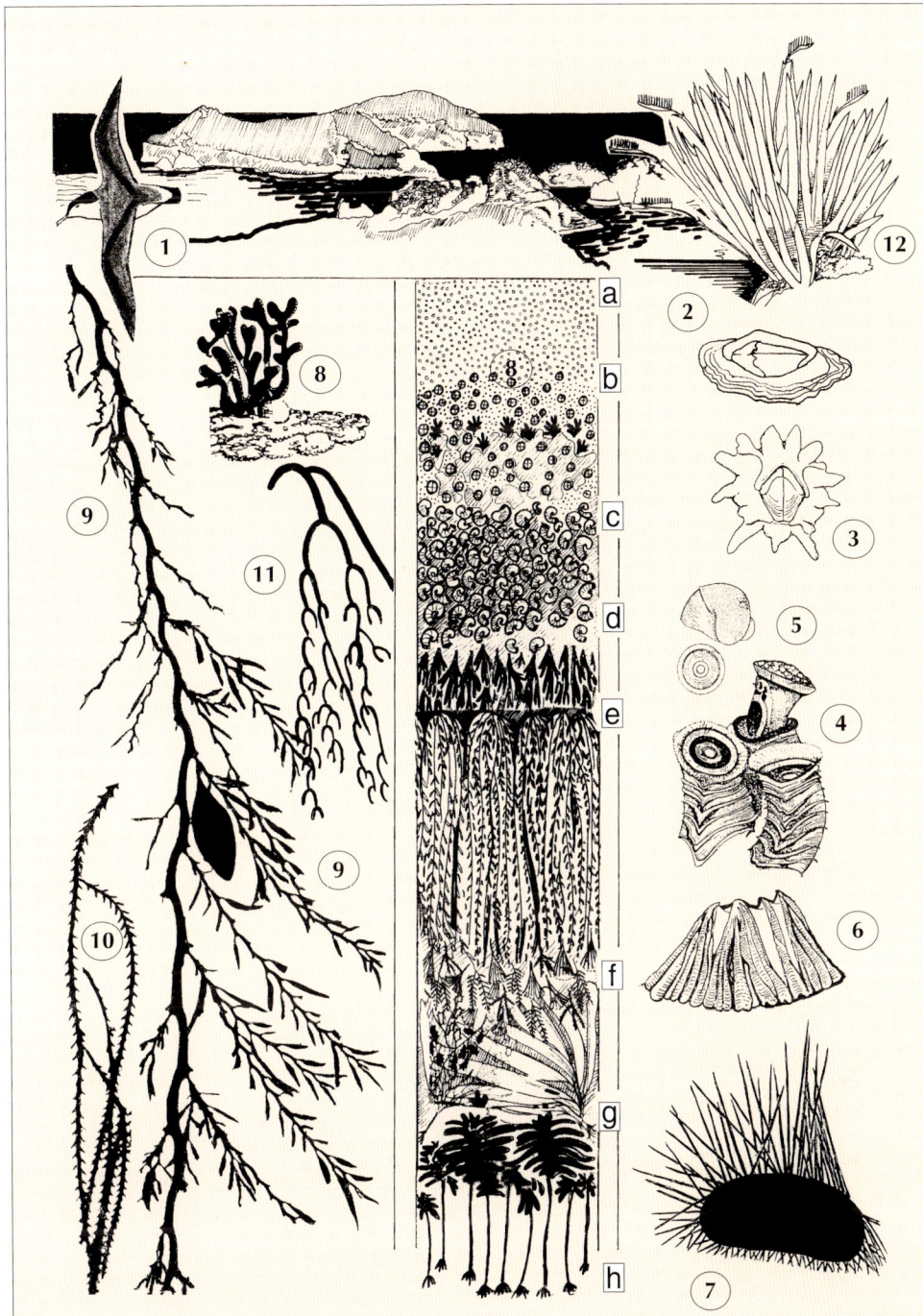

Fig. 10.7 Zoned shore at the Poor Knights Islands

a–b *Chamaesipho brunnea;* b–c *Epopella plicata* and *Apophlaea sinclairii;* c–d *Dendropoma lamellosa;* d–e *Xiphophora chondrophylla;* e–f *Carpophyllum angustifolium;* f–g *Lessonia variegata, Pterocladiella, Landsburgia* and *Melanthalia;* g–h *Ecklonia radiata*

1 Buller's shearwater, *Puffinus bulleri* 2 *Chamaesipho brunnea* 3 *Epopella plicata* (exposed form) 4 *Dendropoma lamellosa* 5 the same, operculum and embryonic shell 6 *Megabalanus tintinnabulum linzei* 7 *Centrostephanus rodgersi* 8 *Apophlaea sinclairii* 9 *Carpophyllum angustifolium* 10 *Osmundaria colensoi* 11 *Xiphophora chondrophylla* 12 Poor Knights 'lily', *Xeronema callistemon.*

The Poor Knights

Lying 24 km north-east of Whangarei Heads, the Poor Knights are a cluster of rhyolitic islands of Miocene age. Of the two main islands, Tawhiti Rahi, 130 ha and 240 m above sea level, is separated by a small passage from Aorangi, a conical peak rising to 254 m.

Pigs left on Aorangi by former Maori inhabitants were found in 1924 by Oliver and Hamilton to have caused huge devastation, nearly exterminating the flax snail *Placostylus hongii*, and forcing back the tuatara and eight petrel species to the cliff-faces. The last of the pigs was shot in 1936. The most remarkable of the cliff plants, unique to this group and Hen (Taranga) Island, is the red-stamened *Xeronema callistemon*, the Poor Knights 'lily'. Its close affinity with a second species in New Caledonia, along with the presence of karaka, puka and *Placostylus*, are all evidence for a Melanesian connection of the Poor Knights.

Of the breeding seabirds, the fluttering shearwater, *Puffinus gavia*, is securely established at higher levels. Lower down are the common diving petrel, *Pelecanoides urinatrix*, the grey-faced petrel, *Pterodroma macroptera*, and the rare Pycroft's petrel, *Pterodroma pycrofti*. Burrowing on the forest floor are Buller's shearwater, *Puffinus bulleri*, and also the fairy prion, *Pachyptila turtur*, resorting to tree-climbing to launch into the sea.

The littoral ecology of the Poor Knights has a combination of features unusual or unique. The islands are a sanctuary with restricted landing, and the shoreline and its surrounds out to 1 km from the coast were proclaimed a marine reserve in 1981. With deep visibility unexcelled, the subtidal cliffs constitute New Zealand's most acclaimed diving grounds.

Shore zonation

With its oceanic clarity and the high regularity of the zones, the Poor Knights were chosen for the pioneering intertidal study by Cranwell and Moore (1937), an early classic of shore description. Their simple schema (Fig. 1.2) affords historically the first recognition of important zoning differences in relation to shade, slope and wave exposure.

In the **upper eulittoral**, *Chamaesipho brunnea* is heavily dominant. It was here that this barnacle was first recognised as a '*Chthamalus* species' to be later fully described by Lucy Moore in 1944. *C. columna* has a reduced importance on these exposed shores. In the **middle eulittoral** it mingles with the much larger surf barnacle, *Epopella plicata*. Through the same zone runs a reddish band of mature *Apophlaea sinclairii* with upright reproductive branches. Neither of the northern mainland mussels *Xenostrobus pulex* or *Perna canaliculus* is to be seen. The tubeworm *Pomatoceros caeruleus* — if present at all — never forms a recognisable subzone. Instead, the **lower eulittoral** is heavily encrusted with a rough-cast of corallines. As on high-exposed shores in general, there is little sign of an algal turf; the corallines instead are present as thin films, nodular crusts and jointed branches, the last often seizing up immovably.

This calcareous base is richly studded with the vermetid gastropod *Dendropoma (Novastoa) lamellosa*, either fully coralline-embedded or revealing the pale mauve body whorl with its sharp, annular lamellae. These massed tubes form a strong band round most of the islands except in highest exposure. The operculum has a long insertion plug and is topped with a dome of coralline algae. The head and foot, quickly retracted on disturbance, are black and yellow, with the mantle edge scarlet. Using a mucus trap and to a smaller extent ciliary feeding, *Dendropoma* is clearly the molluscan analogue of the serpulid worms. Its presence on our east coast islands, south to the Chathams, is a distinctive subtropical feature.

The algal zonation at the Poor Knights resembles that of the northern mainland, though *Pachymenia* would appear to be lacking and *Gigartina alveata* reduced to a thin stubble. A fringe of *Xiphophora chondrophylla* (Fig. 10.8 shows its local branching form) precedes the curtain of *Carpophyllum angustfolium*, the only species of its genus present. In the wave-zone the red algal understorey includes *Pterocladia lucida*, *Pterocladiella capillacea*, *Melanthalia abscissa*, and *Osmundaria colensoi*. Beyond the tidal zone, the palm-like kelp *Lessonia variegata* first appears, with tall, long-stiped *Ecklonia radiata* dominating at greater depths, down as far as 18 m. Only on the most exposed outposts does the bull-kelp, *Durvillaea antarctica*, make its stand at the Poor Knights. Around its bases, embedded or encrusted in coralline algae, occurs the large northern-centred barnacle, *Megabalanus tintinnabulum linzei*.

Lottin Point: almost an island

Midway between Cape Runaway and Matakaoa Point, Lottin Point is formed of Cretaceous basalt. Near New Zealand's farthest eastward reach, it draws close to the Kermadec Trench, and with the 200 m depth line only a few tens of kilometres offshore, a small sea-level change would leave it a shelf-edge island. Like the northern offshore islands, Lottin Point has the vermetid *Dendropoma lamellosa* present in an unusual mainland occurrence. Biogeographically, Lottin Point stands as a natural break, with the last southern stand of *Carpophyllum angustifolium*, *Apophlaea sinclairii* and of *Gigartina alveata* (which is present in a distinctive miniature form, growing on *Epopella plicata*). *Cellana denticulata* appears to have its most northern mainland population here (save for a small stand at Te Kaha in the Bay of Plenty). *Durvillaea antarctica* and *Carpophyllum angustifolium*, here at its southerly limit, grow side by side. High on the shore *Nerita atramentosa* abounds.

Fig. 10.8 also shows the zonation of a slope in reduced exposure. The upper half of the eulittoral has dominant *Cellana denticulata*, with *C. ornata* fewer and smaller. The sublittoral fringe begins with a sharp-edged, straight band of *Xiphophora chondrophylla*, below which are the dominant fans of *Zonaria turneriana*, unusually wide ribbons of *Glossophora kunthii* and short straps of *Carpophyllum plumosum*. These are

Fig. 10.8 The zoned shore at Lottin Point

(top) View from above the shoreline, facing Matakaoa Point.

(centre) Shore in moderate exposure: a–b *Chamaesipho columna, Cellana denticulata* and *Nerita atramentosa;* b–c *Xiphophora;* c–d *Pseudoscinaia, Carpophyllum plumosum, Glossophora kunthii* and *Scinaia firma;* d–e *Lessonia variegata*

(right) Shore in high exposure: a–b *Nerita atramentosa;* b–c *Chamaesipho brunnea;* c–d *Gigartina* and *Epopella;* d–e *Corallina* crust and (lower) *Dendropoma lamellosa;* e–f *Durvillaea antarctica* and *Carpophyllum angustifolium.*

1 high eulittoral poll with *Hormosira banksii* showing (left) shallow specimen, (right) deep-attached with epiphytic *Notheia anomala* 2 *Glossophora kunthii* 3 *Zonaria turneriana* 4 *Scinaia firma* 5 *Carpophyllum plumosum* 6 *Cellana denticulata* 7 *Xiphopora chondrophylla* 8 *Nerita atramentosa* 9 *Epopella plicata* with *Gigartina* 10 *Gigartina* aff *alveata* 11 *Dendropoma lamellosum.*

mingled with the pink, smooth and slippery rhodophyte *Scinaia firma*. The dominant sublittoral kelp is *Lessonia variegata*.

With its high ecological and geographic interest, Lottin Point well deserves Marine Reserve status.

Kermadecian Province

The Kermadec Islands form a small volcanic group of Pleistocene to Recent volcanic origin, some 1000 km north-east of Auckland. At the northern outpost of New Zealand's jurisdiction, they are obviously distinct as a separate biotic Province within the warm temperate Northern New Zealand Region. Lying between latitude 29 degrees and 31.5 degrees south, with a summer isotherm of 23°C but an August surface sea temperature falling to a mean of 17°C, the Kermadecian Province clearly falls within the warm temperate rather than the tropical sphere.

The shore faunal endemism of the Kermadecs is high in some groups, including 68 of the 358 molluscs[5] and five of the 13 echinoderms. Nancy Adams finds that the seaweeds would not uphold Lucy Moore's separate algal province, being largely of genera and species common to most of the subtropical Pacific. The phycologist will, however, find many algal families enriched far beyond their New Zealand representation.

Seventeen hermatypic (reef-building) and seven ahermatypic corals are present subtidally around the Kermadecs.[6] Only one coral, the ahermatypic *Coenocyathus brooki*, is endemic to the Kermadec Islands. The rest of the corals are shared with other subtropical south-west Pacific islands. Only *Culicia rubeola* is also found in mainland New Zealand. A similar pattern is seen in the intertidal rock-pool fish with two endemic species, but most are shared with warm temperate to subtropical east Australia, Lord Howe and Norfolk Islands.[7]

The island of **Raoul**, to the north, is still volcanically active, with a mantle of pumiceous tuff over its andesitic mass. With their remoteness and relative youth, the Kermadecs have obtained most of their land flora by random floating and the agency of birds, plant endemism being low as compared with Norfolk and Lord Howe. The vegetation, including subcanopy trees and shrubs, has been badly debased by goats. Even on Raoul, where the forest has suffered least, the dominant pohutukawa (*Metrosideros kermadecensis*) is failing to regenerate.

Zonation

The transect of a steeply falling tidal shore (Fig. 10.9) is based on a study at **Meyer Island**, off Raoul. The **littoral fringe** is virtually devoid of littorines, though there have been sparse recordings of *Austrolittorina antipoda*, and of the tropical Pacific *Nodilittorina millegrana*. The upper half of the **upper eulittoral** is spangled with the white or pink-tinted barnacle *Tesseropora rosea*. Black *Nerita atramentosa* are common all over the upper shore, as also at Norfolk Island and in northern New Zealand. The tropical *Nerita plicata* turns up only sparsely.

The limpets of the upper eulittoral include the smaller patellid *Cellana craticulata* and the pulmonate *Siphonaria cauleyensis*. The thaid predator in this zone is *Neothais smithii*, shared with New Zealand's Three Kings. In rock crevices, Bob Creese has discovered the presence of the small brooding *Chiton themeropsis*.

In the **middle eulittoral** and towards low water the small pools are lined with the fans of the brown alga, *Padina fraseri*. Higher up, the pools may contain the mussel *Modiolus auriculatus* and the crevice-nestling *Isognomon* sp. Deeper pools are fringed with short *Sargassum tahitense*, *S. cristaefolium* flourishes intertidally in surge.

The **lower half of the eulittoral** is marked by a change from barnacles to pink coralline paint that supports around low water the larger barnacle *Megabalanus tintinnabulum linzei*. Largest too of all Pacific limpets is the remarkable endemic *Scutellastra kermadecensis*, that could be called the icon of Kermadec shores. Ivory coloured and with the margin orange-tinted, these heavy, low-pitched shells, often scarred by attachment of smaller ones, may reach 100 mm long. Bob Creese has found *S. kermadecensis* to undergo sex-change, with all individuals below 60 mm males and enough of the older age-classes then becoming female to result in a balanced sex ratio.

From the pink zone to well beyond low water, the Kermadec algae are small but profuse, most being under 15 cm tall and set close to the rock, generally in multi-species turfs. Vegetative proliferation by growth of stolons mitigates the toll by fishes, echinoderms and gastropods, allowing recovery after apical portions have been grazed away. Green algae abound near low water, including prolific *Caulerpa racemosa*, which entirely replaces *Corallina* in shade. *Boodlea composita* forms a soft turf, and *Cladophoropsis herpestica* is present in a harsher green stubble. Dense mats of *Caulerpa webbiana* and *C. racemosa* proliferate over the sand around the rocks. Regularly associated with *Caulerpa* are the cell-sucking *Arthessa* and *Oxynoe*, both primitive shelled sacoglossans (Fig. 10.9).

The browsing aplysioid *Dolabrifera brazieri* crops the succulent algae around low tide. Ranging Pacific-wide and well established in northern New Zealand, this sea-hare is low-built, with a broad sole for secure prehension in fast-moving water. Its mossy or bottle-green colour is marbled with brown, mauve and white. The

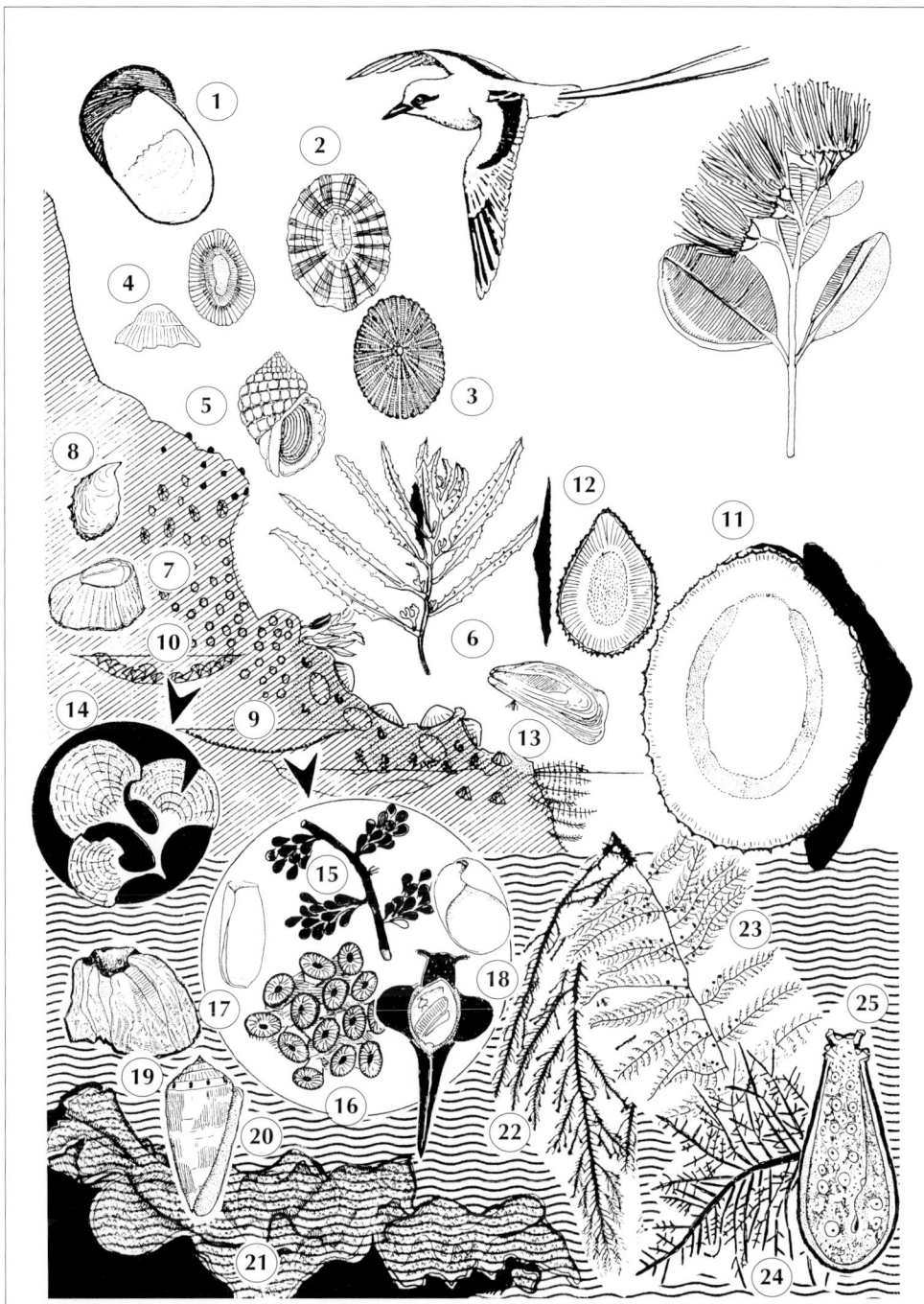

Fig. 10.9 Kermadec Islands: zoned shore at Meyer Island, off Raoul
(top) red-tailed tropicbird, *Phaethon rubricauda* and **(right)** pohutukawa, *Metrosideros kermadecensis*
1 *Nerita atramentosa* 2, 3 *Cellana craticulata* 4 *Siphonaria cauleyensis* 5 *Neothais smithi* 6 *Sargassum tahitense*
7 *Tesseropora rosea* 8 *Isognomon* sp. 9 pool with *Caulerpa* 10 pool with *Sargassum* 11 *Scutellastra kermadecensis*, adult,
with profile 12 the same, immature 13 *Modiolus auriculatus* 14 *Padina fraseri* 15 *Caulerpa racemosa* 16 *Palythoa caesia*
17 *Arthessa* 18 *Oxynoe*, animal and shell 19 *Megabalanus tintinnabulum linzei* 20 *Conus lischkeanus* 21 hermatypic
corals, *Turbinaria frondens, Montastrea curta* 22 *Delisea pulchra* 23 *Asparagopsis taxiformis*
24 *Pterocladiella capillacea* 25 *Dolabrifera brazieri*.

parapodia are more fully fused than in the kindred *Bur-satella* (Fig. 6.9), from which *Dolabrifera* also differs in retaining its chitinous internal shell. Locomotion is almost leech-like (Fig. 10.9) with a narrow forward extension leaving the sole emplaced, to be then brought forward to complete each 'step'. There are in addition

two large *Aplysia* species, *A. dactylomela* and *A. extra-ordinaria*, to be seen by snorkelling, while *A. parvula* turns up in rock pools.

The largest of the algae of the sublittoral fringe is the golden brown *Sargassum tahitense*. Seasonally variable, it produces basal fronds in March, to become long and fertile from September to November. The widespread tropical sargassoid *Turbinaria* seems to be lacking. Characteristic warm-water brown algae include *Hydroclathrus clathratus* (as well as its cosmopolitan relative *Colpomenia sinuosa*). The warm-water Dictyotales are well represented, with nine genera. *Padina fraseri* occurs in pools and also on limpets. *Distromiun skottsbergii* forms a turf on open coasts and *D. didymothrix* lives on stones and coral rubble. *Dictyota intermedia* is abundant subtidally, while *D. bartayresiana* and *Lobophora variegata* are found in low level pools on rock and rubble.

A wide diversity of rhodophytes, including crustose corallines, accounts altogether for over half the algal list.[8] Foliose and filamentous reds were found to cover more than a quarter of the surface. They include such warm-water entities as *Asparagopsis taxiformis*, *Martensia fragilis*, four species of *Galaxaura*, *Liagora harveyana*, *L. farinosa*, and species of *Taenioma nanum*, *Spyridia filamentosa* and *Wrangelia penicillata*. Beyond

low water are *Pterocladiella capillacea*, *Delisea pulchra*, *Gelidium longipes* and *Champia parvula*, with *Plocamia* and *Polysiphonia* species especially common down to 9 m.

In the 1980s, New Zealand biologists David Schiel, Michael Kingsford and Howard Choat studied the ecology of the subtidal grazers. In the first 2–3 m, giant *Patella* (21/sq m) covered some boulders completely. Of the echinoids *Heliocidaris tuberculata* was the most abundant, with sparser *Tripneustes gratilla* and sporadic *Phyllacanthus imperialis* and *Centrostephanus rodgersi*, as well as a few *Diadema* on large table corals. The coral-predating crown of thorns star, *Acanthaster planci*, was occasionally present.

Predatory gastropods of the subtidal include *Neothais palmeri* (known also from the Poor Knights), *Conus lischkeanus* and the nudibranch *Phyllidiella pustulosa* beset with bright orange mammillae. The turbinid *Angarina delphinus* and the small cowry *Cypraea moneta* are also subtidal.

Fig. 10.10 Chatham Islands: the high-exposed shore at Nelly's Nook
a–b adlittoral with *Olearia* and *Muehlenbeckia*; b–c supralittoral zone with lichens, and *Enteromorpha* and *Austrolittorina*; c–d *Chamaesipho brunnea*; d–e pink coralline veneer, with *Pachymenia lusoria*; e–f *Durvillaea* band
1 *Durvillaea chathamensis* 2 *D. antarctica* 3 *Dendropoma lamellosa* 4 *Pachymenia lusoria*
5 broad-billed prion, *Pachyptila vittata* 6 *Olearia traversii*. (Based on notes from Euan Young and Alison Davis.)

Although 17 hermatypic corals live around the Kermedecs, they have not developed into any continuous reef. They grow as scattered colonies, mostly encrusting (such as *Leptastrea, Montipora, Turbinaria, Pavona, Hydnophora, Cyphastrea, Goniastrea*), but some form branching colonies (*Pocillopora damicornis*), vasiform colonies (*Turbinaria frondens*) or submassive colonies (*Montastrea curta*). Around Raoul Island hermatypic corals form 20–40% of encrusting rock-cover at 1–6 m depth with decreasing cover down to about 30 m.

After prolonged advocacy from conservation groups, the Kermadecs were finally in 1990 gazetted under the New Zealand Marine Reserves Act. The resulting zone is now one of the world's largest protected marine habitats, equalling in area the 'no-take' zone of Queensland's Great Barrier Marine Park. With a radius of 12 nautical miles, the Kermadec Reserve covers 7350 sq km. As New Zealand's only true subtropical habitat, the Kermadecs could have been considered far enough from the mainland to have escaped heavy commercial depredation. Already, however, the spotted black grouper, *Epinephalus daemelii*, had come under threat. A large species up to 1.2 m long, it was by its very friendliness and natural curiosity so vulnerable as to have been fished out already in Fiji, and its stock severely reduced at Lord Howe Island.

Southern New Zealand region: Moriorian (Chathams) Province

The **Chatham Islands** lie some 900 km east of Banks Peninsula on the eastern end of the submarine Chatham Rise. These islands are largely composed of schists, overlain by Cretaceous and Tertiary volcanic rocks and shallow marine limestones. Islands of varying size have probably been present throughout the Tertiary. The two largest, Chatham and Pitt Islands, have a land area of 200,000 ha (a fifth of it lake or saline lagoon), with two smaller islands, Mangere and South East (or Rangatira) lying close alongside Pitt.

Assessment of the biogeographical status of the Chathams is founded largely on the results of the DSIR marine expedition of 1954, as reviewed by George Knox in 1957.[9] Based on the Mollusca, Harold Finlay had in 1927 set up the separate Moriorian Province. Figures for species endemism differ widely between groups. The molluscs — taxonomically well-worked — show an endemism of 15%, and the sponges 23% (or five out of the 23 collected). On the other hand there is no endemism in the 18 brachyuran crabs, and little in the algae, fishes, echinoderms and foraminifera.

The terrestrial endemics at the Chathams include 10% of all the vascular plants, including the notable giant forget-me-not, *Myosotidium hortensia* (Fig. 10.12), the giant *Leptinella featherstonii* and a spaniard *Aciphylla dieffenbachii*. Nine of the 11 surviving forest birds are specifically or subspecifically distinct, the best known today being the recently salvaged black robin.

As well as by the former clearance of natural vegetation, the Chathams have suffered more than other islands from heavy commercial depredation. The 'crayfish boom' of the 1970s, wrote Bill Ballantine, 'was conducted with the same speed, waste and carefree ignorance as a gold-rush'. Similar assaults have happened to scallops, paua, kina and other saleable species.

Lying across the subtropical convergence where a warmer, more saline water-mass meets with cold, less saline subantarctic waters, the Chathams enjoy an equable climate on a summer isotherm of 19°C, mean annual air temperature of 11°C and a sea temperature of 13°C. The littoral biota reveals an odd blend of northern and southern elements. Cold water algae such as *Apophlaea lyallii* and *Adenocystis utricularis* grow alongside northern *Carpophyllum plumosum*. *Pachymenia lusoria*, once considered a southern species, is evidently conspecific with populations New Zealand-wide. The warm-water *Chamaesipho brunnea* is the single common barnacle, and the zoning vermetid *Dendropoma lamellosa* here makes its southernmost stand.

The intertidal fauna has more in common with the mainland than we shall find in the far southern islands. The greatest difference is in the absences of numerous mainland species from failure of larval dispersal. Notably lacking on the Chathams are the barnacles *Chamaesipho columna, Austrominius modestus, Calantica spinosa*, and the tubeworm *Pomatoceros caeruleus*. Grapsid crabs will be looked for in vain, as well as the elsewhere abundant *Petrolisthes elongatus*. Missing too are such common mainland molluscs as *Chiton glaucus, Plaxiphora obtecta, Eudoxochiton nobilis, Turbo smaragdus, Nerita atramentosa, Scutus antipodes* and *Benhamina obliquata*. The single patellid limpet is *Cellana strigilis*, with its local subspecies *chathamensis*.

Shore zonation

The small **South East Island**, the site of field studies by biologists from the University of Auckland, has kept its pristine coastal vegetation, followed downshore by rich intertidal sequences.[*]

Around **Nelly's Nook** at the island's north-west point, the low-sloping shore is girded with two species of *Durvillaea*. The mainland *D. antarctica* is accompanied here with *D. chathamensis*, a species peculiar to the

---

[*] A debt is owed to Euan Young and Alison Davis for access to their field data, specimens and photographs, on which much of this description is built.

Fig 10.11 Zoned shore at the Landing Place, South East Island, Chatham Islands

a–b *Chamaesipho brunnea* and **(lower)** *Apophlaea lyallii*; b–c *Epopella plicata, Catenellopsis, Adenocystis* and *Ulva*; c–d corallines and *Xiphophora gladiata*; d–e *Cystophora scalaris* and (lower) *Carpophyllum plumosum* form *quercifolium*; e–f *Marginariella urvilliana* and *Lessonia variegata*

1 *Chamaesipho brunnea* 2 *Apophlaea lyallii* 3 *Adenocystis utricularis* 4 *Cellana strigilis chathamensis* 5 *Catenellopsis oligarthra* 6 *Modelia granosa* 7 *Margarella fulminata* 8 *Cantharidus opalus cannoni* 9 *Xiphophora gladiata* 10 *Carpophyllum plumosum* form *quercifolium* 11 *Landsburgia myricifolia* 12 *Marginariella urvilliana* 13 shore plover, *Thinornis novaeseelandiae.*

Chathams. No more than 3 m long, this second bull-kelp grows seaward of *D. antarctica*, at or just beyond low water. Unlike *D. antarctica*, the lamina has no honeycomb tissue. The blade expanding from a stipe, no more than 18 cm long, is broad and thin, with the margins sometimes undulated. It breaks distally into flattened thongs.

The coastal scene at Nelly's Nook (Fig. 10.10) begins with an **adlittoral** vegetation of low, wind-contoured *Olearia traversii*, fronted by a tangle of pohuehue (*Muehlenbeckia complexa*). Nesting here are the white-faced storm petrel, the Chatham petrel and the broad-billed prion.

The shelving **supralittoral zone** carries lichens (whitish, grey-green and yellow in downward order) along with a rich mix of shore succulents in rock-cracks and fissures. In the **littoral fringe**, swathes of green *Enteromorpha* grow in seepage, with both the periwin-

kles *Austrolittorina antipoda* and *A. cincta* on the surrounding rock.

The **upper eulittoral** carries the barnacle *Chamaesipho brunnea*, and becomes seasonally green-tinged with *Cladophora, Ulva* and *Enteromorpha*. Beyond the barnacle belt, a continuous pink and white band encircles the **lower eulittoral**, with crustose lithophyllum and smooth pink veneer. Turfs of *Corallina officinalis* along with the tails and tufts of *Haliptilon roseum* dominate here. Other algae are the seasonal *Splachnidium rugosum, Adenocystis utricularis* and *Glossophora kunthii*. *Xiphophora gladiata* grows as fringes in low level pools. The single mussel species is the navy blue *Mytilus edulis aoteanus*. The vermetid gastropod *Dendropoma (Novastoa) lamellosa* is found embedded in the lithophyllum crust.

The Chatham Islands limpet, *Cellana strigilis chathamensis*, is large and solid, coarsely ribbed, with the interior yellowish brown. Two small acmaeid limpets are common. *Notoacmea parviconoidea* and coralline-crusted *Radiacmea inconspicua* are often found on the backs of *Haliotis iris*. In the absence of *Turbo smaragdus*, the common turban shell is the heavily pink-encrusted *Modelia granosa*.

The common trochid browsing on algae in tidal pools is the prettily marked Chatham Islands endemic *Margarella fulminata*. *Trochus viridis* and *Micrelenchus dilatatus* are regularly to be found on brown algal fronds, along with *Calliostoma punctulata* and deeper offshore *C. tigris*. The large, handsomely marbled *Cantharidus opalus*, in its local subspecies *cannoni*, lives upon bull-kelp.

Chatham shores have not only the three paua, *Haliotis iris*, *H. australis* and *H. virginea morioria*, but a good representation of fissurellid limpets. *Montfortula rugosa* lives in eulittoral pools, *Tugali elegans* and *Emarginula striatula* are common under boulders, with *Monodilepas skinneri* offshore. *Incisura rosea* and the small fissurellid *Incisura lytteltonensis* occur in the washings of fine red algae. The Family Scissurellidae, also with primitively spiral shells and short pallial slit, offers as many as four minute species of *Sinezona*.

The lowest margin of the pink zone is dull red, with a band of *Pachymenia lusoria*. Immediately below is the *Durvillaea* that runs as a prominent deep orange zone, just above the white line of wave-break.

The **Landing Place** on South East Island enjoys more shelter than Nelly's Nook, with a lower-sloped, shelving shore. Fig. 10.11 brings out its differences from the exposed *Durvillaea* shore. The **upper eulittoral** is clad with *Chamaesipho brunnea* along with *Cellana strigilis chathamensis* and *Mytilus edulis aoteanus*. Common algae are *Catenellopsis oligartlra*, *Adenocystis utricularis*, *Apophlaea lyallii* and *Ulva lactuca*. The *Gigartina* species at the Chathams include *G. decipiens*, *G. ancist-*

roclada, *G. livida*, *G. clavifera*, *G. atropurpurea*, and *G. lanceata*.

The lithophyllum-crusted **lower eulittoral** has *Halopteris funicularis* and *Haliptilon roseum*, while the brown algae of the **sublittoral fringe** are preceded by the pelmet of *Xiphophora gladiata*. Below this come *Carpophyllum plumosum* in its oak-leaved variety *quercifolium* (Fig. 10.11) characteristic of the Chathams. *Glossophora kunthii* is plentiful, with *Marginariella boryana* in the immediate sublittoral and — a little offshore — large groves of *Lessonia*, represented by the Chathams species *L. tholiformis*. A distinctive Chathams endemic is *Landsburgia myricifolia*, with narrow, finely serrate leaves.

The level cliff-tops some 30 m above these shores have their still unspoiled botanical treasures. Amid a pink flower garden of *Disphyma australe*, with the endemic *Hebe chathamensis*, and the large flowered *Geranium traversi*, there are bushes of the azure-blue giant forget-me-not, *Myosotidium hortensia* (ineptly called the Chatham Island 'lily'; Fig. 10.12). With it, or on the salt meadow below the cliffs, grows the endemic (and unprickly) spaniard, *Aciphylla traversii*.

The beach-heads at the Landing Place and Thinornis Bay, facing east, are the haunt of the rare Chatham Islands plover (*Thinornis novaeseelandiae*), nesting among boulders and in the salt meadow. Further back, a wiry tangle of pohuehue (*Muehlenbeckia complexa*) has taken back possession of the former paddocks. Beneath it the soil is riddled with the collapsible burrows of penguins and at least eight petrel species. Today — in all — 15 species of tube-noses (petrels and their allies) resort to the Chathams to breed

Fig 10.12 Giant forget-me-not
Giant forget-me-not *Myosotidium hortensia* (Chatham Island 'lily').

Cold temperate New Zealand: Cookian Province

As part of the South Pacific cold temperate, the **Southern New Zealand Region** embraces on the mainland the Cookian Province, and for the outlying islands the Provinces now designated Moriorian and Antipodean.

In the **Cookian Province**, as over this whole cool Region, the shore is most visibly characterised by the bull-kelps *Durvillaea*, fringing all exposed coasts and in the south pressing increasingly into moderate shelter. Harbour mouths and channels with strong currents, but free from heavy waves, are fringed with the bladderkelp, *Macrocystis pyrifera*, most luxuriant towards the south, where it is associated with the long-stalked tunicate *Pyura pachydermatina*. While *Durvillaea antarctica* has its far outposts at exposed points in the Aupourian Province, *D. willana* is entirely Cookian, only just reaching a few southern points in the North Island. With one

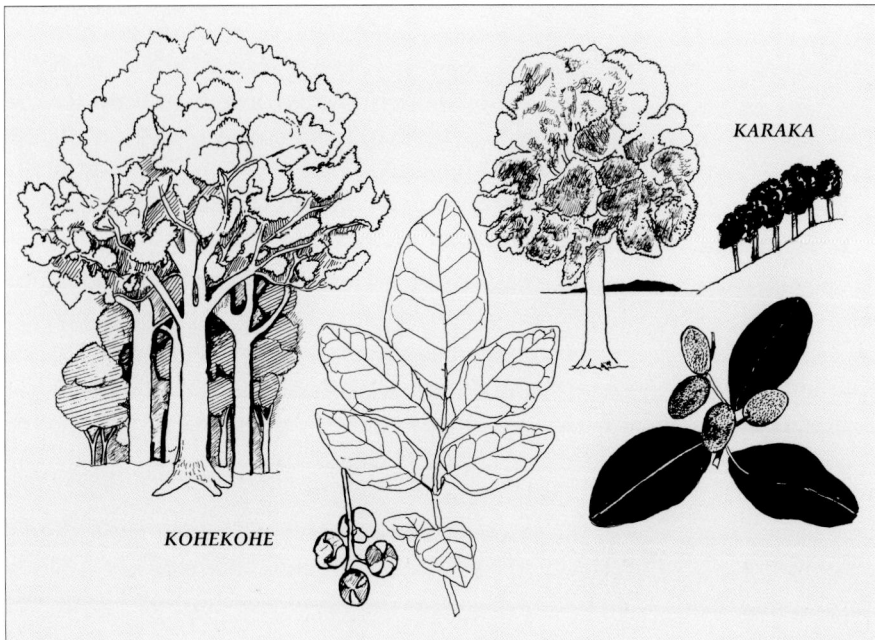

Fig. 10.13 Karaka and kohekohe
Karaka, *Corynocarpus laevigatus*, and kohekohe, *Dysoxylum spectabile*.

of its species (*C. angustifolium*) dropped out in the Cookian, and the other three progressively reduced, *Carpophyllum* now loses its dominance to *Cystophora*, with all of its species commoner in the south and three exclusive to the South Island. Neither *Marginariella* nor *Desmarestia* reaches north of Wellington. So too *Apophlaea lyallii*, *Adenocystis utricularis* and a goodly number of the *Gigartina* species all belong to the southern part of the South Island. Of the smaller kelps, *Ecklonia radiata* — while still present on the southern mainland — becomes in the south entirely subordinate to *Lessonia*.

In the sheltered **middle eulittoral**, blue and ribbed mussels entirely replace *Crassostrea*; though there is a small middle eulittoral rock oyster found from Dunedin south. The cold-water *Ostrea lutaria* grows larger in the south. The green-lipped mussel, *Perna canaliculus*, with its starfish *Stichaster australis* and the thaid *Lepsithais lacunosus*, march typically with bull-kelp on open southern shores.

The **adlittoral** coastal forest in the Cookian Province is far less diverse than in the north. Groves of kohekohe (*Dysoxylum spectabile*) (Fig. 10.13) flourish at Waikanae and Paraparaumu, north of Wellington. Puriri and pohutukawa hardly reach to Cook Strait. Karaka (Fig. 10.13) is still a notable coastal tree, and increasingly important in the South Island is ngaio, *Myoporum laetum* (Fig. 10.14), notably on the Marlborough and North Canterbury coast, with the cliff daisy, *Pachystegia insignis*, and rauhuia, *Linum monogynum*. In the high-exposed far south ngaio is replaced by muttonbird scrub, *Senecio reinoldii*.

The cold south

At Stewart Island and the South Island's southern extreme a significant subantarctic influence has come to bear. For his cold *Forsterian Province* (included here with the *Cookian Province*) Powell listed only 444 Mollusca, compared with 602 for his *Cookian* and 665 for the *Aupourian Province*. Lower diversity is compensated by the vast numbers of individuals of certain southern species. Distinctive of Otago and Southland as well as Stewart Island and the Snares is the locally common limpet, *Cellana strigilis redimiculum*. For the southern algae we have already cited *Apophlaea lyallii*, *Lenormandia chauvinii*, *Adenocystis utricularis* and *Scytothamnus fasciculatus* and *Pachymenia lusoria*, as well as peak status of the *Durvillaea* species.

The molluscs have a notable subantarctic element in the small bivalves of the families **Gaimardiidae** (*Gaimardia*), **Cyamiidae** (*Kidderia* and *Perricrina*), **Philobryidae** (*Philobrya*), and the kelp-browsing trochids *Margarella*, along with the siphonariid limpet *Kerguelenella stewartiana*.

As the third and smallest of our main islands,

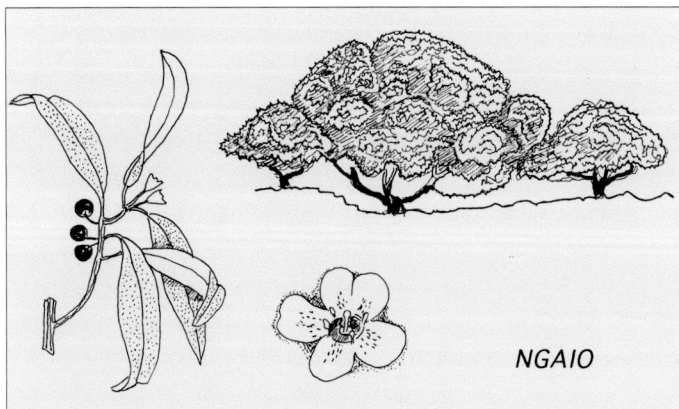

Fig. 10.14 Ngaio, *Myoporum laetum*

Fig. 10.15 Coastal islet near Oban, Stewart Island
1 *Senecio reinoldii* 2 *Metrosideros umbellata* 3 *Dracophyllum longifolium* 4 small *Dacrydium cupressinum* 5 high-tidal fringe of white, then tar-black lichen followed below by *Porphyra*.

Stewart Island (Rakiura) lies across Foveaux Strait 24 km from Bluff, washed by the Southland Current with a flow from the Subtropical Convergence mixing with cold subantarctic water. Temperatures average a little higher than in Southland, but with the plant species mainly hardy to cold.

The coastline is in all some 750 km round, deeply indented on the east, with a profusion of offshore islets known as 'nuggets' in the bays. Human settlement is confined to Oban, with the rest of the coast in its pristine state. The shore character is set by the adlittoral zone of coastal bush that overhangs and shades the supralittoral. The almost grey foliage of muttonbird scrub (*Senecio reinoldii*) contrasts with the dull mauve-green backdrop of *Fuchsia excorticata* and mingles with the light bronze of *Dracophyllum longifolium*. In the summer the seaboard's high colour is given by the scarlet flowering southern rata, *Metrosideros umbellata*.

The Snares

Consisting of Main and Broughton Islands, and a cluster of off-lying stacks, the Snares are composed of granite with a peat mantle averaging 2.5 m deep. Westerly storms lash the coast and the want of safe anchorage has left the islands uniquely undisturbed and immune from introduced mammals.

With 20 species of flowering plants, the chief forest cover down to the cliff-margin is of the tree daisies, blue-grey *Olearia lyallii* and bright green *Brachyglottis stewartiae*, both tending to lie prostrate on the peat. The coastal *Hebe elliptica*, seen already in the South and Stewart Islands, here forms impenetrable thickets, and rhubarb-like *Stilbocarpa robusta* is found in cleared patches along the margin.

Thirteen seabird species nest in millions in burrows and crevices, including prions, cape petrels, diving petrels, gulls and terns. Best known and most numerous is the muttonbird or sooty shearwater, *Puffinus griseus*.

Breeding nowhere else in the world is the Snares crested penguin, *Eudyptes robustus*, today numbering

Fig. 10.16 Snares Islands
Looking landward from the intertidal, with Snares crested penguin, *Eudyptes robustus*, and *Brachyglottis stewartiae* with shrub layer of *Hebe elliptica*.

30,000–50,000 in 130 rookeries scattered through the forest down to the open bedrock of the shore. Three land birds with Snares subspecies, the black tomtit, fernbird and snipe, come right to the shore.

Shore zonation is exposed to high wave action, with a continuous curtain of *Durvillaea antarctica* descending to low water. Prominent above this is a red zone of chiefly *Gigartina lanceata* and also the green finger *Codium gracile*. Next back is a pink-painted coralline zone, and higher up the pale lichen *Pertusaria graphica*, which can be slippery when wet. *Austrolittorina cincta* is present but *A. antipoda* lacking. Barnacles are scarce but the surf species *Epopella plicata* reaches the Snares. The local limpet is *Cellana strigilis* in its subspecies *flemingi* along with *Notoacmea pileopsis sturnus*. *Siphonaria australis* and *Kerguelenella stewartiana* are both present. The algal-browsing trochid is *Margarella antipoda rosea*, very variable in colour.

Antipodean Province

This south-lying province of the New Zealand Cold Temperate Region comprises the distant subantarctic island groups, all administered by New Zealand. The four island groups lie around the oceanward southern and eastern margin of the Campbell Plateau, which has been submerged since the Eocene, for at least 40 million years. Thus none of these volcanic islands has ever been connected by land or shallow water to mainland New Zealand. With increasing distance from the mainland the islands are: Auckland (300 km distant), Campbell (500 km), Bounty (800 km), and Antipodes (850 km).

Around all these islands the coasts are surf-beaten and devoid of any real shelter. As well as their species peculiar to New Zealand, they have some wide-ranging subantarctic entities, especially algae. Campbell and Auckland Islands possess the three kelps *Durvillaea antarctica*, *Macrocystis pyrifera* and *Lessonia brevifolia*. The warmer water *Ecklonia radiata* reaches only to the Snares. Campbell, Auckland and Antipodes Islands have *Scytosiphon lomentaria*, *Adenocystis utricularis*, *Scytothamnus fasciculatus* — all lacking at Bounty Island. Mainland algae unknown from these southern islands include *Hormosira banksii*, *Glossophora kunthii*, *Zonaria turneriana*, *Splachnidium rugosum*, *Colpomenia sinuosa* and *Apophlaea lyallii*. *Marginariella urvilliana* reaches to the Auckland and Campbell Islands, while *Xiphophora gladiata* (shared with Tasmania) and *Pachymenia lusoria* are recorded from all the southern islands but Bounty.

Of the bull-kelps, *Durvillaea willana* is endemic to the mainland and Stewart Island, but *D. antarctica* is shared with southern Chile, Patagonia, the Falklands and New Georgia. Antipodes Island has a still unnamed *Durvillaea* with thick heavy fronds and lateral blades, perforated with holes, and forming underwater forests.

The southern limits of animal species must often result from accidents and hazards of colonisation, as with the absence of all mussels on the Antipodes and with the presence on Bounty as well as on Campbell Island of the mainland ribbed mussel, *Aulacomya maoriana*. The species lists contract with distance from the mainland, with only 223 molluscs listed by Powell for the whole Province.

The *Cellana* limpets, belonging to a genus more typical of warmer waters, have a single widely flung southern species, *Cellana strigilis*, with geographical subspecies *C. s. strigilis* at Auckland and Campbell Islands, *C. s. redimiculum* from Stewart Island north to Kaikoura, *C. s. chathamensis* on the Chatham Islands, *C. s. flemingi* on the Snares, *C. s. bollonsi* at the Antipodes, and *C. s. oliveri* at the Bounty Islands. The *Kerguelenella* species are peculiarly cold-water siphonariids, with *K. flemingi* at Auckland and Campbell Islands while Bounty, the Snares and Stewart Island have *K. stewartiana*. The patellid limpet *Nacella (Patinigera) terroris* dwells on low tidal rocks at the Auckland and Campbell Islands, but is absent on the Bounty and Antipodes. The same species reaches to the cool temperate of western South America. The common low tidal chiton at Campbell, Auckland and the Antipodes, is *Plaxiphora aurata*.

Campbell Island

High and rugged, following complex volcanism and glaciation, Campbell Island lies at 52 degrees south, being 120 sq km in land area. From the east it is deep-incised by the narrow Perseverance Harbour. Though this inlet enjoys some measure of shelter, the driving seas still ensure a high-energy zoning pattern.

Debased to a tussock cover after an ill-starred grazing lease dating from last century, Campbell Island has kept enough of its rare 'megaherbs' to remain a botanist's treasure trove today. Among the tussock, the liliaceous *Bulbinella rossi* forms a lush ground-cover down to the top of the **supralittoral zone**. Conspicuous near the shore are the 'Macquarie Island cabbage', *Stilbocarpa polaris* (Araliaceae), with big foliage like a lime-green rhubarb; *Anisotome latifolia* (Apiaceae) with pink male flower heads and deeply cut leaves; and the two *Pleurophyllum* daisies, purple-flowered *P. speciosum* and *P. criniferum* with apetalate flower heads.

For the following account of the shores, a debt is owed to Kim Westerskov, who visted each of the southern islands in the 1980s, and also to Margaret Morley's molluscan study in the mid-1990s and Robert Morton's shore observations in 1988.

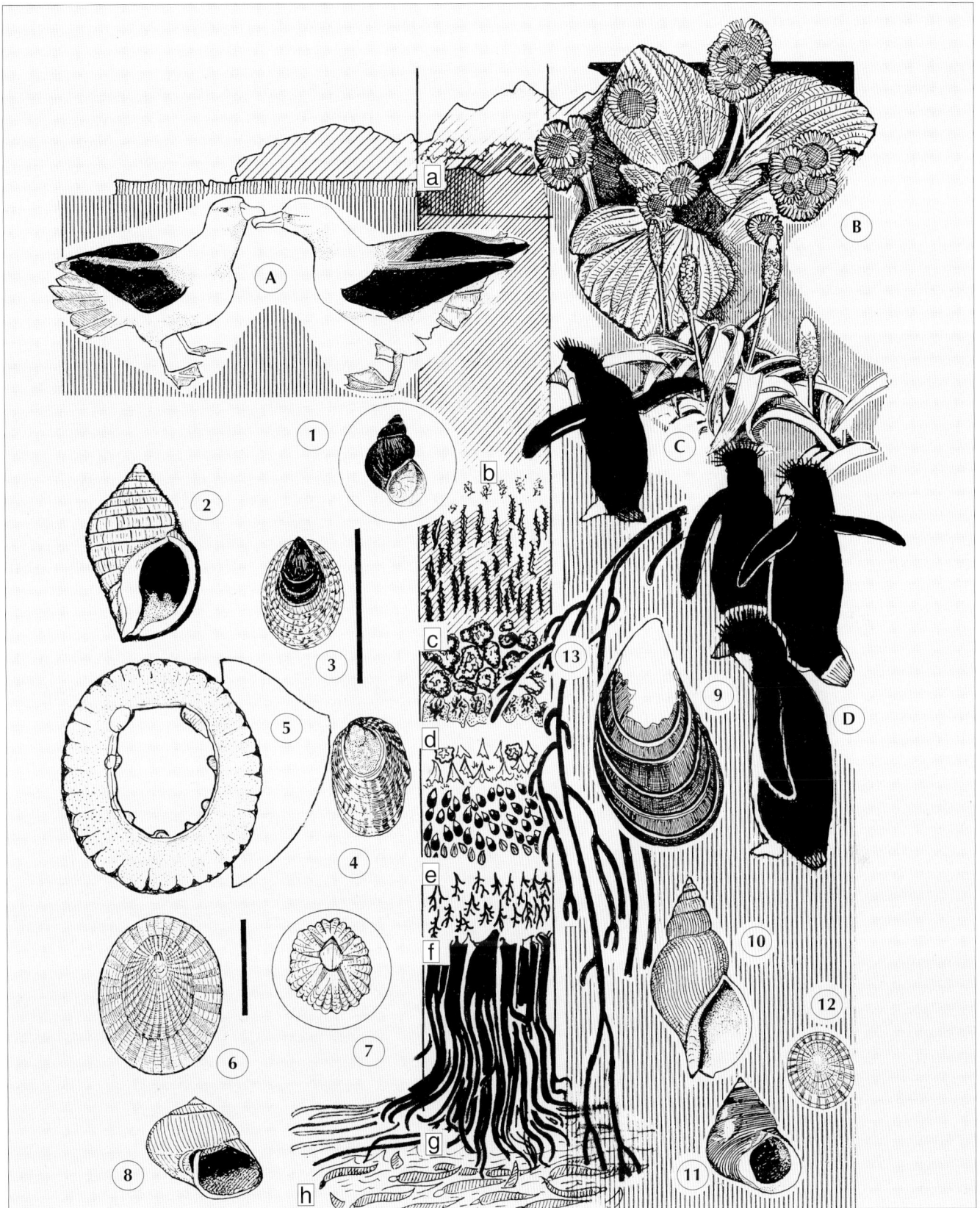

Fig. 10.17 Campbell Island: zoned shore in Perseverance Harbour

a–b black lichen *Verrucaria*; b–c *Porphyra*; c–d *Stictosiphonia*; d–e *Ulva, Mytilus* and *Aulacomya*; e–f *Xiphophora*; f–g *Durvillaea*; g–h *Macrocystis*

A. southern royal albatross, *Diomedea epomophora*; B. *Pleurophyllum speciosum*; C. *Bulbinella rossi*; D. rockhopper penguin *Eudyptes chrysocome*

1 *Laevilitorina antipoda* 2 *Lepsithais lacunosus* 3 *Notoacmea pileopsis sturnus* 4 *Kerguelenella innominata* 5 *Cellana strigilis strigilis* 6 *Nacella (Patinigera) terroris*. 7 *Notomegabalanus campbelli* 8 *Margarella antipoda* 9 *Mytilus edulis aoteanus* 10 *Pareuthria campbelli* 11 *Cantharidus capillaceus* 12 *Actinoleuca campbelli* 13 *Xiphophora gladiata*.

The shore depicted (Fig. 10.17) is a high-exposed *Durvillaea*-girded face at **Boyack Point**, on Perseverance Harbour's north side. At 10 m above the horizontal margin of the bull-kelp, the black lichen *Verrucaria* runs as a primary division round the shore. Above it, swathes of yellow and green-grey lichens reach up together to the first land-plants, with *Bulbinella* now fully returned after grazing. A little below the lichens grows high level *Hildenbrandia leccanellieri*.

The familiar mainland littorines are wanting, but have been replaced under stones and in high level algal tufts by the abundance of tiny *Laevilitorina antipoda*. There are no high level barnacles, but a single low tidal species, *Notomegabalanus campbelli*, lives among *Durvillaea* holdfasts.

The **upper eulittoral** is given over to algae, principally *Porphyra columbina*, *Stictosiphonia arbuscula* and *Apophlaea lyallii*. Around mid-tide these are enriched with green ulvoids, *Codium fragile* and *C. convolutum*, and red *Pachymenia lusoria*, as well as many small tufted and filmy rhodophytes.

The zoning mussels are large-growing *Mytilus edulis aoteanus* and some *Aulacomya maoriana*. Below these begin the algae of the *sublittoral fringe*. A skirt of *Xiphophora gladiata* generally first appears; but in some sites dark red *Gigartina pachymenioides* abuts straight on to the long fall of *Durvillaea antarctica*. The other Gigartinaceae are *G. divaricata*, *G. tuberculosa* and *G. circumcincta*. There is also the cold-water *Iridaea audouinella*. In the sublittoral grow abundant *Lessonia brevifolia* and *Macrocystis pyrifera*.

The most visible gastropods are the limpets. High up among *Stictosiphonia* is *Notoacmea pileopsis sturnus*, intermingled with its narrow elliptical form *subantarctica*. The siphonariid *Kerguelenella innominata* runs through the **middle eulittoral**; below it, down to *Durvillaea*, appears the solid, high-profiled *Cellana strigilis strigilis*. The cold-water *Nacella (Patinigera) terroris* — thinner-shelled and polished — is common on rocks at low tide.

The small, low-pitched trochid *Margarella antipoda* abounds under loose stones and on algae. Both *Durvillaea* and *Lessonia* are grazed by the greyish black *Cantharidus capillaceus*. On its shell is often the small acmaeid *Actinoleuca campbelli*. The common chitons are *Plaxiphora aurata campbelli*, *Onithochiton neglectus* f. *opiniosus* in kelp holdfasts and *Ischnochiton circumvallatus* under stones. The endemic cominellid at Campbell and Antipodes Islands is *Pareuthria campbelli* and the cold-water thaid *Lepsithais lacunosus* is shared with the South and Stewart Islands.

Both the leopard seal (*Hydrurga leptonyx*) and the southern elephant seal (*Mirounga leonina*) (Fig. 10.18), come ashore at Campbell Island. The latter will lie like a heavy bolster near the heaving kelp beds, or rear its snouted head and chest out of the tussock. Of the abounding birdlife, the most numerous species is the rockhopper penguin, *Eudyptes chrysocome*, with breeding numbers estimated at 2–3 million. The yellow-eyed penguin (*Megadyptes antipodes*) has a large breeding colony in Perseverance Harbour. Campbell Island is notable too as the nesting ground of the southern royal albatross, *Diomedea epomophora*. Black-browed and grey-headed mollymawks also breed in thousands; the sooty albatross and giant petrel have smaller colonies. Very tolerant of human approach is the Campbell Island shag, *Leucocarbo campbelli*.

Subantarctic Region

Beyond the Antipodean Province, still part of the South Pacific Cold Temperate Region, we pass to a separate **Subantarctic Region**. This is the only Region made up entirely of small and widely separated islands, and in the broad southern ocean it comes throughout the year under the **West Wind Drift**. Surface temperatures show only minor seasonal fluctuations, from 2°C in winter to 5°C in summer.

Closer to Australia but once included in the New Zealand Antipodean Province, Australia's **Macquarie Island** is accorded its own status as a Province. A second

Fig. 10.18 Macquarie Island
Steep cliff face is extended by low-sloping zoned shore.
1 southern elephant seal, *Mirounga leonina* 2 king penguin, *Aptenodytes patagonicus*.

Of special New Zealand interest is the survival of two ancestral members of our paleo-austral gastropod Family Struthiolariidae. *Perissodonta mirabilis* at Kerguelen and *P. georgiana* at South Georgia are from near the family's point of origin, probably from Cretaceous aporrhaids resembling fossils found in New Zealand. In past times, with a cold temperate climate (and land flora) in Antarctica, this family enjoyed a circum-antarctic range, with species of *Struthiolarella* common in the late Tertiary of South America.

Province takes in **Kerguelen** with the associated islands in the southern Indian Ocean, 2000 km north of the Antarctic continent. This whole far-flung Region possesses many genera and species in common, with the all-year-round West Wind Drift an important disperser.

The Subantarctic Region has relations with the Antarctic shelf, but perhaps more strongly to the cold temperate parts of the southern hemisphere. Kerguelen Province possesses, for example, four penguins found elsewhere only to the west (Falklands, South Georgia, South Shetlands) but not on the Antarctic continent, and nine species of its marine algae are shared with the tip of South America.

This whole Region has a highly endemic cold temperate biota, probably accumulating gradually from clockwise transportation by the West Wind Drift. The chief source has probably been South America and the Falklands, and through the Kerguelen Province a number of species have been carried west to Macquarie Island.

Macquarie Island

Midway between Tasmania and the Antarctic continent, at 54 degrees south, Macquarie Island has a subantarctic air temperature range between 11°C and -10°C. Strong gusting winds of 100 knots predominate, with heavy seas and maximal exposure all the year round. Sea temperatures range between 7°C and 3°C. The Island's shore ecology was surveyed by Kenny and Haysom.[10] Zonation is strikingly simplified, possessing neither littorines nor barnacles. A lichen **supralittoral zone** is followed by a **littoral fringe** with thick *Porphyra umbilicalis*, accompanied by *Rhizoclonium*, *Prasiola* and *Iridaea boryana*. Below these the **upper eulittoral** can be said to begin, with a 'bare zone' of closely cropped *Nothogenia fastigiata* and *Spongomorpha pacifica*, along with the grazing gastropods *Kerguelenella lateralis*, *Macquariella hamiltoni* and *Nacella macquariensis*. The **lower eulittoral** forms a 'red zone' with a *Rhodymenia* species dominant. The small bivalves *Kidderia bicolor* and *Gaimardia trapezina coccinea* abound here.

The single kelp of the **sublittoral fringe** is *Durvillaea antarctica*, with big holdfasts up to 40 cm across. Where kelp is absent, the coralline pink surface carries the asteroid *Anasterias suteri* and the holothurian *Pseudopsolus macquariensis*. Extensive beds of *Macrocystis pyrifera* lie at 10–20 m depth offshore.

Antarctic Region

The broad **West Wind Drift** flows in a band occupying some 20 degrees of latitude, though narrowing to only 650 km between South Shetlands and Cape Horn. Within this band, the **Antarctic Convergence** forms a circumpolar zone where cold surface water sinks. South of this, a band between the convergence and the Antarctic mainland forms a cold, low-salinity layer, some 100–250 m deep, much of it with pack-ice, and expanding and contracting seasonally. All this, to the edge of the continent, comes under the influence of an irregular **East Wind Drift**. The February isotherm is 1°C, with a winter temperature of -2°C.

With the sinking of cold water the temperature and salinity at the continent edge have become fairly uniform, from the surface down to 400 m. Species thus tend to have broad depth tolerances, and the normal depth range of what we may still call the 'shore fauna' can extend as deep as 500 m.

McMurdo Sound

On the Antarctic continent intertidal shore communities as we know them can scarcely exist. There are neither littorines nor mussels, and the only barnacles are those large species attached to whales. At most, there are a few pools between tides, apparently barren, but in fact supporting gastropods, amphipods, small pycnogonids and nemerteans. All these life-forms are found to increase in size and number subtidally. For, in contrast to the poverty between tides, the sublittoral communities are now complex and diverse. Biomass increases as the effect of ice-scour lessens, ultimately falling off as the shelf continues to deepen.

Investigation of the sublittoral calls for an intrepid descent below the ice. From October through November, after winter dark but before high summer, with the solid ice still safe to walk on, a manhole can be cut with an auger, allowing a descent to be made below ice 2 m thick. Under this, it is then 20 m to the rock platform at the bottom. Here, the saline water does not freeze. With the seabed snow-white, pastel-tinted with orange or yellow, a fragile multiplicity of shapes live undisturbed by waves, light or storms.

Here must be the strangest, most other-worldly realm with which a naturalist could come close up. Few ever will, but a fascinating impression near to first-hand can be gained from a Wild South television documentary film[11] by an Auckland biologist, Chris Battershill and his party, or in Kelly Tarlton's Antarctic Experience and Underwater World on Auckland's Waterfront, entitled *Under the Ice* — in giving an illumination of a whole habitat that only a fortunate and intrepid few will ever be able to glimpse for themselves.

An excellent account of the sublittoral ecology at McMurdo Sound, near New Zealand's Antarctic scientific base, has been given by Picken. Three depth zones

Fig. 10.19 Some Antarctic arthropods
1 *Serolis*, dorsal and lateral views 2 *Glyptonotus* 3 a giant pycnogonid.

are to be recognised, according to the effects of depth, scour and anchor-ice disturbance.

Down to 15 m, there is a bare zone briefly recolonised when ice-free by the red or purple star *Odonaster validus* and the urchin *Sterechinus neumayeri*, together with the 2 m-long nemertean *Parborlasia corrugata*, and the giant isopod *Glyptonotus*. The last-named is one of two notable sorts of crustacean that in the Antarctic system entirely supplant crabs as scavengers and fine deposit feeders. The more flattened isopod *Serolis polita* has the ancient look of a trilobite as it rests upon or crawls over the silt.

From 15–33 m, still vulnerable to anchor-ice, is a second zone, of soft corals, anemones, hydroids and ascidians. It is in the third zone, below 33 m, that the sponges, a leading feature of the antarctic sublittoral, come into their own. The majority belong to the glass-sponges of the Order **Hexactinellida**. Large and tubular, they are supported by an exquisite skeletal lattice, fashioned into a lacework out of six-rayed siliceous spicules but lacking in horny spongin. Occupying up to 55% of the ground-cover, they live here well above the much greater depths to which they are confined in other

oceans. Their vertical and arborescent growth provides the substrate for the anemones, hydroids, bryozoans, polychaetes and bivalves in which this third zone also abounds.

Dayton and his colleagues have studied the predation dynamics of the sponge community at McMurdo Sound. The circum-antarctic sponge *Mycale acerta* would probably overgrow many other sessile species, were its dominance not checked by predation from the starfishes *Perknaster fuscus* and *Acodontaster conspicuus*. Along with the lemon dorid sea-slug *Austrodoris macmurdensis*, *Acodontaster* grazes the sponges *Rossella racovitzae* and *Scolymastra joubini*, while larval and young *Acodontaster* are themselves held in check by the star *Odontaster validus*.

In sheltered bays down to 30 m muddy sand replaces rock, and its epifauna of the amphipod *Glyptonotus* and the nemertean *Parborlasia* subsists on decaying organisms. *Serolis* creeps over the sea-bed, while rotting algae provide the substrate for the weird and delicately moving Pycnogonida that pick their way over fronds and small stones to feed on hydroids and anemones. The largest 'sea-spider', *Cecolopora australis* is one of three

10-legged pycnogonids, while *Dodecolopoda mawsoni* is included in a dozen 12-legged species unique to the Antarctic. The extra legs beyond eight come from modification of the pedipalpi and chelicerae.

The molluscs of the Antarctic have thin, poorly calcified shells. A common member of the epifauna is the spasmodically 'swimming' cold-water scallop *Adamussium colbecki*. The largest infaunal bivalve is *Laternula elliptica*, and there are huge numbers of the small protobranch bivalve *Yoldia eightsiu*, along with the tiny *Mysella charcoti*. Important also in the benthos is the file shell *Limatula hodgsoni*. Where algae can exist, the cold-water limpet *Nacella* provides 70% of the animal biomass. The largest molluscan carnivore is a fine, thinly flanged *Trophon*.

Biogeography

The antarctic benthos reveals broadly three geographic components. First, there is a relict fauna left from the continent prior to glaciation; second, a fauna from the deep basins of the Pacific, Indian and Atlantic Oceans; third, a range of shallow-water derivatives from families in South America and Australasia, giving a high proportion of antarctic endemic species. In reverse, New Zealand long ago took from once-temperate Antarctica parts of its present-day biota: including the podocarps, *Nothofagus*, and *Fuchsia*, and in the sea the ancient gastropods *Struthiolaria*.

Physiology

The only intertidal organism regularly exposed to temperatures below freezing point, the limpet *Nacella* can wrap itself in a thick, viscid mucus cocoon and survive temperatures of -20°C for two hours. More common than such 'resistance adaptation' there are the many instances of 'low capacity adaptation' to cold. With low temperatures and shortage of food, the whole tempo of life is slowed. Sponges may grow to be centuries old. *Nacella* can live for 100 years. Though growth patterns are seasonal, slower growing invertebrates may attain greater size than those faster growing. Thus the ascidian *Paramolgula* can reach up to 33 cm long. Giantism seems only possible in organisms that do not require large supplies of calcium for shells. With food scarce, necrophagy is widespread among gastropods, holothurians, ophiuroids, echinoids and nemerteans. Few molluscs or echinoderms can risk the hazards of pelagic larva dispersal. Direct development with large yolky eggs is the almost invariable rule.

CHAPTER ELEVEN
Boulder Shores

Nowhere else between tides shall we find communities as diverse and rewarding as under boulders. For most of us, our first understandings of the riches of the shore have probably been derived from turning over stones. My own dawning realisation about what I can now call taxonomy, ecology and ethology seems to have been stirred by the search for different sorts of crabs. Here — as it could seem — was a life-form pre-eminently designed for existence under boulders. Without at first knowing any names to assign

them, it had soon become predictable where the different sorts would turn up and how they would react and behave. There was the old sense of wrestling with some strange, hitherto unknown, beings, unwilling to let them go until they would tell us their names.

Yet despite such primal appeal, the boulder habitat has been consistently neglected down to the present day, in even the best works published on shores. With all their mastery of the visible panorama, the Stephensons rarely seem to have turned over boulders. Lesser students have

Fig. 11.1 Crabs as under-boulders forms
1 *Heterozius rotundifrons* 2 *Petrolisthes elongatus* 3 *Eurynolambrus australis* 4 *Leptograpsus variegatus* 5 *Plagusia chabrus*.
(From Michael Miller.)

nearly always been daunted by the very richness of these communities. Only recently — in a few advanced countries — have a good number of their species been named or their taxonomic groups adequately understood.

At the outset, it becomes clear what strong affinities the boulder communities have with the sublittoral. They harbour *sciaphilic* or dark-loving species that we would expect to find nowhere else between the tides. Yet by a paradox often remarked upon, these species so seldom visible are generally the most brightly coloured. It is as if colour were an elemental property of living things that needed to be obscured or camouflaged against predators only in lighted places.

Under-boulder communities are in short the privileged outposts of the subtidal translocated up to the intertidal in dark places. They could be compared with the lines of bright flowers from an English woodland sent out in protected salients into the meadows along shaded hedgerows.

We shall thus often find the communities of the undersides of boulders richer by an order of magnitude than those upon open surfaces at the same level. They form a special domain where the constraints of light, temperature and desiccation have been relaxed. Their limiting factors are no longer *climatic* but *biotic*, related to competition or predation by other species.

Orderly zonation will be found to exist under boulders just as out in the open. But like the zones of the subtidal, these boulder patterns are because of their very elaboration harder to delineate. Clean-cut zones with single-species dominance do not often exist. Instead we shall find complex mosaics with a species-composition widely variable and often unpredictable.

With all its resemblances to the wider subtidal, the boulder habitat has, however, a distinctness of its own. First, the space available is narrow and vertically compressed. Luxuriant growth will never occur with the same profusion as in the sublittoral. Fig. 14.1 suggests — for several groups — the way related life-forms have become reduced and simplified to fit their bodies to this confined space.

Physical properties

Boulder and cobble beaches are unlike any other hard shores in their substrate being mobile and shifting. They are also capable of quantitative description using the same physical measurements as a sedimentologist would apply to a sand beach or a soft tidal flat. We can thus begin with the size and shape of their constituent particles — for boulder beaches are still essentially particulate, even though their units are relatively huge, even — it may be — up to a metre's diameter. Their properties derive from the character of the bedrock, and to the con-

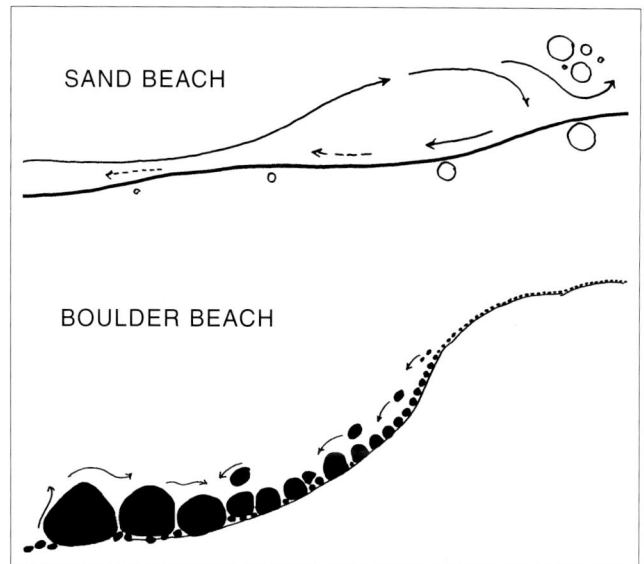

Fig. 11.2 Particle sorting across a sand beach and boulder beach

tinuing action of moving water on the blocks this rockmass has yielded.

Boulder beaches range from stable expanses of low pitch to steep ramps kept constantly mobile by the waves. On shores of low wave energy, the accumulated

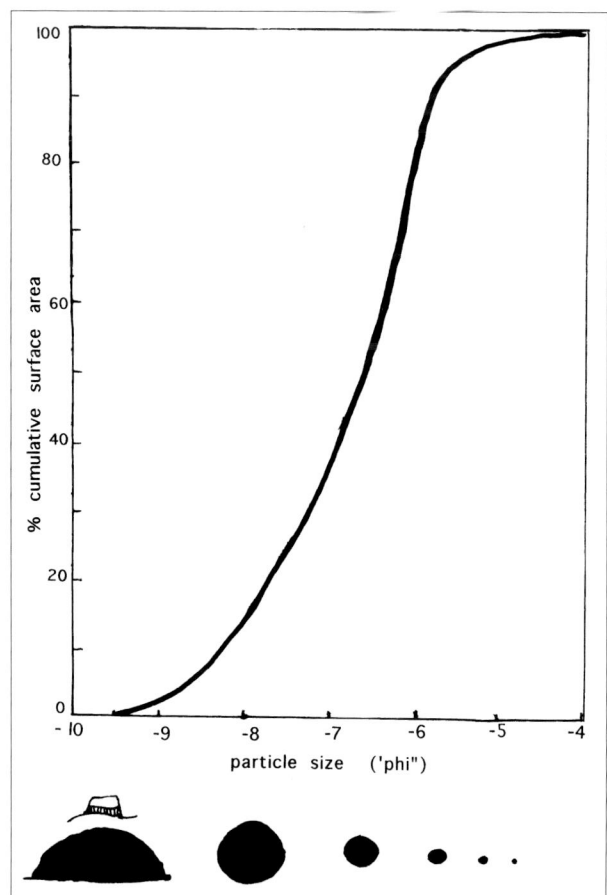

Fig. 11.3 Boulder sorting
Cumulative curve for boulder distribution at a higher mid-level of the beach at Orere Point, south of Auckland.

boulders long remain irregular in shape, either lying on a solid undermass, or bedded in silt or sand. Upon high energy shores, in contrast, they eventually become rounded and wave-arranged. In the technical term such boulders are *well-sorted*, being distributed in a size gradient with the largest ones lowest on the shore, and reducing in diameter from the bottom of the beach upwards. Boulders of high energy shores that have been smoothed by wave-dragging tend to become flat beneath. Those carried down by rivers or regularly overturned at wave-break become smooth and near-spherical.

Boulders, cobbles and pebbles behave differently under wave action from the particles of a sand beach. Moist sand is cohesive, with the property of capillarity, and the close bedding together of the grains makes them more resistant to sheering forces. Transport of sand grains up the beach is effected by *saltation* under the action of waves, with the whole beach ultimately coming to equilibrium at a low angle of slope. Boulders, cobbles and pebbles come to rest at a higher angle of slope. Thrown up the shore by strong waves, they slide or roll by gravity until they are at some spot able to reach stability.

A sand or boulder beach can best be physically described by its overall distribution of particle size. For a given locality we thus need to ascertain the size-composition within a series of samples taken in a transect from the top to the bottom of the shore. With sand, particle sizes are measured by passing each sample through a set of sieves, numerically arranged according to the *Wentworth Scale*. This has a geometrical progression with the mean diameter in each size class twice that of the preceding class. Size is conventionally expressed in *phi* (Ø) units, with *phi* representing the *minus log to the base of 2 of the diameter in millimetres*. Thus, for grain sizes larger than 1 mm, *phi* will have a negative value, with an increasing positive value for grains falling below 1 mm.

With the sediments of sandy and muddy shores, it is convenient to plot size distribution as the percentage by weight of each successive size-class in the sample. But since weight increases cubically with the linear size of the particles, massive boulders and cobbles would distort the shape of such a cumulative curve. While weight is important when a sand particle gravitates down the shore, the wave force causing its upshore saltation will be proportional to its exposed surface area, hence having a squared relation to diameter. With boulders, surface area is additionally important as representing the extent of available settling space.

In plotting cumulative curves for boulder and cobble size, the average diameter was used to derive the surface area. The number of boulders in a sample was multi-

Wentworth Grain Size Scale

Median diameter (mm)	Phi Ø	Size class name	
1024	-10	Boulder	
>256	-8	Large cobble	
128	-7	Small cobble	
64	-6	Very large pebble	
32	-5	Large pebble	Gravel
16	-4	Medium pebble	
8	-3	Small pebble	
4	-2	Granule	
2	-1	Very coarse sand	
1	0	Coarse sand	
0.5	1	Medium sand	Sand
0.25	2	Fine sand	
0.125	3	Very fine sand	
0.063	4	Silt	Mud
<0.004	8	Clay	

plied by this value, to arrive at the aggregate surface area for each size-class. All the classes were next summed to give the total surface area of the sample, with the surface area in each size-class then expressed as a percentage. The cumulative percentage surface areas in each size-class were finally plotted against grain size (Ø).

Shape

The *shape* of a boulder, which is its most important property as a habitat, results both from the rock's geological nature (*lithology*) and its subsequent weathering history. For any boulder a stage is reached when abrasion has reduced its sharp angles until it can be described as *smooth*. Usually expressed as *roundness*, this is a more useful property than true *sphericity*, which is based on a ratio of three dimensions, much influenced by the material's first unweathered shape. In contrast, *roundness*, being the smoothing by obliteration of emergent features, is the important measure of weathering.

Blocks of mudstone and softer sandstone seldom become rounded, since they rapidly crumble on drying out. In contrast hard rocks such as greywacke, and prismatic basalt or andesite, break up along their joints to yield finally well-rounded boulders. With scoriaceous basalt, the surface pitting will never allow complete smoothing.

Various attempts have been made to quantify round-

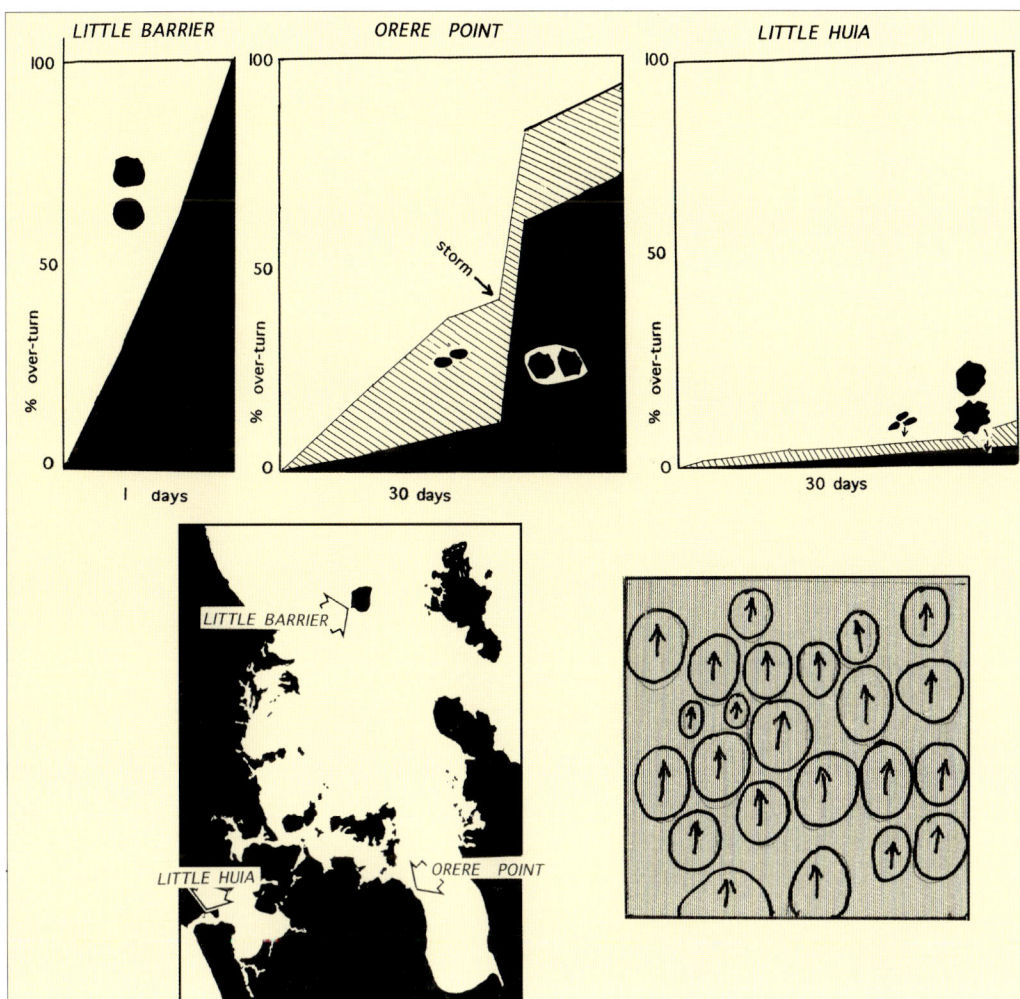

Fig. 11.4 Cumulative wave-overturn of boulders and cobbles at Little Barrier Island (Hauturu), Orere Point and Little Huia
Comparison is given for three Auckland shores: high exposed, moderately exposed and strongly sheltered. Boulders (**black**) and small cobbles (**cross-lined**).
(**bottom left**) Location map and (**bottom right**) boulder quadrat with upper surfaces initially arrow marked.

ness, by measuring the curvature of the sharpest corner, or calculating the average radius of curvature of all corners divided by the radius of the largest inscribed circle. So tedious are these in practice that it is preferable to fall back on a standard pictorial scale.

Stability

A boulder's ability to lie stable, which largely determines its habitat value, will arise both from its size and shape. On the most mobile boulders, sessile animals are lacking. Such shores become the preserve of fast-moving crabs, amphipods and isopods, able to accommodate to the constant shifting of boulders by moving about freely in their interspaces. The longer a boulder is stable, the better the opportunity for sessile communities to build up towards their ultimate climax.

Stability is obviously a function of wave force as well as of size and shape. For any strength of wave there will be boulders too large to move. Such stable boulders, unless embedded in sediment, can be expected to carry a full settled community. At the other extreme, small, wave-rounded cobbles, where stability may not last above one tidal cycle, will carry no sessile fauna at all.

Supply

A boulder's properties, including its hardness and shape, ultimately go back to the parent bedrock. Greywacke, basalt or andesite often become jointed into squarish blocks. Chemical decay first leaches out the minerals along the joint planes, and the wave assault then releases blocks that can be turned by mutual abrasion into spherical boulders. Rounding happens at rates proportional to size and hence initial mobility. Relatively small pieces can be so repeatedly turned over by waves as to become as spherical as those rolled in rivers. Larger boulders will tend to remain broad-based and more stable under wave action, with their tops and sides becoming smooth and dome-shaped by abrasion from smaller pieces.

The most prolific boulder sources are where cliffs with a hard parent rock are being cut back, as for example, with Dunedin's basalt formation at St Clair (Fig. 9.25). Rock debris may be enclosed in a softer matrix as in the cliffs behind the boulder beaches of Taranaki. The andesitic cone of Mt Taranaki is surrounded by a ring plain of laharic breccia extending out to the cliffs. From one such cliff lahar, dated at 7000 years old, the splendid boulder beach at Opunake (Fig. 11.12) is still being

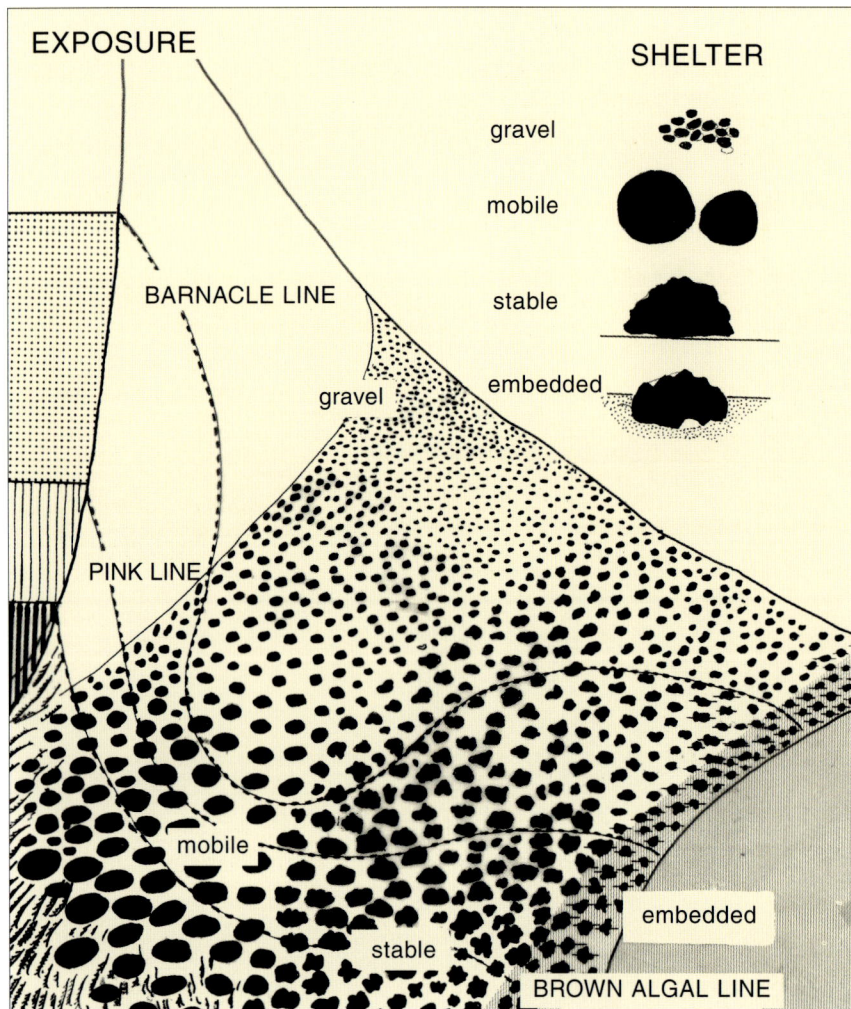

Fig. 11.5 Distribution of gravel types from exposure to shelter
The barnacle-line, pink-line and brown algal-line are projected as from a steep bedrock surface (**left**).
The areas are shown for mobile, stable and embedded boulders and for cobble and pebble gravel.
(**top right**) Characteristic size and shape of each type.

fed. Caroline Bay, Timaru, is another classic site with soft eroding cliffs, built of loess over a basalt boulder bed.

Some cliff formations never yield spherical boulders. The sandstone of the east North Island shores thus weather into slabs that, after release by bio-erosion, may round off at the corners, but retain their flat shape. If kept moist, as below the water table between tides, the softer sedimentary rocks may remain intact. Otherwise they will fracture or crumble as they dry out in the warm air.

Hard greywacke boulders round off well, being worn smooth by constant abrasion against their neighbours. Two alternative shapes result, spherical caused by rolling, and flat and 'platy' after predominant sliding. High on the shore, the weathering of boulders, as of andesite or rhyolite, may be assisted by unequal heating and cooling. A surface 'skin' exfoliates as the outer layer is more rapidly cooled.

The boulder spectrum

A scheme is presented in Fig. 11.5 for the distribution of boulders, cobbles and pebbles across intertidal shores with differences of wave energy.

The most exposed shores that can retain loose cover (around exposure class 5) have smooth boulders, which are **mobile** except at the largest sizes. The mean particle size reduces from low water up to the beach crest, which is a berm covered with small loose cobbles and pebbles. Continued mobility results in a *graded*, size-sorted profile.

Towards low water mark, the largest particles of a boulder beach may be blocks up to a metre across, permanently stable and smoothed by abrasion from smaller moving boulders. High beach ramps, with boulders below and cobbles and pebbles above, girdle many of our offshore islands, as will be described for the landing place at Little Barrier Island (Hauturu) (Fig. 11.11).

At lower exposure levels (classes 4 and 3) such general mobility is no longer possible. **Stable** boulders here lie in extended flats at a lower angle of slope. These are generally less regular in shape, though some wave-smoothing occurs towards low water, and they typically remain for years at a time without overturn. The under-boulder space generally receives a laminar current flow sufficient to maintain a rich sessile fauna.

Towards the sheltered end of the spectrum (exposure classes 2 and l) the boulders are likely to become **embed-ded**, with their bases in silty sand or even impacted in

EXPOSURE ⟶ **SHELTER**

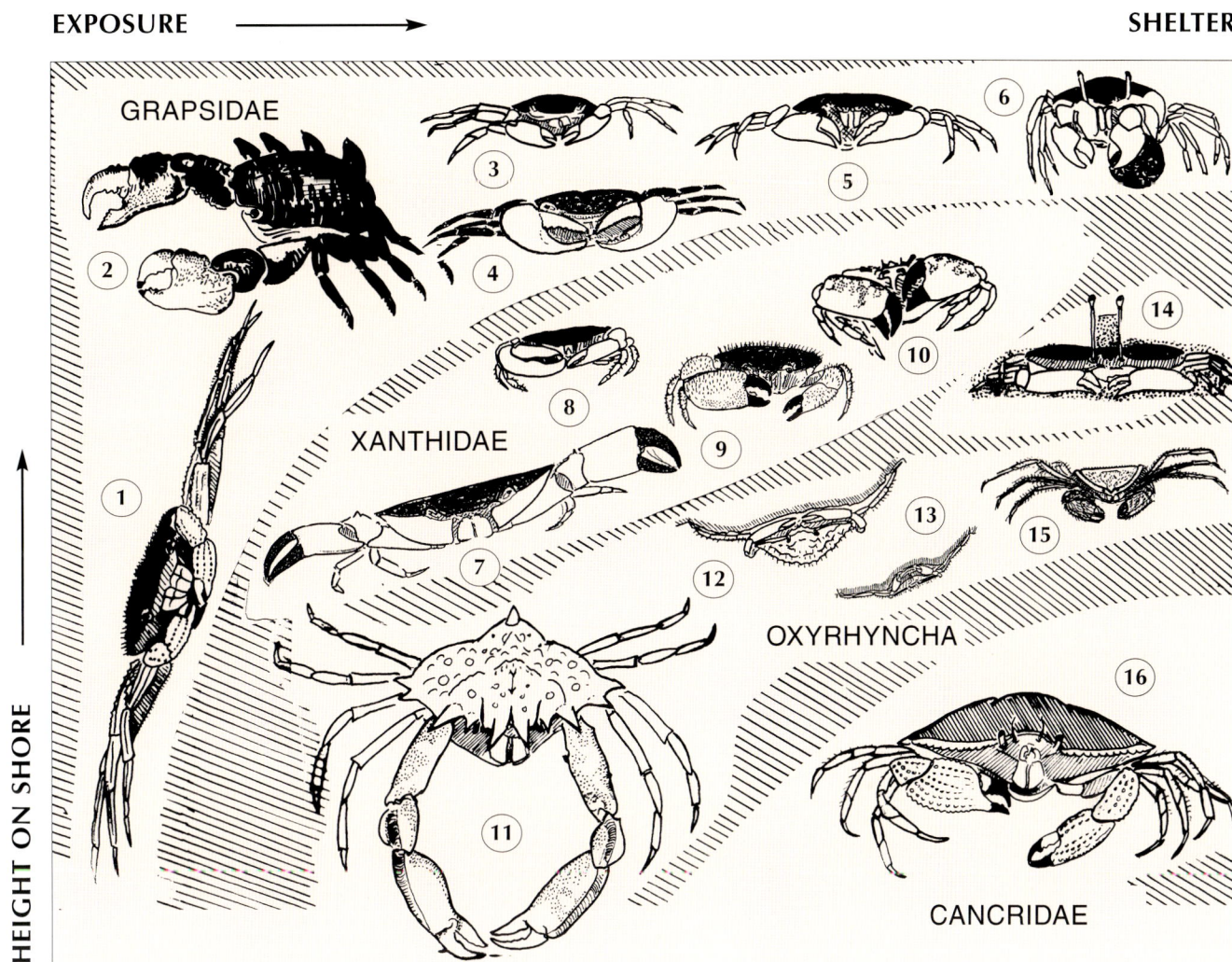

Fig. 11.6 Schema of crab distribution under boulders and in adjacent sediments, in relation to height on the shore and exposure/shelter

1 *Plagusia chabrus* 2 *Leptograpsus variegatus* 3 *Cyclograpsus lavauxi* 4 *Hemigrapsus edwardsi* 5 *H. crenulatus*
6 *Helice crassa* 7 *Ozius truncatus* 8 *Heterozius rotundifrons* 9 *Pilumnus lumpinus* 10 *Pilumnopeus serratifrons*
11 *Notomithrax peronii* 12 *N. minor* 13 *Neohymenicus pubescens* 14 *Macrophthalmus hirtipes* 15 *Halicarcinus whitei*
16 *Cancer novaezelandiae.*

low-oxygenated mud. Their undersurfaces are thus rendered black and anaerobic, with their living communities shifted to the sides of the boulder, just above the line of sediment.

Where boulders, pebbles and gravel have been supplied by rivers and transported by long-shore currents, smaller particles accompanying them will be comminuted to grain sizes that the currents can carry away. The residual beach material remains concentrated between tide marks, with its larger particles continuously sorted by wave action.

Finer **gravel** beaches, as found — for example — in front of Napier and around most of Hawke Bay, are constructed of smaller particles, pebbles or large granules, worn smooth by abrasion. Their steep profile is broken into several ramps, each slightly concave on its wave-exposed face. Often impressive in extent and ver-

tical reach, such beaches are difficult to walk upon and almost desolate of life. No animals can settle on the surface, and few can withstand destruction by particles moving against each other at every wave-fall. Water percolates rapidly through the beach, leaving its top surface hot and dry. Between the particles the wide moist interspaces present in a boulder beach are lacking. Neither is there the continuous water table that supports the rich interstitial communities of a sand beach. The crests of gravel beaches, well above high water mark, are often left stable enough to support their own small, permanent vegetation.

Crabs

Crabs are outstandingly the animals of boulder beaches. Their adaptive pattern is pre-eminently suited to life

207

under loose cover; with all the variations played on this theme, there are crabs equipped for life under boulders virtually anywhere on the shore, from high water to low and from exposure to shelter.

The true crabs (**Brachyura**) are decapod Crustacea that have become grounded by losing the swimming power of the abdomen or 'tail'. The one-piece carapace, that covers the head and thorax, is wider than long, flat-topped and carried close to the ground. The walking legs are splayed sideways for running, or shortened for a slow, more ponderous gait. The body is essentially rebuilt as a wedge to thrust backwards into a narrow space, typically underneath a boulder.

With these changes the front that would otherwise be left vulnerable is now guarded by the strong chelae on the modified first walking limbs. These pincers, the keynote of the crab design, are used not only defensively, but also for sexual display, and in male fighting rituals. Recent studies have revealed how finely the cheliped dentition can be adapted to particular diets.

No longer needed for exploring forwards, the antennae are almost vestigial; with the antennules likewise small. The eye-stalks are set wide apart in deep facets; in some crabs they are long and erectile for periscopic vision. The third maxillipeds close like flat doors over the inner mouthparts.

With such a successful stereotype, the crabs have far outnumbered all the other decapod crustacea. Compared with the tropics and subtropics, New Zealand has only a short list of crabs. But there are species enough, from several families, to occupy — without much competitive overlap — the whole niche continuum displayed in Fig. 11.6.

Unlike cool temperate Europe, or much of Pacific America, New Zealand is warm enough for a fair range of the **Grapsidae**. This is a large and enterprising family, spanning the whole intertidal range on high-energy shores. In the tropics they spread freely above the tides, though never — it would seem — reaching beyond low water. On cool temperate shores, it is the Family **Cancridae** that takes the lead, as with the crabs of Pacific Canada (Chapter 19). In warmer regions the **Xanthidae** catch up in numbers with the grapsids, but never show the same above-tidal, even terrestrial, capacity.

The grapsids, in contrast with both the cancrids and the xanthids, are long-legged and fast-running. The carapace is square-fronted, and the chelae — though larger in the males — are always symmetrical. New Zealand has seven species of shore grapsids, each with its rather well marked territory (Fig. 11.6).

High in the supralittoral, under boulders briefly wetted at high spring tides, live the two small *Cyclograpsus* species. Swift and lightly poised, they run into cover at any disturbance, and under wave action retreat to the deep interspaces of the boulder beach. Common throughout New Zealand, *Cyclograpsus lavauxi* is polished reddish brown, or speckled with mauve and stone grey. *Cyclograpsus insularum* differs only slightly in carapace shape (Fig. 11.10), and is duller than *C. lavauxi*, grey-brown, mauve or light pink. The two also differ slightly but clearly in their niche requirements. Marjorie Bacon in 1970 found from 21 populations that *C. insularum* was slightly less tolerant of extremes of salinity, desiccation and temperature, living significantly lower but still overlapping with *C. lavauxi*. More susceptible to under-boulder silt, *C. insularum* is less widespread and restricted to cleaner shores. Bacon found it absent when the sediment had more than 18% of silt, while *C. lavauxi* could tolerate up to 53%.

The large *Leptograpsus variegatus* is familiar on rocks at the top of high-energy shores. The adults are glossy black, tinted with purple, and cream underneath, while younger crabs are grey-speckled, confined mostly to boulders. This is by far the most agile of our shore crabs, scuttling over the rock, springing by leg-flexions or holding tight with the leg tips and spined tarsi. Widespread in the north, *Leptograpsus* cuts out short of Otago. On the Poor Knights, it can virtually live as a land crab, being a hyperosmotic regulator and predating by night on flax snails, *Placostylus*.

The swift and slender *Plagusia chabrus* is the low tidal counterpart of *Leptograpsus*. Brick-red and finely clad with golden hairs, it has the carapace front toothed and deeply incised. *Plagusia* runs sideways or glides between brown algae. With the flood tide, it comes out to run about just below the water-line, borne up and using the vibrating limbs to part-swim, or actively picking over the corallines with the chelae.

The two New Zealand species of *Hemigrapsus* have shorter legs but are still relatively agile. *Hemigrapsus edwardsi* is found under clean middle eulittoral boulders in moderate shelter, and *H. crenulatus* beneath boulders on silt. *Hemigrapsus edwardsi* is a handsome crab, up to 3 cm in carapace width, purplish black marbled or greyish brown when young. *H. crenulatus* is smaller, with the carapace dull greyish green, sometimes blotched with orange brown, dirty white or buff. The underneath is greyish white, purple speckled round the limb bases. The legs are fringed with dense hairs.

These congeneric species have quite wide adaptive differences, the more interesting in that the two *Hemigrapsus* of western North America form another such ecologically contrasting pair (Chapter 19). There *H. nudus* is the clean-shore species, and *H. oregonensis* its muddy shore counterpart. *H. nudus*, like our *H. edwardsi*, will not shift from well-oxygenated water and was found to be poorly equipped to exclude silt from its respiratory apertures. *H. oregonensis* has these openings

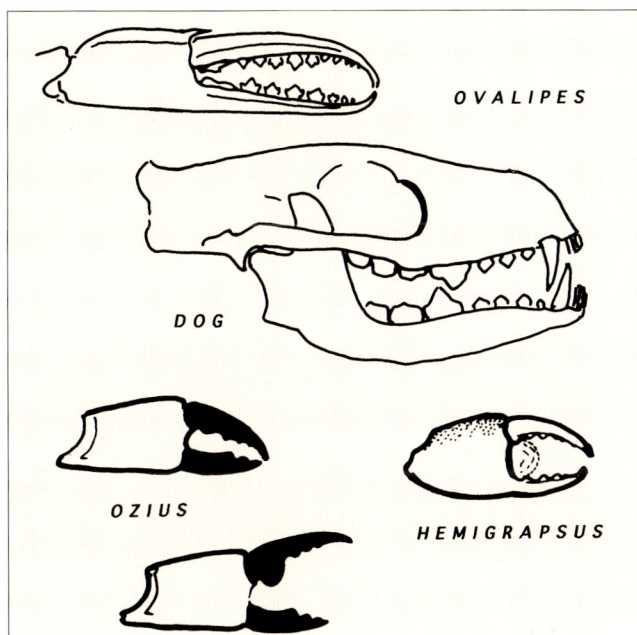

Fig. 11.7 Crab dentition
Ovalipes dactyl as compared to a dog's jaw.
Ozius truncatus chela closed and **(bottom)** open for 'nutcracker' to emerge. *Hemigrapsus* with chitinised tips.

protected by dense bristles, and — like *H. crenulatus* — was found to survive well in low-oxygenated water with fine mud added. There the survival time for *H. nudus* was in contrast cut by a half, and its gills become plastered with mud, where those of *H. oregonensis* remained clean. The habitat adaptations of the two local hemigrapsids could offer a rewarding study.

While *Hemigrapsus crenulatus* lives where cobbles rest in mud, our last shore grapsid *Helice crassa* is usually remote from rocks. Recognised by its long, erectile eye-stalks and the dull orange chelae of the male, it makes warrens in stiff, consolidated mud, often above high water. *H. crassa* is New Zealand's only foretaste of the vast high-tidal radiation of both the Grapsidae and the warm-water Ocypodidae that we shall find in the tropical Indo-Pacific (Fig. 22.30). New Zealand has a single ocypodid species in *Macrophthalmus hirtipes* (Fig. 11.6), with longer eye-stalks than *Helice*, and burrowing in far more fluid mud, as on *Zostera* flats.

With their higher reach upshore, the grapsids have developed both increased resistance to desiccation and enhanced tolerance of actual water loss. They also become increasingly tolerant of intermittent high temperatures and the osmotic effects of changing salinity. Most species are effectively amphibious. All except *Plagusia* are able to pump water out of their branchial chambers to be re-oxygenated, then circulated once more over the gills. From its exit near the orbit, this exhalant flow runs sideways through a reticulum of grooves over the cheek, kept moist by a felt of fine hairs.

Helice and *Cyclograpsus* were found to pump continuously, *Leptograpsus* and *Hemigrapsus* only at intervals, and *Plagusia* hardly at all.

The **Xanthidae** have four New Zealand species (the pebble crab, *Heterozius rotundifrons*, being now placed in the nearly allied Family Bellidae). Xanthids differ notably from grapsids in being round-fronted and slower moving, with short walking legs. The chelae are black-fingered and in the male highly unequal. *Pilumnopeus serratifrons* (Fig. 11.6) lives under silty stones, being the xanthid counterpart of *Hemigrapsus crenulatus*. The finely hairy *Pilumnus lumpinus* is found under low tidal silty boulders (Fig. 11.6). The coarser-whiskered *P. novaezelandiae*, with cinnabar chelae, is a more secretive crab, found under stable boulders, in crevices or in *Ecklonia* holdfasts. This last species has an abbreviated development, with the larvae retained under the tail of the female until they moult into tiny juvenile crabs.

Our largest and fastest xanthid crab, *Ozius truncatus* is iron-grey and rust-tinged, living under clean mid-tidal boulders, somewhat lower than *Hemigrapsus edwardsi*. It has something of the grapsoid alertness and when cornered responds by brandishing and attacking with its chelae.

The pebble crab, *Heterozius rotundifrons*, lives among smooth boulders, often several together. Broad and bow-fronted, the carapace is like a greenish grey pebble. The female has both chelae slender, but in the male the right one is much enlarged and of an attractive avocado green. On disturbance, the walking legs are brought in tight and the crab holds itself immobile. The male can double its face-on width by holding the large cheliped outstretched.

The Family **Portunidae**, known from the paddles on their fourth legs as 'swimming crabs', have only a few New Zealand species. The fast, backward-burrowing *Ovalipes catharus* is common, however, and evidently increasing in low tidal silty sand. *Liocarcinus corrugatus* lives in subtidal shell gravel, occasionally appearing in the sublittoral fringe at a very low spring tide (Fig. 16.6).

In New Zealand our only example of the Family **Cancridae** is *Cancer novaezelandiae*, living at the reef fringe, low or subtidally, where it submerges in fine sand. The rust-brown carapace, growing to 8 cm across, is crenate like a pie-crust round its wide front, and extends sideways over the spindly legs.

Feeding and dietary

Crabs were long assumed simply to be wide-ranging omnivores or scavengers. Detailed study of their habits and dentition has lately alerted us to their special sorts of dietary.

The paddle crab *Ovalipes* does not live on rocky shores, but it must be cited first for its example of well differentiated dentition. The finger of the chela is like a dog's jaw in miniature. The strong tips do duty as canine teeth, employed to chip at the shell margin of a bivalve held against the chest, generally the tuatua (*Paphies subtriangulatum*). The 'premolars' form serrated triangles, one of which is like a large *carnassial* tooth, used for shearing through the adductor muscles, after the slender pincers have been forced in between the valves. Meat is then removed with the whole chela. In large crabs the carnassials tend to be lost, and the operative teeth are now rounded molars able to crack a whole tuatua held inside the chela.

If *Ovalipes* has a dog's dentition, *Ozius* in contrast has blunt molars like a bear's, serving to crack *Turbo* and *Melagraphia* shells, including those occupied by hermit crabs. The shell lip is first chipped away with the very strong 'incisor' tips. When the thick fingers are fully open, a rounded mamilla at the back is positioned for use as a nutcracker on the body of the shell. In Australia, Chilton and Bull have studied predation by *Ozius truncatus* on *Nerita*, *Bembicium* and *Austrocochlea*. Offered all sizes, crabs preferred the smallest. Hence the predominance of small snails, below 10 mm, living in an upper shore refuge, free of predatory crabs.

The diet of *Leptograpsus variegatus* ranges widely. Flea mussels, *Xenostrobus pulex*, are consumed, after crushing their thin shells. The barnacle *Chamaesipho columna* can be scraped from the rock using the chitinised tips of the chelae like a rodent's incisors. At night, the chelae can be inserted under the rim of a mobile *Cellana* limpet or *Sypharochiton*, to wrench it off the rock. *Hemigrapsus edwardsi*, also with chitin-tipped fingers, was found to feed chiefly on *Lepsiella*. *H. crenulatus* is by contrast microphagous, combing off with the mouthparts fine material caught in the hairy tufts of the chelae. Also suspected of particle-feeding is the tiny *Neohymenicus pubescens*, living under silty boulders (Fig. 11.6). Hardly ever mobile, it holds its setose mouthparts open, and particles cleansed from the antennules are regularly passed to the mouth.

Spider and pillbox crabs

The boulder crabs so far described belong to the largest division **Brachyrhyncha**, but do not exhaust the whole tribe **Brachyura**. A second division, the **Oxyrhyncha**, contains those forms broadly known as spider crabs. Here the carapace is triangular, narrowed to a rostrum in front, with the orbits indistinct. The crabs of the Family **Majidae** are convex and high-built, deliberate in gait and not primarily adapted for life under boulders. Our two largest species, *Notomithrax peronii* and *N. ursus*, are most at home among the fronds of *Carpophyllum*. Both are known as decorating crabs, being clad with curved springy hairs that serve to attach camouflaging algae. The smaller *N. minor* is more of a boulder crab (Fig. 11.6), being short-haired and silty grey, with small attached sponges, ascidians and hydroids.

The ungainly but prettily marked crab *Eurynolambrus australis* is found under stones on clean, moderately sheltered shores. From its wide, triangular carapace it used to be classed as a lambrid, but its juvenile shape clearly shows it to be a maiid. The flesh-pink carapace shelves out over short walking legs, marbled in red and gold. The slender chelae are purple.

The second oxyrhynchid family, the **Hymenosomatidae**, is a group well represented in New Zealand, comprising small circular crabs, with a rimmed carapace: hence their common name 'pillbox crabs'. All have frail, slender legs, and some are at home in algae, as *Halicarcinus planatus* or on sea-grass flats (*Halicarcinus whitei*). The smallest species is *Neohymenicus pubescens*, finely hairy and very sedentary, living under silty boulders or holdfasts. *Halicarcinus innominatus* is found among green-lipped mussels (Fig. 5.7) and with fouling species on wharves. *Elamena producta* lives

Fig. 11.8 Porcellanid crabs
(left) *Petrolisthes elongatus* **(right)** feeding action of *Porcellana longicornis*
1 reduced fourth pereiopod 2 antennules 3 antennae 4 third maxilliped 5 second maxilliped 6 left chela.
(From E.A.T. Nicol.)

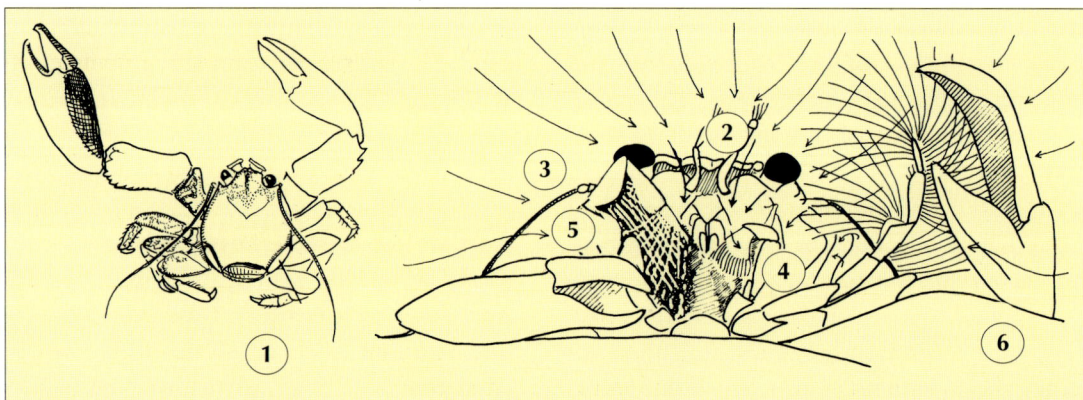

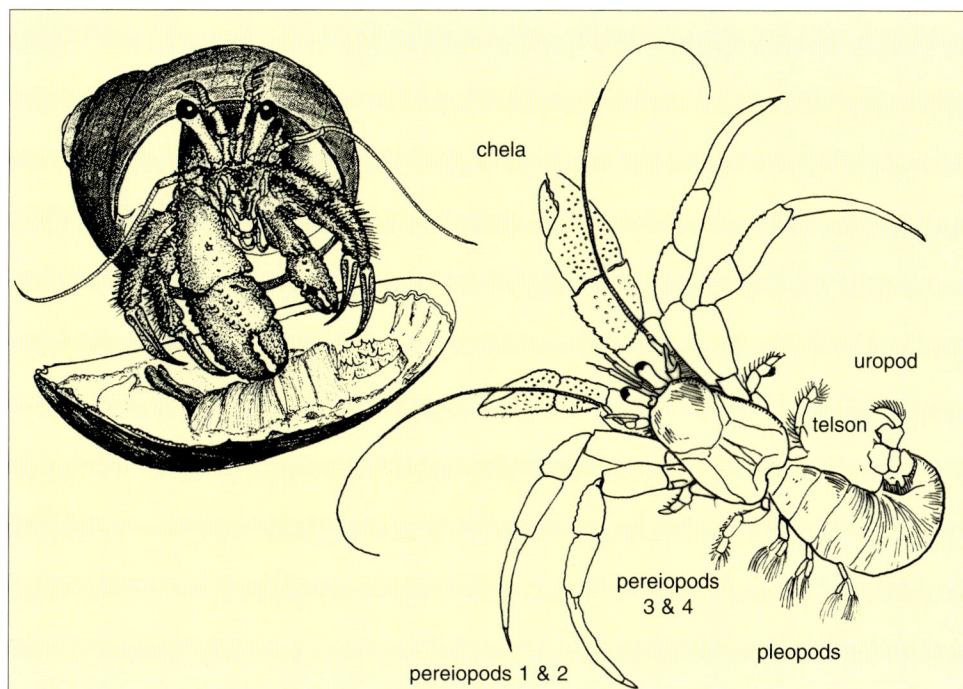

**Fig. 11.9 Hermit crab,
Pagurus novizelandiae
(left)** attacking the meat of a
chiton
(right) removed from shell.

(By courtesy John Walsby.)

under clean stones, sometimes as a commensal of the paua, *Haliotis*.

The peculiar crab *Petalomera wilsoni* is our only intertidal member of the sponge-carrying **Dromiidae**. Found under sheltered boulders at low water, it is told at a glance by the last two pairs of thoracic legs small and turned up over the carapace, usually to hold in place a 'hat' of sponge or ascidian. The New Zealand species is hatless, however, bare of camouflage but finely haired on carapace and limbs.

'Half-crabs'

The crab initiative has been seized not only by the Brachyura, but several times among other crustaceans. Even the small group of the Tanaidacea — relatives of the isopods — have evolved a minute 'hermit crab', *Pagurapseudes*, living in old rissoid shells (Fig. 13.18). Within the Decapoda, an important crab radiation is shown by the squat lobsters and the porcelain or 'half-crabs', **Galatheidea**. These are traditionally classed in the rather mixed bag of the Anomura — near the hermit crabs. The Family **Galatheidae** have obvious links with lobster-like ancestors. In the krill *Munida*, washed ashore on southern beaches, and the small squat lobsters *Galathea* not found intertidally in New Zealand, the abdomen is still elongate and carried permanently as a tail behind the carapace.

The flat porcelain crabs belong to the Family **Porcellanidae**. *Petrolisthes elongatus* is our most ubiquitous and numerous under-stones crab. Almost completely crab-like, it still retains some ancestral features as a 'half-crab'. First there are the long antennae, and — visible on turning over — a broad abdomen lying loose to

the carapace. With its full tail-fan, the abdomen can be flapped in short bursts of swimming. The fifth thoracic limbs are reduced to splints, kept folded in the branchial chambers and tipped with a small brush that is regularly used to cleanse the gills.

Half-crabs are efficient filter feeders, using their large, mobile third maxillipeds to intercept particles from the current drawn towards the respiratory openings. Fringes of silky pinnate-branched hairs on the two last joints overcross to form a straining mesh. The maxillipeds, singly or together, are waved like semaphore flags. This action, charming to observe, can be elicited by adding to the water a little meat juice. The collected particles are combed off by setae on the second maxillipeds and passed to the inner mouthparts.

On the continuum suggested in Fig. 11.6, *Petrolisthes* can be fitted into a broad middle space, coinciding in range with several of the Brachyura. As a microphagous feeder it offers small competitive overlap with the rest, and its territory spans the whole eulittoral, wherever it can scuttle between boulders to cling for filtering, usually upside down on the rock ceiling.

Hermit crabs

Classed also with the Anomura are the hermit crabs **Paguridae**, adapted to use empty gastropod shells as cover for the soft abdomen. The front half of the carapace is still hard-calcified, and the strong second and third walking legs (pereiopods) are used in the scuttling gait, to drag the shell along. The chelae are heavily built, with the right one (or in some forms both) serving as an operculum to close the shell. The third and fourth pereiopods are small and blunt, used for holding the

211

body to the columella of the shell. The uropods form a hook with a similar function. The female uses two or three slender pleopods to carry the egg clusters. The antennae are long, and the antennules shorter, keeping up a flicking motion as they monitor the ingoing water current.

The common tide-pool hermit is *Pagurus novizelandiae* (Fig. 11.9), using chiefly small cerithiid shells when young, then transferring to topshells and cat's-eyes, and finally to *Cominella* or even larger whelks. It is not only an active scavenger or carnivore, but with its filtering maxillipeds samples bottom deposits.

The hermit crabs are rich in species offshore, still imperfectly known. One of the larger subtidal hermits, *Paguristes pilosus*, was found to strain fine particles of food from the respiratory current by special antennary hairs. The antennae are moved in wide orbits, singly or in unison, in the path of the approaching current. Their basal joints can be switched and locked, so that the brush is swept across the current both backward and forward. Particles are removed by combing the antennae through the second and third maxillipeds.

A small subtidal hermit has its first abode in a gastropod shell, normally a turritellid, overlaid with a stony bryozoan, *Smittoidea maunganuiensis*. This grows out from the shell as a spiral tube, ultimately occupied by the hermit. The two symmetric chelae dovetail to form an operculum.

The zoned boulder beach
Coromandel Peninsula

A chain of small boulder beaches fronts the Firth of Thames north from Colville, scalloped out near stream mouths where the cliff-line topped with pohutukawa dips to the shore. A full beach slope comprises several tiers from the berm at road level down to the stable boulders at low water. The middle of the beach consists of mobile boulders and cobbles, grading up to pebbles on the beach crest, all smoothed and wave-sorted. Beach material has either been wrenched straight from the cliff by undercutting, or brought down already well rounded by fast streams.

The different beaches take their character from the diverse rock types of the Peninsula. With some outcrops of old greywacke, most of the Coromandel axis is built of Miocene andesite, rhyolite and minor basalt.

Te Hope (Fig. 11.10), sometimes called Moehau Bay, has a coastal outcrop of white-flecked 'Coromandel granite', more properly a dioritic gabbro. Each of its boulder terraces is gently sloped, almost level, with its drop-face slightly concave. The smallest pebbles and gravel are lifted by storm waves, to reach the berm and

consolidate there. Larger boulders and cobbles are dropped lower on the shore, or roll back before coming to rest. The result is a finely graded profile.

As wave runoff is lost rapidly by percolation between boulders, the dry surface warms up, becoming inhospitable to larval settlement and having little sessile life. A large boulder also absorbs and retains considerable heat. As its outer skin cools faster than the interior, it may often split off or exfoliate, like a rind.

On a boulder beach the **littoral fringe** becomes much extended as compared with steep bedrock. With its lower angle of slope, the dry beach heated by the sun is deprived of the cooling effects of evaporation. Hence, with no steep wave-beaten face, the bare littoral fringe reaches further downshore. Below this, the normally wide **eulittoral** is in turn pushed lower, as an abbreviated strip with *Chamaesipho columna* before the onset of *Carpophyllum maschalocarpum* on the stable boulders of the **sublittoral fringe**.

Looking up the shore from below, some of the boulder beaches would appear to have an entirely bare eulittoral. The reverse view, looking downshore upon the shaded boulder faces, offers an example of 'aspectation', with each boulder encrusted by the reddish brown *Apophlaea sinclairii*, suggestive of dried, congealed blood.

Vegetation

Sun-exposed boulder beaches show a **supralittoral zone** of lichens: grey-green, then yellow and lowest down (in the **littoral fringe**) black. At Te Hope, much of the beach crest is pohutukawa-shaded and given added stability by the roots of old trees. Where humus builds up between its cobbles, a blanket of scrambling pohuehue (*Muehlenbeckia complexa*) takes hold of the stable berm. This is the food plant for larvae of the common copper (*Lycaena salustius*), a familiar low-flying butterfly of boulder beaches. Shore convolvulus or nihinihi (*Calystegia soldanella*) creeps out in front, with polished heart-shaped leaves, and pale mauve or pink flowers. New Zealand spinach (*Tetragonia trigyna*) may be found in slightly moister stretches. Commonly creeping over bare rocks is the narrow-leafed remuremu, *Selliera radicans* (Family Goodeniaceae), with its small flowers like one-sided white stars. On a few islands of the Hauraki Gulf, as at Little Barrier, the native cucumber *Sicyos australis* may still survive (Fig. 11.11). It once grew freely on Motuihe.

Fauna

The deeper interspaces of the **supralittoral zone**, permanently moist and cool, have their characteristic faunule, zoned by depth as well as distance down the shore. At high water spring tide (HWST) and above, the largest animals are the skinks of the genus *Oligosoma*. The

Fig. 11.10 Boulder beach at Te Hope, Coromandel Peninsula
1 aspectation of *Apophlaea sinclairii*, towards the shaded face of a boulder 2 *Anisolabis littorea* 3 *Marinula filholi*
4 *Suterilla neozelandica* 5 *Nerita atramentosa* 6 shore skink, *Oligosoma smithi* 7 *Ligia novaezelandiae* 8 *Betaeopsis*
aequimanus 9 *Heterozius rotundifrons* 10 *Cyclograpsus lavauxi* 11 *C. insularum* 12 *Leptograpsus variegatus*.
(bottom right) overlapping ranges of the two *Cyclograpsus* (with carapace outlines) on an upper shore.
C. lavauxi (10) **(diagonal-hatched)**; *C. insularum* (11) **(stippled)**.(From Marjorie Bacon.)

common skink, *O. nigriplantare*, is the usual inhabitant of this zone in the South Island, southern North Island and Chatham Islands. Around the northern half of the North Island, the shore skink, *O. smithi*, is most frequently found among seaweed wrack on boulder beaches. More patchy and found only north of Gisborne, mostly on offshore islands, is the egg-laying skink

O. suteri, in high density colonies just above high water spring tide on the Coromandel Peninsula.

These smooth, swift lizards move by undulations of the trunk and tail, with the limbs no more than little struts to take purchase. They are scavengers living on dead organisms washed up or on kelp flies.

Boulder beach skinks hunt out small crustaceans,

notably the isopod *Ligia novaezelandiae*. This is a relative of the terrestrial woodlice, turning and twisting as it fast retreats from the light. A voracious scavenger and sometimes a cannibal, *Ligia* comes out to feed nocturnally at ebb tide.

Around high water neap (HWN) lives a common crustacean, the boulder shrimp, *Betaeopsis aequimanus*. Bluish black to chestnut, it has a lighter mid-dorsal streak. *Betaeopsis* does not swim but walks or makes tail-leaps like an amphipod when disturbed. The chelae are small and equal, and the short antennae thick and divergent.

As swift as *Ligia* or the lizard is the wingless seashore earwig, *Anisolabis littorea*. The smooth body, with its strong forceps behind, slithers between cobbles in high beach debris. Shining brown hoppers of the genus *Orchestia* hop about freely when disturbed from cover. These belong to a terrestrial family of the Order Amphipoda, well represented in New Zealand. Beetles, especially wingless staphylinids, as well as spiders and mites, contribute to the predatory and scavenging force around high spring tide. So also in some places may the beach centipede, *Tuoba xylophaga*.

On wet surfaces around 10 cm deep lives the small, horn-coloured prosobranch snail, *Suterilla neozelandica*, of the supralittoral Family Assimineidae. Thin and globular, only 2 mm tall, *Suterilla* is strictly limited in vertical range. With a fine-toothed radula it picks up fragments of organic debris or wave-lodged diatoms.

Between high neap and spring, down to half a metre deep, a paste of decaying brown algae often chokes up the inter-boulder spaces. This provides fodder for *Marinula filholi*, one of the primitive pulmonate snails of the Family Ellobiidae. The pointed, ovoid shell is 5–6 mm tall and reddish brown, with a white, strong-toothed aperture. There is no operculum and — as lung-breathers with the mantle cavity devoid of a gill — the Ellobiidae ventilate through a small, rhythmically opening aperture, the pneumatostome.

Feeding on the same decayed algae may be found a few of the topshell *Diloma nigerrima* (see Fig. 11.13), a species less abundant on northern coasts. Dipteran flies hasten the algal decomposition, with the hairy kelp-fly, *Chaetocoelopa littoralis*, occurring at times in immense droves, and often found as white maggots or dark cylindrical pupae.

Downwards from the **littoral fringe** the boulders have their clear sequence of grapsid crabs. In the fringe proper, with the zoning represented in Fig. 11.10, lives the widespread *Cyclograpsus lavauxi*, overlapped slightly lower down by *C. insularum*. *Leptograpsus variegatus* is common under eulittoral boulders, both dark adults and grey-speckled young, while lowest downshore comes the red *Plagusia chabrus*, fast-moving among brown algae.

At the middle shore is to be found the alert and aggressive *Ozius truncatus*, the only New Zealand xanthid of mobile boulders. In contrast with it is the smaller, grey *Heterozius rotundifrons* that when uncovered stays immobile like a flat pebble.

Little Barrier (Hauturu)

This small elevated island in the entrance to the Hauraki Gulf is the slightly eroded remnant of a 1.5 million-year-old dacite and andesitic stratovolcano. The steep inner portion was part of the composite cone which is surrounded by remaining segments of the ring plain of laharic breccia. Sheer andesitic cliffs are footed with talus, from which waves remove boulders to be spread laterally by long-shore currents. East of the landing place at Te Titoki Point there is a massive supply of stream-rounded boulders. With an open sea angle of 137 degrees, and an average wave fetch of 50 km to the south, wave refraction brings this point into the 'High Exposed' category (index 6).

The boulder shore is highly mobile, both from seasonal shifts by waves, and from long-shore currents. The low growl of the moving boulders follows each receding wave, and even in relative calm the smaller cobbles rattle continuously down the steep slope. Checks of marked boulders on calm days at the places of greatest mobility showed a total overturn after only two high tides.

A study of boulder dynamics at Hauturu showed a cyclical stability of beach profiles. Long-shore boulder transport brings seasonal shifts according to the direction of the prevailing winds and their generated waves. This affects boulders to a depth of 2 m, which are transported within a rather narrow range of diameters. There is a significant correlation between a boulder's size and its rate of drift. Larger boulders travel further by their inertia of motion than smaller ones, which are immobilised earlier by drag. With a given beach-slope and wave regime, there will be boulders so large that they are only infrequently dislodged. The greatest movement will be of those boulders just below the weight threshold for stability.

At Little Barrier Island a bare and dry **littoral fringe** is prolonged well above mean high water of neap tides (MHWN). At Te Titoki Point, periwinkles are totally lacking, but the broad-based, more wave-stable *Nerita atramentosa* is abundant. Limpets, though ideally shaped to resist waves, are disadvantaged by frequent overturn and danger of crushing. *Cellana ornata* is thus absent from its normal high level. Barnacles are also wanting.

Up to the highest splash level, *Nerita atramentosa* dominates, both above and below the boulders, along with the topshells *Diloma zelandica* and — further down the shore — *Melagraphia aethiops*. In decaying

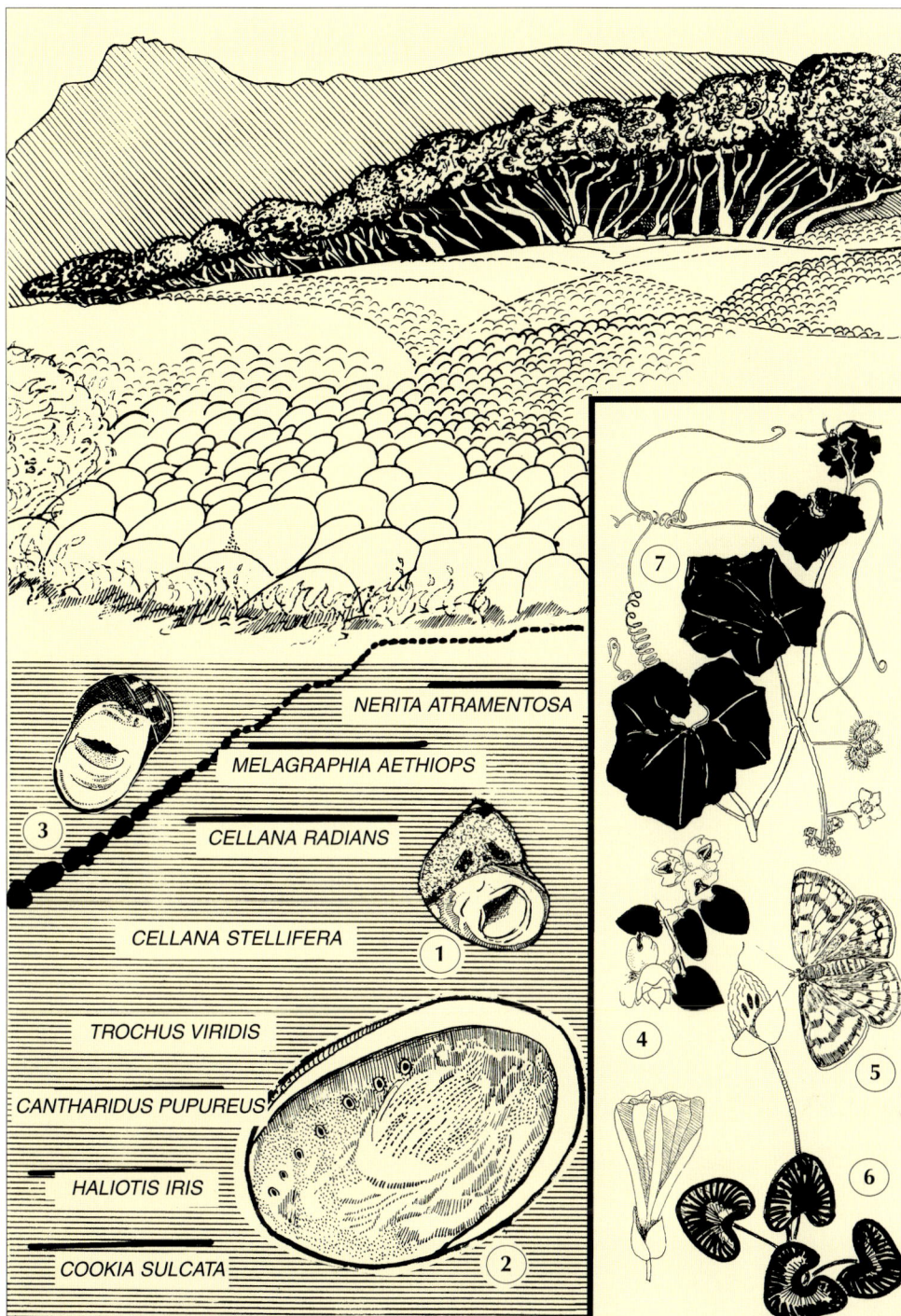

Fig. 11.11 Mobile boulder beach, Little Barrier Island: landing place
(bottom left) Ranges of gastropods across the seaward slope, **(right)** plants of the gravel berm.
1 *Melagraphia aethiops* (coralline encrusted) 2 *Haliotis iris* 3 *Nerita atramentosa* 4 *Muehlenbeckia complexa*
5 *Lycaena salustius* 6 *Calystegia soldanella*, with seed capsule and (left) flower 7 *Sicyos australis*.
The shore is backed with pohutukawa.

algal wrack high up the shore live occasional *Diloma nigerrima*. Confined under boulders, even by night, is the very light-sensitive *Diloma bicanaliculata*.

On a calm day after dusk *Nerita* and *Diloma zelandica* are found all over the boulders. When wave-force increases they move to the lee sides, or eventually close the operculum and fall between boulders. The visible surface is thus left bare; at the first low water after

a storm, only a small percentage of snails have yet returned. At the next low tide, however, the full density is restored, with the surface once more black-spattered.

Melagraphia aethiops does not so readily drop between boulders, and is confined to the pink **sublittoral fringe**. Along with low level *Cellana radians*, *Melagraphia* carries — most unusually — a coralline crust. Under the boulders the browsing gastropods are also

215

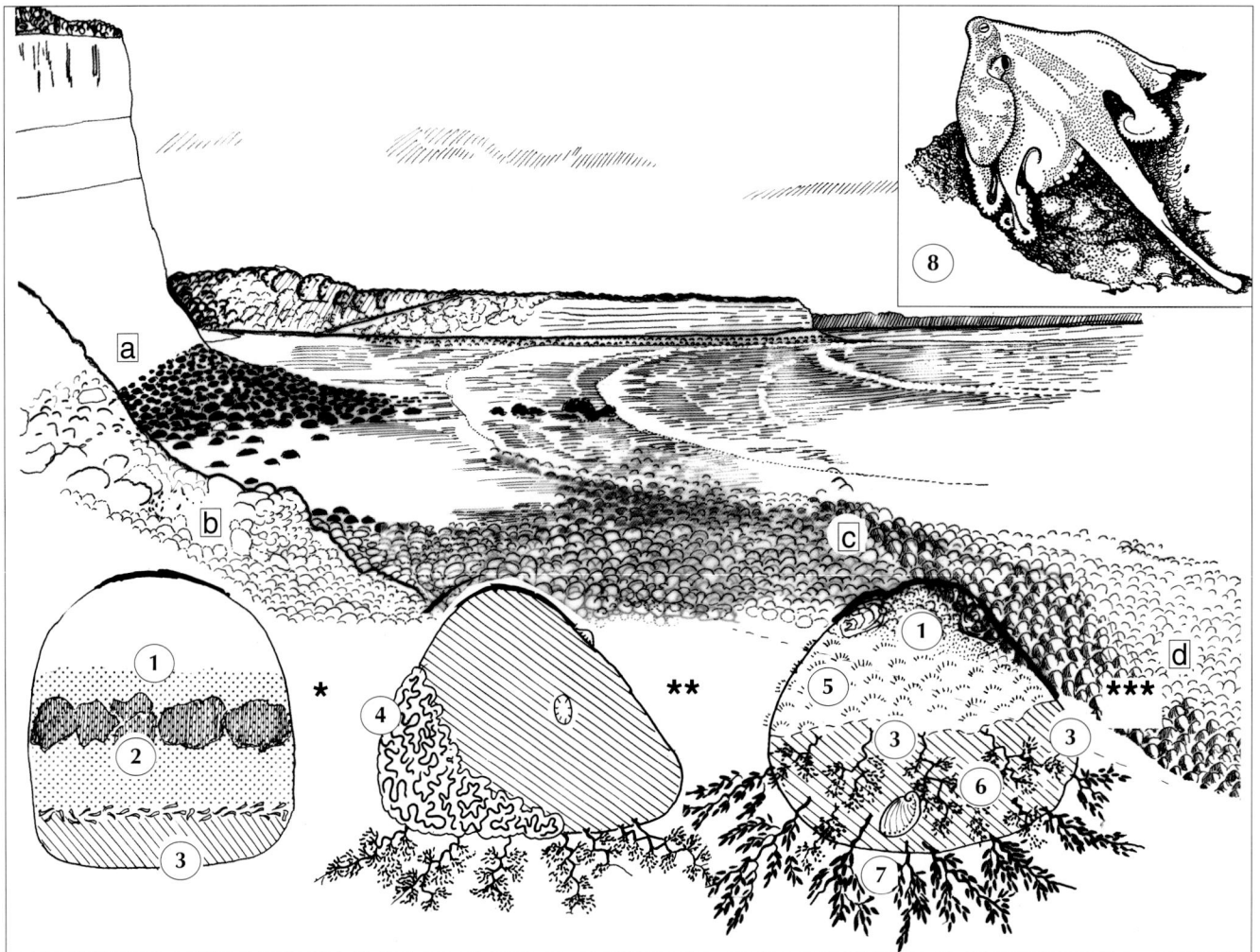

Fig. 11.12 Boulder beach at Opunake, south Taranaki
The boulder beach, looking south, is divisible into:
a–b littoral fringe; b–c eulittoral zone (boulder detail, (*));
c–d sublittoral fringe, with boulder detail (**) and (***)
1 *Chamaesipho columna* 2 *Apophlaea sinclairii* 3 pink
coralline veneer 4 *Codium convolutum* 5 coralline turf
6 *Cystophora torulosa* 7 *Carpophyllum maschalocarpum*
8 *Pinnoctopus cordiformis*.

pink-crusted like iced cake. These include *Cookia sulcata* (the close-allied *Turbo smaragdus* is lacking in high exposure), *Trochus viridis*, and — just a little lower — the black-footed paua, *Haliotis iris*, and the yellow-footed, *H. australis*, along with *Cantharidus purpureus* and the starred limpet, *Cellana stellifera*.

Lowest on the shore there are large, stable boulders, with their pink veneer visible at each draw-back of the waves. These boulders of the **sublittoral fringe** show aspectation according to wave action and sun. To seawards they carry streamers of *Glossophora kunthii*, then — next down — longer tresses of *Carpophyllum plumosum* and *Cystophora retroflexa*. At the base of these large browns spreads a red algal turf, with *Chondria macrocarpa*, *Laurencia* species, and filamentous *Dipterosiphonia* and *Heterosiphonia*. To leeward, the stable boulders are pink-painted, with their highest parts bleached white in summer.

Taranaki: Opunake

South of New Plymouth the coastline is built of sheer cliffs cut through only by the streams flowing radially from Mt Taranaki. Smooth, abraded boulders have been transported by streams or torn by waves from the laharic breccia that forms the cliffs.

The wide, spectacular boulder flat revealed at low tide at **Opunake** comes under the impact of onshore south-westerlies. Looking across its full 200 m expanse, there are several changes of hue revealing the distinct zones.

Above high water spring (HWS) is a steep ramp of sun-warmed, often mobile boulders, with both the periwinkles *Austrolittorina cincta* and *A. antipoda*. In a second zone, descending almost to extreme low water of neap tides (ELWN), the boulders are clean on top, with *Nerita atramentosa*, here nearing its most southerly foothold on the west coast.

The octopus (*Pinnoctopus cordiformis*) may lurk concealed under ledges or stable boulders. By far the largest of New Zealand molluscs, offshore specimens

Fig. 11.13 Boulder beach north of St Clair, Dunedin
1 *Diloma nigerrima*, aggregated between high level boulders 2 *Stictosiphonia arbuscula* with **(right)** thallus detail 3 *Porphyra columbina* 4 coralline paint with *Pachymenia lusoria* 5 *Durvillaea antarctica* 6 *Jania* sp. 7 coralline turf 8 *Codium fragile* 9 *Cladhymenia oblongifolia* 10 *Diloma nigerrima* 11 the same, from beneath 12 *D. arida* 13 *D. coracina*.

may have a metre's arm-spread with a body 20 cm long. A competent funnel-swimmer, the octopus will make no immediate run, witholding its cloud of ink, and using form and colour camouflage until hard harrassed.

In the **sublittoral fringe**, the stable boulders present a sharp change of colour. The seaward faces are painted pale pink with basal *Corallina*, often blotched with tan *Ralfsia* or black *Hildenbrandia* or topped with coralline turf. Their landward faces give a nice example of aspectation, with bottle-green *Codium convolutum* away from the sea in contrast with the pink exposed to full surf. Round the boulder bases are well-grown

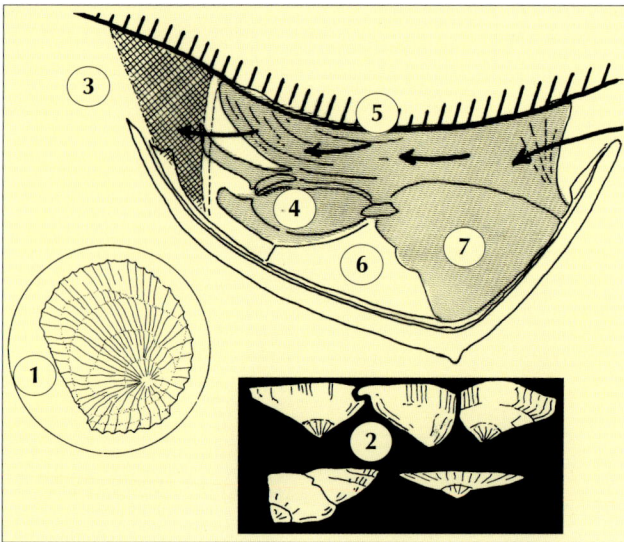

Fig. 11.14 *Gadinia conica*
1 shell, from above, 2 variant shapes of shell, in profile
(top) animal in feeding posture, showing water currents
3 mucus net 4 buccal mass 5 foot 6 mantle cavity,
7 visceral mass.

Cystophora torulosa. In Taranaki this is the predominant brown alga down to extreme low water of spring tides (ELWS) where pure beds of *Carpophyllum maschalocarpum* exclude it. On the large cobbles where surf breaks at low spring tides, the calcareous algae deepen in colour and increase in variety. They come to overspread the boulders completely, welding them into a pavement over which clear water constantly washes.

Otago: St Clair

Close to Dunedin city, at the south end of St Clair beach, the sand is cut off by a cliff-head, which is the mother rock of a fine boulder beach. These cliffs are part of Dunedin's Miocene volcanic outpourings. Rows of prismatic columns are derived from successive basalt flows. Their lava is jointed into blocks that dislodge to become finally smoothed by wave-dragging and mutual abrasion.

The boulders of the upper and lower levels are of very different size and mobility. In the horizontal lower stretch, those of metre-size are flat-based and pyramidal, and support a complex algal zonation. The two bull-kelps, *Durvillaea antarctica* and *D. willana*, are both present, the former with its serpentine thongs visible through white foam, or cast landwards by waves to be draped over the boulders upshore.

Boulders out in front of the kelp are continuously wave-washed, and carry a wealth of smaller algae, including corallines, green finger *Codium fragile* and many delicate, membranous reds, most notably *Cladhymenia oblongifolia*. Inside the kelp-line, rounded boulder tops are festooned with winter and spring *Porphyra*,

or covered with fleecy *Stictosiphonia*. Their sides carry the barnacles *Chamaesipho columna* and *Epopella plicata*. Lower down, in a calmer moat, the boulders display up to their mid level a fringe of *Ulva*, *Scytosiphon lomentaria*, *Scytothamnus fascicularis*, *Bryopsis plumosa*, *Halopteris* sp. and long-stemmed *Streblocladia glomerulata*. Small grazing gastropods take shelter underneath. The rounded, conical limpet *Notoacmea badia* (10 mm across) is peculiar to Otago rock pools. Eroded externally, it is rimmed inside with purple. Under boulders higher up lives the thin, flat *Atalacmea multilinea*. Common too among the algae is the prettily marked topshell *Cantharidella tesselata*.

Above this horizontal stretch a steep, well-sorted slope rises for 3 m, with cobbles of decreasing size up to the mobile gravel at the beach crest. There is no barnacle zone, and the single periwinkle is *Austrolittorina antipoda*, clustered in high sheltered pockets. In cool spots between stable boulders are clusters of the limpet *Notoacmea pileopsis sturnus*. In the south *Leptograpsus variegatus* is absent, leaving *Cyclograpsus lavauxi* as the only upper shore crab.

The lower part of this slope is dominated by topshells. Far commoner than in the north is the near-conical *Diloma arida*, dull black to mauve and flecked with white. On boulders in pockets of sand lives the small, concave-based *D. coracina*. Highest of all, grazing on the paste of decaying *Durvillaea*, are huge numbers of *Diloma nigerrima*, depressed-globose, and glossy black as if rubbed with graphite. These snails sometimes aggregate in thousands to fill the whole space between adjacent metre-sized boulders. The common thaid predator on grazing gastropods is the southern *Lepsithais lacunosus* (Fig. 10.17).

In total dark between squarish boulders lives the pure white pulmonate limpet *Gadinia conica*, found in suitable sites in the north but reaching its highest numbers in the South Island. A single colony may have up to a thousand limpets, immobile and massed together to take a near-hexagonal form. The Family **Gadiniidae** has a handful of species through the temperate world, all confined to dark spaces in strong current flow. They live upon surfaces clean of any algae and the radula is never used to abrade. Instead, food is brought by water turbulence. Oriented precisely to the flow, *Gadinia* hangs upside down, lowering the shell to admit a current from behind. Mucus glands on the mantle and foot secrete a curtain in front of the head, which balloons out with the water pressure behind. Mucus strands serve as 'preventers' to hold this net in, and strained off phytoplankton is held until the mesh is loaded. The net is then grasped by the oral lobes to be hauled into the mouth with the delicate teeth of the small radula. No cilia are involved, nor can feeding be elicited in still water.

CHAPTER TWELVE
Life-forms Under Boulders

The communities living under boulders are effectively protected from high and fluctuating temperatures and the direct effect of waves. Except for a narrow zone of reflection under the boulder margin, light is excluded and plants are altogether lacking. At the line where pink coralline paint cuts out just inside the margin, a threshold has been crossed into a wholly different 'biome' or biological realm, that is held in possession by animals alone.

Dark terrestrial communities in contrast still have their plants of a sort, if we may so regard the *saprobes* (bacteria and fungi), that subsist not by *production* through photosynthesis but by the *reduction* of dead and decaying tissues of other organisms. Under boulders between tides, however, except for a few sheltered pockets, the dark spaces are swept by currents strong enough to ensure that the products of decay will not for long accumulate.

By far the greatest biomass of the dark under-boulder community is to be accounted for by filter-feeding animals that are sessile or permanently fastened down. The constant trend is for the separate individuals to become miniaturised and aggregated into colonial patterns with hundreds of units held together in varying degrees of interdependence. Their food is harvested from the plankton drawn in from the water around, and ultimately from the ocean system at large. With all their diversity of current-driving machinery, a constant water flow is set up, often aided by eddies and turbulence resulting from the configuration of the colony itself.

The sessile filter-feeders themselves form a living substrate in turn grazed by carnivores, with the same dependence on herbivores as have the latter upon plants. A concentrated protein pasture is thus available on site to predators specialised to attack and ingest it. Its exploiting needs only a fraction of the metabolic effort that would be involved in the pursuit of mobile prey.

The second tier of the under-boulder community is thus composed of slow-moving predators, very often food-specific. Foremost among these diet-specialists are the gastropods, ranging from small shelled prosobranchs to reach their zenith of colour and specialisation among the nudibranch slugs. The two other leading carnivorous groups represented under boulders are the crabs, already reviewed in Chapter 11, and the starfish or asteroids.

The filtering life-forms

Three major groupings of sessile ciliary feeders must first be introduced before embarking on our survey of communities under boulders. Two of these — the ascidians and in particular the bryozoans — tend to be most characteristic of high-energy settings in the sublittoral fringe, on current-swept rock-faces, and on shell gravel on continental shelves beyond diveable depths. The third, the sponges, come to their zenith with shelter from strong water movement, and flourish best at sublittoral depths where wave effects and currents are not too severe.

Bryozoa (Ectoprocta)

Cheilostomata

The most numerous and successful of the Bryozoa belong to the large Order Cheilostomata. They can appear in many shapes, from small bushes to coral-like concretions, but under boulders or on algae their most familiar colonial form is of one-layered sheets, built up of box-like zooids, each about half a millimetre to a millimetre in length. Zooids are actually body walls and inside each box is a polypide, made up of feeding tentacles (the lophophore) and a gut. Minute though it may be, each of these zooidal units is a well-organised coelomate metazoan, with complete alimentary, nervous and muscular systems, such as no sponge or cnidarian can aspire to.

The bryozoan polypide at rest is contained completely within its zooid walls. Only when the lophophore is everted from the box does the real mode of life become apparent. The tentacle circlet is beautiful to watch with a binocular microscope, readily enough expanding to its funnel shape, but retracted at the slightest shock. The water currents are centripetal, entering the circlet from above and moving towards the mouth at the bottom of the funnel. Food particles are retained in the funnel while the water currents are driven out between the tentacles by strong lateral cilia.

Some bryozoans are transparent enough in life to show the whole digestive tract at work. The oesophagus can be seen to open into a globular stomach, with a thimble-shaped pylorus leading to the intestine. A rod of compacted food can be watched as it rotates in the pylorus while projecting into the stomach. After a spell, the rod moves into the rectum prior to its expulsion as a faecal pellet. The anus opens at the lophophore base, and **outside** it (hence the alternative name of Ectoprocta for the phylum).

The fertilised eggs of some bryozoans develop into free-swimming larvae (**cyphonautes**). In most others, larger yolky eggs are retained in a brood chamber (ovicell) where they develop into benthic larvae that emerge to found new colonies generally not far from the parent. Once established, cheilostome colonies grow by asexual budding. At the colony margin, the three proximal facets of a zooid are in contact with neighbours, and the three distal are left exposed to produce buds.

The Cheilostomata are widely represented on boulders and algae. A first example can be *Membranipora membranacea*, forming a white lace on the kelp *Ecklonia* (Fig. 13.11). The zooids are lightly calcified and mostly long and rectangular. The frontal surface is only a thin membrane. *Membranipora*, in the Suborder **Malacostegina**, hence belongs to a group of cheilostomes collectively referred to as 'anascans', where the lophophore is everted by hydrostatic pressure generated by parietal muscles pulling down the frontal membrane.

The polypide so thinly covered is left vulnerable, and needs protection by partly over-roofing, which must not, however, impede the hydrostatics. In *Electra* (also found on algae) the frontal surface is proximally calcified and the membrane reduced to an oval, flanked by protective spines. In some genera (*Micropora*, *Steginoporella*) a calcareous shelf forms beneath the frontal membrane, leaving small spaces for the parietal muscles to operate.

In the major Suborder **Ascophora**, there is a complete, calcified roof, either overarching the frontal membrane (*Figularia*, *Celleporaria*) or replacing it, in which case the hydrostatic function is served by an underlying **ascus sac** opening at the anterior end. The sac is dilated by the pull of its parietal muscles, drawing in water to exert pressure upon the body cavity and so evert the lophophore. In *Figularia* and other **cribrimorphs** spines cross the membrane-like rafters or fuse into a sieve-like shield.

In many ascophorans the orifice is subdivided, with separate openings for the lophophore and the ascus. In *Schizoporella* the orifice narrows to a V-shaped notch, which in *Celleporella* and *Watersipora* is a rounded

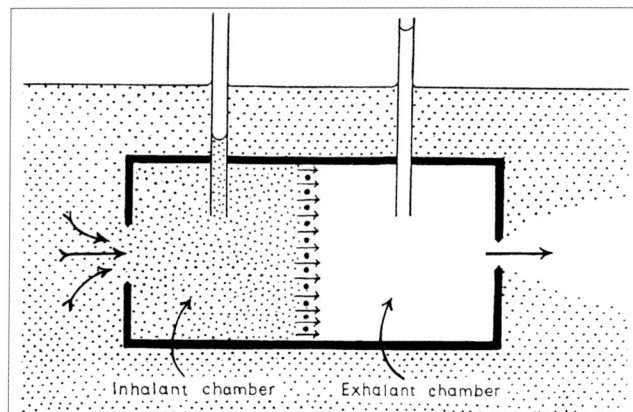

Fig. 12.1 Model of a filtering system in a closed space
The inhalant and exhalant chambers are divided by a screen, with ciliary beat providing the filtering pump and a meshwork forming the straining system.
(From M.J.S. Rudwick.)

sinus. In all microporellids (*Microporella*, *Fenestrulina*) the sinus is separated from the orifice as an ascopore.

The details of pores and ornamentation are important for classification. So too is the polymorphism of the zooids. As well as the **autozooids** (feeding zooids) already described, there are smaller **heterozooids**, of two main sorts, the **avicularia** and **vibracula**. The avicularium has no functional polypide, but carries a movable mandible (a modified operculum) like a parrot's beak, opened and closed by its own muscles. In a vibraculum, the mandible is long, narrow and whip-like.

The origin of separate avicularia is revealed by the genus *Steginoporella*, where there are two sorts of functional autozooid, one with a larger operculum and strong opening muscles. In *Crassimarginatella*, the avicularium with its enlarged operculum still retains a functional polypide. In many flustrids, like New Zealand *Gregarinidra*, smaller **vicarious avicularia** are simply inserted in a linear series of autozooids, and may replace them. In *Micropora* these are smaller and interzooidal (wedged in between the autozooids) and in *Escharina* they are adventitious, inserted on the front of autozooids. In *Bugula* the adventitious avicularia are jointed and stalked, with muscles giving them free mobility. As Charles Darwin wrote: 'They curiously resemble the head and beak of a vulture in miniature seated on a neck and capable of movement.'

The most typical cheilostomes under boulders or on kelp holdfasts are encrusting ascophorans. As well as their microscopic structure, most of these have also some useful features for spotting in the field. *Eurystomella* is red, *Escharoides* and *Smittoidea maunganuiensis* bright orange or pink, recognised by their glistening lustre when fresh. Others are white, such as *Microporella* and *Fenestrulina*, the latter with convex zooids studding the colony. Transparent-buff *Galeopsis*

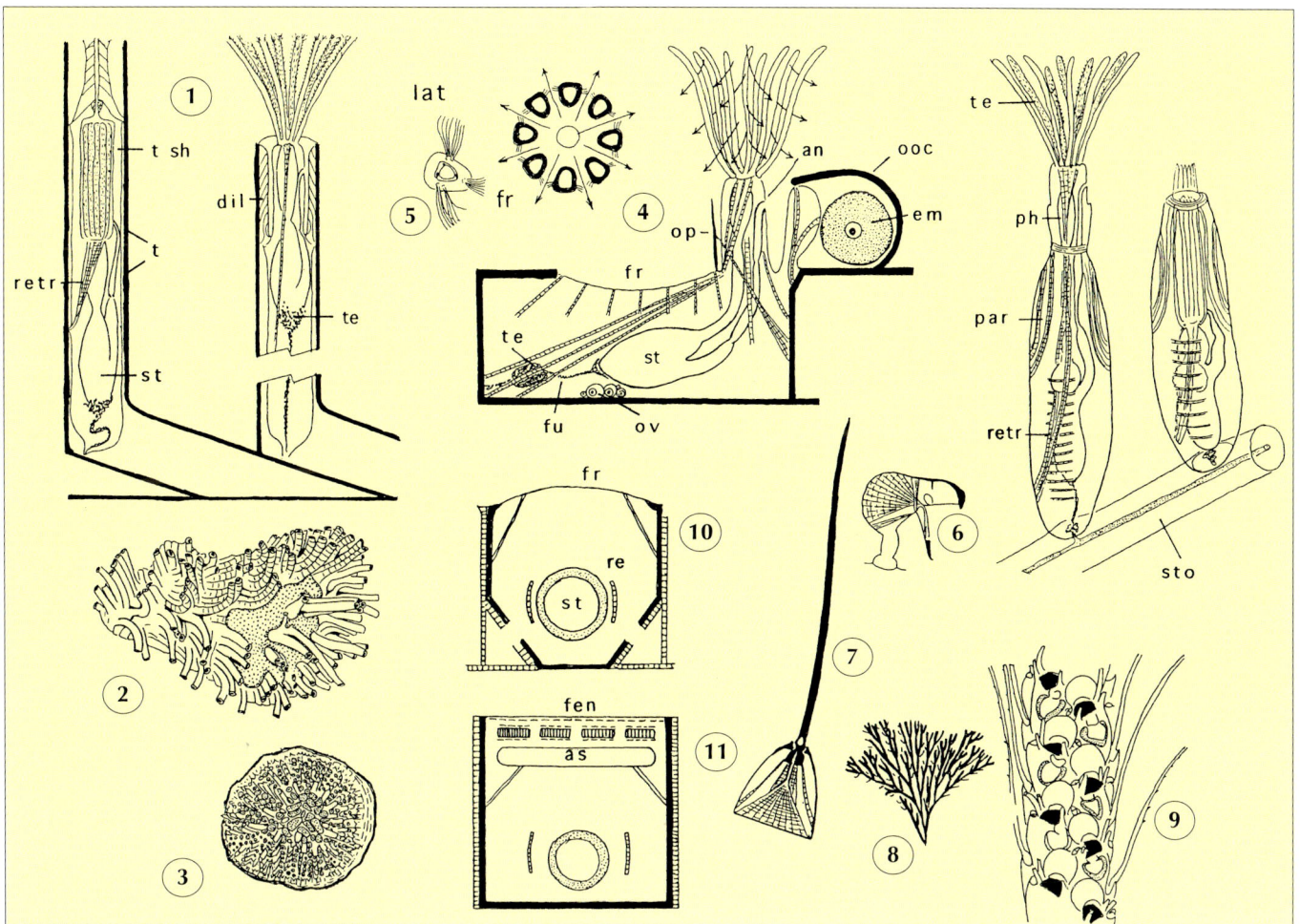

Fig. 12.2 The structure of Bryozoa

Cyclostomata (top left) 1 *Tubulipora* zooid with retracted and **(top right)** expanded (retr) retractor muscle (te) testis (st) stomach (t sh) tentacle sheath (dil) dilator muscle (t) tube wall 2 *Tubulipora* colony 3 *Disporella* colony.

Cheilostomata (centre) (an) anascan zooid (fr) frontal membrane (ov) ovary (te) testis (fu) funiculus (op) operculum (ooc) ooecium (em) embryo (st) stomach (an) anus (as) ascus sac.

4 transverse section through the tentacles, above the mouth 5 section of a tentacle with (fr) frontal and (lat) lateral cilia 6 an avicularium 7 a vibraculum 8 *Caberea zelandica*, branching colony with (9) detail of zooids 10 section of anascan (fr) frontal membrane (st) stomach (re) retractor muscle 11 section of ascophoran with fenestrate roof.

Ctenostomata (top right) zooids expanded and retracted: (par) parietal muscle (ph) pharynx (te) tentacles (retr) retractor muscle.

has flask-shaped zooids with short necks in its encrusting phase (before becoming erect), while *Crepidacantha* is striking for its long needle-like spines horizontally fringing the zooid apertures.

Anascans, with their zooid frontals largely uncalcified, grow either in crusts or upright bushes. The lace-work of *Membranipora* on *Ecklonia* (Membraniporidae) can develop jointed zooids so the colony can bend with the frond. In the related Family Electridae is the silvery-white *Electra pilosa*, in finely spinose colonies on low-water red algae. The species of Steginoporellidae form encrusting or finger-like latticed frameworks with their thickened zooidal walls.

Bushy anascans are typified by the Family **Bugulidae**, with narrow zooids set in two to six series along branches. All (except the common burgundy-coloured fouling species *Bugula neritina*) have well-developed bird's-head avicularia. Dull brown *B. flabellata* is somewhat fan-shaped with three to six serial branches. *B. neritina* and *B. stolonifera* are biserial, the latter forming greyish tufts. *Caberea* (**Candidae**) is fan-like, with somewhat flattened zooids on the front side of the branch and long tube-like vibracular chambers on the back with saw-toothed setae. *Caberea zelandica* and *C. rostrata* grow in orange tufts in crevices and overhangs, on wharves, and *Ecklonia* holdfasts. Related *Tricellaria* is common on wharf piles, as straggling bushes attached by rhizoids. Species of **Cellariidae** are jointed, with cylindrical internodes of well-calcified zooids.

The **Beaniidae** are buguloid anascans that have a

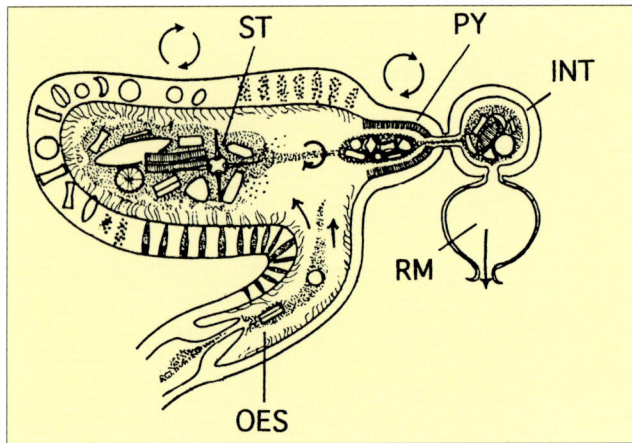

Fig. 12.3 Gut of an ectoproct, Thalamoporella
Diatoms are carried in from the oesophagus (OES) and rotated in the stomach (ST) by the faecal pellet in the pylorus (PY). Each pellet afterwards passes to the intestine (IN) to be discharged by the rectum (RM).

prostrate form, forming loose rhizoid-attached colonies under boulders and on pilings. The stand-apart zooids have tubular links and are frequently covered by incurving spines.

There are two smaller living orders of Bryozoa, both of Paleozoic origin and predating the Cheilostomata (which originated in the Jurassic). In one order, the **Ctenostomata**, the colonies are fleshy or gelatinous, and uncalcified. These will be dealt with as part of the shade community of wharf piles, where they may flourish in summer (Fig. 14.1).

Cyclostomata

This order stands even further apart from the rest of the Bryozoa, having a long ancestry but only moderate success. They retain some primitive features, elsewhere found only in fossil groups, as in their cylindrical zooids, with circular aperture and their lack of opercula. Eversion of the lophophore does not depend on distortion of the body wall, and the coeloms of adjacent zooids are separated by shared walls.

Cyclostomes will regularly be found under current-swept boulders. A favoured habitat is also on the fronds of *Sargassum*. First, the branching *Crisia* species are typified by their diminutive bushy form with the zooids long, erect and curved outward at the apertures. Where the branches fork, swollen balloon-like female zooids contain the embryos. *Hastingsia* species break into club-like branches built up of parallel zooids with their openings like a fine honeycomb. A different design is shown by *Telopora*, with its tubular zooids diverging radially from a pedestal. *Disporella* and *Favosipora* species form small white plate-like colonies, the zooids in the former in spoke-like radial rows.

Ascidians (sea-squirts)

At a first glance the ascidians might seem to display little advance beyond the bryozoans. Some could even be mistaken for sponges. Yet they in fact belong to the highest of the animal phyla, the Chordata, and begin their life history with a briefly mobile stage in the so-called tadpole. The muscular tail is early cast off, as the larva ceases swimming and fixes itself to the bottom. By chordate standards this could be called a degeneration. But from an ascidian standpoint, it is the higher vertebrates that could be looked upon as 'sea-squirts manqués'. Rather than maturing to the sessile stage, they have made the mobile larva prematurely reproductive. From such a beginning, with *neoteny* or 'escape from specialisation' all the vertebrates — including ourselves — have proceeded, by retaining and enhancing the locomotor ability conferred by the muscular tail.

In becoming sedentary the ascidian adult has lost its tail, brain rudiment and elementary sense organs, while the pharynx has became highly enlarged for filter feeding (Fig. 2.9). The tadpole — with the mouth still imperforate — was not yet concerned with feeding but with site-finding and settlement. Unlike any higher chordate, many ascidians have resorted to asexual reproduction, leading to complexly patterned colonies.

From inspection of external characters, the primary division of the Class **Ascidiacea** is hardly self-explanatory. The two Orders **Enterogona** and **Pleurogona** each include both *simple* and *compound* (or colonial) forms. Their prime difference lies in the position of the ovary and testis: in the Enterogona these ramify over the gut, but in the Pleurogona they lie in the atrial lining.

Simple ascidians

Within the Order **Enterogona**, the Suborder **Phlebobranchiata** contains the well known simple ascidians, *Ascidia*, *Ciona* and *Corella*. The test is cartilage-like and translucent, and the soft body comes cleanly out of it on incision. The oral siphon is fringed with eight lobes, the atrial with five or six. Commonest under boulders is the ovate, red-siphoned *Corella eumyota*. *Ascidia aspersa* has a more quadrangular test, without siphonal colour. *Ciona intestinalis* is cylindrical and pendent from its base, with the siphons tipped in yellow.

All the rest of the simple ascidians are placed in the Order **Pleurogona**, with leathery tests fixed by the long side, and with both siphons directed upward. Slips of muscle strongly secure the body within the test. The **Styelidae** have the oral tentacles (at the base of the branchial siphon) unbranched, the siphonal apertures four-lobed or square, while the branchial sac has four large folds at either side. The styelid species (Fig. 12.4)

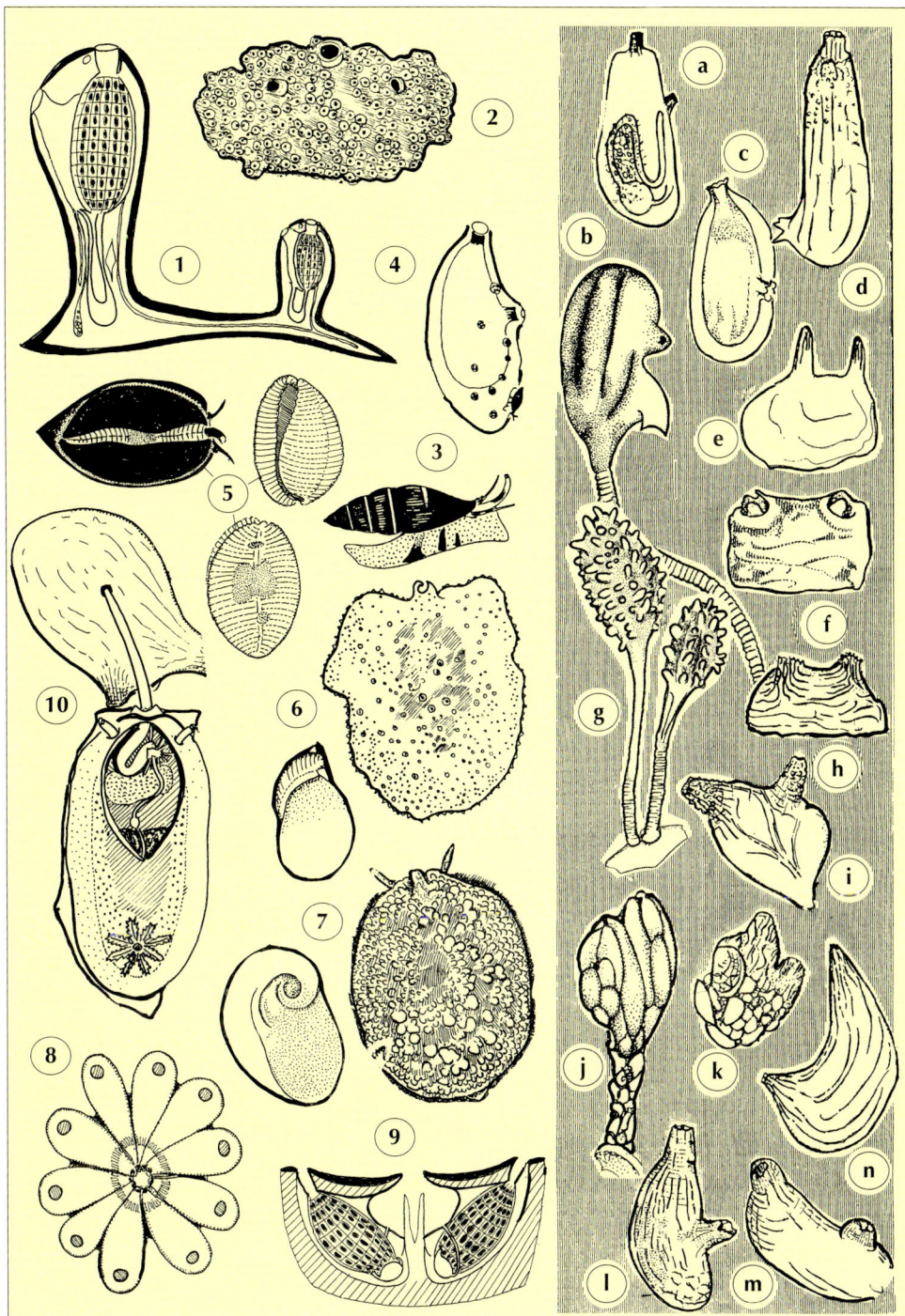

Fig. 12.4 Ascidians and their predators

(a) *Ascidia aspersa* (b) *Pyura pachydermatina* (c) *Corella eumyota* (d) *Ciona intestinalis* (e) *Pyura subuculata* (f) *Pyura pachydermatina* (g) *P. spinosissima* (h) *P. cancellata* (i) *Microcosmus squamiger* (j) *Styela plicata* (k) *Pyura carnea* (l) *Asterocarpa coerulea* (m) *A. humilis* (n) *Cnemidocarpa bicornuta*.

1 structure of *Clavelina* zooids on stolon 2 surface of a *Didemnum* colony with separate branchial and large common cloacal openings 3 *Austromitra rubiginosa* with (4) egg capsules in *Corella* test 5 *Trivia merces* **(left)** animal, **(right)** two views of shell 6 *Lamellaria ophione*, shell and animal 7 *L. cerebroides*, shell and animal 8 detail of rosette of *Botryllus schlosseri*, with (9) zooids in vertical section 10 *Dendrodoris citrina*, dissected to show fore-gut with proboscis inserted in *Microcosmus*.

are distinguished by test shape and hue, siphonal colour, and the form and arrangement of gonads.

The **Pyuridae** have in contrast branching oral tentacles, and six or more folds to the branchial sac. The test is gnarled or wrinkled, often covered with shaggy growths or silt. These ascidians commonly cluster under ledges or overhangs, on wooden piles or in holdfasts.

223

Compound ascidians

The compound ascidians are as colourful as the sponges, but contribute far less biomass to the encrusting community. The individual zooids show a progressive reduction in size, with the eventual loss of their independence by incorporation in a colonial test. In the **Clavelinidae**, submergence of the zooids has hardly begun, and separate individuals a centimetre tall stand up from stolons running in a common base. The *Pycnoclavella* zooids are translucent and beautifully green or blue opalescent.

In the **Holozoidae** small zooids have become embedded in a continuous test. *Sigillina australis* has no patterned 'systems', and the individuals open separately, but in *Distaplia* they form regular series alongside the shared cloacal canals. In the **Polycitoridae** (*Eudistomia*, *Cystoides*) the zooids are completely absorbed in the test, and arranged in incipient circles with the atrial aperture at the centre. The **Polyclinidae** form thick encrusting sheets or cushions, containing the attenuate zooids having the gonads prolonged behind the gut-loop. *Aplidium* is the largest, most diverse genus, with rounded heads or — in the New Zealand-wide *Aplidium phortax* — translucent crusts on wharf piles and overhangs. In the most advanced family, the **Didemnidae**, the colonies are flat and encrusting, with minute zooids in complex systems around the wide cloacal spaces that occupy much of the interior of the colony. *Didemnum* crusts are impregnated with microscopic limy spicules, but *Diplosoma* lacks these.

All the above compound ascidians belong to the Suborder **Aplousibranchia** in the Order **Enterogona**, with budding from a stolon at the back of the branchial sac. Compound forms have also evolved within the **Pleurogona** in two subfamilies of the **Styelidae**. First, the **Polyzoinae** are miniaturised styelids showing progressive simplification, with separate individuals standing up from a common basal test. The **Botryllinae** are far more advanced, having patterned systems with the zooids in *Botryllus* opening to a common cloaca in circular and radial systems, or in long double rows.

Sponges: Porifera

Even at the dawn of the Cambrian, the sponges must already have been long in existence. Evidently the earliest well-organised filter-feeders, they are also the first animals we can properly call multicellular. Rather than call them Metazoa, it is customary to place the sponges in a lower grade, the Parazoa. Most sponges are highly variable in shape, better regarded as patterned colonies than as finished, unified organisms. They lack both the nervous system or directed sets of muscles that co-ordinate and integrate the life of the Metazoa.

Yet the sponges do show many aspects of unity. They

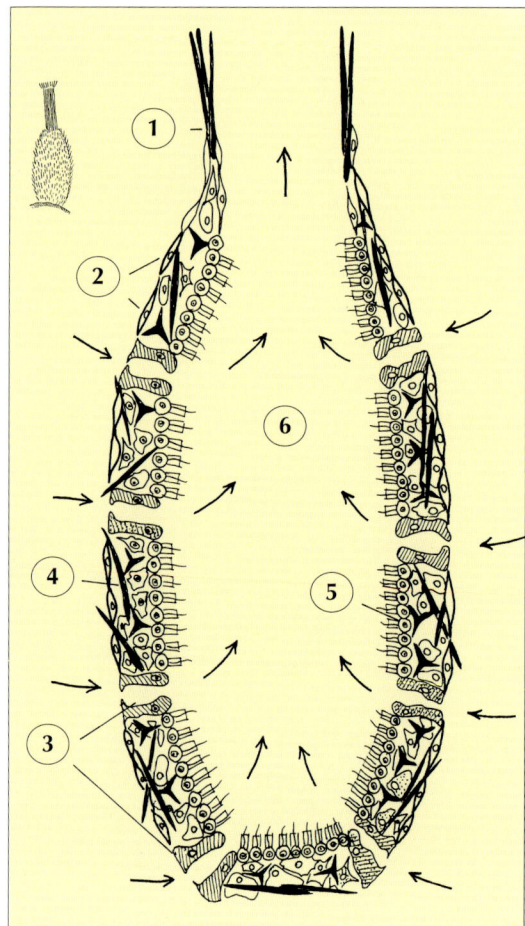

Fig. 12.5 A calcareous sponge
Simple asconid sponge in longitudinal section to show structure and current flow.
1 osculum 2 ectoderm cell 3 pinacocyte or pore-cell of ostium 4 spicule 5 flagellated cell or choanocyte
6 flagellated cavity.

have a controlled, even though plastic, growth strategy, and the whole colony maintains well-directed current systems. Water is diffusely taken in through numerous pores in the body wall (*ostia*); exhalant currents leave the sponge through the larger openings of the chimneys called *oscula*. The inhalant and exhalant channels communicate by narrow passages through which currents are driven by the flagellated collar-cells known as *choanocytes*. Fig. 12.5 shows a sponge's simple histology, with *choanocytes*, mobile *amoebocytes*, *pinacocytes* or pipe-like pore-cells, and the *sclerocytes* that secrete the skeletal spicules.

The single flagellum of each choanocyte is encircled by a protoplasmic collar with an ultra-structure of micro-villi. It is these that ingest microscopic particles, which are passed down the collar by protoplasmic streaming and phagocytosed at the base. Intracellular digestion begins in vacuoles in the choanocytes, with the contents then passed to amoebocytes to complete digestion and distribute the products through the tissues.

Seemingly vulnerable to predators or bacteria, the

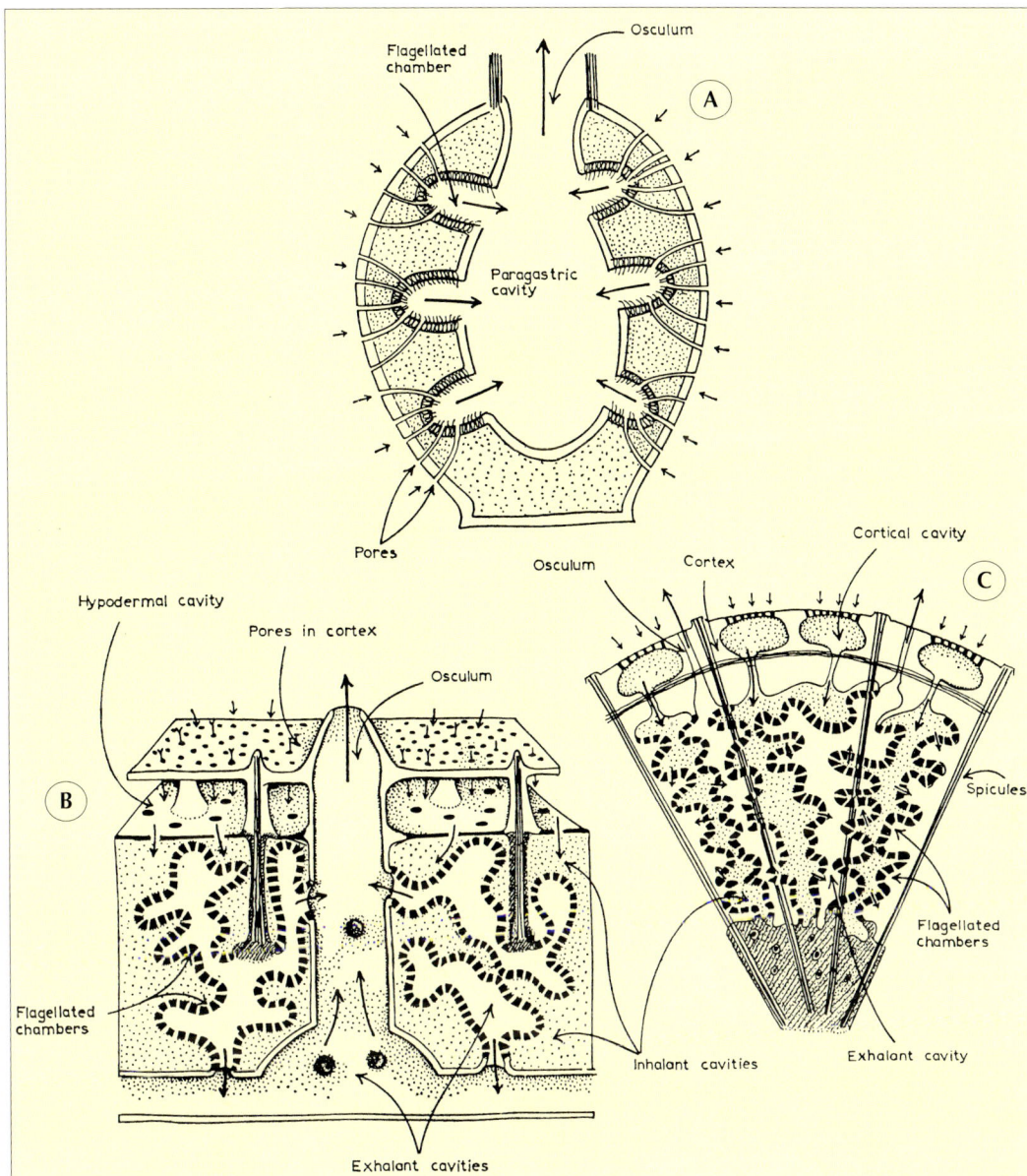

Fig. 12.6 The morphology and water currents of sponges

A. A calcareous sponge of the syconid type. B. A halichondrine sponge.

C. A hadromerine sponge. (B and C adapted from Miner.)

sponges have in a long history evolved their own set of defences. Their glassy spicules are abrasive, their elastic connective tissue is tough and unyielding, and their smell and taste (signalled by warning colours) often disgusting. In addition, the sponges have acquired distinctive chemical compounds, with antiseptic properties, comparable to antibiotics. Some of these are today being investigated for new pharmaceuticals.

The University of Auckland has been a recent centre of research on the Porifera, and the best modern introduction to their classification, biology and physiology is Patricia Bergquist's *Sponges*.

Sponge identification has traditionally relied on rough but serviceable field characters, including colour, texture, spicule density and colony shapes. Final deter-

minations must, however, depend on the microscopic forms of the spicules, which have now been well enough described in New Zealand sponges for serious students to be able to use them. The subkingdom and Phylum **Porifera** is divided into three classes, **Calcarea, Demospongiae** and **Hexactinellida.**

The least likely sponges to be seen by ordinary students are the **Hexactinellida**, which have elaborate glass-like skeleta of six-rayed siliceous spicules, and altogether lack *spongin*, a horny substance chemically akin to silk. With the soft tissues removed, the latticed cylinder of the Venus' flower-basket (*Euplectella*) has a singular beauty of design. Hexactinellids belong typically to deep water; and can be glimpsed in relatively shallow shore communities only in Antarctica.

Most remote from all the rest are the sponges of the Class **Calcarea**, with their spicules of calcium carbonate. These are small, simply organised sponges, generally white. A single vase or tube opens distally by an exhalant osculum, while the inhalant ostia are scattered all over the wall. In the most elementary construction, of the *Ascon* type, the internal cavity is itself lined with choanocytes. This lining is separated from the outside wall by a loose intermediate layer with spicules and sclerocytes. In sponges of the *Sycon* type, the choanocytes are removed into thimble-shaped radial tubes, marked from the outside as low papillae, each pierced by its ostium. The third degree of complexity is shown in the *Leucon* construction, with a complex canal arrangement more like the Demospongiae.

Class Demospongiae

These are the largest, most widespread and colourful of the sponges, having siliceous spicules normally mixed with fibres of spongin. Two large subclasses are recognised: the **Ceractinomorpha** and **Tetractinomorpha**, along with a third smaller one, the **Homoscleromorpha**.

Subclass Ceractinomorpha

Here the *megascleres* are one-, two- (rarely four-) rayed, and the *microscleres* generally sigmoid or chelate, never asteroid. The *spongin* may consolidate to form a primarily horny skeleton. In the *keratose* families, with spicules lacking, this constitutes the total support. Reproduction is typically *viviparous*, with *parenchymella* larvae incubated within the parent colony.

• Order Dictyoceratida

Compressible and rubbery sponges, sometimes toughened with collagen as well as spongin. These include: *Ircinia*, often tubular and stalked, with many collagen filaments; and foetid-smelling *Dysidea*, less tough and massive.

• Order Dendroceratida

Encrusting or erect-branching sponges, rubbery with a fibrous skeleton. *Darwinella* is crust-forming, with conulose surface, usually bright-coloured. *Dendrilla* is upright, with rubbery dendritic skeleton, slimy when living.

• Order Haplosclerida

A springy reticulate skeleton of spongin may envelop the small spicules; but fleshy tissue is sparse. *Haliclona* species encrust intertidal boulders. *H. clathrata* and *H. heterofibrosa* have tall pale mauve cones with wide oscula. *H. petrosioides* is a smooth transparent crust

with canals converging on low oscula. The branched *Callyspongia* species live subtidally.

• Order Poeciloderida

A large, diverse order with spongin reduced to a basal crust, and spicules of many different forms. Red encrusting sponges include: soft velvety *Microciona;* rough spiculose *Plocamia;* deep red *Ophlitaspongia;* yellow to pink *Adocia;* and the dull yellow *Tetrapocillon* is soft, almost pulpy.

• Order Halichondrida

Lightly built sponges, with progressive reduction of spongin. In crumb o' bread *Halichondria* and *Hymeniacidon* the skeleton has little keratin, and the spicules have only a slight regularity. Good field characters are the crumbling texture, and the detachable dermal crust supported by pillars of spicules and roofing a wide hypodermal cavity. From this cavity opens a mesh of deeper inhalant spaces, communicating by flagellated chambers with a basal exhalant space that sends up funnel-like oscula (see Fig. 12.6).

Subclass Tetractinomorpha

In this division, the large spicules (*megascleres*) are of two forms, *tetraxonid* (with four equal rays) and *monaxonid* (a single ray ornamented at the ends). The *microscleres* are chiefly star-shaped (*asteroid*), but *sigmas* and *rhaphides* may occur. Where *spongin* is wanting, the megascleres become clustered in radial, axial or plumose patterns. A strong cortex is often developed. Reproduction is typically *oviparous* with the larvae free-swimming.

• Order Axinellida

Branching red or orange sponges, rough from projecting spicules, and fibrous with axial concentration of spongin. *Axinella* and *Phakellia* are subtidal. *Raspailia* species are intertidal and subtidal, with radial spicules outside the strong axis.

• Order Hadromerida

Massive and crusting sponges with coherent structure, little spongin but strong spicules. A firm outer cortex, and an inner *choanosome* permeated by flagellated canals (Fig. 12.6). This order includes the firm, spherical *Aaptos*, *Suberites* and *Tethya*, also massive *Polymastia* and eroding *Cliona celata*.

• Order Astrophorida

Massive sponges generally of deep water, but *Ancorina* and *Geodia* are low tidal. Rough and sand-papery outside, with internal spicules radial.

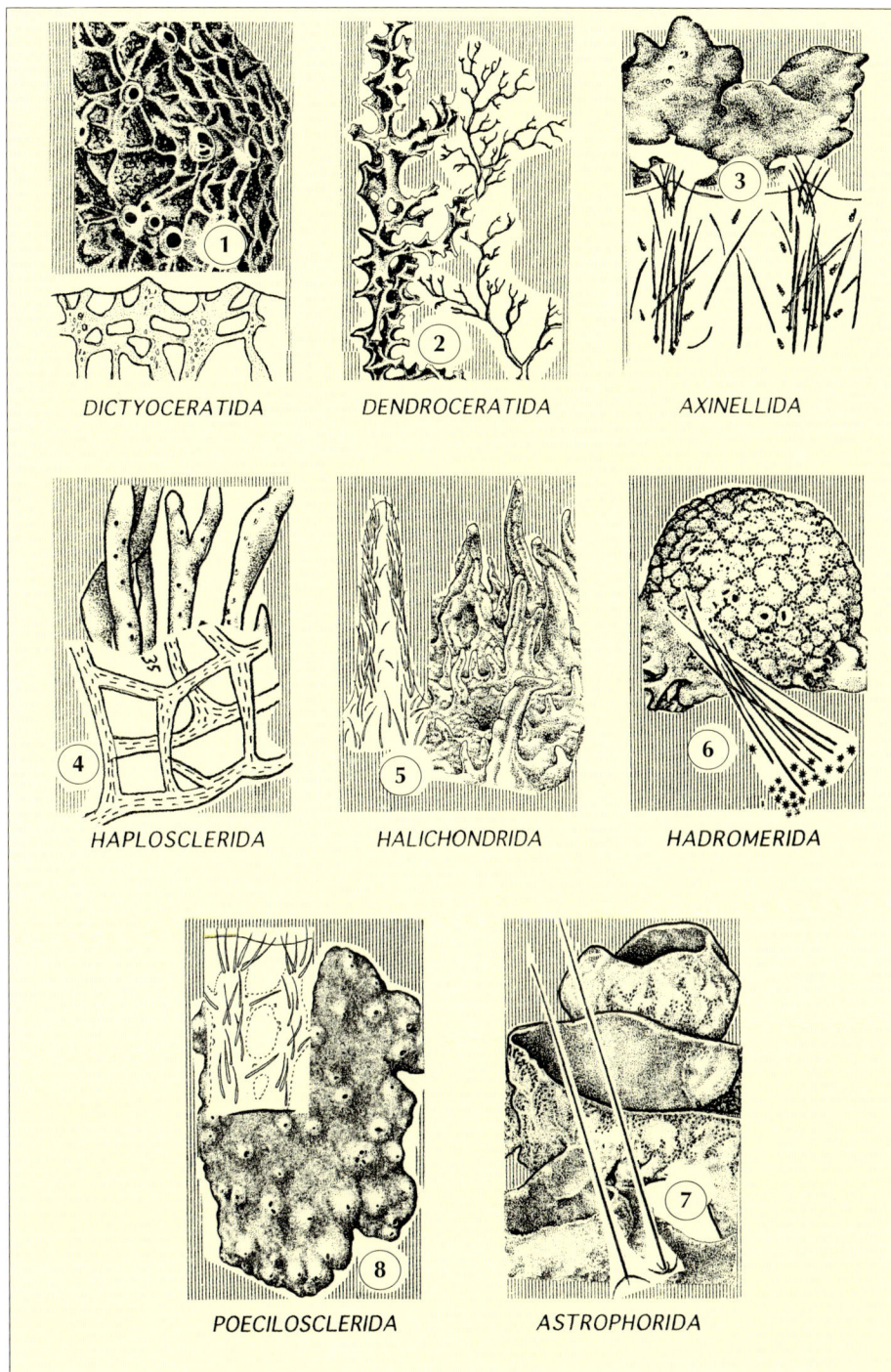

Fig. 12.7 The orders of Demospongiae
The sketches of colony form are accompanied by skeletal detail, based on the illustrations of Leigh sponges by Vivienne Ward. The orders with their species examples are 1 *Ircinia novaezelandiae* 2 *Dendrilla rosea* 3 *Raspailia agminata* 4 *Callyspongia ramosa* 5 *Ciocalypta* sp. 6 *Tethya aurantium* 7 *Axociella* sp. 8 *Ancorina alata*.

Subclass Homoscleromorpha

Contains the thin, jelly-like sponges of the Family **Plakinidae**. These are thinly encrusting, with jelly-like lobes, and a skeleton of fibrillar collagen but no mineral content.

Sponges widely replace the algae subtidally and around low water wherever illumination is low. Here they bring two availing qualities: first, their preference for reduced light including tolerance of water turbid with silt, second, their prolonged and ultimately massive growth. The sponges are initially at a disadvantage against faster growing tubeworms, bryozoans and ascidians, but will eventually outstrip them in quiet waters, under stable boulders, or where low-illumined sponge gardens can supersede the coralline algae. Their tolerance of sediment allows sponges to spread where the

227

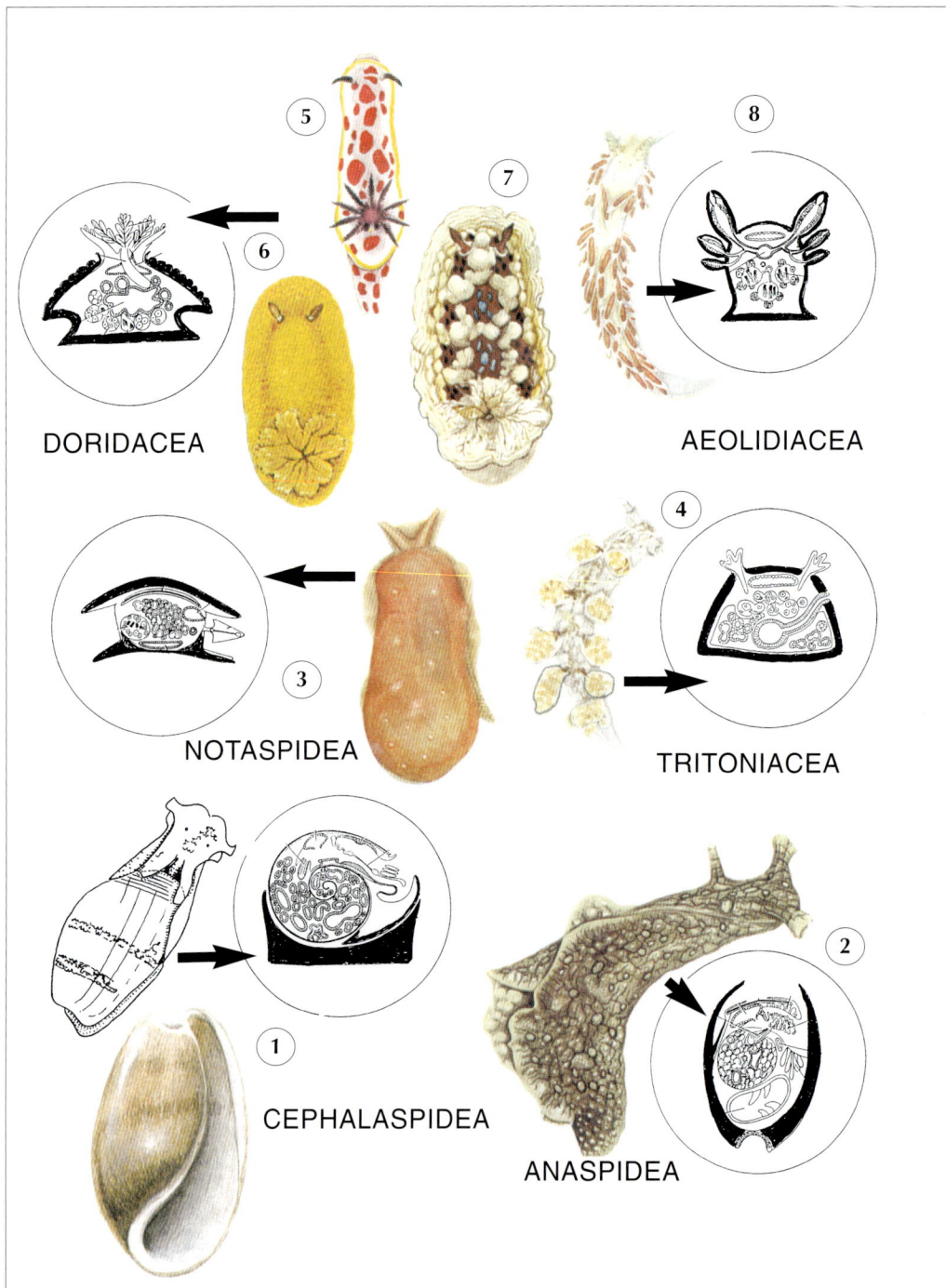

Fig. 12.8 Opisthobranch range and structure
Six orders are represented by the animal in dorsal view, along with cross sections showing from below upwards the progressive reduction and loss of spiral form and pallial cavity.
[TECTIBRANCHIA] **Cephalaspidea** 1 *Bulla quoyii* **Anaspidea** 2 A*plysia dactylomela* **Notaspidea** 3 *Bouvieria aurantiaca*
[NUDIBRANCHIA] **Tritoniacea** 4 *Doto pita* **Doridacea** 5 *Ceratosoma amoena* 6 *Archidoris wellingtonensis* 7 *Dendrodoris gemmacea* **Aeolidiacea** 8 *Coryphellina* sp. (Colour illustrations from Michael Miller.)

edge of the continental shelf is lightly covered with sand. Between tides, sponges can withstand heavy silting on coralline flats, as with *Hymeniacidion perleve*, *Suberites perfectus* and *Aaptos confertus*. On exposed shores bright *Polymastia granulosa* flourishes in clean water around bull-kelp, where the white reflecting foam cuts down the light.

Grazing predators
Opisthobranchs

Of the three gastropod subclasses, the **Opisthobranchia** is the smallest, but its members are by far the most visually attractive. They are also the most adaptively enterprising in their loss of their shell and spiral pattern and the making over of the body to new styles of life. The most primitive opisthobranchs, still with their spiral shells, subsist mainly on plants, grazing on their detritus as we have seen with the Cephalaspidea, or browsing their live tissues, as in the Anaspidea. The Sacoglossa have developed the special habit of sucking the cell fluids of green algae.

The higher opisthobranchs, collectively referred to as nudibranchs, are carnivores with specialist diets. They are cast among the main predators of sessile invertebrates. Their largest grouping, that of the dorids, contains mostly sponge-feeders, but some live upon bryozoans and ascidians. Second, the aeolids and most of the dendronotids rely for their food upon cnidarians. Few indeed of the big sessile invertebrate groups are immune from some form of nudibranch predation.

Pleurobranchs

This first group are not yet quite 'nudibranchs', since they preserve at least a vestige of the shell. They have been aptly called the 'side-gilled' slugs, all retaining a plume-like gill under the mantle or *notum*, on the right side. The nearest to the ancestral form is the chinaman's hat slug, *Umbraculum umbraculum*, that has kept its large external shell covering the whole dorsal surface. In nearly all the rest the shell is reduced to a thin internal plate. Above the mouth the pleurobranchs carry a broad oral veil. Beneath this diverge paired tentacles, rolled into incomplete tubes, which are the chemo-sensors or *rhinophores*.

In the New Zealand north, four pleurobranchs turn up, somewhat unpredictably, under low tidal boulders, with good water movement. The lemon coloured *Berthellina citrina* has been found to graze on the sponges *Demira*, *Halichondria*, *Iophon* and *Tetrapocillon*, in that order of preference. *Berthella ornata* (milk-white and blotched with wine-red) and *B. medietas* (pock-marked and dull orange brown) both take bryozoans and ascidians as well as sponges. The brown speckled *Pleurobranchaea maculata* is more of an opportunist, eating anemones, mobile annelids and molluscs. All the pleurobranchs have a wide, rather unspecialised radula with numerous scalpel-like teeth, and chitinous jaws.

Nudibranchs

The dorids are the most numerous and familiar of the nudibranchs. They are recognised by the shift of the anus to a posterior-dorsal site, encircled by the gill plumes, while the solid paired rhinophores spring up through the notum in front. Bilateral symmetry is now externally complete, except for the genital aperture still on the right side. With a good number of exceptions the main dorid diet is of sponges.

The daringly bright colours of the dorids are generally held to be *aposematic*, giving warning of noxious or distasteful properties. Their veracity may be cautiously tested by touching a dorid with the tongue. Certain colours may, however, serve for camouflage, as with the dull, yellow-ochre *Archidoris wellingtonensis*, feeding on the sponges *Halichondria* and *Hymeniacidon*. Of a clearer yellow is the smaller sponge-predator *Doriopsis flabellifera*, with a fan of gill-pinnules retractible under a skin-flap. This slug can flatten itself to the surface so as often to pass unobserved. Also coloured like its prey is the small scarlet *Rostanga muscula* (up to 20 mm) found to feed on red sponges, in the order of preference *Holoplocamia novizelanicum*, *Ophlitaspongia seriata*, *Microciona coccinea*.

The larger dorid *Alloiodoris lanuginata* (confined like *Rostanga* to the North Island), is reddish brown to ashen grey, white speckled, and lightly spiculose, giving off a faint aromatic odour. The rhinophores are yellow and the gills greenish. Another sponge-feeder, *Aphelodoris luctuosa*, is common between tides in the north, being smooth and slippery, and cream-coloured with rather distinct chocolate bands.

Some of the most graceful nudibranchs belong to the **Chromodorididae**, famous — as the name suggests — for their bright colour patterns. Their slender tongue shape was denoted by their former name 'glossodorids'. Much more abundant in the warmer Pacific, these have been compared to other dorids as butterflies to moths. Their rhinophores and gills are often startlingly coloured, and predators can be put on notice by twitching the gills or flapping the mantle edge. New Zealand's two chromodorid species form a contrasting pair. *Ceratosoma amoena* is vividly flamboyant with its patches of vermilion on yellow and the gills and rhinophores picked out in magenta. *Chromodoris aureomarginata* is modest white, adorned only with a thin gold border.

One of the smallest dorids is the 4 mm-long *Okadaia cinnabarea*, to be recognised as a scarlet streak under clean boulders with spirorbids (Fig. 13.16). The radula bores a neat hole in the posterior part of the worm's tube; the proboscis is then inserted to engulf the prey, which is secured by the elongate radular teeth.

The Family **Dendrodorididae** includes suctorial dorids that have lost both their radula and jaws. The chrome-yellow *Dendrodoris citrina* is our commonest intertidal nudibranch (Fig. 12.4) reaching far into sheltered harbours. It feeds on the ascidian *Microcosmus squamiger*, opening a hole in the test with a salivary secretion, then inserting the narrow proboscis. Of the habits of the two related species we still know all too little. *Dendrodoris nigra* is jet-black with vermilion margin, while *D. denisoni* is elegant without rival among our nudibranchs, having the notum thrown into grey-brown folds, carrying big lozenges of peacock blue.

Cladobranchs

A second great cluster of the nudibranchs, distinct from the dorids, is grouped as the Cladobranchia from the development of dorsal outgrowths. It consists of the dendronotids and aeolids, both specialist predators on various kinds of cnidarians. The evolution of the **Aeolidoidea** has been intimately involved with that of their prey. These are the most attenuate and agile of nudibranchs, typically found twining among anemones and hydroids, normally in fair current flow. Most are small, and often well camouflaged against their polyps.

Through their whole saga, the aeolids have stayed faithful to Cnidaria. The *Glaucus* species are pelagic, having followed their host cnidarians up to the ocean surface. Here they share the vivid blue of the by-the-wind sailor *Velella* or its relative *Porpita*. They rely on these cnidarians for food and transport, rasping away the host tissues until only the chitinous float remains. In strong contrast to this pelagic venture, the interstitial fauna of coarse sand beaches includes a miniaturised aeolid *Pseudovermis*, attacking tiny hydroid polyps (in the northern hemisphere *Psammohydra* and *Halammohydra*) attached to single sand grains.

Aeolids, however, belong primarily to hard shores between and beyond the tides. Their club-like outgrowths (*cerata*) are entered by branches of the digestive gland, which store the still unexploded nematocysts derived from the prey. These stinging bodies find their way into the epithelium of the cnidosacs that open at the tips of the cerata. They are held available as the slug's prime defence against attackers.

The largest New Zealand aeolid, *Jason mirabilis*, reaches 60 mm long and lives subtidally on the metre-high trees of the hydroid *Solanderia ericopsis* (Fig. 16.12). On ships, wharves and the ropes of buoys, we may find *Flabellina albomarginata* (25 mm) feeding on *Tubularia* and *Sarsia eximia*, and yellow and green *Cuthona scintillans* attacking *Tubularia* polyps. The orange and red *Phidiana milleri* lives on hydroid colonies in tidal pools.

The *Aeolidiella* species, with the cerata not clumped but in oblique rows, specialise on anemones rather than hydroids. The small nurse anemone, *Cricophorus nutrix* (Fig. 3.7), living in the axils of the brown algae, *Cystophora* and *Carpophyllum*, has as its predator an exquisite aeoliid, *Berghia australis*, kelp-brown and white-speckled, with the blunt cerata light blue, tipped with crimson and white.

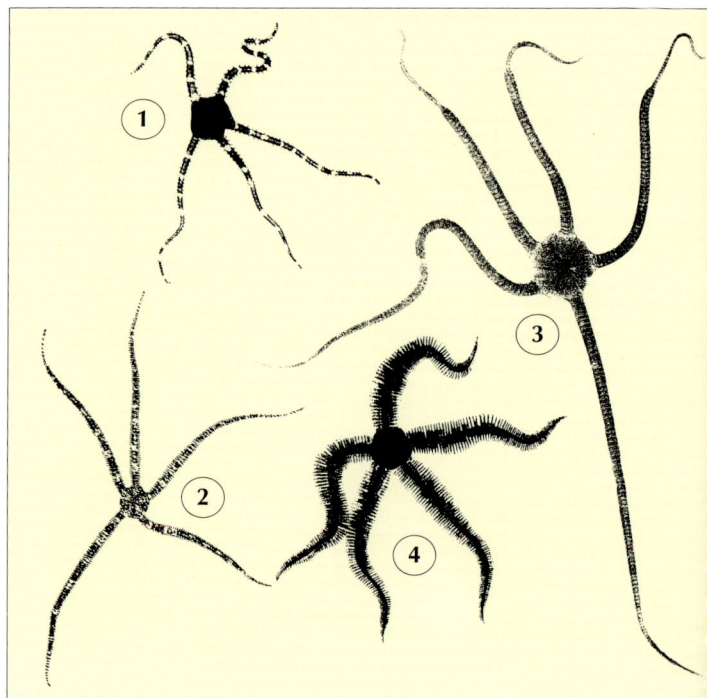

Fig. 12.9 Brittle stars (Ophiuroidea)
1 *Ophiomyxa brevirima* (soft-skinned sand-star)
2 *Ophionereis fasciata* (mottled sand-star)
3 *Ophiopsammus maculata* (snake tail) 4 *Ophiopteris antipodum* (oar sand-star). (From Michael Miller.)

The cladobranchs include also the **Dendronotoidea**, with short gill tufts along the dorsal margin. They have neither the dorid gill circlet nor the aeolid cerata with nematocysts. The rhinophores have a basal sleeve, and — unlike the dorids — the anus remains lateral. The large apricot to vermilion *Tritonia incerta* is — like its genus worldwide — a predator of alcyoniids. Tissue chunks of *Alcyonium aurantiacum* are rasped away with the large, strongly jawed buccal bulb.

The smallest dendronotaceans are the pale yellow *Doto* species, with blunt appendages. The delicate *Doto pita* lives in tide pools, feeding on the thecate hydroids *Obelia*, *Orthopyxis*, *Clytia* and *Sertularella*.

Carnivores and deposit feeders
Echinodermata

The five living classes of the echinoderms have an undoubted unity — in their five-radiate organisation, tube-feet, water vascular system and dermal spines. But, like the asteroids and echinoids already described, each of the classes has its own highly characteristic design.

Asteroidea
The small cushion star, *Patiriella regularis*, and the large *Stichaster australis* (Fig. 5.6) have been introduced as

asteroids well at home on the open surface. Other members of this class come fully into their own under boulders and beyond the tides. As well as *Patiriella*, there is a second asterinid living under boulders in the handsome web-star, *Stegnaster inflatus* (Fig. 3.6).

The Family **Asteriidae** comprises prickly skinned stars with more than five arms, flexible and much longer than the disc. *Astrostole scabra*, the largest New Zealand star, reaches 50 cm across, with (usually) seven steel-blue or grey-brown arms. The tube-feet are bright orange, flanked with long ambulacral spines. Much commoner around low water is *Coscinasterias muricata*, some 12 cm across and with up to 11 arms, varying with arm loss and regeneration. The aboral surface is rust-red to grey, the tube-feet cream. Both upper surface and ambulacral margins are heavily spined. *Coscinasterias* clings tightly under stones and preys on gastropods and bivalves, especially oysters. Commoner in the South Island is the five-armed *Sclerasterias mollis*, bright orange with five longitudinal bands of yellow on each arm, corresponding to the rows of long spines set in the skin like inverted tacks.

The incubating star *Anasterias suteri* (up to 15 cm across) is brownish grey to green, being found at kelp level from Banks Peninsula south. Noted for the brooding of the young, the female places her body over the eggs and carries the small stars underneath for several months.

As well as the large mauve or ochre *Stichaster australis*, the **Stichasteriidae** includes the small 'dividing stars' *Allostichaster*. These have the arms in two sets, resulting from reproduction by fission, with each resultant half growing a new set of smaller daughter arms. *A. polyplax*, the four-and-four star, bluish to buff, is common throughout New Zealand, with *A. insignis*, the three-and-three star, orange or purple, confined further south.

The biscuit stars belong to the **Goniasteridae** — a family far more diverse in the warmer Pacific than in New Zealand. Our main low tidal species, found among stones from Napier south, is the attractive jewel star, *Pentagonaster pulchellus*, red, orange or purple-grey. The rigid disc has short, bluntly rounded arms, with both the oral and aboral margins equipped with wide plates. On the aboral surface there is a mosaic of close-fitting plates.

Ophiuroidea

In all their essentials the brittle stars are highly uniform. The small rounded disc carries five flexibly jointed arms, with no open ambulacral grooves as in the starfish; nor are the pointed tube-feet employed for locomotion. Instead, the brittle stars move by sinuous arm-flexions in the horizontal plane. The digestive cavity is a simple sac with no anus. The madreporite and the five genital pores open upon the oral surface.

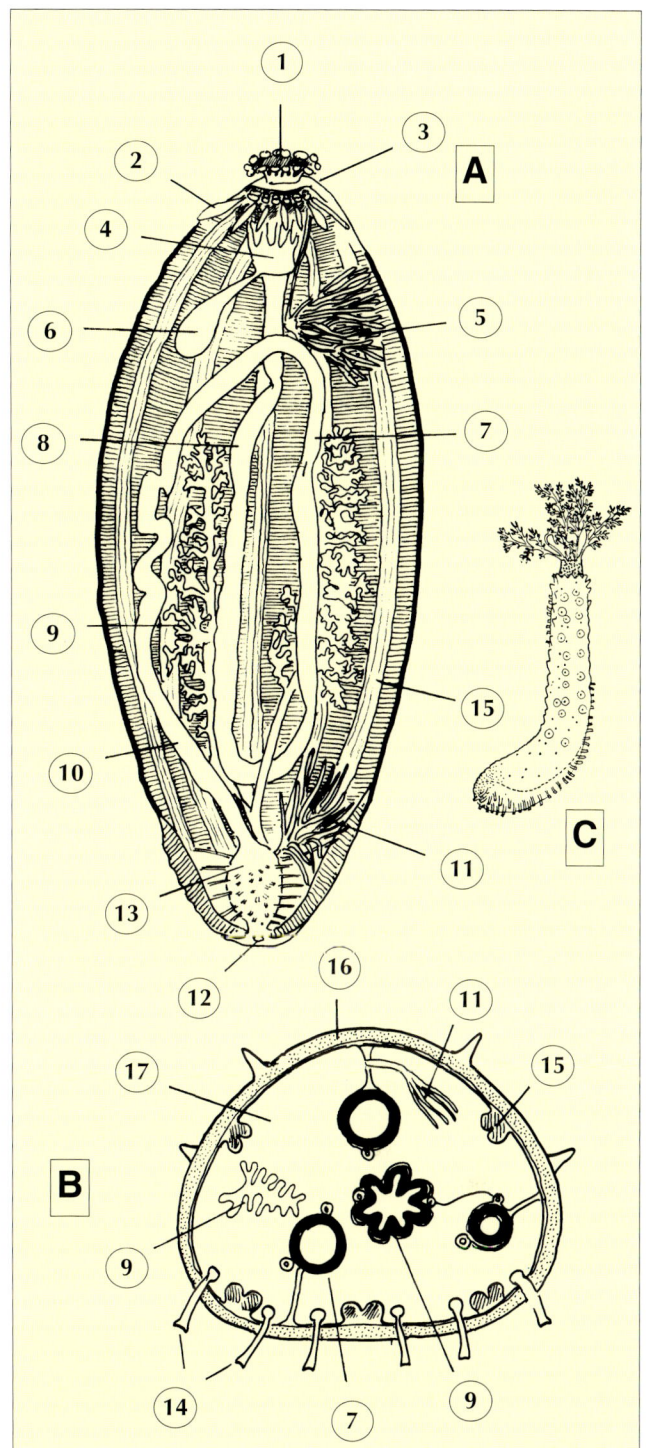

Fig. 12.10 Structure of Holothurioidea

A. *Stichopus* internal structure
B. The same in cross-section.
C. *Australocnus calcareus*, with branched oral podia.
1 oral tube-feet 2 buccal ampullae 3 genital opening 4 water ring canal 5 gonad 6 polian vesicle 7 digestive tract 8 respiratory tube 9 longitudinal muscle 10 rectum 11 cuvierian organ 12 cloacal opening 13 cloaca 14 tube-feet 15 longitudinal muscle 16 circular muscle 17 coelome.

Despite their basic similarity the ophiuroids show much versatility in feeding. The long-armed snake-star, *Ophionereis fasciata*, found on sand under clean boulders, captures suspended food in a screen between the arm-spines, made of mucus secreted from glands inside the spines. The mucus with its collected particles is wiped off by the tube-feet that pass it towards the mouth. The smaller *Ophiactis resiliens*, ensconced in crevices and holdfasts, filters in the same way, with the arms oriented to the water currents.

As well as mucus-netting, long-armed brittle stars have other feeding techniques such as sweeping the surface film with the arm-tips, arm-looping, and deposit-browsing.

Under the same boulders as *Ophionereis* is often found the jet-black oar-star, *Ophiopteris antipodum*. The arms are fringed by long, flat spines, adhering to the substrate by mucus secreted from the knobbed tube-feet. Pressing the oral surface against the rock, *Ophiopteris* scrapes off encrusting bryozoans with the jaws. In addition it can filter-feed by mucus-screening.

The *Amphiura* brittle stars, that burrow in soft flats, have converted filter feeding to deposit grazing. From a small disc, long sinuous arms reach up individually or in pairs through the sand. Their waving tips catch a harvest of particles from the surface, to be carried down by tube-feet to the mouth. One or more arms may be retained below the sand and looped for 'elbow feeding' on larger particles.

Our smallest brittle star, *Amphipholis squamata* (Fig. 8.5), could lie on a five-cent piece, and lives in silty sand at the base of *Corallina*. With its short arms it is able to select individual diatoms.

Finally there are carnivorous snake-stars, including our largest species, *Ophiopsammus maculata*. The strong arms are rounded in section, up to 15 cm long, with reduced arm-spines close-pressed like scales. *Ophiopsammus* takes purchase with the arms as it glides between stones and algae at low water. Items of animal food are looped with a tapered arm-tip and handed direct to the jaws.

Holothurioidea

Where the brittle stars have perfected a flat design with multi-radia movement, the holothurians or sea cucumbers have become cylindrical by stretching out from mouth to anus. They have thus evolved towards a bilateral, 'dorsi-ventral' symmetry, and a flat undersurface used for progression is developed from three of the five inter-ambulacra.

The last traces of a skeleton are retained in the microscopic plates and spicules embedded in the skin. The old five-radiate symmetry is preserved in the five ambulacral lines running along the body. The mouth is encircled with special oral podia which take up fine deposits from the substrate. One by one these feet are inserted into the mouth, then pulled back like sucking jam from the fingers. The anus, at the posterior (original aboral) pole, takes in water into a branched diverticulum of the intestine, known as the *respiratory tree*. Gaseous exchange is thus possible with the coelomic fluid.

New Zealand's most familiar intertidal holothurian, *Stichopus mollis*, grows to 20 cm long. Commonest onshore in the south, it is found under low tidal stones and in sandy pools, often detectable by its cylindrical castings of fine sediment. The body is mottled grey or brown, studded above with pointed tubercles. The lighter undersurface has tube-feet not in ambulacral rows but diffuse. Like all the Order **Aspidochirota**, *Stichopus* has 20 podia in the oral circlet, each with its terminal adhesive disc. Handled alive, it will defensively auto-eviscerate, casting out through the anus the whole digestive tract. These lost parts will rapidly regenerate over a few weeks.

The Order **Dendrochirota** is typified by the small cucumbers of the **Cucumariidae**, having 10 branched oral podia. *Australocnus calcareus* (Fig. 12.10) is a little over 1 cm long, stiff and chalky white from its microscopic 'buttons' and scattered with minute scarlet tubercles. Both this species and the slightly larger *Squamocnus brevidentis* occur throughout New Zealand in concavities and holdfasts. In rock crevices in the South Island can be found a small, plump cucumber, *Psolidiella nigra*, up to 2 cm long, taking permanent grip by a flattened 'sole'.

Finally, in the Order **Apoda**, there are holothurians entirely worm-like, with the ambulacra devoid of tube-feet. The reddish speckled *Taenigyrus dunedinensis* (up to 5 cm long), burrows in sand pockets on rocky shores, while the larger *T. dendyi* is common in protected low tidal flats. The integument is studded with microscopic hooks and wheels. *Kolostoneura novaezelandiae* (8 cm), found on rocky shores throughout New Zealand, is pale pink, with its granulated skin devoid of wheels or hooks.

CHAPTER THIRTEEN

Dark Communities

Though we have left behind us the long banding of the visible shore, we will find on the undersurfaces and sides of boulders community patterns that are still clearly zoned.

On the most irregular boulders, like pitted scoria, it would be pointless to look for such order. But a regular boulder, flat-based and stable, will have its concentric bands we may call *zonules*. These will run from the lower pole of the boulder to its well-illumined upper pole. They will, moreover, have a clear correspondence with the broad zones of the surrounding shore. A boulder 25 cm high may present a microcosm of the zonation of a whole shore.

The surface extent of sessile or encrusting species may be plotted against the height of the boulder on the shore (Fig. 13.4). Position on the boulder is shown by a contrived 'latitude' from 0 at the top to 180 on the undersurface. From their regular slope it is apparent that organisms affected by desiccation and illumination can come progressively higher on boulders lower down the shore. This effect is most marked for barnacles, oysters and encrusting red algae. For bryozoans, most susceptible to light, the trend is nearer to horizontal.

At the lowest reaches, there is more to the story. The under-boulder pattern diversifies and enriches, with a majority of species never found in the light. Temperature is here kept low and stable and evaporation mitigated, though still operative at the edge. We begin to glimpse under boulders an assemblage unlike any in the visible shore. This is the same sciaphilic or dark-loving community that we shall find (in Chapter 16) most extensive on the broad faces of subtidal slopes. A unity of composition links up all these dark communities, including those of the tropics, in the inclusive term *general sublittoral hypobion*.

Secluded sites under boulders are privileged enclaves for a community confined in its full richness to the subtidal. Such upward salients we have compared to English woodland flowers reaching under shaded hedgerows across meadows. Or they are like small stands of tropical rainforest translocated to moist spots in drier savanna. In both tropical forests and coral reefs, with all their shaded space, community complexity and biological diversity can reach a zenith. Lighted communities, of grassland or in the visible zones between tides, would then be seen as lower diversity extrusions from this primal wealth, adapted to survive in a harsher, more demanding climate.

The sciaphilic community, as the shore's rich 'underface', will be described in detailed description in this chapter. With the greater number of species able to settle, in the mitigating of harsh climate, the boulder community grows up with a degree of opportunism, with a 'lottery effect' more like that of the subtidal. At the extreme we shall expect no regular species patterns, only a diverse mosaic.

But these dark or 'sciaphilic' communities share some strong similarities. They are most often dominated in the temperate by the **bryozoans**, **ascidians** and **sponges**, where a boulder habitat runs through an 'ecological succession' these three sessile groups appear and partly supplant each other in that order in time. As well as their high diversity, a feature never so far fully accounted for is their bright colour. It is as if colour were a wide property of life, at once beautiful and gratuitous: in a dark world evolution will not act to suppress it, since it could be near invisible and its effect selectively neutral.

The boulder microclimate

Surface temperature, illumination and evaporation were measured on spherical boulders half a metre across. John Walsby made concrete boulders marked with a 'latitude' from 0 at the top to 180 below. To determine water loss, glass capillary tubes were put in holes at different latitudes, and the contraction of their water column recorded as surface evaporation took place from a 1 cm circle of moist filter-paper. Evaporation varied markedly with the weather. Two curves are shown in Fig. 13.2, one for a sunny day without wind, and one for a night with light breeze. The night figures show evaporation to be highest in the narrow corridor between the boulder edge and the ground. With perceptible windflow this space presents climatic hazards.

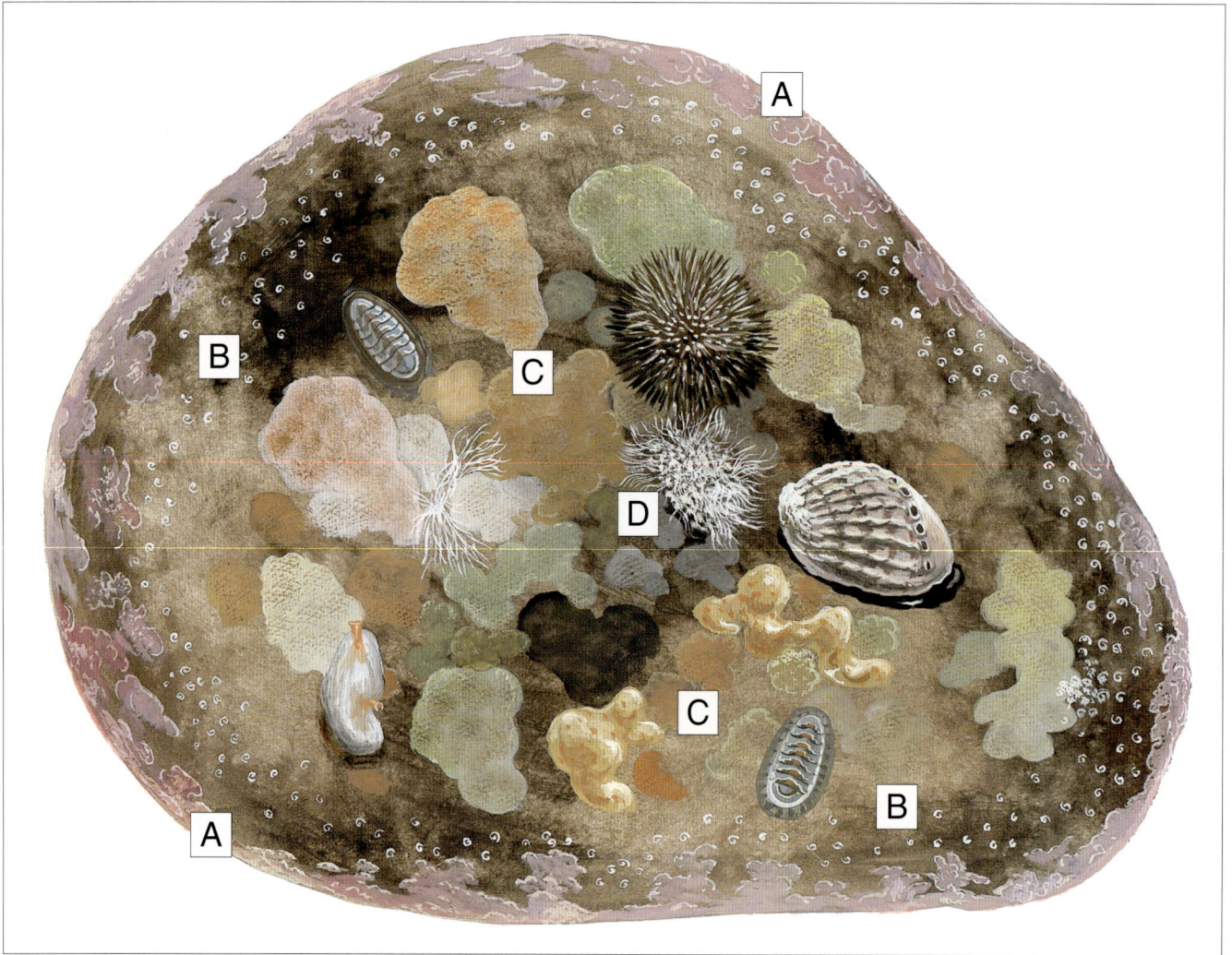

Fig. 13.1 Undersurface of a bryozoan and tubeworm dominated boulder in clean water outside Whangarei Heads.
The succesive zonules are (A) shaded sub-marginal coralline crust (B) next inward zonule of cheilostome bryozoans with
(C) simple and compound ascidians (D) central zonal of filogranid tubeworms. Associated mobile animals are
Sypharochiton, Haliotis and *Evechinus*.
(Painting by Ron Cometti.)

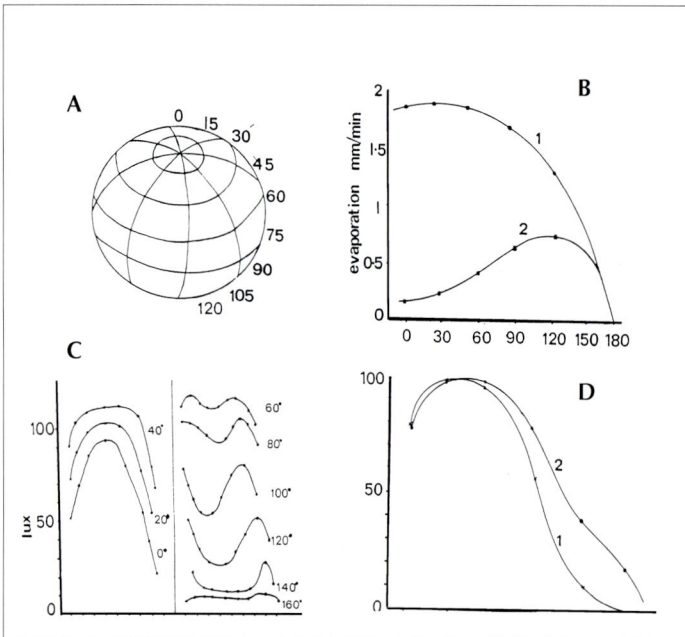

Fig. 13.2 The microclimate of boulder surfaces
A. Artificial boulder cast in concrete, showing portion
of inscribed 'latitude' (0 to 120).
B. Evaporation in relation to latitude (1) on a sunny day
with no breeze, and (2) at night with a light breeze,
showing the 'venturi' effect in the shaded overhang.
C. Variation of illumination with latitude, during hours
of daylight.
D. Illumination of boulder (1) resting on a low-reflecting
rock surface and (2) on a high-reflecting surface of
white sand.
(Courtesy of John Walsby)

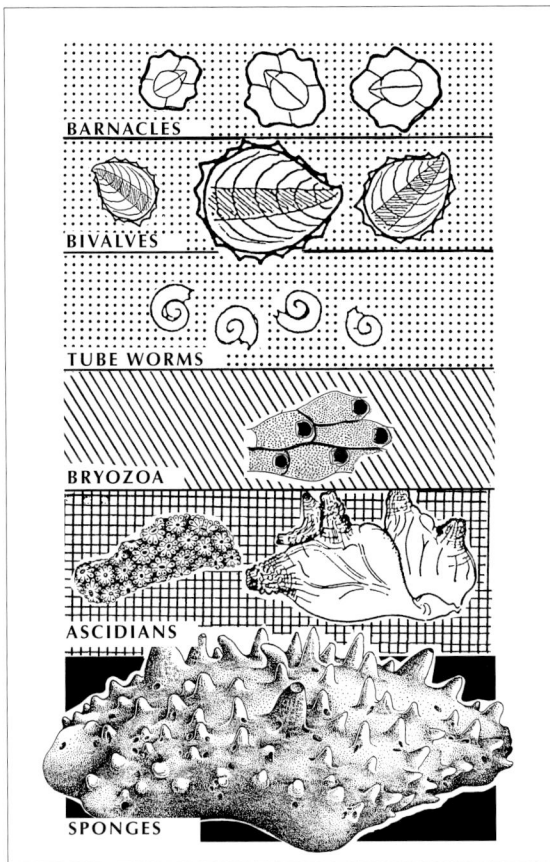

Fig. 13.3 The life-forms of filter feeders down the shore.

Variation of surface illumination with the sun's position through the hours of daylight is also shown. For the lower reaches of the boulder, illumination is highest in the early morning and late afternoon, with the sun low in the sky. Different illumination is received by boulders resting on high-reflecting sand and low-reflecting rock surfaces.

With species in the sublittoral fringe sensitive to desiccation, survival could be critically affected by the time of day of the lowest spring tides. With light-shunning species under boulders, hazards from illumination would be greater where low spring tides fall early or late in the day rather than when the sun is high in the sky. On a normal shore, where boulders are not isolated and light-exposed but sheltered by adjacent rocks or under algae, such potential hazards will seldom be realised.

The detailed positioning of species in boulder zonules is clearly **adaptive**. Thus, just inside the margin, spirorbids, securely protected by their shells against desiccation during low tides, can take feeding advantage of the fast current through the narrow corridor at the edge. The 'worm-shell' *Stephopoma* is their molluscan equivalent.

Bryozoan colonies, next inward, are better secluded from light or evaporation than tubeworms; and their level expanse of zooids takes maximum feeding advantage of the laminar current flow. Black-pigmented

Watersipora has a greater tolerance of reflected light than other Bryozoa. In winter it can spread out widely around the boulder sides, or even over the top.

Only the largest boulders are found to carry a predominant sponge load. Latest to appear, sponges are generally located towards the centre beneath boulders. They reach the edge, or dominate the whole surface, only where currents are slight, the water turbid, or the boulder is shaded by tresses of brown algae on the top and sides. At their full growth, sponges are of massive bulk and able to spread and fuse to neighbouring boulders.

Ordering on the shore

The filter feeders living under boulders as well as on open surfaces could be arranged in several meaningful sorts of classification. At the outset, they might be *structurally* ranked according to their degree of protection by a shell or hard integument, and the extent to which their units have become aggregated into colonies. Thus considered, they would fall into the following order.

Structure

I   *with a hard shelly covering; sometimes aggregated but never colonial, i.e., each individual remains distinct*
(a) Cirripedia: barnacles; (b) Polychaeta, Sedentaria: chiefly serpulid tubeworms; (c) Mollusca: Bivalvia, and a few Gastropoda

II   *with thin, part-calcified boxes, aggregated into colonies. Individuals remain distinct, with little tissue continuity.*
Bryozoa: Cheilostomata and Cyclostomata

III   (a) *with no hard cover, but a protective test of tunicin; individuals distinct*
Simple Ascidiacea: sea-squirts
(b) *colonial and with much tissue continuity, though individuals still more or less distinct*
Compound Ascidiacea: compound ascidians

IV   *with no tunic, but there may be an outer spiculose cortex; individuals never distinct; colony a tissue continuum*
Porifera sponges.

Shore occurrence

It is obvious that these life-forms, each adapted for a particular strategy, are far from fortuitously arranged across the shore. In their biological characteristics, they can be placed in a ranking according in some ways with their morphological character.
1   First, the groups may be ranked by their niche-width or **ubiquity**, as distinct from narrow localisation.

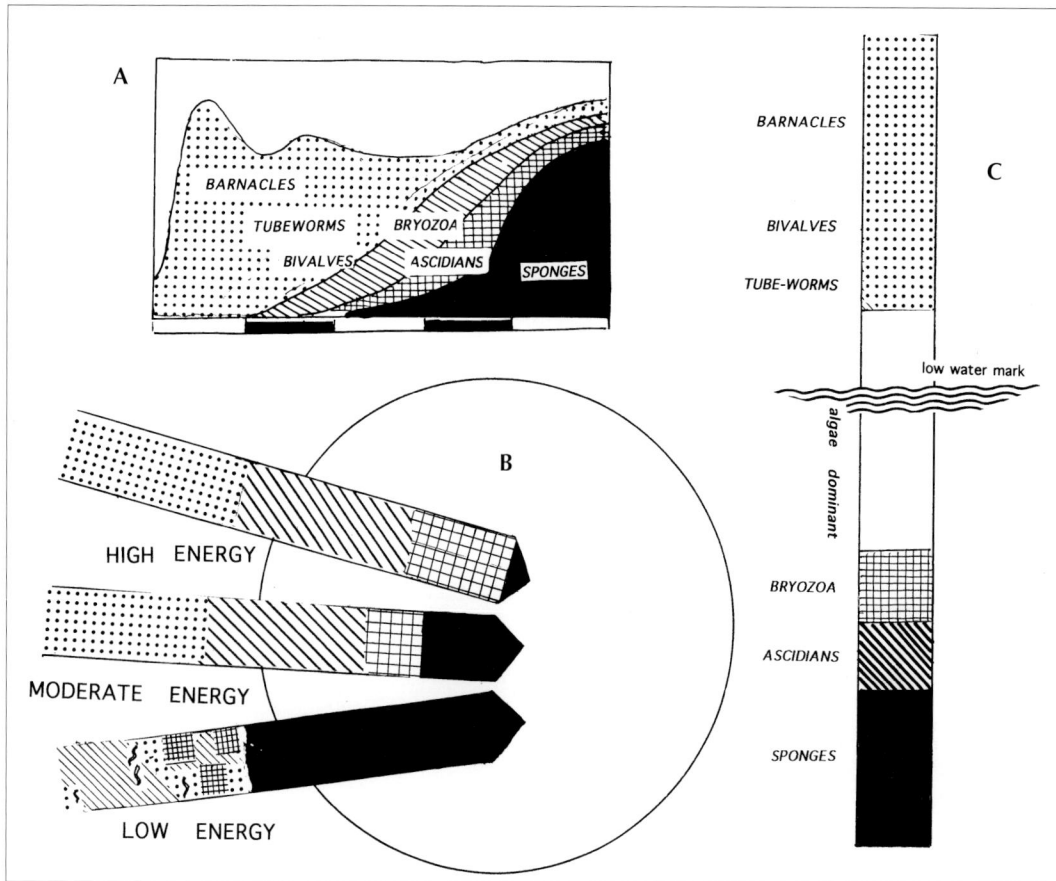

Fig. 13.4 The succession of life-forms under boulders

A. Temporal order of settlement and establishment over four years.

B. Arrangement under boulder, from periphery towards centre, on low, moderate and high energy shores.

C. Succession on the same life-forms on the lighted (upper) surface, from the intertidal, into the sublittoral zone.

(For shading conventions, see Fig. 13.3.)

Barnacles are clearly the most widely competent, able to live almost everywhere on a shore. But they no doubt first evolved for high-level open surfaces, and though regularly appearing under boulders, they are seldom found dominant there. A somewhat narrower range — but still extensive in the light or dark from the middle eulittoral down — is enjoyed by *bivalves* and *tubeworms*.

Bryozoans, *ascidians* and *sponges*, by contrast, are scarcely found anywhere else but in the dark community under boulders or in the subtidal. Rather exceptional is the success of a few zoning ascidians on open rock.

2 Second, the groups can be ranked in the **vertical order** in which they each appear on open surfaces. Through the upper two-thirds of the eulittoral there are *successively barnacles, bivalves* and *tubeworms*. The sequence is then broken where the lowest part of the intertidal is yielded to the algae (Fig. 13.4); but when animal dominance is resumed subtidally, it is again in the order *bryozoans, ascidians,* and finally *sponges*.

Strategies

The same groups show a comparable ranking in their **biological strategies,** under which term we may include rate of growth, reproduction and pace of life. Thus most of the *barnacles* are small in biomass and have extended breeding seasons. Fast-growing and short-lived, they are opportunists quick to colonise empty space, being thus — in the ecologist's term — examples of '*r*' *strategists*. The same can be said of the *tubeworms, Pomatoceros* and spirorbids, settling early on bare surfaces, as also of bivalve spat, notably of *Xenostrobus pulex* seizing surfaces recently denuded by sand burial (Fig. 7.17). At the opposite extreme are the *sponges*, which are typically '*k*' *strategists*: being the last to settle, slow in reproduction and growth and evidently immensely long-lived. They ultimately achieve the highest biomass, with their systems imperfectly individualised. Rather than being colonies with reduced individuality, as with the compound ascidians, the sponges are in effect 'pre-individual'. Like the barnacles at the high extreme of the shore, the sponges — in their low level seclusion — hold by default a habitat where few competitors can successfully challenge them.

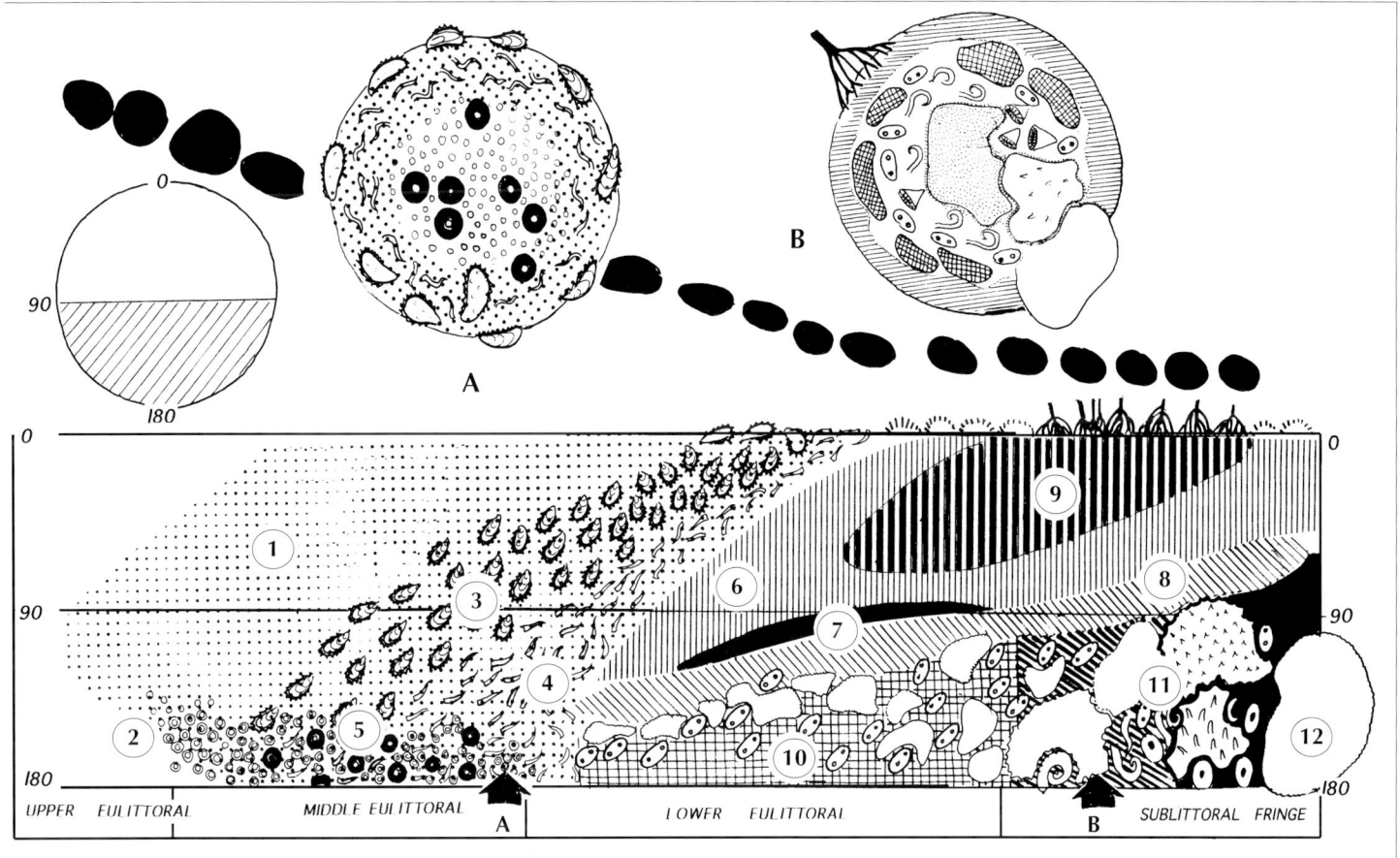

Fig. 13.5 Boulder communities down the shore: Northland and Hauraki Gulf
Community composition is shown in relation to height on the shore (= duration of emersion) and 'latitude' on the boulder.
A and B **(top)** show the undersurface of boulders at the levels marked.
1 *Chamaesipho columna* 2 *Tetraclitella depressa* 3 *Crassostrea glomerata* 4 *Pomatoceros caeruleus* 5 *Actinia tenebrosa*
6 coralline pink veneer and **(top)** turf 7 *Watersipora* 8 other cheilostome bryozoans 9 *Carpophyllum* and *Ecklonia*
10 simple and compound ascidians 11 crusting sponges 12 fusing sponges.

Settlement

We can finally discover among the groups a similar ranking order when we turn to the spatial and temporal aspects of their **settling behaviour** under boulders.

- Where a mature boulder community is space-partitioned the ranking I to IV (above) corresponds rather well to their spatial order of settlement from the part-lit boulder rim to the dark and secluded central sites.
- If we look to the groups that dominate the undersurfaces of boulders at successive levels down the shore, we find a sequence that recalls the vertical order in the zonation column of intertidal and subtidal surfaces.
- The ranking I to IV also corresponds to the temporal order in which the larvae of the groups are found to settle, whether on bare boulders or experimental plates.

Communities under boulders

For boulder habitats we shall use four of the shore types already described for visible zoning in Chapter 4. The account will begin with Class 3 **Mainly Sheltered**, with mobile boulders, as well as large immobile blocks and

sand scour common, but silt only occasional. Shores of Class 4 **Semi-Exposed** have movable boulders in varying degree, some sand scour but no silt. Finally, **Sheltered** (Class 2) and **Highly Sheltered** (Class 1) shores, can have extended boulder fields of any size and with movable sand, silt and finally mud, as they are found increasingly enclosed by land.

Mainly sheltered
Upper eulittoral zone (mobile fauna)

Highest up in the intertidal, boulders are chiefly refuges for mobile animals. Only with the middle eulittoral level do we begin to find sessile communities established, and becoming richer and more complex towards low water. Clean and smooth boulders within the upper eulittoral are generally bare of any sessile community underneath. They are notably a habitat for herbivorous gastropods, especially trochids and limpets, taking shelter during emersion, and to varying degrees coming forth in shade, at night or when immersed.

The commonest of the trochids, *Melagraphia aethiops*, sits visibly on the boulder tops, and with its movements haphazard, evincing no homing behaviour. The other topshells appear to return after feeding to an under-boulder site, though never so exactly as the homing limpets. In the north, the commonest topshell under stones is *Diloma zelandica*, joined increasingly by *D. arida* towards the south. The beaded, sharp-keeled *D. bicanaliculata* is New Zealand-wide.

Two small acmaeid limpets live under clean boulders. Both are fragile and difficult to dislodge without damaging their shells. *Atalacmea fragilis* (15 mm long) is especially thin and almost flat, marked with concentric green lines and emerald-centred interior. *Notoacmea elongata* is smaller (5 mm), with fine reticulate marking.

The behaviour of these species is shown in Fig. 13.6, as they move out to graze the film of algae on the boulder sides. When the tide is in, *Diloma zelandica* and *Atalacmea fragilis* can make feeding migrations both by day and night. *D. zelandica* will also emerge by day in overcast weather where the tide is out but the rock surface wet. *D. bicanaliculata* and *Notoacmea elongata* take sorties only at night, but the northern *Nerita atramentosa* is more tolerant than the trochids or limpets to heat and evaporation, and grazes actively on wet rock at night and by day, only ceasing when the surface is entirely dry.

Where a boulder retains its water film underneath, the minute rissoids *Pisinna impressa* and *P. zosterophila* and the near-rissoid *Eatoniella olivacea* can all live, moving about actively to pick up diatoms. Held to smooth rock by the surface tension of the water film is sometimes found the tusk shell *Caecum digitulum*.

Two common chitons complete the tally of grazers: *Sypharochiton pelliserpentis*, in its uneroded *sinclairi* form, and our most familiar under-stones chiton, *Chiton glaucus*. This species is clean and smooth-shelled, varying from dark bottle green or bluish-green to olive, rust-brown or straw-coloured, often with zigzag streaks.

The typical crabs under clean eulittoral boulders are the dark-mottled *Hemigrapsus edwardsi*, the aggressive xanthid *Ozius truncatus* and the far more quiescent, pebble-like *Heterozius rotundifrons*. The 'half-crab' *Petrolisthes elongatus* is usually commoner than all the rest together, clinging upside down to the boulder, and scuttling away when disturbed.

Under flat slabs resting on sand in the upper eulittoral huge numbers of amphipods can be found. The commonest are generally *Hyale* species, jack-knifing on their sides in the water film or taking off to swim freely. Almost as versatile are the squadrons of small isopods, *Isocladus armatus* and *I. spiculatus*, variously camouflaged against shelly sand and with the males dorsally spined. Low-built and broad-based, they walk or swim through water or fluid sand, with their abdominal pleopods driving a current backwards.

Smooth boulders are also the home of polyclad flatworms, adhering by their mucus-coated ventral surface. The leaf-like *Leptoplana* is greyish brown, up to 2 cm long, and crawls by a rippling wave of the whole body, assisted by cilia working in the secreted mucus film. A larger *Stylochus* species is thick and dorsally opaque, irregular when at rest, but in motion broadly oval. Beige-speckled above, it is translucent underneath, revealing the complex branching of the gut and reproductive system. Marginal eyes extend all round the edge, and there are clusters of antero-dorsal, central and lateral eyes.

The olive rockfish, *Acanthoclinus fuscus*, belongs to a genus peculiar to New Zealand and east Australia, and well adapted for life under mid-tidal boulders. Greyish brown and slippery, this fish is like a short eel, growing to 25 cm offshore, but seldom half that length in the intertidal. The head is camouflaged with a white badger stripe, best seen in the young. The long dorsal and anal fins have spines with soft sensory tips, used during progression under stones. The rockfish is virtually amphibious, with the gill chamber protected by a membranous fold to cut down evaporation during low tides. It feeds at night on small crabs, worms and other fishes, being at times cannibalistic. The male stays on guard during breeding, irrigating a gelatinous ball of eggs by tail-fanning.

The small sucker fishes of the Family **Gobiesocidae** are also adapted for life among boulders. Smooth, broad-based and scaleless, with the dorsal and anal fins reduced, they swim spasmodically with flicks of the tail. As they come to rest, or when a wave breaks over a pool, they attach to stones by an adhesive cup, formed from the ventral fins and the lower rays of the pectorals. Our common species under eulittoral boulders is the lumpfish, *Trachelochismus pinnulatus*, with a blunt snout, and the upper surface marbled with broken stripes on a pale ground.

Middle eulittoral (sessile barnacles, tubeworms and anemones)

It is at this level of the shore that boulders first develop a recognisable sessile community underneath. In the middle eulittoral this is a rather simple assemblage, typically composed of white-shelled barnacles, tubeworms and small oysters, with bright red anemones as high spots. The most individually numerous barnacle is generally the fast-growing opportunist *Austrominius modestus*. But the barnacle most characteristic under middle

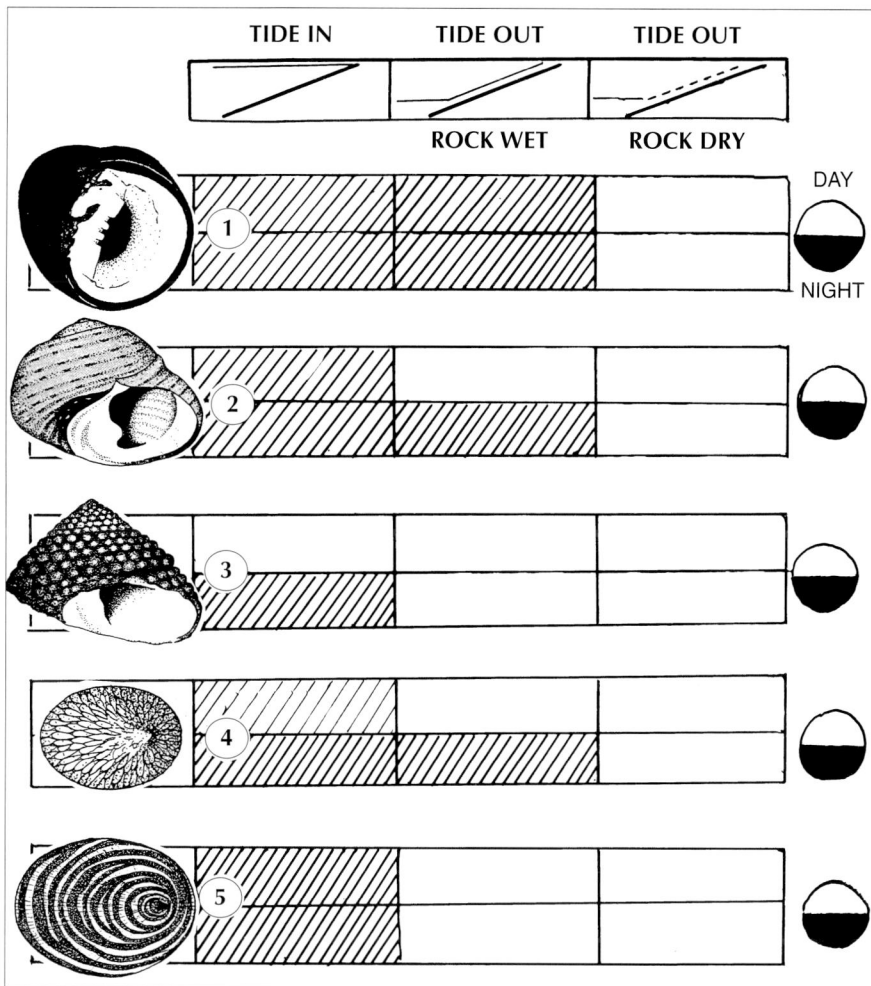

Fig. 13.6 Activity of five grazing gastropods under boulders
Activity is indicated by crosshatching with the tide in, and with the rock wet or dry when the tide is out. Division is shown between day and night.
1 *Nerita atramentosa* 2 *Diloma zelandica* 3 *Diloma bicanaliculata* 4 *Notoacmea elongata* 5 *Atalacmea fragilis*.

(From Michael Barker.)

eulittoral boulders is the flat, wafer-shaped *Tetraclitella depressa*. Fawn to white and up to 1.5 cm across, the shell has a four-radiate sculpture; most of its wide base is occupied by a honeycomb of air spaces. *Tetraclitella* confines itself to eulittoral sites in extreme shade, in caves, crevices or under boulders.

In the same confined spaces as *Tetraclitella*, specimens of the primitive stalked barnacle, *Calantica spinosa*, can occasionally be found. The peduncle is compressed against the rock, and the capitulum faces outward (Fig. 13.7) with its cirri passively deployed in the accelerated current under the boulder's edge.

Of the serpulid tubeworms, the keeled *Pomatoceros caeruleus* mingles with smaller *Hydroides elegans*, with smooth tubes circular in section. Spirorbid tubes are often numerous in the broad edge zone of the boulder. Small rock oysters, *Crassostrea glomerata*, settle numerously under boulders, but will not grow to maturity. The young shells generally have a chevron of dull purple and the smallest, most recently settled are covered with dark channelled spines.

The deep red anemone *Actinia tenebrosa* is a shade species (as its specific name — of the *tenebrae* or gloomy dark — brings out). It prefers surfaces emersed for several hours between tides; red *Actinia* species — through-

out the temperate world — are thus the denizens of dimly lit middle eulittoral sites, including cave-walls. The column is dark and hemispheric when contracted, and lightens to rose colour as it extends. The tentacles and oral disc are paler, with marginal tubercles or *acrorhagi* tinted with blue.

The chief food of this anemone appears to be the mobile amphipods caught by the tentacle-spread. At mid-tidal level, its feeding must disclose a wet-and-dry periodicity. With specimens able to be collected at regular intervals through the cycle, the imposed tidal rhythm of digestion could be a rewarding study.

In moist rock fissures and in pools, as well as under boulders, lives a second species of anemone, *Isactinia olivacea*, varying through emerald green, olive, pale yellow to brown. The tentacles are brown or green and the oral disc lighter coloured. The palest specimens are found towards low water, part-embedded and fixing shell-sand to the column.

A search should be made in silty sand near coralline turf for the elegantly sculpted *Epitonium jukesianum*. This slender gastropod feeds by inserting its proboscis into the column wall of *Isactinia olivacea*. *Epitonium tenellum* is found similarly associated with the anemone *Anthopleura aureoradiata*.

239

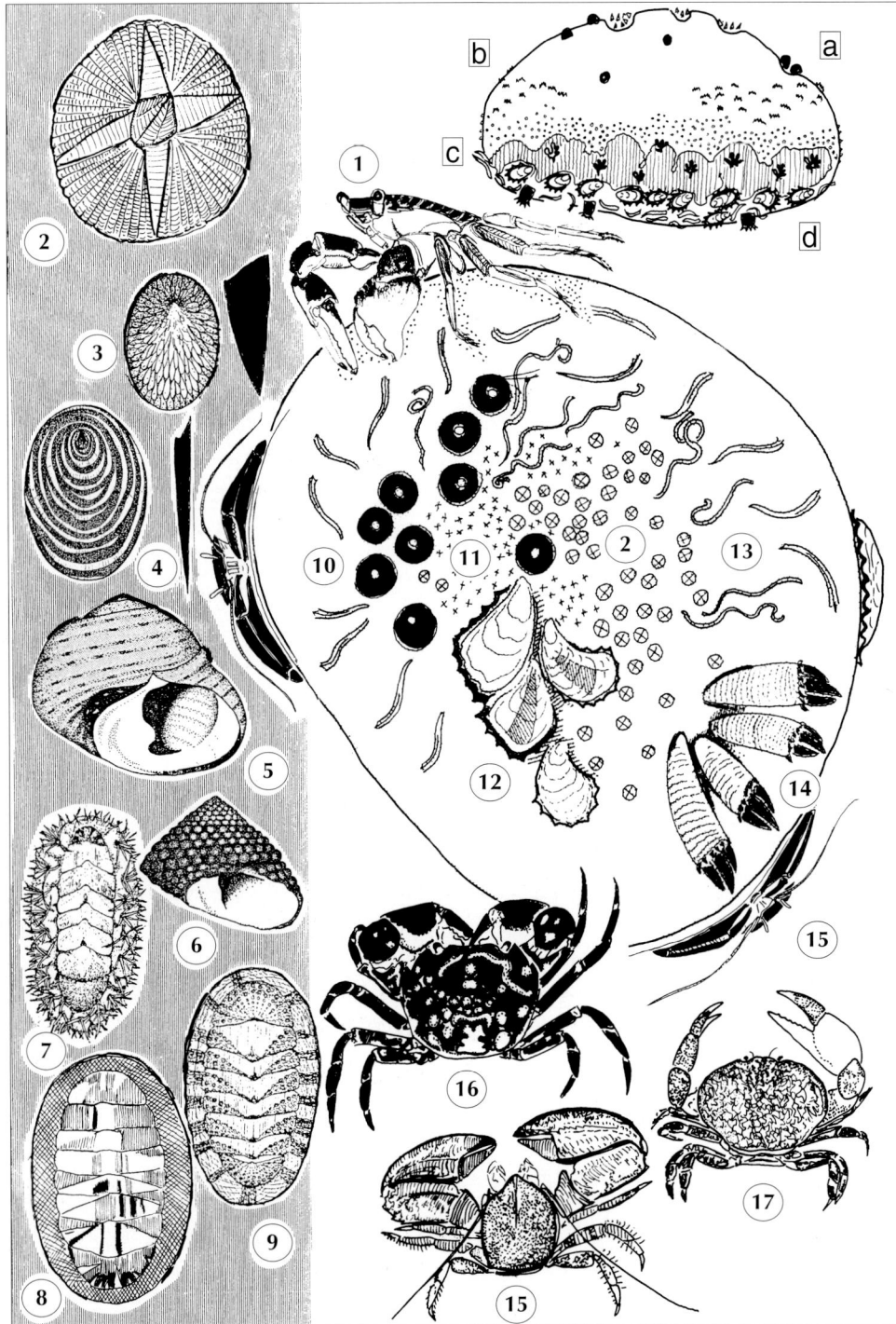

Fig. 13.7 A middle eulittoral boulder: Whangarei Heads

a–b *Lichina, Nerita, Austrolittorina;* b–c *Chamaesipho columna, Apophlaea sinclairii;* c–d *Crassostrea glomerata, Pomatoceros, Actinia tenebrosa*

1 *Leptograpsus variegatus* 2 *Tetraclitella depressa* 3 *Notoacmea elongata* 4 *Atalacmea fragilis* 5 *Diloma zelandica* 6 *D. bicanaliculata* 7 *Acanthochitona zelandica* 8 *Chiton glaucus* 9 *Sypharochiton pelliserpentis* 10 *Actinia tenebrosa* 11 *Austrominius modestus* 12 *Crassostrea glomerata* 13 *Hydroides norvegicus* 14 *Calantica spinosa* 15 *Petrolisthes elongatus* 16 *Hemigrapsus edwardsi* 17 *Heterozius rotundifrons.*

Fig. 13.8 A lower eulittoral boulder: Milford, Auckland

a–b *Chamaesipho columna* with *Crassostrea;* b–c *Corallina-Hormosira* and (**underneath**) *Codium convolutum;*
c–d *Watersipora;* d–e unlighted community
1 *Beania plurispinosa* 2 *Watersipora subtorquata* 3 *Taron dubius* 4 *Herpetopoma bella* 5 *Thoristella oppressa* 6 *Scutus antipodes* 7 the same, shell, and immature animal 8 *Cominella virgata* 9 *Buccinulum vittatum* 10 *B. linea*
11 *Paratrophon quoyi* 12 *Neohymenicus pubescens* 13 *Notomithrax minor* 14 *Ozius truncatus* 15 *Pilumnus lumpinus.*

Lower eulittoral (bryozoans and tubeworms)

At the coralline turf level, marking the lower eulittoral, stable boulders get a generous addition of bryozoans, which are the next big group to become dominant downshore. The wide-based, circular boulder shown (Fig. 13.8) from **Milford**, Auckland, was covered on top with *Corallina*. Below this, inside the margin, it is shown encircled with shaded *Codium convolutum*. A strong spread of *Pomatoceros caeruleus* can alternatively reach some distance beneath the overhang, in turn to be superseded in some spots by the tubes *Stephopoma roseum*, a worm-like gastropod liable to be mistaken for a serpulid (Fig. 13.18).

The next two circles are composed of ectoproct bryozoans, with two species more resistant to periodic emersion than most of their group. Closest to the light, in the outer zonule, there is the jet-black *Watersipora subtorquata*, an immigrant species from east Australia in the late 1950s, that for some 25 years enjoyed a bonanza on sheltered northern shores, before declining in number. Still typical on the shaded overhang of any suitable boulder, and reaching into darkness further in, *Watersipora* forms curling black sheets, edged in brick-red, often so brittle as to break off at a touch. Settling next below the tubeworms, this bryozoan always remains clear of silt. Its black lamellae may overgrow the crusts of *Pomatoceros*, sometimes so rapidly that live tubeworms and even *Sypharochiton* may be found beneath it. In winter, evidently shielded from light by its dark pigment, *Watersipora* can fast grow upwards as a flimsy canopy over *Codium convolutum* and even *Corallina* turf.

Next away from the light is the second ectoproct zone, of beige *Beania plurispinosa*. Only thinly calcified, *Beania* belongs not with the hard crustose bryozoans but is a prostrate member of the bush-like Anasca. The zooecia, not in contact but interlinked by tubular connections, are held down by rhizoids, and can be pulled away as loose flaps. Colonies are ornamented on top with close-lying, microscopic spines. Obviously tolerant of some silt, *Beania* species can reach well outside the clean sublittoral fringe, where most cheilostomes come into their own.

In the lower eulittoral of the north, several other cheilostomes are regularly to be found under boulders notably *Rhynchozoon larreyi*, *Eurystomella foraminigera*, and *Steginoporella neozelanica* form *typica*.

Several kinds of ascidian usually grow alongside the Bryozoa, or more centrally on the boulder base. Common under almost every boulder is *Asterocarpa coerulea*, smoothly rounded, with the thick test pearly white and suffused with dull blue, with the siphon linings deep blue. The familiar *Corella eumyota* has in contrast a translucent test, brittle and cartilage-like, broadly fixed by one side, and with the siphon tips bright red. *Cnemidocarpa bicornuta* has a wrinkled, brown or orange test, with a sunken-cheeked appearance. The leading compound ascidian is *Didemnum candidum*, in small off-white crusts somewhat like putty.

The prevailing sponge colours are those of red *Microciona coccinea* and buff *Halichondria* species. Towards the boulder edges grow the small tangerine golf-balls of *Tethya aurantium*. In cloudy water, quite high in the eulittoral, the sides of boulders may carry thick crusts of the crumbling, dull orange *Halichondria moorei*.

Lower eulittoral boulders have their good complement of crabs. *Hemigrapsus edwardsi*, mostly encountered higher up, generally gives place here to the wide-brandishing xanthid *Ozius truncatus*. As further upshore, the filter-feeding *Petrolisthes elongatus* outnumbers all the rest. There are in addition several species rather distinctive of lower eulittoral boulders from quiet shores. The smallest of its genus, the decorator crab, *Notomithrax minor*, clings to the rock, with its silty carapace garnished with flaps of *Beania*, halichondriid sponges or small compound ascidians. Concealed even flatter against the rock is our smallest crab, the virtually immobile *Neohymenicus pubescens* (Fig. 11.6).

The xanthid crab *Pilumnus lumpinus*, finely hairy and often silt-covered, is frequently found under these boulders. The chelae are pearly white with black fingers, whereas the coarser haired *Pilumnus novaezelandiae* has cinnabar pink chelae, and commonly hides in crevices or immures itself in *Ecklonia* holdfasts. The third xanthid, *Pilumnopeus serratifrons*, is more characteristic of boulders in still muddier sites (Fig. 11.6).

A number of gastropod species first come into prominence at this lower eulittoral level, to continue into the sublittoral fringe. The largest is the jet-black fissurellid *Scutus antipodes*, reaching up to 8 cm long. The vernacular name 'ducksbill limpet' is derived from the shell plate, concealed by integument, and roofing the mantle cavity in front. The body is rounded and muscular, with its broad foot adhering to a bare scar. *Scutus* appears to 'home' to this site after browsing on succulent green and red algae, and young *Hormosira*. With its size and conspicuous form *Scutus* was the first and only shore invertebrate to be painted by Sydney Parkinson, at the Bay of Islands on Cook's first visit.

More securely attached, under boulders at the same level, lives a second fissurellid, *Tugali elegans*, grazing chiefly on sponges. The yellowish grey mottled foot and epipodium extend out far beyond the oval shell, but in the much smaller species, *T. suteri*, the soft parts can be entirely shell-contained.

Two small topshells are to be found regularly in the

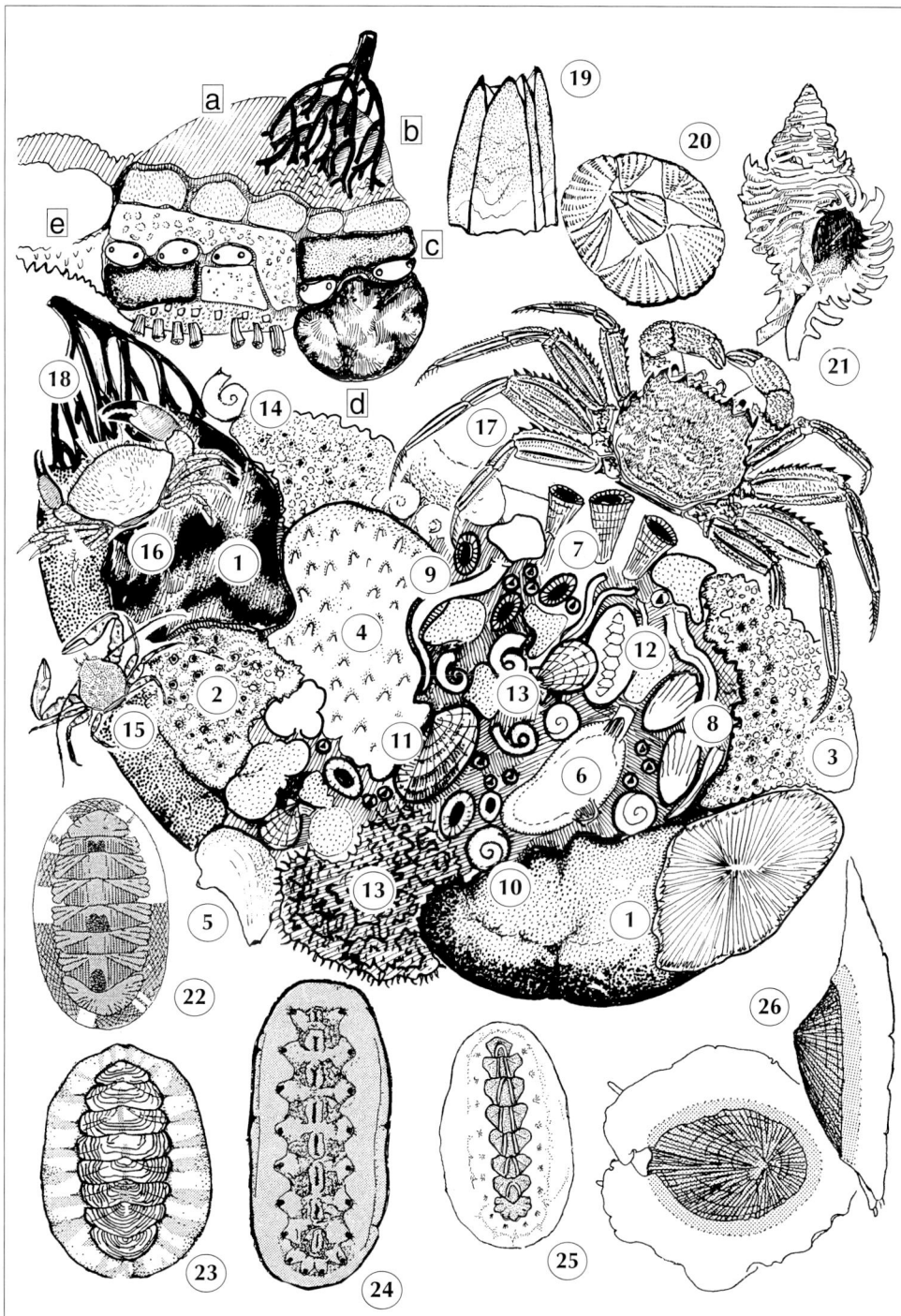

Fig. 13.9 A sublittoral fringe/zone boulder: Whangarei Heads

a–b coralline crust with *Ecklonia;* b–c bryozoans; c–d ascidians, sponges, cup corals; e fusing sponge (*Cliona celata*) between boulders

1 *Ancorina alata* 2 *Cliona celata* 3 *Spongia officinalis* 4 *Halichondria* sp. 5 *Cnemidocarpa bicornuta* 6 *Corella eumyota* 7 *Monomyces rubrum* 8 *Maoricrypta costata* 9 *Galeolaria hystrix* 10 *Sigapatella novaezelandiae* 11 *Barbatia novaezealandiae* 12 *Acanthochitona violacea* 13 *Chlamys zelandiae* 14 *Serpulorbis zelandicus* 15 *Pterolisthes novaezealandiae* 16 *Pilumnus lumpinus* 17 *Plagusia chabrus* 18 *Ecklonia* holdfast 19 *Notobalanus vestitus* 20 *Balanus trigonus* 21 *Muricopsis octogonus* 22 *Rhyssoplax aerea* 23 *Onitochiton neglectus* 24 *Cryptoconchus porosus* 25 *Rhyssoplax violaceus* 26 *Tugali elegans*.

lower eulittoral, the 'stepped' *Thoristella oppressa*, and the rounded *Herpetopoma bella*, to be detected at once by its nearly invariable covering of red sponge.

Of the neogastropod predators, *Taron dubius* is our only common member of the mainly tropical **Fasciolariidae**. Overlaid with coralline crust, its conspicuous feature is the family character of a scarlet foot. *Taron* commonly haunts the sides of boulders near the tube-

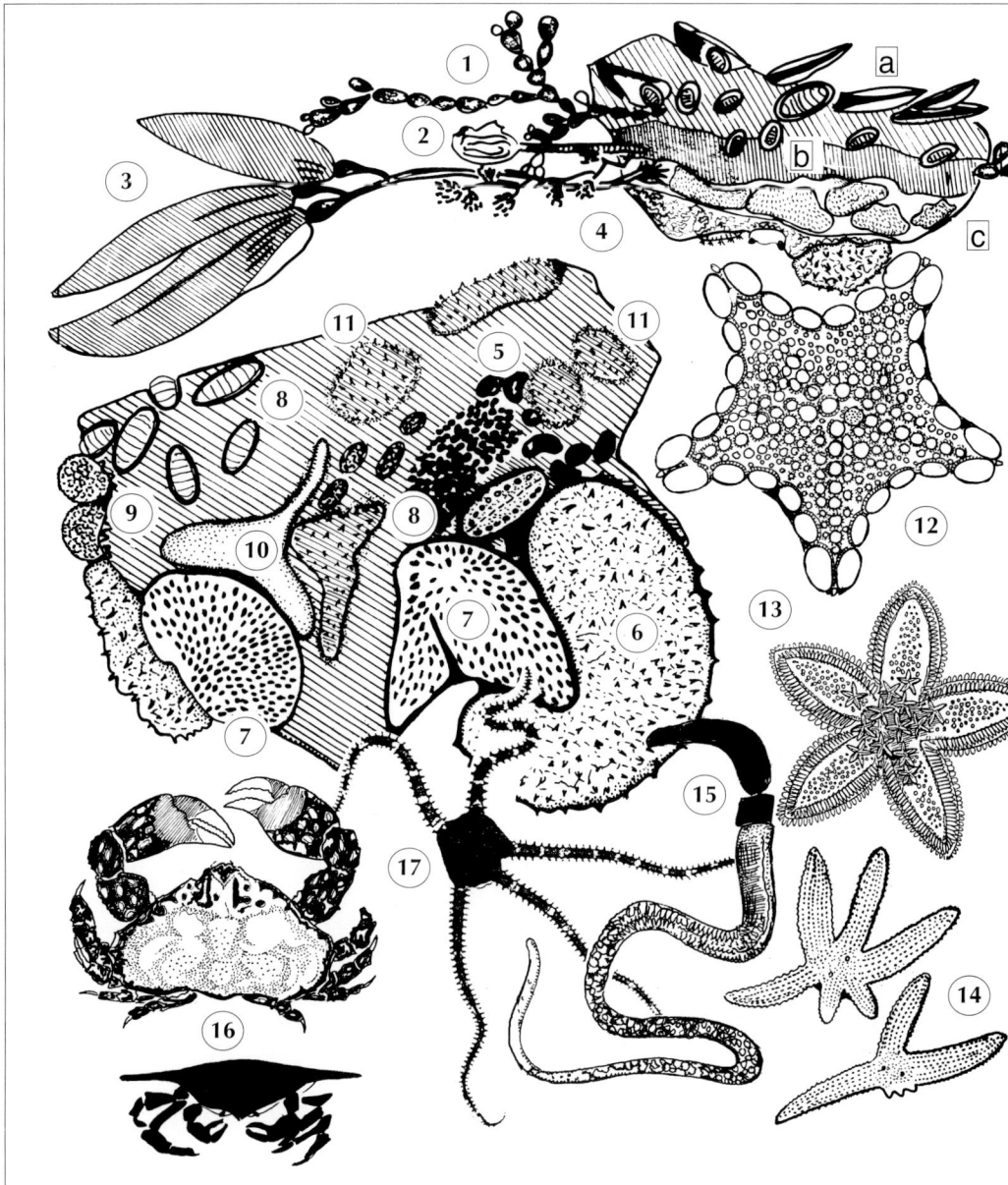

Fig. 13.10 Lower eulittoral (top) and sublittoral fringe boulders: Broad Bay, Otago Harbour

a–b chitons and *Ostrea lutaria;* b–c bryozoans, ascidians and sponges

1 *Hormosira banksii* 2 *Pyura pachydermatina* 3 small *Macrocystis pyrifera* (pool form) 4 *Cystophora scalaris* 5 *Aplidium* sp. 6 *Ophlitaspongia reticulata* 7 clathrate keratose sponge 8 *Oscarella* 9 *Tethya aurantium* 10 *Didemnum* sp. 11 keratose sponge 12 *Pentagonaster pulchellus* 13 *Anasterias suteri* 14 *Allostichaster insignis* 15 *Saccoglossus otagoensis* 16 *Eurynolambrus australis*, with (**bottom**) view from in front 17 *Ophiomyxa brevirima.*

worms and *Stephopoma* upon which it has been found to prey. The Family **Buccinulidae** is represented by *Cominella virgata*, and the *Buccinulum* species, *B. linea* and — in the north — *B. vittatum*, thought to be wholly or in part grazers of bryozoans.

Sublittoral fringe (sponge boulders)

On the most stable boulders of the sheltered sublittoral fringe and the adjacent sublittoral zone, the sciaphilic community achieves its greatest maturity and longevity

and its highest biomass. Broadly to be called *Sponge Boulders*, these are the habitats best looked for in open embayments with no more than light wave action. In the north, the reef fringes of the Hauraki Gulf from Narrow Neck to the marine reserve of Long Bay are rich in such boulder cover. The chosen examples (Fig. 13.1) are from High Island, in the mouth of the Whangarei Harbour, fronting **Taurikura Bay**.

On a calm day, along these quiet stretches hardly a ripple breaks the surface. At the lowest tides, heavy tresses of *Carpophyllum* lie slack above the water-line; shiny heads of *Ecklonia radiata* may be exposed to the

Fig. 13.11 The fauna of *Ecklonia* and *Sargassum*: east Northland
(top left) the blade of *Ecklonia radiata* **(bottom left)** *Ecklonia* holdfast **(right)** *Sargassum sinclairii*.
1 spirorbid tubeworms 2 *Membranipora membranacea* 3 same, enlargement of colony 4 stipe 5 *Amphiroidea media*
6 *Plakarthrium typicum* 7 *Batedotea elongata* 8 *Planes minutus* 9 *Stylochoplana* sp. 10 *Balanus trigonus* on *Sargassum*
vesicle 11 *Bugula neritina* 12 *Smittoidea maunganuiensis* 13 *Hastingsia* n. sp. 14 *Bugula flabellata* 15 *Sargassum* vesicle
with spirobids, *Hydroides* and *Lichenopora* 16 *Tenagodus weldii* 17 *Corella eumyota* 18 *Sycon* 19 sabellid *Branchiomma
curta* 20 *Hiatella arctica* 21 *Petrolisthes novaezelandiae* 22 *Australocnus*.

air for up to an hour. Over the coralline turf around low water small marine caddises, *Philanisus plebeius*, feebly take to the wing. Redolent of old sponge communities, the heaviest boulders, overturned for perhaps the first time in at least a decade, give off the sour aroma of carbon disulphide.

At High Island the andesite boulders are round-angled, with the base roughly flat. They lie stably bed-

ded, ultimately welded into a pavement by the living crusts that fuse between them until no glimpse of bare rock remains. On the top, beneath *Carpophyllum* or *Ecklonia*, the rough-cast of coralline spreads like a cement. In the space beneath, still open to currents, undisturbed communities have gone on enriching with the years.

Such closed strongholds should not be thoughtlessly

broken open though some intrusion can in conscience be justified for serious study. With such an intent a few large boulders can be prised apart, with the wrenching of their sponges; or smaller loose ones — waist-deep at low water — can be lifted out by their *Ecklonia* stipes. The deepest boulders carry brachiopods and cup corals, with their vivid scarlet becoming apparent for the first time as they are overturned. Total sacrifice of boulders can be avoided by shortening their exposure to the light, and restoring them carefully to their resting sites.

The sponges — with their long protracted life — have laid hold of the major part of the under-space. Those that spread out to attach boulders together may be designated the 'fusing sponges'. In the north these are contributed from three main species. First, nearly every boulder carries the heavy-lobed *Ancorina alata*, steel-grey with a firm sandpapery skin broken off, it is found to be densely spiculose and pale yellow inside.

The second fusing species, *Ophlitaspongia reticulata*, is by contrast rubbery and compressible, altogether devoid of spicules but pervaded with elastic fibres. The tough, shiny brown skin is studded with pointed villi bearing the oscula. It is chiefly this species — along with *Ancorina* — that emits the heavy sponge aroma. The third fusing sponge is the chrome yellow *Cliona celata*. This species begins as small, eroding pustules that will join up massively to bind neighbouring boulders together or swell out from kelp holdfasts. A fourth fusing species, usually subtidal but sometimes seen onshore, is *Geodia regina*, matt and opaque white outside, with a brown interior.

Far more diverse are the 'crusting sponges', of lower profile and smaller biomass. The wide-spreading *Halichondria* species are the lightest in texture, crumbling like bread as in *H. panicea*. They contain virtually no elastic spongin or sharp spicules. Another common halichondrine is the yellowish brown *Adocia venustina*. In all the species the exhalant oscula open from tall chimney pots. The related *Haliclona* sponges spread flush with the surface, as in the dull orange *H. stelliderma*, with clear radial canals coverging on low oscula. *Timea aurantiaca* forms pale yellow crusts, netted with a pale surface relief.

The red crusting sponges are low-contoured and generally smooth. *Suberites axinelloides* is dull orange, velvet to the touch. The soft-textured *Microciona coccinea* is like red baize, following closely the surface relief of the rock. Of similar hue, *Plocamia novizelanicum* is distinguished by its finely spiculose surface.

The orange golf-ball sponge, *Tethya aurantium*, and the smoother, flesh-coloured *T. australis* avoid the undersurface of boulders to grow higher up the sides.

The thick, soft crusts of *Tetrapocillon novaezealan-*

diae can be mistaken for nothing else. They are dull yellow in hue, and semi-fluid beneath a smooth skin, like congealed mustard with its surface smoke-filmed. Even less substantial, little more than lobules of soft grey jelly, are the so-called 'slime sponges' *Oscarella*, with their skeletal structure reduced to the ultimate minimum and spicules eliminated.

Finally, among the **Demospongiae**, there are those strong-textured sponges, sometimes described as 'keratose', constructed of elastic fibres and lacking spicules. The fusing *Ophlitaspongia reticulata* (close to the traditional bath-sponge) is the most massive of these. Spread more thinly over the surface grows a *Darwinella* species of dull verdigris colour. *Dysidea* sp. is yellowish white, like a rubbery plastic studded with sharp villi, while the rough-textured *Ircinia fasciculata* is dirty yellow brown sprinkled with darker pustules.

Distinct from all the siliceous sponges are the small vases and tubes belonging to the Class **Calcarea**. These include firstly the *Clathrina* species, simple cylinders springing up from creeping stolons, with choanocytes lining the whole central cavity. Second and a little more complex are the vase-shaped sponges of the syconid type, such as *Sycon ornatum*, having the choanocytes withdrawn into radial tubes opening from the central space (see Fig. 13.11).

Bryozoa

Though present on almost every sponge boulder, the bryozoans are far less profuse than on higher energy shores. They are generally detectable by their file-like hardness to the touch, but some of the commonest are only weakly calcified as with the flattened laminae of *Beania* (Fig. 13.16). Familiar species are the beige *B. discodermiae*, orange *B. magellanica*, and *B. plurispinosa*, covered with long overlapping spines.

Ascidians

The ascidians generally occupy more space than the bryozoans. The common simple sea-squirts are the large *Corella eumyota*, with cartilaginous test and red siphons; the blue-suffused *Asterocarpa coerulea*; and sometimes the brown-tinted *A. humilis*.

Compound ascidians offer less variety than under wave-swept 'bryozoan' boulders, but may include the bright heads of polyclinid species, or fleshy spreads as in *Aplidium phortax*. The most frequent of all the compound ascidians are likely to be the patches of *Didemnum* species, apricot, beige, mauve or most commonly chalky white.

Tubeworms

The most prolific of the serpulids, such as *Filograna* and the spirorbids, prefer current-swept boulders. A com-

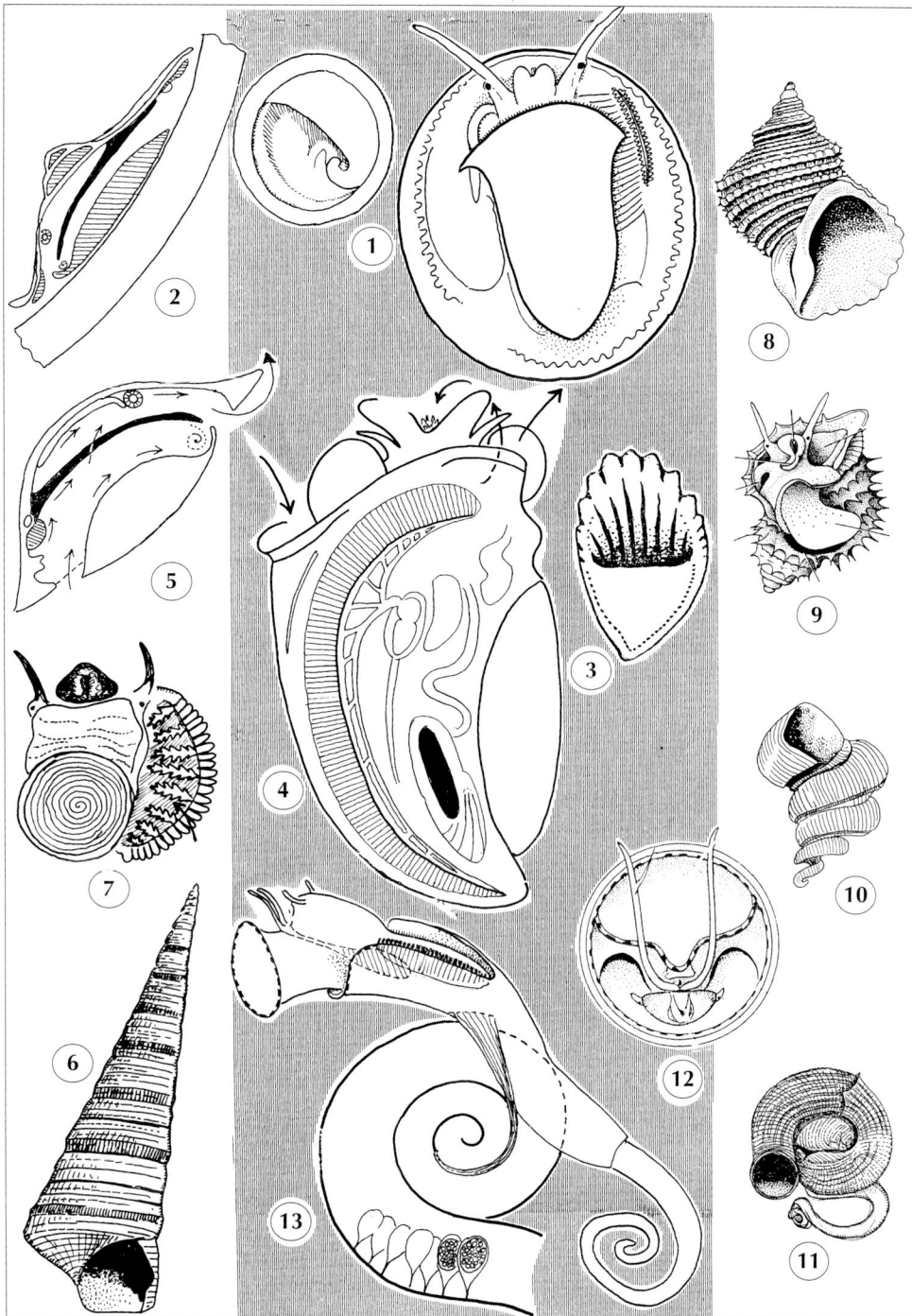

Fig. 13.12 Ciliary feeding and related gastropods
1 *Sigapatella tenuis* **(left)** shell and **(right)** animal viewed from below 2 the same, cross section of pallial cavity
3 *Maoricrypta costata*, shell from below 4 *Maoricrypta costata*, animal removed from shell, viewed dorsally 5 *Maoricrypta*, cross-section of pallial cavity 6 *Maoricolpus roseus*, shell 7 the same, foot and entry to pallial cavity 8 *Trichosirius inornatus* 9, the same, animal 10 *Tenagodus weldii* 11 *Serpulorbis zelandicus*, shell 12 the same, head and foot 13 the same, part removed to show egg capsules shell attached.

mon species in quieter waters is the large, double-keeled *Galeolaria hystrix*, with its white tube, pointed up in scarlet. The operculum is spined like a porcupine (hence the specific name). *Spirobranchus latiscapus* is named from the wide helix of its gill plume. The operculum has a winged stalk, and is built like a five-tiered tiara, marginally spined. *Neovermilia sphaeropomata* builds an

unkeeled solitary tube with an incurved spine at the aperture. The gills are whorled in a double crown, and the globular, hyaline operculum is carried on a long stalk. A second *Neovermilia* with a funnelled operculum may form white clumps in kelp holdfasts. In a third species, found in twos and threes under boulders, the operculum is spherical, with a calcified top.

Cirripedia

On crowded sponge boulders the operculate barnacles claim only a minor space. Throughout New Zealand *Notobalanus vestitus* is, however, a regular species at low tide; its sharp-pointed cone, pink at the valve-tips, is covered by a pale yellow epidermis. The cosmopolitan *Balanus trigonus*, brought by ship, is now the commonest barnacle of the sublittoral and its fringe in the Hauraki Gulf and through east Northland. The aperture is triangular, and the column plates pink with white ribs.

Anthozoa

Two sorts of ahermatypic coral reach up to the sublittoral fringe. Their anemone-like polyps sit in skeletal cups, divided up by radial partitions or *septa*. *Culicia rubeola* has short, cylindrical cups up to 5 mm across and with 24 septa, linked up by horizontal tubular rhizomes, also calcified. The polyps vary from delicate pink to pale green. The solitary 'fan coral', *Monomyces rubrum*, is larger and rather variably shaped. Near low water, the cups are generally short and near-cylindrical, but deeper down they flatten and expand like a fan, attached by a pedicel. The oral disc is brilliant orange, with a whorl of pale yellow tentacles.

Kelp holdfasts

The holdfasts of *Ecklonia radiata* offer a natural extension of the shelter of the boulder, most noted for their wealth of polychaete worms. The most visible are generally the tube-living **Sabellidae**. As distinct from serpulid tubes, these are flexible and leathery and close elastically when the animal is retracted. The branchial crown is spectacular, with the feathered pinnae widely deployed or flashing back upon disturbance. The sabellids *Megalomma suspiciens* develop special eyes near the filament tips. Others, such as the small *Branchiomma curta*, are devoid of ocelli, but have the pinnae brightly patterned.

Common serpulids in holdfasts include an unnamed *Neovermilia* with its tube-clumps giving refuge to small interstitial worms; and a tightly spiralled *Hydroides*. The large terebellid *Terebella plagiostoma* builds membranous, sand-studded tubes among the holdfast branches. A small *Terebella*, bright red and cream and encased only in a mucus sheath, is found both in holdfasts and crevices.

The errant polychaetes in kelp holdfasts include the long grey *Marphysa depressa*, with its antero-dorsal 'blood gills', the vigorous greenish blue *Perinereis ambylodonta*, and the smaller *Nereis falcaria*. Among the scale-bearing polynoid worms may be found the small *Lepidonotus polychromus*, sometimes *L. jacksoni*, or even the broadly oval sea-mouse *Euphione*, reaching to 3–4 cm long.

The smallest and most abundant worms in holdfasts are the miniaturised **Syllidae**. Seldom above 2 cm long, they live in mucilaginous grooves in deeper reaches of the holdfast branches. Under a strong lens syllids present clear identifying marks. The head bears four eyes, three tentacles and a pair of large unjointed palps. The peristomium carries two pairs of cirri, while the parapodial cirri are usually long and slender. All syllids have a tubular pharynx, often visible through the body wall, that can be shot out to impale small prey with its sharp denticles.

Several crab species are almost certain to be found in *Ecklonia* holdfasts. The coarsely haired *Pilumnus novaezelandiae* can become virtually imprisoned by the crampons as it grows, while *Pilumnus lumpinus* is more frequent under lower eulittoral boulders (Fig. 11.6). The decorating crab, *Notomithrax minor*, and the tiny pillbox crab, *Neohymenicus pubescens*, are both common in holdfasts as is also *Petrocheles spinosus*, a small mauve-brown porcellanid with long, slender chelipeds.

Ascidians sometimes cluster together to fill much of the holdfast space. This is a favourite habitat for *Corella eumyota*, *Pyura carnea*, *P. rugata*, and *Styela plicata*.

The brittle star *Ophiactis resiliens* deploys its arms from narrow cavities. The small (2 cm) cucumber *Australocnus calcareus* is common under *Ecklonia*, with the tube-feet of the 'sole' taking firm grip. Its white integument, firm with spicules, is sprinkled with scarlet tubercles on the upper side.

Woody holdfasts are bored by at least two species, the small and often misshapen bivalve *Hiatella arctica* and the kelp-eroding isopod *Phycolimnoria insegnis*.

Brachiopods

The sponge boulders at Whangarei Heads shelter two very different sorts of Brachiopoda. The coral red lampshell, *Calloria inconspicua*, has hinged valves (Fig. 16.16) and attaches to the rock, or to others of its kind, by a strong peduncle. On the same boulders may be found also the hingeless *Novocrania huttoni*, with the valves held together only by a complex of muscles. *Novocrania* has evolved so far in a limpet direction that until its top valve is prised off the rock with a strong knife, there is little hint it is a brachiopod at all. The lower valve, technically 'dorsal' (see Fig. 16.16), is transparent and fused to the rock, with the mantle that lines it carrying the gonadial canals. Lying within the upper (properly 'ventral') valve is the spiralled feeding organ or *lophophore*.

Mollusca
Chitons

Under the shaded boulder edges can be found several sorts of chitons, some of which — like the *Tugali* limpets (Fig. 13.9) — have evidently moved to a diet of sponges.

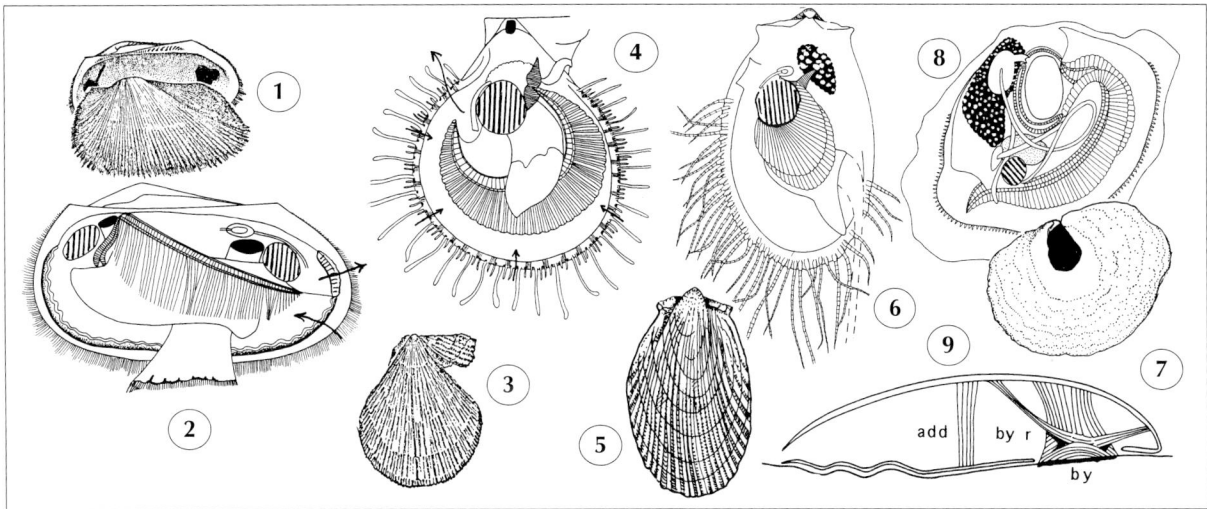

Fig. 13.13 Byssus-attached and mobile bivalves
Barbatia novaezealandiae, 1 shell 2 pallial organs;
Chlamys zelandiae, 3 shell 4 pallial organs; 5 *Limatula maoria* 6 the same, pallial organs; *Anomia trigonopsis* 7 shell 8 pallial organs 9 section showing byssus and musculature (by) byssus cable (by r) byssus retractors (add) adductor muscle.

Below: Fig. 13.14 Bivalves: the mussel form
(inset left) veliger larva with symmetrical fore and aft shell 1 *Crenella radina*, small and sublittoral; *Trimusculus barbatus* 2 shell 3 pallial organs 4 *Modiolus areolatus* 5 *Modiolarca impacta*; *Zelithophaga truncata* 6 shell *in situ* in soft rock 7 pallial organs 8 *Perna canaliculus*, pallial organs.
(bottom left) comparable evolution in Carditidae, from 9 *Venericardia purpurata* to 10 *Cardita aoteana* shell and 11 pallial organs.

The orange-brown *Crypotoconchus porosus* has a thick, rubbery girdle, leaving the shell valves visible only by a mid-dorsal slit. On removal they reveal the attractive blue that has won them the name of sea-butterflies. The close-related *Acanthochitona violacea* has a wide, orange girdle and narrow, purplish brown valves.

A common species under boulders, though never — like *Ischnochiton* or *Chiton glaucus* — gregarious, is *Onithochiton neglectus*. The soft, scaleless girdle is dull pink, roughened with microscopic spicules, and the valves are prettily marbled with green and brown crescents. Our largest chiton, *Eudoxochiton nobilis*, is sometimes to be found at the sides of sponge boulders; it grows up to 8 cm long, with the girdle black-stubbled and the wide valves coralline-painted.

Gastropods
Several of the under-boulder gastropods have become near-immobile, and resorted like bivalves to ciliary feed-

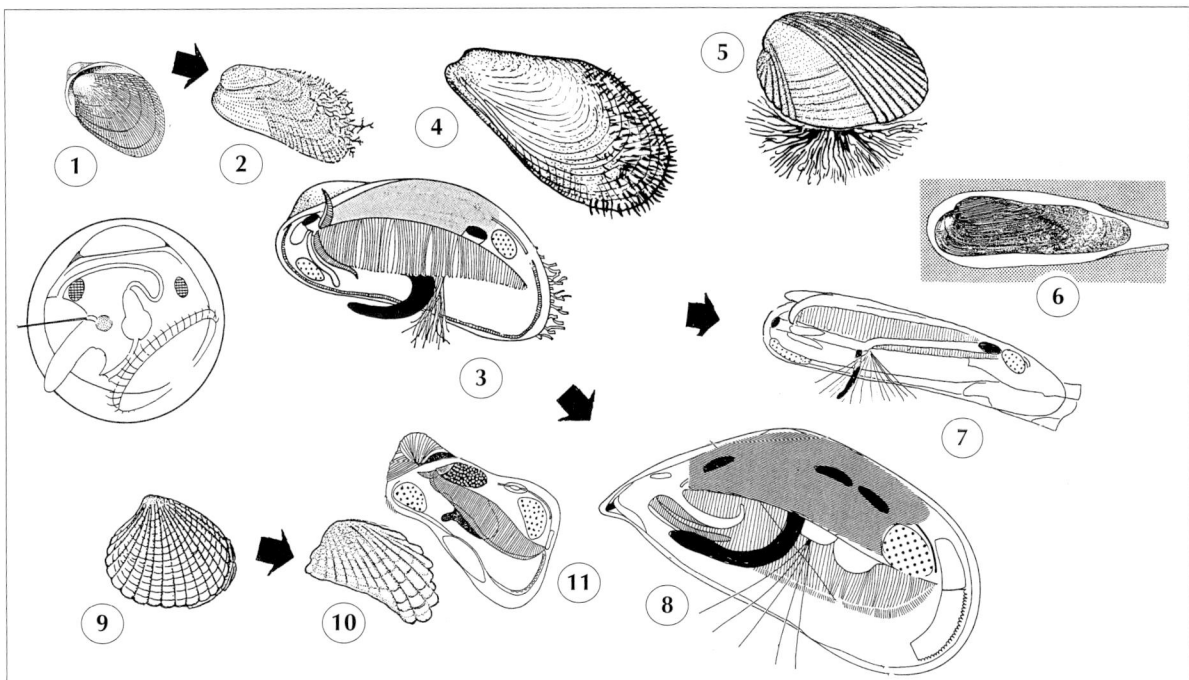

ing. The gill has in this process become highly modified, with its filaments narrow and linear (though never reflected to form a double lamina as in bivalves). At their end-point of evolution these gastropods have in effect become limpets, the same transformation that in this community has overtaken both a bivalve and a brachiopod.

Early in this series is the small, turreted *Trichosirius inornatus*, common under sponge boulders at Whangarei Heads. This snail is still mobile and spirally shelled, but the long, grooved proboscis and the enlarged gill herald the start of ciliary feeding. In the Family **Calyptraeidae** the saucer limpet, *Sigapatella novaezelandiae*, has kept its short spiral coil while another calyptraeid, the subtidal *Sigapatella tenuis*, still remains freely mobile (Fig. 13.12). A more finished limpet is the slipper shell, *Maoricrypta costata*, where a cross-septum forms a shelf separating the pallial cavity from the foot. The smaller slipper shell, *M. monoxyla*, can take several shapes according to situation. The normal convex form is found under boulders, while curved and arched individuals live attached to *Turbo smaragdus* on *Corallina* turf. A flat, or even upwardly concave, form *M. sodalis* lives in the mouths of empty whelk shells.

In all these filter feeders, the pallial cavity is divided into inhalant and exhalant chambers by the large gill. Strong lateral cilia pass water between the gill filaments from a lower left to an upper right chamber. Strained particles are conveyed by frontal cilia to the tips of the filaments, to be formed into a mucus string and conveyed to the mouth. Portions of this string are regularly pulled off by the small radula.

All the Calyptraeidae are protandrous hermaphrodites, changing from smaller males through an intermediate bisexual stage to mature females. They generally associate in sex pairs, and in *Maoricrypta* a small male settles on the right side of the female's shell, a strategic site to reach the vagina with the extensile penis. The once-male then develops female features as it grows, and is in turn settled on by a new young male. Egg capsules are brooded by the female under the shell in front of the foot.

The plankton-feeding gastropods of the **Vermetidae** — as we have earlier seen (Fig. 2.8) — are not gill-filterers but mucus trap feeders. Unlike *Dendropoma lamellosa* on the open shore (Fig. 10.7), the *Serpulorbis* species under boulders are semi-solitary and lack opercula. On disturbance the animal darts far back into the long tube. The eggs are incubated in tear drop-shaped capsules attached inside the tube along a dorsal slit in the mantle of the female.

The Family **Siliquariidae** at first sight resemble the vermetids, but are in fact filterers like the calyptraeids.

They belong to a separate lineage, evidently derived from the mobile Turritellidae scattered among small boulders near sand, which had already developed ciliary feeding. *Stephopoma roseum* is a common species in the north, forming dense clusters under moderate energy boulders (Fig. 13.18). The operculum is fringed with long, branched bristles, evidently serving as strainers to exclude larger particles when the disc is held part-closed against the rim of the tube.

A close relative is the elegantly coiled cork-screw shell, *Tenagodus weldii*, embedding in the sponge *Cliona celata*. The spiral shell is incised with an exhalant slit along the right side, and the opercula are capped with red sponge, studding the yellow *Cliona* like cherries in a cake.

Bivalves

Most of the bivalves beneath sponge boulders are byssus-fixed. Their most ancient family, the **Arcidae** (ark shells) is well typified by the purplish brown *Barbatia novaezelandiae*, roughly rectangular, with radial ribs and a beard of periostracal hairs. The shell is held upright, with byssal threads fused into a tough membrane along the keel of the foot.

The mussels (**Mytilidae**) come from a lineage almost as old as the ark shells. *Mytilus* and *Perna*, that we have encountered on zoned shores, are advanced mussels streamlined to resist the waves. We have to look under boulders for the survivors of the earliest mussels, more oval and still with large beaks or *umbones*. Our largest species is the fringed mussel, *Modiolus areolatus*, 3–4 cm long and more often found subtidally, with the shiny brown epidermis drawn out into coarse hairs. The much smaller *Trimusculus barbatus* (10 mm), often found in holdfasts, produces an epidermal beard. The nut mussel, *Modiolarca impacta*, forms dense colonies, with shells wrapped in a nest of byssus fibres, or sunken into ascidian tests. Immune from the action of waves or currents, the shell retains an early unspecialised shape.

The earliest of the scallop family (**Pectinidae**) are also byssus-fixed, represented under boulders by the small fan shell, *Chlamys zelandiae*. Radially ribbed and rose-pink, purple, orange or yellow, the shell lies over on its left side, with the byssus emerging through a notch in that valve.

The saddle oysters (**Anomiidae**) have in effect turned themselves into bivalved limpets. Our commonest species, *Anomia trigonopsis*, has its honey-coloured left valve contoured to the rock. The thin right valve underneath is pale green, with a deep embayment for a thick calcified byssus cable. The retractor muscles of the byssus now serve to adduct the shell closely to the rock, and the true adductor muscle, like the right valve, plays a reduced role.

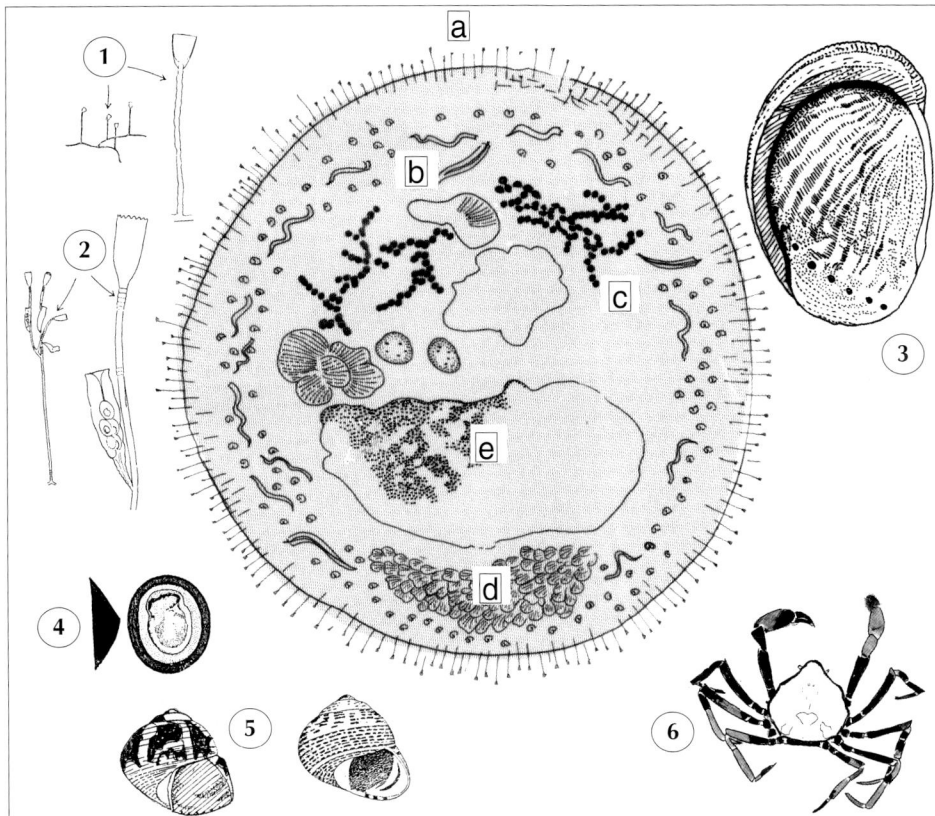

Fig. 13.15 A high energy boulder: St Clair, Dunedin
(a) fringe of thecate hydroids (*Orthopyxis* and *Clytia*) in the zone of fast flow (b) *Galeolaria, Pomatoceros* and *Hydroides*
(c) *Alloeocarpa minuta* (d) *Beania*, and (e) sponges *Tetrapocillon* and *Halichondria*.
1 *Orthopyxis integra* 2 *Clytia hemisphaerica* 3 *Haliotis australis* 4 *Notoacmea badia* 5 *Cantharidella tesselata* 6 *Elamena producta*.

Higher exposure

On the most exposed shores carrying loose boulder cover the encrusting community is thin and transient. With constant overturn, as for example at **Little Barrier**, the boulders are mostly pink, with the algal paint that can extend right round a boulder when each of its aspects is periodically turned to the light. Animals briefly appearing may include spirorbids, small *Balanus trigonus*, polyzoan patches, heads of a polycitorid ascidian, even a light settlement of the sponges *Tetrapocillon* and *Halichondria*, as well as the chitons *Chiton glaucus* and *Onithochiton neglectus*.

Rounded boulders at **St Clair**, Dunedin (Fig. 13.15), have progressed towards a more stable but still early community. The painted coralline margin carries a silvery fringe of hydroids (*Orthopyxis* and *Clytia*) in the zone of fast flow under the boulder edge. In from the edge are the tubeworms, spirorbids, *Hydroides* and young *Galeolaria*. Next in turn comes a bryozoan circlet, with *Beania*, and more heavily calcified species in thin pink and orange crusts. The centre is occupied by compound ascidians *Didemnum candidum*, and long chains of the small, cherry red *Alloeocarpa minuta*, a 'social' rather than a 'compound' species.

Sublittoral fringe (bryozoan boulders)

High-energy coasts with shelving reefs may carry boulders that by shape, or secure lodgment, can avoid frequent overturn. Swept beneath by strong laminar currents, they are mostly kept free of sediment; though some may rest in dips and depressions on a floor of shell-sand.

Such boulders vary in shape according to their parent rock and weathering history. **Smuggler's Bay**, Whangarei Heads is strewn with boulders of andesite or dacite, rounded on top but flat-based and relatively stable. The prolonged intertidal stretch at **Opunake**, South Taranaki (Fig. 11.12) has large boulders approaching spherical, resulting from erosion of laharic breccia in the cliffs.

At **Goat Island Bay**, Leigh (Fig. 13.18), as on many shores in the outer Hauraki Gulf, flat slabs have been eroded from the sandstone strata of the Waitemata Formation. Lodged in pebbly or sandy channels, and secure from overturn, these are smooth on top and bottom with the angles erosion-rounded.

At Smuggler's Bay the boulder sides are clad with the clean-water brown algae *Carpophyllum plumosum* and

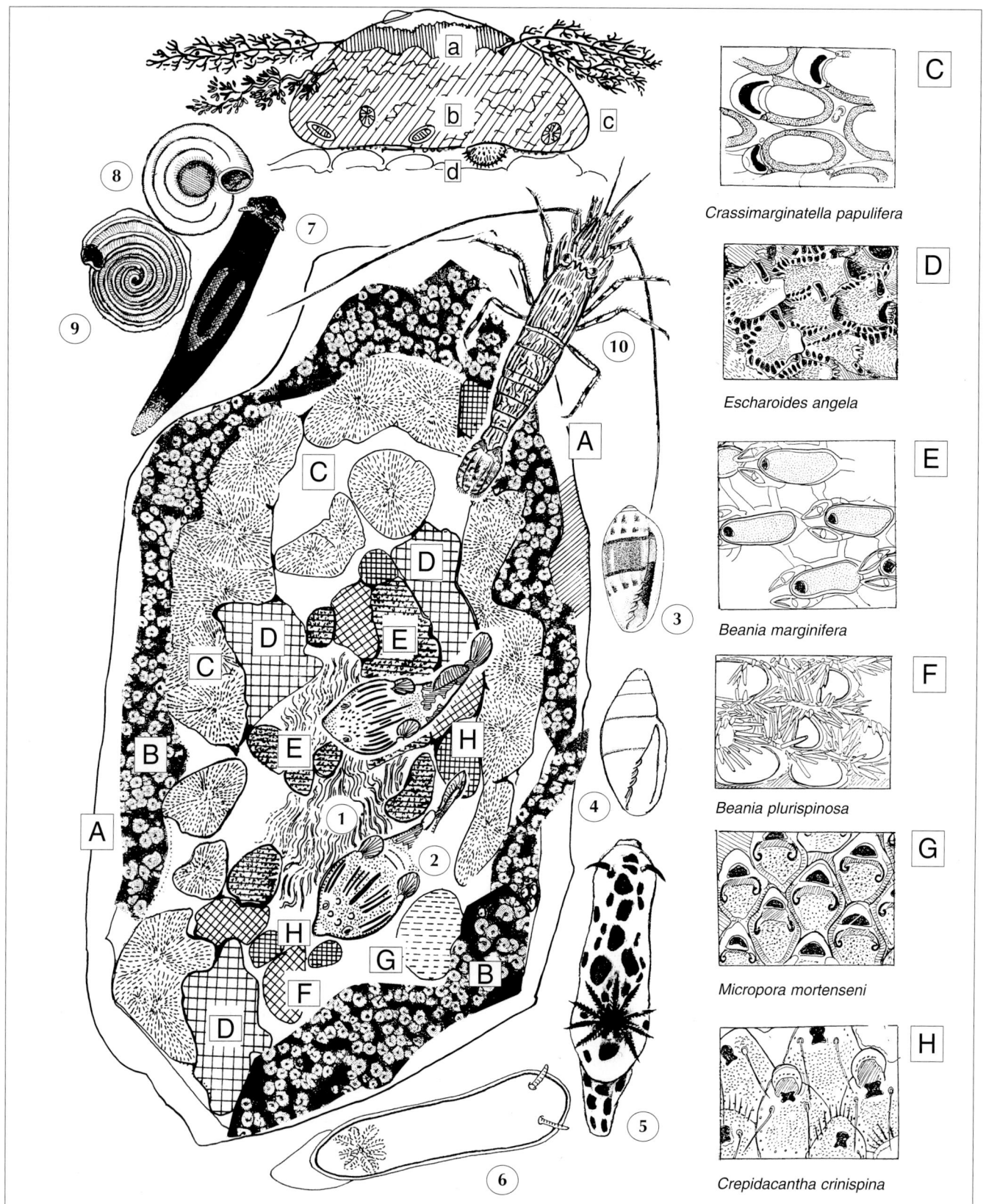

Fig. 13.16 Goat Island Bay, Leigh: a stable bryozoan boulder, with strong laminar flow

(top) side view of boulder with (a) *Hildenbrandia*, dark top (b) coralline veneer, with limpets (c) *Cystophora torulosa* (d) pink submarginal zone of spirorbids (middle) under-surface of boulder showing the leading encrusting species: (A) submarginal shaded zonule of pink coralline crust (B) submarginal zonule of spirorbid words (C–H) the six leading cheilostome bryozoans, shown also in zooid detail on right: (C) *Crassimarginatella papulifera* (D) *Escharoides angela* (E) *Beania marginifera* (F) *B. plurispinosa* (G) *Micropora mortenseni* (H) *Crepidacantha crinispina*. (1–10) other under-boulder species 1 crust of filogranid worm tubes 2 *Trachelochismus crenulatus* 3 *Haloginella mustelina* 4 *Volvarinella cairoma* 5 *Ceratosoma amoena* 6 *Chromodoris aureomarginata* 7 *Okadaia cinnabarea* 8 *Simplaria ovata* 9 *Neodexiospira pseudocorrugata* 10 *Alope spinifrons*. (Bryozoan after D.P. Gordon)

Occupation of space by sessile and sedentary cryptic organisms under 100 boulders on Echinoderm Reef Flat at Goat Island Bay

	No. of species	Ubiquity (% occurrence on 100 boulders)	% of total animal cover	% of surface covered
Bryozoa	35	73	56	14
Polychaeta	7	97	14	4
Porifera	13	42	8	2
Mollusca	7	45	6	1.6
Cirripedia	5	60	5	1.6
Cnidaria	18	29	5	1.3
Ascidiacea	16	46	3.5	1.0
Protozoa	3	40	0.2	0.06

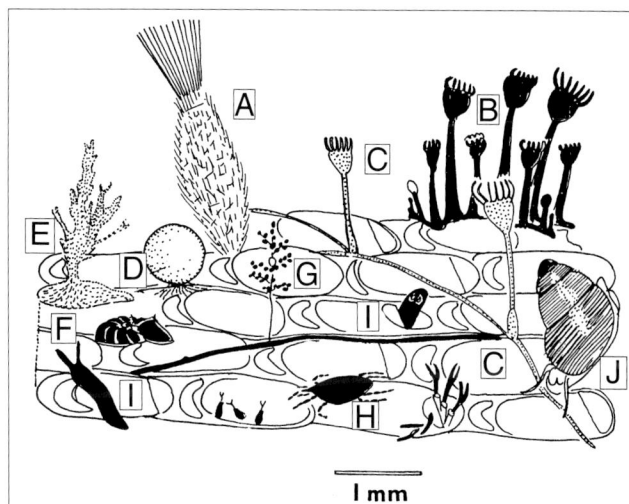

Fig. 13.17 The epifaunal micro-community on dead *Cheilostome zooecia*

(A) *Sycon* sponge (B) *Pedicellina* (C) *Barensia* (entoprocts) (D) (E) (F) three Foraminiferan species (G) *Zoothamnion* (Ciliata) (H) marine acarine mite (I) Tubellarians (J) gastropod *Eatoniella*.

(From D.P. Gordon.)

Cystophora torulosa. High-energy boulders are thinly crusted on top, as with dark red *Hildenbrandia* and tawny *Ralfsia*. Beneath these, or in places of shade coming right over the boulder top, is a pink algal veneer. Under the dimly lit edge, the coralline crust may carry the star limpet, *Cellana stellifera*.

On their undersides high-energy boulders are markedly different from the sponge boulders we shall find on secluded shores. They can be easily lifted up, being never held together by lithophylla or fusing sponges. The encrusting sponges form a minor element. Instead, the bryozoans are peculiarly rich, under constant water movement, and with medium-term stability and freedom from siltation. Boulders like these seldom get overturned by waves, but can be left so by thoughtless human foragers. The sad results are easy to recognise, with the pink under-margin bleached white and the animal community killed by light and sloughed away.

On a principal cover of bryozoans and ascidians, certain larger organisms draw attention. Clean boulder shores have long been ransacked for paua (*Haliotis*) and kina (*Evechinus*). Goat Island Bay, with its abundance of echinoids, asteroids and ophiuroids, has been reprieved by the Marine Reserve, and has come again to deserve its name, the Echinoderm Reef. The strong return of traditional species, like kina and paua, has been a vindication of the new regime.

Beneath the boulders the first glimpse can come as a delight. Pale pastel shades of the bryozoans contrast with high points of colour from ascidians, sponges and nudibranchs. Compared with sponge boulders these are more lightly loaded, leaving much of the bare rock surface still apparent.

The zoning pattern is also more regular and concentric. First, the pink coralline margin reaches under the edge as far as dimly reflected light allows. The next zonule is densely composed of spirorbid tubeworms. These are par excellence the small, quick settling and

fast growing organisms of clean, moving water. Their helical tubes attach in vast numbers both to shaded rocks and smooth kelp blades. A living spirorbid can be a fascinating object of study with a binocular microscope, displaying in miniature all the workings of the ciliated serpulid crown. Only five pinnae are present, with a sixth modified as an opercular plug, that in many species holds the eggs.

The spirorbids from Goat Island Bay and Northland have been well described, with their shells, setae and opercula, by Peter Vine.[1] The common species under boulders are *Neodexiospira pseudocorrugata* and *Simplaria ovata*.

The familiar barnacle in northern waters settling in smaller numbers with the spirorbids or wider scattered under the boulder is *Balanus trigonus* (Fig. 13.9).

Next inward from the spirorbids is a zonule of cheilostome Bryozoa essentially continuous, but generally broken into patches with up to half a dozen main species, and a host of minor ones. The Echinoderm Reef Flat at Goat Island Bay is highly suited for bryozoan growth. Sixty-four species have been recorded by Dennis Gordon as living there. Their favourable conditions are optimal wave exposure, with moderately stable boulders and slabs resting on bedrock and the clarity or reduced turbidity of the water. For boulders at Goat Island Bay, Dennis Gordon has calculated the cover by each of the major groups, with the Bryozoa easily in the lead.

The bryozoans can with a little practice be identified from their colour, colony size and shape, as well as

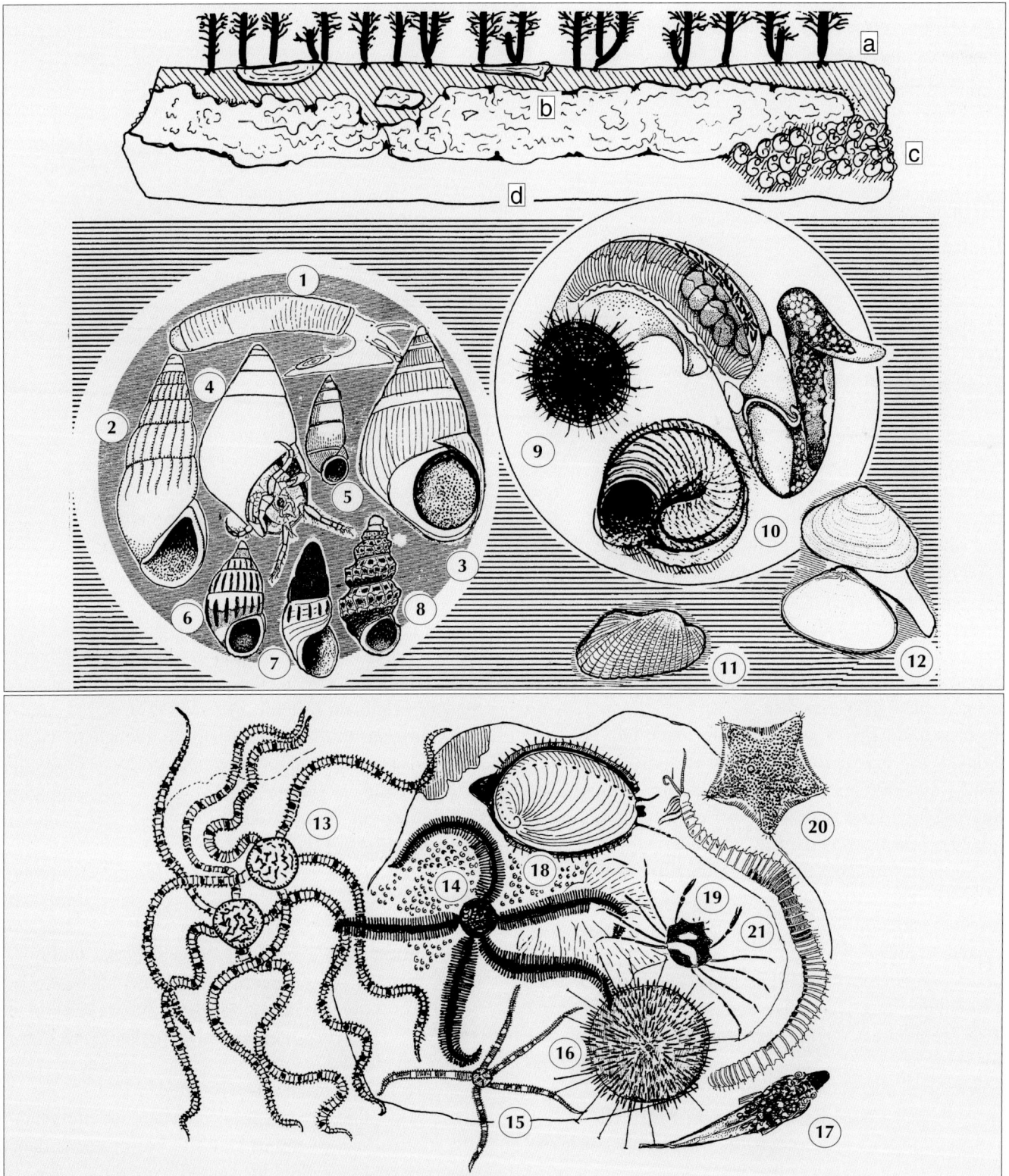

Fig. 13.18 Goat Island Bay, Leigh: clean boulder community of lower eulittoral and sublittoral fringe

(top) side view of boulder showing (a) *Carpophyllum plumosum*, cut off short at water level of pool (b) a crustose coralline alga (c) *Stephopoma roseum* (d) under-boulder community.

(centre) small molluscs of smooth undersurface of small slabs and cobbles. 1 *Caecum digitulum* 2 *Rissoina chathamensis* 3 *Eatoniella olivacea* 4 *Eatoniella* with a *Pagurapseudes* 'hermit crab' 5 *Amphithalmus hedleyi* 6 *Pisinna impressa* 7 *Fictonoba carnosa* 8 *Merelina lyalliana* 9 *Stephopoma roseum*, animal removed from shell, showing opercular ornament 10 *Stephopoma roseum*, shell 11 *Arca sociella* 12 *Borniola reniformis* some mobile and sedentary species beneath boulders in the sublittoral fringe 13 *Ophionereis fasciata* on clean sand beneath boulders 14 *Ophiopteris antipodum* 15 *Ophiactis resiliens* 16 *Evechinus chloroticus* 17 *Trachelochismus pinnulatus* 18 *Haliotis iris* 19 *Elamena producta* 20 *Patiriella regularis* 21 *Dodecaceria* sp.

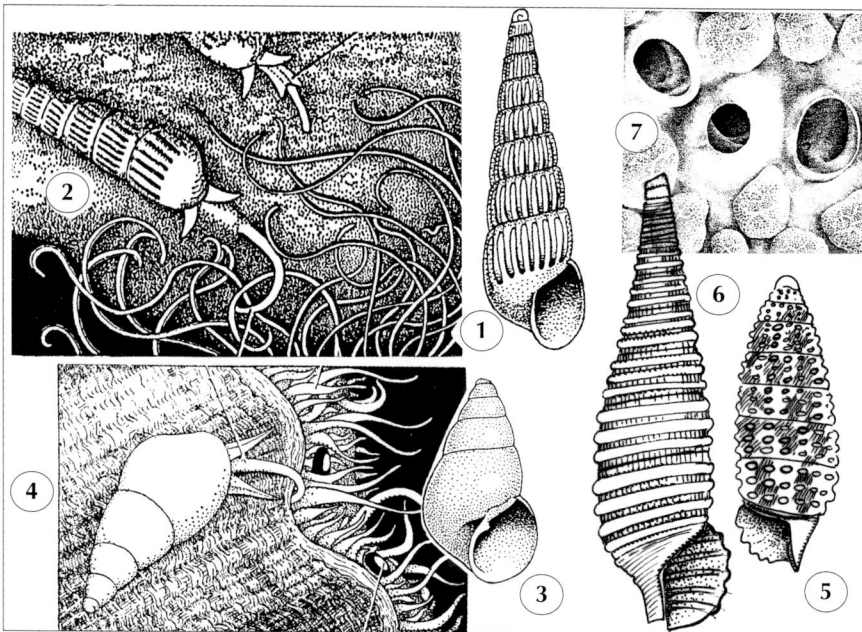

Fig. 13.19 Minute gastropod predators: suctorial and sponge-grazing
1 *Chemnitzea zealandica* 2 a British *Turbonilla, in situ* on a cirratulid worm (after Fretter and Graham) 3 *Odostomia takapunaensis* 4 a British *Odostomia, in situ* at the mantle edge of a scallop (after Fretter and Graham) 5 *Sagenotriphora ampulla* 6 *Seila cincta* with (7) its habitat sponge, *Cliona celata*.

Polycitoridae
Cystoides dellichiajei firm gelatinous sheets with rounded edges, lemon or brown.

Polyclinidae
Aplidium phortax, heavy crusts or stalked clusters, translucent yellowish to grey.
A. benhami, cushions or knobs on sides of rocks or bases of algae; translucent with bright orange zooids.
A. notti, thick mats or white sheets, with pink zooids.
Synoicum kuranui, crimson gelatinous sheets, may be sand-encrusted.

Didemnidae
Didemnum densum, parent colonies surrounded by discrete daughters; opaque pink or fawn.
D. cf. *candidum*, opaque white, fawn, orange or mauve.
Leptoclinides marmoreus, encrusting mats, ridged and humped, mottled grey, with white streaks following the zooid systems.

surface texture. Confirmation, down to species level, can be made with a binocular microscope. Gordon has listed in their order of abundance the 17 principal species of Bryozoa recorded from Goat Island Bay. The first of these, *Beania plurispinosa* was found to occupy the greatest area. It was followed by the total of spirorbid worms, and then in turn by *Steginoporella magnifica* and *Crassimarginatella papulifera*, and by the vermiform gastropod *Stephopoma roseum*. The bryozoans next in order were *Crepidacantha crinspina*, *Escharoides angela*, *Micropora mortenseni*, *Beania magellanica*, *Galeopsis polyporus*, *Eurystomella foraminigera*, *Diaperoecia* sp.,

Chaperia granulosa, *Fenestrulina* cf. *disjuncta*, *Odentionella cyclops*, *Exochella tricuspis*, *Bitectipora cincta*, *Hippopodinella adpressa*, *Calloporina angustipora*.

The zooecial structure of six species is shown in Fig. 13.16. For full taxonomic detail, the reader should consult Dennis Gordon's monographs.[2]

The two other bryozoan orders are more sparingly represented. In the Cyclostomata, with their rows of calcified tubular zooecia, Goat Island Bay has a *Diaperoecia* species up to 8 mm across the base, and smaller *Disporella novaehollandiae*, like tiny inverted mushrooms, with radiating zooecia. In *Eurystrotos* the small zooecia stand up like prickles. *Telopora* has tubular branches spreading from a pedestal, while *Tubulipora* and *Platonea* colonies form radiate spreads with dichotomous branching. The Order Ctenostomata is represented by an uncalcified *Bowerbankia*, forming a carpet pile of upright tubes rising from prostrate stolons.

At a miniature scale, Dennis Gordon has illustrated the epifauna living on empty cheilostome zooecia. As well as ciliate Protozoa taking shelter, there are at least four sorts of Foraminifera (*Dendronina*, *Gromia*, *Trochulina* and *Shepheardella*) together with a minute *Sycon* sponge, an halicarid mite, a prolecithophoran flatworm and the snail *Eatoniella olivacea*.

The small nodding animals known as Entoprocta were once classified with the bryozoans. Typified here by *Pedicillina whiteleggei* and a *Barensia* species, this small phylum (with one species in fresh water) has notable differences from the ectoprocts. Their cup-shaped zooecia are mounted on contractile stalks, and the tentacles curl up rather than retract. The water currents pass inward towards the centre of the tentacle spread. The anus opens *inside* the circlet (hence the name Entoprocta).

Towards the centre of the boulder base, surrounded

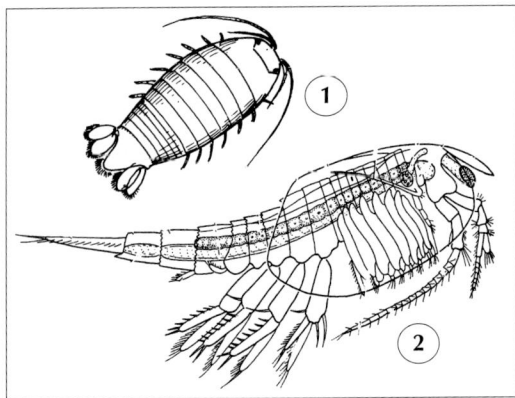

Fig. 13.20 Sand and boulder crustaceans
1 *Cirolana arcuata* 2 *Nebalia bipes.*

by the zone of Bryozoa, grow ascidians, sponges and serpulid worms. After the spirorbids, the most numerous tubeworms are *Filograna implexa*, forming a thin filigree away from the light or building up into strong ridges. The live worms occupy only the termini of the tubes, and new tubes arise by branching off from the parents. Their small branchial crowns stand out in bright red.

Larger serpulids include *Pomatoceros coeruleus*, *Hydroides elegans*, *Neovermilia sphaeropomatus* and *Galeolaria hystrix*. Common as well are the hard-cemented sand tubes, half-round in section, of the sabellariid *Paridanthyrsus quadricornus*.

The vermetid-like gastropod *Stephopoma roseum* (Fig. 13.18) is convergent in habit with the tubeworms, in clusters of translucent or faintly pink shells. The bristled opercula are easily apparent as they partly open. At Goat Island Bay, *Stephopoma* was found to rank fifth as a space occupier, preceded only by spirorbids and the three leading bryozoans.

The familiar sea-squirts of current-swept boulders are *Corella eumyota*, *Cnemidocarpa bicornuta*, *Asterocarpa coerulea* and *A. humilis*. Belonging like *Asterocarpa* to the Family Styelidae is the colonial ascidian, *Alloeocarpa minuta*, forming chains of flame-red zooids, 3 mm across.

The principal compound ascidians at Goat Island Bay can be tabulated by their families:

On bryozoan boulders the sponges are mostly of low profile. Encrusting *Pseudosuberites sulcatus* is common at Goat Island Bay, with *Timea aurantiaca* and *Adocia parietaloides*, along with red crusts of both *Plocamia novizelanicum* and *Microciona coccinea*. Orange *Tethya aurantium* or flesh-pink *T. australis* are attached at the boulder sides. Thin lattices are formed by the elastic *Chelonaplysilla violacea*. On most clean boulders white calcareous sponges will also feature, either as single cylinders of *Sycon ciliata* or small tubes of *Clathrina coriacea* from a creeping stolon.

Mobile species

Current-washed boulder shores, with their high diversity of molluscs, crustaceans and echinoderms, have traditionally been rich collecting places, whether for seafood or science. In the newly founded Marine Reserves — including the first one at Goat Island Bay — the boulders are protected from any sort of human pillage.

Small cling-fishes (Family **Gobiesocidae**) are regularly to be found under high energy boulders, attaching with their suction cup in strong currents. *Diplocrepis puniceus* is orange, while the blunt-snouted *Trachelochismus pinnulatus* has broken bands of brown or yellow upon a pink ground. The smallest of this family, averaging only 3 cm long, is *Dellichthys morelandi*, living as a commensal beneath *Evechinus chloroticus*. A scatter of electric blue spots cover its back and head.

Crustacea

The two largest crabs of the sublittoral fringe, the fast *Plagusia chabrus* and the slower *Notomithrax ursus*, live not beneath the boulders but among their brown algae, especially the bushy *Carpophyllum plumosum*. Confined under the boulders can be found the ungainly *Eurynolambrus australis*. With its wide dull pink carapace, this crab can be turned over to reveal its violet chelae and short legs splashed in red and gold. A much smaller crab under clean boulders is the dainty *Elamena producta*, sometimes commensal under the rim of a *Haliotis*. The smooth polygonal carapace is orange, with the rest of the body and legs black or wine-red.

The first crustacean to be encountered under current-swept boulders could be the rather heavily built shrimp *Alope spinifrons*. Sometimes gregarious, this species clings to the surface or walks about slowly when a boulder is turned, before taking off with a short leap. *Alope* is glassy and translucent, longitudinally striped with red and green. The chelae are tiny, but the male has very long third maxillipeds. There is a supraorbital spine as long as the eye.

Various 'worms'

As eye-catching as the nudibranchs are some of the flatworms and ribbon worms, so little studied in New Zealand that some common species still lack specific names. First and most lowly are the marine flatworms of the Order **Polycladida** (Phylum Platyhelminthes). Pride of place must go to the pseudoceratid species, up to 3 cm long. One is deep imperial purple or almost black, another wears the doctoral colours of crimson, edged with gold. Both swim by undulating their thin margins. So also does the oval *Thysanozoon brocchii*, grey with dense, pointed papillae. A species of leptoplanid found

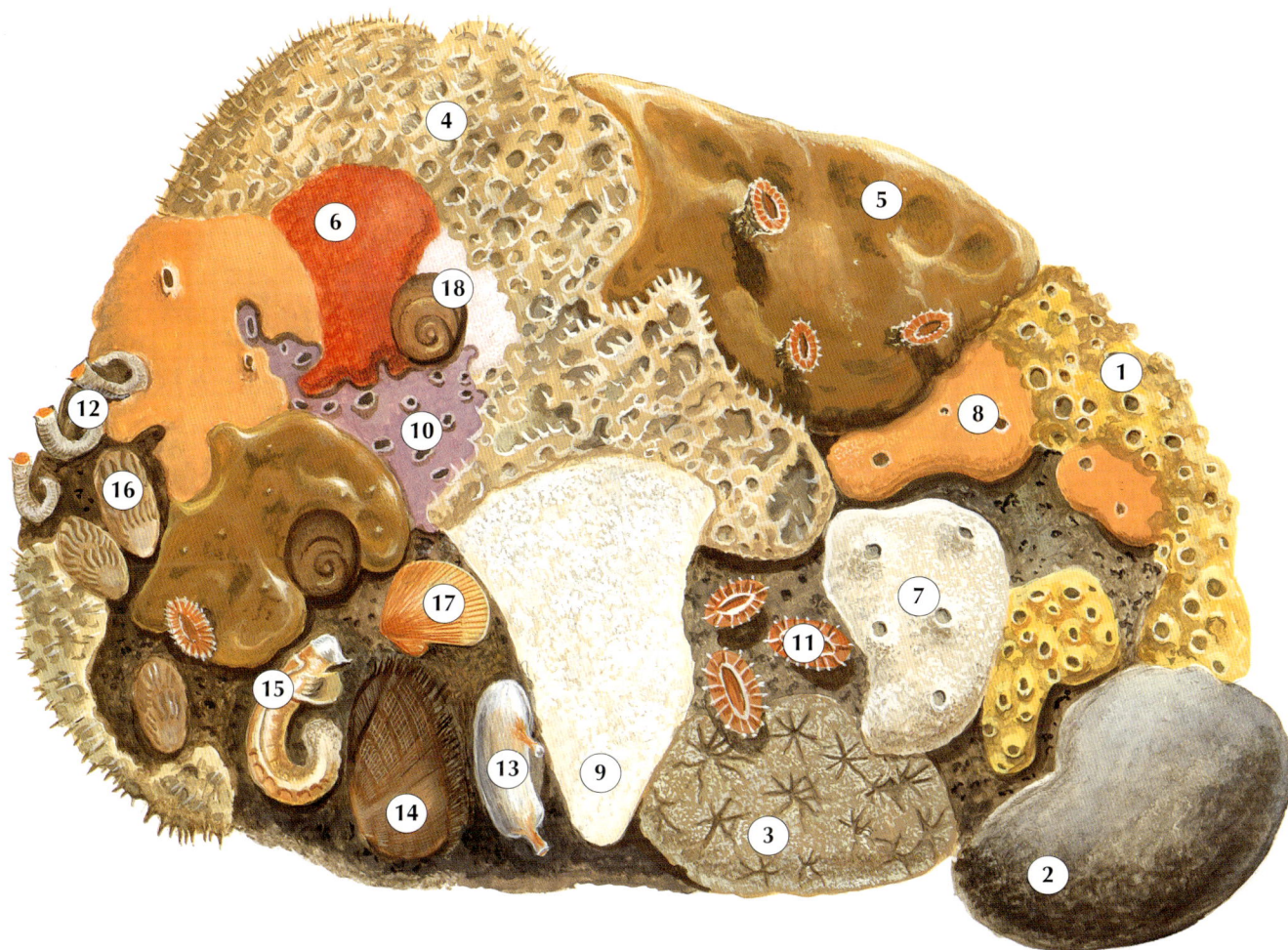

Fig. 13.21 Undersurface of a sponge boulder, Whangarei Harbour

1 *Cliona celata* 2 *Ancorina alata* 3 *Haliclona* sp. 4 *Ophlitaspongia reticulata* 5 *Tetrapocillon novaezealandiae* 6 *Microciona coccinea* 7, 8 *Halichondria* sp. 9 *Didemnum* sp. 10 a halichondriid 11 *Monomyces rubrum* 12 *Serpulorbis* sp. 13 *Corella eumyota* 14 *Barbatia novaezealandiae* 15 *Galeolaria hystrix* 16 *Maoricrypta costata* 17 *Chlamys zelandiae* 18 *Sigapatella novaezelandiae.* (Painting by Ron Cometti.)

under clean, low tidal boulders, is translucent white, light orange-netted and with mobile dorsal tentacles. A narrower bodied *Eurylepta* species is rust-speckled on a pale ground. The smallest and perhaps most charming of low-tidal polyclads is the milk-white, grey-speckled *Stylochoplana*, attaching to *Ecklonia*, and breaking into periodic bursts of swimming.

Like the flatworms, the ribbon worms (**Nemertea**) first attract notice by their colours. A *Tubulanus* species, 5 cm long, is reddish brown with a median dorsal orange stripe, and regular hoops of yellowish white. Beneath stones or twining in kelp holdfasts lives a slender nemertean, more than 25 cm long, and apricot-coloured with a white line at the sides. Under clean low

tidal stones there is a familiar species of *Amphiporus* brown above and steel-grey below, with a white head bar and pale forward tip. Our largest under-stones nemertean is a dull orange *Cerebratulus*, reaching 1.5 cm wide and 25 cm at full length.

Of the errant polychaete worms, one of the most colourful is a rich green *Acrocirrus trisectus* (Family **Cirratulidae**) growing to 15 cm long with each segment ringed in lime yellow. When picked up it extrudes an intense yellow dye. The long tentacular cirri are in *Acrocirrus* restricted to the four pairs of gills and two palps at the anterior end. These freely curl in, then spread out, with their ciliated grooves applied to the ground to pick up food particles for passage to the mouth.

Mollusca

The largest browsers of algae on high energy shores are the paua or ear-shells (**Haliotidae**), feeding most of all on succulent reds. Among gastropods they are notably fast-moving. The foot produces a double muscular wave, alternating on right and left, which makes for agility in advancing and turning. In the subtidal a paua will freely ascend to the top of a kelp. They can also take tight hold of the rock, and keeping sensory contact by a skirt of epipodial tentacles round the base of the foot.

The large *Haliotis iris* (up to 12 cm) is black-footed, with the shell peristome a flat rim, and the interior brilliantly blue and green. The yellow-footed paua, *Haliotis australis* (up to 9 cm), is corrugated, with a pink and silvery iridescence, and the peristome continuing inside up the spire. The much smaller *Haliotis virginea* is more brightly patterned in green and pink.

The prosobranchs belonging to Neogastropoda are carnivores and usually diet-specialists. The *Buccinulum* species are the boulder counterparts of the more visible *Cominella*, and are thought to be bryozoan grazers. The largest (up to 3 cm) is *B. linea*, with widely spiral lines. Most numerous is the small *B. vittatum*, spirally ridged and purple and brown banded in the north and in the south smoother, and more distantly lined. The small fasciolariid *Taron dubius* (Fig. 13.8) attacks tubeworms and *Stephopoma*.

The elegant little margin snails (**Marginellidae**) have three species that need mention. The dark-banded *Volvarinella cairoma* and the white urn-shaped *Sinuginella pygmea* are found in coralline turf as well as under stones. Confined to clean boulders in the north is the exquisite *Haloginella mustelina*, like polished mahogany with chestnut and white banding.

In the same places live two small snails of the **Columbellidae**: *Zemitrella chaova* and *Paxula paxillus*, both fusiform and dark brown. The latter is often found on brown algae with the brooding anemone *Cricophorus nutrix* (Fig. 3.7). To the **Mitridae** belongs the similar, but ribbed, *Austromitra rubiginosa*, an ascidian-feeder that lays its egg capsules in the tests of *Corella eumyota* (Fig. 12.4).

Allied to the cowries is the small *Trivia merces*, with the white, pink-blotched shell finely ribbed, covered in life by the dark mantle. Like the **Triviidae** the world over, it is assumed to graze on compound ascidians, implanting its eggs in their tests.

The **Lamellariidae** are also ascidian specialists, at first view slug-like, but easily recognised as prosobranchs from the glassy internal shell, head and mantle cavity. The southern *Lamellaria cerebroides* (4–6 cm long) has soft yellowish mammillae, while the smaller *L. ophione* is smooth and blotched with greyish white.

Smaller molluscs

Clean cobbles up to the middle eulittoral, freely scattered with spirorbids, yield a wealth of small molluscs. The bivalves include (Fig. 13.18) the ark shell, *Arca sociella* (confined to east Northland), a diminutive bearded mussel, *Trimusculus barbatus*, the kidney-shaped *Borniola reniformis* (Erycinidae) and the small, strongly ribbed *Cardita brookesi*. All are byssus-attached.

The most abundant gastropod is the rissoid-like *Eatoniella olivacea*. Some of its empty shells are occupied by the miniature 'hermit-crab', *Pagurapseudes*, distantly related to the isopods. Smaller still are the under-stones rissoids: sculptured *Merelina* and smooth *Pisinna* species and — still more minute — the seed-like *Anabathron hedleyi* and *Amphithalmus semen*. The tusk shell, *Caecum digitulum* (Fig. 13.18), is found under the same smooth cobbles. Larger than the true rissoids, the axially ribbed *Rissoina chathamensis* lives under boulders on coarse sand, where it gleans live Foraminifera.

The slender-spired **Cerithiopsidae** and **Triphoridae** both feed on siliceous sponges by thrusting the proboscis into their ostia, or breaking into the tissues direct. The Cerithiopsidae have a normal, right-handed coil, with the common species *Alipta crenistria*, on the sponge *Raspailia*, and *Seila cincta* with *Cliona*. The Triphoridae are by contrast sinistrally coiled with aperture to the left, as in *Sagenotriphora ampulla* and *Metaxia exaltata*, grazing on encrusting sponges.

The **Pyramidellidae** (sometimes classed with shelled opisthobranchs and evidently remote from the rest of the prosobranchs) are tiny ectoparasites found living only by searching their known hosts. Most of New Zealand's species were described from dead shells in sand, and some go back to the Pliocene as micro-fossils. Pyramidellids extract the body fluids from many sorts of invertebrate host, specialising on echinoderms, coelenterates, sedentary polychaetes or molluscs. The capillary proboscis carries a salivary duct and sucking tube, with hard chitinous stylets derived from the jaws, the radula being entirely lost. The pyramidellid genera are each distinctive in shape and ornament. *Odostomia* is short and smooth, with an apertural tooth, *Eulimella* tall and unsculpted. In *Chemnitzea* axial ribs may extend over the base, while in *Striacana*, they stop on the body whorl to leave the base smooth.

In close shelter

On higher energy boulder shores, fine particles are scarce or altogether swept away. In increasing shelter, clean sand may accumulate in the depressions where flat or spherical boulders have found stability. Flat slabs may lie

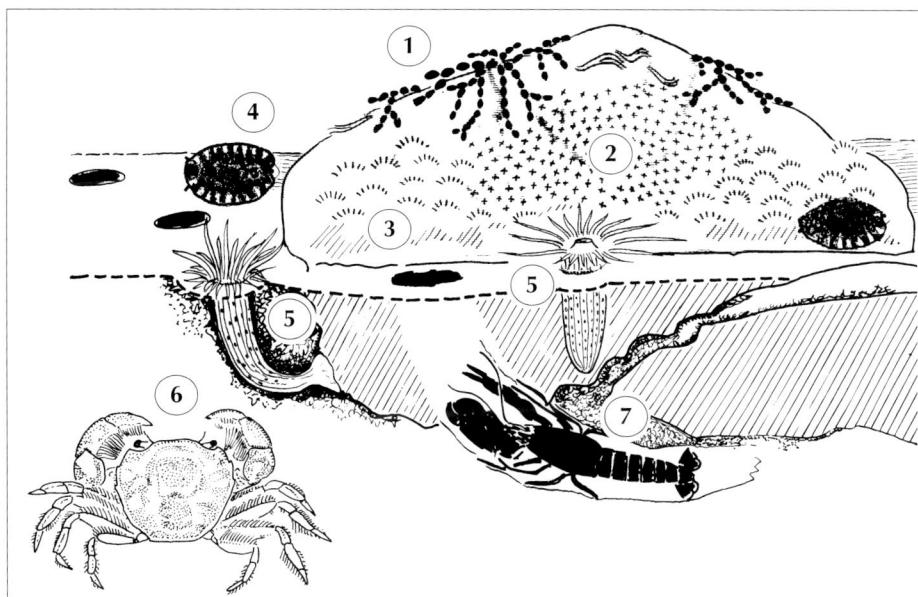

Fig. 13.22 Hauraki Gulf: mid-eulittoral boulder, part-buried in silty sand
1 *Hormosira banksii*
2 *Austrominius modestus*
3 coralline turf 4 *Onchidella nigricans* 5 *Anthopleura aureoradiata* 6 *Hemigrapsus crenulatus* 7 *Alpheus richardsoni*.

upon clean sand so that the organisms beneath will live in burrows closed off by the boulder from above.

On sediments

Rock slabs lying upon sand may thus be lifted up to reveal the pinkish brown polychaete *Platynereis australis*, gliding through its superficial galleries. Where there is more silt or a stiffening with clay, the euniciid *Marphysa depressa* may also appear, grey with red blood-gills.

In ultimate shelter, blacker anaerobic sediments offer a habitat for the cirratulid *Timarete anchylochaeta*. The same places, with the remnants of algal decay, might also be searched for the pallid white *Nebalia*, a small 'phyllocaridan' shrimp with a bivalved carapace, being one of the most primitive and isolated of the malacostracan Crustacea.

Clean, fine-grained sand is the milieu of the smooth, streamlined isopod, a *Cirolana* species. With the beat of their pleopods these swim rapidly in the groundwater or bulldoze with the smooth head through fluid sand, jetting out particles behind by the action of the pleopodal limbs.

Boulders deeper-immersed in muddy sand carry few or no attached animals underneath. But around their periphery they may have a ring of the greyish brown anemone *Anthopleura aureoradiata* attached just beneath the sand level. The tentacle circlets open out flush with the surface, displaying the oral disc radially streaked with brown and gold. This anemone is known to incorporate in its mesenteries photosynthetic algal symbionts (*zooxanthellae*), though their role is not yet fully understood.

Over the surrounding soft mud, the rubbery pulmonate slug *Onchidella nigricans* creeps about freely.

Galleries roofed by a boulder may be occupied by the snapping shrimps, *Alpheus richardsoni*, betraying their presence by a sound like hailstones from the clicking of the enlarged left chelae. Lowest down on the shore, the anomuran *Upogebia hirtifrons* is sometimes found under stones in the silty sand. Up to 5 cm long, this shrimp is pale orange with large subchelate first legs. Irrigation is kept up by the abdominal pleopods, with food particles intercepted by nets of crossed setae on the first walking legs.

Beneath silt-lying boulders the common crab is *Hemigrapsus crenulatus*, varying from grey green to buff or orange brown. It relies for food on the fine particles gathered by the tufts inside the 'palms' of the chelae. A similar habit is found in New Zealand's only ocypodid crab, *Macrophthalmus hirtipes*, which lives not under boulders but among the sea-grass *Zostera*. The flat pillbox crab, *Halicarcinus cooki*, with pilose carapace and long fragile legs, prefers the same semi-fluid mud.

On black sand

In the sublittoral fringe of low energy shores some boulders rest in sand rendered black and anaerobic by organic decay. Overshaded by *Carpophyllum flexuosum* or *Ecklonia radiata* their communities of sponges and compound ascidians have shifted up to the sides. Underneath there are only a few species tolerant of anoxic sediments. The boulder periphery (Fig. 13.23) may have half a dozen milk-white anemones, *Isoparactis ferax*. Mostly sand-buried, the peduncle attaches to the rock, and the brown-rayed oral disc displays slender tentacles at the surface.

Three kinds of polychaete worms live under the same boulders, with their feeding tentacles directed radially outward through the surrounding sediment. The tere-

Fig. 13.23 Milford Reef, Auckland: a boulder on anaerobic sediment

(a) *Corallina* turf (b) *Carpophyllum flexuosum* (c) ascidians and sponges (d) black sediment with anemone *Isoparactis ferax*. 1 *Isoparactis ferax* 2 cirratulid, *Timarete anchylochaeta* 3 *Terebella plagiostoma* 4 *Leptochiton inquinatus* 5 *Ischnochiton maorianus* 6 *Flabelligera affinis* 7 *Penion sulcatus* 8 *Cominella quoyana* 9 *Upogebia danai* 10 the same, in burrow roofed by boulder 11 *Cancer novaezelandiae* 12 *Flabelligera affinis*, detail of head.

bellid *Terebella plagiostoma* attaches to the rock in its shell-studded parchment tube. *Timarete anchylochaeta*, a bronze-coloured cirratulid, has no secreted covering; its blood-red feeding and respiratory tentacles radiate out like those of *Terebella* towards the boulder edge.

The **Chlorhaemidae** are a small family of polychaetes,

'green-blooded' from the presence of the pigment *chlorocruorin*. *Flabelligera affinis* is found on black, anoxic surfaces, with its yellowish green body wrapped in flaccid mucilage like a bag of jelly. The worm progresses by peristalsis with the help of setae projecting from this sheath. The head segments carry a cage of golden setae

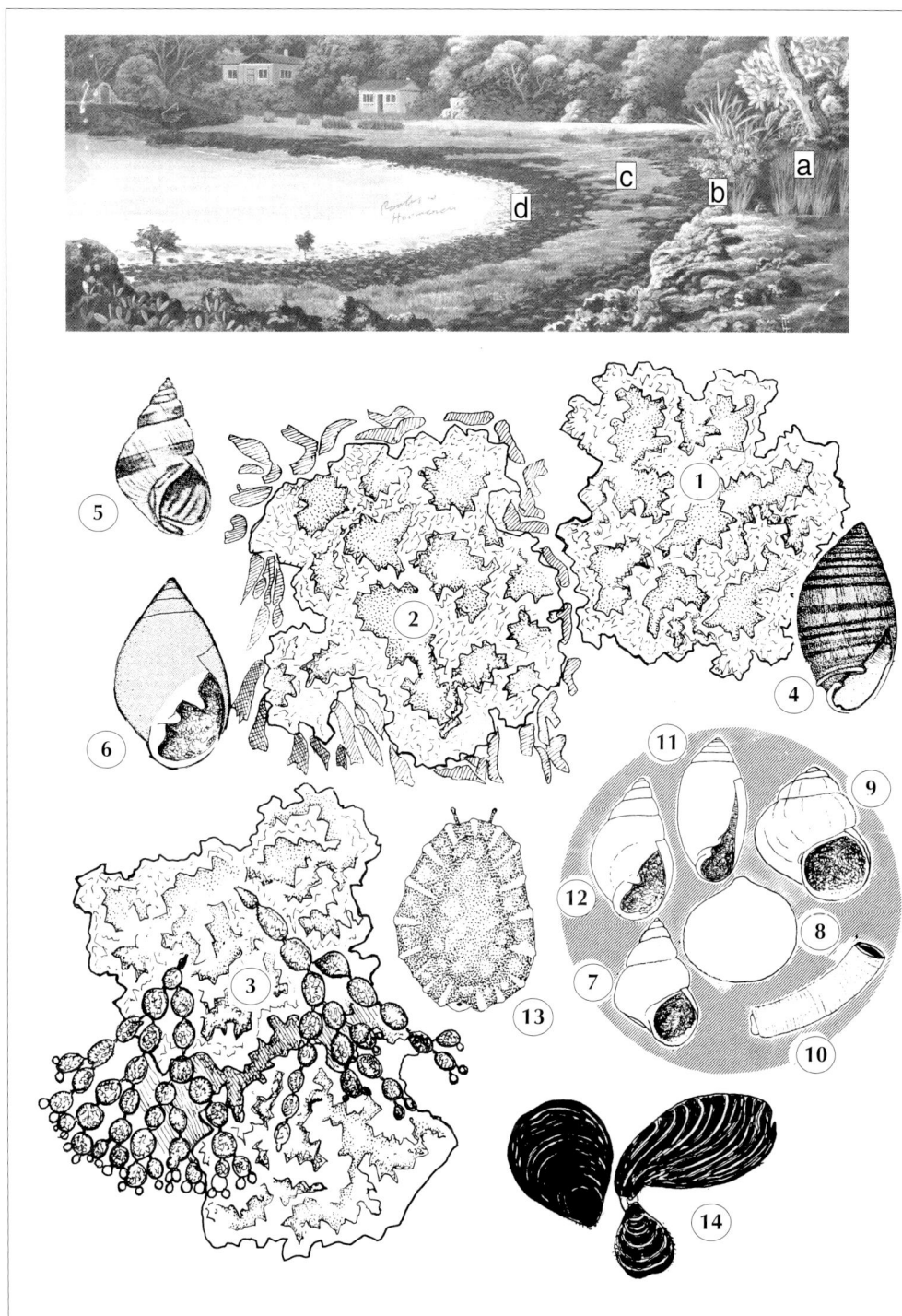

Fig. 13.24 Rangitoto Island, Auckland: scoriaceous boulders lying over mud

(above) communities of the sheltered shore (painting by Ron Cometti)

a–b flax and *Leptocarpus*; b–c *Sarcocornia* belt; c–d eulittoral stretch with boulders on mud

(below) three successive boulders and their species (1–14) 1 supralittoral boulder 2 mid-eulittoral boulder in decaying algal wrack 3 lower eulittoral boulder with *Hormosira banksii* 4 *Ophicardelus costellaris* 5 *Austrolittorina antipoda* 6 *Marinula filholi* 7 *Assiminea vulgaris* 8 *Lasaea maoria* 9 *Suterilla neozelandica* 10 *Caecum digitulum* 11 *Microtralia occidentalis* 12 *Leuconopsis obsoleta* 13 *Onchidella nigricans* 14 *Calloria inconspicua*.

enclosing a collar with four rows of tentacular gills, and two grooved palps. Gill currents bring in fine particles to be caught in the mucus lining of the bristle cage, from which the palps carry streams of food to the mouth. Detritus may also be directly sucked in by the mouth.

Chitons are not only adapted like limpets for strong water flow; some species may protectively withdraw beneath boulders into sediment. *Ischnochiton maorianus* has the photoreceptor 'shell-eyes' well developed and is remarkable for the high speed at which it makes

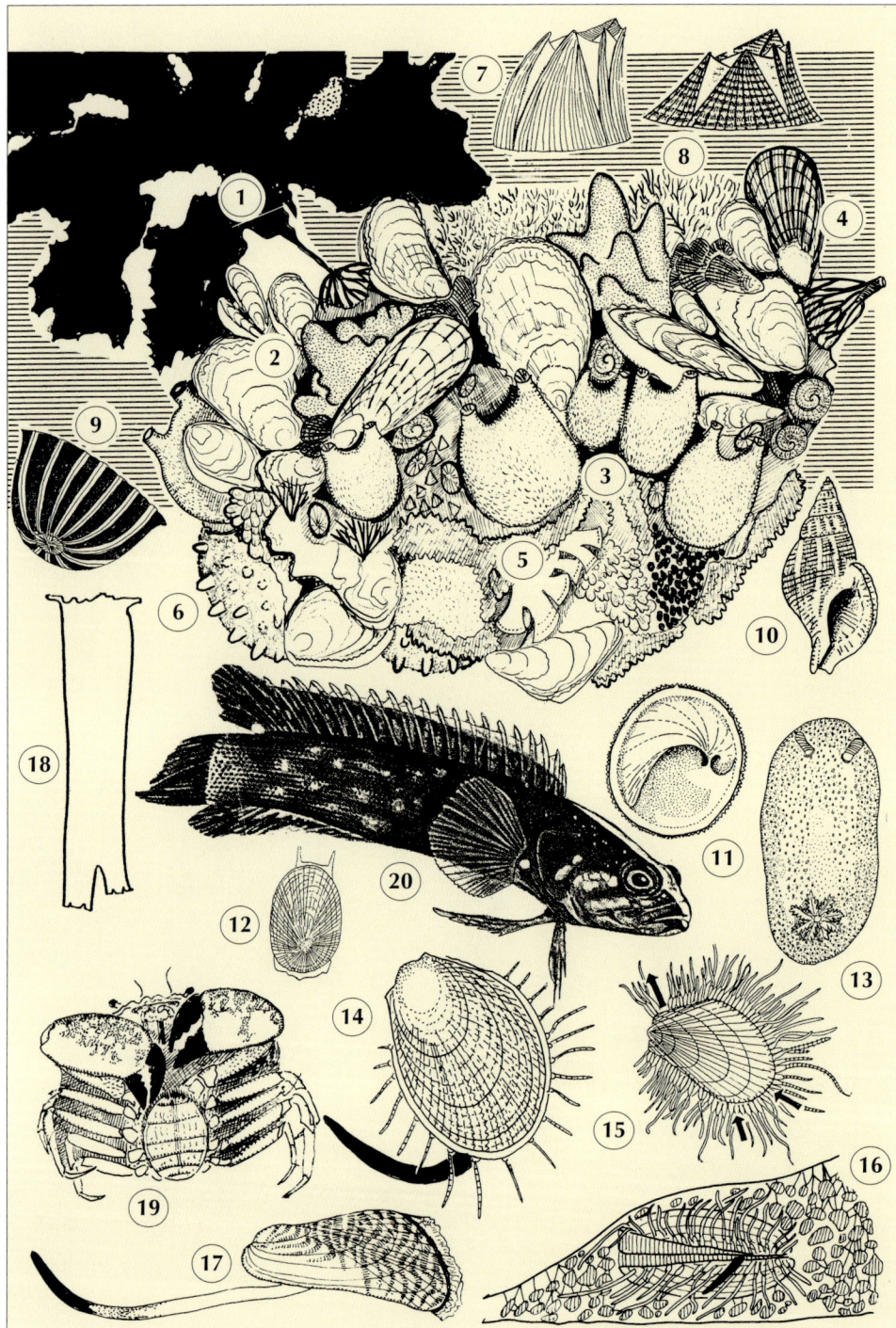

Fig. 13.25 Black Reef (Te Tokaroa Reef), Westmere, Auckland: boulder community in high shelter
1 *Ecklonia radiata* (high-sheltered form) 2 *Ostrea lutaria* 3 *Microcosmus squamiger* 4 *Perna canaliculus*
5 *Gregarinidra* sp. 6 *Suberites perfectus* 7, 8 *Balanus amphitrite* 9 *Haliplanella lineata* 10 *Buccinulum vittatum*
11 *Sigapatella novaezelandiae* 12 *Tugali suteri* 13 *Dendrodoris citrina* 14 *Limaria orientalis* 15 the same, swimming and
16, in nest of byssus threads and sand particles 17 *Musculista senhousia* 18 *Ciona intestinalis* 19 *Pilumnopeus serratifrons*
20 *Acanthoclinus fuscus*.

for the dark when a rock is taken up, turning quickly or — in the strongest light — detaching from the rock to fold up double. Long and narrowly oval, this chiton is dirty grey-green to metallic blue when adult, but immature specimens can be pink with a black girdle, or dark with a median white streak. *Ischnochiton maorianus* has

been shown to be highly photosensitive and positively geotactic. It has a lunar periodicity and appears on the upper surface only in low moonlight. With its poor resistance to desiccation, it never exposes itself to the air.

The smallest New Zealand chiton, *Leptochiton inquinatus* (1.5 cm long), is biscuit-brown with a red spot at each valve apex. It has the same preference as *Ischnochiton* for withdrawing under a boulder into sediment. Belonging to the most primitive chiton family, **Lepidopleuridae**, this easily available species could well repay morphological and behavioural study.

One chiton, the subtidal *Pseudotonicia cuneata*, can move with ease through sand and shell debris. The foot is narrow and the girdle thin and flexible, elevated at any point to improvise an inhalant aperture, or to expose the gills along one side.

A crab near sedimented boulders, less common at Auckland in recent years, is *Cancer novaezelandiae* (Fig. 13.23), living where it can submerge into fine sand, with only the back exposed. The wide, rust-red carapace, up to 9 cm across, expands over the slender legs, and is bow-fronted and crenellated like a pie-crust around the edge. *Cancer* is a slow crab. It holds the legs tightly beneath the chest when picked up, and makes no attempt at escape. Cockles and thin-shelled *Ostrea lutaria* are opened by chipping the edges until a chela can be inserted.

The whelk *Penion sulcatus* may be found in the same fine sublittoral sand. Presumed to feed on the bivalve *Dosina zelandica*, it has also been found with its proboscis inserted into the rock-borings of pholads.

Basalt on mud

The community under smooth, un-encrusted boulders of the littoral fringe has been well shown by the high-ramped boulder beaches of the Coromandel Peninsula (Fig. 11.10). A rather different regime is found where boulders of pitted basalt lava come to rest on a soft substrate in close shelter. The high tidal lava field at **Rangitoto Island**, Auckland (Fig. 13.24), was the site of a classic early study by Powell in the 1930s. Blocks of scoriaceous basalt lie permanently in a matrix of decaying algae, mixed with a sticky organic mud. The surrounding wrack gives a continuing protection from high temperature and desiccation.

The small gastropods gleaning detritus under high level boulders belong — as we have already seen — to two principal families, the prosobranch **Assimineidae** and the pulmonate **Ellobiidae**. Both show their complete species range at Rangitoto.

The four ellobiid snails have their own niche differences. Highest on the shore under basalt or among the succulent glasswort (*Sarcocornia quinqueflora*) occurs *Ophicardelus costellaris*. Numerous *Austrolittorina antipoda* live on the boulder tops. *Ophicardelus* also creeps about freely under driftwood and halophytic plants, feeding upon organic deposits or green filamentous algae such as *Rhizoclonium*. Its ovoid shell reaches 10 mm tall, grey and brown-banded, with the aperture

toothed as in Fig. 13.24. A little lower on the shore, the smaller *Marinula filholi*, with reddish brown shell and the animal pure white, grazes upon the paste of decaying algae.

Two small, thin-shelled ellobiids, only 3 mm long, live deeper down, in pits in the boulders below the mudline. Both are white and permanently confined to the dark, *Leuconopsis obsoleta* being urn-shaped and *Microtralia occidentalis* more cylindrical. The latter was first discovered in New Zealand on this shore, and was first named by Powell *Rangitotoa insularis*.

Under cleaner stones may be found the two species of Assimineidae, *Suterilla neozelandica* (Fig. 13.24) and its smaller relative *Assiminea vulgaris*. In similar places lives the minute tusk shell *Caecum digitulum* (Fig. 13.18) a horn-shaped gastropod with a tubular shell only 2 mm long, which is pulled forward as the foot advances in steps. Diatoms and morsels of similar size are gleaned individually with the radula as the proboscis searches the ground. In greater numbers, clustering in hollows under basalt blocks, is one of our smallest bivalves, *Lasaea maoria*, attaching with a single byssus thread, or creeping by extension of the narrow foot.

Finally, at Rangitoto may be found in open spaces under mud-submerged boulders, its typical habitat of total dark, the pale red, pedunculate brachiopod *Calloria inconspicua*.

In extreme shelter

In land-locked stretches, as at **Westmere** in the upper Waitemata Harbour (Fig. 13.25), the intertidal platform is over-laid with mud, and boulder-based communities can hardly be said to exist. Instead sessile animals may themselves build up diverse clumps, starting from a single oyster shell (*Ostrea lutaria*) with other shells fused on. An early recruit is the sea-squirt *Microcosmus squamiger*, and sponges finally establish round the sides and on top. With waves or currents negligible, these stable oyster-ascidian aggregates collect a complex of attached species raised just above the sediment line.

From the interspaces, the steel-blue polychaete worm *Perinereis amblyodonta* reacts vigorously on disturbance. The dark green *Eulalia microphylla* threads a slower course, coming out to feed on shaded surfaces. At deeper levels live the euniciid *Marphysa depressa*, bright-coloured terebellids and finally cirratulids lying in pockets of sediment.

Sedentary molluscs include the saucer limpet, *Sigapatella novaezelandiae*, and wafer-like saddle oysters, *Anomia trigonopsis*. The silt-tolerant chitons are *Acanthochitona zelandica*, *Ischnochiton maorianus* and *Leptochiton inquinatus*.

263

The bryozoans *Watersipora subtorquata* and *Bugula flabellata* are found on almost every cluster. Common too may be the fleshy lobules of the ctenostome *Gregarinidra*. As well as *Microcosmus*, the simple ascidians regularly found include *Corella eumyota* and *Asterocarpa coerulea*. The fleshy compound ascidian *Aplidium phortax* invests the bases of oysters; cushions of *Hypsistozoa* and rosettes of *Botryllus* can exist close to the silt level.

A few sorts of gastropod thrive exceedingly in silty habitats under extreme shelter. The ducksbill limpet, *Scutus antipodes*, and the cat's-eye, *Turbo smaragdus*, reach exceptional size. Carnivorous snails include *Cominella virgata*, *C. adspersa*, with — lowest down — *Taron dubius*, *Buccinulum vittatum* and *Muricopsis octogonus*. The pale yellow nudibranch, *Dendrodoris citrina*, probably a regular predator on *Microcosmus*, is particularly common at Westmere.

Slow to move when disturbed is the crab *Pilumnopeus serratifrons*, a xanthid with a bald, ill-favoured look, holding the chelipeds to the chest when picked up. The chelae are coloured burnt sienna with the dark fixed finger slightly deflexed. *Hemigrapsus crenulatus* sometimes occurs near silt with *Pilumnopeus*. Under cleaner rocks near the middle shore black-marbled *Hemigrapsus edwardsi* often turn up.

On high-silted reefs, the sponges call for special notice, flourishing in turbid waters wherever the algae fail with loss of light. The soft masses of the crumb o' bread sponge, *Halichondria moorei*, lend their orange-pink colouring to shaded sides and overhangs. The general grey is relieved by the scarlet of *Microciona* and *Plocamia* and the yellow of *Halichondria* and *Haliclona*. Some open coast sponges are equally abundant in quiet waters, especially the cherry-coloured knobs and mounds of *Crella incrustans*. Globular sponges, orange *Tethya aurantium* and chocolate *Suberites perfectus* appear at every hand. Dull purplish hemispheres of *Aaptos confertus* grow under shade.

CHAPTER FOURTEEN
Overhangs, Wharves and Caves

Upon every shore where boulder cover can lie stable, we find the shade-loving fauna called 'sciaphilic'. On the richest shores, with wave action sufficient to remove silt and bring food currents to filter feeders, there is a foretaste of the diversity and bright colour widely spread out beyond low water. The living space under the boulders, though rich in species, is, however, a restricted one. Its narrow spaces are washed by quite strong laminar currents, sweeping beneath exposed rock slabs, or a more feeble ebb and flow in quieter waters. Conditions are ideally suited for such groups as barnacles and spirorbid polychaetes, as well as bryozoans, compound ascidians and sponges.

We find sciaphilic communities made up of much larger life-forms where shaded space is more widely open. Such habitats include vertical or overhung rock-faces on open coasts, with constant movement of clean water and virtual freedom from sediment. They may be referred to as **open shade** habitats, having important differences from the constricted spaces under boulders.

First, where some light is still admitted, certain algae,

particularly the encrusting reds, can find their place within the shaded zonation. Secondly, with their wider available space, most of the major life-forms seen on boulders can be given more ample expression (see Fig. 14.1). No longer held back to a crustose spread swept by laminar currents, they can grow taller, with tentacle systems able to exploit a wider feeding orbit.

Fig. 14.1 Some life-forms of a wharf stringer and under boulders

(a) red sponges *Raspailia topsenti* (**top**) and *Microciona coccinea* (**bottom**); (b) ctenostome bryozoans *Zoobotryon* (**top**) and *Alcyonidium* (**bottom**); (c) anascan cheilostome bryozoans *Caberea* (**top**) and *Beania* (**bottom**); (d) majid crabs *Notomithrax peronii* (**top**) and *N. minor* (**bottom**); (e) athecate hydroids *Pennaria* (**top**) and *Hydractinia* (**bottom**); (f) compound ascidians *Aplidium* (**top**) and *Didemnum* (**bottom**); (g) simple ascidians *Ciona intestinalis* (**top**) and *Pyura rugata* (**bottom**).

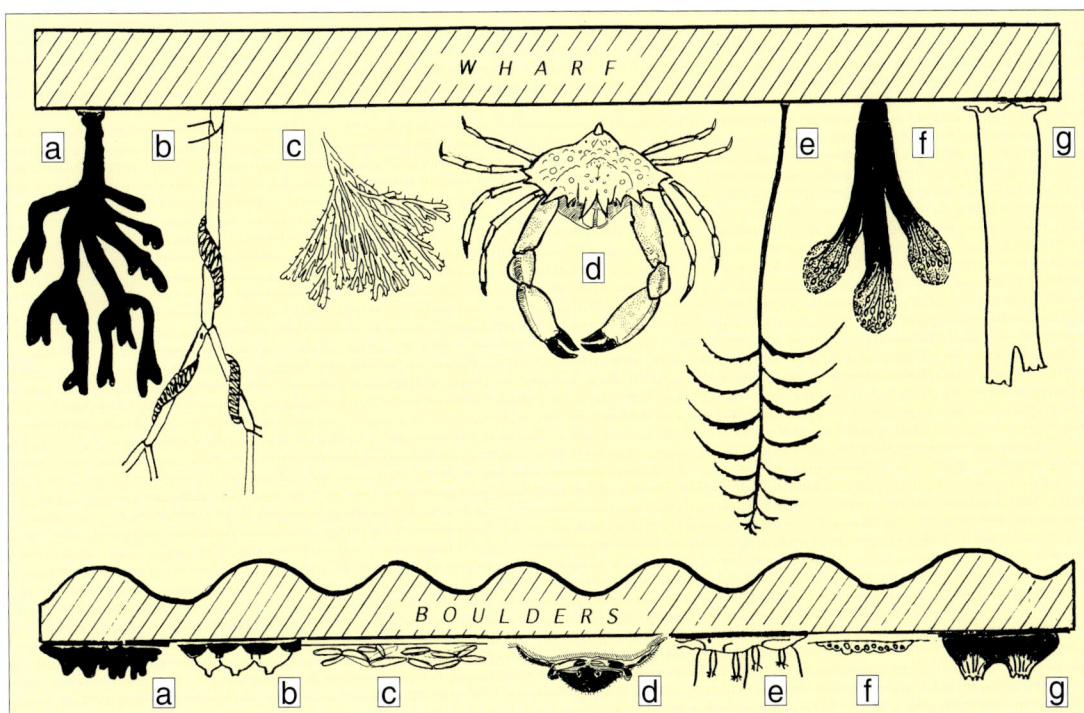

Fig. 14.2 Cnidarians and sponges in overhangs and caves
An intertidal rock-face in moderate shelter is shaded by an overarching ledge, and from the eulittoral downward the shaded surface is rich in cnidarians, ascidians and sponges.

Zonation beneath overhang
a–b upper and middle eulittoral overhang with *Actinia tenebrosa*; b–c mid eulittoral *Actinothoe albocincta* and *Spirobranchus cariniferus*, with some *Corynactis australis*; c–d lower eulittoral with *Aplidium phortax* and *Alcyonium aurantiacum*; d–e upper and lower sublittoral fringe with ascidians and sponges.
1 *Actinia tenebrosa* 2 *Diadumene neozelanica* 3 *Actinothoe albocincta* 4 *Corynactis australis* 5, 6 *Culicea rubeola*
7 *Isocradactis magna* 8 *Oulactis muscosa* 9 *Tethya aurantium* 10 *Cliona celata*. (By courtesy of Michael Miller.)

Greater space has allowed special prominence to the cnidarian polyps, both the hydrozoans and also the anemones and their kindred. Cnidarians feed essentially like other sessile life-forms, by entrapping moving organisms from the water. But instead of relying on slow ciliary currents they are efficient hunters, micro-carnivores capturing with their tentacles relatively larger items of zooplankton. Glancing against a moving tentacle, a small crustacean or larval fish is disabled by batteries of stinging *nematocysts*, borne in the *cnidoblast* cells. Special nematocyst threads also function in adhesion and lasso-type prehension.

Eulittoral overhang

We may show first the shaded surge overhangs, as illustrated for **Piha**, west Auckland (Fig. 14.2). Here the swaying curtains of brown algae have given place to bright sponges, fully visible each time a wave draws back. The orange and yellow sponges are of the three sorts by now familiar to us. Furthest from light there are clusters of orange golf-balls, *Tethya aurantium*, and also the smoother, flesh-pink *T. australis*.

Scatters of the yellow eroding sponge *Cliona celata* finally coalesce into heavy crusts. Tangerine *Polymastia granulosa* forms thick mammillose sheets, while the heavy rinds of *P. fusca* are chocolate brown with a greenish over-tinge.

The walk up a surge gully at low tide will reveal the zoning patterns of full shade. In the more open reaches, a grotto of red algae may first be seen. With greater overhang, all the algae but the coralline overlay will have disappeared, to be replaced by colourful encrusting animals, first anemones, next downward ascidians, and lowest of all sponges.

At the lowest level in the sublittoral fringe the rock surface may often have become recessed into concavities like miniature caves. The zonation spread out on their walls can be lit with a torch to show species almost identical with those on sublittoral fringe boulders. On the cave floor, there may be fixed boulders in effect turned upside down with the zonules normally seen underneath now convexly displayed on the boulder tops.

The zoned interior of such a small cave was shown at Whangarei Heads. Lowest down or over the floor are the massive steel-grey sponges *Ancorina alata*. Next appears a zone of pale compound ascidian heads (*Aplidium*) among orange golf-ball sponges (*Tethya*). Above this comes a crust of the red sponge *Microciona*, broken by the cup coral *Culicia rubeola*.

At the next level up, on boulders or cave walls, is the black bryozoan *Watersipora subtorquata*. Beyond this, at middle eulittoral level, the cave roof is clad with rock oysters, *Crassostrea glomerata*. At spots where enough light can enter, *Watersipora* gives place to pink coralline paint, with *Pomatoceros* and occasionally the alcyonacean *Clavularia*.

On the low tidal surface we have called open shade, two important groups are rich and diverse. The opistobranch molluscs have already been described, with their feeding habits, in Chapter 12. The anthozoan cnidarians can now be reviewed in a little more detail.

Anthozoa

Aside from the Subclass **Hexacorallia**, which contains the anemones, the Anthozoa include the colonial polyp-systems of the Subclass **Octocorallia**. In the first order, the Alcyonacea, the small polyps are embedded in a coenenchyme of stiff mesogloea, permeated with calcareous spicules. These invariably have eight tentacles fringed with pinnae, with the internal cavity much simpler than in anemones, divided only by eight folds or *mesenteries*. The soft coral *Alcyonium aurantiacum*, traditionally called 'dead men's fingers', lives below the anemone level and among ascidians. The club-shaped branches are firm but compressible, and white, star-like polyps stud their orange surface.

In still deeper shade or under boulders a smaller octocoral, *Clavularia*, may spread over the surface in violet to brown patches, up to 10 cm across. Their stolons, no more than a millimetre thick, give off small polyps up to 5 mm long, with pinnate tentacles.

Anemones

The sea anemone of the Subclass **Actiniaria** have a broad oral disc sharply differentiated from the column and fringed with hollow tentacles that may number several hundreds. Anemone colour patterns are strikingly beautiful and, as Alan Stephenson showed in his finely illustrated Ray Society monograph, must be relied upon for identification. The slit-like mouth is the radial body's only bilateral element. At either end it has a ciliated channel or siphonoglyph, carrying a continuous water current into the coelenteric cavity.

The anemone's whole system of symmetry is based primarily on the order of the mesenteries. No longer eight simple partitions as in Alcyonaria, these now form a complex of radial folds partly dividing the coelenteron. Six primary mesenteries are the largest, with their inner edges attached to the tubular stomodaeum. The rest of the margin is thickened to a mesenteric filament, carrying ciliated and glandular cells and nematocysts. The edges are sometimes drawn out into whip-like filaments or 'acontia', loaded with nematocysts like the tentacles. These are extruded by way of the mouth or through pores in the column called 'cinclides'. Corresponding to each primary mesentery, a large primary tentacle arises from the disc. Between the primary mesenteries is a cycle of six secondary mesenteries, with a corresponding cycle of tentacles. To either side of these are members of a tertiary cycle of mesentery pairs and tentacles, generally with quaternary and even smaller and more numerous lower cycles inserted in turn.

The longest emersed anemone, pendent under shaded ledges at about half-tide, is the red *Actinia tenebrosa*, as at home here as in caves or under mid-tidal boulders. At

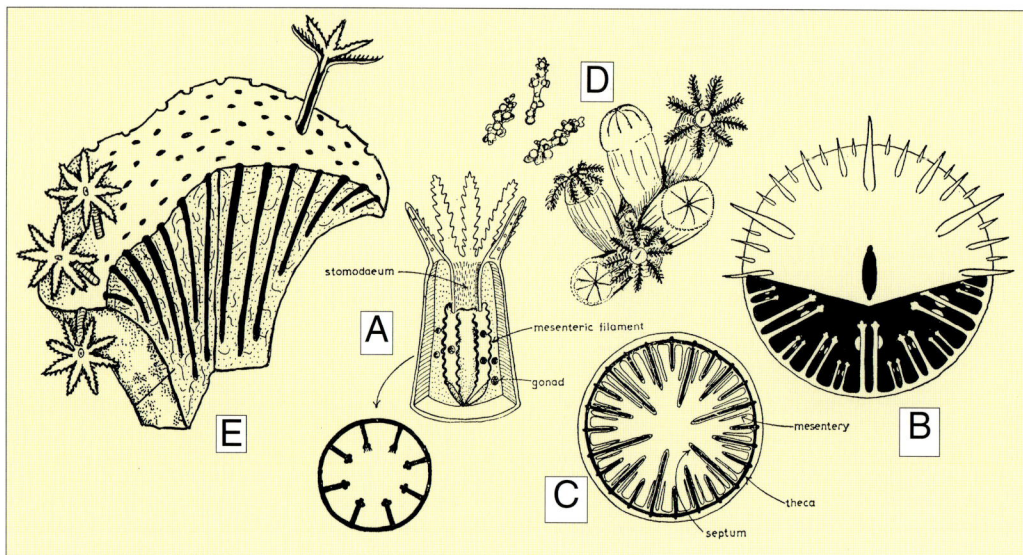

Fig. 14.3 Structure of Anthozoa
Diagrams of the arrangement of tentacles and mesenteries in:
A. an alcyonarian polyp
B. an actiniarian anemone with oral disc part-removed to reveal mesenteries
C. septa of calcareous theca, shown black, with intervening mesenteries, in a solitary coral of the *Scleractinia*
D. an intertidal species of *Clavularia* with a group of spicules highly magnified
E. portion of a colony of *Alcyonium*, with small polyps studding a stiff coenenchyme.

a lower level live the two anemones of deeper shade, *Actinothoe albocincta* and *Diadumene neozelanica*. Typical throughout New Zealand beneath overhangs, both continue subtidally. *Actinothoe* is brown or orange, with the column striped in pale green; the red oral disc is fringed with white tentacles. The column is pierced with small holes (*cinclides*) from which can emerge long, nematocyst-bearing filaments (*acontia*) produced from the edges of the mesenteries. These may also appear through the mouth. *Diadumene* is more closely striped in bluish grey and white, with deep brown tentacles around a bright orange disc, circled with an inner ring of much larger 'catch tentacles'.

An attractive actinian, seen only on open coasts, is the jewel anemone, *Corynactis australis*. This is a gregarious species, clustering in sheets under ledges and reproducing by fission, so that imperfectly separated polyps are often found. The column is rose-red to salmon pink, occasionally paler. The transparent tentacles, each tipped with a brown knob, increase in length towards the margin. The column is set on a calcareous basal plate, like the beginnings of a cup; some specialists would accordingly classify *Corynactis* near the corals.

The cup corals are represented on shaded surfaces by *Culicea rubeola*, with cylindrical corallites, 4.5 mm in diameter, linked together by thick, calcified rhizomes.

Towards low water we may find the largest anemone, *Isocradactis magna*, generally part-buried in fissures with clean sand. The disc may reach up to 10 cm across, lightly coloured in brown, olive, cerise or purplish blue, with the tentacles in the same, or contrasting, shades. The edge of the disc is thrown into four undulant lobes, surrounded with a white ruff of finely branched acrorhagi. The column is buff or cream, covered with small warts, often with chips of shell sand attached. A related species, *Oulactis muscosa*, is smaller and lives chiefly under sublittoral fringe boulders. The column is yellowish white, with soft warts (verrucae) and the ruby oral disc is radially splashed with brown, white and gold. Short, blunt tentacles are arranged in four cycles (12, 12, 24, 48), the inner ones mottled and the outer rose-pink. Orange marginal tubercles surround the disc.

Wharf piles

New space and freedom for growth are provided beneath wharves, where wooden or concrete piles and stringers can carry rich and diverse low-tidal communities. Such constructed sites are ideal too for experimental studies of settlement and succession. They offer plane surfaces where gradations of tides, illumination, waves and currents can be far more simply contrived than on an irregular rocky shore. Wharf piles moreover form islands where a species chosen for observation can easily be promoted or erased. Plates can be attached with a range of different settling surfaces, and grazers or predators excluded by cages of wire mesh.

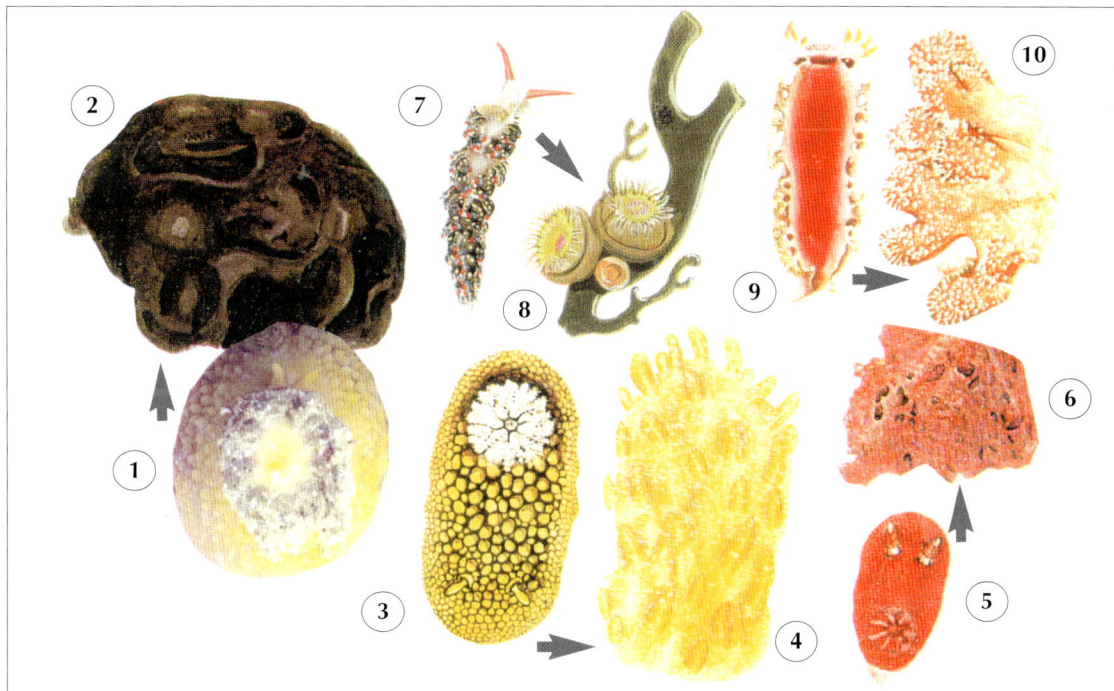

Fig. 14.4 Some opistobranchs and their dietary
1 *Umbraculum umbraculum* with 2 sponge *Ancorina alata* 3 *Archidoris wellingtonensis* with 4 sponge *Polymastia granulosa* 5 *Rostanga muscula* with 6 sponge *Crella incrustans* 7 *Berghia australis* with 8 anemone *Cricophorus nutrix* in axial branch of *Cystophora* 9 *Tritonia incerta* on 10 *Alcyonium aurantiacum*. (By courtesy of Michael Miller.)

Each of the life-forms described on boulder shores can gain more luxuriant expression on wharves. Hydroids that would form a short pile under boulders (e.g. *Hydractinia*) are here exchanged for the dense beards of sertulariids (longest in *Amphisbetia* on mussels), the plumes of *Pennaria* or the long, soft polyps of *Tubularia*. Flat-spreading cheilostome Bryozoa can give place to the bushy *Bugula* and *Tricellaria catalinensis-aria*, or to the long stolons of *Zoobotryon* and *Amathia* among the Ctenostomata. Sponges such as *Raspailia topsenti* change from a flat crustose to an elaborate branching growth. The flat sea-squirts *Ascidia* and *Corella* are now replaced by plump upright tests of *Microcosmus* or the pendent tubes of *Ciona*.

Though small flat crabs — *Petrolisthes* and *Neohymenicus* — are still there, the large crabs of wharf piles are higher built, like *Notomithrax peronii*, with stilt legs and slow gait, picking its way over the living substrate. The one grapsid is likely to be the low tidal *Plagusia chabrus*, running swiftly or taking momentary hold by sharp claws.

With their spacious settling surfaces, fast currents and shaded microclimate, wharves have — within living memory — provided a luxuriant sublittoral fringe at the heart of Auckland city. The old ferry wharf at **Devonport**, with a 3 m range of spring tides, stood in moving water by today's standards relatively clean. In the 1940s the sublittoral fringe was golden green with hydroid-bearded mussels, pale yellow *Aplidium*, and the widely ranging red, mauve and orange of sponges. With its host of small mobile species, this community could be inspected and even reached for sampling from the steam-ferry as it drew alongside at a low spring tide. Lower parts of the wharf could be walked upon from the ferry's lower deck — a frequent exit for the more intrepid passenger, occasionally a biologist. In the generation before scuba equipment, this was a special way of accessing the rich shaded communities at low tide. In 1948 the Devonport base commander had enrolled in person as a part-time marine biology student; access all through this habitat was then smartly laid on by a long six-oared naval rowing boat.

Some of Devonport's richness had already gone by 1962 when the illustrations first used in Morton and Miller were drawn up during Robin Harger's masterate study. We have kept them here not only as recording the site of Miller's opisthobranch studies, but as showing what is still to be found on the diminishing number of wooden wharves of the quieter northern coastline.

The zonation of piles from the old wooden wharf at Devonport is presented in Fig. 14.5. In a **littoral fringe** with *Enteromorpha* and cyanobacteria, littorines are lacking, along with all the other grazing snails. The only topshell is *Calliostoma pellucidum* that has turned to

269

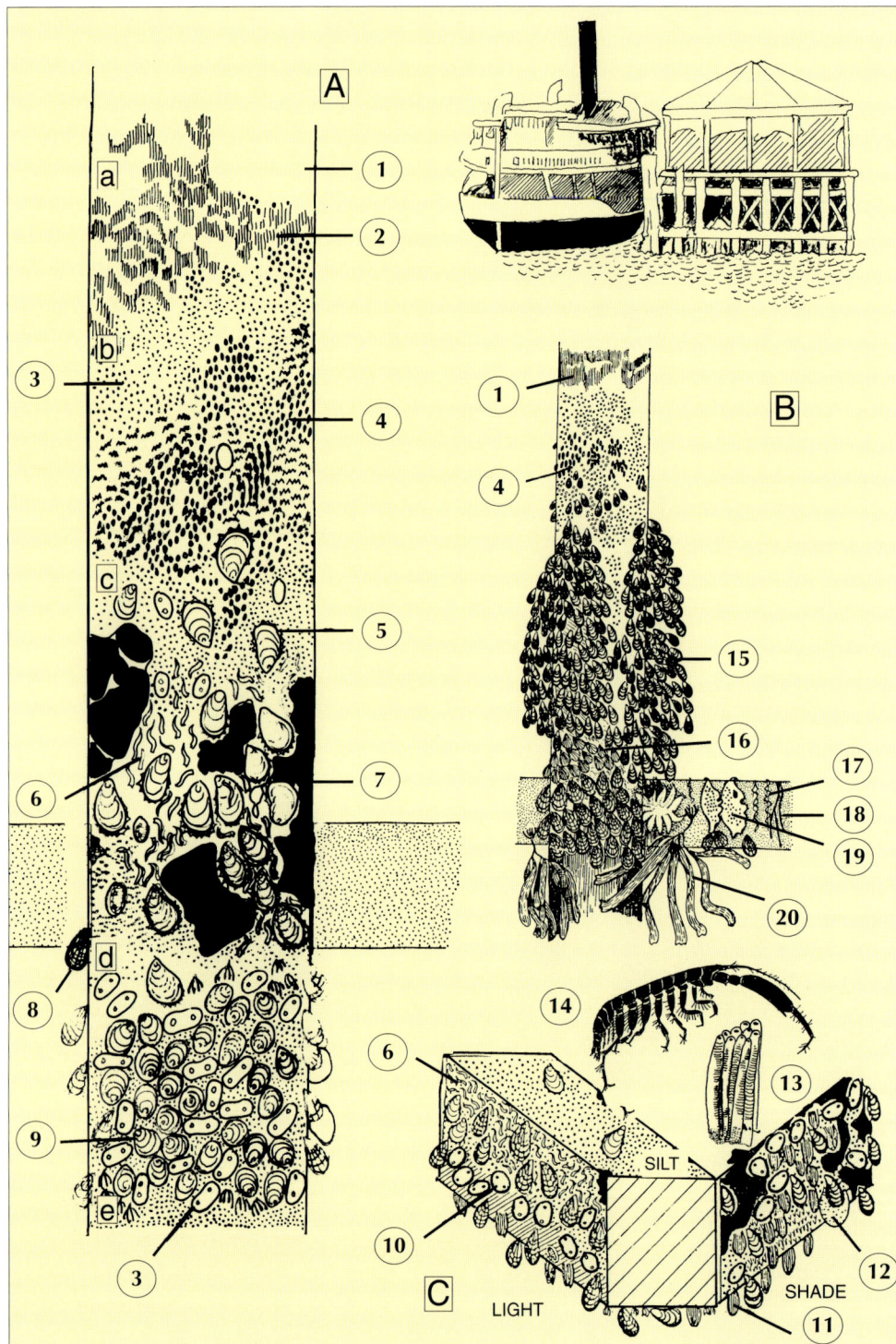

Fig. 14.5 Zonation on Auckland and Wellington wharf piles

These formations are shown, before recession, in the greater richness of 1960. (A) wooden wharf pile at Devonport, Auckland Harbour a–b black lichens and *Enteromorpha*; b–c *Austrominius*; c–d *Crassostrea glomerata*, *Pomatoceros caeruleus* and *Watersipora subtorquata*; d–e *Ostrea lutaria*, *Microscosmus squamiger* and *Perna canaliculus*. (B) wharf pile at Eastbourne, Port Nicholson, Wellington. (C) Devonport wharf pile at low water, showing the effect of ospectation on horizontal and on high- and low-shaded vertical surfaces.

1 blue-green algae 2 *Enteromorpha* 3 *Austrominius modestus* 4 *Xenostrobus pulex* 5 *Saccostrea glomerata* 6 *Spirobranchus caniniferus* 7 *Watersipora arcuata* 8 *Perna canaliculus* 9 *Ostrea angasi* 10 *Microcosmus kura* 11 *Diadumene neozelanica* 12 *Aplidium phortax* 13 *Corophium* tubes 14 *Corophium* animal 15 *Mytilus edulis aoteanus* 16 *Aulacomya maoriana* 17 ulvoid species 18 *Scytosiphon* 19 *Gigartina* 20 *Lessonia variegata*.

sponge-feeding. The pervasive barnacle is the opportunistic *Austrominius modestus*. Where this species and *Chamaesipho columna* occurred together as on the rocks nearby, *Austrominius*, being larger, became the preferred prey of the thaid *Lepsiella scobina*. With this gastropod absent from wharves, *Austrominius* was secure from predation, and can grow faster to oust its smaller competitor *Chamaesipho*. In the past two

Fig. 14.6 Devonport wharf (1960): a community built up just below low tide on a settlement plate immersed for 30 months
1 *Ecklonia radiata*
2 *Microcosmus squamiger*
3 *Perna canaliculus*
4 *Corophium* tubes 5 *Bugula neritina* 6 *Botryllus schlosseri*
7 *Amphisbetia bispinosa*
8 *Aplidium phortax*
9 *Notomithrax peronii*.

decades, *Lepsiella* has disappeared from the rocks around Devonport and most of the rest of the Waitemata Harbour as a result of poisoning by TBT antifouling paint.

In the **mid-eulittoral**, *Austrominius* yields some of its space to the flea mussel, *Xenostrobus pulex*, and the rock oyster, *Crassostrea glomerata*. Lower down the barnacle patches intersect with a mosaic of *Watersipora* and *Pomatoceros*. *Austrominius* reaches lower than would *Chamaesipho*, settling freely beyond low water neap (LWN). The individual lifespan seems limited to a month or two, with predaton from cushion stars *Patiriella* or the flatworm *Stylochus*.

In the **sublittoral fringe**, the 'big three' are the low-tidal flat oyster (*Ostrea lutaria*), the green-lipped mussel (*Perna canaliculus*) and the sea-squirt *Microcosmus squamiger*. At the lighted front of the piles, *Ecklonia radiata* reaches maturity in a couple of years; round its base are the mussels, pale green and brown-rayed where sheltered from full light. Among these primary growth forms are festoons of bushy anascan bryozoans. Three common *Bugula* species — all apparently ship-borne and cosmopolitan — are *B. neritina*, *B. flabellata* and *B. stolonifera*. There are also crisp tufts of *Tricellaria* and fans of *Caberea rostrata*. The flat, prostrate bushes of *Beania* are also common.

The abundant compound ascidians are the translucent yellow *Aplidium phortax*, and the rosettes of *Botrylloides leachi* and *Botryllus schlosseri*. Prominent sponges at Devonport include yellow *Cliona celata* and brown *Haliclona heterofibrosa*. Of the red sponges, dull *Raspailia agminata* and brighter *Plocamia* and *Microciona* compete for space with well grown *Crella incrustans*. In deep shade grow the orange golf-balls *Tethya aurantium* and the brown tennis balls of *Suberites perfectus*.

Vertical and horizontal surfaces, in light and shade, present aspectation effects. The sides of the low-tidal stringers are clad with oysters, mussels and *Microcosmus*. On their flat tops there may be pure *Austrominius*; or a layer of settled silt, permeated with the tubes of a small amphipod, *Corophium* species, long and straight, with the second antennae unusually large. The gnathopods are reduced and non-chelate, food being intercepted by the mouthparts from a current driven through the tube by the beating of the pleopods.

Under the horizontals grow the large pendulous species: with the amorphous shapes of *Aplidium phortax*, tubular *Ciona intestinalis*, and the anemones *Actinothoe albocincta* and *Diadumene neozelanica*.

Bryozoans and their predators

Along with their hydroids and anemones, wharves are favoured resorts for the soft, uncalcified bryozoans of the Order **Ctenostomata**. These are an ancient group,

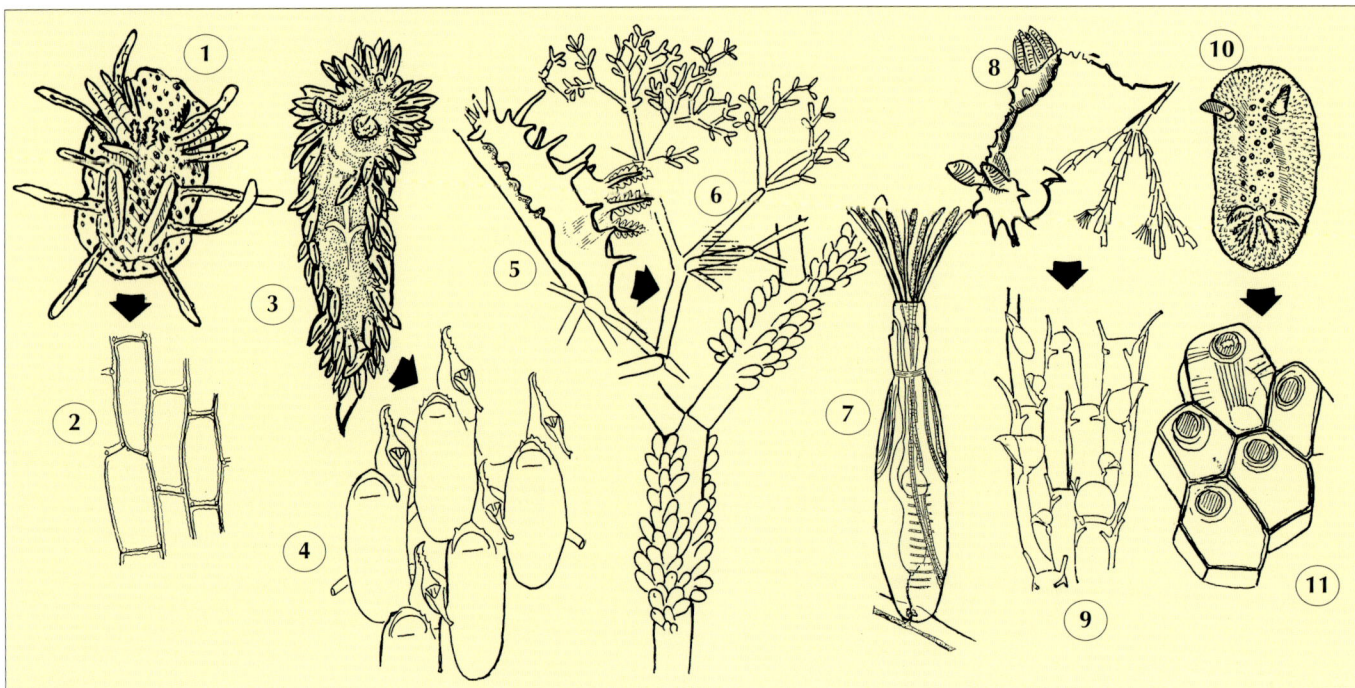

Fig. 14.7 Bryozoans and their opisthobranch predators
1 *Okenia plana* with 2 *Membranipora membranacea*
3 *Caldukia rubiginosa* on 4 *Beania* 5 *Trapania rudmani*
with 6 *Zoobotryon* 7 *Zoobotryon*, zooid structure
8 *Polycera hedgpethi* on 9 *Bugula;* 10 *Acanthodoris
mollicella* with 11 *Alcyonidium*. (From Michael Miller.)

stemming from a Paleozoic ancestry, and thus far older
than the cheilostomes which are of Cretaceous origin
and — by bryozoan standards — 'young'. Some of the
Ctenostomata are gelatinous and stolonate, others firm
and fleshy, but a hard skeleton, ooecia and avicularia are
all lacking. The zooid structure is illustrated in Fig. 12.2.

The largest of the New Zealand ctenostomes is
Zoobotryon verticillatum, one of the Family **Vesicularii-
idae**. The translucent stolon — up to as much as a metre
long — is formed of much-attenuate zooids with
branches at each internode. Short, normal zooids are
wound upon the stolon in a spiral chain. In *Amathia dis-
tans* by contrast, the colonies form erect tangled stems,
dichotomously branched, with spirals of tubular zooecia
just below each fork. The epizoic *Bowerbankia* species
have tubular zooecia set on a creeping stolon.

In the **Averrilliidae** the ovoid zooecia are joined in
pairs to a long, creeping stolon. Oppositely branching
Aeverrillia forms tufts on algae, hydroids and other bryozoans.

The **Nolellidae** are small ctenostomes with opaque
zooids. Short zooecia spring directly from the stolon.
Anguinella palmata, presumably ship-borne, has been
found in Auckland and Nelson Harbours.

The two families **Flustrellidridae** and **Alcyoniidae** are
non-stolonate. In the first, *Elzerina binderi* has zooecia

embedded in a gelatinous layer, with apertures bordered
with spines. Two *Alcyonidium* species close to the European ones have been widely reported round New
Zealand coasts. The gelatinous crust has transparent
hexagonal zooecia, without spines.

Where the dorids feed chiefly on sponges and the
aeoliids on cnidarians, there are some miniaturised
dorids specialising on bryozoans. The small, suctorial
Goniodorididae have a narrow radula with large,
hooked lateral teeth. *Okenia plana*, with dorsal
appendages spotted in lilac or brown, lives upon *Membranipora* on *Ecklonia* fronds. The related *O. lucida*
sucks the fluids of *Zoobotryon*.

The small red *Acanthodoris molicella* (Family **Onchi-
dorididae**) feeds suctorially on fleshy *Alcyonidium* by a
muscular pharynx pump.

The wharf bryozoan, *Bugula neritina*, is attacked by
a small species of *Polycera hedgpethi*, black and gold-
speckled with long side horns.

In the smaller group **Arminacea**, the **Janolidae** resemble the aeoliids in their long cerata, but have no
cnidosacs. The small, flat *Caldukia rubiginosa* feeds
upon *Beania*, not by sucking, but by tearing open the
zooecia and rending them with the radula.

Cnidaria: Hydrozoa

With their wide tentacle sweep, the polyps of the Class
Hydrozoa (known as hydroids) are micro-carnivores,
capturing the larger zooplankton as successfully as ciliary
feeders take the smaller zooplankton and phytoplankton.

The full life history shows two very distinct phases:
the fixed *polyp*, usually colonial, and the *medusa*, a
small jellyfish bearing the sex-organs and generally free-

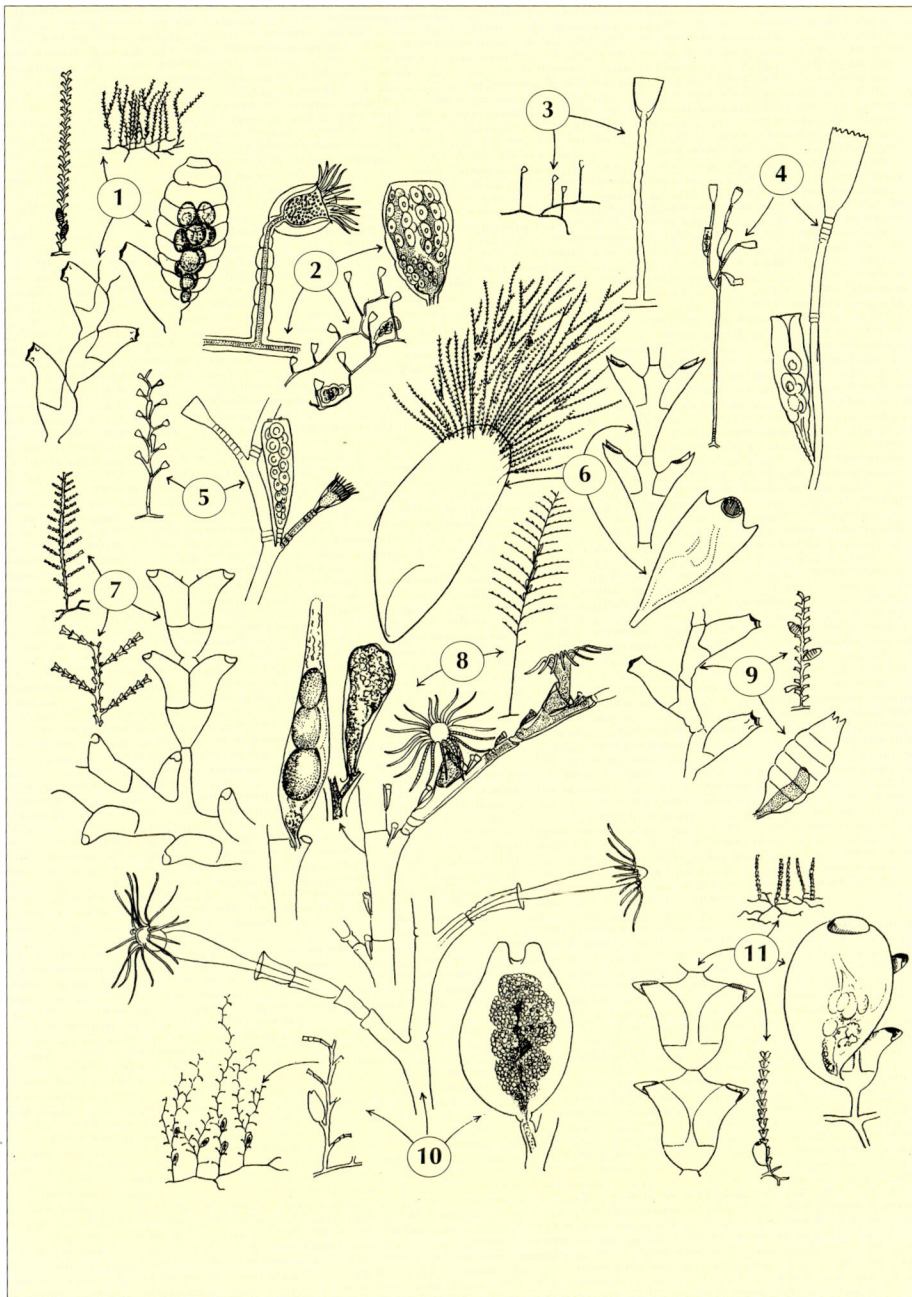

Fig. 14.8 Some Auckland thecate hydroids
Habit sketches and hydrothecae and gonothecae variously enlarged. (After Michael Miller.) 1 *Sertularella crassiuscula* (on *Carpophyllum*) 2 *Orthopyxis crenata* (on *Gigartina marginifera*) 3 *Orthopyxis integra* 4 *Clytia hemisphaerica* 5 *Obelia longissima* 6 *Amphisebetia bispinosa* (on *Perna* shells) 7 *Sertularia unguiculata* (under low tidal ledges) 8 *Plumularia setacea* 9 *Sertularella simplex* (on *Perna* shells) 10 *Halecium delicatulum* (on *Carpophyllum* and wharf piles) 11 *Amphisbetia minima* (on *Carpophyllum*).
(From Morton and Miller.)

swimming. The hydroid colony is built up of branching stolons, giving rise to upright polyps on separate stalks, or to groups of polyps from a common main stalk (*hydrocaulus*). The whole axis (*coenosarc*) is covered by a secreted cuticle (*perisarc*).

In the fixed generation the Hydroida are polymorphic, with different sorts of polyp serving distinct roles. Most numerous are the *trophozooiids* (feeding polyps) complete with mouth and gastral cavity. The *gonozooids* are the reproductive polyps, whose function is to bud off the alternate generation of *medusae*. In the full life cycle these become free-living, to develop male or female gonads and reproduce sexually. The resulting larvae settle to form a new polypoid colony. Many hydroids have in addition protective polyps (*dactylozooids, tentaculozooids*) that have lost the mouth and consist of little but a long nematocyst-studded tentacle.

The Class Hydrozoa is divided into (1) the **Leptothecata** (thecate hydroids)[1] where the horny perisarc invests the trophozooids with protective cups (*hydrothecae*), and (2) the **Anthoathecata** (athecate hydroids) in which the perisarc stops short at the base of the naked polyp, which is left naked. The leptothecate medusae (*Leptomedusae*) are umbrella-shaped, have statocysts and carry the gonads on the radial canals. Those of the anthoathecates (*Anthomedusae*) are bell-shaped, with ocelli but no statocysts, and the gonads on the mouth-tube or *manubrium*.

Leptothecata

The least specialised among the thecate hydrozoans belong to the Family **Campanulariidae,** where hydranths with a globular hypostome are enclosed in wide bell-

Fig. 14.9 Some athecate hydroids
1 *Pennaria disticha* 2 *Ectopleura crocea* with centre polyp carrying gonothecae 3 *Hydractinia parvispina* 4 *Hybocodon prolifer* 5 *Stomatoea* sp. 6 *Coryne pusilla* 7 *Turritopsis* sp. (From Michael Miller.)

shaped hydrothecae, and the gonothecae form narrow mouthed vases. The most familiar are the *Obelia* species with their hydrothecae on upright stems with regular sympodial branching. The medusae are saucer-shaped with more than eight marginal tentacles.

In the *Clytia* species single cups with the rim toothed arise direct from a stolon. The gonothecae are ringed, and the medusae globular with only four tentacles. In *Orthopyxis*, cups with thickened walls are also put up direct from the stolon. *Silicularia* has the cups bilaterally symmetrical, and so thickened that polyp cannot fully retract.

In the **Plumulariidae** the colonies carry their hydrothecae on only one side of the pinnate branches (hydrocladia). *Plumularia* has the cups fused close to the axis with accessory cups (nematothecae) holding mouthless polyps with stinging cells.

In the **Sertulariidae**, the largest family of the calyptoblasts, the hydrothecae are curved sessile tubes with operculum of one to four flaps, set in close series so as to give the stalk a serrate appearance. *Sertularia* species have the cups opposite, *Sertularella* and *Symplectoscyphus* alternate. Sertulariid colonies are frequently long

and densely branched, dragged from offshore, washed up in drift, or — like *Amphisbetia bispinosa* — found attached to mussels at low water. *A. fasciculata* forms beards over a metre long.

Anthoathecata

Among the athecates the shortened cups are best developed in the branched colonies of the **Bougainvilliidae**, which are not unlike the *Obelia* species, but with conical hypostome and the gonophores pear-shaped. The much-branched *Bougainvillea muscus* is found in New Zealand both intertidally and on algal drift. Also collected from drift are the *Eudendrium* species, with the perisarc finely ringed throughout the colony by stopping short at the hydranth bases.

The **Hydractiniidae**, as typified by *Hydractinia parvispina*, are much shorter and crustose, occurring under rocks or on shells that often contain hermit crabs. They cover the surface with a plate-like perisarc putting up tiny trophozooids and longer, mouthless dactylozoids.

In the Family **Tubulariidae** the hydranths are large

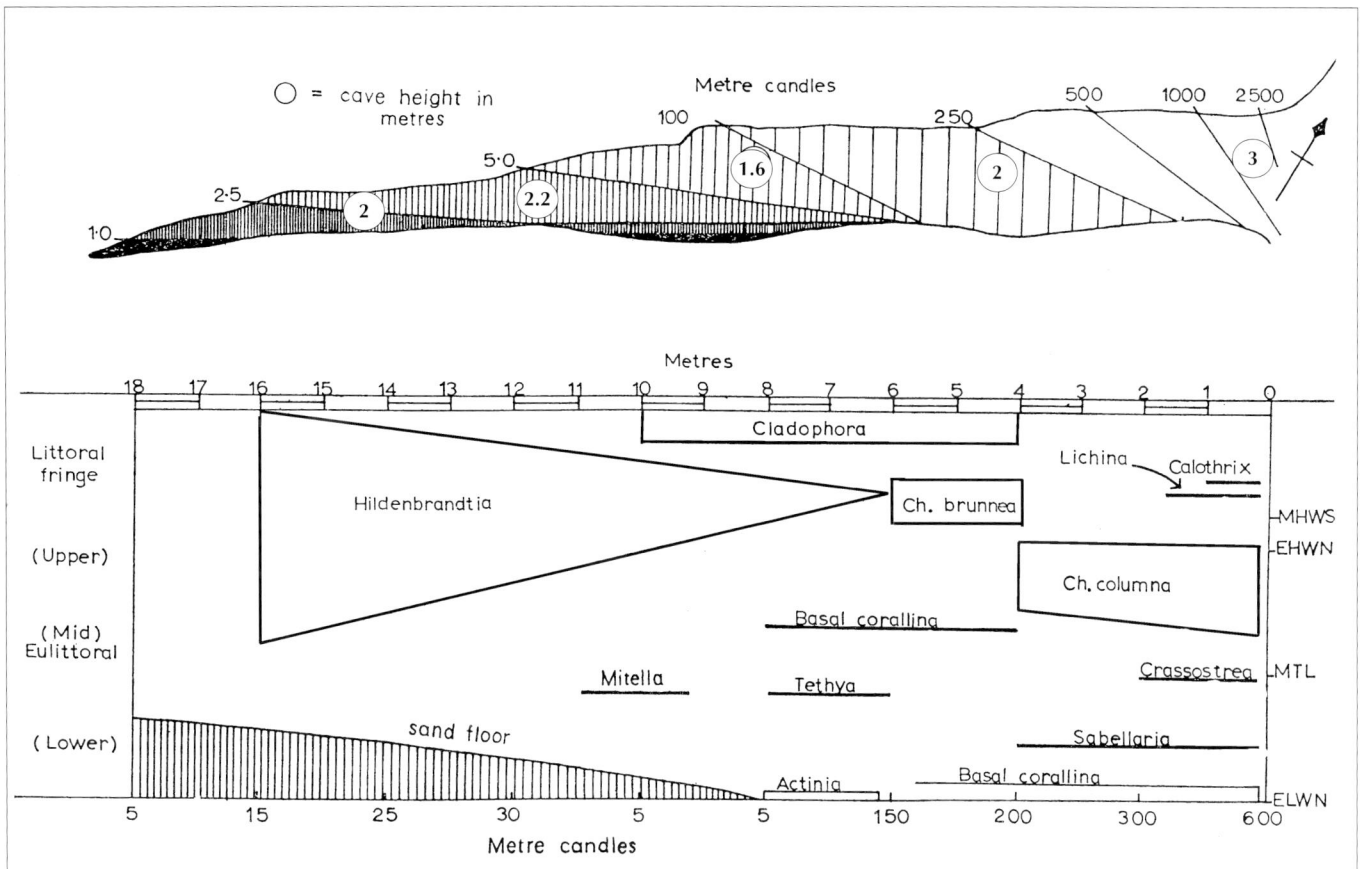

Fig. 14.10 Cave-wall zonation at Red Beach, Whangaparaoa, Auckland

(top) Ground plan of cave showing penetration of indirect daylight, in metre candles, with height of cave roof in metres, shown in circles.

(bottom) Schematic diagram of zonation pattern on south-east wall, illuminated by indirect daylight (summer).
Vertical scale 2 x horizontal.

and pink, with a proximal and a distal ring of filiform tentacles. The most familiar New Zealand species is the large *Ectopleura crocea*, growing on the lower reaches of wharf piles and other shaded subtidal situations. Its soft stems form a tangled colony with upstanding polyps, broad-belled or flask-shaped. The spherical gonophores cluster densely above the proximal circlet of tentacle. The medusae are highly simplified and non-swimming, but set free a sexually produced polypoid larva, the 'actinula' which on early settlement gives rise to a new colony. The related Tubulariidae have long, unbranched stems bearing single hydranths with two distal whorls of tentacles. They produce free medusae. *Hybocodon prolifer* is common in the sublittoral fringe of exposed coasts.

The athecate Family **Pennariidae** has both filiform and capitate (knobbed) tentacles. In *Pennaria disticha*, found as long, feather-like colonies, especially common under Auckland wharves, the pinnate branches bear large, pink hydranths with 7–12 filiform basal tentacles, and 9–14 capitate tentacles. Spherical gonophores are attached above the proximal tentacle circle.

The **Clavidae** have filiform tentacles not in a ring but scattered over the tubular hydrant. The chief intertidal species is *Turritopsis nutricula*, with branched single hydranths erect from a creeping stolon. In the **Corynidae** the scattered capitate tentacles have strongly swollen tips as in *Coryne pusilla*, sometimes found in seaweed drift and in the branched intertidal *Sarsia eximia*.

Cave communities

Narrowly open to the sea, intertidal caves can extend as fissures along a fault-line or other zone of weakness in the rock, for up to 50 m or more into a cliff-face. Floored with sand or boulder scree, but with walls clearly zoned, they can be followed inwards at low tide from a low-lit entrance to total dark.

Cave wall zonation sees the expansion into an open vertical sequence of the patterns we have first met with in concentric zonules under boulders. Caves are the longest and ultimately the darkest of secluded habitats. They have some close resemblances to upper eulittoral boulder beaches, but with the species there restricted to the cold and dark undersurfaces now spread out over a whole rock-face.

For sessile or immobile species, a food supply is able to be filtered from fast currents driven up narrow caves as the tide returns. Such water movement is well conducive to fast current feeders such as stalked barnacles

275

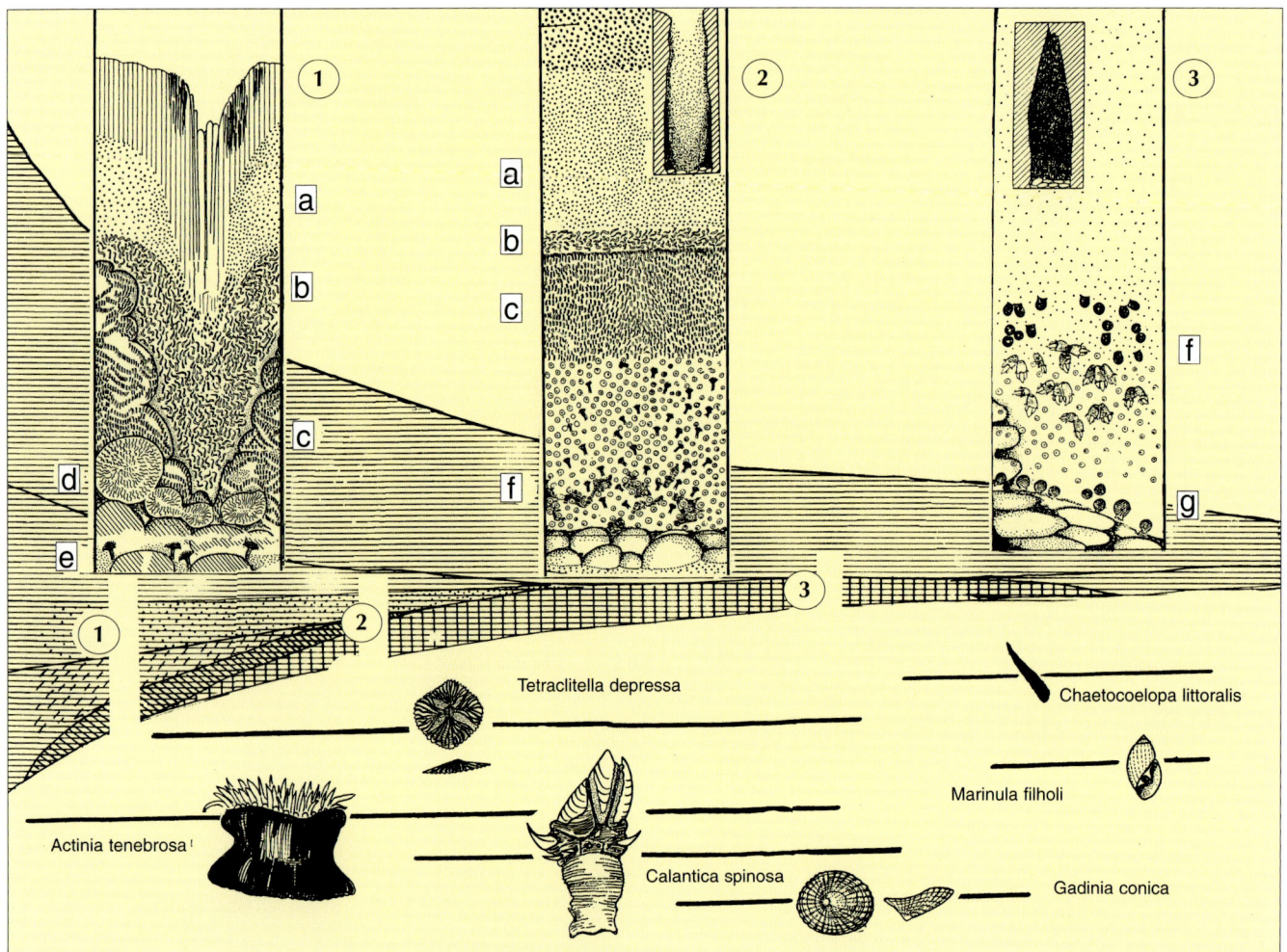

Fig. 14.11 Piha, Auckland West Coast: cave wall zonation

1   shaded cave mouth with freshwater seepage
2   cave wall in strong shade just inside cave mouth, and part-open above **(inset, sectional view)**
3   region of total darkness, near head of cave, with sectional view **(inset)**.

Zoning sequences

(a) *Chamaesipho columna* (b) *Pomatoceros caeruleus* (c) *Neosabellaria kaiparaensis* (d) plaque-forming tube-worm (*Serpula* sp.) (e) terebellid worm (f) *Tetraclitella depressa* with sabellid worm and in (3) with *Actinia tenebrosa* and *Calantica spinosa* (g) *Tethya aurantium*.

(bottom) Extent of occurrence of six common species within the cave depth.

and *Gadinia*. Photosynthesis being entirely lacking, the only plant food evidently derives from the decay of algae washed in.

The illumination and the horizontal and vertical ranges of plants and animals in a cave at **Red Beach**, Whangaparaoa, are shown in Fig. 14.10, from the findings of Vivienne Cassie. Towards the entrance, each species is more or less tidally zoned around its 'normal' level. The gradient of indirect daylight, of more influence than the direct sun, finally leads to the loss of the tidal order. Species become arranged instead by threshold light requirements. At the entrance are two light-preferring algae, the blue-green *Calothrix scopulorum*, and the brown *Ralfsia verrucosa* as well as *Lichina confinis*, all ceasing at c. 400 m candles. They are succeeded, between 250 and 100 m candles, by a range of species

with wide light tolerance, yet able to flourish in shade, including the barnacles *Chamaesipho columna* and *Epopella plicata*, with *Crassostrea glomerata*, *Xenostrobus pulex* and *Pomatoceros caeruleus*, as well as coralline crust and *Stictosiphonia arbuscula*. Still more shade is needed by *Neosabellaria kaiparaensis*, *Actinia tenebrosa*, and the encrusting alga *Peyssonnelia rugosa*. *Hildenbrandia* can grow in light intensities as low as 5 m candles. Below 250 m candles, other algae may still continue, such as green *Cladophora*, and blue-green *Nodularia harveyana* and *Oscillatoria nigro-viridis*. Red *Ptilothamnion* and *Audouinella* can also flourish in this low light.

For a cave at **North Piha**, Auckland west coast, Fig. 14.11 illustrates the transition from the shaded mouth to total darkness. With the cave wall still narrowly open to direct light, the dominant lower eulittoral barnacle is the wafer-flat *Tetraclitella depressa*, typical of extreme shade, or total dark as under middle eulittoral boulders (Fig. 13.7). This zone is studded with the tubes of a terebellid worm, probably a *Lanice*. Above *Tetraclitella* the cave wall has a stubble of *Neosabellaria* tubes not built into crusts. At a higher level there is the last sprinkling of *Chamaesipho columna*.

At the light level of the last *Tetraclitella*, clusters of the stalked barnacle *Calantica spinosa* appear. As a regular indweller of caves and crevices with fast currents (sometimes mixed or replaced with *C. villosa*) this is the most primitive of all our shore barnacles. The capitulum is covered with two pairs of pointed terga and scuta, and the median carinal plate. Round the base of the major plates are many accessory ones. *Calantica* is not — like most barnacles — an active sweeper but uses its stiff cirri as passive strainers in moving water. They are withdrawn at intervals and the food caught on their setae passed to the mouth.

With *Calantica* in near-darkness are the red anemone *Actinia tenebrosa* and the golf-ball sponge *Tethya aurantium*. There is also the white pulmonate limpet *Gadinia conica*, always confined to the darkest places. Small numbers cluster in crevices, or patches on a cave roof. The shell can be raised to allow entry of a filter current to the secreted mucus net (Fig. 11.14).

In the blind end of the cave at Piha, around high water, decaying *Durvillaea* wrack gives off its sweetish odour. Grazing on this or crawling openly on the cave walls, the ellobiid snail *Marinula filholi* abounds. Maggots of the hairy kelp fly, *Chaetocoelopa littoralis*, are teemingly common. The isopod *Ligia novaezelandiae* runs about freely over the rock, with small *Leptograpsus variegatus* further down. In the same places the limpet *Notoacmea elongata* is found on the open surface. We thus find repeated on the dark cave walls the same community as encountered under boulders on a high tidal beach.

CHAPTER FIFTEEN
Communities and Succession

The concept of the *community* is far older than the word *ecology* itself, having been in use for rocky shores since different animal and plant species were first found regularly co-existing. In 1877 the naturalist Mobius had applied the term *'biocoenosis'* to the species assembled together on clusters of oyster shells. Correspondingly the physical space in which the organisms lived came to be referred to the *'biotope'*.

From even earlier times special names had been assigned to the particular zones regularly observed across the shore. By 1832 Audouin and Milne-Edwards had recognised five such zones, naming them: (1) *Balanus* (2) *seaweed* (3) *coralline* (4) *laminarian* and (5) *oyster*.

Time was to see the invention of an ascending hierarchy of names — as in the North American fashion of ecology in the 1920s — denoting each of the levels of formation and subformation within the large and inclusive *biome*. Alan Stephenson was roundly to dismiss the accumulated apparatus of associations and fasciations as 'the positively terrifying outburst of terminology referring to communities'.

In these pages we have felt the same hesitation to employ even the term *community* until we could reach some conviction about its real nature and status. The first pitfall of its undiscerning use could be to infer some clear-cut, discernible entity, with its species boundaries regularly expected to coincide.

It is true we could point to examples of closed communities in just this sense. Some of the most obvious could be human-contrived ones, like a camembert cheese in a small box presenting both biotope and biocoenosis together. The mature cheese has resulted from setting in train certain biochemical processes. In the right microclimate a succession of saprophyte species (bacteria and fungi) decompose the lactose, protein and fat of milk. With a precision unusual in the wild, the community moves predictably to a *climax*. As in most unlighted situations, its course is based on the breakdown of a complex organisation, with the freeing of its constituent materials to be recycled as nutrients.

The cheese is only one example of a universal process wherever there are saprophytes at work, whether it be in treating sewage, maturing silage or compost, or making wine or cheese. In the last two, the outcome happens to be also humanly gratifying. In every case, hand in hand with saprophytic *reduction*, there is the *production* or building of new tissue by the activity of the saprophytic organisms.

It is, however, with communities of *production*, based on the primary activity of photosynthesis, that the elementary textbooks of ecology generally begin. Here, the green plants rank as *primary producers*, with herbivorous animals as *primary consumers*. Then follow tier by tier those animal species, designated *secondary*, *tertiary* and *quaternary consumers*. It is such ascending pyramids or *ecosystems* that in trophic ecology are designated *communities*. A flow exists of materials and hence of energy passing up the pyramid. The body size of the species in the successive tiers tends to increase, but the numbers of individuals and also the total biomass are progressively reduced.

Clearly the word *community* is here being preempted for the ecosystems familiar to first year biology students. It will soon be obvious that the *biocoenoses* or collections of species we have been finding between tides cannot always be interpreted as self-contained ecosystems. A mussel or barnacle 'community', as we might speak of it on the shore, has within its occupied space no built-in support system. Its primary resources come from the plankton, itself dependent on the currents and upwelling of nutrients in a system ultimately as large as the ocean. Further, because waves are constantly sweeping away from rocky slopes the products of decay, the saprophytic processes of reduction are mostly shifted to sites remote from the rocky intertidal scene.

Do 'communities' exist?

Visual recognition of a community becomes more difficult when we have to rely on sampling the sediments of the seafloor with a grab or filtering the water column with a plankton-net. Here we are using the statistics of association to find what aggregations of species are in

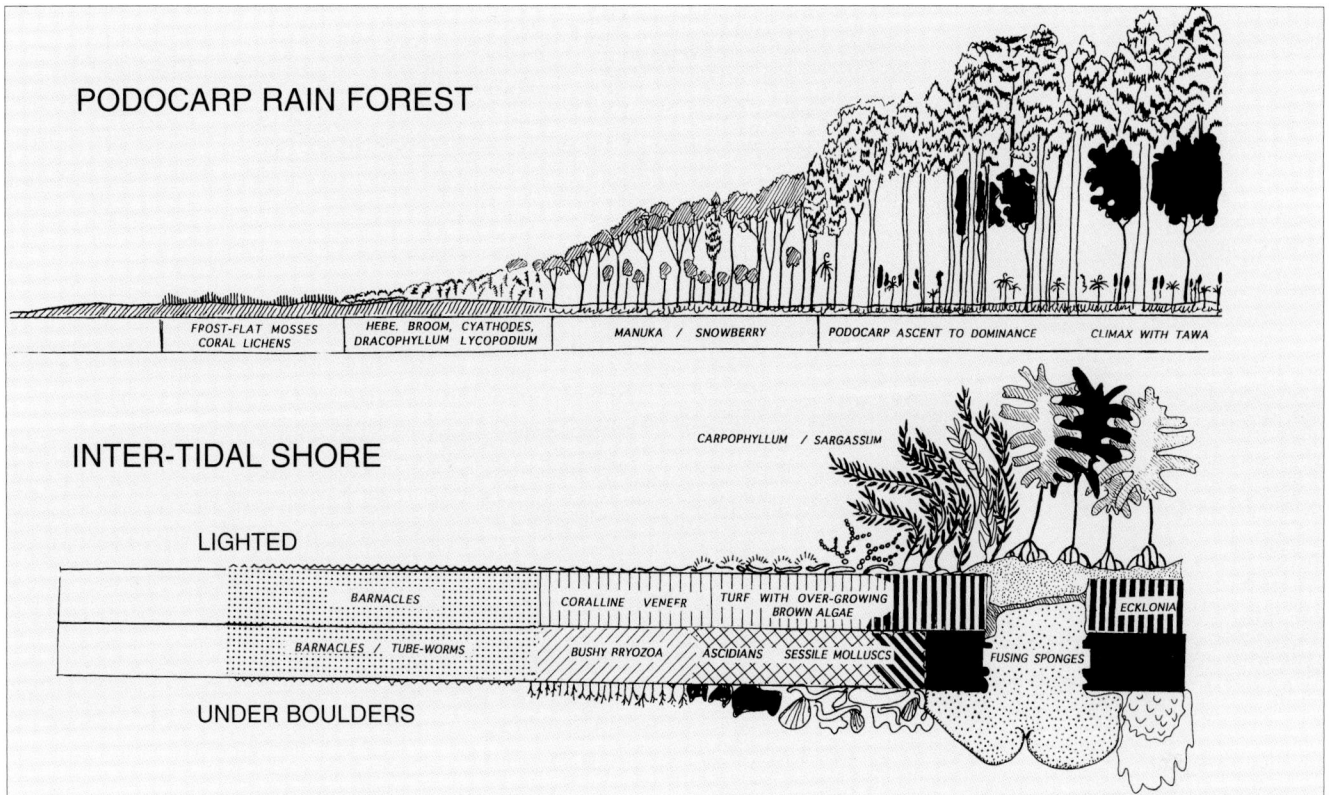

Fig. 15.1 Ecological succession
The course of a terrestrial succession up to a forest climax (**top**) is compared with succession on the shore, on lighted surfaces (**top panel**) and under boulders (**bottom panel**).

fact existing together. We have no grounds as yet for invoking causal explanations. The first notion of the community is just that certain species predictably occur together.

From early in the twentieth century a very different perception of the community had grown up, again with a North American school of ecologists. The terrestrial community was to be envisaged as a natural entity or 'super-organism', to be validly compared with a cell or an individual in the possession of its own metabolism, ontogeny, growth, senescence and death.

Such an advanced organismal view includes enough of the mystical to draw sharp scepticism, even a doubt as to whether the community existed at all. One such reaction was the doctrine of *species individuality*, put forward by the plant ecologist Gleason in 1926. The trees and shrubs in a woodland, far from being organised into a vegetational community, were held to lack any real affiliation with each other. Natural selection would instead tend to lead towards the *separation* of different species territories.

Associations of plants would thus be no more than coincidental. Their distributions might in places overlap, but these could be expected to dissociate and diverge by the operation of selection. Thus, any observed pattern of vegetation would be the result of fortuitous immigration of seeds and the equally fluctuating environment and microclimate. No two species would be quite alike in distribution, and species along an ecological gradient would be found to intergrade continuously rather than forming discrete zones.

No doubt most of the species associated together on rocky shores can also have discordant distributions, with their boundaries individual rather than coincident. Far more, however, than terrestrial plants, rocky shore organisms are normally narrowly distributed, with their cut-offs sharp and sudden. Not only are their ranges compressed, but their very limits of survival may be stringently drawn. All these facts on the shore lead towards species concentration in what we have perceived as strict zones. For many sessile species, gregarious larval settlement sharpens up zonal boundaries in a way that seed distribution in plants cannot match.

A botanist confining his attention to the canopy species from which terrestrial plant communities are named might expectably find these species acquiring more individual distributions as they diverged selectively. Where species were making similar demands on resources, competition would produce differences in their mean adaptive positions. They would thus evolve to occupy different spaces. We would here be finding '*beta diversity*', measured by the rate of change of the species along an ecological gradient.

If it were widely true that selection leads not towards but away from territorial overlap, a body blow would

279

have been given to the community idea. On the shore, however, we do repeatedly find clusters of species more or less coincident in range, suggesting some measure of co-adaptation. Far from competitive exclusion there is more often biotic interdependence with the reliance of many species on the nearby availability of another living organism. Such dependence may involve food specificity, the various grades of commensalism, or bio-camouflage.

Instead of *beta diversity* as between adjacent territories, we are finding high *alpha diversity* among species occupying the same territory. Such diversity reaches a plateau where a substrate or *biotope* is itself living, contributed by the bodies of plants or animals. Community complexity promotes increased stability, with positive feedbacks favouring still greater alpha diversity.

Ecological succession

Ecological succession is to be seen as a nearly universal fact of nature. Bare ground, rock or soil does not remain empty, and in a given area changes follow each other directionally, as the physical environment is modified by its existing tenants so as to prepare the way for successors.

This changing composition through time has been classically illustrated from the major vegetational communities on land. But it is obviously a property manifested also in the intertidal community, on a far shorter time-scale (Fig. 15.1).

The *climax* stage of a rainforest may require centuries to attain. New Zealand's central North Island forests, for example, have arisen by a progression from frost-flats and low tussock, through shrub-like *Dracophyllum* and *Kunzia/Leptospermum* (tea-tree) to a multi-stratal community with podocarp giants, leading eventually — as some would predict — to an ultimate hardwood climax in tawa.

The series of events as listed in a standard text-book of terrestrial plant ecology[1] can be lifted with scarcely a change of wording, to the course of change on the shore, exemplified on boulders and wharf piles:

Succession on the shore
1 Lighted intertidal
Castor Bay, Auckland

Settlement experiments were initiated by placing rock slabs, split from prismatic basalt, half a metre square, as settling surfaces just beyond low water. Over 24 months, a succession was enacted (Fig. 15.2), passing from bare rock to (1) the barnacle *Austrominius modestus* with small serpulid tubeworms (2) *Corallina officinalis* (3) brown algae. This order in *time* turns out to be the same as the order in *space* of the *white*, *pink* and *brown* zones we have already recognised in proceeding down the shore.

In the *upper eulittoral*, colonisation of settling surfaces came to an early halt at the first or barnacle stage, with *Austrominius modestus* at times superseded by the slower arriving *Chamaesipho columna*. In the *middle eulittoral*, the white barnacle zone was retained, but generally became enriched with rock oysters (*Crassostrea*) and tubeworms (*Pomatoceros*). In the *sublittoral fringe*, the succession advanced to the pink of *Corallina* paint. This proceeded to a full-grown coralline turf, entered also by *Hormosira banksii* towards the end of the first year. Lowest downshore, in the *upper sublittoral*, just beyond low water, the succession advanced to the brown zone of *Carpophyllum* and *Sargassum*, with the entry — beyond the second year — of juvenile *Ecklonia*.

Goat Island Bay, Leigh

Settlement slabs placed just beyond low water, under higher wave energy than at Castor Bay, were observed over two years. On both the light and dark surfaces succession began with a barnacle spat-fall: *Austrominius modestus* on top and small *Balanus trigonus* beneath. On the undersurface, the tubeworm *Hydroides elegans* also appeared. The two communities thereafter diverged. The lighted surface acquired a cover of coralline paint, at first upon the shells of still-living *Austrominius*, then thickening to cover all the barnacles save for a slit where the cirri could for a while continue

---

i   The *substrate* becomes progressively altered and developed from original bare soil or rock.

ii the *biotic strata* become higher, and more massive and differentiated.

iii *productivity* (= rate of formation of organic matter) increases as the structure of the community elaborates.

iv *microclimate* is increasingly moderated as height and complexity increase, and becomes increasingly determined by the characteristics of the community itself.

v   Species *diversity* increases from earlier to later stages (though sometimes declining before climax).

vi populations rise and fall, *replacing* one another along a time gradient, just as do stable populations along an ecological gradient.

vii the rate of such replacement slows down as *smaller and short-lived* species are replaced by *larger and long-lived* forms.

viii *stability*, as well as *diversity*, increases as the succession advances. After earlier stages that are very unstable and fast replaced, the climax community is usually stable, dominated by a few long-lived species, which maintain their composition, showing oscillation but no longer directional change.

CASTOR BAY

JFMAMJJASONDJFMAMJJASON

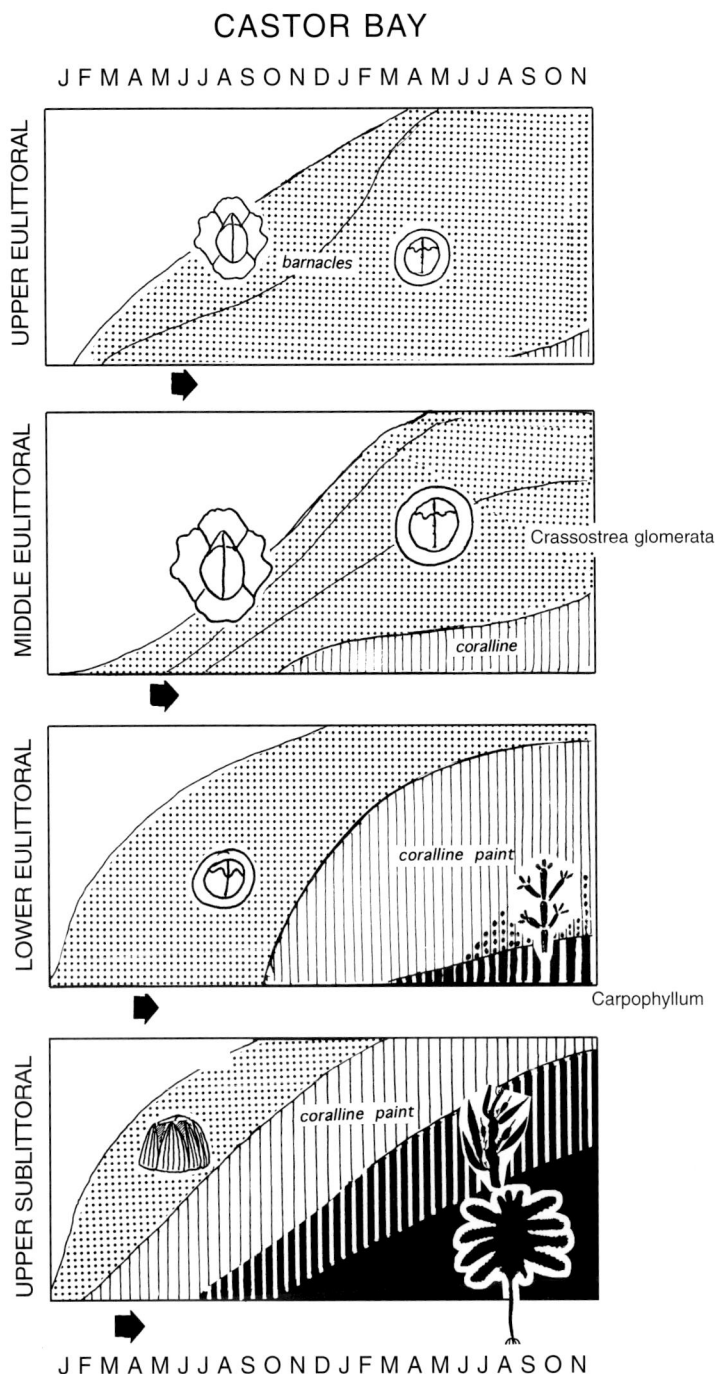

UPPER EULITTORAL

barnacles

MIDDLE EULITTORAL

Crassostrea glomerata

coralline

LOWER EULITTORAL

coralline paint

Carpophyllum

UPPER SUBLITTORAL

coralline paint

JFMAMJJASONDJFMAMJJASON

Fig. 15.2 Settlement at Castor Bay, Auckland
Occupation of space on the lighted surface of settlement plates is shown over two years for three levels in the intertidal and for the upper sublittoral.

to emerge. Full coralline turf was to spring up after 12 months. The brown algae made their appearance at 15 months, first with *Colpomenia* and the red *Liagora harveyana*. *Sargassum scabridum* and *Carpophyllum plumosum* were ultimately dominant.

2 Dark intertidal and subtidal

For the dark community underneath the settling slabs, the succession took the following course: (1) bare (2)

barnacles and tubeworms, (3) bryozoans, (4) compound and simple ascidians with sponges, with further tubeworms and sessile molluscs, (5) encrusting sponges, (6) fusing sponges (see Fig. 15.4).

This basic under-boulder scenario was found to be run through more rapidly, and to be carried to a more advanced stage, with increasing distance down the shore. Boulders at low water acquire bryozoans and ascidians in little more than a year. Fusing sponges — the climax achieved only at low water and developed best in shelter — never became established in under three years. On the middle shore succession reached no more than the bryozoan stage, while at the highest settled levels it was arrested at barnacles and tubeworms (*Hydroides*), with occasional small *Crassostrea* and *Pomatoceros*.

The different sizes of boulders at a given shore level have an important influence on natural settlement. It was found to be the larger ones that soonest attained the more advanced stages. Community development is thus a reflection of boulder stability. Completion of the scenario can be set back by the storm-overturn of a boulder. On the new lower surface, the scenario goes back to the beginning. The previous shade communities, now exposed to the light, died and slough off.

The state of colonisation of a boulder with the ratios of the stages 1 to 6 can be represented by histograms showing the percentage cover achieved by each type. The progress of a given boulder in the scenario will then depend on the following factors:
- Boulder mobility, depending on size, shape and weight.
- Position on the shore which — as well as its effects on mobility — will determine the length of filtering time.
- Water clarity, as related to wave energy, with effects on community growth and diversity, as for example in the differences between Goat Island and Castor Bay (Fig. 15.3).

Goat Island Bay, Leigh

For Goat Island Bay, Leigh, Fig. 15.4 shows the percentage of the boulders of different sizes reaching successive stages of the scenario for three levels in the eulittoral zone. None of these eulittoral boulders has reached the final stage of colonisation, but for equivalent boulder size, those lower on the shore carry later stages. The prospect of a boulder of given size reaching an advanced stage is thus enhanced by its lowness on the shore.

Under wharves

Some of the causal factors in succession can be well brought out by the fortunes of species on low-tidal settlement plates. Fig. 15.5 presents the course of events

281

Fig. 15.3 Settlement on dark surfaces
The occupation of space is shown over two years for the undersurfaces of settlement plates at Castor Bay **(top)** and Goat Island Bay **(right)**.

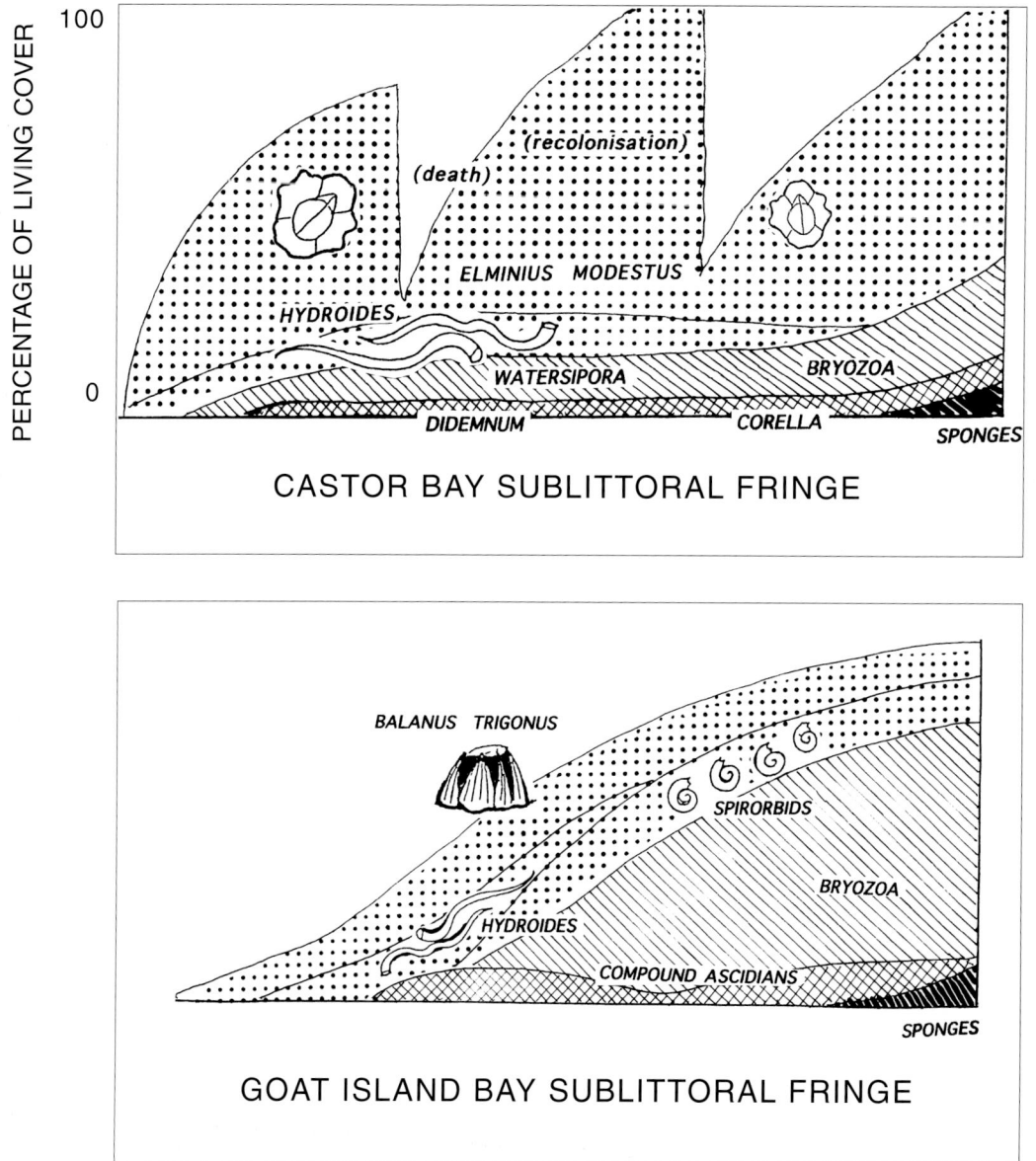

PERCENTAGE OF LIVING COVER

100

0

HYDROIDES

(death)

(recolonisation)

ELMINIUS MODESTUS

WATERSIPORA

DIDEMNUM

BRYOZOA

CORELLA

SPONGES

CASTOR BAY SUBLITTORAL FRINGE

BALANUS TRIGONUS

SPIRORBIDS

BRYOZOA

HYDROIDES

COMPOUND ASCIDIANS

SPONGES

GOAT ISLAND BAY SUBLITTORAL FRINGE

observed by Robin Harger in a pioneering study of Devonport wharf, Auckland, in 1962. The first species to appear on a bare surface was usually the opportunist barnacle *Austrominius modestus*. Capable of year-round colonisation, the barnacle was sometimes overtaken in early spring by the filamentous brown alga *Ectocarpus*. This brief showing of algae was soon grazed off by parore (*Girella tricuspidata*), and *Austrominius* would then succeed, often with the hydroid *Tubularia* settling on its shells. At times there would arrive a load of a second hydroid, *Pennaria*. The barnacles so smothered, or with their feeding impeded, dropped off after a short spell, carrying their hydroids with them. Alternatively, moribund barnacles could be eaten out of their shells by the flatworm *Stylochus zanzibaricus*, while *Tubularia* was removed by its nudibranch predators.

Following any of these events, fresh barnacle spat could opportunistically move in. But at suitable seasons cheilostome bryozoa were eventually able to seize the

initiative, no doubt helped by the ability of their avicularia to remove the alighting barnacle cyprids. The bushy cheilostomes *Bugula*, *Caberea* and *Tricellaria catalinensisaria* would hence become established. Tubeworms, later hydroids and saddle oysters *Anomia* contributed to the rising diversity.

After about four months, incursions of compound ascidians arrived in July, establishing heads of *Aplidium*, sheets of *Didemnum* or *Diplosoma* and the rosettes of *Botryllus*. By August simple ascidians had begun to follow, heralded by *Corella eumyota* and *Asterocarpa coerulea*.

The long-enduring climax community is dominated by the simple ascidian *Microcosmus squamiger*, with the co-dominants in the mussel *Perna canaliculus* and — where access to light is sufficient — the kelp *Ecklonia radiata*. Each requires a prepared surface to settle. *Ecklonia* succeeds if it has the chance to move in while arborescent Bryozoa are still there, but it will avoid

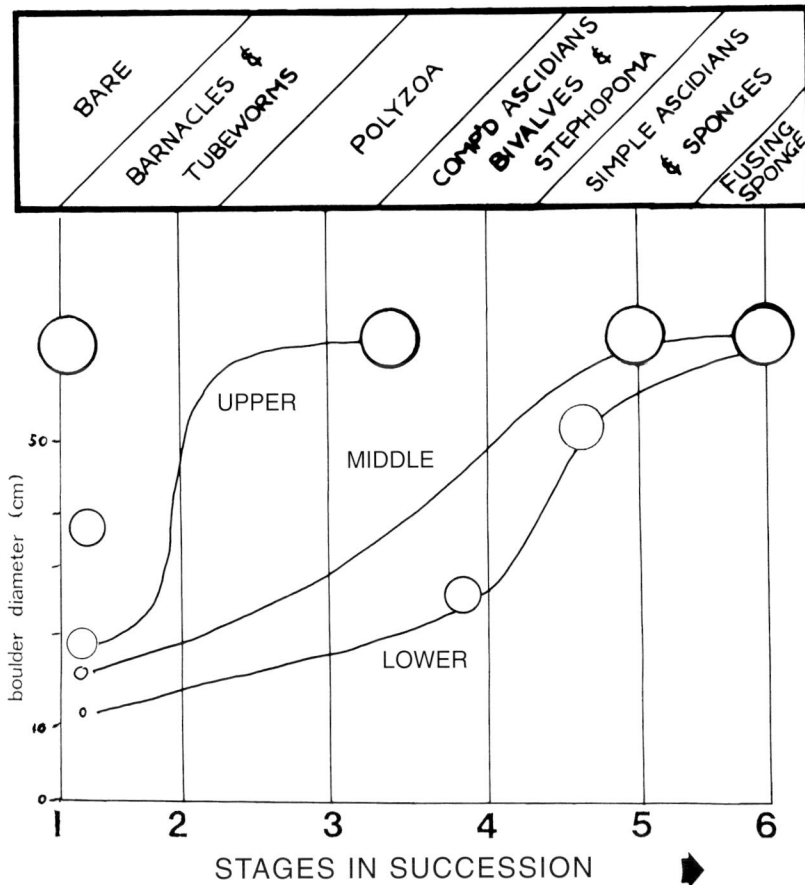

Fig. 15.4 Succession and boulder size across the shore

At Goat Island Bay, Leigh, the course of succession is shown, with the climax attained, for increasing boulder sizes at high, middle and low levels in the eulittoral zone. None of these eulittoral boulders has reached the final stage of colonisation, but for equivalent boulder size, those lower on the shore carry later stages.

(Based on John Walsby.)

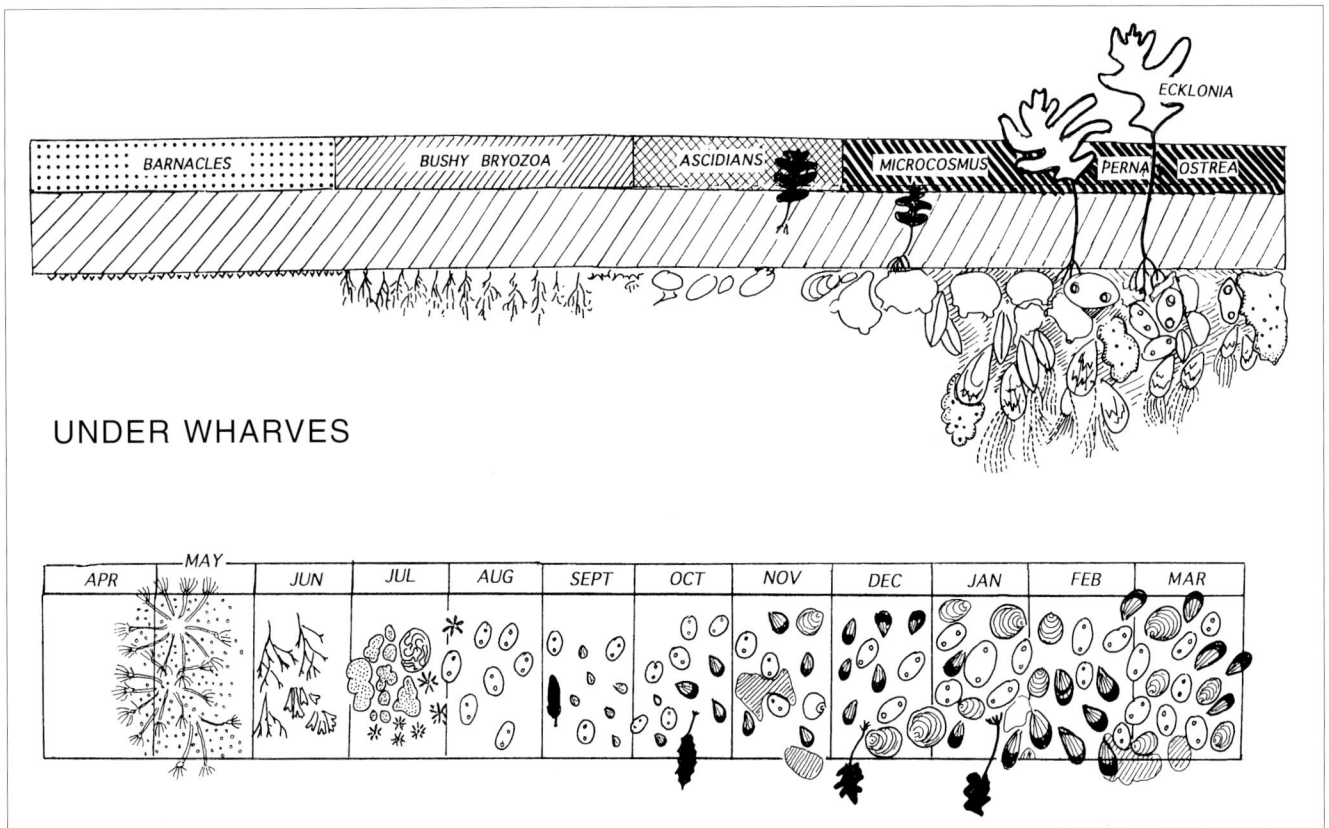

Below: Fig. 15.5 Ecological succession under a wharf at Devonport

(top) The course of succession over three years (1962–65) is shown for a low tidal stringer on Devonport Wharf, Waitemata Harbour. **(bottom)** Events on low tidal plates over 12 months at Devonport. (From the data of Robin Harger.)

283

plates already carrying ascidians. Mussels need a fibrous substrate of Bryozoa or filamentous algae. Cultivated mussels are found to settle best on frayed nylon ropes.

The path of succession will depend not only on the structure of the habitat but on the surrounding community and the kinds of larvae produced. The direction can be switched by the arrival of novelties, as on the hulls or in the bilge of ships. *Watersipora*, the bugulid bryozoans and the cosmopolitan ascidians *Ciona* and *Ascidia* evidently reached New Zealand in this way. In reverse, our barnacle *Austrominius modestus* reached Europe during the Second World War and has acquired a temperate-wide distribution with the increase in speed and frequency of shipping. Important too will be the season at which a species can produce settling larvae or propagules. It may thus enter aggressively at a propitious time, with effect on the immediate course of events, where the prior requirement in the physical and biological structure of the surface has been satisfied.

The final outcome is thus in effect a lottery. Unpredictability increases, with added chances and hazards further down the shore, given the greater diversity, greater complexity and higher elaboration of the living or non-living substrate. Higher on the shore, in earlier stages of succession, or on a bare initial surface, it is climatic factors that will exercise tighter constraints, with only one or a few species of high importance.

As well as succession there are also short-term community changes that are seasonal or cyclical. In New Zealand there is the regular rise and decline of the small winter and early spring algae *Porphyra*, *Scytosiphon* and *Petalonia* (Fig. 15.6).

Dominance

Under a climate-determined regime, communities (in the upper or middle eulittoral) tend to be narrow-based, and their composition predictable. They are thus said to have a high degree of *dominance*, with the importance of one or a few species. In many-species communities we find a more opportunistic regime, less predictable, and essentially under biological control. At the top of the shore, as with barnacles, desiccation will control the upward limits, while biotic factors such as competition or predation are likely to set the boundary for the downward limits.

The nature of succession

As one of the community's inherent properties, several models have been suggested to account for ecological succession. With a **facilitation** model, succession is held to be assisted by the remains of previous species. Thus brachiopod colonisation needs the presence of dead barnacles; mussel settlement is facilitated by the previous foundation of bushy Bryozoa; the sporelings of brown algae require a ground-base of coralline paint.

Under the **tolerance** model, the earlier colonisers modify the environment to make it less suitable for continued recruitment of their species, but the modification has little direct influence on the later species. Thus, in a forest, new recruitment of podocarps is inhibited under the shade of their existing canopy. To new colonists, the residents show a passive response or tolerance. With an **inhibition** model we have species early in a succession tending to restrict the colonisation or growth of later species.

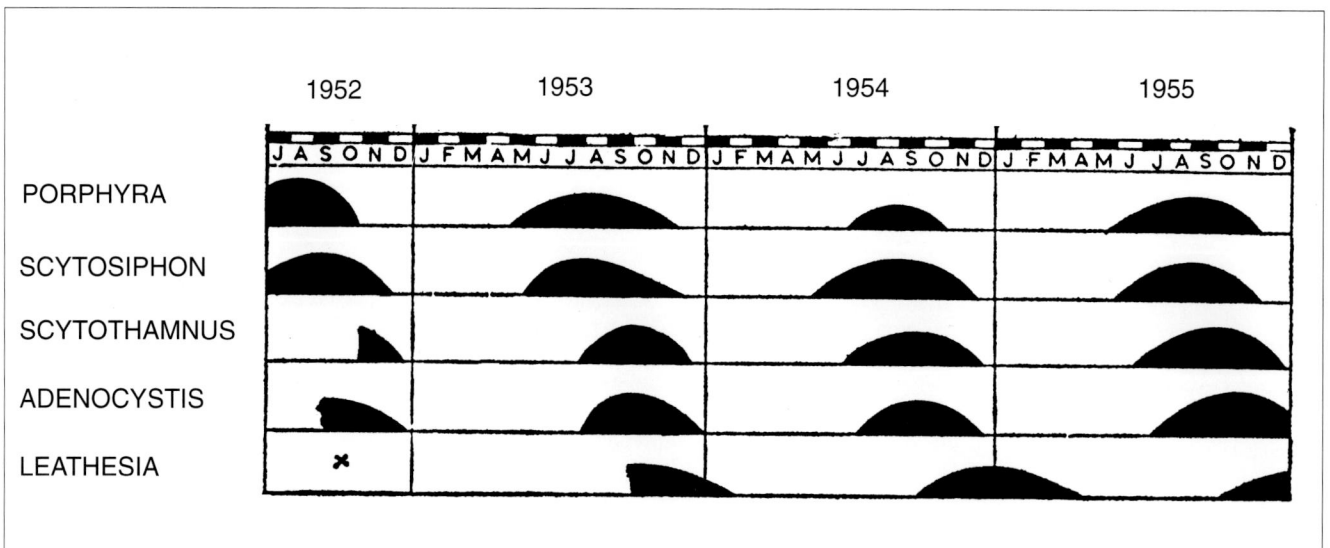

Fig. 15.6 Seasonal succession of algae

The seasons of occurrence of five annual algae (*Porphyra columbina*, *Scytosiphon lomentaria*, *Scytothamnus fasciculatus*, *Adenocystis utricularis* and *Leathesia difformis*) are shown for Portobello. (From Betty Batham.)

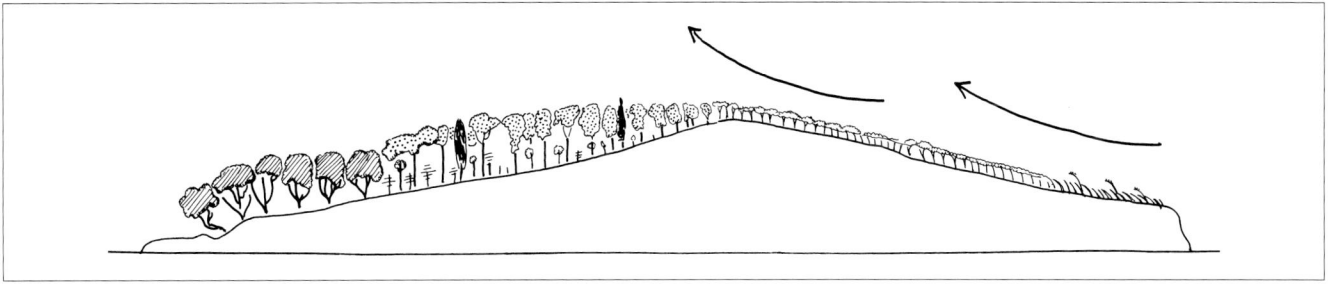

Fig. 15.7 Goat Island from shelter to windward

The path to climax

In land vegetation, the climax of a succession can have a complex causation. Over the one general area, local differences of shelter, soil and terrain may lead to a mosaic of different climax types, including patches of pre-climax stages. A *poly-climax* thus develops, with the influence of slope, aspectation, drainage, wind strength or soil type. In a comparable way, we have seen in sub-tidal communities how the straight and predictable zonation of the intertidal has been replaced by a mosaic resulting from local differences of light.

Terrestrial succession may often be arrested by the effect of microclimate at one of its pre-climax stages. Fig. 15.7 shows the exposed and sheltered aspects of **Goat Island**, Leigh. Over the south-facing slope, away from winds and wave exposure, there is a *climax* of northern coastal forest (kohekohe, mahoe, houpara, kawakawa) with cliff-hanging pohutukawa. Where wind sweeps the seaward slope, there remains a *pre-climax* stage with a low two-metre canopy, of mapou, manuka, and — at the seaward edge — karo and harakeke. With stronger winds bringing spray, the cliff-top is held with a part-succulent salt meadow.

The shore as a multi-climax

The intertidal shore has a steep ecocline, and with a strong gradient of ecological conditions — in the broad sense *climate*. The poly-climax mosaic that results here takes on a step-wise linear expression. The full sequence could be said to comprise the zones we have repeatedly found: a *black* zone (blue-green algae with periwinkles), a *white* zone (barnacles with oysters and tubeworms), a *pink* zone (coralline algae), with the ultimate climax a *brown* zone of large ochrophytan algae and their subordinate species, around low water.

With this interpretation, the zonation of the inter-tidal presents us with a single community, having its development arrested at various pre-climax states, according to tidal level and hence the constraints of microclimate. Different potential dominants may gain expression in different situations. The community shows plasticity as well as stability, with alternative expressions never realised together at a single section of time. Succession may be expressed not only temporally but in a linear sequence down the shore.

Equilibrium of climax communities

Biological interactions as well as patchiness of climate are also responsible for spatial mosaics varying in dominance over narrow areas. The model shown in Fig. 15.8 represents neighbourhood stability as seen on rocky shores, among fouling communities and sand and mud-flats. Type A represents a climax community by the black ball with the side of the valley as the scale of perturbation. Whichever way the community is disturbed it will always return to a single species dominance with stable equilibrium. In Type B, the species are locally stable in Stage 1 but a slight perturbation may send it into a different species dominance in Stage 2. A larger perturbation would be required to change the state from 2 to 1 or 3.

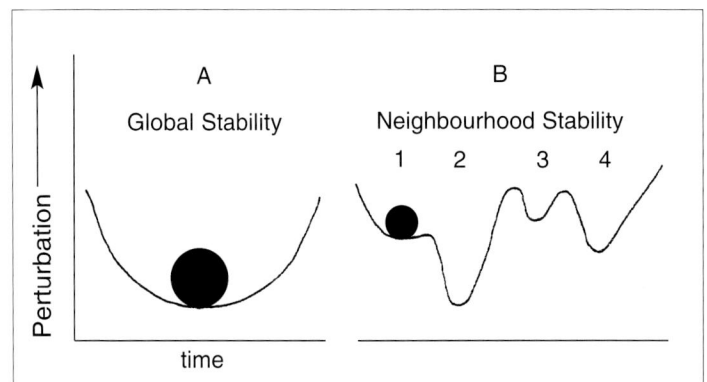

Fig. 15.8 Equilibrium of climax communities

CHAPTER SIXTEEN
Beyond the Tides

In the decade of the 1960s, marine biology was still mainly engrossed in the intertidal. *The New Zealand Sea Shore* (1968) acknowledged that 'our picture has generally been cut off by the increasingly arbitrary line of low tide'. Nor did the subtidal chapter added to the 1972 edition offer more than a brief tail-piece. Just as in some degree today, such imbalance had to be justified — historically if not logically — from Morton and Miller's prime range and experience.

The shore between tides has still the advantage of being immediately accessible at trifling cost. Its study can be both concentrated and comparative, both in place and through time with regular observation over months and even years. This is moreover the only part of the shore that a majority of naturalists will see. Since Aristotle, the shore has been the theatre where study of animals is likely to begin.

Yet in three decades the coming of scuba diving has changed all our realisations about the sea. With many who would not profess to be scientists, its influence has brought a new conscience in our treatment of coasts, and has been a driving force in the creation of marine reserves. Most of all, scuba has brought home to us the unbroken continuity of the intertidal with the subtidal. The traditional cut-off at low water mark is not only being transgressed. In a biological sense, it has been shown hardly to exist.

It is the intertidal, rhythmically exposed to the atmosphere, that is now seen as a specialised margin. Rigorously zoned under its own climatic stress, it is but the prelude to the extended sequence beyond the tides. The steep cliff-fall continues subtidally until the continental shelf cuts it off. With light diminishing and water movement ultimately stilled, this whole rock-face presents its own zonation, of a kind far more spacious and diverse.

As the first dive will make clear, the prime determinant of subtidal zoning is *illumination*. Light passing through water diminishes in quantity and alters in spectral quality. This fact governs the pattern of life. The algal community first becomes enriched, not only with the forest of the browns, but in the underworld of small reds at the lighted levels below wave-base. With continued loss of light, the algae are reduced to extinction. With the new space available and the lifting of other climatic constraints the fauna enriches. This is most true of those sessile *sciaphilic* groups we have already met with between tides. It is the under-boulder communities that have already given us in advance a foretaste of the subtidal shore.

The quantity of illumination at a given depth is controlled by shelter and local topography. The value of a shore for diving is diminished to the extent that underwater visibility is reduced by suspended particles carried down from the land. Thus, in its attraction for divers, the Hauraki Gulf can be divided into a clear 'Outer Gulf', reaching in its extended sense far up the east Northland coast, and a more clouded 'Inner Gulf' including the Firth of Thames and the approaches to the Waitemata Harbour.

Fig. 4.11 shows the successive contraction of the cliff-slope and the depth zones lying above the sediments of the level seafloor. From the Poor Knights to Mimiwhangata, Hen and Chickens and Goat Island, these are the good diving grounds of Northland and the outer Gulf. These stand in contrast with Tiri Tiri Island, Castor Bay, and finally Northcote Point, in the inner confines of the Waitemata Harbour.

Just as the build-up of a sand beach may cut short the sequence of the rocky intertidal, the sediments of the continental shelf truncate the full column of subtidal zonation. It is on offshore islands, with minimal sediment runoff and deepest transparency, that we can best appreciate all three divisions of the sublittoral zone.

We shall first describe the communities of the subtidal cliffs of east Northland, chiefly as typified at **Goat Island**, just outside Cape Rodney.

Divisions of the sublittoral

First, the subtidal forests of brown algae constitute the **upper sublittoral**, reaching up without a break into the **sublittoral fringe**. In the north — as we have seen — the fringe is composed of *Carpophyllum* species, and to the south increasingly of *Cystophora* and *Marginariella*.

Beyond low water the fringe first continues into a thicket of subtidal fucoid alga. Lower down these are soon replaced by pure groves of kelp.

The upper sublittoral and its fringe are unified by the encrusting of coralline algae beneath the shade of the tall browns. This pink veneer begins just above the sublittoral fringe, with the coralline turf in the lower third of the eulittoral zone. Beyond low water it continues essentially without a break to the ultimate depths possible for photosynthesis.

In the 5 m or more above wave-base, water movement continues strong through the upper sublittoral zone, but is somewhat diminished or broken under the kelp canopy, that also casts a deep shade over the territory beneath.

The canopy can in some places experience the major disturbance of biotic *perturbation*. At Goat Island the continuity of the sublittoral fringe with the beds of *Ecklonia* offshore can thus be broken. As the alternative to kelp the upper sublittoral develops a grazed meadow created by activity of the kina, *Evechinus chloroticus*, in pushing back the boundaries of the forest.

The second division, the **middle sublittoral** zone, can be taken to begin after the last of the brown algae, and with the disappearance in quantity of the pink veneer. Not only does illumination fall below photosynthesis level. Water movement is markedly reduced beyond wave-base, though long-shore currents may persist in some places. This middle zone takes its whole character from the dominance of the sponges, whose biomass and diversity increase with depth. The kelp brown and coralline pink now give place to a bright mosaic. The colours of Demospongiae range from scarlet, orange, chrome yellow, maroon, pink and mauve to cream, pure white or jet-black. The beauty of this sponge assemblage can be properly captured only by camera flash, or by bringing samples up to the full light where they are normally never viewed. More exquisite in detail and finer in scale is the subsidiary wealth of ascidians and bryozoans.

At Goat Island, as on most of the mainland, the third division, the **lower sublittoral**, will hardly be realised. Instead and at the deepest, fine sand is likely to reach right up to the base of the sponges. Where sediment remains thin, less than 10 cm deep, a different wealth of sponges ensues. Globular, bowl-shaped and finger sponges grow up through the sand; others — under a very thin sand cover — spread over the rock. The gently sloped platform is in effect a sand-strewn prolongation of the middle sublittoral. Off Goat Island, divers have designated this the **Sponge Garden**.

On the steep faces of most of the middle sublittoral, above the Garden, sponges are still subject to long-shore currents. With their most luxuriant growth subdued,

they form a low profile of heavy crusts. It is in the lower sublittoral, with water movement ultimately subdued, that the sponges grow long and tubular, often branching like upstanding candelabra. But their absolute dominance now comes under challenge. In this sub-zone, though with no sharp demarcation from the middle sublittoral, the sponges are being overtaken by cnidarians of the Class **Anthozoa**. First come the planed surfaces of beige and puce sea-fans or gorgonians, in a few places mingled with pink *Oculina* coral. At the greatest depths we shall come upon the trees and shrubs of *Antipathes*, pallid white though — from their skeletal axis — ineptly called 'black corals'.

Here we have reached the furthest depth that the scuba diver can safely explore. *Antipathes* were already known in east Northland, when by an intriguing surprise they turned up at much shallower depths in Fiordland. Here — in diminished illumination — normally deep communities have moved close to the surface.

Determinants of zoning

Insolation (warming by the sun) and desiccation cease at low tide. Water movement, still significant on high-energy shores, is hardly effective below wave base, generally — on an open coast — at approximately 20 m. The quantum of illumination is controlled not only by depth and transparency but by all the small-scale accidents of topography. Far from a smooth continuity, the subtidal cliffs are complexly broken with ledges, caves, fissures and overhangs. A long vista of a single hue or dominated by one species would be unimaginable. Such uniform zones with extended visibilty are the characteristic of the intertidal.

On nearly every sublittoral surface, a plethora of species contribute to a dappled mosaic. Instead of broad, predictable zones we find a patchwork of colours critically dependent on the angle and amount of lightfall, that is, on whether the slope be low-pitched, vertical or overhung.

After their experience of the relatively simple layout of zones between tides, the first students to venture into the subtidal, at the end of the 1960s, were looking for physical determinants — then assumed to be microclimatic — for the complex patterns they were discovering. Beginning with illumination, the pioneer study at Goat Island by Tony Ayling — then an undergraduate — focused on the critical importance of *depth* plotted against *angle of slope*. As determinants of illumination, depth and slope were measured on a subtidal cliff down to 25 m. The resulting plot of the organisms no longer looked like an actual picture, as with the exposure sequences from between tides. The diagram (Fig. 16.2) had instead become abstract.

OUTER HAURAKI GULF INNER HAURAKI GULF

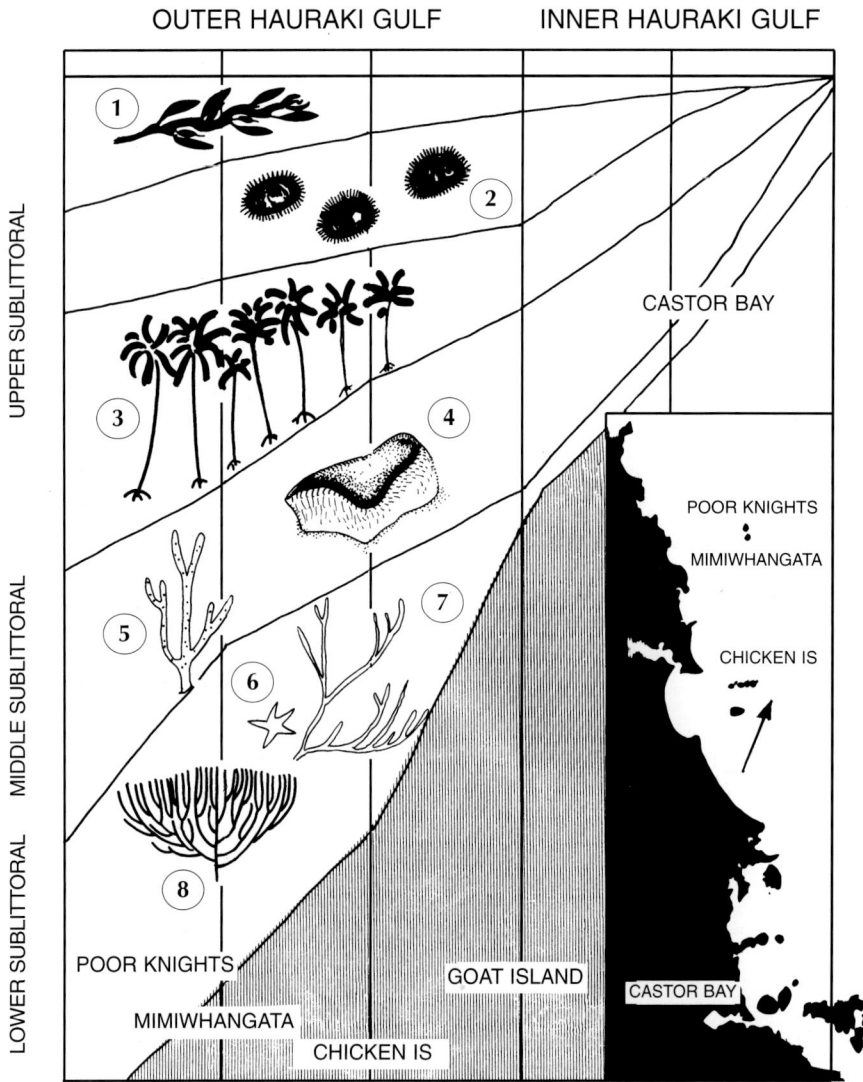

Fig. 16.1 Zoning of sublittoral rock-faces in the Hauraki Gulf and east Northland

The key species shown are
1 *Carpophyllum maschalocarpum*
2 *Evechinus chloroticus* 3 *Ecklonia radiata* 4 black sponge, *Ancorina alata* 5 finger sponge, *Callyspongia* 6 starfish, *Asterodiscides* 7 sponge, *Iophon*, 8 gorgonian.

(From a diagram by Roger Grace.)

At shallow depths with a low slope, sponges such as yellow *Tedania* were found to dominate. Increase of slope favoured ascidians, especially *Cnemidocarpa bicornuta*. With increasing depths at moderate slope, the anemone *Actinothoe albocincta* became common. Further increase both in depth and slope brought a mixture of encrusting species, with the ascidian *Didemnum candidum* most prominent. The greatest depths and darkest underhangs were reserved for a community dominated by the brachiopod *Calloria inconspicua*.

Tony Ayling has shown the influence of depth and slope on the standing crop of the main groups of organisms.

In the early 1970s, many students were to become sceptical of zonation beyond low water, at least as they had been taught to look for it in the intertidal. The broken and mosaic occupation of space, with the difficulty of taking in broad vistas, had made it harder to visualise wide-extending zones.

It began to be questioned whether the zoning concept — useful enough in the rigorous climate of the intertidal — was appropriate to the species-rich subtidal, obvi-

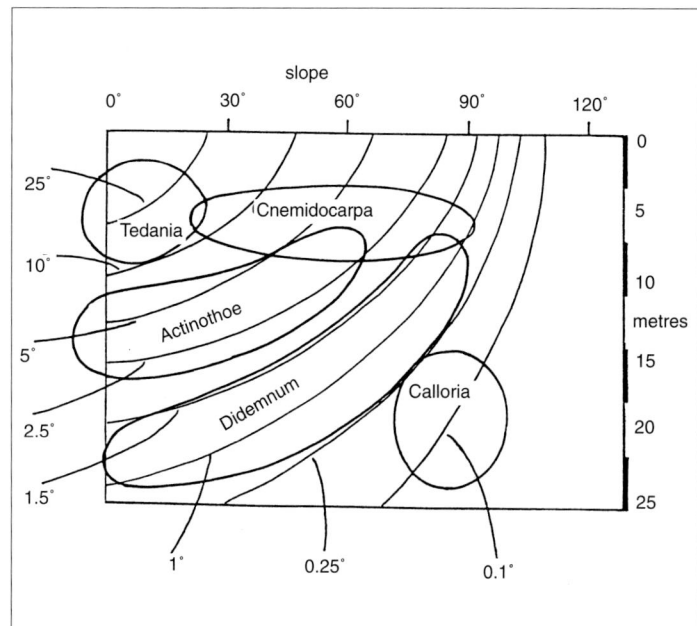

Fig. 16.2 Distribution of organisms in relation to light intensity with depth and angle of slope

The isophotic lines express the percentage of surface illumination. (From Tony Ayling.)

ously colonised with so much diversity and opportunism. At least, subtidal 'zones' — if the term were to be used at all — had to have far greater vertical extent and more internal subdivision. Further, their edges could be complexly intermeshed. Free from the straight-line regularity of the intertidal, the living pattern had to depend more on the accidents of micro-topography with their small-scale effects on illumination.

Moreover, it was soon to become doubtful if climatic constraints, including illumination, could offer a full explanatory frame for species distribution on the shore. The claim of *The New Zealand Sea Shore* (second edition, 1972) that 'we can use a continuum with vertical height, water movement and illumination, to express the position of any plant or animal from the splash-line to the continental shelf, and from open surfaces to deep caves and boulder beaches' might be not only over-ambitious, but even untrue.

No climatic parameters — even light and water movement — could by themselves offer a complete explanation. The spatial distribution of sublittoral species was found to have a new element of unpredictability. This arose from the *lottery effect*, by which species settled opportunistically, as the biotic pressures of other organisms in the time and place allowed.

In 1976 Tony Ayling completed an important account of the role of biological disturbance in the organisation of subtidal encrusting communities in temperate waters.

Several categories of questions derive from the fact that precise predictions of community development and organisation cannot be made in the marine subtidal region. Firstly, is it in fact worthwhile to attempt exercises such as this in an environment where variability and unpredictability are the rule rather than the exception? It is possible to identify a series of mechanisms and pathways, but trying to view organisms as a part of these processes and pathways achieves little. Instead an attempt should be made to look at the consequences on the level of the individual of some of the unique features of many marine animals. How do the different species cope with special problems associated with high fecundity, recruitment, variability, unpredictability in the provision of living space, and unpredictable disturbance from which there is no escape in large size? An individual cannot know in advance the precise array of competing disturbing organisms it will encounter, and must have the flexibility to handle this uncertainty.[1]

Biotic disturbance by the intervention of 'perturber species', rather than being an occasional event, came to be accepted as the rule. Through the years 1975–85 this was brought to light as research went on at Goat Island

Bay and the Poor Knights, rich locations that were, in this same period, to be proclaimed New Zealand's first and second marine reserves.

At Goat Island Bay, the site of the University of Auckland's Marine Laboratory since 1961, the first Reserve was created by Order-in-Council under The Marine Reserves Act 1972 on 28 October 1975. In the area from high water to 40 chains offshore between Cape Rodney and Okakari Point, and taking in Goat Island, all species are protected from commercial or any other exploitation. The Reserve is freely open to visitors, and by a firm consensus of divers and local residents, spear-gunning and leisure fishing have been given up without regret. The University's collecting for teaching was already being done elsewhere, and the taking of specimens for research requires permission from the Reserve's management body, now the Department of Conservation.

In 1981, the (then) Department of Lands and Survey published in three sheets, at a scale of 1:2000, a map of the subtidal habitats of the new Reserve. Perhaps the most detailed record of its kind for any subtidal area, this is based upon information from thousands of individual dives between 1975 and 1979, supplemented with echo-sounding and aerial photography.

The section of this map around Goat Island (Fig. 16.3) reveals with clarity a pattern of zoning for the sublittoral, however different in scale and regularity from that of the intertidal. The formations recognised are:

Upper sublittoral zone
- Shallow, broken rock, 1–12 m, with mixed brown algae on peaks and moderate *Evechinus*.
- Rock-flats 1–12 m, *Evechinus* abundant, *Ecklonia* sparse or absent.
- Dense *Ecklonia* forest 5–18 m, *Evechinus* absent.

Middle sublittoral zone
- Deep reefs, 15–20 m, few *Evechinus* or *Ecklonia*; sponges, ascidians and brachiopods abundant.
- Sponge flats, 12–20 m, sediment-covered rocks with massive and finger sponges abundant.
- Cobbles.
- Sand and gravel.

Upper sublittoral

A short way offshore the upper sublittoral zone becomes transformed from the fucoid margin that links it to the sublittoral fringe into a forest of kelp. Where light penetration is deep, as off Goat Island, *Ecklonia* can reach downwards to 18 m. With its landward margin of *Carpophyllum* reaching up to be emersed at a lower than average tide, we are reminded of the biological insignif-

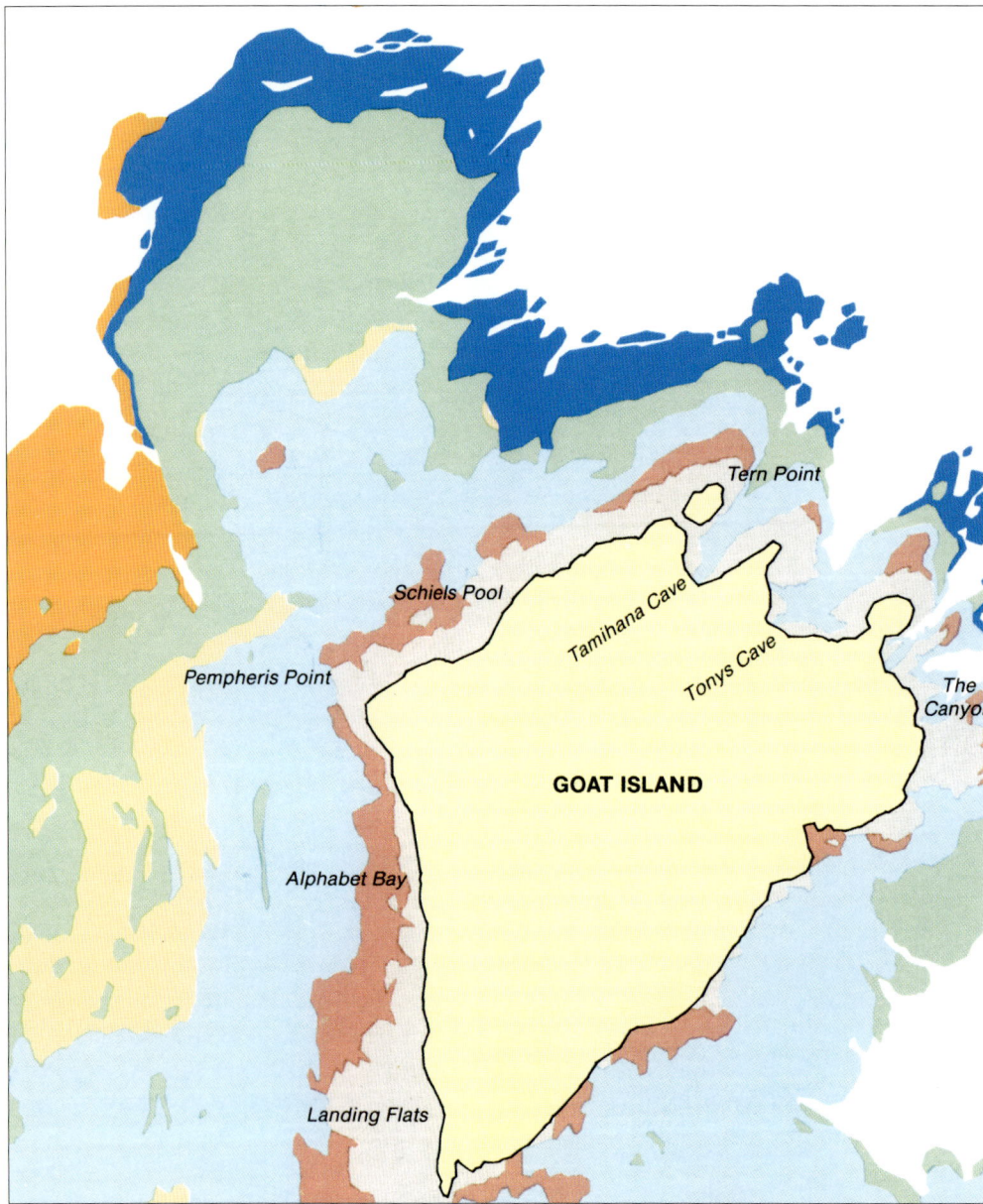

Fig. 16.3 Offshore habitats at Goat Island Marine Reserve

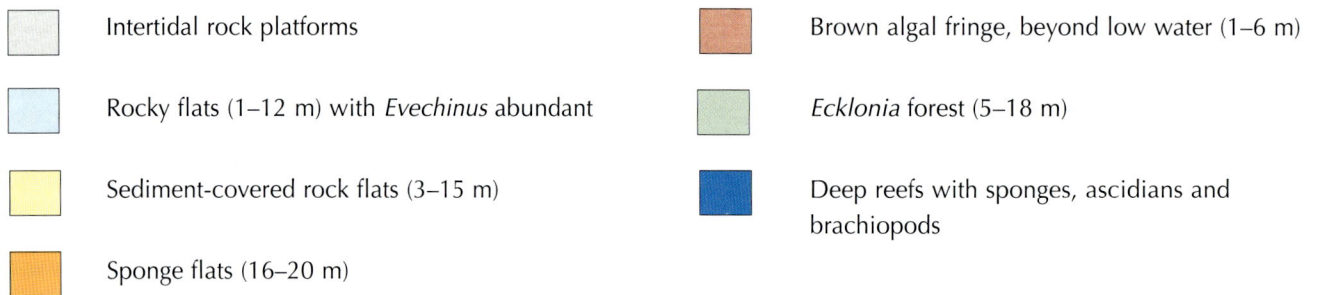

Map labels: Tern Point, Schiels Pool, Tamihana Cave, Tonys Cave, Pempheris Point, The Canyon, GOAT ISLAND, Alphabet Bay, Landing Flats

Legend:
- Intertidal rock platforms
- Rocky flats (1–12 m) with *Evechinus* abundant
- Sediment-covered rock flats (3–15 m)
- Sponge flats (16–20 m)
- Brown algal fringe, beyond low water (1–6 m)
- *Ecklonia* forest (5–18 m)
- Deep reefs with sponges, ascidians and brachiopods

icance of the line of low water. It can indeed be difficult to decide where — on a particular day — it happens to fall.[2]

The kelp forests of the upper sublittoral are dominated in New Zealand by one of the three laminarians: *Ecklonia radiata*, *Lessonia variegata* or *Macrocystis pyrifera*. On cold west coasts of the great continents, the sublittoral zone, up to its tidal fringe, is composed almost purely of Laminariales. In Australasia, a fucoid element is added, with *Carpophyllum*, *Sargassum* and *Marginariella* the leading New Zealand genera of sublit-

toral fucoids. *Cystophora* species on the whole attain less importance below the intertidal. The *Durvillaea* species — conventionally called 'bull-kelps' — are in fact giant Fucales, specialised for heavy wave action. Though *Durvillaea willana* is sited just beyond low water, the essential bull-kelp territory may be said to be 'onshore'.

The brown algae of the upper sublittoral descend to relatively shallow depths. At 20 m, light intensity is reduced — depending on turbidity — to between 0.1% and 10% of the surface value. The algal forests vary in

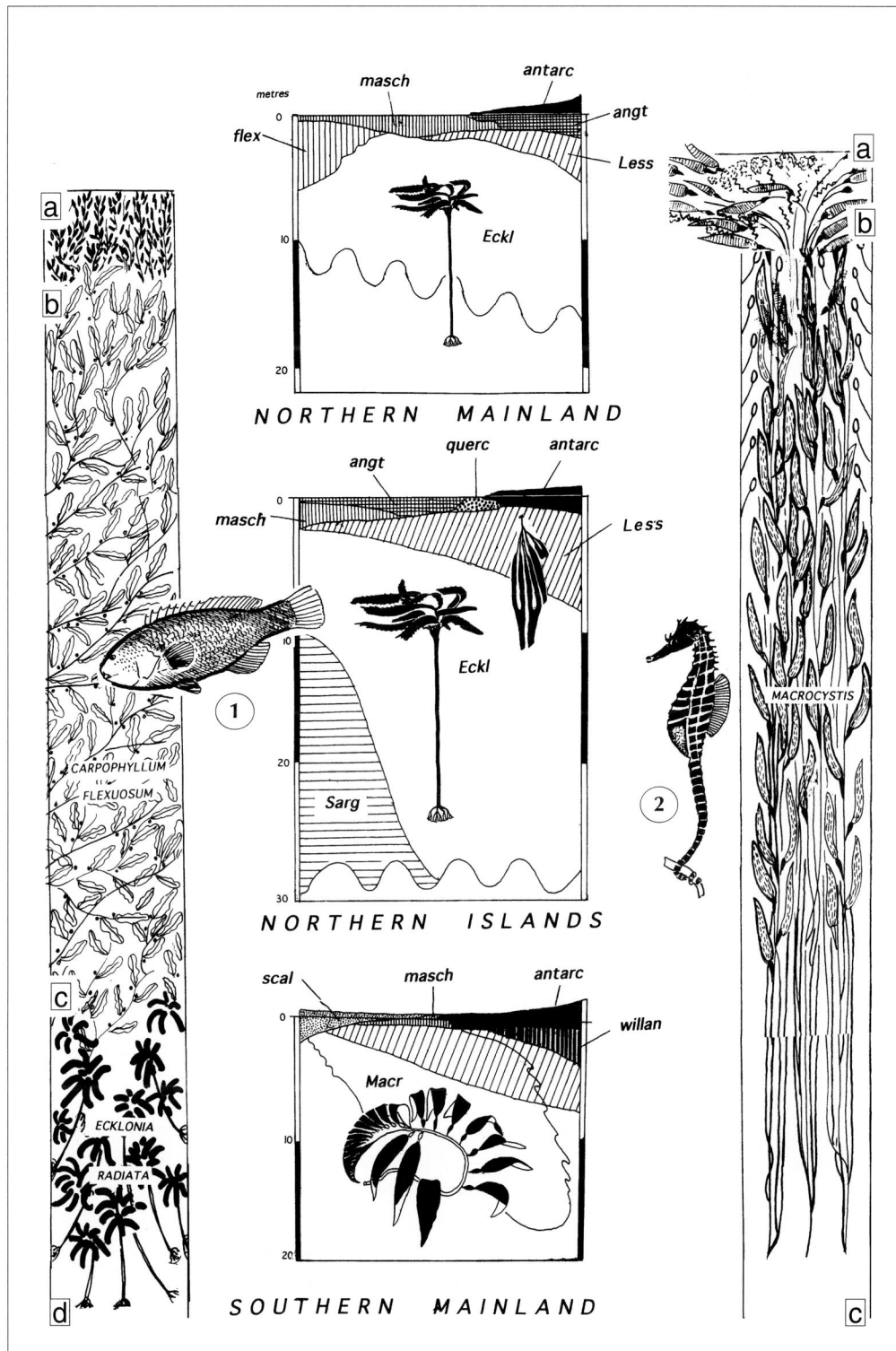

Fig. 16.4 The large brown algae of the upper sublittoral zone
(left) northern depth zonation (sheltered) a–b *Carpophyllum maschalocarpum;* b–c *C. flexuosum;* c–d *Ecklonia radiata*
(right) southern depth zonation (sheltered) a–b *Cystophora scalaris;* b–c *Macrocystis pyrifera*
(centre) Distribution of larger brown algae, from shelter (left) to exposure (right) for northern mainland, northern offshore islands and southern mainland. (Generalised diagrams from Tony Ayling.)
(angt) *Carpophyllum angustifolium* (antarc) *Durvillaea antarctica* (Eckl) *Ecklonia radiata* (flex) *Carpophyllum flexuosum* (querc) *Landsburgia quercifolia* (Less) *Lessonia variegata* (Macr) *Macrocystis pyrifera* (masch) *Carpophyllum maschalocarpum* (Sarg) *Sargassum* spp. (scal) *Cystophora scalaris* (willan) *Durvillaea willana.*
1 banded wrasse, *Notolabrus fucicola* 2 seahorse, *Hippocampus abdominalis.*

composition not only with waves and light, but with change of mean sea temperature from north to south.

In the north, under close shelter, the dominant fucoid down to 10 m is *Carpophyllum flexuosum*. Not only in Auckland and east Northland, but as far south as Tasman Bay (Fig. 16.4), this is the indicator species for quiet waters, presenting a tangled, almost impenetrable thicket from 2 m down, with its foliage usually filmed with silt. Typical animals of this community are the holothurian *Stichopus mollis*, picking up fine detritus and strewing the bottom with castings; the attractively coloured web-star, *Stegnaster inflatus*, and the kina, *Evechinus chloroticus*, in depressions and crevices.

Porae (*Nemadactylus douglasii*) are generally the abundant fish among *Carpophyllum flexuosum* in the north, sucking animals out of the mud or brittle stars from crevices with soft, protrusible lips. Seahorses (*Hippocampus abdominalis*) twine with their prehensile tails. Seeking out small animal prey among the foliage are the paketi, *Notolabrus celidotus*, and the banded wrasse, *N. fucicola*.

Carpophyllum maschalocarpum favours clearer waters in only moderate shelter. It does not reach as deep as *C. flexuosum*, being generally limited to the upper 2 m, with its darker tresses swirling with the waves. At little more than a metre long, its fronds are shorter and tougher than in the very lax *C. flexuosum*; it prefers a rock-base clean or only lightly silted. Common sheltering fish are parore, *Girella tricuspidata*, and the butterfish, *Odax pullus*.

The basal coralline crust carries its large grazing force in the gastropods *Cantharidus purpureus*, *Trochus viridis*, *Cookia sulcata* and *Cellana stellifera*, the last our only subtidal patellid.

The most exposed coasts of east Northland and the Bay of Plenty see *Carpophyllum maschalocarpum* replaced by a third species *C. angustifolium*, distinguished by tough, hardly breakable fronds, sparse, elliptic leaves and general absence of bladders. Even more than the rest of its genus (including *C. plumosum*) *C. angustifolium* is a 'fringe' species, concentrated in the wave-break line, and generally replaced subtidally by *C. maschalocarpum* or directly by *Lessonia variegata*.

The three kelp species contribute little to the sublittoral fringe (though a special growth form of *Ecklonia* moves inshore in turbid secluded waters). The fringe is characteristically fucoid territory, with leafy Carpophylla dominant in the north, giving place in the south to *Cystophora* or to *Durvillaea* on open southern coasts, and the most exposed outposts in the north.

The sublittoral forest round most of the North Island is of *Ecklonia radiata*. Beyond the tides this kelp grows much taller than its marginal stragglers above low water, and the inshore stipe length of c. 0.7 m more than doubles. Its wide blades form a closed canopy, with 7–10 plants to a square metre. The quiet under-space has slow water movement and light intensity no more than 3% of that outside.

On more exposed northern shores can be found *Lessonia variegata*, hardly ever emersed — even briefly — at a low tide. This kelp tends to intervene between *Carpophyllum angustifolium* and *Ecklonia* (which is always the deepest northern kelp). The stipe is shorter and thinner than in *Ecklonia* but more woody. The fronds form narrow straps sometimes lacerated against the rock below. The *Lessonia* forest is not a quiet one, and its lower canopy is open and straggling. Light is not so markedly reduced as under *Ecklonia* and at the exposed extreme the coralline pavement is heavy and gnarled.

In southern New Zealand the sublittoral forest looks wholly different. At shallow depths a fringe of *Carpophyllum maschalocarpum* may remain, or — in more shelter — a subtidal extension of *Cystophora*, chiefly *C. scalaris*. In moderate or high exposure, there are the bull kelps. Thongs of *Durvillaea antarctica* are flung right up to the intertidal or trail back into white surge. Further out, weighed down by their side-paddles, are the arching stipes of *D. willana*.

Below *Durvillaea*, beginning quite shallow in full to moderate exposure, grows *Lessonia variegata*, towards shelter overlapping with the bladder-kelp *Macrocystis pyrifera*. In high exposure *Lessonia* will hold the greater depths on its own. In strong currents rather than waves, the bladder-kelp has undisputed sway, with its intertwined cables striking up from 20 m down, like pylons towards the light.

The tangle of *Macrocystis* at the mouths of southern harbours may form an obstruction to small craft. On more exposed coasts, bladder-kelp continues abundant at greater depths. From the leaves photosynthesising high up, metabolites can be translocated to the base, where tissue production is going on. New fronds grow up from just above the holdfast at half a metre a day, lasting for five to six months, then dying back to be replaced by others. Under the canopy light may be reduced by up to 98%. Though not wave-stressed, the leaves and branch tips that top the soaring groves receive constant exchange of water.

Beneath or adjacent to the bladder-kelp, the rocks may be covered with sea-tulips, *Pyura pachydermatina*, on stalks up to 1 m long. Loaded with epiphytic algae, *Pyura* also harbours at its base a wealth of other tunicates and sponges. In Otago, 20 ascidian species have been recorded from a *Macrocystis* forest.

Small islands

In east Northland, at the edge of the continental shelf,

small islands such as Piercy, Poor Knights and Moko-hinaus, receive full ocean swell. Wave refraction ensures that from whatever direction the wind blows, the whole island has a coast of sustained high energy. In these oceanic waters, algae can function at twice their main-land depth. *Carpophyllum maschalocarpum* and even *C. angustifolium* are now relegated to the least exposed sites. In deeper water, *Lessonia* dominates in relative shelter. On exposed promontories the cold-water *Durvillaea antarctica* can oust *Carpophyllum angusti-folium* at high levels, giving place directly to *Lessonia* below.

On these offshore sites the fucoid *Landsburgia quer-cifolia* comes best into its own. Found on the exposed mainland from North Cape to Wellington and in the north of the South Island, it forms forests only on outer islands, including the Chathams.

Below *Lessonia*, as can be well seen at the Poor Knights, *Ecklonia* becomes dominant down to 45 m. In shelter, and from 16–45 m, *Sargassum scabridum* may often occur; at the Three Kings, this forest has a spe-cially interesting composition, of *Sargassum johnsonii* and *Perisporochnus regalis* (Fig. 10.6).

Goat Island upper sublittoral

The whole upper sublittoral zone is unified by its basement crust of coralline paint. Where conditions are right it will carry a forest of pure *Ecklonia*. Around low water at Goat Island, there is a fringe of *Carpophyllum maschalocarpum*, sometimes with *C. flexuosum* dominant at 3–6 m. *Cystophora retroflexa* and *C. torulosa* sometimes replace *Carpophyllum*, sometimes mixed with *Sargassum scabridum*. The under-layer is of *Melanthalia abscissa*, and *Pterocladia lucida*, *Pterocladiella capillacea* and *Osmundaria colensoi*.

On the steepest rock slopes the brown algae are largely replaced by animal communities. Down to 11 m, Goat Island has a sponge/ascidian formation, with brick-red *Hymedesmia* (*Tedania*) and *Cnemidocarpa bicornuta*. From the pink basal crust spring up fronds of *Corallina officinalis* and several *Jania* species. The glob-ular sponge *Tethya australis* is often found, along with the tufted bryozoans, *Bugula*.

From 11–18 m, rock slopes tend to be dominated by the anemone *Actinothoe albocincta*, in stands of more than 1000/sq m. Needing water movement to feed, they close down tight in calm weather.

Ecklonia forests

Though *Ecklonia* can reach up into *Carpophyllum* groves, its proper domain is below the major water movement. With no real strength against wave attack, it comes above wave-base only on low energy coasts,

where lowered illumination permits. The holdfast is weak under stress, with *Ecklonia* generally the first alga to be thrown up by storms. Its range is thus influenced by severity of wave action, illumination and — very importantly — by the activity of *Evechinus chloroticus*, a species with profound disturbance effects on the *Ecklonia* community.

At Goat Island, *Ecklonia* forests become continuous at about 5 m, and can descend to 18 m. At upper levels the stipes are no more than 60 cm long but the deeper groves, from 15 m down, at the Marine Reserve, are made up of plants reaching 1.5 m tall and at least five years old. Shorter plants in densities of about 60/sq m form a closed canopy; whereas the taller *Ecklonia* — at densities of 5–10/sq m — are easier to move through.

The smallest *Ecklonia* are those being recruited at the margin. Juveniles first appear in June, and by Octo-ber may number thousands per square metre, being added to as long as ground-space affords, right through to March. The primary lamina grows rapidly from the meristem, where the blade joins the stipe. Small lateral pinnae soon develop, increasing markedly in a year. Growth is fastest in spring and summer, slowest in win-ter. After the first year, the stipe markedly extends, and the blade becomes heavy with side pinnae. Growth goes on vigorously for three years, with a full span of six years or more. Reproduction begins in the second year, and with spore release, the wrinkled fronds lighten from brown to yellow.

The blade and ribbons of *Ecklonia* have their own faunule, notably of silvery cheilostome Bryozoa. Most typical are the laceworks of *Membranipora mem-branacea*, with straight rectangular zooecia, hinged to bend slightly with the swaying blade. *Electra pilosa* is the second common bryozoan, bristling with minute spines. The compound ascidian *Botryllus schlosseri* spreads its small rosettes on smooth *Ecklonia* blades, and white spirorbid tubeworms settle on the wrinkled laminae. Thecate hydroids flourish on kelp, notably *Obelia longissima, O. geniculata, Halecium corrugatis-simum, Clytia hemisphaerica* and *Sertularella simplex*, along with their attendant predators, aeoliid slugs and minute *Doto*.

Ecklonia has its full assemblage of browsing gas-tropods. The leading three are the topshell, *Trochus viridis*, the cat's-eye, *Turbo smaragdus*, and half-grown *Cookia sulcata*. The most exquisite are the *Cantharidus* topshells, living on kelp like rissoids on more delicate algae. The largest is *Cantharidus opalus* (up to 4 cm), marbled with purplish red, with the mouth peacock iri-descent. *C. purpureus* (2.5 cm) is dull greenish pink in life, but dries to a dull rose-pink; its flat spiral ribbing is finely cross-sculptured.

The two smaller cantharid browsers are *Micre-*

Fig. 16.5 The community of the upper sublittoral zone at Goat Island
a–b shallow sublittoral, with *Carpophyllum;* b–c grazed *Evechinus* meadow; c–d *Ecklonia* forest
1 *Evechinus chloroticus* 2 mature *Ecklonia* canopy 3 *Cookia sulcata* 4 *Micrelenchus sanguineus* 5 *Trochus viridis*
6 *Cantharidus purpureus* 7 red moki, *Cheilodactylus spectabilis* 8 butterfish, *Odax pullus* 9 snapper, *Pagurus auratus*.

lenchus dilatatus, bronze to pink with rounded aperture, and *M. sanguineus*, greenish with blood red splashes over its spiral ribbing, or tessellated with white. All these topshells move actively with the conical shell hanging from a close-gripping foot or hauled forward over the swaying plant. A dislodged snail can hang fast by the viscid mucus of the foot.

Ecklonia are freely browsed and nibbled by fish, most commonly the butterfish, *Odax pullus*. Adult males have a distinctive look with the dorsal and anal fins elongated back to the tail tip and used in courtship. The young are shorter finned with white dashes from snout to tail. Like a wrasse, *Odax* has the teeth fused into a beak. Only the tender central part of each frond is taken, sucked flat against the open mouth. One bite can neatly remove a disc, which is cut up by sharp pharyngeal teeth before swallowing. Older kelp can be left tattered with large areas dead or dying.

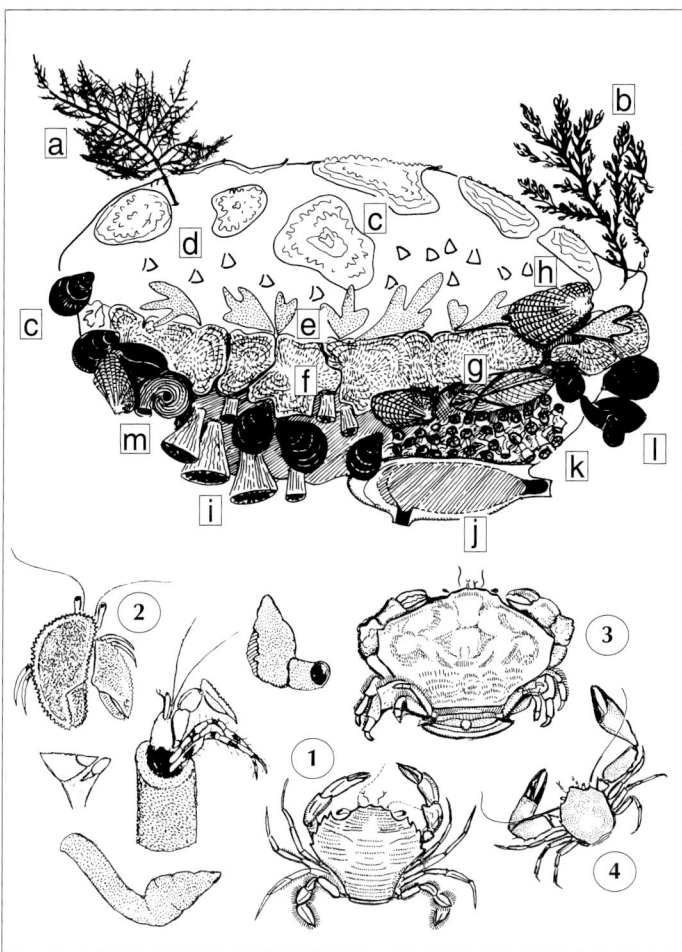

Fig. 16.6 Small boulder community in grazed upper sublittoral zone: east Northland
(a) *Pterocladiella capillacea* (b) *Plocamia costatum*
(c) patches of encrusting coralline (d) *Balanus trigonus*
(e) *Rhodymenia leptophylla* (f) cheilostome bryozoans
(g) *Barbatia novaezealandiae* (h) *Cardita brookesi*
(i) *Monomyces rubrum* (j) *Corella eumyota* (k) *Culicia rubeola* (l) *Calloria inconspicua* (m) *Serpulorbis zelandicus*
Crabs: 1 *Liocarcinus corrugatus* 2 a hermet with a bryozoan extension to its gastropod shell 3 *Petalomera wilsoni* 4 *Petrolisthes novaezelandiae*.

Lying generally between *Carpophyllum* above, and intact *Ecklonia* forest below, the productive open meadows run down from 3–12 m. They are pink with coralline and studded with dark kina in their pits. Where secure from collecting, the urchins can reach 40/sq m, with an average of 8/sq m. They initially clear the rock, then maintain a short-cropped meadow that — left alone — would soon return to *Ecklonia*.

Kina clear kelp only from the edges of forest stands. Their mobility to forage is interrupted close to low water by breaking waves, and the few kelp that can withstand wave movement are left immune from urchins.

As in any forest, tree-fall leaves a light gap, where ground productivity increases. The new sporelings and filamentous algae will then be grazed by urchins and gastropods, so that the full underworld of red algae is held back, though up to 10% of the rock may have coralline turf, and transient patches of succulent reds.

The bottom is covered by small pink-crusted boulders, often with the yellow sponge *Cliona celata*. Grazing gastropods abound, not themselves clearing the kelp, but well able to keep the ground bare. Five common species (*Cantharidus purpureus*, *Trochus viridis*, *Micrelenchus sanguineus*, *Cookia sulcata* and *Cellana stellifera*) can together reach 50/sq m.

Deforested flats are lacking in shelter, but several fish come out of the kelp to graze, at Goat Island mainly snapper, goatfish, paketi and blue cod. In summer the snapper take a high toll of gastropods, bivalves, crabs and small urchins. Old paketi emerge to feed on crabs, hermits, small mussels and amphipods. The goatfish, *Upeneichthys lineatus*, explores the bottom with sensitive chin barbels; its fine teeth seize amphipods and small crabs. The blue cod, *Parapercis colias*, stalks smaller fish or takes up whole crabs and hermits to be crushed with the strong pharyngeal teeth.

Dynamics

Without kina grazing, a closed forest of *Ecklonia* would directly abut with *Carpophyllum*. In keeping open a

Fish feed also from the rich community in the *Ecklonia* holdfasts (see Fig. 16.14). Most typical at Goat Island is the red moki, *Cheilodactylus spectabilis*, conspicuous by its red-brown stripes and fins, set against a pale ground. With their fleshy lips, moki suck up sediments with small invertebrates from both turf and holdfasts. Linda Leum gave a good account of their behaviour around Leigh, feeding non-selectively on crabs and smaller crustaceans, polychaetes, urchins and brittle stars.

With low lighting under the canopy, the productivity of the floor is comparatively small. Advantage goes to filter feeders living in the holdfasts, drawing food from outside into their neighbourhood.

Kina flats
There are large areas of the upper sublittoral that *Ecklonia* fails to hold. The kelp has given way to low meadows that, like grazed pastures on land, have their own productivity higher than the long-standing biomass of the forest. At the forest edge the urchin *Evechinus chloroticus* can get good access to *Ecklonia*. The stipe is then gnawed right through just above the holdfast, with several urchins converging to consume the fallen plant. At the forest's upper bounds, *Ecklonia* would be spreading by juvenile recruitment if *Evechinus* were not holding it in check.

productive pasture the urchin is (like the starfish *Stichaster* that removes mussels) a 'keystone species', promoting greater diversity. Perturbations by a grazer or carnivore change the community's final state, switching it to an alternative subclimax.

The dynamics of the upper sublittoral at Goat Island were studied from the 1970s by Howard Choat and his students. Removal of *Ecklonia* was found to promote small red and brown algae, soon to be urchin-grazed. Only in spring and early summer, with higher nutrients and stronger light, could algal growth keep pace with grazing.

In shallow areas there is also gastropod grazing, by topshells, turbans and star limpets and also by the paua species, *Haliotis iris*, *H. australis* and *H. virginea*. All these, with their preference for flat surfaces, can finely crop the meadow, clearing the small algae left behind after an urchin's coarse grazing.

A grazed surface keeps its pink veneer, over-filmed with sporelings and filamentous algae, that if left alone would lead to a second-growth forest. The primary surface off Goat Island was found to be shared by crustose coralline algae (65–75%); simple and compound ascidians (0.1–3%) and encrusting sponges (19–24%). Temporary space-holders (*Cnemidocarpa*, *Calloria* and many sponges) are short-lived and seasonal; being opportunists they vary in performance from year to year.

As well as their 10% permanent holding, succulent reds make augmented flushes in spring. The yellow sponge *Cliona* is a regular variant on the algal pink, its yellow pustules erode the calcite of dead shells, serpulid tubes, or *Lithophyllum*, and join up in massive crusts. The interstices develop a rich microfauna.

On an encrusted boulder from the productive flats, as compared with one from the sublittoral fringe, algal paint has been reduced to a few patches at the top. The rest is animal-encrusted; or with security from urchin-grazing, a red algal fringe of *Plocamia* and *Rhodymenia* can develop.

Evechinus thus releases wide settling space. Few sponges can withstand its attack, but *Hymedesmia lundbecki* can tolerate moderate grazing, and persists in small clusters even when grazed heavily. It is suspected to be toxic to *Evechinus*. The fast-growing red sponge *Microciona coccinea* can also thrive under such disturbance.

When grazing had removed most *Hymedesmia*, the advantage was found to pass to *Cliona*, with minute fragments surviving under coralline crusts, building up to 20% of cover.

When urchins were excluded with stainless steel cages, *Lyngbya* filaments and turfing algae soon appeared, like long grass and scrub on a pasture with cattle removed. Jointed corallines were soon springing

up from their crustose base. But 15 months saw no return of *Ecklonia*.

With kina as the prime disturber, no single equilibrium point can be as reliably predicted as in the intertidal. Community structure at any one time can only be explained by historical events. Neither the pathways nor the end-point are fixed. Particular happenings (settlement, disturbance, predation, extremes of climate) act as switches. The result is a 'lottery effect'.

The second disturber is the trigger fish or leatherjacket, *Parika scaber*, feeding on encrusting animals. One of our commonest subtidal fishes, it achieved portraiture by Sydney Parkinson on Cook's first voyage. The body is rigid and diamond-shaped, with an ensheathed dorsal spine, and a shorter one buried in the skin of the belly. With the tail-fin folded down, the trigger fish slowly advances or reverses with the dorsal and anal fins, keeping the tail-fin folded down. Ayling stresses the large independently swivelling goggle eyes that sit on top of the head, with the whole unlikely body encased in a thick skin with sandpaper-like scales.

Parika is our only fish to take two-thirds of its food as sponges and ascidians, but it can also consume hydroids, barnacles, seaweeds, bryozoans, jellyfish and even lunge into a fortress of urchin spines. The small mouth is chisel-toothed, with jaw leverage giving immense pressure at the cutting edge. Schools of leatherjackets moving over the bottom with heads down and tails aloft have been likened to a field of strange plants.

At peak numbers trigger fish were estimated to take 35,000 bites annually from each square metre of surface. Low densities produced no visible effect, but at 42 per 500 sq m the change was dramatic. They released 9% of the ground-space, removing all or part of the large sponges *Ancorina*, *Cliona* and *Polymastia*. Their major effect was to prevent ascidian, bryozoan and sponge recruitment by removing newly settled individuals.

The leatherjacket courts and spawns from June until October. Between December and April the minute greenish young settle out of the plankton to shelter in *Ecklonia* until four months old and 10 cm long. They reach 20 cm in the first year and are adult at 25–35 cm before the end of the second. They then cease to grow, although they may live six years.

A third biological perturbation could result from the episodic bacterial infection of sponges, degrading large colonies by rotting and releasing significant free space.

With biotic intervention, multiple-stand configurations are the rule. At Waterfall Flats, Goat Island, a disturbance-terrain of encrusting coralline, *Cliona* and tufting algae, resulted from four years of high urchin and gastropod density. On the adjacent flats there was dense *Ecklonia* forest with few urchins, but a high population of leatherjackets and gastropods. A third expanse was dominated by the *Hymedesmia* sponge, and coralline turf with *Microciona* and *Cnemidocarpa*.

Middle sublittoral

At Goat Island, the last reach of the subtidal cliff before the sand-strewn continental shelf is known among divers as the 'Deep Reef'. Composed of large tumbled boulders at the base of the steep face, it is preceded above by a last attenuate fringe of *Ecklonia; Evechinus* is scarcely to be seen.

We have passed to a world of new colours. Pink crusts of coralline and lithophylla have given place to a mosaic of encrusting animals, diverse and multi-hued. On an immense scale it is like turning over a rich boulder. With delighted realisation, the diver has entered into a realm set off in all its novelty from anything seen above.

Sponges now cover around two thirds of the surface. Coralline paint is hardly seen anymore, though a few low-lit red algae remain, commonly *Plocamia costatum* and *Rhodymenia leptophylla*. The basic industry is no longer one of photosynthesis nor grazing, but of filter feeding: reliant on the harvest of plankton and dead remains raining down from the lighted productive zone above.

As well as sponges, the bryozoans and ascidians figure largely. There is usually a broad depth gradation in colour, with the shallowest levels reddish brown, tawny or grey, breaking next into the brighter chrome, pink and tangerine. Then — towards the deepest reaches of the Deep Reef — the background pales to predominantly white, but always spangled with high points of scarlet.

Currents are still sufficient — at or below wave-base — to keep the surface clean of loose particles. Though most sponges are adapted to cope with sediment, this proficiency becomes more important in the sand-strewn Sponge Garden further out.

In its sponge communities most of all, the middle sublittoral enjoys a new freedom of growth. With the slowing of water movement the sponge shapes can run to far more variety, as will again be more evident when we reach the quiet, little-disturbed Sponge Garden.

First, and accounting for the greatest biomass on the Deep Reef, are some heavy sponges already glimpsed under boulders at low tide. *Ancorina alata* is nearly everywhere the most visible, steel grey and like shark-skin to the touch, but yellow and densely spiculose inside. Far more than in the sublittoral fringe, it now grows in heavy lobes, or hollows out into deep bowls. In vivid contrast is the yellow *Cliona celata*, beginning as pustules among calcareous sediments, and finally breaking the surface to fuse into wide crusts.

Other massive sponges, rough-skinned and dense with spicules, belong — with *Ancorina* — to the Order Astrophorida. *Stelletta conulosa* is slate-grey and triangular in section, like a loaf with a smooth strip, pimpled with low conules. The orange *Stelletta crater* forms a massive cup and the maroon *S. maori* is also cup-like. Yet another cup sponge is the hard and stony *Geodia regina*, generally living in faster currents higher up, often among *Ecklonia* holdfasts.

The hadromerine sponges are here in plenty, firm and spiculose, and chiefly spherical or hemispheric. The grey-brown tennis balls of *Aaptos confertus*, and orange golf-balls of *Tethya aurantium*, are familiar from clouded waters at low tide but grow larger here. The smooth, flesh-coloured *Tethya australis* is most favoured by fast currents. Tangerine *Polymastia granulosa* grows as bright hemispheres, putting up thick papillae.

In the quieter water, the finger sponges for the first time come into prominence. Conspicuous by its bright yellow is *Iophon minor* with its upright branches irregular and anastomosing. Orange red *Raspailia topsenti* produces large, blunt fingers mounted on a small stalk. Dull mauve, slender-fingered *Callyspongia ramosa* is one of the commonest subtidal sponges, frequently cast up on beaches after storms.

Of smaller biomass, yet prominent by their colours in archways and passages, are the horny sponges (Dendroceratida), that lack any spiculose skeleton. Yellow *Darwinella oxeata* spreads in rubbery, yellow sheets, beset with sharp conules. *Chelonaplysilla violacea* is dark purple, also finely prickled. The salmon pink *Dendrilla* species are tree-like, with chunky, sharp-angled branches, elastic and slimy to the touch. The black *Ircinia novaezelandiae* forms tough, compressible crusts, with a reticulate and conulose surface.

Towards the deeper shade, the prevailing colour changes to cream or white, with numerous bright points of red. The clean tubes of the serpulid worm *Galeolaria hystrix* are painted with scarlet. *Filograna implexa* tubes form a spreading filigree, or are bound into crusts and bundles, inter-fusing and putting out small crowns of scarlet tentacles. The smallest and most prolific tube-worms are the tiny coiled spirorbids, scattered over every available surface.

The red brachiopod *Calloria inconspicua* attaches by its peduncle to the rock or other shells, and forming wide spreads over the steepest, most shaded surfaces. As an ancient survivor, this lampshell finds its refuge in the darkest places, in suitable sites reaching up to the secluded intertidal.

The same dark preference is shown by *Monomyces rubrum*, our largest cup coral, conspicuous by its bright orange oral disc. Under stones at low water, the corallite is short and cylindrical, but in the sublittoral the well-

Fig. 16.7 'Deep reef' community of the middle sublittoral off Goat Island, east Northland

a–b *Ecklonia* at its lowest reach, with finger sponges

b–c zone of massive and branched sponge predominance

c–d paler zone with red high spots (*Monomyces*, *Calloria* and serpulid worms)

1 *Ecklonia radiata* 2 leatherjacket, *Parika scaber* 3 *Ancorina alata* 4 *Ophlitaspongia reticulata* 5 yellow finger sponge, *Iophon*
6 serpulid tubeworms 7 *Monomyces rubrum* 8 *Calloria inconspicua* 9 *Ircinia* sp. 10 *Aaptos confertus* 11 *Iophon minor*
12 *Ircinia novaezealandiae* 13 *Axinella* sp. 14 *Dendrilla* sp. 15 vase sponge, *Leucettusa* 16 spirorbid tubeworm
17 *Galeolaria hystrix* 18 *Filograna implexa* 19 *Hydroides elegans* 20 *Elzerina binderi* 21 *Hastingsia* sp. 22 *Jasus edwardsii*.

grown cup becomes fan-shaped, with a narrow pedicel.

Finally, there are the calcareous sponges, pure white or occasionally pink-tinted. Confined to the most shaded spots beneath ledges or boulders, most are frag-ile and easily crushed. A common species, *Leucettusa lancifer*, forms small, wide-mouthed urns, joined by a common base (Fig. 16.7).

The bryozoans of the Deep Reef form mostly

encrusting sheets, but they have increasing freedom of lamellar growth in quiet water. The most conspicuous is perhaps *Steginoporella neozelanica*, with upstanding fingers, orange brown and bright at the tips. On its stalks often grows the erect and frondose *Beania bilaminata*, beige or yellowish brown.

Compound ascidians are far more diverse in the sublittoral than under low tidal boulders. Important crustose species include the sand-impregnated *Ritterella arenosa*, and the white or pale pastel *Didemnum* species. Additional Didemnidae are the tawny, translucent sheets of *Diplosoma listerianum*, and the flask-shaped colonies of ochre-coloured *Polysyncraton chondrilla*.

Short-stalked ovate heads, often massed in colonies, are the common growth form of the Polyclinidae. *Pseudodistoma aureum* forms firm plates thrown into flattened yellow lobes. *P. novaezelandiae* presents a mass of transparent heads with brilliant orange zooids. Some of the polyclinids are sand-impregnated, such as the crimson *Synoicum kuranui*, and the lobate sheets of *Aplidium scabellum*, out towards the Sponge Garden. The most attractive *Aplidium* still evidently lacks a name; its large, frosted white hemispheres are faceted with windows of the red zooid systems.

The Family Clavelinidae produces stalked heads, each being a single zooid. *Euclavella claviformis* is colourless except for a bright blue hyperbranchial groove showing transparently. *Pycnoclavella kottae*, from east Northland, has pale orange, yellow-stalked heads.

The large foragers below the kelp-line include the numerous fish, nibbling or fossicking from the living crust, to which we will later return.

In common repute, the best known forager is the spiny crayfish or rock lobster, *Jasus edwardsii*. The first sign of a squadron of crayfish under a ledge or in a fissure is generally a forest of antennae all aquiver. Spiny crayfish stay withdrawn by day, clinging vertically or upside down, but come out at night to walk on tip-toe, or hold their position as if by treading water with the paddles of the abdomen. *Jasus* takes all manner of dead or living prey, from urchins and tubeworms to fish and even oysters, which are opened by breaking the thin edge of the shell.

Alister McDiarmid's research, showing the upsurge of the spiny crayfish since the gazetting of the Goat Island Marine Reserve, is the classic reference work for its breeding and population biology.

The Sponge Garden

Beyond the Deep Reef the sponges gain their final monopoly of the shore. From the foot of the middle sublittoral at Goat Island a gently inclined platform extends outward, from 15–22 m deep. Thinly covered with white sand, this slopes off eventually into the deep sediments of the continental shelf.

The inner reaches of this flat form the Sponge Garden, with the sand shallow enough to allow sponges to grow through it from the rock below. We have seen in the intertidal how the sponge *Hymeniacidon perleve* can spring up where sand collects in *Corallina* turf. In the Sponge Garden where sand reaches 5 cm deep, the sponges have little else to compete with. Some of the larger ones are estimated to have reached a century old. Growing slowly and rarely succumbing to predators, they change little from year to year. Wide formations of sponges must have remained stable through prolonged stretches of time. Low tidal shores could even have looked like this at the opening of the Cambrian era.

The panorama of miniature trees beyond the influence of light or strong currents has an outlandish, otherworldly look. Its uniform cover is interrupted only where a sand river winds through, or a natural gutter is filled too deep to allow sponges to grow.

Sponges flourish well in clouded water, with primitive but effective ciliary systems to take up the rain of plankton, and to dispose of unwanted sediment. The shapes already met with — vases and bowls — are all here, but the landscape takes its character from the finger sponges, striking up through a hand's depth of sand, to 30 cm or more in height. The more massive cups, vases, bowls, clubs or spheres grow only where the sand cover is light. There is a further range of encrusting sponges, to be seen only by sweeping sand away, to expose their colour and texture.

The branching sponges include mauve-brown *Callyspongia ramosa*, perpendicular to gentle currents, orange-red *Raspailia topsenti*, and apricot-grey *Chondropsis kirkii* with thick fingers, often growing up in crests or lamellae. A new species of *Pararhaphoxya* has pointed fingers with some of its branches anastomosed. *Axinella* species are abundant, one being orange, with flexible fingers in a flat plane, another yellowish, soft and slimy, with knobbly fingers. The dull yellow *Homaxinella erecta* has a single upright branch, while orange-red *Phakellia dendyi* is broad and fan-like.

The massive sponges include the widespread *Ancorina alata*, *Stelletta crater*, *S. maori* and *Cliona* species. A grey, cup-shaped *Ircinia* reaches 20 cm high on its stalk, having a hard skeleton of spongin with collagen fibres.

Many spherical forms are already familiar, including *Aaptos confertus*, *Tethya australis*, *Suberites perfectus* and *Polymastia fusca*. There is also a shaggy hemispherical *Cinachyra*, yellow-grey and compressible, showing long projecting spicules when sand is removed.

The part-hidden crusts include a cream *Polymastia*,

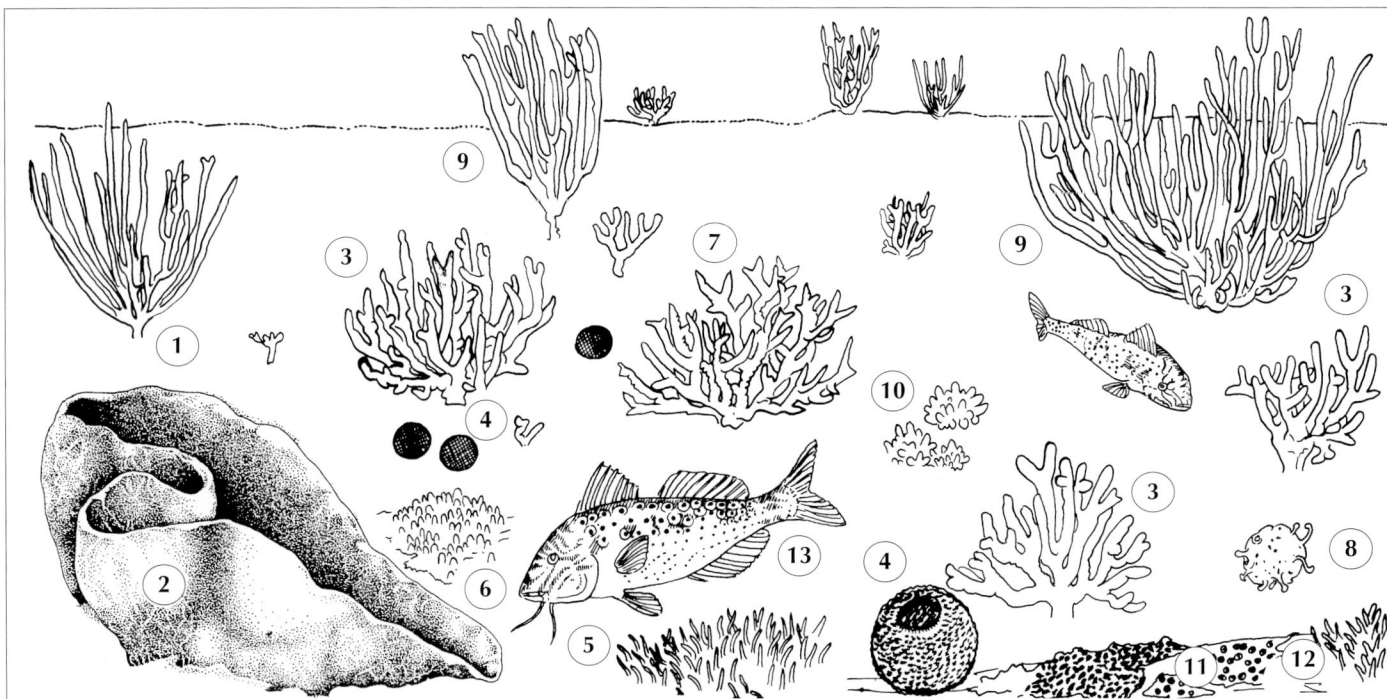

Fig. 16.8 Goat Island, Leigh, east Northland: a prospect of the Sponge Garden
1 *Axinella* sp. 2 *Ancorina alata* 3 *Raspailia topsenti* 4 *Aaptos confertus* 5 *Polymastia agglutinans* 6 *Biemna* sp.
7 *Pararhaphoxya pulchra* 8 *Tethya australis* 9 *Callyspongia ramosa* 10 *Polymastia granulosa* 11 *Ophlitaspongia reticulata*
12 *Hymedesmia* sp. 13 goatfish (red mullet), *Upeneichthys lineatus*.
(By courtesy of John Walsby and Vivienne Ward.)

with papillae from a basal mat in medium to coarse sediment, whereas *P. granulosa* and *P. hirsuta* are only lightly sand-tolerant. A yellow halichondrid, a species of *Ciocalypta*, puts up firm papillae bearing the oscula. There is finally a white lithistid sponge, a species of *Monanthus*, brittle to the touch, with corrugated turrets springing from a thick base. This sponge has interlocking desmaspicules, a remnant of earlier times when lithistid sponges dominated the seas.

Towards the lower sublittoral

The Sponge Garden of the middle sublittoral marks the diving limit at Goat Island. With the subtidal cliffs cut off by the shelf, no room is left for the cnidarians (gorgonians and antipatharians) that typically belong to the *lower sublittoral*. For these we must move to those deeper reaches of Northland at Mimiwhangata and the Poor Knights.

Mimiwhangata

Gorgonian sea-fans were known in the north only from the Poor Knights, until the discovery of diving grounds at Mimiwhangata, near Helena Bay, south of the Bay of Islands. The history of Mimiwhangata is a portent of good hope. In the 1960s a coastal farm had been acquired by New Zealand Breweries Ltd for a tourist resort. When the quality of its beaches and offshore reefs was realised, the Chairman, Sir Geoffrey Roberts, commissioned a Report by three marine biologists, Bill Ballantine, Roger Grace and Wade Doak. As a result the Company gifted Mimiwhangata to a private trust as a reserve, today managed by the Department of Conservation.

The Report completed in 1973 revealed the richness of the subtidal. Down to 20 m, *Ecklonia* forests covered every slope that was not too steep or shaded. Further out was a more barren rock platform, cut by winding canyons up to 5 m wide and dropping to 25 m deep. Under tidal currents and wave surge, their twin faces were clad in an amazing array of filter feeders: tall candelabra of *Raspailia* sponges, hydroid trees of *Solanderia* and scarlet branches of the sponge *Darwinella gardineri*. At ground level were myriads of gem-like bryozoans, compound ascidians, jewel anemones and finely sculptured calcareous sponges.

The sandy slope from the canyon mouths to seaward was next to be explored. Echo-sounding detected rock formations a metre above the sand; by dropping a grapnel-line on these, the divers descended to 35–50 m, the maximum safe depth for scuba.

The rock stacks were dissected with trenches and covered with a delicate primnoid gorgonian, up to 54/sq m. Mingled with these sea-fans were exquisite assemblages of sponges, regular and symmetrical, forming bright red antlers; yellow candelabras; golden lattices, large grey goblets and lilac vases. With these grew delicate alcyonaceans, and zoanthids like yellow daisies investing dead fans. Here and there lay the saffron yellow starfish, *Knightaster bakeri*, formerly known only from the Poor Knights and other island sites.

A third zone at 50 m deep was kept bare by sand abrasion, but an elevated stack was found with the branching ivory coral, *Oculina virgosa*, first discovered in New Zealand by Tarlton and Doak at 50 m deep on the West King. Its thickets reach 20 cm high and under water look sombre brown, but are bright pink when brought to the surface. The corallites are slightly protuberant corallites and spirally arranged.

Lower sublittoral Anthozoa

In all the Class Anthozoa, the medusoid phase is wanting. Their achievement has been the elaboration of the polyps. The largest, and most elaborate polyps are the sea anemones, **Actinaria** — that have remained solitary, with many species between the tides. For the rest, the anthozoans are nearly all colonial, living — with a few exceptions — well beyond low water.

The **Anthozoa** fall into two subclasses, **Octocorallia** and **Hexacorallia**. The first, through all their colonial diversity, share one point of recognition. The polyp has always eight pinnate tentacles. Its internal cavity is divided by eight mesenteries, with their thickened margins — like the tentacles — armed with cnidoblast cells carrying nematocysts. The gonads are never external as in Hydrozoa, but lie inside upon the mesenteries.

The Subclass **Alcyonacea** consists of the 'soft corals' typified in New Zealand by *Alcyonium aurantiacum*. Under their old name 'dead men's fingers' the thick branches are built of a fleshy coenenchyme, derived from the *mesogloea*. Their surface is studded with star-like polyps with their distal parts (*anthocodia*) protruding, and their gastrodermal cavities with cross-connections (*solenia*). Calcareous spicules secreted by the mesogloea are scattered through the coenenchyme.

The gorgonian sea-fans also in the Octocorallia branch in one plane, typically at right angles to the current flow. Their axial skeleton is made of horny proteinaceous material, *gorgonin*. The main trunk is fastened by a basal plate and the side-branches run more or less parallel to it, sometimes cross-connected. Only in the quietest waters do gorgonians branch multi-directionally to become bushy.

In the primnoids the branch axis resembles a thin black wire, covered with the interlocking calcified beads that house the small polyps. The 'red coral', *Corallium*, has a hard calcified axis carrying tall, wide-spaced polyps.

The sea-pens (Order **Pennatulacea**) are found subti-

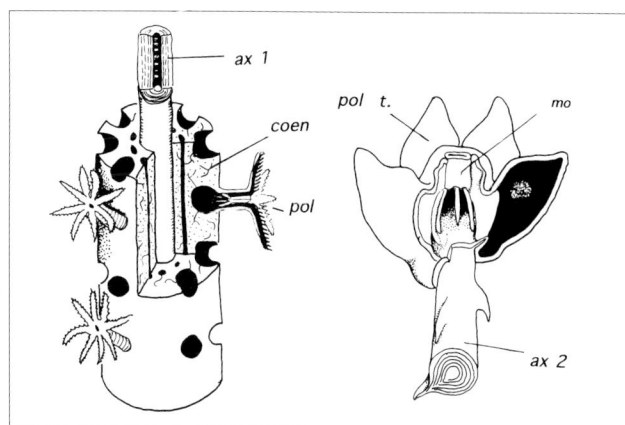

Fig. 16.9 Structure of Anthozoans
(left) an alcyoniid soft coral, **(right)** a gorgonian (ax 1) horny axial skeleton (ax 2) skeleton of antipatharian (coen) coenenchyme (mo) mouth (pol) polyp (pol t) polyp tentacle.

dally in one or two favoured spots such as the southern fiords. The colony has a fleshy axis developed from an attenuate original polyp. Its two rows of side branches carry close ranks of feeding polyps (*asutozooids*), and also special polyps called *siphonozooids*, adapted to maintain a water current across the colony.

The Subclass **Hexacorallia** has also a wide range of form. Setting aside the large, solitary polyps of the anemones (Order **Actinaria**), the rest are smaller and in varying degrees colonial. All the Hexacorallia agree in having the tentacles (and the mesenteries) in multiples of six, generally in several circlets.

The Order **Scleractinia** will be met with in their full diversity as tropical reef corals (Chapter 20). Their polyps lie in calcareous cups or *corallites*, making up the massive colony known as the *corallum*. New Zealand's several cup corals are mainly solitary, with no dependence like the reef corals on symbiotic algae. Of our 127 scleractinian species, only 11 have been found in depths less than 50 m around mainland New Zealand.[3]

Monomyces rubrum and *Culicia rubeola* can extend into the sublittoral fringe, beneath boulders. Of the rest, *Oculina virgosa* was first discovered at 50 m off the West King. The bright orange *Caryophyllia profunda* was first found by Wade Doak, at 57 m, off Twelve Fathom Pinnacle at the Poor Knights. The salad green, cave-dwelling *Tethocyathus cylindraceus* has recently been discovered at only 11 m, in the recesses of Rikoriko Cave at the Poor Knights.

The anemone-like polyps of the Order **Zoanthiniaria** lack basal discs. Instead they are united into a colony by a thick sheet of coenenchyme. The tentacles number 24 or more, in a single marginal circle. Easily peeled off, the zoanthids are typically epizoic, on sponges, hydroids and especially gorgonians that are eventually smothered and killed. A cream zoanthid attaches to the bare-stripped branches of *Primnoides*, as if to replace the original polyps. A second species has been likened to brilliant yellow daisies, raised on slender columns.

Rather remote from other Hexacorallia are the black corals of the Order **Antipatharia**. Their branching colonies form snow-white bushes with a black horny axis secreted from the polyp bases. The pale polyps so completely cover the axis as to belie the name 'black coral'. Unlike the gorgonian polyps, they derive no protection from the skeletal axis. Each has six short, non-retractile tentacles that sweep the water for the rain of plankton. Two ciliated grooves lead a water current down the gullet.

The first requirement of the black corals *Integra* is evidently a low light level, confined as they are between 35–60 m. This seems more important to the larvae, that settle in caves or in places of deep shade. Adult branches brought up to only a couple of metres deep have been found actually to grow towards a light source. The optimum current speed for black corals is about 2 knots, with neither sediment, nor risk of abrasion to the soft, lightly attached polyps. Where the currents are weak, *Integra* are bushy, but in stronger currents the tree becomes tall and laterally oriented.

Wade Doak has described first sighting of a black coral at a horizontal visibility of 50 m.

Arising from a rock pedestal below, silhouetted against the deep blue of the continental shelf, was a prodigious black coral tree. The depth gauges read 60 m at its arm-thick base. The feathery undulating tips were 6 m above. While flash bulbs popped, the divers literally climbed up and down the iron-hard branches, incredulous that such a tree could be real.[4]

Finally, the Order **Cerianthatia** is another isolated group of Hexacorallia, having large, solitary polyps like tube-dwelling anemones. They were first observed in New Zealand by night-divers in Fiordland and at the Poor Knights. *Cerianthus* forms a tapered column implanted in sediments like a taproot, enclosed in a parchment-like calcareous secretion. The numerous finely pointed tentacles are expanded only at night, to be quickly withdrawn at the slightest disturbance.

Poor Knights Islands

Influenced by subtropical oceanic waters from the north, the Poor Knights have been the scene of important underwater exploration since the 1960s.

Where horizontal visibility at Goat Island has a maximum of 15 m, averaging 5–7 m, the subtidal cliffs at the Poor Knights can offer underwater vistas of 27–30 m, sometimes reaching 50 m. Comparable illumination

reaches at least three times deeper than at Goat Island.

The Poor Knights shores drop off with steep cliffs down to 50 m, then either form a ledge or fall deeper at a lower angle of slope. There are a few sites with a straight vertical drop to 100 m. Wave refraction results in high exposure all round the islands, but on near-vertical faces much of the wave force is lost through reflection. Below 15 m the water is relatively calm, and beyond 35 m movement is almost lacking.

In clear water, the kelp forest of the **upper sublittoral** descends to 35 m or a little more, with *Ecklonia radiata* replacing *Lessonia variegata* at 3–10 m, depending on wave energy. Under their shade erect cheilostomatous bryozoans grow to some 10 cm tall: a bright green *Bugula*, an orange *Emma* and the reddish brown *Margaretta barbata*. The thecate hydroids *Lytocarpia*, *Aglaophenia* and *Symplectoscyphus* contribute to the same community. Encrusting sponges such as *Tedania* are common, and below 10 m the jewel anemone, *Corynactis australis*, becomes prominent.

Below the kelp follow the **middle** and **lower sublittoral** zones. While the line cannot be precise, the predominance of gorgonians and the appearance of black corals signal the change from middle to lower. The sessile biomass increases with depth as does the ratio of erect to encrusting sponges. Pink algal paint continues patchily beyond *Ecklonia* into the middle sublittoral, still forming 20% by area of the living cover at 75 m depth.

The sponges in the middle sublittoral have attained their most elaborate quiet-water growth. Outstanding at the Poor Knights are the splendid funnels of *Vagocia imperalis*, the yellow candelabras of *Iophon proximum* and the orange-red branches of a *Desmacidon* species. The white *Iophon laevistylus* has a lower, spreading habit, with wide tubes and prominent oscula partly closed by membranous rims. Some shallow water red sponges are still present, such as *Raspailia topsenti* and *Crella incrustans*, far more luxuriant than in disturbed water. A large orange tethyid is studded with spiny conical mamillae. The deep yellow *Aplysilla sulfuria* and lighter yellow *Cliona celata* are much in evidence. Darker places are reserved for clusters of the white calcareous sponge *Leucettusa lancifer*. Bryozoans are still varied and numerous in the middle sublittoral. The largest and most exquisite is the brittle network of the 'lace-coral', *Hippellozoon novaezelandiae*.

Wade Doak has memorably described this zone at the Poor Knights as it merges into the lower sublittoral. With the water above acting as a light-filter, this ultimate calm is supremely the region of sponges that prefer below 1% of normal sunlight:

...in the quiet, dim depths as competing life-forms diminish, sponges gradually dominate the rock slope

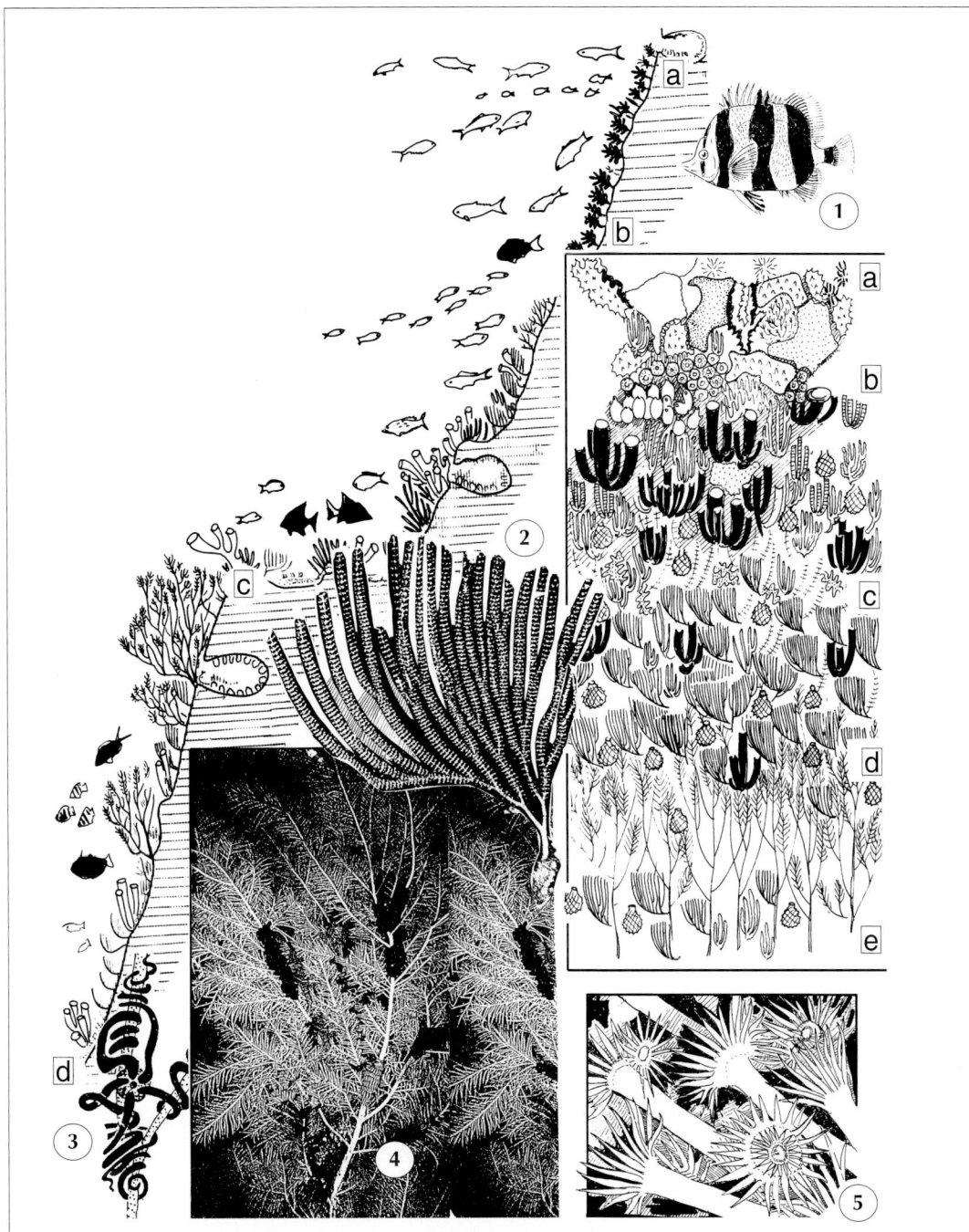

Fig. 16.10 Poor Knights Islands: profile of subtidal cliffs from upper to lower sublittoral zone
a–b upper sublittoral with *Ecklonia;* b–c middle sublittoral with sponges and gorgonians; c–d lower sublittoral, with black corals and gorgonians
(right) detail of lower sublittoral zone
a–b low and crustose sponges; b–c tubular and candelabra sponges; c–d zone of gorgonians; d–e deep level with black corals
1 Lord Howe coralfish, *Amphichaetodon howensis* 2 gorgonian 3 *Astrobrachion constrictum* 4 *Antipathes* sp. 5 a yellow zoanthid. (With assistance from Wade Doak.)

with encrusting varieties, lemon yellow, violet or bright orange-red. Further along the wall stand graceful erect sponges like candelabra, salmon pink funnel shapes, organ pipes, yellow golf-balls and — where scorpion fish lie camouflaged — massive dark blue crenellated cups, like eroded battlements. At these depths the hydroids too have developed, from the stunted one inch growth at 20 feet to dainty filamentous plumularians up to four inches long. They festoon the slope, along with the fragile lolly-pink reteporan bryozoan known as lace coral [*Hippellozoon novaezelandiae*] and the pink, delicately beaded gorgonian fans of the Primnoides family, the gilled cup corals and the fleshy alcyonacean corals in pastel hues. The divers, falling gently down through shoals of friendly demoiselles, land on a shining white beach. From the cliff-foot, they gaze up the fantasy staircase, its multitudinous life-forms in silver-rimmed silhouette, to where the *Ecklonia* forest looms in dark knots, merging in the haze of distance, just beneath the living quicksilver of the surface. A poetry of drowned colours, frozen sound and sublime weightlessness invades the mind at these depths.[5]

In the lower sublittoral, where cnidarians dominate, there are large plumularian hydrozoans 15 cm tall in still water, and graceful gorgons or sea-fans, set vertically across the current. Commonest at the Poor Knights is a pink, parallel-branched primnoid, entering the zonation from 40 m down. A yellow or buff sea-fan, one of the family Gorgoniidae, with branches freely anastomosing, is confined to deep recesses and overhangs.

Grazers and carnivores
The predatory gastropods of the sublittoral appear to be largely independent of depth and limited only by food. The trumpet shell, *Ranella australasia*, preys upon the ascidian *Cnemidocarpa bicornuta* and the anemone *Actinothoe albocincta*. Subtidal *Dicathais orbita* also takes both. The turban shell, *Cookia sulcata*, however, browses on algae only down to 8 m. The larger top-shells, *Calliostoma tigris* and *C. punctulata*, have turned to sponge-grazing, the first on *Tedania* in shallow depths, the other upon *Ancorina*. In the middle sublittoral, two common sponge predators are the nudibranchs *Aphelodoris luctuosa* and *Ceratosoma amoena*.

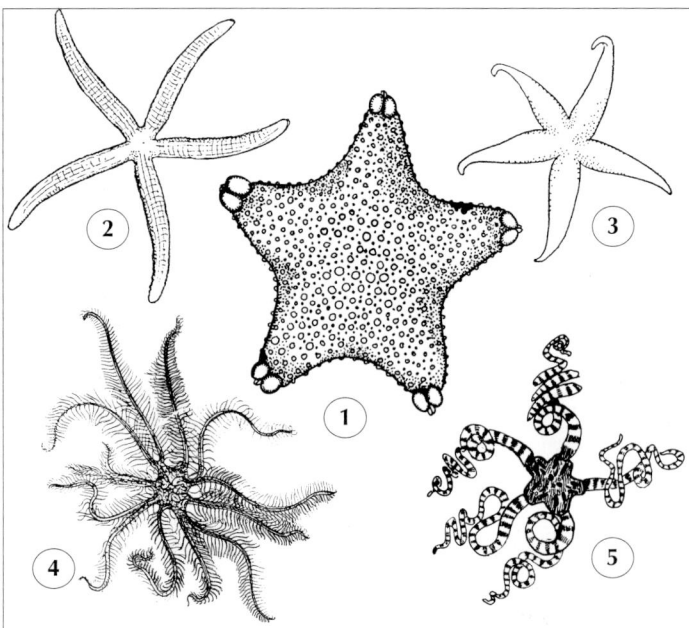

Fig. 16.11 Some echinoderms of the sublittoral zone
1 *Asterodiscides truncatus* 2 *Ophidiaster kermadecensis* 3 *Henricia* sp. 4 *Cerolia spanoschistum* 5 *Astroceras elegans*.

Tropical influence
The forms mentioned above are familiar New Zealand species up to and around low water. So too are the leading sublittoral starfishes: *Coscinasterias muricata*, preying on urchins, and the less numerous *Astrostole scabra*, taking the sea cucumber *Stichopus mollis*.

In greater depths, a new tropical element is revealed at the Poor Knights, not so far met with on the mainland. The south-flowing drift of subtropical water that joins the East Auckland Current to wash the Knights raises their summer surface temperature to as much as 25°C.

Mollusca
Two widespread Pacific cowries are now established offshore in Northland: *Cypraea cernica*, honey coloured with white spots, and the larger *C. vitellus*, also pale-spotted. *C. caputserpentis* has been more recently reported. There are two cowry relatives with a more specialised dietary. *Phenacovolva wakayamaensis* has been found at the Poor Knights, living on gorgonians at 50 m. The white shell flushed with rose is drawn out narrowly at either end. The smaller cap-shaped *Pedicularia pacifica* lives off the Three Kings, clinging to red *Corallum*.

Mesogastropod carnivores recently arrived include a large naticid, *Polinices tawhitirahi*, a bursid, *Bursa verrucosa*, and a frog-shell, *Tutufa bufo*. The delicate pink dawn murex, *Prototyphis eos*, today found intertidally in Northland, was previously known only as a Pliocene fossil. A newer tropical muricid is *Morula palmeri*, found subtidally at the Poor Knights and also at the Kermadecs.

Echinoderms
Two feather stars or sea lilies of the Class **Crinoidea**, *Cenolia spanoschistum* and *Argyrometra mortenseni* are found at the Poor Knights and other Northland sites, and *Oxycomanthus plectrophorum* lives in the special habitats of Fiordland. These are our last remnant of the stalked or sessile crinoids that had reached their zenith by the end of the Paleozoic. No crinoid in New Zealand lives between tides, and the long-stalked sea-lilies belong to deep water. The comanthids lose their stalked attachment early in life, taking temporary hold by claw-like cirri at their base. They can spasmodically swim by the leisured arm-flexions.

The viscera of a crinoid lie within the calyx, a cup with the mouth and anus both on its soft upper surface or *tegmen*. Pinnate arms spring from the rim (12–20 in *Cenola*) sensitively curling and extending at a touch. Their pinnules channel into an ambulacral groove leading down each arm to the mouth. Small zooplankton is caught by direct contact with the waving arm, to be narcotised by poison glands on the pinnules. Particles are carried to the mouth by cilia and suckerless tube-feet.

Plankton is taken also by multicoloured snake-stars (**Ophiuroidea**) that coil so tightly around gorgonians and black corals as to defy removal without damage to both. Clambering about like spiders, they use the part-extended arms to filter in the normal ophiuroid mode. The three species at the Poor Knights are *Astroceras elegans*, *Astrobrachion constrictum* and *Astrothorax waitei*.

The most colourful asteroid of the northern sublittoral is the resplendent fire-brick star, *Asterodiscides truncatus*. The thick chrome yellow arms are spangled with vermilion and smaller purple tubercles. Two starfish with tropical affinities are a bright yellow *Henricia* and the serpent starfish *Ophidiaster kermadecensis*, with cylindrical, yellow and brown mottled arms. Both are unselective grazers like *Patiriella*.

In the middle sublittoral, *Evechinus* is replaced by the large purplish black *Centrostephanus rodgersi*, with sometimes one to a square metre. The heavy spines are hollow and coarsely serrated (Fig. 10.7). The bright red, long-spined *Diadema palmeri* is found occasionally at the Poor Knights; it possesses venomous spine glands like its tropical relatives, and is negatively phototactic, living securely under ledges. Two other echinoids frequently found at the Poor Knights are *Heliocidaris tuberculatus* on rock, and among subtidal boulder and cobble gravel the big heart urchin, *Brissus gigas*.

Sea caves and overhangs

Just as the undersides of boulders introduce a new order of diversity between tides, the richest subtidal habitats are in dark sea caves, altogether protected from wave action. The steep subtidal cliffs present a wealth of caves, tunnels and archways. Life here is stratified — as under boulders — in accordance with the fall of illumination. Thus, with a boulder on a small scale, or a cave on an expanded one, we can find translated to higher levels, salients of a fauna that on an open surface would only exist much further down.

The subtidal caves and overhangs of the Poor Knights and east Northland have been well described by Tony Ayling. The simplest shaded habitats are vertical faces, with incidental light reduced to 5% of the horizontal, too low for most algal growth. The next stage, in effect a very shallow cave, is an overhung rock-face, with only reflected illumination amounting to 1% of the horizontal. The current flow is, however, unimpaired, allowing a diversity of filter feeders, freed from space-competition by algae. With a half metre quadrat in 12 m depth off Goat Island, Ayling found six such species on

the horizontal, with 24 on a vertical face and 28 under a moderate overhang.

The red brachiopod *Calloria inconspicua* is usually dominant in deep shade at up to 1200/sq m. Crowded among them are bryozoans, ascidians and encrusting sponges. Here also grows our largest, most tree-like hydroid, *Solanderia ericopsis*, with its profusely branching fan turned across the current to ensnare zooplankton. A specialist predator, *Jason mirabilis* lives on these trees, being the largest of New Zealand's — if not the world's — aeoliids. Lavender pink, with milk-white cerata, this opisthobranch reaches 60 mm long.[6]

In those deeper cracks and fissures that Ayling termed *mini-caves*, water movement is buffered, but some efficient filterers — bryozoans and thin encrusting sponges — can survive on the reduced food supply. The two corals *Monomyces rubrum* and *Culicia rubeola* also live here. Mini-caves are shelters as well for mobile predators, including the large snake-star *Ophiopsammus maculata*. Small and inquisitive blennies dart out or goggle from cracks and holes, along with the larger and aggressive scorpion fish. Spiny crayfish prefer these narrow retreats to open caves, lining up in rows along cracks with feelers wary for intruders.

The larger *caves* are blind alleys tapering to a narrow end, and finally to total dark. Currents can still enter with little obstruction, bringing rich food supply to most of the cave wall. But towards the 'empty quarter' at the far end, the fauna ultimately diminishes from reduced food supply, and the rock is finally left bare. *Tunnels* differ from caves in having more than one entrance, with vigorous water flow. There is no empty quarter, and the walls are flushed by a steady current producing dense growth right through.

On steep outer rock slopes, as in Fig. 16.10, the living community is complexly stratified: (1) closest to the rock is a mosaic of encrusting sponges, ascidians and bryozoans; (2) above this is an *inner meadow*, rich in feather-like and fan-shaped hydroids and foliose bryozoans; (3) an *outer meadow* is dominated by a still unnamed purple gorgonian, a metre in height, turned across the current flow. Dead gorgonians sometimes carry yellow zoanthids. The salmon pink coral *Oculina virgosa* is rare at the Poor Knights, but commoner under overhangs further north. Finger sponges — mauve *Callyspongia ramosa* and yellow *Iophon laevistylus* — are important in the outer meadow too.

Within a sea cave, communities replace each other as light intensity falls. First, the cave entrance is essentially like the overhang just described, though as the current slows, finger sponges may be lost and gorgonians reduced to a single stem. Next inward, the outer meadow is entirely lost, but colourful encrusting sponges and compound ascidians remain conspicuous,

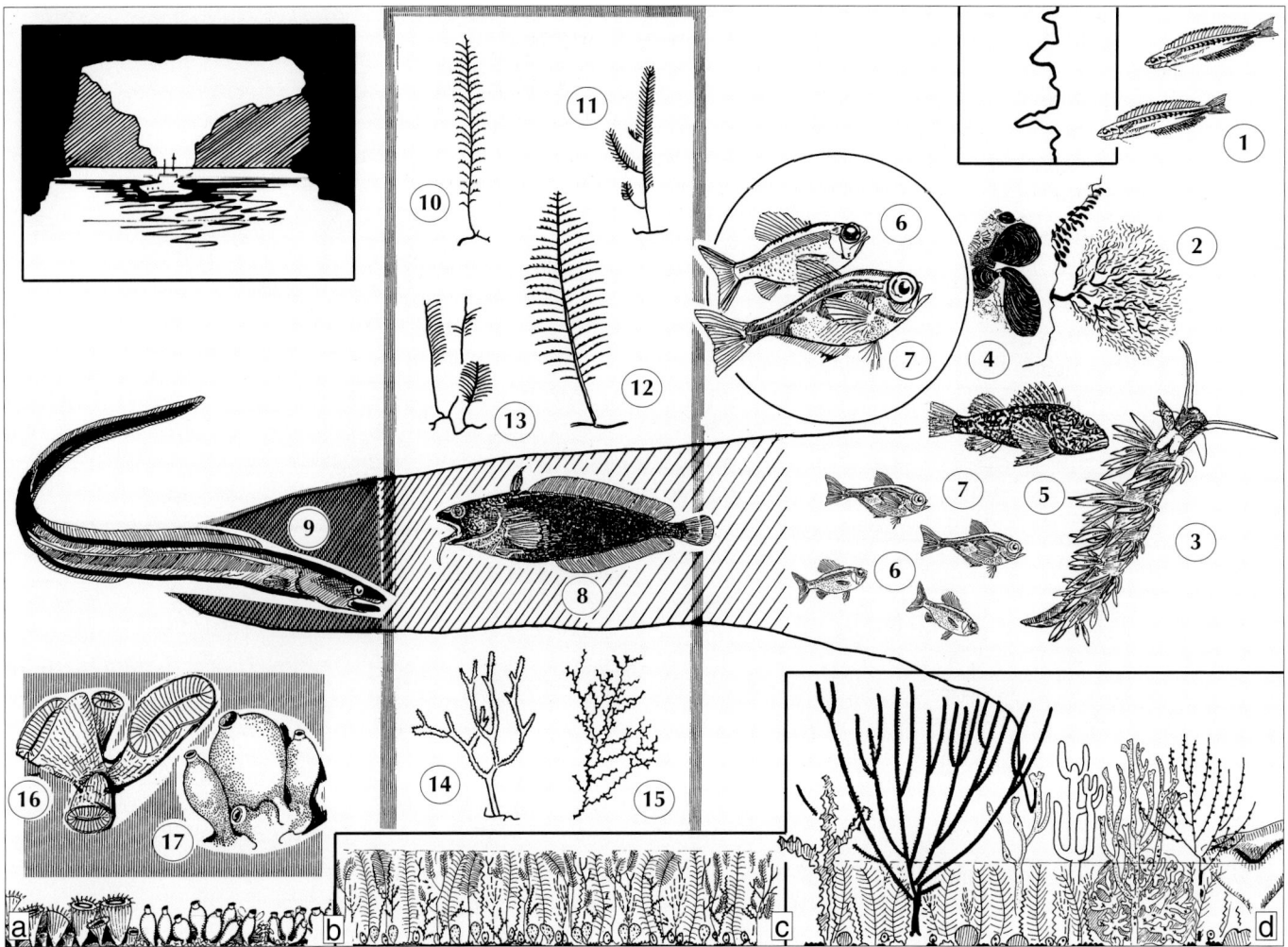

Fig. 16.12 Sea caves and overhangs in Northland
(top right) small niches with 1 mimic blenny, *Plagiotremus tapeinosoma*, shaded overhang with 2 *Solanderia ericopsis*
3 *Jason mirabilis* 4 *Calloria inconspicua* 5 *Scorpaena cardinalis*.
(centre) blind sea cave, with succession of fishes towards dark 6 *Optivus elongatus* 7 bigeye, *Pempheris adspersis* 8 rock
cod, *Lotella rhacinus* 9 northern conger eel, *Conger wilsoni*.
(bottom) from dark end **(left)** to light cave meadows with three levels (see text): a–b, b–c, c–d.
10 *Plumularia setacea* 11 *Aglaophenia laxa* 12 *Sertularia unguiculata* 13 *Lytocarpia incisa* 14 *Symplectoscyphus*
subarticulatus 15 *S. johnstoni* 16 *Monomyces rubrum* 17 *Leucettusa* sp. (Based on the findings of Tony Ayling.)

together with hydroid fans and bryozoans. There are vivid splashes, from reds and pinks, orange and yellow, to occasional green or blue, or a range of deep purples, mauves, pure white or jet-black. Yellow zoanthids encrust some of the sponges. Clusters of transparent, blue-highlighted ascidians, *Pycnoclavella*, add their jewel-like effects.

In the quiet, unlighted waters furthest back in the cave, there is an increase of bare rock. Pale calcareous sponges, bryozoans and some simple ascidians are common. *Monomyces rubrum* may be finally the only species left, but there is sometimes the green cup coral *Tethocyathus cylindraceus*, and a scatter of ascidians.

Caves, overhangs and tunnels are more important habitats in the islands of the continental shelf than on the mainland. Plankton-rich waters and subtropical temperatures bring communities distinct from the mainland. Northland's chain of islands may thus be a 'gateway' for rich subtropical recruitment. Adequate protection of habitats is necessary to prevent it being closed.

Coastal fishes

Though all fishes are in relative terms mobile, those of shallow inshore waters have intimate ties with the bottom. Most have a well-defined territory, and dependence on particular foods. Some fishes are most active by day, some by night. Others again are in evidence at any time. Just as with birds, they have their own flocking and schooling behaviour, moving about in mixed or single-species formations.

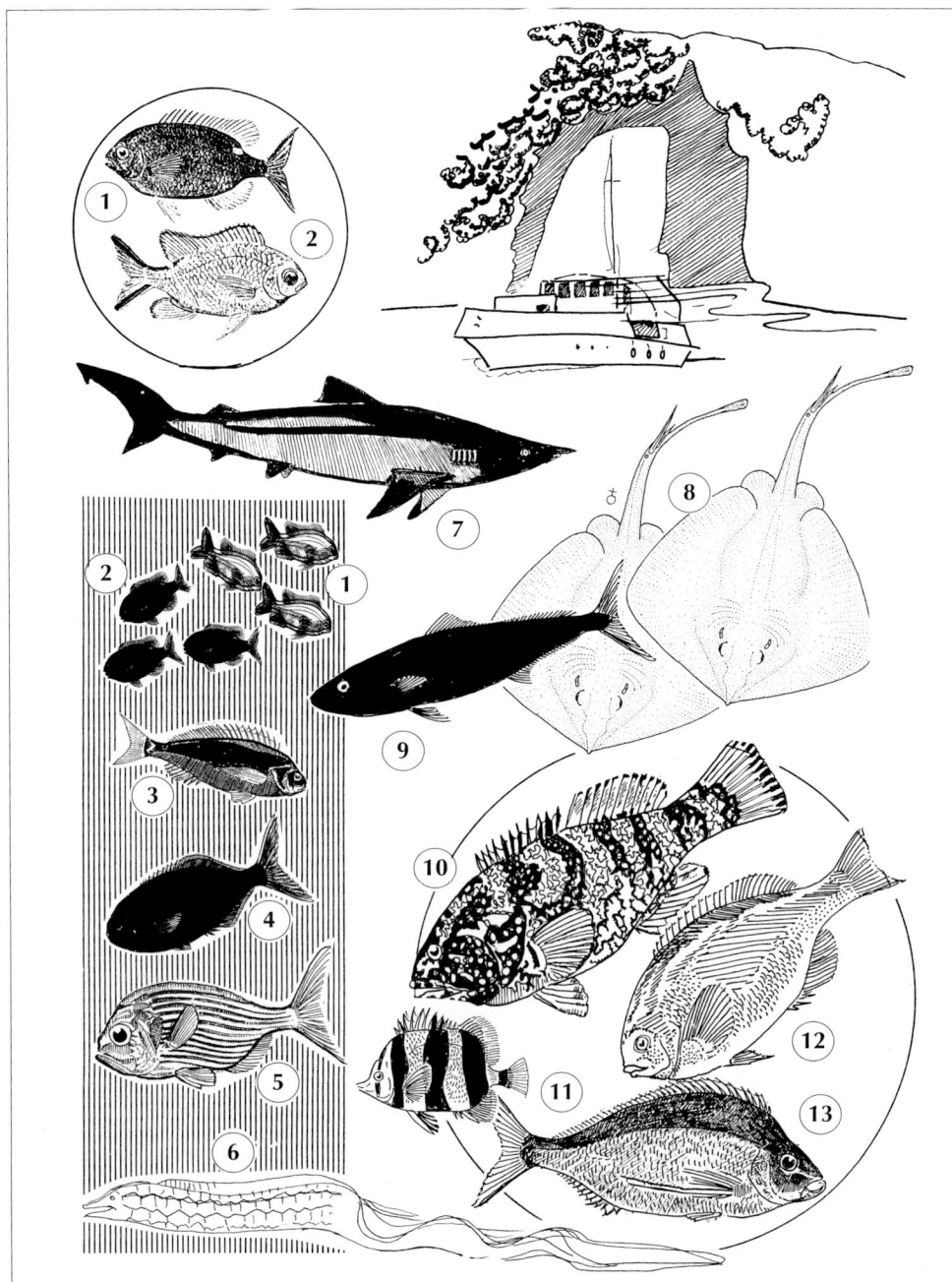

Fig. 16.13 Vertical stratification of fishes at The Archway, Poor Knights Islands
1 single spot demoiselle, *Chromis hypsilepis* 2 yellow demoiselle, *Chromis fumea* 3 pink mao mao, *Caprodon longimanus*
4 blue mao mao, *Scorpis violaceus* 5 golden snapper, *Centroberyx affinis* 6 mosaic moray, *Enchelycore ramosa* 7 bronze
whaler shark, *Carcharhinus brachyurus* 8 short-tailed stingray, *Dasyatis brevicaudatus* 9 kingfish, *Seriola lalandi* 10 spotted
black grouper, *Epinephalus daemelii* 11 Lord Howe coralfish, *Amphichaetodon howensis* 12 blue drummer, *Girella cyanea*
13 porae, *Nemadactylus douglasii.* (Based on the description by Tony Ayling.)

Our knowledge of the *feeding guilds* of northern fish, built up since the 1960s, has been brought together in Tony Ayling and Geoff Cox's *Guide to the Sea Fishes of New Zealand* (1982). Fig. 16.13 is constructed from Ayling's description of a descent down the steep wall of a great submarine archway at the Poor Knights.

From the surface, moving down through the clear blue, first seen are the schools of the plankton-feeders: demoiselles and blue mao mao:

...swimming and weaving, hindering the view by their sheer numbers. Looking up, a veritable storm of fish is silhouetted against the light, slanting down from the surface, crossed by the occasional roving kingfish on the lookout for an unwary schoolfish. Squadrons of stingrays like huge delta-winged bombers soar among the nonchalant plankton-eaters, unusual behaviour for a species that is usually a sand bottom dweller.[7]

The small plankton-feeding fish are a tropical element confined to the North Island and mostly to east Northland. The golden snapper, *Centroberyx affinis* (Berycidae), keeps to rocks and caves by day but disperses at night to take large plankton with its fine-toothed jaws. Pink mao mao (*Caprodon longimanus*), blue mao mao (*Scorpis violaceus*) and trevally (*Pseudocaranx dentex*) often school together. The blue mao mao — one of the drummer Family Kyphosidae — is an open water plankton feeder, and 'their bright blue foreheads break from the water as their schools gulp down euphausiid shrimps'. The pink mao mao is a plankton-eating serranid, one of the grouper family, forming loose schools off rocky headlands between North and East Capes. Another plankton-taking serranid is the butterfly perch, *Caesioperca lepidoptera*.

The bigeye, *Pempheris adspersis*, and the slender roughy, *Optivus elongatus*, shelter in crevices by day and come out to gulp the rich plankton at night. Mid-water fish pursuing active food thus range from the mao mao, seizing minute plankton, to John dory (*Zeus faber*) with telescopic jaws, sidling up to take small fish by their lightning extension.

Schools of the different plankton feeders are stratified with depth: the demoiselle (*Chromis dispilus*) from the surface to 15 m, and the blue mao mao down to 6 m. From 16 to 30 m the pink mao mao and the butterfly perch associate. From 30 to 50 m the splendid perch, *Callanthias australis*, is dominant, with golden snapper below 40 m.

On the rocky bottom, porae (*Nemadactylus douglasii*) feeds with their thick lips on animals sucked from sediment. From the broken rock-faces scorpion fish (*Scorpaena cardinalis*) snatch and swallow active prey. Still more aggressive are the lurking moray eels of the Muraenidae, a tropical family reaching the North Island's east coast. The yellow moray (*Gymnothorax prasinus*) has reached the mainland, pouncing on crabs and shrimps, also blennies and young scorpion fish. Four other moray species are confined to offshore islands.

At the Poor Knights stable sex-pairs of the Lord Howe Island coral fish (*Amphichaetodon howensis*), may also be found, New Zealand's only chaetodontid or butterfly fish. Standing out by its black and yellow striping, it has a family habit of snout-probing for crustaceans or tubeworms, or snipping off single polyps with its small teeth.

Boulders at the tunnel bottoms offer retreats for larger fish, such as the spotted black grouper, *Epinephelus daemelii*, and the marble-fish (*Aplodactylus arctidens*) with its relative the notch-head marble-fish (*A. etheridgii*) from Norfolk and Lord Howe Island.

The drummers (Kyphosidae) include some of our relatively few herbivorous fish. At the Poor Knights, the brilliant blue drummer (*Girella cyanea*) replaces the silver

Fig. 16.14 Fish distribution on subtidal slope at Goat Island, near Leigh

1 garfish, *Hyporhamphus ihi* 2 parore, *Girella tricuspidata* 3 banded wrasse, *Notolabrus fucicola* 4 marble-fish, *Aplodactylus arctidens* 5 red moki, *Cheilodactylus spectabilis* 6 blue mao mao, *Scorpis violaceus* 7 leatherjacket, *Parika scaber* 8 snapper, *Pagurus auratus* 9 scarlet wrasse, *Pseudolabrus miles* 10 kingfish, *Seriola lalandi* 11 demoiselle, *Chromis dispilus* 12 butterfly perch, *Caesioperca lepidoptera* 13 goatfish, *Upeneichthys lineatus* 14 moray eel, *Gymnothorax prasinus* 15 John dory, *Zeus faber* 16 blue cod, *Parapercis colias*. The brown algae are **(top)** *Carpophyllum maschalocarpum* and **(left)** *Ecklonia radiata*. (With acknowledgement to Barry Russell.)

drummer (*Kyphosus sydneyanus*). Both use their small but powerful mouths, with close-set teeth, to cut tender algae from the rock surface, like their familiar cousin the parore (*Girella tricuspidata*) that schools in estuaries.

Where large boulders rest on an uneven bottom with several entrances there may lurk red moki (*Cheilodactylus spectabilis*), and the hiwihiwi or kelp-fish (*Chironemus marmoratus*) — in shape and habits like a scorpion fish — together with the silver drummer and the two sorts of marble-fish.

Cleaning symbiosis is a widespread tropical relationship, now well known from a few locations in northern New Zealand. Small crimson cleanerfish (*Suezichthys aylingi*) and juvenile Sandager's wrasse (*Coris sandageri*) can be seen going over demoiselles and pink mao mao to pick off parasites and unhealthy tissue. Less than 15 cm long, the cleaners are reddish brown, advertised with a long white stripe. A dozen or more customer fish may be seen waiting for service.

There is a crustacean cleaner species now established in New Zealand, the tropical banded shrimp, *Stenopus hispidus*, with red and white striped chelae signalling its presence with the six long white antennal flagella.

Fiordland

The fiords of Southland are unique habitats where normally deep communities can flourish at relatively shallow levels. Until recently it has been their very remoteness that safeguarded the fiord habitats from exploitation or destruction. Today the stricter protection of the Fiordland coast has become urgent. Proposals to export fresh water in super-tankers could overturn a fiord's surface layer and spell disaster for the zoned communities. There are even ambitions for harvesting black coral, now a protected species. Irreparable harm might be allowed to happen, with only a handful of New Zealanders yet realising what is at stake. Yet these isolated and beautiful habitats could today be accorded the secure status of marine reserves without any detriment to commercial interests yet established. Only two small areas in the fiords currently have marine reserve status.

The coast of Fiordland extends over 200 km from Milford Sound to Preservation Inlet, and the longest inlet, Doubtful Sound, reaches inland for 30 km. Each fiord has originated as a U-shaped glacial valley, with its steep, near vertical sides carved in glaciations, the last one ending just 12,000–15,000 years ago, ending around 20,000 BP. Their lower reaches have today been drowned to a depth of 100–400 m, to form long basins. Each is closed by a shallow sill marking the seaward extent once reached by glacier ice. Today the sill cuts off the water mass of low salinity in the fiord's deep interior

from the normally saline water outside.

The fiords region has a high rainfall, up to 7000 mm/yr, with the spring and summer snow-melt added. Soil erosion is effectively halted by the dense bush and its ground vegetation layer holding back runoff above the sheer sides. The entering water thus carries little sediment, but picks up leaves and organic detritus, along with dissolved nutrients. A freshwater layer about 3 m deep, coloured like weak tea, thus floats on the fiord surface. With the absence of waves or swells this layer, as it flows seawards, never fully mixes with the deeper water. But since some of the water is carried out from the top of the saline mass, the upper 20–40 m has a saline counter-current flowing in. Below this, the saline water, protected by the sill, has remained virtually isolated for years. Dissolved oxygen is lost through animal respiration, while hydrogen sulphide is produced by sulphate reduction. So at relatively shallow depths the sediments resemble those of 1000 m bathyal habitats with their special fauna of heart urchins, tubeworms, bivalves and scaphopods.

Scientists from the (then) New Zealand Oceanographic Institute began active study of the Fiordland ecology and hydrology in 1980. To the leader of the biological survey, Ken Grange, we are indebted for a detailed description of Central Long Sound, Preservation Inlet.

In the transect shown from this location (Fig. 16.15), a steep drop to 36 m is interrupted by a gently sloped, sand-covered ledge at 25–30 m. Within the low salinity zone *Ulva lactuca* and *Hormosira banksii* grow in the upper 2 m, followed by the mussels *Mytilus edulis aoteanus* and *Aulacomya maoriana*. The upper 3 m has patches of the barnacles *Chamaesipho columna* and *Austrominius modestus*. The sea urchin, *Evechinus chloroticus*, feeds on algal drift dislodged from immediately above, and the star *Coscinasterias muricata* is found at the base of a shelf at 6 m.

A steep rock wall then continues from 6 to 40 m carrying two intergrading zones. The shelf beginning at 6 m is covered with silty shell gravel, supporting the bivalves *Tucetona laticostata*, *Venericardia purpurata* and *Chlamys gemmulata*, as well as the gastropod *Maoricolpus roseus* and the sea cucumber, *Stichopus mollis*.

Down to 9 m the rock-face carries little more than a thin coralline veneer, with the echinoderms *Coscinasterias muricata*, *Sclerasterias mollis* and *Pseudechinus huttoni*. From 10–15 m runs a dense zone of serpulid tubeworms with mixed clumps of *Pomatoceros caeruleus*, *P. terraenovae* and *Protula bispiralis*, and a fourth species, *Galeolaria hystrix*, entering lower down. The grazing molluscs include the handsome trochid *Calliostoma tigris* and the turbinid circular-saw shell, *Astraea heliotropium*. The aeoliid slug *Jason mirabilis*

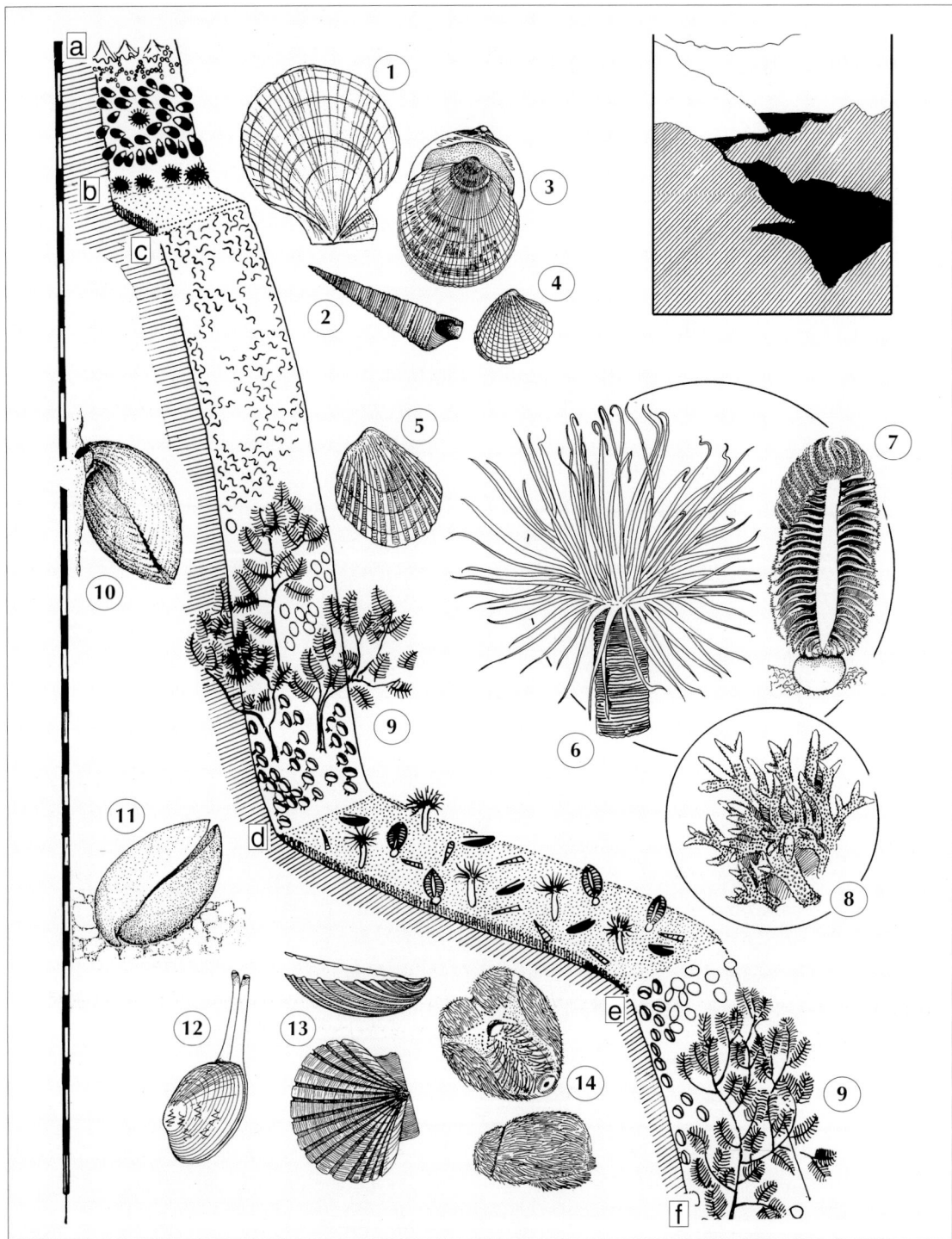

Fig. 16.15 Sublittoral zonation of the fiord wall at Central Long Sound, Preservation Inlet

a–b low salinity zone with *Ulva*, mussels and *Evechinus;* b–c sandy shelf with *Chlamys, Tucetona, Maoricolpus* and *Venericardia;* c–d long cliff-fall, with tubeworms and **(bottom)** brachiopods and *Integra;* d–e sandy shelf with *Cerianthus, Pennatula, Pecten, Echinocardium, Ruditapes;* e–f lowest slope with brachiopods and *Integra*

1 *Chlamys gemmulata* 2 *Maoricolpus roseus* 3 *Tucetona laticostata* 4 *Venericardia purpurata* 5 *Lima zealandica* 6 *Cerianthus* 7 *Sarcophyllum bollonsi* 8 *Errina novaezelandiae* 9 *Integra fiordensis* 10 *Magasella sanguinea* 11 *Neothyris lenticularis* 12 *Ruditapes largillierti* 13 *Pecten novaezelandiae* 14 *Echinocardium cordatum*. (Based on the description by Ken Grange.)

and a small species of *Doto* feed on the large colonies of the hydroid *Lafoea dumosa*.

At about 12 m brachiopods become numerous. Their mode of life has been studied in detail by the National Geographic Society's Expedition to Fiordland in 1977 and 1979, led by Joyce Richardson. In the region at large, eight species have been discovered, from three of the five orders of living Brachiopoda.

While all the species attach at settlement to rock or sand grains, muscular adjustment of the pedicel allows freedom to orient to water currents. The nature of the ground determines whether adults will stay attached or lie free. Both *Magasella sanguinea* and *Calloria inconspicua* are able to lie free on sediment or attach to rock, the first showing a preference for free-lying, the second for attachment. *Neothyris lenticularis* is also best adapted for free-lying, while *Notosaria nigricans* is always fastened down.

On the transect illustrated, the highest brachiopods are *Liothyrella neozelanica*, *Notosaria nigricans*, *Magasella sanguinea* and *Calloria inconspicua*. These are often mixed with tubeworm clumps and with the bivalve *Lima zealandica*.

Between 25 and 30 m there follows a gentle sand-slope with a substrate entirely different from the rock-face. This enclave is dominated by the tube anemone, *Cerianthus* (Fig. 16.15), with a dense spread of pale filiform tentacles, orange mouth and parchment tube up to 10 cm tall. The molluscs here include *Maoricolpus roseus* and the bivalves *Pecten novaezelandiae*, *Chlamys gemmulata*, *Atrina zelandica*, and *Ruditapes largillierti*. *Neothyris lenticularis* lies free and the heart urchin, *Echinocardium cordatum*, burrows into the sand.

Anthozoa

From 15 m downward, the serpulids are lost and brachiopods dominate. Mixed with them are cnidarians, with the most complete range of Anthozoa yet found in New Zealand, and especially remarkable at such shallow depth. Seven orders in all of the Anthozoa are exemplified, including cerianthids on the shelf. Largest and most conspicuous are the thickets of the 'black coral', *Integra fiordensis*, dividing into terminal twigs beset with small white polyps. The tallest trees reach 2 m. This is the shallowest habitat so far known for black corals anywhere in New Zealand. They have their own two sorts of tightly twined brittle star: dark red or striped *Astrobrachion constrictum* and brown and white-spotted *Astroceras elegans*.

Of lower profile, up to 30 cm, is the orange yellow gorgonian *Acanthogorgia* species, the ecological equivalent of the sea-fans at the Poor Knights. There are also freely branching colonies of orange *Alcyonium aurantiacum* and a related white species. The jewel anemone *Corynactis australis* is widespread, while there are three scleractinian corals, the solitary *Monomyces rubrum*, *Desmophyllum dianthus* and *Caryophyllia profunda*.

Perhaps the most interesting discovery has been of a *Telesto* species, the first New Zealand recording from the Suborder Telestacea of the Subclass Octocorallia.

The final prize from this rich assemblage of Anthozoa is one of the Order **Pennatulacea**, the sea-pen *Sarcophyllum bollonsi* (Fig. 16.15), a species that normally occurs in much deeper waters on the continental shelf.

Hydroida

As well as such a range of anthozoans, the fiord slopes have revealed in much less than its usual depth a 'hydrocoral' of the Family Stylasteridae, in the Class Hydrozoa. *Errina dendyi* grows in fragile fans, with circular gastropores and smaller more numerous dactylopores concentrated on the terminal branches. Fiordland specimens are large and delicate, red or orange to pale yellow and white, and sometimes attached to brachiopod valves. In the deepest waters on submarine ridges and sea mounts away from all land derived sediments, New Zealand has in fact the largest stylasterid fauna in the world, totalling some 59 living species.[8]

Brachiopoda

With their bivalved shells, the brachiopods at first sight resemble molluscs. But the likeness is no more than superficial, with the shell valves not left and right, but dorsal and ventral; while compared to the peak working efficiency of the bivalves, the internal organisation of a brachiopod has aspects that seem archaic.

The surviving brachiopods belong to some of the most ancient invertebrate lineages still living. *Lingula* — not found living in New Zealand but common in tropical Pacific sandflats — has come down from the Ordovician, and is said to be the oldest animal genus still extant. The brachiopod phylum had its two classes fully evolved at the opening of the Cambrian, and the more primitive **Inarticulata** was already in decline before the Silurian. The **Articulata** had reached their high noon by the Devonian and Carboniferous and by the Permian they too had started to regress. Only a few persistent lines have come down to us today, mostly in secluded habitats beyond low water. New Zealand has an unusual number of living species, with one or two, including an inarticulate, in the intertidal.

Brachiopods of both classes have a bivalved shell. In the Articulata the dorsal and ventral valves are hinged together, and an internal shell skeleton supports the spiral feeding apparatus or *lophophore*. The intestine ends blindly without an anus. In the Inarticulata the hinge is lacking, nor is there a lophophoral internal skeleton, but the gut opens by an anus.

The brachiopod lophophore is enclosed in the cavity between the two sheets of the mantle. Compared with the ctenidium of the lamellibranch and their quick acting adductors, the design of the brachiopod pallial cavity seems both complex and also antiquated. The two spiral coils of the lophophore are fringed with mobile filaments, broadly like the organ of the same name in bryozoans and *Phoronis*. The filaments have lateral cilia, drawing water currents between them into the mantle cavity. Frontal cilia carry particles to food

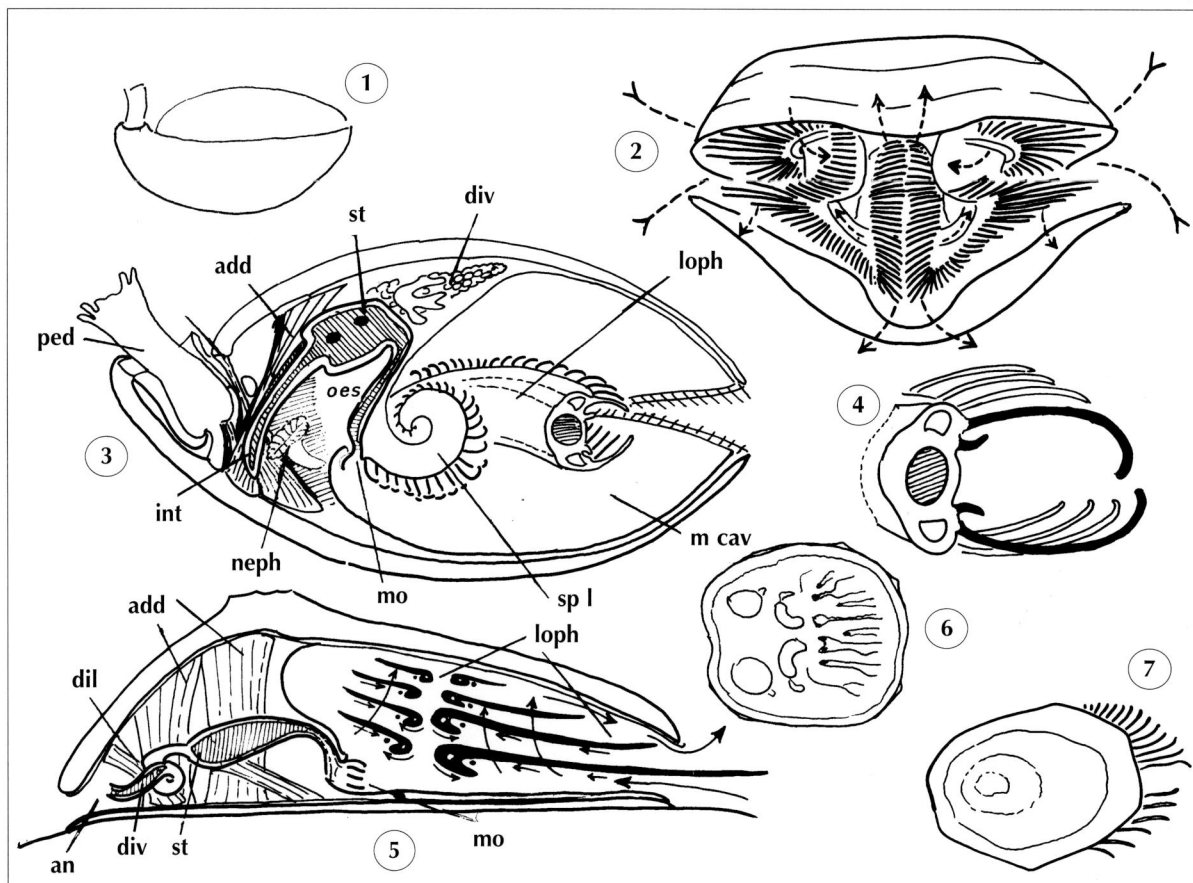

Fig. 16.16 Brachiopod morphology
1 articulate brachiopod, *Calloria inconspicua*, with shell and peduncle; 2 the same from in front, showing lophophore and direction of currents; 3 internal structure: (add) adductor muscle (dil) dilator muscle (div) digestive diverticulum (int) blind intestine (loph) side-arm lophophore (m cav) mantle cavity (mo) mouth (neph) nephridiumoes oesophagus (ped) pedicle (sp l) spiral part of lophophore (st) stomach; 4 cross-section of lophophore, side-arm enlarged; 5 inarticulate brachiopod, *Novocrania huttoni*. Internal structure (add) adductor muscle (an) anus (dil) dilator muscle (div) digestive diverticulum (loph) side-arm of lophophore (mo) mouth (st) stomach; 6 lower valve with muscle scars and gonads seen from within; 7 upper valve with lophophore filaments emergent.

grooves around the lophophore base, from where they are then conveyed to the mouth.

The shell skeleton that in the Articulata supports the lophophore is attached to the inside of the dorsal valve. This feeding apparatus becomes complexly folded and spiralled, with effects on the current pathways. *Calloria inconspicua* (Fig. 16.16) has two lateral inhalant paths and a median exhalant one. The upper bar of the lophophore is coiled towards the middle line. In *Notosaria nigricans* a spiralling cone extends to either side. The feeding of *Neothyris lenticularis* has been described in detail by Joyce Richardson.

In the Inarticulata the valves are joined only by their complex set of muscles, and the lophophore lacks a skeletal support (Fig. 16.16). A highly modified inarticulate, *Novocrania huttoni*, is fairly common under boulders at low tide in east Northland.

The causes of the long brachiopod decline has been

widely speculated upon. From evidence of low oxygen levels, some in the Permian, some have inferred a shortfall in phytoplankton production in the Permian, to the advantage of the more 'modern' bivalve molluscs over less efficient filterers like the brachiopods, bryozoans, corals and crinoids. In their relict habitats today, the Brachiopoda could be getting along by dint of their reduced size and low metabolic rate. Their more 'leaky' feeding apparatus could be of actual advantage in sedimented places where the more efficient uptake of the bivalve gill would collect more waste than the rejection mechanisms could handle. The superior retentive power of the bivalve gill, with its coalescence of filaments and screens of latero-frontal cilia, can hardly be doubted. Fused mantle edges and extensible siphons too have allowed the bivalves to evolve through infaunal niches altogether closed to Articulata.

PART TWO
THE PACIFIC RIM

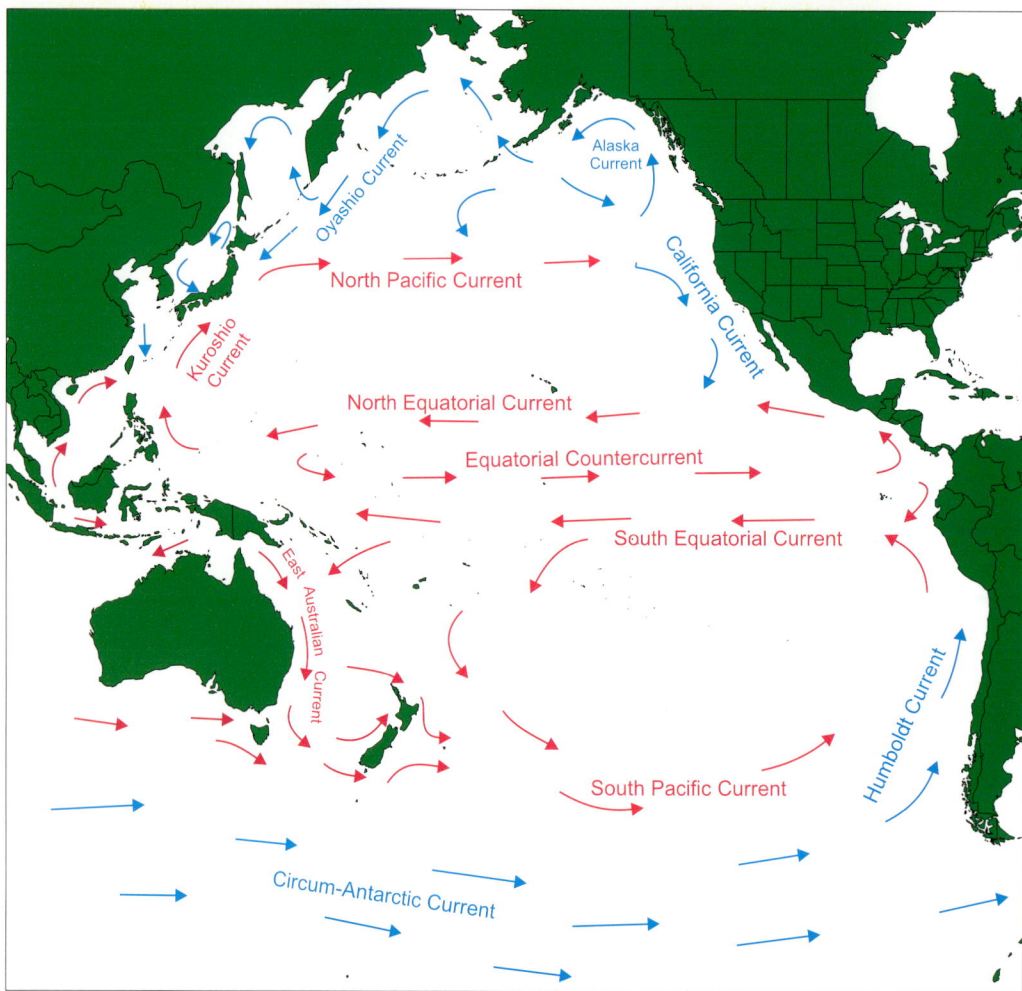

Fig 17.1 Current map of Pacific Ocean

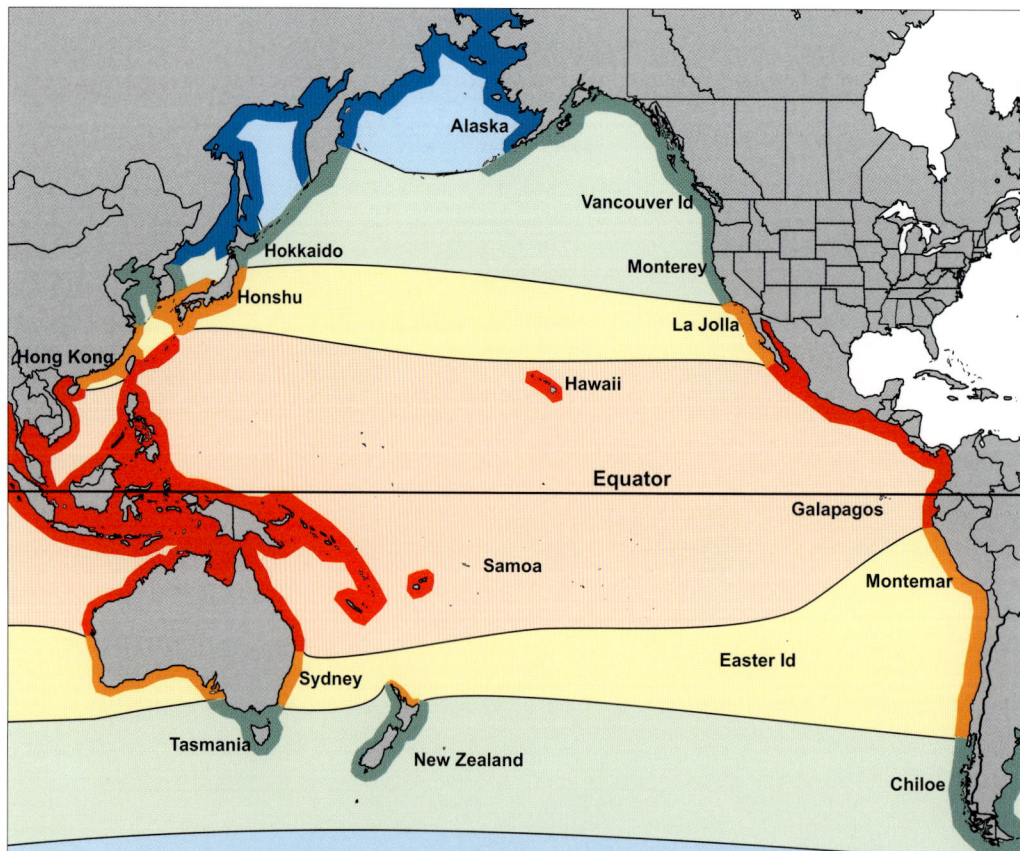

Fig. 17.2 Climate-related marine regions of the Pacific

CHAPTER SEVENTEEN
East Australia

The wider Pacific

It was appropriate to begin this story in New Zealand. No other Pacific country shows so compactly quite such a spectrum of shore types. Its North and South Island coasts are twice the length of the United States' Pacific seaboard, with far more diversity of terrain. Not only does New Zealand's mainland span from warm to cool temperate. The island extremes run from subtropical at the Kermadecs to subantarctic shores in the remote south.

Long-attenuate from north to south, New Zealand also shows contrasts from east to west. In the zone of the south-west anti-trades, the western seaboard compares with the cold kelp coasts of North and South America. Save for Tasmania, Australia has no cold west coast, but to the east both its warm and cool temperate invite comparison with New Zealand.

The currents map of the Pacific (Fig. 17.1) well explains New Zealand's west-east disparity. West-facing coasts, under the impact of continuing onshore winds, are washed by north-flowing cool ocean currents. These wave-pounded shores are favoured by rich nutrient upwelling from offshore. New Zealand's cool **Westland Current** is hence comparable with the larger **Alaska Current** of west North America and the **Humboldt Current** of South America, as well as with the **Benguela Current** of western South Africa.

On the continental west coasts, the temperate/tropical boundary is thus moved to lower latitudes, in South America to as low as 4 degrees south, with the Humboldt penguin reaching to the equator. The truly tropical seaboard here becomes confined within only 30 degrees of latitude. By contrast, on eastern continental coasts washed by warm currents the warm temperate is greatly extended. The influence of the North Atlantic *Gulf Stream*, as far as Cape Cod, is the classic example. On the Asian Pacific seaboard we shall similarly find the warm temperate prolonged into higher latitudes and subtropical shores reaching to the southern island chain of Japan.

For southern Africa the Stephensons, in their classic papers from 1936–47, first showed in detail the biological impact of such a west-east temperature contrast. On the east coast of Natal there are tropical populations that persist southwards to between Durban and East London. Warm-temperate species characterise the south coast to the tip of the Cape Peninsula; with cold-loving populations characterising the west coast from the Cape to Walvis Bay.

New Zealand's eastern seaboard, with the warm *Aupourian* and cool *Cookian Provinces*, has its closest affinities with east Australia. The *Aupourian Province* is influenced also by the **East Auckland Current** flowing from the north-east and contributing to the **East Auckland Current** south to East Cape and to a **West Auckland Current** for a shorter and variable distance down the west Northland coast.

The New Zealand fauna does not closely reflect this tropical connection, receiving only a minority of eurythermal tropical species from the Tonga-Samoa area. Even for the *Kermadecian Province* the main affinity is with New Zealand and eastern Australia. The east North Island mainland gives much evidence of the Australian relation. Powell in 1940 first showed the large percentage of *Peronian* (New South Wales) Mollusca brought to New Zealand by the *Tasman Front* of warm water. H.B. Fell in 1949 pointed to a large recent migration of east Australian echinoderms by the same route, while the largest element of our fish fauna from Northland to East Cape is shared with eastern Australia.

Temperate Australia

Our progress around the Pacific Rim can first begin with the long sequence of hard shores of eastern Australia, having many affinities to New Zealand, in the sharing of the same species or others closely related.

Southern Australian region:
Peronian Province

Warm-temperate New South Wales has a majority of long, straight shores directly confronting the Tasman

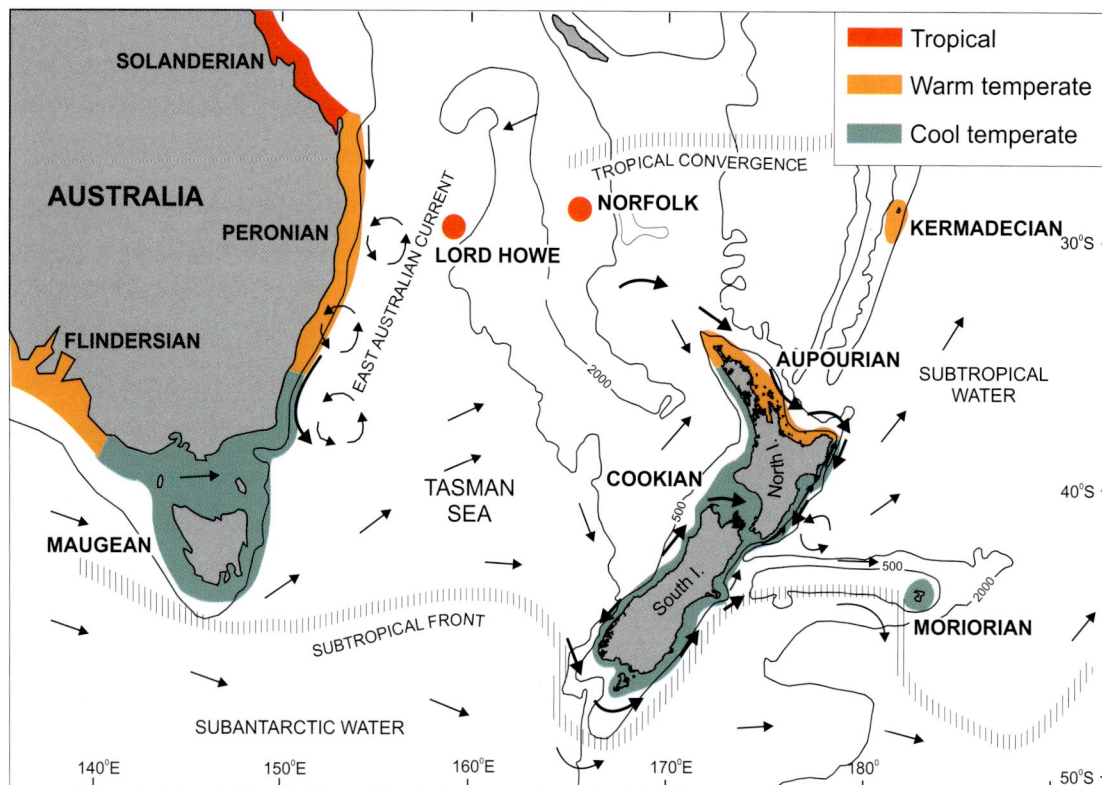

Fig. 17.3 The marine biogeographical provinces of New Zealand and East Australia

Sea, broken from Sydney north by just a few long, deep-embayed estuaries. Sydney's straight ocean coasts are carved out from the iron-brown Hawkesbury Sandstone of Permian or Triassic age. The cliff-lines are fronted with reefs on carved platforms, gently sloped to low water, or with high tidal benches subjected to constant wave crash, as at the Sydney Heads.

Long Reef at **Collaroy** is today a coastal reserve, though indiscriminate gathering of sea-foods in the past has much impoverished it. This reef is the *locus classicus* of Elizabeth Pope's pioneering shore paper that first brought home to me the idea of the integrated study of a whole habitat complex.

Right along Sydney's ocean shores, horizontal zones stand out clearly. In the **supralittoral zone** the brown sedimentary rocks lack most of the lichen colour of New Zealand greywacke or the hard granites of Victoria. The **littoral fringe**, on bare rock above the barnacle-line, has always the three common open coast periwinkles. *Nodilittorina unifasciata*, temperate-wide through Australasia, is rounder-whorled than New Zealand's cognate species *antipodum*. *Nodilittorina pyramidalis* is a warm-water and tropical species, in Australia located highest upshore. The third and very typical Australian littorine is *Bembicium nanum*, varying from tall-conical to a wave-exposed form as low as a limpet, sitting in its scar. The shell's oblique zebra-bands are generally much corroded.

All through New South Wales the **eulittoral zone** maintains a clear three-fold division. The **upper eulittoral** is invariably a barnacle territory, with generally two species. *Chamaesipho tasmanica*, already familiar from New Zealand, commonly shares this band with a second barnacle, *Chthamalus antennatus*, widely distributed through the warm South Pacific, and distinguished from *Chamaesipho* by its six-piece column.

In the lower half of the barnacle territory, on shores of moderate or high exposure, two larger species form the constant markers of the east Australian **middle eulittoral**. The more numerous is generally the pale tetraclitid *Tesseropora rosea*, pink-flushed when young and often imparting this tint to a whole shore band. Mingled with it is a second barnacle, *Catomerus polymerus*, to be picked out at once by its extra whorls of small basal plates. This is one of the few coasts where such a link with the primitive operculates having supernumerary plates still survives. Both *Tesseropora* and *Catomerus* prefer vertical faces in strong surf, and spread highest upshore on slopes under constant surge.

As in New Zealand, the lower band of the middle eulittoral is occupied by a serpulid tubeworm, but the Australian *Galeolaria caespitosa* is much more general and prominent than its New Zealand counterpart *Pomatoceros*. Forming a white zone standing out from a far distance, its tightly packed tubes build up the foundation crusts into convex ridges and mounds in the formation commonly known as 'Sydney coral'.

The **lower eulittoral** begins with a dark band sharply abutting upon the white, with a single closely massed species, the large ascidian *Pyura stolonifera*, familiarly called the 'cunjevoi'. The *Pyura* band forms a conspicuous segment of New South Wales' open shores, but extends only thinly into Victoria. Comparable species occur in warm temperate South Africa and Chile, but New Zealand has no such massive ascidian zone; only in the cold south does the stalked sea tulip, *Pyura pachydermatina*, appear sublittorally. At Sydney, the whole cunjevoi band stands out in dull reddish brown. Each broad-based individual, some 7–10 cm across, carries at the top two conical siphons, invested by a thick, tough tunic (recalling the former specific name *praeputialis*). Firm to walk over at low tide, cunjevoi make their presence known by sending up thin, sporadic exhalant jets.

The lower edge of the cunjevoi zone, sometimes intermingled with a patchwork of *Corallina* turf that can also extend subtidally, may be equated with the bottom of the eulittoral zone. Below this, the **sublittoral fringe** begins, with its heavy blanket of brown algae fully revealed at low tide, but being constantly splashed by the break of surge. Around Sydney, this is typically a sargassoid zone. Its most widespread and largest-leaved species is *Sargassum fallax*, with thick, slightly ribbed basal leaves and long festoons of smaller leaves with fertile axillary branchlets.

Shore platforms

Towards the platform edge, the eulittoral pools become fringed with *Hormosira banksii*, or their shallow floors may be bright green with well-lit *Enteromorpha*. Where rock stacks rise out of sand, branching straps of the green alga *Caulerpa filiformis* are often conspicuous.

Ending to seaward with its belt of cunjevoi and pinkish grey *Corallina*, the eulittoral zone drops to the **sublittoral fringe**. Towards its lowest edge, the eulittoral is generally enriched with algal pools. The deepest and more shaded are densely choked with *Sargassum* species. Shallow and rather tepid pools are floored with warm-water brown algae of the Dictyotales, such as *Padina pavonia*, with its dainty fan-like thalli scrolled at the edges, and the small-branched *Lobophora variegata*. The branched ribbons of *Dictyota dichotoma* are thin and smooth, common in well-lit shallows. They contrast with the transversely wrinkled fronds of *Dilophus marginatus*.

Beyond the cunjevoi and corallines, *Sargassum* becomes dense and continuous as the platform drops off to wave-break. Beneath the shade of *Sargassum*, there is a shining pink coralline paint, and the tufting *Corallina*

officinalis, diversified with *Jania* species and coarser-jointed *Cheilosporum* and *Amphiroa* species. Here too are scattered *Codium fragile*, *C. convolutum*, as well as *Laurencia* tufts. For the detailed treatment of the low tidal algae, recourse should be made to the comprehensive handbook of Brian Womersley.[1]

Heads of the small kelp *Ecklonia radiata* appear in the lowest reaches of the sublittoral fringe, under wave-break, as well as subtidally. This is unmistakably the same kelp common through northern New Zealand, though showing regional differences, as in the roughening of the blade and ribbons with pointed papillae (see Fig. 17.5), and in the shape of pool specimens (Fig. 17.7). For New South Wales, *Ecklonia radiata* is the single species of laminarian.

Exposure and shelter

This clear-cut sequence, constant over long stretches of open coast, is modified under extreme exposure or shelter. In the most exposed places, the *Galeolaria* zone is lost and the cunjevoi strip reduced almost to vanishing. The whole middle shore is left barnacle-dominated, with the high-level *Chamaesipho* mostly giving place to an upward spread of *Tesseropora rosea*, intermingled as always with *Catomerus*. Among cunjevoi or replacing them may be a scatter of the large barnacle, *Austromegabalanus nigrescens*, recognised by the rich blue lining revealed when the operculum is open.

On high-energy shores, the coralline crust reaches down to a wine-red zone of *Pterocladiella capillacea*. Unlike New Zealand or Victoria and Tasmania, New South Wales has no fucoid brown algae really distinctive of high exposure. To the south, long-leafed tresses of *Phyllospora comosa*, however, stand up well to surge, often with *Ecklonia radiata* under more exposure than this kelp would normally tolerate in New Zealand.

On the most sheltered coasts of New South Wales, lying in deep-indented harbours, the zonation is much transformed. The barnacles *Chamaesipho tasmanica* and *Chthamalus antennatus* give place to the opportunist *Austrominius modestus*, that arrived originally from New Zealand. Through the middle of the eulittoral zone runs a strong band of the rock oyster *Saccostrea glomerata*, today augmented or often replaced with the now Pacific-wide *Crassostrea gigas*. Both oyster species are farmed in the quiet waters of the George Estuary. Beneath the oyster zone appear the bladdered tresses of *Hormosira banksii*, often followed by a band of the mussel *Trichomya hirsuta*, with pointed apex and hairy epidermis.

Next below, the sublittoral fringe is sometimes made up of pure *Ecklonia radiata*, though *Sargassum* species may to varying extents reach far into the harbours. Two gastropod predators at low level are *Dicathais orbita*

317

Fig. 17.4 A New South Wales shore in moderate exposure

(Based on Balmoral, Sydney Harbour, looking out towards the North and Middle Heads.) a–b *Chthamalus antennatus;* b–c *Tesseropora, Catomerus* and *Galeolaria;* c–d *Pyura* and corallines; d–e *Sargassum* and *Ecklonia*

1 *Nodilittorina pyramidalis* 2 *Nodilittorina unifasciata* 3 *Bembicium nanum* 4 *Nerita atramentosa* 5 *Chthamalus antennatus* 6 *Tesseropora rosea* 7 *Catomerus polymerus* 8 *Morula marginalba* 9 *Notoacmea petterdi* 10 *Patelloida latistrigata* 11 *Cellana tramoserica* 12 *Montfortula rugosa* 13 *Turbo torquatus* 14 *Turbo undulatus* 15 *Astralium tentoriformis* 16 *Padina pavonia* 17 *Lobophora variegata* 18 *Sargassum* sp. 19 *Sargassum fallax*.

and *Cabestana spengleri*, both shared with New Zealand. The common oyster borer in estuaries is the small thaid *Bedeva paivae*.

Gastropods and chitons

Though harbour-enclosed, the rocky shore of Balmoral (Fig. 17.4) faces out directly to the Sydney Heads, and receives high energy from waves that frequently roll into the Harbour. This shore well illustrates — in the long vertical drop to the shelving platform of the lower eulittoral — the clear-cut ranges of the barnacles, limpets and other gastropods.

The highest littorine, *Nodilittorina pyramidalis*, is restricted to the sun-warmed tops of the rock stacks. The

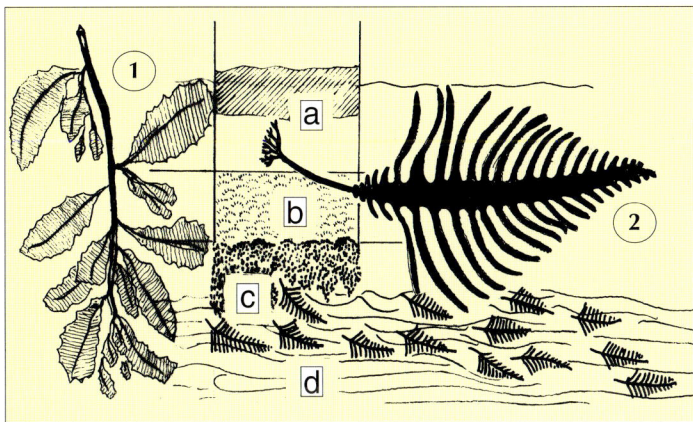

Fig. 17.5 The lower shore at Seal Rock, north of Sydney
(a) *Tesseropora* and **(beneath it)** *Galeolaria* (b) *Corallina*
(c) *Sargassum* (d) sublittoral channel with narrow-bladed
Ecklonia 1 *Sargassum fallax* 2 *Ecklonia radiata*.

**Fig. 17.6 A sheltered New South Wales shore at
Chinaman's Bay, Sydney Harbour**
a–b *Littorina*; b–c *Chthamalus antennatus* and *Cellana*;
c–d *Saccostrea* and **(beneath it)** *Galeolaria*; d–e *Pyura*;
e–f *Ulva*, *Petalonia* and *Sargassum*
1 *Chthamalus antennatus* 2 *Austrocochlea constricta*
3 *Austrocochlea concamerata* 4 *Bedeva paivae* 5 *Petalonia
fascia* 6 *Sargassum* sp. 7 *Cabestana spengleri* 8 *Dicathais
orbita* 9 *Saccostrea glomerata* 10 *Galeolaria caespitosa*
11 *Trichomya hirsuta* 12 *Pyura stolonifera* 13 *Pyura*,
dissected to show pharynx lying in atrial cavity.

black snail, *Nerita atramentosa*, familiar in northern
New Zealand, is tolerably common at the same level.
Nodilittorina unifasciata reaches well down into the
upper reach of the barnacle *Chthamalus antennatus*.
The variably shaped littorine, *Bembicium nanum*,
extends down to the middle eulittoral, which is clad
with *Tesseropora rosea* and *Catomerus polymerus*.

The most important east Australian trochids are the
two species of *Austrocochlea*. The sharply ovoid *A.
obtusa* with its dark axial striping is commoner on open
surfaces. *A. concamerata*, which is more depressed and
with nodulose spiral ribs, generally lives under boulders
or in crevices. Both these trochids most abound on
broad intertidal platforms as at Long Reef or Balmoral.

Over the same stretches will be found the turban
shell *Turbo torquatus*, with a deep umbilicus and white,
shelly operculum carrying two spiral lamellae. *Turbo
undulatus* is a smaller shell (2–3 cm), light and dark
green-banded and with a smooth operculum. The largest
of the turbans ranging north into tropical waters is
Turbo militaris. Quite distinct in shape is *Astralium
tentoriformis*, a tall conical turbinid with a flat base,
chiefly found in pools. The interior is pearly and
nacreous and the shelly operculum is flushed with pink.

Through the Sydney intertidal, a sequence of limpets
fill a succession of niches largely comparable with those
on New Zealand shores. Highest up the small limpet
Notoacmea petterdi (the counterpart of New Zealand's

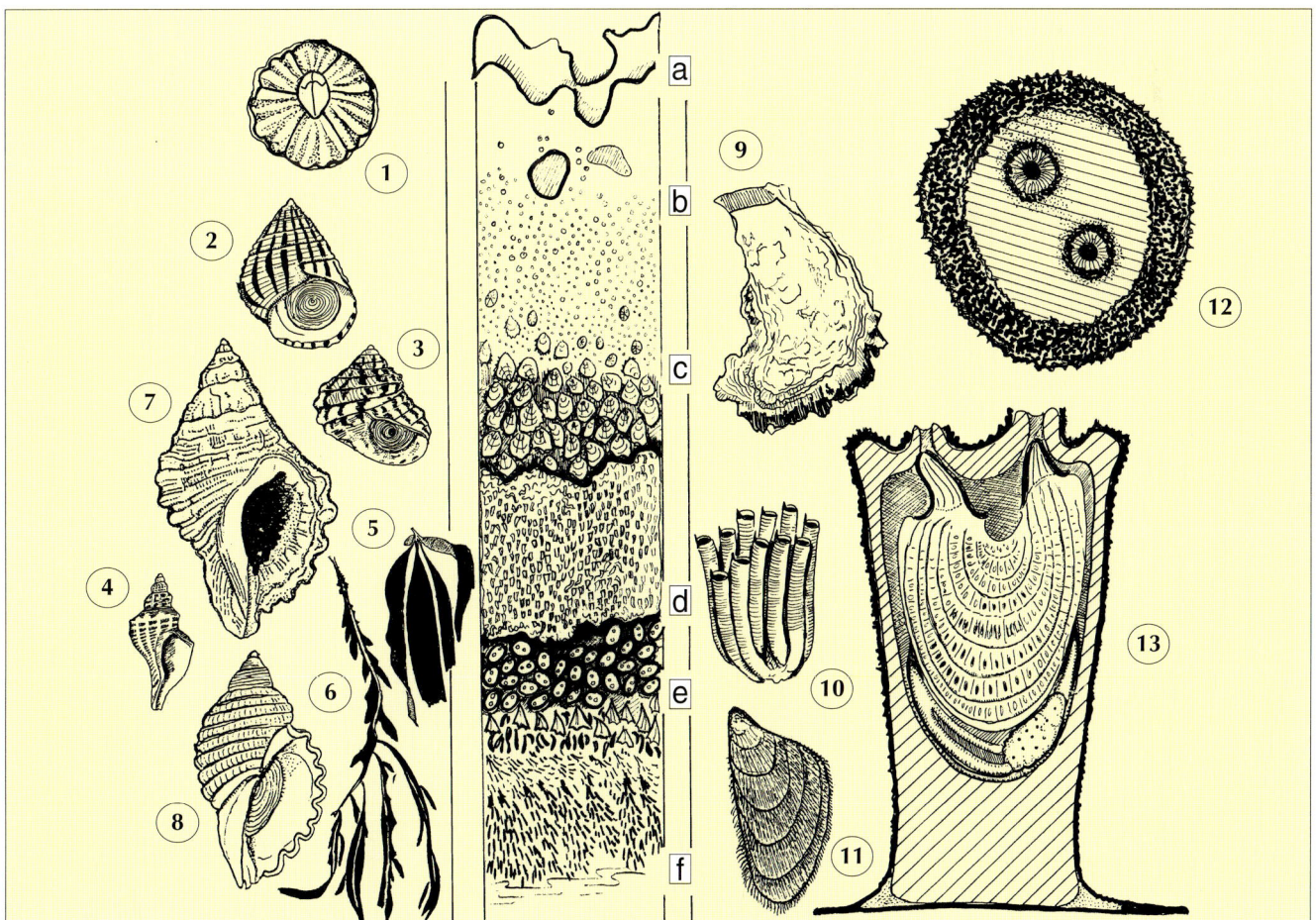

319

N. pileopsis) clusters on shaded vertical faces. *Patelloida latistrigata*, irregular in shape and often eroded, comes next below, to extend down through the barnacle zone.

The prosobranch limpets begin on the low tidal platform, reaching highest in pools within the eulittoral zone. The most widely familiar of the Australian limpets is *Cellana tramoserica*, variably coloured in orange, pink or white, with some 36 radial ribs, and the space between them darker. A few of the ribs are accentuated as dark streaks. This limpet excavates home-scars over the whole platform. On the lowest intertidal rocks permanently wet with surge, lives *Scutellastra chapmani*, a small limpet only 2 cm long, and sometimes stellate, always with eight strong radial ribs.

Two pulmonate limpets are common in New South Wales: *Siphonaria denticulata*, larger and depressed, with strong radial ribs, lives on the lower platform, while the smaller *S. funiculata*, conical and finer-ribbed, is confined to vertical faces within the barnacle zone. The pulmonate slug *Onchidella patelloides* grazes over platforms and shaded surfaces as in New Zealand.

East Australia has also a zoning fissurellid, the high-built *Montfortula rugosa*. Found in only a few parts of New Zealand (Fig. 7.29), this limpet is common at Sydney on open faces from the barnacles down to its prime sites among *Galeolaria* and cunjevoi. The shell is pale greenish and fine-ribbed, marked by a vestigial shell-slit in front, continued as a furrow inside.

The familiar middle shore chiton *Sypharochiton pelliserpentis* has the same range within the barnacle zone as in New Zealand. Below mid-tide levels occur two larger chitons, *Onithochiton quercinus*, with greenish brown valves, often eroded, and a broad, smooth girdle, and *Plaxiphora albida*, also eroded, but orange-footed and with the girdle beset with minute spicules.

The common thaid whelks at Sydney are the eulittoral black oyster borer, *Morula marginalba* and — near low water — *Dicathais orbita*, found singly or in clusters in deep pools or crevices. The last named has in New South Wales deep ridges and grooves (being called the cart-rut shell). Elsewhere it varies widely from the smooth '*textiliosa*' to be found on southern coasts to the bluntly tuberculated '*aegrota*' in Western Australia.

Biogeography

The whole eastern seaboard of Australia from Port Phillip to Cape York runs through more than 1500 km and traverses 30 degrees of latitude. The sea temperatures (winter to summer) range from 7–18°C in the south to 20–25°C in the extreme north. There is thus presented an unbroken sequence of shores from cool temperate to tropical. As well as both warm and cool temperate provinces, comparable with those of New Zealand, almost half this prolonged coastline falls within the *Indo-Polynesian Province* of the *Indo-West Pacific Tropical Region*. In the final chapter, on the coral-girded Pacific, this Province — with an area and extent approached by no other — will be treated in detail.

The seminal paper by Bennett and Pope (1953) divided Australia's warm temperate coasts into a south-western *Flindersian Province* (the Perth-Fremantle coast and South Australia with western Victoria as far as Cape Otway), and a south-eastern *Peronian Province*, typified in our description of Sydney shores (Fig. 17.4) and running from Sandy Cape on Fraser Island, Queensland, to Bermagui in southern New South Wales. The *Flindersian* (south-western Australian) and the *Peronian* (south-eastern Australian) *Provinces* were grouped together by Briggs (1974) in a warm temperate *South Australian Region*.

The east to west continuity of the Region is broken by the cool temperate shores of south-east Australia. Bennett and Pope saw these as a cool *Maugean Province* extending from Cape Otway in west Victoria to Bermagui and taking in Tasmania as well. In sea temperature and basic zonation these shores align well with southern New Zealand.

For extended detail of the intertidal animals and plants, particularly in the south-east, the reader should turn to Isobel Bennett's revised edition (1980) of the classic *Australian Seashores* of Dakin, Bennett and Pope (1952), invaluable not only for its illustrations of species, but for its broad treatment of ecological principles and coastal forms.

South Australia: Flindersian Province

Technically outside the Pacific, we have included here an exposed Flindersian shore, on the **Fleurieu Peninsula**, south of Adelaide. West of Cape Jervis from Port Elliot to Victor Harbour, the shores confront high wave action from the Southern Ocean, but have their main affinities with warm temperate New South Wales, rather than Victoria and Tasmania.

Granite Rock, **Victor Harbour** (Fig. 17.7) is an old dome of slow-weathering granite. High wave action elevates the vertical zonation to at least three times the tidal range. On the seaward face and smooth top the bush cover of the **adlittoral zone** is pushed upward, with the shore communities just below it drenched with splash and spray. High in the **supralittoral zone** the dark granite is lichen-rich, with grey *Buellia* and the bright orange of *Caloplaca*, conspicuous from far off. The

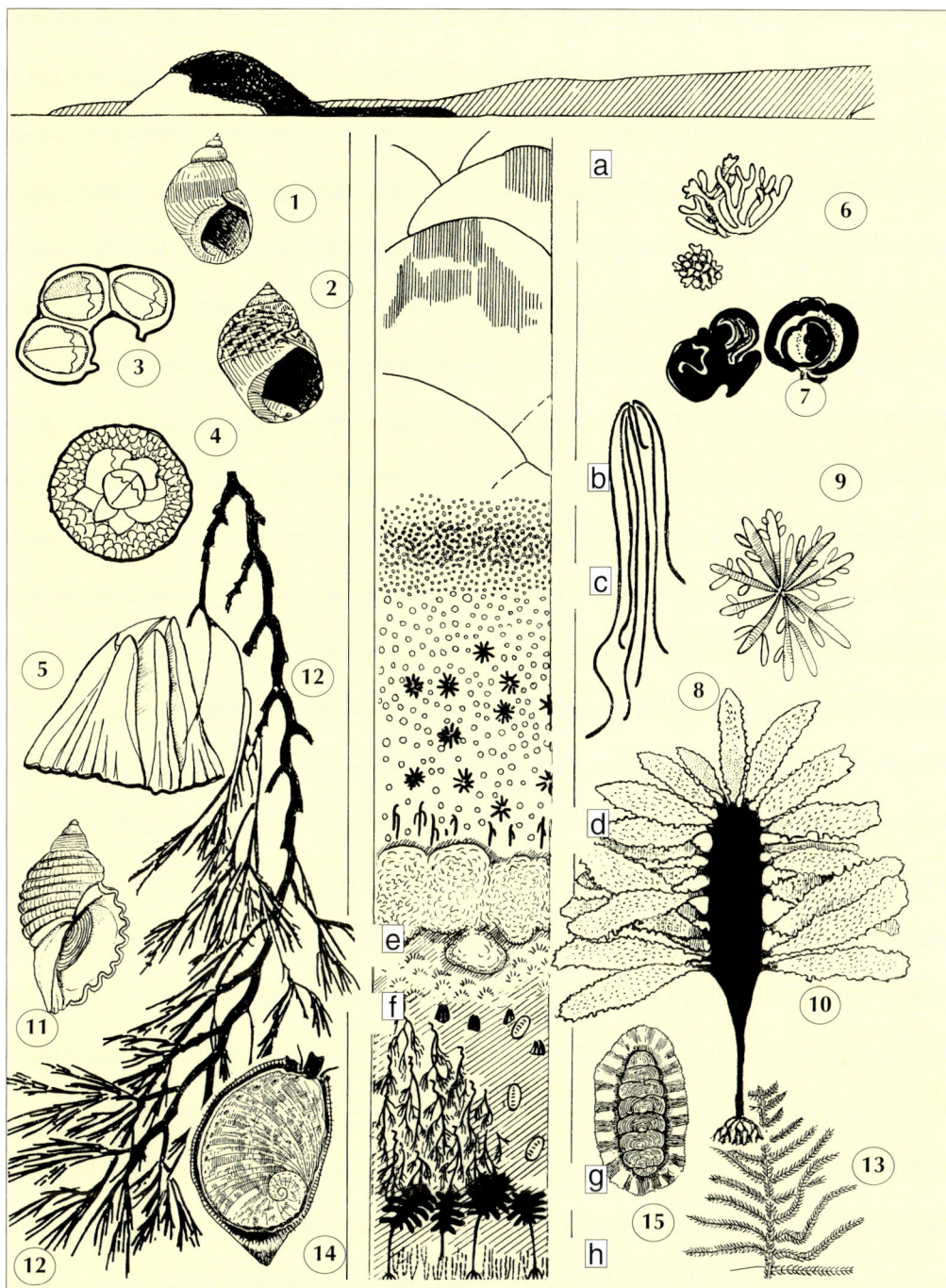

Fig. 17.7 An exposed shore at Victor Harbour, Fleurieu Peninsula, South Australia

a–b *Caloplaca, Calothrix* and *Littorina;* b–c *Chamaesipho tasmanica* and *Chathamalus antennatus;* c–d *Catomerus polymerus* and *Splachnidium rugosum;* d–e *Galeolaria caespitosa;* e–f *Corallina;* f–g *Cystophora;* g–h *Ecklonia* and *Caulerpa*

1 *Nodilittorina unifasciata* 2 *Nodilittorina praetermissa* 3 *Chamaesipho tasmanica* 4 *Catomerus polymerus*

5 *Austromegabalanus nigrescens* 6 *Lichina confinis* 7 *Anabaina* sp. 8 *Nemalion helminthoides* 9 *Splachnidium rugosum*

10 *Ecklonia radiata* 11 *Dicathais orbita* 12 *Cystophora intermedia* 13 *Caulerpa brownii* 14 *Haliotis rubra* 15 *Plaxiphora albida.*

littoral fringe is clad with *Verrucaria* or *Lichina*, or with the cyanobacteria *Calothrix fasciculatus*, shining black when wet. At this same level, or moving down among the barnacles in hot weather, live the periwinkles *Nodilittorina unifasciata* and — somewhat lower — *N. praetermissa*. The last named is absent from New South Wales, but is found in Victoria and Tasmania (Fig. 17.11).

As in New South Wales, the **upper eulittoral** is dominated by the barnacle *Chamaesipho tasmanica*, often intermixed with small *Chthamalus antennatus* The **mid-eulittoral** is sprinkled with larger *Catomerus polymerus*, with patches of the mussel *Brachidontes rostratus* appearing only in calmer places. Low in the mid-eulittoral white *Galeolaria caespitosa* encrusts the rock. Amid the barnacles are the limpets *Cellana*

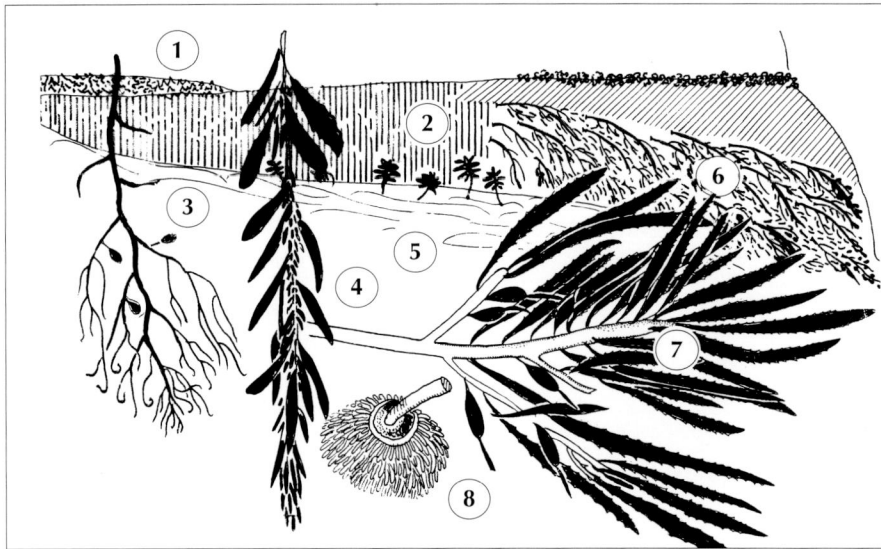

Fig. 17.8 The sublittoral fringe, near Bermagui, southern New South Wales
1 finger *Codium* 2 algal curtain, with 3 *Cystophora* and 4 *Sargassum* 5 channel with *Nemalion helminthoides* on sand and *Ecklonia radiata* 6 *Phyllospora comosa*, on wave-exposed face, with detail of 7 leaves and vesicles and 8 holdfast.

tramoserica, *Patelloida latistrigata*, and *Siphonaria diemenensis*, as well as the predatory whelk *Dicathais orbita*. The algae at mid-eulittoral level are seasonal and desiccation-adapted: the mucilage-filled tubes of *Splachnidium rugosum* and the thick mucous cords of *Nemalion helminthoides*; along with the blue-green vesicles of *Rivularia firma* and patches of *Isactis plana*.

The **lower eulittoral** is occupied by a turf of *Haliptilon roseum* and *C. officinalis*, occasionally with the large barnacle, *Austromegabalanus nigrescens*. Subordinate algae include *Laurencia filiformis*, *Wrangelia plumosa*, and *Dasya clavigera*. *Plaxiphora* chitons are common. *Hormosira banksii* gains entry only in the most shaded local spots.

From the **sublittoral fringe**, *Corallina* turf may continue to 1–2 m beyond low water. In high exposure it is, however, widely replaced by *Cystophora intermedia*, well adapted by its tough axis and pinnules to resist strong wave action. Other *Cystophora* species, *Ecklonia* and species of *Sargassum*, with *Caulerpa brownii* as an understorey, come in only where conditions are calmer. This same zone may also contain the stalked sea squirt *Pyura pachydermatina*.

Tasmanian region

The warm temperate *Peronian Province* had its boundary with the traditional *Maugean Province* at around Bermagui, in southern New South Wales. Today the coast south of Bermagui is placed by John Briggs within a cool temperate *Tasmanian Region* having a single Province taking in the entire seaboard of Victoria and Tasmania, with its western boundary set at Robe, in South Australia. Its own indigenous biota is relatively small, but is augmented with numerous incursions from the warm temperate north.

The little known west coast of Tasmania — still difficult to access by road — is directly exposed to the West Wind Drift, while cold currents sweep up the eastern shores to Bass Strait. Victoria shows to a greater extent than Tasmania an overlap of warm-water species.

Victoria

The intertidal zonation of Victoria differs clearly at first sighting from that of the warm *Peronian Province*, presenting a cool temperate picture basically comparable with southern New Zealand.

The journey south from Sydney by the Princes Highway gives access to rocky coasts marking the progressive change towards a colder regime. Rewarding stops could be the narrow-benched breccia shore of **Bombo**, then the clean granitic slope of **Bawley Point**, with cunjevoi now scarce, and *Cystophora* species first becoming common. At **Bermagui** the brown alga *Phyllospora comosa* first appears in quantity and the bull-kelp, *Durvillaea potatorum*, can frequently be found washed up onshore.

Phyllospora comosa soon becomes the common fucoid of the sublittoral fringe, beginning sparsely in the north around Port Macquarie and reaching from Victoria to Robe in South Australia. The branching, cylindrical stipe can grow to more than a metre long, with a discoid holdfast of compressed fingers. The long, elliptical leaflets are sparsely toothed and interspersed with ovoid vesicles, tipped with a small leaflet.

At **Merimbula**, *Phyllospora* and several species of *Cystophora* abound, while at **Bastion Point**, Mallacoota, just over the Victorian border, the bull-kelp, *Durvillaea potatorum*, has attained its full zone, and the bladder-kelp, *Macrocystis integrifolia*, makes its appearance in the upper sublittoral. **Cape Conran**, the last outpost of rocky coast, shows us two intertidal mussels well-established, a band of *Brachidontes rostratus* (reaching furthest north to Tuross Heads) and good patches of *Xenostrobus pulex*.

Numerous warm-water species have concurrently fallen out. The common grapsid shore crabs of New South Wales, *Leptograpsus variegatus* and *Pachygrapsus transversus*, are replaced in Victoria by *Brachynotus octodentatus* and *Paragrapsus quadridentatus*. Several mollusca are lost too, including the turban *Turbo torquatus* and several chitons, such as *Sypharochiton pelliserpentis*, *Cryptoplax mystica*, *Ischnochiton lentiginosus* and *I. smaragdinus*.

In land-locked estuaries south from the New South Wales border, the cool temperate oceanic patterns are interrupted by enclaves of a warmer regime. As far south as Port Phillip rock oysters, mangroves and black snails (*Nerita atramentosa*) make their final stand.

Victoria's open coasts are mostly built of hard rocks, typified in the basalts of Phillip Island, or the old, wave-smoothed granites of Wilson's Promontory. Once an island, this southernmost outpost of the Australian mainland forms a dome rising more than 600 m. Well-vegetated on top, it faces the constant wave attack of the south-westerlies.

The zonation pictured in Fig. 17.9 is based upon the open shore at **Lorne**, west of Port Phillip. A **littoral fringe** with the black lichens *Verrucaria maura* and *Lichina confinis* stands out darkly in a way seldom seen on softer bedrock. The commoner periwinkle, *Nodilittorina unifasciata*, ranges highest, with a second Victorian species, *Nodilittorina praetermissa*, further down in the fringe. *Nodilittorina pyramidalis* is lacking south of Bastion Point, New South Wales.

In the upper stretch of the **eulittoral**, *Chamaesipho tasmanica* stands alone, but east of Cape Otway it shares possession with warm temperate *Chthamalus antennatus*. *Catomerus polymerus* continues from warmer coasts, but *Tesseropora rosea* is found in Victoria only in the far east. The lower reach of the barnacle zone seasonally develops a band of the brown alga *Splachnidium rugosum*.

The salient feature of Victorian shores is the black cordon of mussels cutting across the eulittoral. Patches of the large ribbed mussel, *Brachidontes subramosus*, join up in a continuous belt essentially all round the coast. The much smaller flea mussel, *Xenostrobus pulex*, runs — as in New Zealand — at a high level through the barnacles of the **upper eulittoral**, sometimes in a distinct black band but elsewhere merging with the top of *Brachidontes*. Rather than forming their own band, scattered *Catomerus* barnacles may settle on *Brachidontes*. Below the mussels the white band of the serpulid *Galeolaria caespitosa* is broadly constant on open coasts right through Victoria. As in New South Wales, *Hormosira banksii* has only a narrow vertical range, but on rock platforms it may form wide carpets.

In the **lower eulittoral**, the cunjevoi, *Pyura*

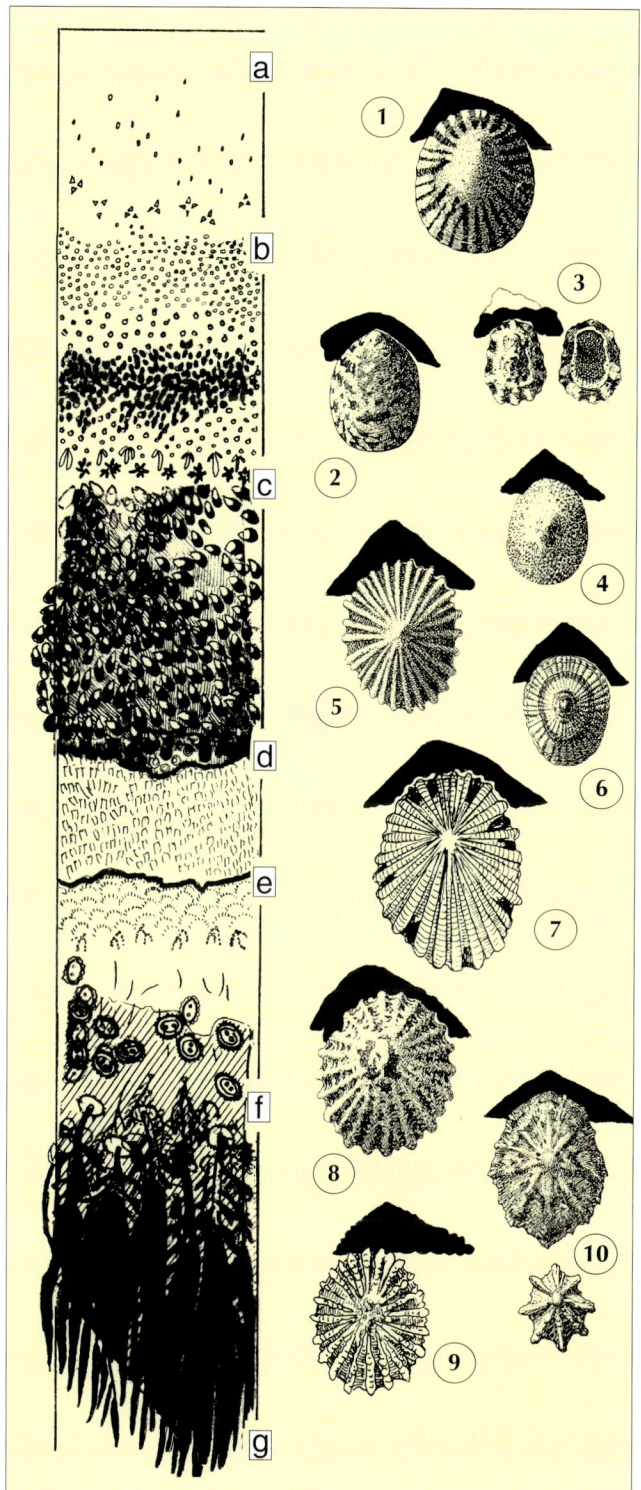

Fig. 17.9 A high-exposed shore at Lorne, Victoria
a–b littorines; b–c *Chthamalus antennatus*; c–d *Catomerus polymerus* and mussels *Xenostrobus* and *Brachidontes*; d–e *Galeolaria caespitosa*; e–f *Corallina* and *Pyura stolonifera*; f–g *Durvillaea potatorum* and *Macrocystis*

Limpets with approximate vertical distribution:
1 *Notoacmea petterdi* 2 *Notoacmea mayi* 3 *Patelloida latistrigata* 4 *Notoacmea flammea* 5 *Siphonaria diemenensis* 6 *Siphonaria tasmanica* 7 *Cellana tramoserica* 8 *Patelloida alticostata* 9 *Scutellastra peronii* 10 *Scutellastra chapmani*.

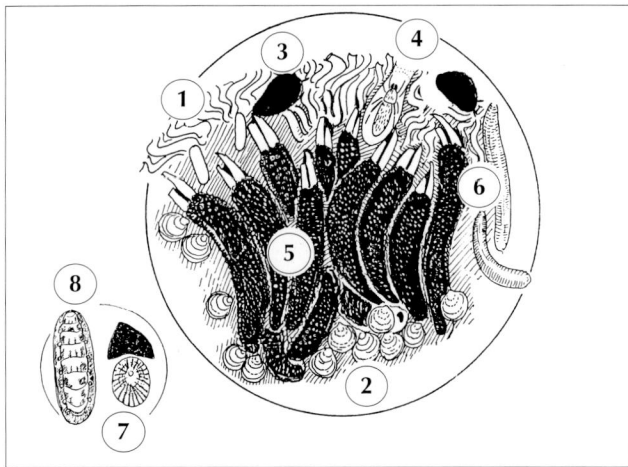

Fig. 17.10 The community among serpulid tubes
1 *Galeolaria caespitosa* 2 *Lasaea australis* 3 *Onchidella patelloides* 4 *Desis crosslandi* 5 *Ibla quadrivalvis*
6 *Phascolosoma noduliferum* 7 *Scutellastra chapmani*
8 *Acanthochitona retrojecta.* (Based on Bennett).

stolonifera, occurs only spasmodically at a few places east of Cape Otway, as at Lorne and Wilson's Promontory. Pink *Lithophyllum* provides the chief rock cover, studded with *Plaxiphora albida*, up to 10 cm long, which is Victoria's only large surface chiton.

Limpets

The principal Victorian limpets, including the siphonariids, are shown in Fig. 17.9. Found on intertidal platforms, as one of Australia's commonest shells, the nacellid *Cellana tramoserica* has the same range as in New South Wales. In Victoria and Tasmania the larger *Cellana solida* shares this habitat, being broader ribbed and prominently orange-red margined. A third limpet, *Scutellastra peronii*, is common only at low levels, from *Pyura* down to the holdfasts of *Durvillaea*; it is marginally scalloped and variable, with crenulations corresponding to channels between the strong radiating ribs. The *Scutellastra chapmani*, sometimes in its stellate form, lives in surge in the lowest intertidal.

The high level acmaeid limpets are *Notoacmea petterdi* and the southern *N. mayi*; the latter is shared with Tasmania, and recognised at once by the apex in advance of the front margin. As at Sydney, *Patelloida latistrigata* occupies steep faces at the next level down in the upper eulittoral. A related species is the small *Notoacmea flammea*. A larger acmaeid is *Patelloida alticostata* — up to 3.5 cm — living on rock platforms at the lowest levels; it is marked by its scalloped edge, and sharp ribs with black cross-lines in their interstices.

The southern siphonariids include *Siphonaria diemenensis*, on lower platforms, larger, flatter and strongly ribbed, and the smaller, conical *S. funiculata* and *S. tasmanica*. The pulmonate slug *Onchidella*

patelloides grazes in the eulittoral, as also in New South Wales and Tasmania.

The common mid-tidal chiton is *Sypharochiton pelliserpentis*, ranging from central New South Wales to Victoria and Tasmania. Below mid-tide it is replaced in Victoria and Tasmania by the large, generally eroded *Plaxiphora albida*.

The interstitial community or 'cryptofauna' among *Galeolaria* tubes is represented in Fig. 17.10. With *Onchidella* grazing upon the surface, a narrow chiton, *Acanthochitona retrojecta*, and a small, high-built limpet, *Notoacmea alta*, penetrate deeper. The silken pockets of the spider *Desis croslandi* open at the top. Wedged between the serpulid tubes are the small, primitive stalked barnacle *Ibla quadrivalvis*, and often the sipunculan worm *Phascolosoma noduliferum*. Around the tube-bases are crowds of the pallid little bivalve *Lasaea australis*.

On sheltered Victorian coasts numerous species of *Cystophora* occur in or beyond the sublittoral fringe. At Wilson's Promontory, *C. intermedia* grows in exposed places among *Pyura* or may even replace *Durvillaea* down to low water. In more protected sites or rock pools may be found *Cystophora torulosa*, *C. paniculata*, *C. siliquosa*, *C. retorta* and *Caulocystis uvifera*. *Austromegabalanus nigrescens* makes a definite band at the upper part of a *Cystophora* zone.

In the sublittoral fringe, below the pink veneer, the bull-kelp, *Durvillaea potatorum*, holds full sway. In essentials this species is close to the New Zealand *D. willana*, with wide basal disc, and side-paddles carried on the stipe. At many sites it is exposed only at low spring tides in calm weather, but with continuous swells its top limit is pushed higher up the shore, and on Victoria's most exposed coasts bull-kelp forms a continuous zone daily emersed.

In more sheltered spots, as at Bastion Point and Sorrento, there are stands of the bladder-kelp, *Macrocystis integrifolia*, growing slightly below the bull-kelp and differing from *M. pyrifera* chiefly in its narrower leaves. In Victoria *Ecklonia radiata* and *Phyllospora comosa* play a secondary role to *Durvillaea*, being relegated to wave-replenished pools or sheltered reef margins. They are altogether lost in high exposure.

Tasmania

Everywhere in Tasmania there is an essential similarity with the shores of cool temperate Victoria, but the island has its special differences from the mainland, as well as variations of its own from exposure to shelter. The westward shores face the highest exposure, with the seas calmer on the more sheltered north coast, while the east coast experiences sub-extreme to moderate exposure.

Fig. 17.11 A high-exposed shore at Bicheno, east Tasmania

a–b orange *Caloplaca* and black *Verrucaria*; b–c *Chthamalus antennatus* and *Chamaesipho tasmanica*; c–d *Brachidontes, Xenostrobus* and *Mytilus galloprovincialis*; d–e *Lithophyllum hyperellum*; e–f *Xiphophora gladiata* and (**beneath**) f–g *Durvillaea potatorum*

1 *Nodilittorina unifasciata* 2 *Nodilittorina praetermissa* 3 *Xenostrobus pulex* 4 *Mytilus galloprovincialis* 5 *Brachidontes rostratus* 6 *Cellana solida* 7 *Dicathais orbita* 8 *Lepsiella vinosa* 9 *Lithophyllum hyperellum* 10 *Durvillaea potatorum* 11 *Macrocystis integrifolia* 12 *Lessonia corrugata* 13 *Xiphophora gladiata*.

The shore sequence shown in Fig. 17.11 is from **Bicheno** on the Tasmanian east coast. As well as black *Lichina confinis*, the **littoral fringe** is noted for a crustose orange lichen that widely lends its colour to the Tasmanian and Victorian upper shore. The barnacles *Chthamalus antennatus* and *Chamaesipho tasmanica* are found as on the mainland, but the former is lost in the extreme south. *Catomerus polymerus* is likewise thinning out to the end of its range. Mussels dominate the middle shore, with a strong band of *Brachidontes rostratus* mixed with *Mytilus galloprovincialis*, and above it small *Xenostrobus pulex*. As in Victoria, there is seasonal *Splachnidium rugosum*.

Below the mussels the exposed shore takes its more

distinctive Tasmanian character. Except in shelter *Galeolaria* has given place to the hard nodules of mustard yellow *Lithophyllum hyperellum*, sculpted as shown in Fig. 17.11. Coralline algal turf is succeeded by a broad belt of *Xiphophora gladiata*, branching into slender whips. *Hormosira banksii* in Tasmania is restricted to places of high shelter.

A reduced band of *Pyura stolonifera* may mark the top of the **sublittoral fringe**, where *Durvillaea potatorum* is strong and dominant. On the north coast, the bull-kelp is found to be replaced by *Cystophora torulosa* and *C. paniculata*. Offshore, beyond *Durvillaea*, as well as *Macrocystis integrifolia*, there is the slender-strapped *Lessonia corrugata*.

Towards the Coral Coast
Subtropical Queensland

Beyond the warm temperate *Peronian Province*, based on New South Wales and southernmost Queensland, the northern half of the east Australian seaboard is subtropical or tropical. The real transition is nowhere sharp or clear-cut. In the 800–1000 km north of Sydney, up to latitude 26 degrees, near Noosa Heads, Queensland, Bennett and Pope could find no place with enough tropical species to give it fully a subtropical rather than warm temperate status. It is at Caloundra that corals, zoanthids and alcyonaceans including *Xenia*, as well as tropical echinoderms and molluscs, begin to set the scene. A province truly to be called *tropical* was thus taken to begin at the southern extreme of the Great Barrier Reef.

South of this, Queensland's southern reach had been traditionally designated as a separate *subtropical* province of Australia, called the *Solanderian*. John Briggs has regarded this coast — like the Great Barrier Reef system — as a southern salient of the most extended of all marine provinces, the *Indo-West Polynesian*, with the winter sea temperature, safely above 20°C, allowing the beginning of reef corals in mainland Queensland.

The off-lying reefs of Queensland are the theme of the excellent and comprehensive *Reader's Digest Book of the Great Barrier Reef* (1984). Leaving the diversity of the coral shores for our final chapter, we shall notice here those features above the coral-line that distinguish the Queensland mainland from the warm temperate *Peronian Province*.

Bennett and Pope cite tropical species reaching sporadically into New South Wales. The coral *Plesiastrea versipora* and the zoanthid *Palythoa australiensis* extend even as far south as Nambucca Heads. The sea urchins *Echinometra mathaei* and *Diadema setosum* enter northern New South Wales, and *Tripneustes gratilla* may reach south to Sydney during summer.

A composite representation (Fig. 17.12) based on the mainland shores of north Queensland, brings out the many differences from the warm temperate. In the **adlittoral zone** the coastal forest, with its luxuriant shrub and ground layers, reaches almost to the high water mark. No tree is so typical as the screw-pine, *Pandanus tectorius*, with its harsh, spear-shaped and serrated leaves, trunk-struts and thorny main stem and branches.

Over the sun-warmed **supralittoral zone** runs the convolvulacean *Ipomoea pescaprae*, with purple trumpets and thick, shiny leaves. Behind the reef there may be a shrub layer of *Scaevola taccada* (Family Goodeniaceae), with asymmetric white flowers like half-stars, and pale, waxen berries. At ground level the scrambling, blue-flowered *Scaevola calendulacea* often grows. A regular feature of the Queensland adlittoral is also the yellow-flowered *Hibbertia scandens*.

The **littoral fringe** is notably enriched with small gastropods. The Queensland periwinkles are *Nodilittorina trochoides*, continuing southward right through the warm temperate, the small, granulose *Nodilittorina vidua* and the taller, often wavy-marked *Littoraria undulata*. Each of these has a wide tropical range.

On the back-shore, the primitive pulmonates of the Family **Ellobiidae** increase in size and diversity. In the Queensland littoral fringe and supralittoral zone, among litter or under boulders, there are several *Melampus* species. In the far north, *Pythia scarabaeus* lives entirely terrestrially in coastal scrub; it is the size and shape of a plum-stone, with a narrow, tooth-lined aperture. The high-level Family **Neritidae** also diversifies in the tropics. With the warm temperate *Nerita atramentosa* reaching to South Queensland, there is an accession of tropic-wide species: ridged *Nerita plicata*, smooth *N. polita* and, further down-shore, the low, hemispheric *N. albicilla*.

The barnacle-line begins with a scatter of the widespread *Chthamalus antennatus*. *Chamaesipho tasmanica*, along with *Tesseropora rosea* and *Catomerus polymerus*, terminate in South Queensland; and the dominance below *Chthamalus* passes to the tropical *Tetraclita squamosa*, coming around Cape York to continue south. The sculpture is coarsely striate and the column plates are insulated by the air spaces boxed off with trabeculae. The small aperture retards desiccation.

The grapsid crabs of the littoral fringe show a new release of speciation. *Cyclograpsus* species are common under high tidal litter and stones, along with tropical *Sesarminae* and *Metopograpsus*. The free-running *Leptograpsus* of the warm temperate now gives space to brightly painted species of *Grapsus*. An important new

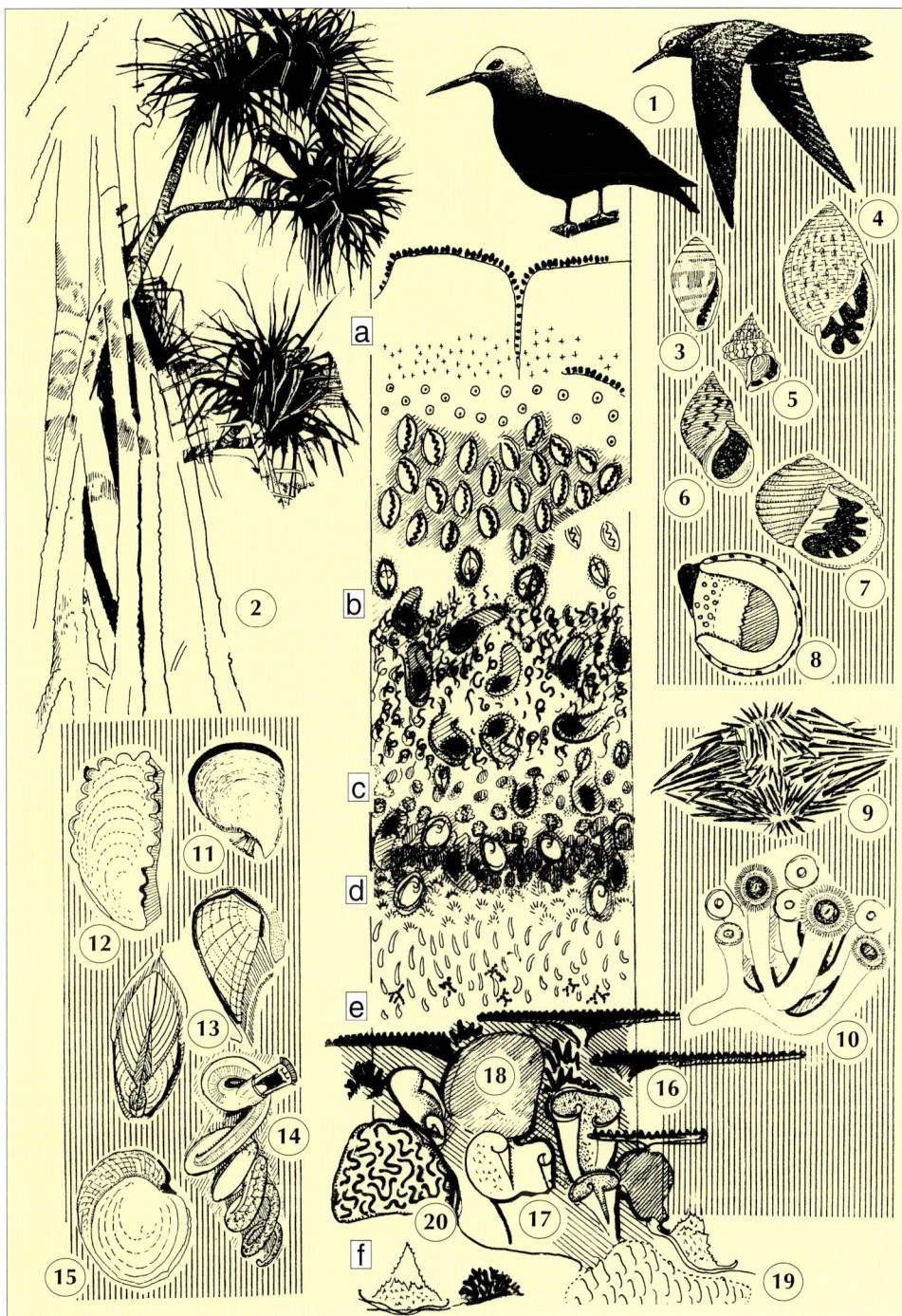

Fig. 17.12 Shore zonation in North Queensland

(A composite picture representative of Queensland tropical shores.) a–b *Chthamalus* and *Isognomon;* b–c *Tetraclita, Crassostrea* and *Arca;* c–d vermetids and *Echinometra;* d–e *Chama,* corallines, green algae and zoanthids; e–f corals 1 black noddy, *Anous minutus* 2 screw-pine, *Pandanus tectorius* 3 *Melampus fasciatus* 4 *Pythia pollex* 5 *Nodilittorina pyramidalis* 6 *Littoraria undulata* 7 *Nerita plicata* 8 *Nerita albicilla* 9 *Echinometra mathaei* 10 *Zoanthus* sp. 11 *Isognomon nuclesis* 12 *Crassostrea mordax* 13 *Arca avellana* 14 vermetid *Dendropoma* removed from shell 15 *Pseudochama exogyrata* reef corals 16 *Acropora* table 17 *Merulina ampliata* 18 *Porites* 19 *Lobophytum expansum* 20 meandroid faviid.

element as we enter the Queensland tropics is the semi-terrestrial hermit crab, *Coenobita rugosus.*

At a high level for a bivalve, rows of the small *Isognomon perna* fix by the byssus in cracks, feeding from inundation with splash above the barnacle-line. On all but the most exposed coasts, rock oysters dominate the **mid-eulittoral,** as typically as mussels in the cool temperate. The Queensland species, built into a close mosaic, is the crenate-edged *Saccostrea cucullata.* Below the oysters, particularly on steep, shaded surfaces, serpulid tubeworms are tropically replaced by vermetid gastropods. Small species of *Dendropoma* with their loose-coiled shells lie crusted or embedded in calcareous algae.

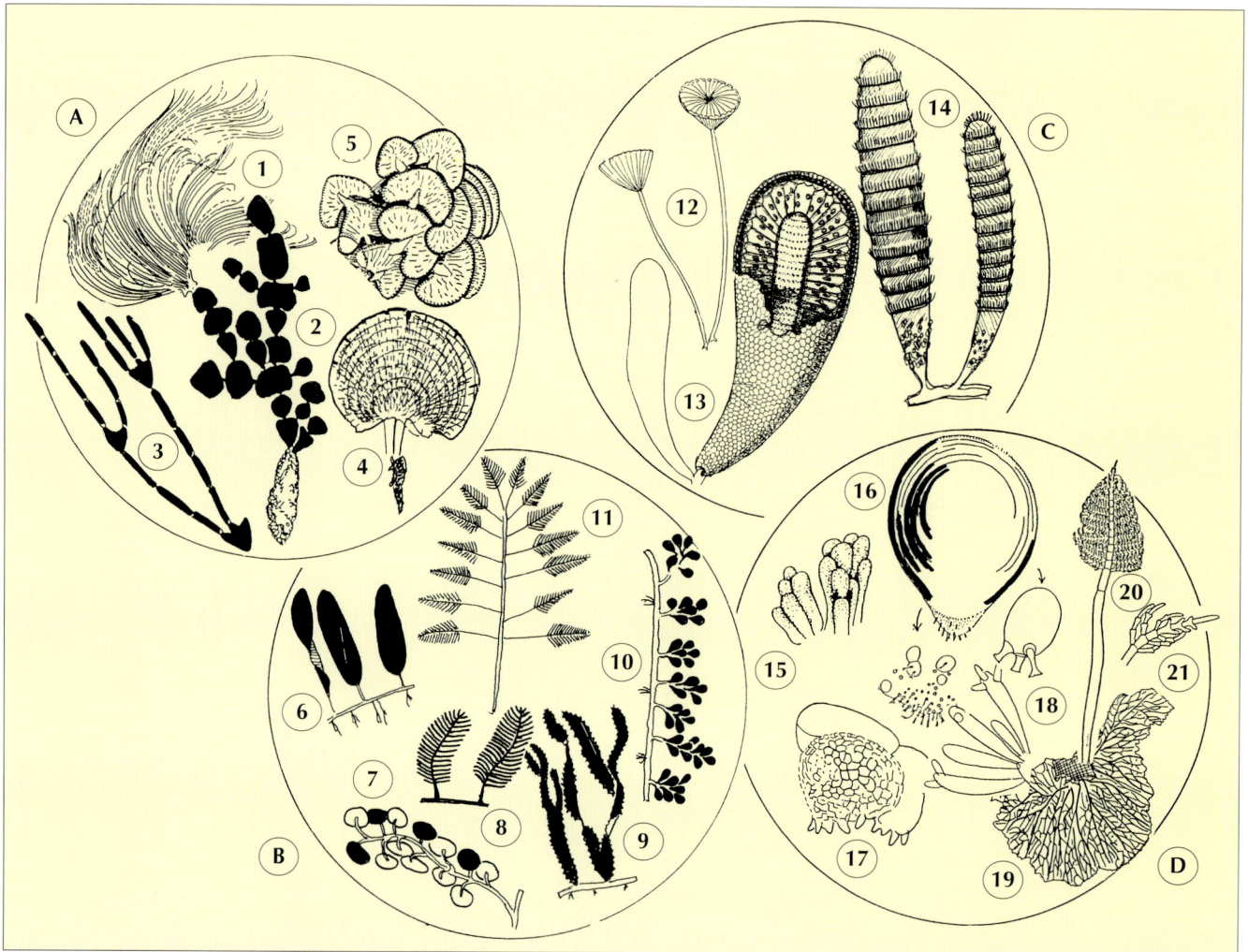

Fig. 17.13 Tropical and subtropical green algae

A **Codiales** 1 *Chlorodesmis fastigiata* 2 *Halimeda macroloba* 3 *Halimeda cylindracea* 4 *Udotea* sp. 5 *Halimeda discoidea*

B **Caulerpales** 6 *Caulerpa brachypus* 7 *Caulerpa peltata* 8 *Caulerpa cupressoides* 9 *Caulerpa serrulata* 10 *Caulerpa racemosa* 11 *Bryopsis plumosa*

C **Dasycladales** 12 *Acetabularia* 13 *Bornetella* 14 *Neomeris*

D **Siphonocladales** 15 *Boergesenia forbesii* 16 *Ventricaria ventricosa* 17 *Dictyosphaeria* 18 *Ernodesmis verticillata* 19 *Microdictyon* 20 *Struvea* 21 *Struvea* detail.

On the upper shore, true limpets are virtually replaced by pulmonates, with two widespread species, the low, strong-ribbed *Siphonaria atra* and the higher pitched *S. normalis*. *Sypharochiton pelliserpentis* cuts out just north of the Queensland border; the common chiton of the upper shore is now the big *Acanthopleura gemmata*, with its broad girdle covered with horny villi. Several species of thaids prey on barnacles or young oysters, the commonest being *Morula granulata*.

Right up to the vermetid level, the rock surface is scarred by the small urchin *Echinometra mathaei*. Ovate-oblong rather than circular, and with short abrading spines, this urchin abounds right through the tropics. Shortly below this level two sorts of bivalves are ubiquitous on tropical shores. The *Chama* species are oyster-like, with the lower valve cup-like and fused to

the surface and the other a convex lid. Their real affinities are with higher bivalves in their possession of a fused mantle with two short siphons. Nestling or rock-abrading among the chamids is an ark shell, *Arca avellana*, byssus-attached and eroding as the shell is rocked to and fro. The spiral umbones like ram's horns at once identify the embedded shell.

The brown algae in the tropics are smaller and less important than in the temperate. In the upper eulittoral tufts of *Ectocarpus* and other filamentous browns grow under frequent splash. Permanent high level pools contain the fan-shaped *Padina* and kindred Dictyotales. Towards low water, in constant surge, the Fucales are represented by rather small Sargassa, or their much modified relatives, *Turbinaria*. The *Sargassum* species of the tropics have great foliage diversity, and with high wave attack, their short thalli become crisp and cartilaginous.

Though smaller in biomass than the reds or browns, the green algae draw early attention on a subtropical shore. *Caulerpa* and *Halimeda* species are increasingly common, along with small *Microdictyon*, *Dictyopsphaeria*, *Bornetella*, and *Bryopsis*.

Dominance of the lower shore passes eventually from the algae to the cnidarians, with spreading sheets of zoanthids, anemones and alcyonacean soft corals. Finally, the growing array of stony corals (**Scleractinia**) sets the stage for the tropical **sublittoral fringe**. In the temperate this fringe is no more than an enriched tail-piece to the tidal shore. In the tropics it is prolonged far to seaward, in the rich, crowning achievement of the 'biotic reef', where both corals and calcareous algae have shared in the work of construction.

Tropical algae

Rhodophyta

From the threshold of the **pink-line** downwards, the crust of calcareous red algae thickens and diversifies. While many soft-fronded rhodophytes remain, the main pre-occupation of the tropical reds has been to deposit calcium carbonate within their cell walls. They thus contribute massively to reef growth, cementing the interspaces of living corals, and overgrowing the dead ones. Some writers have adopted the term 'coral/algal reefs', to recognise the contribution of these algae.

Throughout the tropics, crustose lithophylla are the associates of reef corals, sometimes outweighing the coral in their production of calcite. In the much-quoted Funafuti Island coring, the calcareous algae were found to go down to 300 m. They are evidently geologically ancient, with *Archaeolithothamnion* running back to the Cretaceous and most of our living genera found through the whole Tertiary. *Solenopora* — apparently allied to the Corallinaceae — is of Ordovician to Jurassic age.

Chlorophyta

The green algae have evolved most of their diversity, with such tissue complexity as they attain in warm seas. Their temperate members are mostly very simply structured. Thus *Ulva* is a sheet with two cell layers, or *Monostroma* with one; *Enteromorpha* is tubular, *Cladophora* is branching-filamentous; and *Chaetomorpha* forms chains of large, thick-walled cells.

The tropical green algae stand out most vividly in the sublittoral fringe or on the reef front where pools are protected from direct waves but constantly replenished by surge. A further suite of species grow from rhizomes creeping over sand.

The most advanced of the **Chlorophyta** have their unique mode of forming tissues, with branched tubular filaments that are not cellular but syncytial, partitioned into compartments each with several nuclei. From this feature, they were grouped together in the older works as the Order **Siphonales**.

Four orders of the higher greens are recognised today, each with a separate sub-pattern of organisaton.

In the Order **Caulerpales**, the simplest forms like *Bryopsis* have pinnate fronds, multinucleate but non-septate, from a creeping rhizome. *Caulerpa* is more differentiated, with the syncytium divided up by cross-struts. They display a wide variety of frond structure, fashioned as clubs or peltate heads, bunches of grapes, or cypress or yew leaves (see Fig. 17.13). Some *Caulerpa* may spread over much of the bottom of a lagoon, being some of the few green algae fixing in sand or silt.

The Order **Dasycladales** specialises in whorled systems, sometimes lightly calcified, with rings of branches springing from an original basal vesicle. *Neomeris* is shaped into a club with its whorls of branches densely adpressed to it. In the tiny toadstool-like *Acetabularia*, the vesicle is a slender stalk, with a two-whorled crown of branches at its head.

In the Order **Siphonocladales**, *Struvea* has as its primary vesicle a creeping rhizoid from which rises a loose network like a leaf-skeleton. *Boodlea* has become more densely branched, while in *Microdictyon* (Fig. 17.13) a cluster of one-planed branch systems is gathered into a mossy cushion. In *Ventricaria ventricosa* the large primary vesicle is as large and firm-walled as a ping-pong ball. In *Boergesenia forbesii* the vesicles are elongate and close-aggregated side to side. The pervasive *Dictyosphaeria* species form clusters of small vesicles with thick, cartilage-like walls scattered over the reef edge.

It is in the Order **Codiales** that the green algae develop the most substantial thalli with the branched threads compacted into a pseudo-parenchyma, sometimes fleshy, as we have seen in the temperate *Codium* species (Fig. 7.25). The tropical Codiales have a diverse form range. *Chlorodesmis fastigiata* grows in bright green tresses, generally conspicuous near the reef edge. On soft flats, *Avrainvillea* and *Udotea* put up vertical sheets, sometimes with intersecting flanges, built up of densely interwoven filaments.

The *Halimeda* species, the most widespread tropical greens, have perfected the planar structure, with lime-impregnated segments and dichotomous-branching. *Halimeda* is thus — on a coarser scale — the green equivalent of the corallines, with the segments separated by uncalcified joints. The species found on rock are the smaller segmented *H. discoidea* and *H. opuntia* (Fig. 17.13). *Halimeda macroloba*, with coin-like discs, and *H. cylindracea* with rod-like segments, grow in sediment.

CHAPTER EIGHTEEN

Pacific Asia

Hong Kong

The island of Hong Kong and its formerly styled Mainland Territories lie 320 km south of the Tropic of Cancer off the continental coast of China. With its 1036 sq km in total area, Hong Kong presents a diverse and intricate coastline, about a fifth as long as that of New Zealand's North Island.

To its special geographical position, Hong Kong owes its complex, seasonally changing biology. Landwards it sits at a conjunction of the vast *Palearctic Region* with the tropical *Oriental Region*. Coastally it forms a northward salient of the tropical *Indo-West Pacific* biogeographical province. The shores are rich in tropical species that manage to survive cold winters. In the same places there are temperate algae that come to their maximum under the same winter regime. Hong Kong's mangroves, though dwarfed compared with Malaysia, still muster some eight species. Around the islets of Mirs Bay to the east, there are approximately 50 coral species in 25 genera. Near the northern limit for reef-builders, all avoid low winter air temperatures by confinement to the subtidal.

The ocean currents are the leading determinants of this warm temperate/subtropical mix. Northern and southern water masses impinge on Hong Kong at different seasons, bringing larvae of quite separate origins. The *Kuroshio Current* with warm Pacific water has brought many tropical forms, like the corals and their associates, to the islets to the east. Alternatively, the *Taiwan Current* transports animals and plants from the temperate shores of Japan and east China, while the *Hainan Current* brings species from the South China Sea. During the winter growth of intertidal algae, many temperate herbivores come inshore to feed. With algal die-off towards summer, the more tropical herbivores make their seasonal appearance.

Oceanic water masses from the south-east also support a high standing crop of plankton. The inshore animals breed here, with a single peak of larval production at high summer temperatures.

Hong Kong's coastal regime is also markedly affected by the influx of fresh water from two sources. Heavy monsoon rains (averaging 200 cm/yr) from May to September, appreciably dilute the coastal waters. In addition a huge accession of fresh water from the mouth of the Pearl River, greatest in summer, governs the ecology of the western seaboard. Surface salinities west of Hong Kong may in summer fall to as little as 1–2%.

With all these influences, Hong Kong's shore ecology is marked with contrasts, in the transitions (1) from turbid harbours to clean ocean coasts, and (2) from tropical waters of high salinity in the east, to low-saline warm temperate waters in the west. There are also (3) seasonal climatic changes, from a cool dry winter to hot wet summer, with spring and autumn the periods of transition.

The shore and shallow subtidal communities of Hong Kong, as investigated in the middle 1970s, have been described in detail in *The Sea-shore Ecology of Hong Kong* (1983).

Shore zonation

The basic pattern presented in Fig. 18.3 is from a shore chosen under moderate wave action at **Clear Water Bay** on a far eastward salient of the New Territories. The steep-sloping bedrock is of jointed rhyolite, and its biological zonation is clear-cut and horizontally regular. Like the granite shore of Cape D'Aguilar — the site of Hong Kong Island's University Marine Laboratory — the Bay is washed by clean oceanic water.

The zoned shore with its seasonal algae was drawn in March, but no one month would reveal an all-year 'typical' algal picture for Hong Kong. In December, the brown algae of the sublittoral fringe would have been looked for in vain; by March they have only partly returned. For the upper shore too, continuity through the year will be shown not by its evanescent algae, but far more reliably by its sessile animals.

Below the black *Verrucaria maura*, with coloured lichens out-topping it, the **littoral fringe** has two periwinkle species we have already met with in North Queensland. The higher reaching is the pointed *Nodilittorina trochoides*, and — overlapping it but also

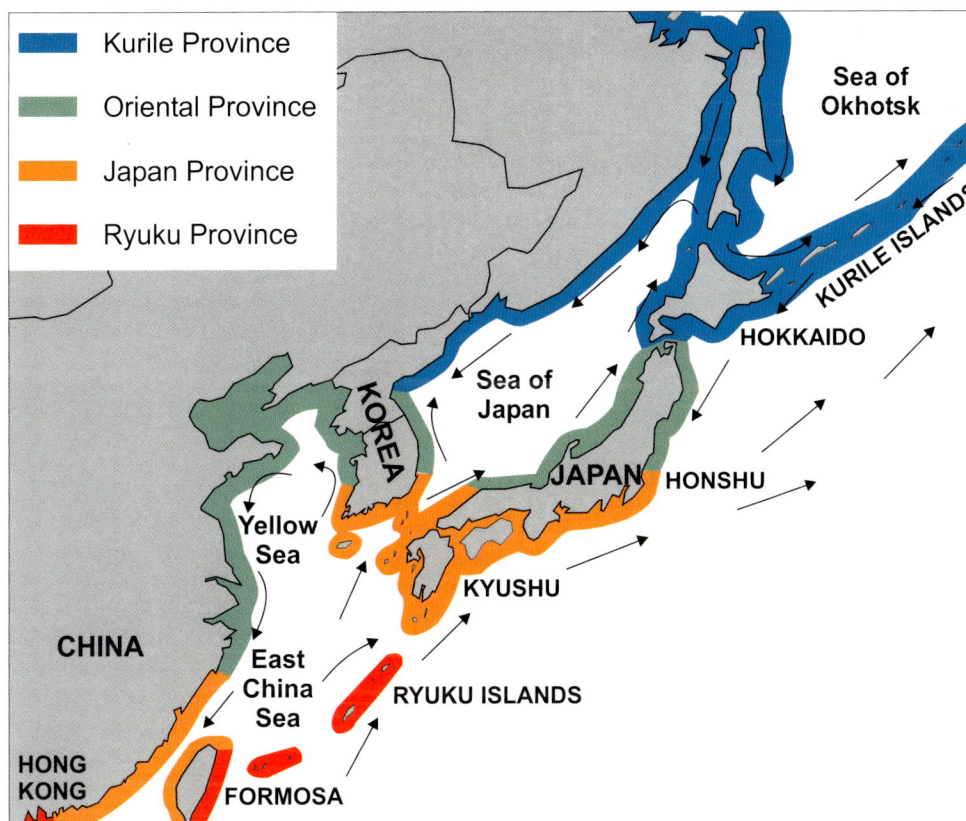

ranging lower — the more rounded *N. millegrana*, overlapping in range from lower down. A band of the high level red algae contains both *Porphyra suborbiculata* and — immediately above it — *Bangia fusco-purpurea*. By March, both had begun to regress.

At the top of the **eulittoral zone**, small barnacles are — as in most of the tropics — sparse here represented by a seasonal scatter of *Chthamalus malayensis*. Larger barnacles belong to the mid-eulittoral, with a band formed of the pan-tropical *Tetraclita squamosa*, that we last saw in Queensland. Its adaptive features are the reduced aperture to retard evaporation, and insulating air-spaces in the column plates. Along rock fissures grows the primitive pedunculate barnacle, *Capitulum mitella*, straining food with short, thick cirri held motionless in the current. In the lowest eulittoral grows a large open coast barnacle, the cosmopolitan *Megabalanus tintinnabulum*.

Scattered within the eulittoral is the mussel *Septifer bilocularis*, distinguished by its thin radial ribbing. Most abundant in higher exposure, it grows in lines like *Capitulum*, or forms fist-sized clumps, a reminder of the mussel sub-zone developed on cool temperate shores. The small *Isognomon perna* wedges into crevices high in the eulittoral.

Towards greater shelter, with the eulittoral barnacles reduced or lost, there is instead a close-paced rock oyster zone of *Saccostrea cucullata*. In extreme shelter *Crassostrea gigas* becomes predominant, still, however, with *S. cucullata* and usually the mytilids *Brachiodontes vari-*

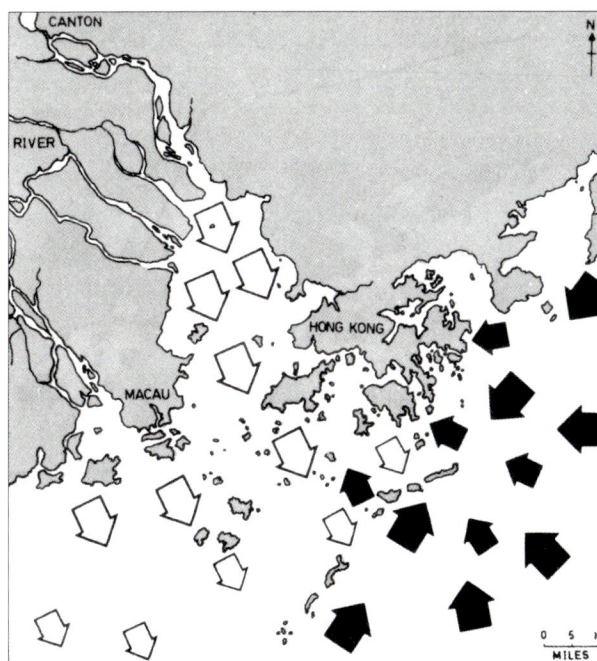

Fig. 18.2 Water masses and ocean currents affecting the coast of Hong Kong
The influence of the Pearl River with freshwater run-off (white arrows) and the oceanic water full of salinity (black arrows).

abilis and *Modiolus auriculatus*, as well as the mussel-like *Trapezium liratum*. In ultimate shade beneath wharf piles, *Balanus amphitrite* becomes the prime barnacle; *Perna viridis* with is long green shells is the predominant

Fig. 18.3 Moderately exposed shore based on Clear Water Bay, Hong Kong
a–b *Verrucaria* and littorines; b–c *Bangia*, *Porphyra* and *Chthamalus*; c–d *Tetraclita*, *Septifer*, *Capitulum* and *Dermonema*; d–e *Gloiopeltis* (**top**) and coralline turf; e–f *Sargassum* and *Anthocidaris*
1 *Nodilittorina millegrana* 2 *Nodilittorina trochoides* 3 *Porphyra suborbiculata* 4 *Sargassum hemiphyllum* 5 *Dermonema frappieri* 6 *Gloiopeltis furcata* 7 *Sargassum patens* 8 *Collisella dorsuosa* 9 *Chthamalus malayensis* 10 *Siphonaria sirius* 11 *Capitulum mitella* 12 *Tetraclita squamosa* 13 *Septifer bilocularis* 14 *Patelloida saccharina* 15 *Scutellastra flexuosa* 16 *Anthocidaris crassispina*.

mussel, accompanied with *Modiolus auriculatus* and *Musculus cupreus*.

Clear Water Bay, in late November-December, reveals a **lower eulittoral** zone and **sublittoral fringe** bare of large algae. A pink veneer of basal *Corallina* is spangled with black, short and thick-spined urchins, *Antho-*

cidaris crassispina. Two commensal crustaceans, the half-crab *Porcellana ornata* and the alpheid shrimp *Athanas dorsalis*, can be found clinging beneath them near the mouth and poaching food. The common low tidal chiton is *Onithochiton hirasei*, while *Acanthopleura japonica* is regularly found higher up in the middle and upper eulittoral. The cut-back summer stems of last season's *Sargassum*, or the beginnings of a new autumn crop, are visible on the cropped surface.

The seasonal algal poverty of brown algae is a surprise to a visitor from temperate shores. Villagers and fishermen on the circuit of Lantau and other islands have been unanimous that there are no salt-water algae to be had there in July, August, September, October, November and December, but that beginning with January or February, they come in with increasing abundance.

At Clear Water Bay the low-tidal algae have become fully restored around mid-March. The autumn-winter band of upshore red algae has meanwhile begun to disintegrate. *Bangia* had reached its height in December, with *Porphyra subarticulata* that began as small pink thalli in mid-November having its maximum in January.

By the end of December the pink sublittoral fringe is displaying young, golden brown *Sargassum* plants. Of the host of sargassoid species in warm south-east Asia, the most important at Hong Kong is *Sargassum hemiphyllum*, with its thick dark brown leaflets, coarsely dentate and lacking a mid-rib. By contrast, *Sargassum patens* produces finely serrate, lightly coloured leaves, distinctly ribbed. *Sargassum hemiphyllum* makes its fastest growth after January until the thongs reach over a metre in April, with vesicles and mature reproductive organs. By May many plants have broken away to form floating masses offshore.

Of the smaller seasonal brown algae higher up the shore, *Colpomenia sinuosa* begins in January, with its maximum in March; while *Petalonia fascia* and *Scytosiphon lomentaria* have already attained half a metre by February. All three are cosmopolitan in warm and temperate seas.

At Clear Water Bay the barnacle band of *Tetraclita squamosa* is invaded in mid-March by the miniature treelets of *Dermonema frappieri* (Helminthocladiaceae), one of the commonest east Asian algae. Next below, scrambling over the mid-eulittoral during March and April, are the mucilaginous branchlets of *Gloiopeltis furcata*, also widespread in east Asia.

Small red and green algae complete the rich spring turf of the lower eulittoral. The common corallines are *Corallina pilulifera*, *C. sessilis* and *Jania ungulata*. Tufts of *Laurencia undulata* and *L. japonica* are recognised by their cartilaginous texture. Large-celled filaments of *Chaetomorpha antennina* stand out in bright green. The feathers of *Bryopsis plumosa* grow in small pools, while *Boodlea composita* and *Struvea anastomosans* put up their fine meshes of anastomosing filaments. Conspicuous on every eulittoral turf are the large-fronded *Ulva lactuca*. In moist or shaded spots a bottle-green scramble of *Gigartina intermedia* may appear. Here too grows

Gelidium amansii, with wine-red pinnate fronds, as large as the closely allied *Pterocladiella*.

At the same level, there are a host of small algae, especially the greens, witnessing to the tropical affinities of Hong Kong. The common species of *Caulerpa* are *C. peltata*, *C. scalpelliformis*, *C. taxifolia*, *C. cupressoides* var. *lycopodium*, and *C. sertularioides*, with each name suggesting a distinctive leaf shape. *Codium* species are common too: *C. intricatum* in small cushions, and the dichotomous *C. cylindricum*. *Dictyosphaeria cavernosa* studs the rock surface, with its small vesicles, aggregated in stiff green pads. *Galaxaura fastigiata* (a tropical rhodophyte) forms stiffish dichotomous-branching fans.

The small turfing algae include *Gelidiopsis variabilis*, springing up among the corallines, and epiphytic *Polysiphonia* growing where surge washes at low water. Found as epiphytes upon *Sargassum* there are the small, cross-septate tubes of *Champia parvula*, and the larger, springy *Hypnea cervicornis* and *H. japonica*, with curling tips to their dichotomous branches.

In the summer algal drift cast up from the subtidal many warm-water browns of the Dictyotales are to be found, including: fans of *Padina*, kidney-shaped fronds of *Lobophora variegata*, stiffer *Zonaria coriacea* and the dichotomous ribbons of *Dictyota*. Common also are the fleshy lattices of *Hydroclathrus clathratus*, close allied to *Colpomenia*.

Gastropods

In warm seas and towards the tropics, the grazing littorines and nerites, the browsing trochids and turbinids, and the thaid carnivores all increase in diversity.

At Hong Kong, the periwinkles (**Littorinidae**) present six species from exposure to shelter (Fig. 18.4). Below *Nodilittorina trochoides* and *N. millegrana*, a small *Peasiella*, low-pitched and sharp-keeled, lives in crevices. Among rock oysters, in bays and harbours, the spirally ridged *Littorina brevicula* becomes dominant. In brackish reaches or on mangroves appears the larger tropical species, *Littoraria scabra*, while the fawn-coloured *L. melanostomum* is restricted to salt meadows and swamps. Close allied to the littorines is the grey and white checkered *Planaxis sulcatus*, common through the Indo-Pacific on hard rocks in the upper eulittoral.

Though most **Trochidae** retire under boulders, *Monodonta australis* is found in the open around mid-tide. In *Gelidium* and *Corallina* the **Turbinidae** appear, with the smaller *Lunella coronata* among turf, and the tall *Astralium rhodostomum* upon painted and encrusting corallines.

At Hong Kong the **Siphonariidae** are common everywhere, without supplanting the true limpets as they virtually do in the full tropics. The small, strong-ribbed *Siphonaria sirius* lives on exposed shores with *S. japon-*

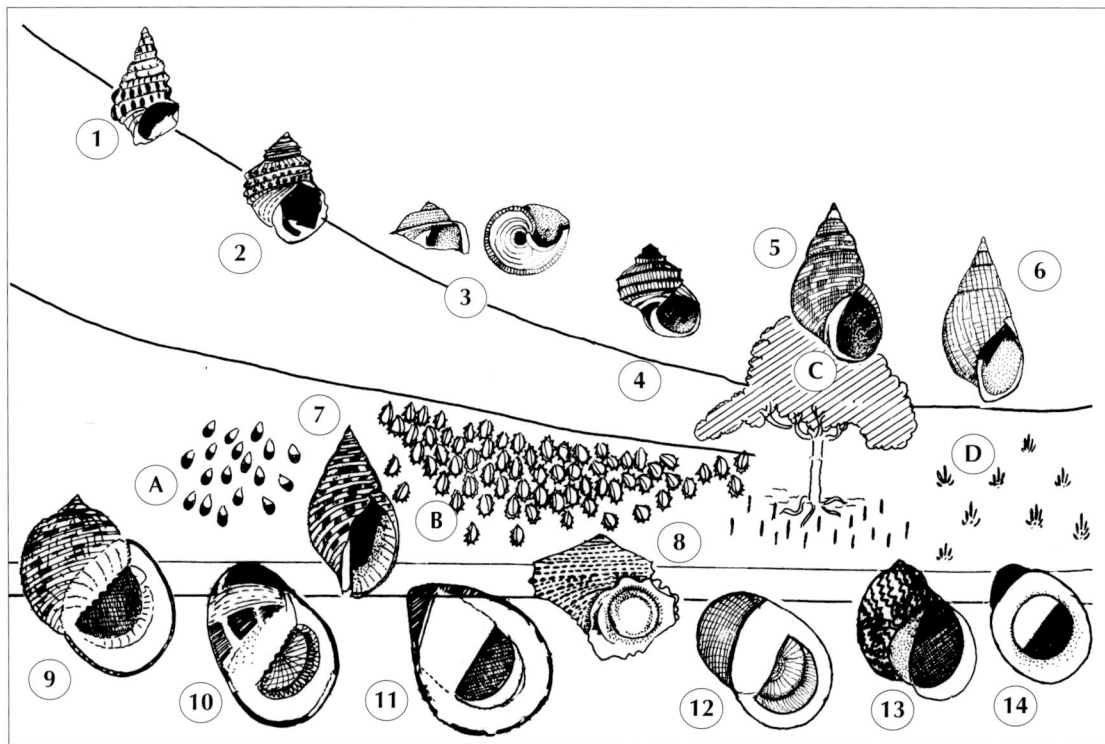

Fig. 18.4 Range of grazing gastropods from exposure to shelter at Hong Kong

A. mussels (*Septifer*) B. *Saccostrea cucullata* C. mangrove, *Avicennia* D. salt meadow

1 *Nodilittorina trochoides* 2 *Nodilittorina millegrana* 3 *Peasiella* sp. 4 *Littorina brevicula* 5 *Littoraria scabra* 6 *Littoraria melanostomum* 7 *Planaxis sulcatus* 8 *Lunella coronata* 9 *Nerita undata* 10 *Nerita polita* 11 *Nerita albicilla* 12 *Nerita yoldii* 13 *Clithon oualaniensis* 14 *Neritina violacea*.

ica commoner in shelter upon oyster shores. The dark-ribbed *S. atra* is found in small pools on open coasts.

The topmost of the **Acmaeidae** on exposed shores is the high-arched *Collisella dorsuosa*, confined to shady, vertical faces. In more shelter, the flatter *Collisella luchuana* and *Notoacmea concinna* replace it just above oyster level. In empty *Crassostrea* shells, along with *Siphonaria japonica*, is found the small *Patelloida pygmaea*. Down at coralline level lives a tropic-wide species, the stellate *Patelloida saccharina*. Of the **Patellidae**, the exposed shore species are *Patella flexuosa* and *Cellana rota*, replaced in shelter by the lighter and flatter *C. toreuma*.

The **Neritidae** are notably successful on all warm shores. Familiar tropic-wide species at Hong Kong include *N. plicata*, *N. undata*, *N. polita* and — in more shelter — *N. chameleon*. *Nerita albicilla* is broader-based, almost limpet-shaped for life under stones. This family ranges widely into estuaries with the *Clithon* species and to salt marshes where *Nerita balteata* and *Neritina violacea* live upon mangroves. With the fresh-water *Septaria* the neritids have evolved a limpet form adapted for boulders in fast streams.

Muricid gastropods, particularly of the thaid sub-family, are the leading carnivores on any shore that offers barnacles, oysters, mussels or herbivorous gas-

tropods. John Taylor made a study of the niche-diversity of 10 muricids feeding in the intertidal and shallow sub-tidal of Tolo Harbour, Hong Kong.

The intertidal prey-species are in vertical order the oyster *Saccostrea cucullata*, the mussel *Brachidontes variabilis* and the barnacle *Balanus trigonus*. Subtidally there is the oyster-like bivalve *Alectryonella plicatula*, with *Spirorbis* tubeworms densely covering the boulders. Deeper still lies a zone of live corals reaching down to 10 m. The highest zoned thaids, *Thais clavigera* and *Morula musiva*, feed mostly on *Saccostrea*, but can also take *Brachidontes*. *Thais* operates by pulling the two valves apart, *Morula* by drilling. *Thais luteostoma* feeds exclusively on *Balanus*, while *Morula margariticola* with a longer depth range takes barnacles, *Spirorbis*, small bivalves and even attacks corals. *Mancinella echinata* and *Ergalatax contractus* feed on barnacles and *Alectryonella*, while *Drupella rugosa* is exclusively a coral predator. The largest muricid, *Chicoreus microphyllus*, feeds by drilling an entry hole in *Alectryonella* and also on other bivalves and barnacles.

Deep Bay and its marshes

The north-western reaches of Hong Kong's New Territories are massively affected by a monsoon rainfall of

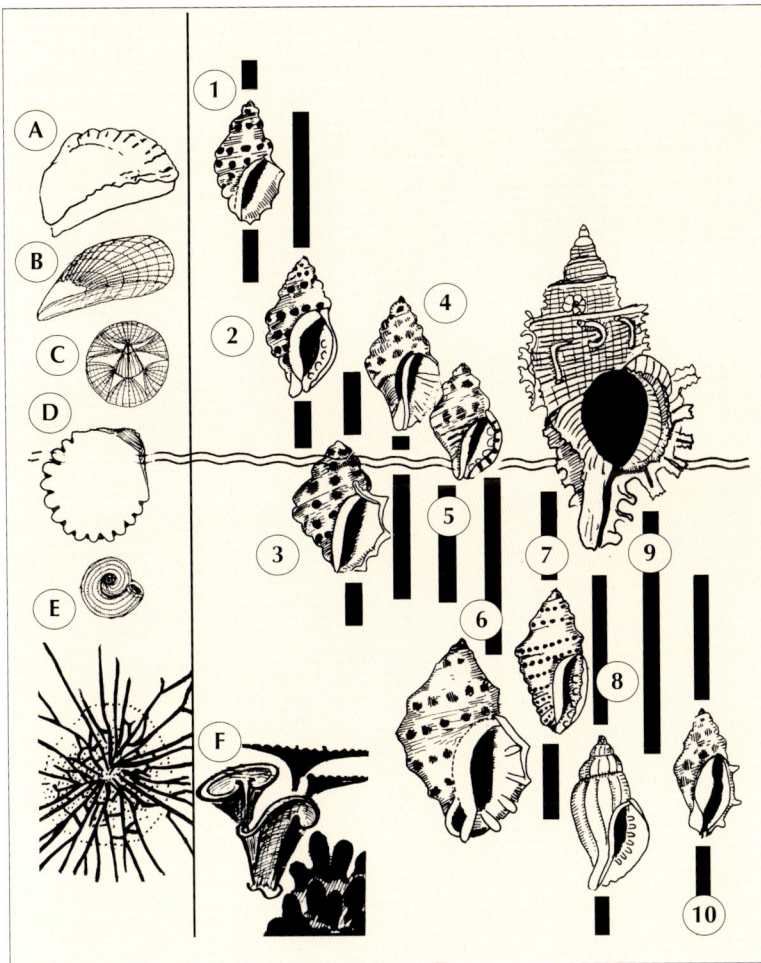

Fig. 18.5 Shore range and foods of predatory muricids at Hong Kong
A. *Saccostrea cucullata* B. *Brachidontes variabilis* C. *Balanus trigonus* D. *Alectryonella plicatula* E. *Spirorbis* F. scleractinian corals.
1 *Thais clavigera* 2 *Morula musiva* 3 *Thais luteostoma* 4 *Thais tissoti* 5 *Morula margariticola* 6 *Mancinella echinata* 7 *Drupella rugosa* 8 *Ergalatax contractus* 9 *Chicoreus microphyllus* 10 *Morula spinosa*.
(From observations by John Taylor, in Tolo Harbour.)

200 cm/yr from May to September. West of Hong Kong the summer outflow from the Pearl River lowers the surface salinity to as little as 1–2%. This incursion of fresh water is visible to the south of Lantau as a brown swathe sharply delineated from the blue oceanic water.

This western sector is essentially estuarine, with mudflats and mangroves rich in *euryhaline* species (tolerant of fluctuations of salinity). Sublittoral corals — important in Mirs Bay to the east — are excluded from the west. In Deep Bay (in fact broad and shallow) the surface and bottom temperatures draw close, with warming and cooling both accelerated.

The communities of Deep Bay fall into three broad types: (1) salt meadows; (2) mangrove-fishpond-paddy-field systems, and (3) open mudflats with oyster beds on natural or artificial cover.

Salt meadows

Strewn with silty sand and pebbles, these dry stretches are dominated by the maritime grass *Zoysia sinica*, superseded in the wet by marsh sedges bordering the constructed paddy-fields. To seaward, these are in turn replaced by mangroves. Among *Zoysia* may grow succulent dicotyledons, chiefly *Limonium sinense* (Plumbaginaceae); with the dwarfed *Scaevola hainensis* (Goodeniaceae), and the sea-blight *Suaeda australis* (Chenopodiaceae).

On the salt meadows the most numerous and colourful animals are the fiddler crabs, *Uca*, belonging to the Family **Ocypodidae**. Small and gregarious, they run about constantly or sidle into a complex of shallow burrows. Their social behaviour is fascinating to watch close up with glasses. The male fiddler has one chela (generally but not always the left) enlarged almost to the bulk of the rest of the body. This claw is bright-coloured and as 'chic' as costume jewellery. Its pattern is highly species-specific, and so is the ritualised arm-waving used in signalling to the females. Two males may also employ their large chelae in formalised jousting.

The male's other chela is tiny (as are both in the female), hardly larger than the claw of a walking leg. Its fingertips are spooned and bristle-edged, for picking up sediment. Particles are swirled round in the basket of mouth appendages to extract the nutrients. The remains are then discarded as sand balls the size of lead shot that accumulate in hundreds around the burrows. For each *Uca* species the straining bristles are finely adapted to the particle size of the sediment. Of the four Hong Kong fiddlers, *Uca lactea*, with the large chela mustard yellow, lives in coarser sand highest on the beach. *Uca vocans*, with the chela mauve and brick-red, prefers the silty terrain a little further down. The highest zoned fiddler, among beach vegetation, is *U. chlorophthalmus*, with the large chela coral red, white-tipped and the carapace

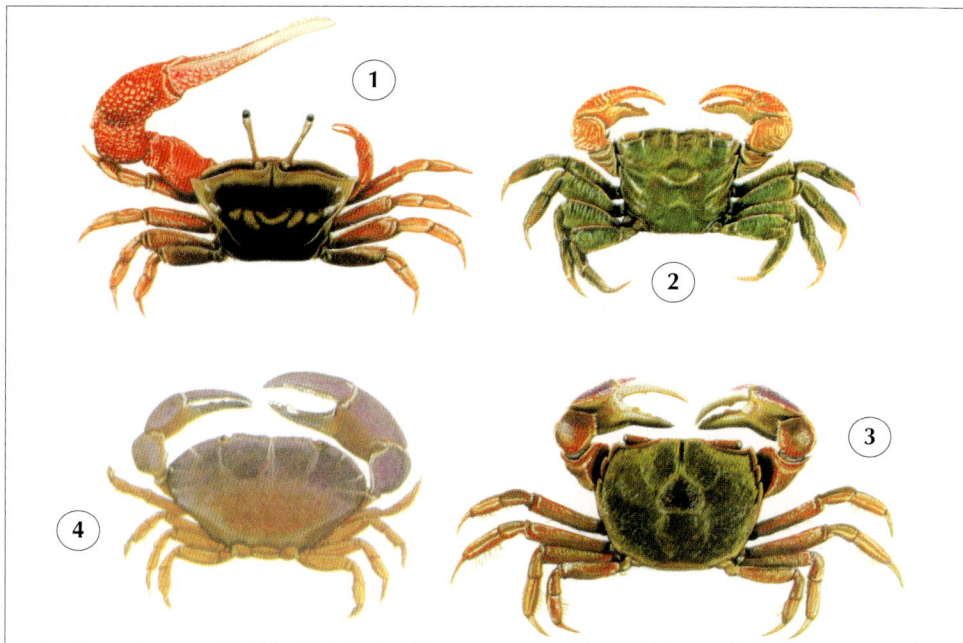

Fig. 18.6 Some crabs of Hong Kong

1 *Uca arcuata*
2 *Chiromanthes bidens*
3 *Chasmagnathus convexum*
4 *Epixanthus frontalis*.

(By courtesy of Karen Phillips.)

jet black. The largest species — living in mangrove mud — is *Uca arcuata*, with the claw deep red and orange and the carapace black with turquoise.

Far less gregarious than the ucids are the small grapsids of the subfamily **Sesarminae**.[1] Some 15 species of Sesarminae have been identified for the Hong Kong upper shore, living mainly under stones and rubble. *Metopograpsus* species may be seen spiralling around mangrove trunks and branches. All are fast runners, with short eye-stalks wide-set at the corners of the cara-

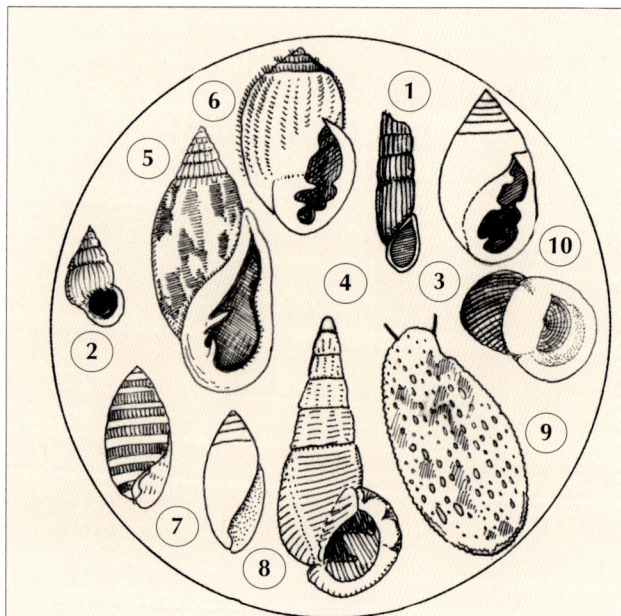

Fig. 18.7 Some gastropods of the Hong Kong salt meadow

1 *Truncatella* sp. 2 *Assiminea violacea* 3 *Nerita yoldii* 4 *Terebralia sulcata* 5 *Ellobium politum* 6 *Cassidula schmackeriana* 7 *Auriculastra subula* 8 *Auriculastra duplicata* 9 *Onchidium* sp. 10 *Cassidula plectotrematoides*.

pace, which is square in *Sesarminae*, and rhomboid in *Metopograpsus*.

The gastropods of salt meadows come from a few highly typical estuarine families. The largest of the prosobranchs is one of the Cerithiidae, *Terebralia sulcata*. Though it contains a gill, the mantle cavity serves also as a lung, with the shell-lip formed into an air-tube that can stay open while the operculum is closed (Fig. 18.7). Like all its family, *Terebralia* grazes the surface detritus. Gelatinous spawn strings are attached to debris, and the eggs hatch to free-swimming larvae.

Aside from the common prosobranchs *Assiminea violacea*, *Littoraria melanostomum*, and the slender-spired *Truncatella*, all the remaining estuarine snails are pulmonates. It is in such tropical salt meadows and mangals that the primitive **Ellobiidae** must have had their genesis. Hong Kong is rich with species, with its polished, urn-shaped *Melampus*, barrel-like *Laemodonta*; smooth spindle-shaped *Auriculastra*; and larger, more sturdy *Cassidula* with the aperture flatly rimmed. The two *Pythia* species are wholly terrestrial, in ground vegetation and on trees. *Pythia fimbriosa* is compressed like a plum stone, and *P. cecellei* is top-shaped, both having a narrowed aperture with protective teeth. The giant of the family, *Ellobium politum*, reaches 7.5 cm high, being found around mangrove trunks in Deep Bay marshes. Only slightly smaller is *Ellobium chinense*.

The Family **Onchidiidae** is represented by two or three species, rubbery skinned, with heavy warts upon the naked skin. *Onchidium verruculatum* occurs on small estuarine boulders.

Mangroves

Hong Kong's mangals are some of the most northerly in Asia but are comparatively rich, with eight species,

Fig. 18.8 A prospect of Deep Bay, Hong Kong, with mangal, fish-ponds and mudflats with oyster cultivation
A. enclosed ponds B. mangroves C. stakes and battens for oyster growing
Seven species of mangroves are represented by their symbols on the mangal **(centre right)** 1 *Acanthus ilicifolius*
2 *Excoecaria agallocha* 3 *Aegiceras corniculatum* 4 *Lumnitzera racemosa* 5 *Avicennia marina* 6 *Bruguiera conjugata*
7 *Kandelia candel.*
(inset top left) The use of the mud-scooter.
(bottom) The fauna of an oyster clump: 8 *Thais carinifera* 9 *Trapezium liratum* 10 *Crassostrea gigas* 11 *Saccostrea cucullata*
12 *Metopograpsus* sp.

leading from landward out to the pioneer mangroves of soft mud. Their backshore associates include the screw-pine, *Pandanus tectorius*, the yellow-flowered *Hibiscus tiliaceus*, the mangrove fern, *Acrostichum aureum*, and the St John's lily, *Crinum asiaticum*. Outside or mingled with *Hibiscus* grow three landward mangroves, the euphorbiacean *Excoecaria agallocha*, the tall, plank-buttressed *Heritiera littoralis* and the holly-like *Acanthus ilicifolius*.

The five seaward mangroves at Deep Bay can be told at a distance by their shape and hue. Furthest out, *Avicennia marina* has its broad, low spread surround of breathing roots (pneumatophores). Almost as far out grows *Kandelia candel*, a small rhizophoracean mangrove with clusters of torpedo-shaped embryos or 'droppers'. Hong Kong has no true *Rhizophora* species, but the tallest of that family, *Bruguiera conjugata*, stands out tree-like above the shrub stratum. Less numerous and a little higher on the shore are the white-lowered *Lumnitzera racemosa*, and *Aegiceras corniculatum*, with its clusters of small banana-shaped embryos.

The Mai Po marshes

As the last considerable spread of mangal and kindred habitats still left, the Mai Po Marshes have a commanding importance for Hong Kong. They are a pre-eminent wildlife area forming an international staging point for migrant shore birds, and in total harbour some 230 of the 370 species of Hong Kong's birds, including 100 found nowhere else in the Territory. They are thus the favourite resort for resident and visiting naturalists.

Down to recent times, ponds and paddy-fields have continued to enhance the diversity and accessibility of bird habitats. As represented in Fig. 18.8 with the fish-ponds and mangroves are true wetlands lying seaward of an inner bund. The ponds are cut up by paths, leading to an outer bund closing them off in turn from the still unaltered mangrove stretch.

In *The Sea-shore Ecology of Hong Kong* we were still able to record that the pond area, like the paddy-fields, is today the habitat of a bare-footed people to whom soft mud and turbid water are elements like the sky and air. The few secure buildings are on the bunds or raised above the fishponds on stilts. These waters are used extensively for fattening fish, which are regularly caught as the tide goes out through the sluice-gates of the ponds. Within the large ponds a dredge is regularly at work, shifting and accumulating loads of bottom mud.

The largest bird assemblage at Mai Po consists of the waders including as passage migrants curlew, whimbrel, eastern bar-tailed godwit and knot. In addition, Kentish or Alexandrine plover, common sandpiper, redshank, dunlin and three sorts of snipe all share the paddy-fields,

marshes and margins of streams. Four heron species and four egrets are found on the flats beyond the mangal. As well as the great bittern, *Botaurus stellaris*, there are three smaller tree bitterns, *Ixobrychus* species, perching in the mangroves. Winter visitors include the black stork, white ibis and two species of spoonbill.

The rails of Mai Po are the banded rail, Chinese water rail and coot. Small grebes swim and dive in the ponds with pheasant-tailed jacana, on the decline since pond conversion. Of the five kingfishers found at Hong Kong, four are regularly to be seen at the marshes.

The chief predators at Mai Po are marsh harriers and osprey. The duck tribe is represented by teal, garganey, widgeon, yellownib, shoveller and white shelduck. Though only wintering in the marshes, black-headed and Sanders gulls and whiskered, gull-billed and little terns are common birds of passage.

Largest of all the marsh birds is the Dalmatian pelican, with the smaller spotted-billed pelican often flying in V-shaped formations or fishing at the edges.

Many passerine or song birds play their part in the mangal. White-eye, chestnut munia, Chinese bulbul and coucal; yellow bellied and brown warblers, are joined seasonally by von Schrenck's and the great reed warbler. Bluethroat can be seen on the bunds in winter. Swallows and yellow wagtail manoeuvre to catch insects just above the floating vegetation.

Mudflats with oyster beds

Oysters have been cultivated at Hong Kong for 700 years. Today's industry located around Deep Bay has come to rely upon the Pacific cup-oyster, *Crassostrea gigas*. The smaller *Saccostrea cucullata* has now been displaced, though it still grows alongside *C. gigas*. The original, more rounded *C. rivularis* grows wild and can also settle on tiles. Unfit for human consumption without microbial depuration, most of today's oyster crop goes into cooked products, including oyster sauce.

On the soft mudflats extending far out from the land margin, villagers are constantly at work, placing out oyster-cultch, and tending it or harvesting. They glide over the semi-fluid surface, as fast as cyclists, on the traditional plank mud-scooters.

Flat stones or tiles have introduced islands of hard benthos onto the mud. Both oyster species attach to the tops, along with the mussel-like bivalve *Trapezium liratum*. Two gastropods are common on the hard cultch: the strong-ribbed *Nerita undata* and the oyster predator *Thais carinifera*, light-shelled with a spinose shoulder.

Byssus-fixed bivalves attach underneath, notably the ribbed mussel, *Brachidontes atratus*, *Trapezium liratum* and the saddle oyster, *Placuna ephippium*. Two sorts of crab abound. The faster is *Metopograpsus messor*, attractive with its eau-de-nil carapace and violet chelae.

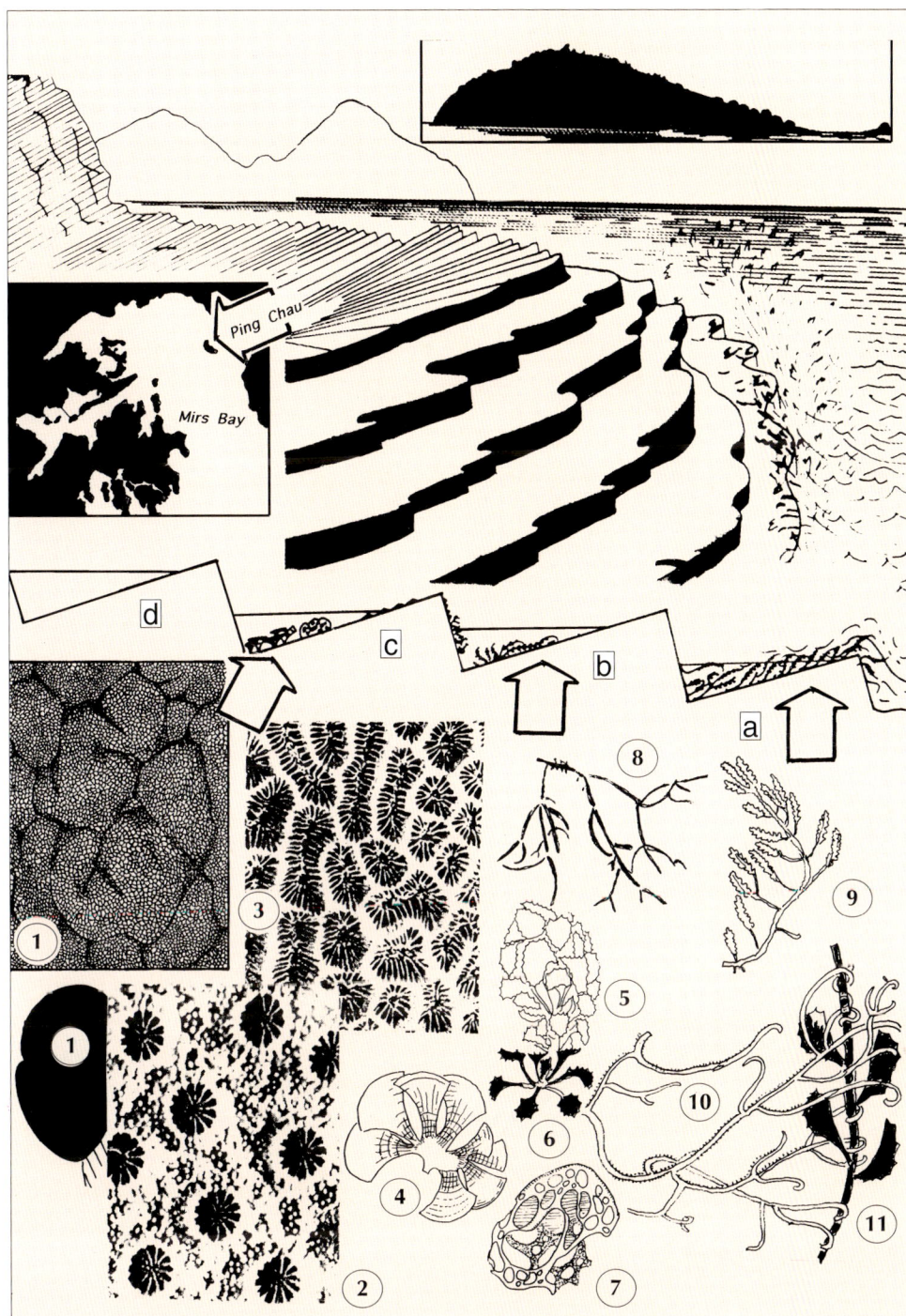

Fig. 18.9 Intertidal terraces and coral pools of Ping Chau Island, Mirs Bay
The expanse of the west-facing shore is illustrated, with the arrangement of successive terraces and pools upshore (a) to (d) formed of metamorphosed Tolo sandstones. The principal animals and plants are shown in the lower half.
1 *Porites lobata* 2 *Goniastrea aspera* 3 *Cyphastrea serailia* 4 *Padina durvillaea* 5 *Turbinaria peltata* 6 *Turbinaria*, reduced summer fronds 7 *Hydroclathrus clathratus* 8 *Gloiopeltis furcata* 9 *Laurencia japonica* 10 *Hypnea musciformis* twining on 11 *Sargassum hemiphyllum*.

The second is the brown and white blotched *Hemigrapsus penicillatus*, catching food particles with tufts of setae on the palms of the chelae. Out on the soft flats, several mussel species sit upright in mats knitted from their byssi: *Musculista senhousia* (recently an immigrant to New Zealand, Fig. 6.5), the black triangular *Modiolus metcalfei* and the shining green *Arcuatula elegans*.

The gastropods *Cerithideopsilla djadjariensis* and C.

cingulata trail over the soft mud. Small *Nassarius* whelks are mobile and abundant on tropical flats. In patches of coarser sand lives the speckled *Notocochlis tigrina*, usually first detected from its sandy niduses or spawn-coils. The trailing gastropod, *Turritella cerea*, is — like its whole family — a ciliary feeder.

Old oysters have their own sessile and penetrant fauna. A small pholad bivalve, *Aspidopholas obtecta*, makes bottle-shaped burrows in the shell, while the outer shell layer is eroded by the spionid polychaete *Polydora*. Far more ultimately destructive is the yellow boring sponge *Cliona celata*.

As well as *Thais*, the cap-shaped gastropod *Capulus yokoyami* lives at the oyster's expense, attaching like a limpet near the margin, with its proboscis gleaning particles crossing the bivalve's mantle edge.

The coral shore of Hong Kong

The island of Ping Chau, lying furthest east under the warm currents washing Mirs Bay, is one of the few places in Hong Kong with intertidal coral. The shores, though unusually rich, conform in the main with the zonation we have seen. But their pattern is expressed by a more horizontal unrolling of what has elsewhere been more vertically compressed on a steep shore.

Metamorphosed Tolo Sandstone forms terraces inclined back from the sea with a short, steep 'strike face' and a long 'dip slope'. A series of intertidal benches extends for a good distance along the island's outer or west-facing coast. Level by level shallow pools are formed, each with a different ecology. On the bottom terrace the pools are surge-swept at low water and kept pink with a *Lithophyllum* crust. By early March the pools and open rock have developed long *Sargassum hemiphyllum*. Twining about the dark *Sargassum* are the golden yellow thongs of the rhodophyte *Hypnea musciformis*. It branches into clinging tendrils and its larger branches have small spinules. Conspicuous also are brown filaments of *Ectocarpus*, and bright green *Caulerpa peltata*.

The second terrace up, also breached by surge, contains *Sargassum* with *Hypnea*, while above the water-line at low tide the dip-slope carries small dichotomous tufts of *Chaetangium* and on the strike face *Gloiopeltis furcata* is dominant from February on.

On the third step the pools warm up to become tepid between tides. Limpets abound on the rocks: *Cellana toreuma* and *C. grata*, as well as *Siphonaria japonica*, *S. sirius* and *S. atra*. The brown algae are those of quiet water rather than surge. *Turbinaria peltata* replaces its relative *Sargassum*. Fans of *Padina* are common, with golden brown vesicles of *Colpomenia sinuosa* and its tropical relative *Hydroclathrus* perforated with large holes as an open network. At the bottom of the pools are large attached bivalves, *Chama dunkeri*, and the black, thick-spined urchin *Anthocidaris crassispina*. The steep strike face, dry at low water, carries the stalked barnacle *Capitulum mitella*. As well as *Chama*, there are small pearl oysters, *Pinctada margaritifera*, and a large warm tropical oyster not previously noted *Hyotissa hyotis* (Family Gryphaeidae). Flat coils of the vermetid

gastropod *Serpulorbis imbricatus* are common.

It is in these pools also that corals grow, forming low brown or mauve plaques. Foremost in high mid-tidal pools and resistant to moderate turbidity are the fawn porous crusts of *Porites lobata* (Fig. 20.12). Faviid corals are also important, notably the small-cupped *Cyphastrea serailia* and the meandrine brain coral *Platygyra sinensis*. At least eight other faviids were recorded by Paula Scott among the intertidal corals of Hong Kong. The strike face of the coral pools carries the rock oyster, *Saccostrea cucullata*, and the urchin, *Anthocidaris crassispina*. The pools at the higher level of the fourth step are blotched and marbled with black *Hildenbrandia* and grazed by *Nerita albicilla*.

Conservation

In *The Sea-shore Ecology of Hong Kong* the authors described the Territory's coastal communities — surveyed in 1977 — as resting ecologically on a knife-edge. 'The real wonder,' we wrote, 'is perhaps that Hong Kong still has many healthy shores. The conservationist must ask how this came about and how long it may be maintained.'

Around most of Hong Kong's 800 km of coastline a diversity of communities was then still intact, some of them not too remote from their original state. Other parts of the natural coast had been degraded, a warning of disaster that could rapidly compound. Victoria, once the 'Fragrant Harbour' had long served as a water closet, though open at either end and part-flushed by the tides. Smaller, blind inlets were more like uncleaned cesspits, choked with leachate-producing organic wastes and non-degradable plastics. Those parts of the Tolo Harbour's Shatin arm still unreclaimed had been fouled by human, agricultural and industrial effluvia. The inner reaches of this once biologically rich second harbour were already a dead sea.

Beyond these dead shores, there still remained in the 1970s much opportunity of reprieve. Almost extant were the splendid marshes of Mai Po, the coral-fringed islets of Mirs Bay, the unspoiled sand beaches of Lamma and Lantau Islands; and the ocean promontories at Cape D'Aguilar and Clear Water Bay. More has been lost since the 1970s and thus this is a historical account in retrospect. The paradox it displayed, and still does today, owed much to what can be called 'conservation by concentration'. Hong Kong's urban population of 5 million has nearly all been crowded into a small part of its 1000 sq km. With an estimated half million inhabitants in a block of 2 sq km, Kowloon has some of the planet's highest human densities.

To save the periphery by forcing congestion at the centre would only have been an option for the most

Fig. 18.10 A moderately exposed shore at Taiping Point, Qingdao, People's Republic of China

a–b   littorines

b–c   *Chthamalus challengeri*

c–d   *Crassostrea gigas, Xenostrobus atratus* and *Mytilus galloprovincialis*

d–e   *Gelidium* and *Ulva*

e–f   *Sargassum*

1 *Nodilittorina millegrana* 2 *Littorina brevicula*
3 *Chthamalus challengeri* 4 *Xenostrobus atratus*
5 *Mytilus galloprovincialis* 6 *Lunella coronata*
7 *Cellana toreuma* 8 *Patelloida pygmea* 9 *Saccostrea cucullata* shell containing *Siphonaria japonica* and *Patelloida pygmea* 10 *Gelidium divaricatum*
11 *Sargassum pallidum.*

(Based on the study by Morton, B.S.)[4]

ruthless planner's logic. Today, with still mounting population pressure, and all the political uncertainty since 1997, this central confinement, with its saving benefit to the margins, could be about to break wide open.

Continued effluent pollution, human and agricultural, is the first ecological threat. No relief has yet been offered to Victoria Harbour. The death of Tolo Harbour has run its full course, and the rich communities of Hoi Sing Wan (Starfish Bay) described in *The Sea-shore Ecology of Hong Kong* no longer exist. Industrial and farming effluent from this Harbour now poses a danger to the proposed establishment of a marine park at Hoi Ha Wan ('Garden under the Sea'). This is the home of some 30 species of hermatypic corals, already reduced in area and declining in diversity and species density.

Yellow Sea

The north-east coast of the Chinese Republic, bordering the Yellow Sea, falls within the *Oriental Province* of the *North Pacific Cool Temperate Region*. The boundary of the warm temperate is set by Briggs (1974)[2] near Wenchou, so as to exclude the Yellow Sea, but taking in the tip of the Korean Peninsula. The Yellow Sea, however, retains many warm temperate features. With rock oysters and Sargassa its coasts are like the protected enclaves we shall find remaining in many parts of the Sea of Japan.

One of the rather few accounts of China's mainland coast ecology is Brian Morton's recent study of the rocky shores at Taiping Point, Qingdao.[3] The hard granitic shore is wave-exposed, especially during summer, with a cliff-face undercut by waves cresting over offshore dykes at high tide. In a **littoral fringe** washed by spray, the highest periwinkle is the small *Nodilittorina millegrana*. The two other species are *Littoraria articulata* and *Littorina brevicula*. The sea-slater *Ligia exotica* runs about freely at upper levels.

The barnacle of the **upper eulittoral** zone is *Chthamalus challengeri*, with neither *Capitulum* nor *Tetraclita* to be seen. Below *Chthamalus* runs a strip of the alga *Gloiopeltis furcata* with the small black mussel, *Xenostrobus atratus*. The **middle eulittoral** is constituted of a dense band of rock oysters, *Saccostrea cucullata*, invaded by *Xenostrobus* from above and from below by scattered *Mytilus galloprovincialis*, a colder water mussel, as distinct from *Septifer* at Hong Kong. This same mussel is locally cultivated in bunches hung from ropes connected by a matrix of floats.

The lowest oyster levels are covered by dense tufts of *Gelidium divaricatum*, and by *Ulva pertusa*, both being abundant all-year-round algae.

As on most of the coasts of China, the **sublittoral fringe** is *Sargassum*-dominated, here with long swathes of *Sargassum pallidum*, sweeping a surface pink with *Lithothamnion*. On nearby shores as at Da Hei Lan, there is *Sargassum thunbergi*. The sublittoral fringe has two seastars *Asterias rollestoni* and *Asterina pectinifera*. Algal-grazing molluscan herbivores include *Ischnochiton hakodatensis* and *Acanthochitona rubrolineata*. There are also turban shells, *Lunella coronata*, with the topshells *Monodonta labio* and *Tegula rustica* as in Hong Kong. The common limpets in the **lower eulittoral** are *Cellana toreuma*, *Patelloida pygmaea* and *Notoacmea shrencki*.

All this shore assemblage has a general warm temperate character. Qingdao shores thus have a threshold status, with summer occurrence of subtropical algae such as *Padina crassa*, *Dictyota indica* and *Champia parvula*, and their replacement in winter by such cold northern species as *Petalonia debilis*, *Porphyra yezoensis* and *Desmarestia viridis*.

Japan

In size and character Japan has many affinities with New Zealand. Tectonically unstable, both have north-to-south mountain spines upfolded at times not geologically remote. A high proportion of their shores are rocky, with salient capes and headlands, alternating with bays and harbours recently drowned.

Like New Zealand, Japan lies across both a warm and a cool temperate region, while the north and south extremes complete an even wider span. The remote northern shores of the Kurile island chain fall within a high boreal or subarctic region, with pack-ice in winter and laminarian algae held in common with the Bering Strait. To the far south, the Ryukyu chain lies in subtropical waters, and has reef-building corals, tropical algae and marine grasses and mangroves.

Japan's west and east coasts are moreover widely disparate, in a way that has no parallel in New Zealand. Essentially all the western seaboard looks out to a closed inland sea, with low wave action and reduced salinity, along with high land runoff and sedimentation. All this is in contrast with the ocean face looking east to the Pacific.

With such a drawn-out coast, from subarctic to subtropical, and with its differences of west and east, Japan has in total acquired one of the world's richest algal floras. As well as local and regional diversity, there is also a south-north trend that sees the ratio of brown to red algae regularly increased.

The Japanese people have long been engrossed with gathering seafood, ranging from fish, molluscs and crustacea to the signally important marine algae. While the seaboard is on the whole used sustainably, the demands upon it have been alleviated only by new exploitation —

not always prudent or moderate — of the ecosystems deeper offshore.

A high proportion of Japan's shore communities have been brought under surveillance and cultivation. Algal-collecting has seen the continuous use of a renewable resource, mostly in colder northern Honshu and Hokkaido. The advancing of mariculture has been a high preoccupation of biological science. The Imperial Marine Biological Laboratory, at Sagami Bay, founded and inspired by the late Emperor Hirohito, remains a symbol of this commitment.

Not surprisingly, Japanese marine biology has a strong systematic tradition. The abundance of good taxonomic literature contrasts with its paucity for the Asian mainland. Yet Japan would seem not yet to have produced — in a form available to English readers — any definitive account of shore zonation. The Stephensons' *Life Between Tidemarks on Rocky Shores* (1972) had scarcely any information sources for eastern Asia. But there have been papers enough from Japan, rounded out with first-hand reports from colleagues and students, to encourage me to attempt a synoptic account of these shores that I have not yet seen.

The islands of Japan all told lie between latitudes 23 and 45 degrees north, from the tip of the Ryukyu chain in the south to Hokkaido extended by the Kurile chain in the north. From the Philippine Sea the *North Equatorial Current* turns towards Japan as the *Kuroshio Current*. Just south of Kyushu, the southernmost main island, this current divides. One branch, the *Tsushima Current*, passes between Korea and Kyushu, into the Sea of Japan. The other continues east to wash the outer coast of Honshu, reaching as far as Cape Inuboe just north of Sagami Bay. Converging here with the cold *Oyashio Current* washing eastern Honshu from the north, it then takes an abrupt turn across the Pacific to join the *West Wind Drift*.

Biogeography

The broad Pacific-wide designations used in this book accord well enough with the biogeographic division of the Japanese coasts first made by K. Okamura on the basis of their algae. From his initial paper published in 1889, to his matured conclusions of 1926–31, this distinguished author has been saluted as the 'father of Japanese phycology'.

Okamura recognised in Japan three distinct kinds of algal flora: **subtropical** (= our tropical), **temperate** (= warm temperate and low boreal), and **subarctic** (= high boreal), having their correspondence with the three marine biogeographical regions of Briggs (1974) adopted in the present work.

The *North Equatorial Current* turns north at the

Fig. 18.11 East Honshu, Japan: comparison of warm and cold temperate shores

(left) The warm temperate Japan Province, Izu Peninsula, Shizuoka Prefecture.

(right) The cold temperate Oriental Province, Rikuchu Seashore National Park, Iwate Prefecture.

1 *Bangia atropurpurea* 2 *Porphyra* sp. 3 *Gloiopeltis furcata* 4 *Myelophycus simplex* 5 *Ishige okamurae* 6 *Ishige sinicola* 7 *Hizikia fusiformis* 8 *Sargassum thunbergii* 9 *Corallina pilulifera* 10 *Sargassum horneri* 11 *Eisenia bicyclis* 12 *Ecklonia cava* 13 *Chthamalus challengeri* 14 *Nemalion vermiculare* 15 *Septifer bilocularis* 16 *Heterochordaria abietina* 17 *Undaria pinnatifida* 18 *Chondrus yendoi* 19 *Calliarthron yessoense* 20 *Laminaria* sp. 21 *Costaria costata* 22 *Alaria crassifolia*.

(Details based on Chihara, Mitsuo.)[6]

Philippines as the *Kuroshio Current*, cutting through Japan's Ryukyu Islands chain just east of Taiwan. Its influence extends for a short way into the Korean Strait, and up the Pacific coast of Honshu to Cape Inuboe. A general consensus has placed at that point the northern limit of the *Japan Province* of the *North Pacific Warm-Temperate Region*. This embraces the warm ocean coast of Honshu with the coast of southern Shikoku and Kyushu, as far as Hyuga-Oshima, as well as the neck of the Straits of Korea, the tip of the Korean Peninsula and the west side of Taiwan, with the mainland China coast south of the Yellow Sea.

The shelf fauna of Japan south of Tokyo — with the large tropical element brought by the *Kuroshio Current* — has been so often referred to as Tropical-Subtropical, as to obscure the recognition of a Japanese warm temperate region. Only 20% of the fish belong to typical cold-water families, the rest being tropical derivatives. As Briggs wrote:

…the result has been a swamping of the resident fauna to the extent that if one depended on distributional records alone, a tropical fauna would appear to extend clear to Tokyo and in some cases even to Hokkaido.[5]

There remains, however, a strong tally of warm temperate endemics in southern Japan — 28% for both echinoderms (asteroids, ophiuroids and echinoids) and fish; while the rich fauna of 240 brachyuran crabs has 32% of warm temperate endemics, in addition to 53% shared with the Indo-Pacific tropics, with the rest being eurythermal temperate.

The truth is that for Japan the Kuroshio Current has played the role as the Atlantic Gulf Stream, that has brought tropical elements (including a horse-shoe crab and a fiddler crab) right to the southern shore of Cape Cod, Massachusetts.

The rest of the mainland, fronting the Sea of Japan and the Pacific coast of Honshu north of Cape Inuboe, is regarded as part of the *North Pacific Cool-Temperate Region*.

The Sea of Japan receives warm water from the branch of the *Kuroshio Current* passing through the Strait of Korea, to wash the north-west coast of Honshu as the *Tahushima Current*. From the cold water entering the Sea of Japan from the north, the *Linan Current* flows south along the mainland (west) coast to complete a counterclockwise gyre. There is a major faunal break at the Strait of Tsugaru between Honshu and Hokkaido; south of this the four species of Pacific salmon (*Onchorhynchus*), the smelts (Osmeridae) and a large number of the Cottidae do not extend. A similar break occurs at Chongjin on the north Korean coast.

Three Provinces impinge on Japan. The southern division, which is Briggs' *Oriental Province*, takes in the Yellow Sea, the central part of the Sea of Japan and the Pacific coast of Honshu north of Cape Inuboe.

The Sea of Japan is thus biotically diverse, with elements from several geographic groupings. The extreme southern end is warm temperate, invaded as well by many eurythermic tropical species. The cool temperate (low boreal) *Oriental Province* occupies the centre. North of 41–42 degrees north begins the *Kurile Province* with its high boreal biota, washed by cold water and with a predominance of brown algae over the reds.

The *Kurile Province* extends from the northern reaches of the Sea of Japan and Hokkaido along the east side of the Kurile Island chain, as far north as Cape Olyutorsky in the Western Bering Sea.

The Sea of Okhotsk is regarded as a separate *Okhotsk Province*, very cold, with its entire area below 0°C winter surface temperature and an August mean of under 12.8°C. As in the Sea of Japan, there is a counterclockwise gyre, receiving its inflow from the *East Kamchatka Current* coming from the cold Western Bering Sea, and with the main outflow through the middle of the Kurile Islands chain.

Warm and cold temperate

On the Pacific seaboard the transition from warm to cold is as always gradual, but can be taken to hinge upon Cape Inuboe, where the warm northerly flow turns east and a cold influence begins. Okamura's (warm) temperate coast began somewhat further north at Kinkazan, proceeding down south-east Honshu to take in Shikoku and reach Hyuga Oshima on the east coast of Kyushu. Though a few boreal algae persist as far south as Cape Inuboe (*Heterochordaria abietina*, *Pelvetia wrightii* and the offshore *Desmarestia ligulata*), the flora from Kinkazan south is as a whole warm temperate.

Both Provinces lie along clean coasts interspersed with bays and inlets and notably rich in algae. In the warm temperate, *Sargassum* species — continuing from their tropical dominance — still constitute the major brown algae. North of Cape Inuboe, entering the cool temperate, laminarians reach increasing size and diversity and come into their own in the sublittoral fringe and upper sublittoral zone.

Cold temperate (low boreal)

As representative of the cool temperate shores in east Honshu, an intertidal strip is reconstructed from Rikuchu Seashore National Park, north of Kinkazan, based upon the figure and data in Chihara's review of the communities of the algae. Cool temperate brown algae

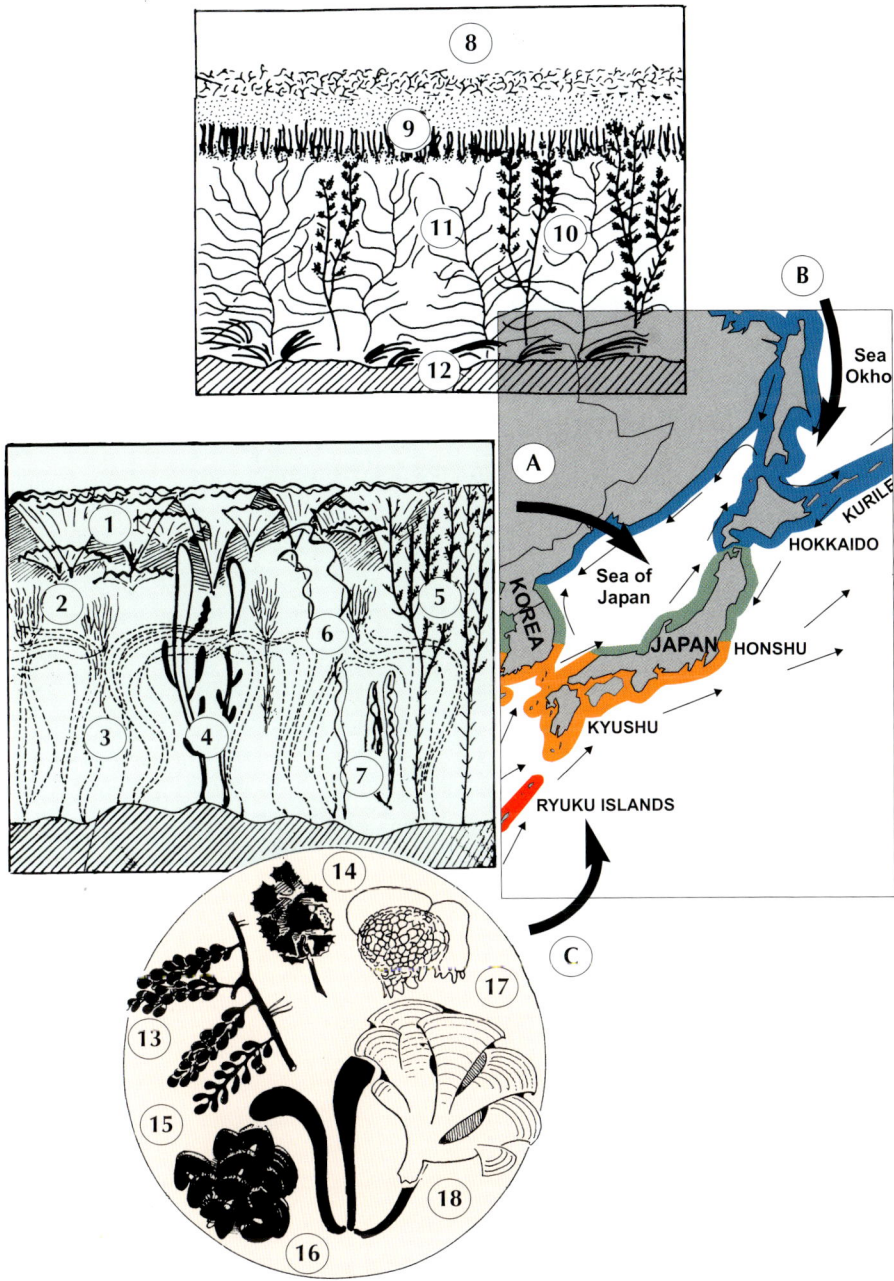

Fig. 18.12 Algal communities of warm, cold and tropical Japan
A. The Sea of Japan; West Honshu.
B. The cold shore of North Hokkaido, facing the Sea of Okhotsk.
C. The tropical islands of Ryukyu.
1 *Monostroma nitidum* 2 *Cladophora lehmanniana* 3 *Scytosiphon lomentaria* 4 *Colpomenia bullosa* 5 *Sargassum thunbergii* 6 *Enterorpha linza* 7 *Enteromorpha compressa* 8 *Gloiopeltis furcata* 9 *Heterochordaria abietina* 10 *Sargassum thunbergii* 11 *Chordaria flagelliformis* 12 *Phyllospadix iwatensis* 13 *Caulerpa racemosa* 14 *Turbinaria turbinata* 15 *Halimeda opuntia* 16 *Bornetella* sp. 17 *Dictyosphaeria cavernosa* 18 *Padina* sp.

predominate, though from south of Kinkazan there are warm-water elements such as *Hizikia fusiformis*, *Sargassum confusum* and *Undaria pinnatifida*.

The vertical sequence at Rikuchu begins at the top with seasonal *Bangia atropurpurea* and *Porphyra suborbiculata*. The periwinkles of the **littoral fringe** are two temperate species, *Littorina sitkana*, and — lower down and entering the eulittoral — *L. squalida*. The barnacle cover consists of the cold-water *Chthamalus dalli*, regular and close-spaced, beginning in the **upper eulittoral**. Over the topmost barnacles there is a scatter of the warm-water red alga *Gloiopeltis furcata*, and also — reminiscent of colder reaches — the bushy cords of *Heterochordaria abietina*. Seasonally the mucilage-covered cords of *Nemalion vermiculare* grow in shaded places.

The **middle eulittoral** includes a mussel zone, with the exposed shore species *Septifer virgatus* or — in greater shelter — *Mytilus galloprovincialis*. Below this a strong red algal turf begins, principally of *Corallina pilulifera*, scattered through with *Chondrus yendoi*, and enriched lower down with other reds, notably *Calliarthron yessoense*. Directly below the barnacles a high-level sargassoid, the narrow-leafed *Hizikia fusiformis* regularly grows.

At Rikuchu, the rich zone of brown algae (**sublittoral fringe** and **upper sublittoral**) has no sargassoids. Instead the laminarians hold sway down to a good depth. Dominant in the fringe are the blade-shaped *Alaria crassifolia*, the long, smooth straps of *Laminaria* and the cross-ribbed sheets of *Costaria costata*. These merge below into a rich zone dominated by *Undaria pinnatifida*, growing on the lowest intertidal rocks and continuing to 5 m beyond low water. The *Undaria* species are cool temperate laminarians, until recently with a restricted east Asian range, centering — in Japan — around the Izu Peninsula. *Undaria pinnatifida*, locally known as 'watasene' is Japan's choicest and highest priced edible brown alga. Its recent appearance in cool temperate New Zealand and Tasmania is a biogeographical event of major interest.

Warm temperate

South of Cape Inuboe, the algal balance has shifted. While the kelps are not lost, their dominance has moved from *Laminaria* to the warm temperate *Eisenia* and *Ecklonia*. The sublittoral zone, from 5–15 m, has dense underwater forests of *Ecklonia cava* and *Eisenia bicyclis*. *Ecklonia* is reported to give its maximum yield in July of 20 kg fresh weight per metre, with the largest plants reaching 3 m long. The *Undaria* species are still common in the lowest intertidal, but the sargassoids are now much more prominent, with *Sargassum horneri*, and — in its higher, middle shore position — *Sargassum thunbergii*. *Hizikia fusiformis* is also abundant in the mid-eulittoral.

Numerous species can be listed as typically warm temperate: *Monostroma nitidum, Chaetomorpha spiralis, Chaetomorpha crassa, Codium latum, Cutleria cylindrica, Padina arborescens, Galaxaura falcata, Serraticardia maxima, Phyllymenia sparsa, Carpopeltis angusta, Callophyllis crispata, Callophyllis japonica* and *Martensia denticulata*.

Warm enclaves

Well beyond the Warm-Temperate Province, enclaves of warm-water species persist in the land-locked bays of the Sea of Japan, and around the Simokita Peninsula and the Asamushi coast at the northern tip of Honshu.

At **Asamushi**, the rocky shore zonation was described by Hoshiai. The littoral fringe has the warm-water periwinkles *Littorina brevicula* and *L. mandshurica*, with the uppermost limpet species *Collisella heroldii*. The highest occurring barnacle in the upper eulittoral is *Chthamalus challengeri*. Reminiscent of subtropical Hong Kong is the band of *Tetraclita squamosa japonica* at the base of the barnacle zone, with occasional clusters of the stalked *Capitulum mitella*. At wave-exposed sites a band of the mussel *Septifer virgatus*, stronger than at Hong Kong, follows the barnacles, being exchanged in deep shelter for *Mytilus galloprovincialis*. *Thais bronni* is the common mussel-eating thaid. In the lower part of the barnacle zone, under stronger shelter, a warm-water enclave of rock oysters may develop (*Crassostrea gigas* at Asamushi and *Saccostrea echinata* at Shimokita). On steeper shaded surfaces, below mussels or oysters, the serpulid tubeworm *Pomatoceros crosslandi* builds up a strong zone, often mingled with a species of *Hydroides*. The small venus shell *Claudiconcha japonica* nestles among the tube-bases.

The algae *Ishige okamurae* and *Hizikia fusiformis* have their standard level on the warm middle shore, with *Gloiopeltis furcata* a little higher up. A common limpet in the eulittoral is *Siphonaria japonica*. The whelks of the *Sargassum* zone are the predatory *Ocenebra japonica* and the scavenging *Nassarius fraterculus*.

The algal turf of the lower eulittoral is composed of *Corallina pilulifera* enriched with *Laurencia* species, *Chondria crassicaulis* and *Symphocladia latiuscula*. *Undaria pinnatifida* occasionally enters the lower reaches.

In the sublittoral fringe sargassoids are the sole or dominant large brown algae. At Asimushi *Sargassum thunbergii* reaches to the middle of the intertidal, followed by *S. hemiphyllum* (Fig. 18.3), down to extreme low water. *S. tortile* is confined to subtidal spots protected from wave action. Among the holdfasts of *S. thunbergii* cluster the bivalves *Modiolus nipponicus* and *Arca boucardi*, the tubeworm (*Galeolaria*) and the anemone *Metridium senile fimbriatum*. Algae associated with *S. hemiphyllum* are *Corallina pilulifera, Dictyota dichotoma, Gymnogongrus flabelliformis* and *Acrosorium yendoi*.

Beyond the algal turf at extreme low water follows an encrusting coralline zone, with its cluster of chiefly warm temperate species: *Metridium* anemones with hydrozoans, the barnacles *Balanus trigonus* and *B. rostratus*, the tubeworms *Hydroides* and *Spirorbis*, and the bivalves *Striostrea circumpicta, Anomia chinensis* and *Arca boucardi*. Also common are the stalked ascidian *Styela clava*, the urchin *Strongylocentrotus nudus* and the seastars *Asterias amurensis* and *Aphelasterias japonica*.

These enclaves of warm temperate and even subtropical species are believed to be the most ancient surviving ecosystems of the east Asian shallow seas. In the west Pacific they evidently represent a very old fauna, established by the first half of the Miocene (25 million years ago) in the central and southern islands of Japan. The warm-water littorines, along with *Tetraclita* and *Septifer*, were in those times widely distributed. Later, as the Pacific became colder, they contracted their range, remaining dominant today only in the protected bays of the Sea of Japan, and a few other localities. Even as far north as the south Kuriles their presence bring enclaves of the warm systems into the lower boreal.

The tropical reaches

The *Kuroshio Current* washes the Ryukyu chain of islands that — with Okinawa and Amami Islands — fall into the *Indo-West Pacific Tropical Region*, with the February (winter) isotherm at 20°C. The northern boundary is placed just north of Amami, with the tropical Region taking in the east coast of Taiwan, then cutting across to the China mainland to take in the coral-girded and mangrove shores of Hong Kong.

Japan's mainland accession of tropical algae is most marked in the southern reaches of the large islands. Already, within the Izu Archipelago, only a little south of Sagami Bay, most of the algae held by Okamura to be subtropical are found co-existing with warm temperate elements. These include *Willeella japonica*, *Microdictyon japonicum*, *Anadyomene wrightii*, *Valonia macrophysa*, *Dictyosphaeria cavernosa*, *Cladophoropsis zollingeri*, *Boodlea coartica*, *Halimeda discoidea*, and *Liagora japonica*, along with *Padina* and *Caulerpa* species.

Through the Ryukyu arc, running south to Okinawa, the algae attain full tropical status. At **Yoron Island** (latitude 27 degrees north), near Okinawa, Tanaka described a barrier reef exposed at low water in a chain of coral blocks about 50 m wide and 6 km long. Sea temperatures for the year run between 19°C and 30°C. Of the 184 marine algae, 45 are greens, 108 reds and only 31 browns. The outer reef-face at wave-break has a tropical algal meadow including *Chlorodesmis fastigiata*, *Cladophoropsis zollingeri*, *Laurencia papillosa*, *Chondrococcus hornemannii*, and *Dictyosphaeria cavernosa*. On the reef flat exposed at low tide can be found *Hypnea nidulans*, *Colpomenia sinuosa*, *Microdictyon okamurai*, *Asparagopsis taxiformis* and a *Liagora* species. Tide pools of the reef flat are rich with species of *Codium*, *Padina*, *Gelidiella*, *Galaxaura*, *Peyssonnelia* and *Plocamium*. The ratios of Chlorophyta (green algae) to Ochrophyta (brown algae) indicate the increase of tropical influence:

Akkeshi, Hokkaido 0.35
Izu Peninsula. 0.6
Yoron Island 1.45
Yonakuni Island, southern part of Ryukyu Chain . . 4.5

Of the 15 species of sea-grasses found in Japan, all the tropical element (*Enhalus acoroides*, *Halophila ovalis*, *Thalassia hemprichii*, *Cymodocea isoetifolia*, *C. rotundata*, *C. serrulata*, *Diplanthera pinnifolia* and *D. uninervis*, as well as the cosmopolitan *Zostera nana*) are found exclusively in the Amami-Ryukyu area.

Japan's enclosed sea

The west-facing seaboard of Japan has its own algal communities, very distinct from those of Honshu's open Pacific coast. As well as the reduced action of waves and strong currents, especially in the quiet stretches of the

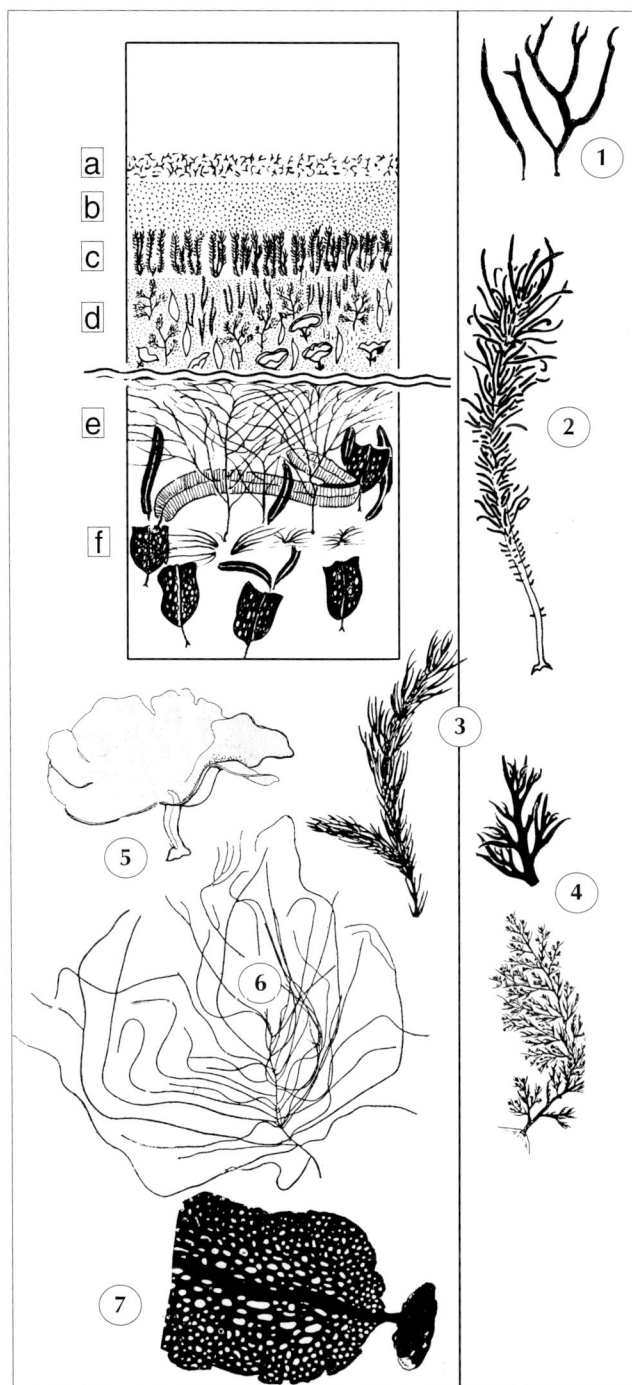

Fig. 18.13 The cold shore of North Hokkaido, facing the Sea of Okhotsk

(a) *Gloiopeltis* band (b) barnacle zone (c) *Heterochordaria abietina* (d) lower eulittoral with red algae (e) sublittoral zone with *Chordaria* and laminarians (f) lower sublittoral zone with *Phyllospadix iwatensis* and *Agarum*.
1 *Gloiopeltis furcata* 2 *Heterochordaria abietina*
3 *Neorhodomela larix* 4 *Odonthalia floccosa* (enlarged branch above) 5 *Constantinea subulifera* 6 *Chordaria flagelliformis* 7 *Agarum cribrosum*.

bay-heads and harbour backwaters, the tidal range is much shorter, and there is both a general lowering of salinity and increased runoff of sediment from the land.

Almost absent from the scene are the golds or rich browns of the laminarians and the olive or wine red of the larger rhodophytes. The leafy Sargassa are also much reduced. Instead, beneath the quiet waters, the algal hues are pallid green and dull yellowish brown. Expanded and membranous rather than wave-robust, such algal thalli lie slack between tides to form heavy swards across much of the emersed shore.

First, in the west, there are large river mouths bringing a strong freshwater influence. These occur on the Pacific coast as well: on the western coasts of Sagami Bay as at Tsuruga Bay and Tosa Bay, the community is dominated by short-turfed reddish brown *Gigartina intermedia* and *Sargassum sagamianum*.

Second, over most of the shoreline of the Inland Sea, there is a subdued flora, primarily of greens and browns. In the upper eulittoral grow wide, pale green sheets of *Monostroma nitidum*. A band of long, straw-coloured *Scytosiphon lomentaria* frequently marks the lower eulittoral. There are in addition *Enteromorpha compressa*, *E. linza*, *Cladophora lehmanniana* and tubular *Colpomenia bullosa*, with long tresses of *Sargassum thunbergii*, small-leaved and yellowish green.

All the above species have good access to the moderately high-saline waters of the outside sea. As salinity is reduced, algal diversity falls, leaving *Ulva pertusa*, *Dictyopteris divaricata*, *Neorhodomela larix*, *Carpopeltis affinis* and *Neodilsea yendoana*. With the lowest salinity, *Ulva pertusa* and *Grateloupia turutusu* are the last species to remain.

From Cape Nomo in the south to the Strait of Tsugaru in the north, the Sea of Japan has — as a whole — a shorter list of temperate algae than the Pacific coast. From its small indigenous element, Okamura concluded that this Sea was formed much later, with species supplied from the Pacific by the Tsushima current. At the beginning of the Pleistocene the Yellow Sea is thought to have been dry land and the Sea of Japan an inland body of water. With its past isolation, a good number of endemic invertebrates have evolved, including four out of 25 stone crab species (Lithodidae) and a number of molluscs and polychaetes. Briggs attributes the richness of this endemic fauna to its derivation from four zoogeographic groups, giving the southern end a warm temperate assemblage, invaded by many eurythermal tropical species.

Japan's cold seas

Falling within the low boreal region, the cold coasts of east Hokkaido, continuing north along the ocean seaboard of the Kurile Islands arc, are bathed by south-flowing currents. Facing the open seas, the typical algal associations are of *Porphyra pseudappressa*, *Pelvetia*

wrightii, *Fucus evanescens* and — appearing around low water mark — *Laminaria longissima*. The marine grass *Phyllospadix iwatensis* grows at and beyond low water.

The coldest Japanese coast is that of north Hokkaido, fronting the Okhotsk Sea. Between January and May, flat-topped ice-drifts float south from Sakhalin in the meridian of 144 degrees east. Emerging up to 2 m above the sea surface, these move up and down with the tides, causing repeated onshore abrasion. The dominant algae are *Chordaria flagelliformis* and *Heterochordaria abietina*. Outside the bays, in strong wave action, *Pelvetia wrightii* and *Fucus evanescens* are lost, but *Heterochordaria* persists, along with *Iridaea cornucopiae*. In north Hokkaido some notable high boreal laminarians begin to appear, including *Agarum cribrosum* and two forms not found in mainland Japan: *Arthrothamnus bifidus* and *Thalassiophyllum clathrum*. As well, there are the cold-water red algae *Neorhodomela larix*, *Iridaea laminarioides*, *Constantinea subulifera* and *Odonthalia floccosa*.

Such cold seas with drift ice present us with a world remote from middle or southern Japan. They introduce us to the diversity of the Pacific kelps, a group that are now briefly reviewed.

To the Arctic Circle

The Asian seaboard of today's *North Pacific Boreal Region* is thought to have been once comparatively warm. An ancient intertidal biota seems to have been widespread during the Miocene, with subtropical littorines (*L. brevicula*), mussels (*Septifer*), oysters (*Crassostrea*) and brown algae (*Sargassum*), now confined to warm protected enclaves. Early in the Miocene, this original community evidently contracted, and in the later Miocene widespread climatic change was to follow. With the onset of cooler conditions in the north Pacific, a range of species began to appear in the southern part of the Bering Sea that are today typical of the Pacific Boreal.

The *Boreal Region* is believed to have first been climatically uniform as far south as 38–39 degrees north. With climate change, a lower boreal became recognisable, distinctly warmer in summer than the higher boreal. The original species either changed their spawning times or survived only over the southern part of their range, forming the basis of today's coastal ecosystems over the whole boreal Pacific. They include the barnacle *Chthamalus dalli*, *Littorina sitkana*, several *Lottia* limpets, the mussel *Mytilus galloprovincialis*, the anemone *Metridium fimbriatum* and the brown alga *Fucus evanescens*.

Late in the Miocene, the hydrologies of coastal Asia

and America began to diverge. With the continued deepening of the south-west part of the Bering Sea, separate sets of high boreal species were able to evolve in Kamchatka and Alaska. Further falls in temperature gave a distinctive character to the West Pacific high and low boreal.

Low boreal

For today's *North Pacific Low Boreal Region*, we have described the **Oriental** province in northern Honshu and Korea. Its open shores are dominated by the following pattern. The periwinkles are *Littorina sitkana* at high level, while in the eulittoral *L. squalida* has filled the lower niche occupied by *L. littorea* in the Atlantic boreal. The mid-eulittoral mussel is *Mytilus galloprovincialis* with *Mytilus coruscans* towards low tide. *Pelvetia wrighti* and *Fucus evanescens* are the common intertidal fucoids. In the sublittoral fringe grows *Laminaria japonica*, with the common urchin *Strongylocentrotus nudus* among it. At 12–16 m the subtidal *Laminaria cichorioides* appears, while the mussels beyond low tide mark are the cold-water *Crenomytilus grayanus* and *Modiolus difficilis*.

High boreal

The **Kurile** province takes in the northern end of the Sea of Japan, Hokkaido, and the Pacific side of the Kurile chain, and reaches up the Kamchatka Peninsula to a boundary with the *Arctic Region* at about latitude 60 degrees north. Surface water temperatures never exceed 8–12°C and in most sectors salinity is high, at 32–33.5%. The shores of this Province are kept cold by the *East Kamchatka Current*, originating in the Bering Sea and flowing down the Japanese coast as far as Cape Inuboe where it turns east to mix with warm Kuroshio water along the Subtropical Front to become the West Wind Drift.

In the high boreal, *Littorina sitkana* is the uppermost periwinkle, succeeded by *L. sitkana*, with *L. squalida* at lowest intertidal level. The barnacles are the small *Chthamalus dalli* and at mid-eulittoral level the much larger *Semibalanus cariosus*. *Mytilus galloprovincialis* is the common intertidal mussel. *Fucus evanescens* remains, but *Pelvetia wrightii* has dropped out.

From the sublittoral fringe to the lowest level still sufficiently lit, the laminarians have unchallenged dominance. *Laminaria bongardiana* occurs at 0.5 m, *Alaria marginata* from 0.3 m down. The giant of all the kelps, prolonged up to 20 m long and as much as a metre wide, is *Alaria fistulosa*, anchored at 6–12 m depth. The hollow mid-rib has a series of chambers cut off by partitions. Further down, at 15–18 m, stones

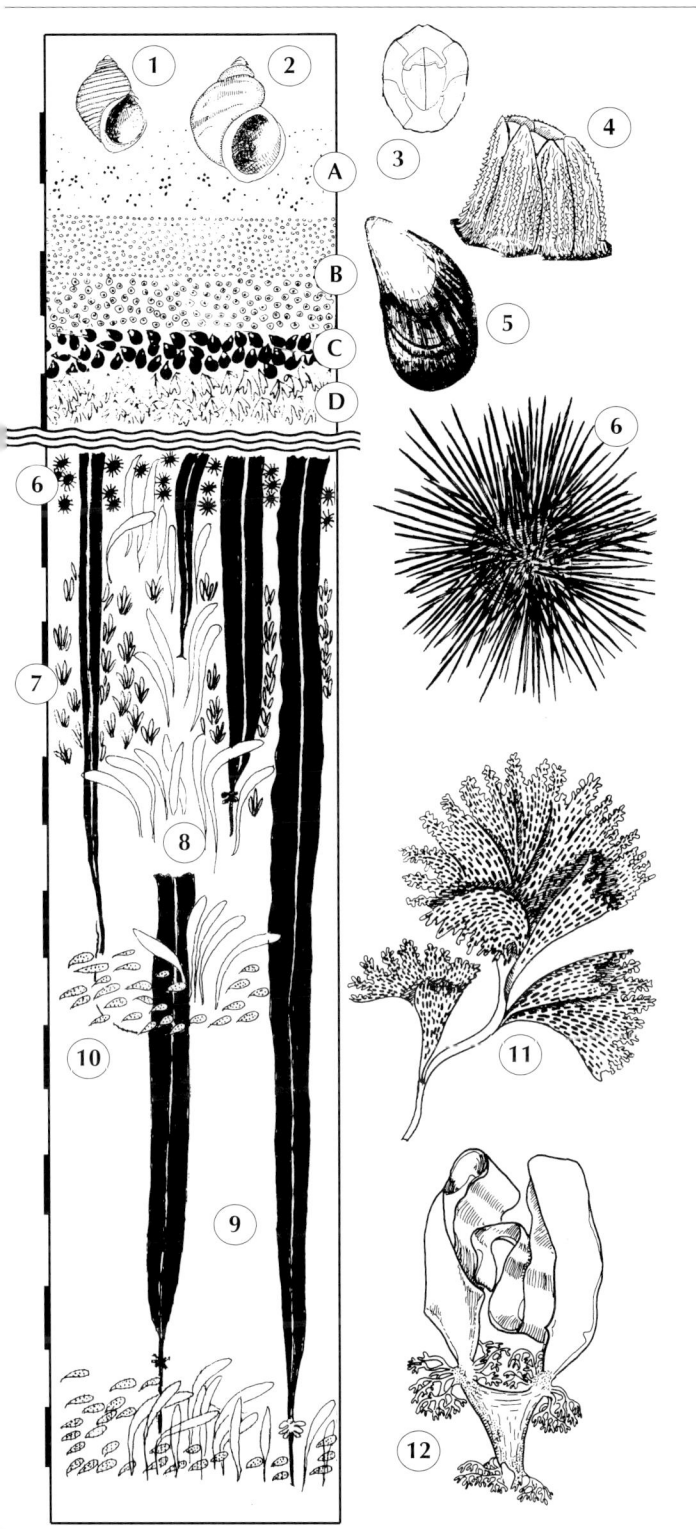

Fig. 18.14 General scheme of zonation of the cold Pacific high-boreal

Intertidal zonation
A. littorines; B. *Chthamalus* and (**bottom**) *Semibalanus*;
C. *Mytilus*; D. *Fucus evanescens*
Sublittoral zonation with *Alaria fistulosa*,
Strongylocentrotus and *Agarum cribrosum*
1 *Littorina sitkana* 2 *Littorina squalida* 3 *Chthamalus dalli*
4 *Semibalanus cariosus* 5 *Mytilus galloprovincialis*
6 *Strongylocentrotus* 7 *Laminaria* sp. 8 *Laminaria longipes* 9 *Alaria fistulosa* 10 *Agarum cribrosum*
11 *Thalassiophyllum clathrum* 12 *Arthrothamnus bifidus*.

carry beds of *Agarum cribrosum*. Here are found also those strange but characteristic cold-water laminarians *Thalassiophyllum clathrum* and *Arthrothamnus bifida*. As well, there are strong carpets of red algae, notably *Ptilota asplenioides*. Here live too the anomiid bivalve *Pododesmus macrochisma*, and the urchin *Strongylocentrotus sachalinicus*, along with *Actinia* anemones, the holothurian *Cucumaria japonica* and rich sponge communities.

The Pacific cold boreal has clear affinities with the Atlantic. In the later Miocene (Golikov and Scarlato)[7] a sea route is thought to have linked the Pacific and Atlantic Oceans across North Canada. Migrant species from the North Pacific could thus pass over to the boreal regimes already established in the north Atlantic. The original Atlantic subspecies so formed were soon to progress to new species making up the greater part of the present-day Atlantic ecosystems. There remains also a minority element from older Atlantic stocks that first emerged on the European coasts of the Sea of Tethys, for example the amphipods (*Gammarus*) and the cods (*Gadus*).

The Pacific migrants now dominant in today's Atlantic include such familiar high boreal species as *Semibalanus balanoides*, *Littorina saxatilis*, *Littorina littorea*, *Mytilus galloprovincialis*, *Littorina obtusatum*, *Ascophyllum nodosum*, *Fucus vesiculosus*, *Fucus serratus*, and *Rhodymenia palmata*. Beyond low water, there are *Alaria esculenta*, *Laminaria saccharina* and in some places *L. digitalis*. All these elements are common today on the coast of Great Britain. They are likewise to be seen on the cold shores of the Bay of Fundy that were — after glaciation — recolonised over the past 5000 years by a high Arctic route.

The Arctic region

Of the ice-bound regime of the Arctic, the Pacific Ocean affords no more than a glimpse, where pack-ice enters the northernmost reaches of the Bering Sea. In the Atlantic, circumpolar seas that can properly be designated Arctic reach as far south as Labrador, Greenland and the north of Iceland. In the north Pacific the winter pack-ice can extend to 60 degrees, bringing the Arctic boundary with the High Boreal to about Nunivak Island off Alaska and to Cape Olyutorsky in Siberia. In summer, North Pacific water flows through the Bering Straits to clear pack-ice from much of the Chukchi Sea, north of Siberia.

The post-Pliocene climatic story, reviewed by Briggs (1974), explains the formation of today's *Arctic Region*. In the late Pliocene (2.5 million years ago) the fall of temperature with the onset of glaciation put an end to the exchange of boreal populations between the Pacific

and the Atlantic. Then or later, boreal/arctic species of Atlantic origin were formed in the Barents Sea, and others of Pacific origin in the Bering Sea. Some of these have mingled with Atlantic and Pacific boreal communities. Others in the Arctic sector have formed associations of their own.

Among today's boreal/Arctic species are the bivalves *Macoma calcarea*, *Nuculana pernula*, and a number of species of *Astarte*, *Musculus*, *Neptunea*, *Buccinum* and *Eunephyta*. Most are found today in the high boreal of the Bay of Fundy.[8]

Regression of the oceans at the beginning of Pleistocene glaciation (1.8 million years ago) may have led to the formation of a self-contained Arctic water mass in the polar basin. Here would have originated the high Arctic species, to enrich the existing boreal/Arctic associations, or — more seldom — to form Arctic associations (as with *Laminaria longicruris*, the amphipod *Anonyx sarsi*, the echinoid *Strongylocentrotus pallidus* and the gastropods *Buccinum hydrophanum*, *Oenopota gigantea*).

Briggs (1974) has argued for an Arctic fauna of comparatively late origin, becoming distinct for two reasons. Some groups such as sponges, bryozoans and amphipods have large numbers of species still evolving in the Arctic. Arctic waters have also served as a haven for relics that survived by avoiding competition with the more advanced temperate forms. Examples are the three monotypic whale genera of the Arctic: the Greenland whale (*Balaena*), the narwhal (*Monodon*) and the white whale (*Delphinapterus*). There are in addition a large number of Arctic/boreal species with a circumpolar distribution.

We have no full description of the zonation of any Arctic hard shore, but Golikov and Scarlato have described two sorts of shorescape. Those that remain ice-free during four to six weeks of a 'hydrological summer' are able to develop a moderate cover of laminarians: *Phyllaria dermatodea*, *Alaria esculenta*, *Laminaria saccharina*, *L. longicruris* and a few red algae such as *Phyllophora*.

Shores ice-covered the year round will have no algae. But there can be *crio-littoral* communities with the amphipod *Gammarus setosus*, and littoral diatoms may develop in the firm soldering ice. On the lower surface of pack-ice in *crio-pelagic* communities *Apherusa glacialis* is the leading amphipod.

On any secure settling ground may be found the bivalve *Hiatella arctica*, a cosmopolitan species reaching to New Zealand (Fig. 13.11) but growing to its largest size in the Arctic. Hard shores will present also the mussel *Musculus corrugatus*, with ascidians, soft corals, sponges, and bryozoa. There are some equally characteristic Arctic species burrowing in sediments, notably *Mya truncata*, *Macoma moesta* and *Yoldia hyperborea*.

The cold-water kelps: Laminariales

In the Pacific boreal region we shall find nearly the whole range of these cold-water giants of the algae. The warm temperate shores have shown us a sublittoral fringe largely captured by the southern Fucales, leaving one or two laminarian species rising to canopied forests beyond low water. It is on the cool temperate west coasts of the great continents that the **Laminariales** come into full pre-eminence in the fringe, all but excluding the Fucales.

On cold boreal shores, both above and beyond low water, the great kelps are unchallenged, needing only freedom from permanent ice, and cold currents maintained by constant upwelling. Their original centres of distribution were evidently around the Arctic and Antarctic, with the northern and southern kelps belonging today to almost wholly different genera.

The Laminariales are far less tolerant than the Fucales of either high illumination or desiccation. Thus in the warm temperate they reach only exceptionally into the intertidal, mostly remaining as dense groves in the sublittoral. *Macrocystis* and *Nereocystis* anchor at considerable depth, and raise their blades to the surface by their spectacular length of stipe.

In the cool temperate of New Zealand and south-east Australia, we have found the bull-kelps (*Durvillaea*) in their relative gianthood around low tide. Properly speaking, these are not kelps, but Fucales that have assumed a laminarian habit and stature.

The large laminarian sporophyte must count as the high point of algal evolution. Though producing hardly any massive tissues, many have a prolonged axis that could be called 'woody'. The support and conducting tissues seen in a land plant are, however, rudimentary. Histological differentiation is very slight, with only a holdfast and a stipe carrying a lamina, sometimes cut into separate blades.

The Laminariales show a marked alternation of generations. The sporophyte carries the sporangia in patches of sori, never enclosed in conceptacles as with Fucales. The small gametophyte has an independent existence, albeit as little more than microscopic filaments with antheridia and oogonia.

The first family, the **Laminariaceae**, are middle-sized kelps, having their focus in the northern hemisphere with their highest diversity in the Pacific. The Atlantic *Laminaria saccharina* forms a simple stipe up to a metre long, fringed with an undulating lamina. In *L. groenlandica* of west America the blade is divided into segments with their own intercalary growth.

Pacific North America has several monotypic genera at the first stage of elaboration. *Cymathere triplicata* shows a discoid holdfast, a stipe only 5 cm long, and a blade up 4 m carrying three narrow folds. In *Pleurophycus gardneri*, there is a flat mid-rib with undulant wings. In *Agarum typicum* the expanded blade is covered with perforations, and in *Costaria costata*, also perforated, it is impressed with five long folds.

Hedophyllum sessile of Pacific America has a divided blade with the fingers of the holdfast (*haptera*) springing straight from its base. From such a lobate form could

Fig. 18.15 Three Alariaceae

1 *Pterygophora* with (2) juvenile; (3, 4, 5) *Eisenia* (immature), with undivided vegetative frond shown in black, with the sporophylls left unshaded 6 *Eisenia*, mature 7 *Alaria*.

have arisen the elaborate subarctic *Thalassiophyllum clathrum*, where the blade is a perforate net complexly rolled in at the edges. In *Arthrothamnus bifidus*, at the same high latitudes, the primary lamina ultimately disappears, except for its decumbent base from which secondary blades may arise.

In the second family, **Lessoniaceae**, unlike the Laminariaceae, splitting extends into the transitional growth-zone between the stipe and the lamina. Thus the secondary blades of *Lessonia* and *Macrocystis* have a meristem and can develop stipes of their own. In *Lessonia*, confined to the southern hemisphere, the young lamina splits at the base to initiate secondary branching. In New Zealand's *L. variegata*, the woody stipe stays short, but in *L. flavicans* of South Africa and *L. nigrescens* of South America it grows into a considerable trunk with repeated side-branching.

The bladder-kelps, *Macrocystis*, are the largest of the perennial Laminariales. Though *M. pyrifera* reaches Pacific North America, the genus is southern-centred. Each lamina undergoes multiple division with each strip developing its bladder and being paid off as a new blade.

On North American coasts with fast tides or high wave action *Macrocystis* is replaced by the giant *Nereocystis luetkeana*, anchored at 10–15 m deep. A buoyant spherical cyst produces ribboned fronds, splitting down to the base. The whole growth, up to 25 m, is achieved in one season. The Californian *Pelagophycus* has a similar structure but with a pair of leaf-bearing branches beyond the vesicle.

The high-exposed *Lessoniopsis littoralis* of Pacific North America forms a monotypic genus with its separate fertile and sterile blades. The former are short-stalked and oval, with the sori on both surfaces. The sterile blades are much longer, and tapered to pointed flails. On the same coasts grows the peculiar 'palm weed', *Postelsia palmaeformis*, a smaller kelp standing upright in the wave-exposed eulittoral.

One of the most aberrant of the Lessoniaceae must be *Dictyoneurum californicum*. From a normal-looking

juvenile, the stipe becomes thin and prostrate, attaching to the rock by marginal hapteres. The lamina splits beyond the transition zone, and the secondary blades attach by short decumbent stipes in the same way as the primary. This is repeated so that the front part of the system advances, with the older part dying. One plant thus becomes a clump.

The third family, **Alariaceae**, keep the vegetative frond un-split, but develop special sporophylls from either the stipe or the blade. The young *Alaria* resembles a simple *Laminaria*, with undulant lamina and short stipe. This carries two fringes of sporophylls, which — as in *A. marginata* — can become relatively long. In the monotypic *Pterygophora*, the stipe is a woody pole with a tuft of large sporophyll blades up to 1.5 m long, carrying basal sori. The primary blade, not much longer than the sporophylls, can itself become fertile.

With *Eisenia*, in California and Japan, the sporophylls appear as marginal pinnae on a primary or apical blade ultimately superseded. The sporophyll margins then roll in and thicken until, with the primary blade worn down, the two first pinnae are carried as arms at the top of the stipe. Growth of new pinnae is transferred to secondary meristems at their tips, while the original top of the stipe goes on growing, to become rigid and flattened.

Undaria, until recently confined to Japan and China, has a pinnate lamina, and the flattened stipe bears the sori on undulant wings. The Californian *Egregia* shows higher elaboration, with long branching straps growing out from the primary blade to reach up to 8 m long. These in turn develop fringes of marginal photosynthetic appendages, serving also as sporophylls, with some distending to form air-bladders.

The chiefly southern genus *Ecklonia* is represented in New Zealand and south-east Australia by *E. radiata* and reaches Japan with *E. cava*. Its spectacular giant is *Ecklonia buccinalis*, the 'bamboo seaweed' in south-west Africa, in habit somewhat like *Nereocystis*, with a hollow erect stalk up to 5 m long.

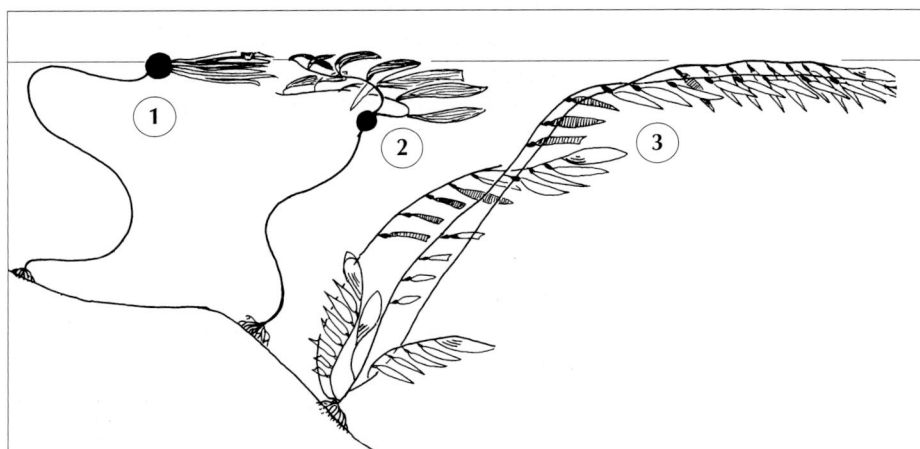

Fig. 18.16 Three large Pacific Laminariales
1 *Nereocystis* 2 *Pelagophycus* 3 *Macrocystis*.

CHAPTER NINETEEN
Pacific America

The whole seaboard from Alaska to Point Conception in California falls into the *Oregonian Province* of the cool temperate *Pacific Boreal Region*. These shores are washed by the *North Pacific Current*, flowing eastwards across the Pacific, then sending one branch south as the cool *California Current* and another north to the Gulf of Alaska. This current's maximal influence is rather brief (September–October); at other times the shores are cooled by deep upwelling (February–September), or washed by the northerly flowing *Davidson Current* (November–February).

This cool seaboard presents us with a rich low boreal biota, having certain resemblances to the Oriental Province of Japan. The shores first to be described, on the ocean-facing coast of Vancouver Island, where I researched and taught in 1977, are the most pristine and scenically distinguished I have yet seen.

Provinces

Biogeographers have argued over the boundary of the *Oregonian* with the high boreal *Aleutian Province* that reaches round the Alaska Peninsula into the Bering Sea. Briggs recognised a well-defined low boreal fauna and flora, and on the evidence of the fishes, echinoderms and marine algae would place the boundary somewhere along the Alaskan panhandle, perhaps as far north as Sitka. Reliance on the molluscs would bring the major faunal break to about the Dixon Entrance, north of the Queen Charlotte Island, making the Oregonian Province co-terminus with British Columbia and leaving all the islands of the Alexander Archipelago within a cold Aleutian Province.

Vancouver Island

The low boreal *Oregonian Province* — with the seaboards of Canada, Washington, Oregon and most of California, will be described at two localities. **Bamfield**, on the western side of Vancouver Island, and — towards the southern extreme — **Pacific Grove**, California.

The Stephensons visited this region in 1961, and

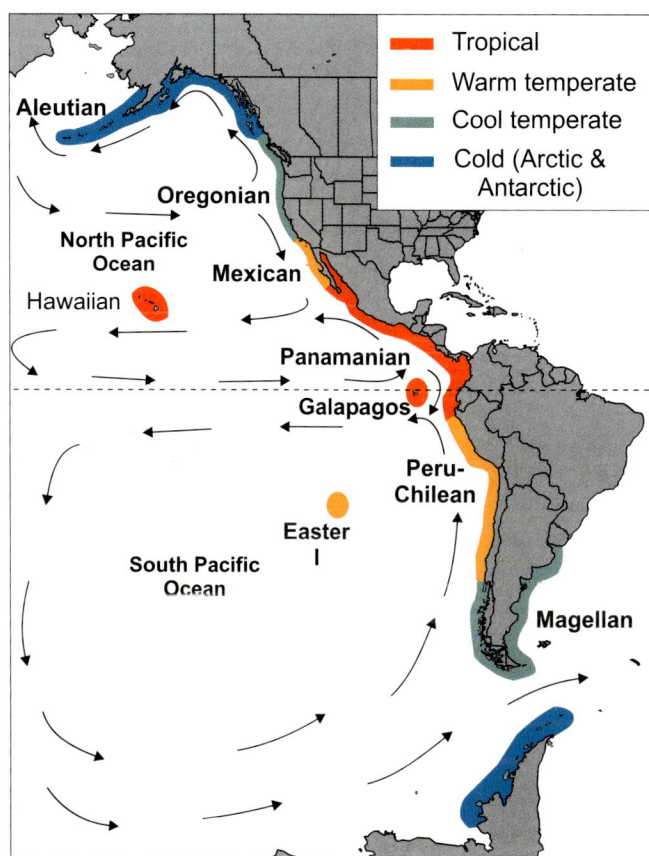

Fig. 19.1 The ocean currents and the biogeographic provinces of the east Pacific

made a detailed study of Pacific Grove. In British Columbia their observations were confined to the land-locked Strait of Georgia washing the sheltered coast of Vancouver Island and the mainland. A careful account of the low energy shore of Brandon Island, compared the zonation on the sheltered and less sheltered side. Prime stress was laid on the disparity of the Strait's whole shore from California's high-exposed Pacific Grove. The ocean face of Vancouver Island the Stephensons were never to see.

Like the Rockies on their grander scale, Vancouver Island has its high north to south spine. The 130 km journey from the ferry terminal at Nanaimo by way of the timber settlement of Port Albernie brings us to the

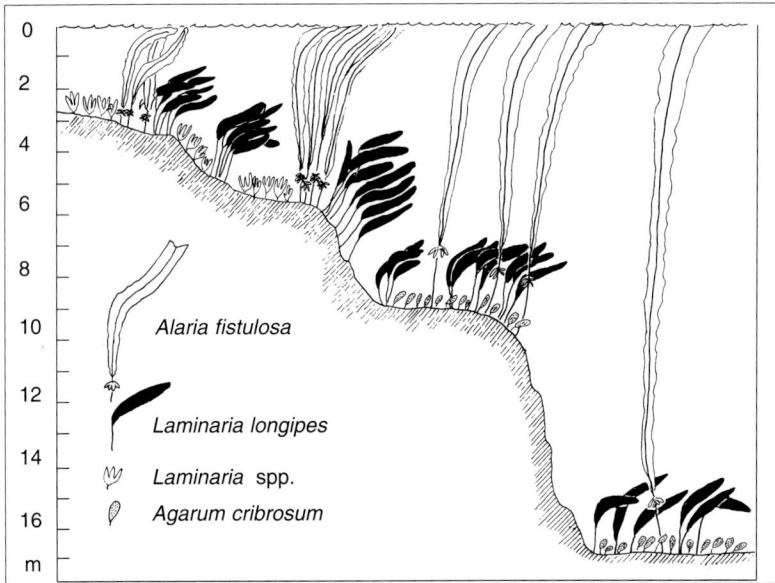

Fig. 19.2 A link with Asia: depth zonation of a kelp community at Amchhitka Island, Alaska

(From V.J. Chapman after Dayton.)

Alaria fistulosa

Laminaria longipes

Laminaria spp.

Agarum cribrosum

still isolated fishing village and old cable station of **Bamfield**, cut off by a spit from the Trevor Channel that forms the southern passage of the Barkley Sound (Fig. 19.3). For long distances north and south of the Sound, a wild, kelp-girded coastline faces the westerlies in the belt of the high 'forties'. With scarcely a break the *taiga* forest dips right to the shore at embayed points, or under strong winds reaches from the cliff-edge to the elevated skyline.

Right through summer, the sea temperature remains as low as 10°C. The tradition is that the Indians never learned to swim. To fall into these waters was to perish. With cold current-set from the north and with offshore upwelling, the outer coast is pre-eminently a domain of the large kelps. Tier upon tier the lower half of the intertidal descends steeply into the sublittoral zone, clad with a couple of dozen laminarian species, with varying resistance to the waves or — for just a few species — to the drying power of the air. More than any other marine province, these low boreal coasts are notable too for their diversity of algal-grazers and browsers: chitons, acmeid limpets, fissurellids and ormers.

At an easy distance from Bamfield, the ocean-exposed promontories on Cape Beale were accessible from a boat on only one calm day in a two months' stay. Here, at the south head of Barkley Sound, tumultuous surge sweeps up to a vertically prolonged littoral fringe. Above the direct effect of the waves, the whole terrain is continuously clouded with spray, and the tree cover has receded towards the skyline above the lichen-tinted supralittoral zone.

Barkley Sound is a squarish bite cutting deeply into Vancouver Island behind Cape Beale. The settlement of Bamfield lies on its southern shore at the confluence of the two arms of the Bamfield Inlet, cut off from the open Sound by its low spit (Fig. 19.3). From the Bamfield entry to Cape Beale the Trevor Channel is sheltered from

the wider Sound by the chain of Edward King, Diana, and Helby Island, with several smaller islets.

Above the landward shore of the Inlet is the Bamfield Marine Laboratory, serving a consortium of the five universities in British Columbia and Alberta. Clustered at their moorings below are the small craft seasonally employed in the lucrative business of salmon-catching, flat-fishing and shrimping.

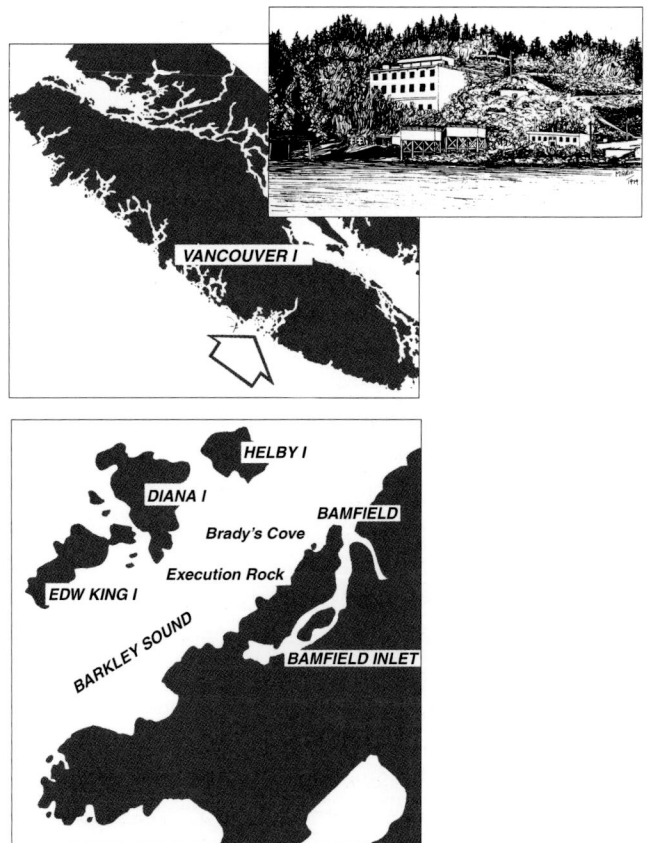

Fig. 19.3 Vancouver Island

Showing location of Barkley Sound and Bamfield, with Marine Laboratory.

The shores of the Trevor Channel outside the Bamfield Spit as at Brady's Cove and Scott's Beach are moderately exposed. Above the surge at Execution Rock, a California sea lion may rear itself on a high promontory or be watched at play amid the floating spheres of *Nereocystis* kelp.

Inside the inlet — by contrast — sombre green *taiga* forest touches down to the littoral fringe with sitka spruce, western hemlock, western white pine, Douglas fir and red cedar. The Inlet's southern arm reaches up to a brackish pond, with the intertidal swathed in *Zostera* grass (Fig. 19.10). Behind this stretches a secluded cedar bog, with sundew, swamp laurel (*Kalmia polifolia*) and cranberry. On drier ground at the edge of the sprucewood grow the yellow aroid 'skunk cabbage', *Lysichiton americanus*, and British Columbia's national flower, the bunchberry, *Cornus canadensis*.

Where forest touches the shore a black bear may wander out to the reef, turning over boulders with a talon to nuzzle for crabs. Raccoons with their predilection for washing come down to the backwaters. In the taiga forest live also pine martens. In these quiet reaches is the vigilant great blue heron, and loons will break the silence with their eerie call. A bald eagle, icon of the wild coasts of Vancouver Island and Puget Sound, may be picked out upon a bare promontory. Above a dead Douglas fir, two courting osprey were watched circling and finally pairing over their high platform nest.

Shore zonation

Brady's Cove, on the outside of Bamfield spit, presents us with a moderately exposed seaboard typical of the Canadian reach of the low boreal Province.

The **supralittoral zone** is darkened with the black lichen, *Verrucaria maura*. In rock cracks with a thin humus flowers grow generously, with glistening yellow *Potentilla anserina* the most widespread. Common too on the rock face are rosettes of *Plantago maritima juncoides*. In gravelly bay-heads can be found white thimble-berry (*Rubus parviflorus*), brick-red hawkweed or Indian paintbrush (*Castillya miniata*) and orange western columbine (*Aquilegia formosa*), along with yellow monkey flower, (*Mimulus guttatus*), alum-root (*Heuchera micrantha*) and fringe-cup (*Tellima grandiflora*).

The **littoral fringe** is marked by two cold-water species of periwinkle, the spiral-ridged *Littorina sitkana* and the checkered *L. scutulata*, both ranging north to Alaska. The barnacles of the **upper eulittoral** may include an uppermost scatter of small, smooth-worn *Chthamalus dalli*. But by far the commonest barnacle is the larger, strongly ribbed *Balanus glandula*, dominant through the upper eulittoral. Lower down, where the

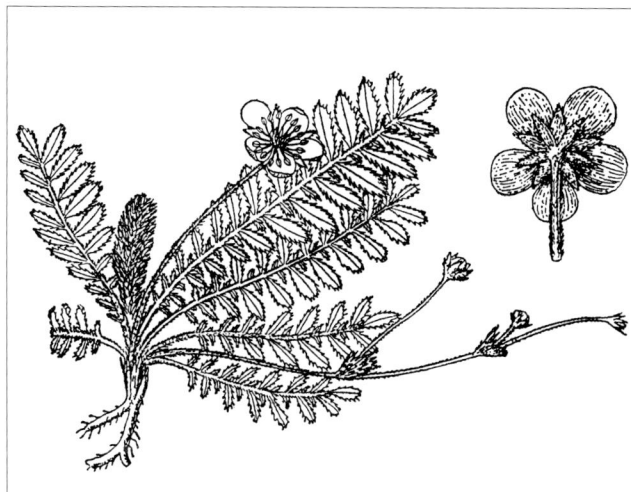

Fig. 19.4 *Potentilla anserina*

eulittoral is mussel-zoned, the much larger barnacle *Semibalanus cariosus* appears on rock or *Mytilus* shells; its ribbed column can reach 50 mm tall, carrying a thatch of spiny scales, often detaching towards the apex.

The **middle eulittoral** of the Pacific boreal has two species of mussels. An upper tier consists of the smaller *Mytilus trossulus*, dull blue with the beaks white-eroded. In all but the most sheltered sites, these intergrade below with the longer *Mytilus californianus*, with its shiny black periostracum. Though commonly overlapping, the two are found to have divergent preferences. *Mytilus trossulus* is optimally adapted to shelter and arrives at suitably clean spots by crawling away more actively from heavy silt. In a mixture with *M. californianus* in artificial clumps, it is *M. trossulus* that will move to the outside. In exposed conditions, with the fine sediment removed by waves, the competitive advantage passes to *M. californianus*. With both the shell and byssus stronger, it grows to more than twice the length of *M. trossulus*. With increased water movement, *M. californianus* eventually outgrows and crushes the smaller species.

Under high wave energy the mussels are invaded by pale patches of the stalked barnacle *Pollicipes polymerus*. This species stands out by its ivory-white capitular plates on the dull red peduncle 5–6 cm long. A regular feature of exposed Oregonian coasts *Pollicipes* enlarges its holdings at the expense of the mussels with increase of surge.

Like most ocean shores with mussels, the American Pacific has its own large asteroid. The orange-brown to purple *Pisaster ochraceus* grows to 30–40 cm across, with five strong arms sculpted with a network of short spines. This star is the original 'keystone species', shown by Robert Paine to open up new settling space by its predation of mussels and *Pollicipes* (Fig. 19.5). Paine's study was later to extend to *Stichaster australis*, the compara-

Fig. 19.5 A moderately exposed shore at Brady's Cove, Barkley Sound, Vancouver Island

a–b *Fucus gardneri* and *Balanus glandula;* b–c *Mytilus* and *Pollicipes;* c–d corallines *Semibalanus cariosus, Anthopleura xanthogrammica* and *Endocladia muricata;* d–e *Hedophyllum* and *Laminaria groenlandica;* e–f *Alaria, Egregia, Phyllospadix scouleri, Laminaria,* and with *Nereocystis* offshore

1 *Littorina sitkana* 2 *Littorina scutulata* 3 *Mytilus trossulus* 4 *Fucus gardneri* 5 *Balanus glandula* 6 *Mytilus californianus*
7 *Pollicipes polymerus* 8 *Nucella canaliculata* 9 *Nucella lamellosa* 10 *Hedophyllum sessile* 11 *Laminaria groenlandica*
12 *Nereocystis luetkeana* (**left**) spherical bladder with fronds, (**right**) juvenile plant 13 *Alaria marginata* 14 *Egregia menziesii.*

Fig. 19.6 A sheltered shore inside Bamfield Spit

a–b littoral fringe shaded by Sitka spruce and with *Verrucaria* and limpets; b–c *Balanus glandula*; c–d *Fucus gardneri* with *Mytilus trossulus*; d–e *Neorhodomela larix*, with *Ulva*, *Cladophora* and *Soranthera*; e–f *Sargassum muticum*
1 *Littorina sitkana*
2 *Tectura persona*
3 *Searlesia dira*
4 *Fucus gardneri*
5 *Balanus glandula*
6 *Neorhodomela larix*
7 *Sargassum muticum*.

ble star on New Zealand's high-energy mussel shores.

Also reminiscent of New Zealand's west coast are the large anemones in pools or with the lower mussels. *Anthopleura xanthogrammica* can grow to 12 cm across the disc with a column twice that length covered with soft warts. The disc and tentacles are pink, yellowish green or lavender, becoming deep green in brighter light, evidently from the presence of algal symbionts. Slightly higher up the shore lives the attractive red and green *Anthopleura elegantissima*.

In the short algal turf of the **lower eulittoral**, corallines are usually sparser than in the warm temperate. Dominance is taken over by red-brown clumps of the non-calcified rhodophyte *Endocladia muricata*. The branchlets are beset with fine spinules visible with a hand lens.

In the **sublittoral fringe**, and far beyond low water as sufficient light is available, the laminarian brown algae hold the dominance. The one intertidal fucoid species on this shore is the small, olive-green *Fucus gardneri*, forming patches breaking into the upper barnacle zone. The simple mid-ribbed fronds are dichotomous, with the tips swollen at sexual maturity. Remote from the large Laminariales round low tide, this high-level species is a remainder of the full suite of fucoid species running right through the eulittoral in the Atlantic boreal.

In place of fucoids, the sublittoral fringe has a larger sequence of laminarians than we have so far described on any Pacific shore. The first tier, becoming briefly emersed even at a low neap tide, is composed of the small *Hedophyllum sessile*, with its thin, often tattered blade attached direct by a branched holdfast. At first oval, it soon becomes incised and lobate, with the surface raised into flat vesicles. Below or mingling with *Hedophyllum* grows *Laminaria groenlandica*, raised on slender stipes like small banners with the smooth blade divided into two or three straps. Around low water of spring tide may appear the second member of this genus, *Laminaria saccharina*, with a long, mid-ribbed frond, undulant along the margin.

Directly below *Laminaria groenlandica* there is a crowded cascade of *Alaria marginata*. Brown or olive in hue, the slender, mid-ribbed lamina can reach 2 m long and 15 cm wide. As in all *Alaria* species, there is a small stipe, fringed with a row of short sporophylls at either side. Far more attenuate than *Alaria* is *Egregia menziesii*, with only the highest plants emersed briefly at low water. Near the base, the *Egregia* thallus divide into long, flat blades, sometimes growing to 10 m long and carrying at either side spatula appendages some of which inflate into vesicles. In a large offshore *Egregia* the blade can reach 25 cm wide.

Among these laminarians, typically mingled with the highest *Egregia*, green shocks of the sea-grass *Phyllospadix scouleri* stand out brightly. Exposed to the full wave-force, this grass reaches a metre long, with the leaves up to 20 mm wide. From the base springs the small inflorescence up to 6 cm long.

The laminarians enlarge and diversify beyond low water. By far the most attenuate is the sublittoral *Nereocystis luetkeana*, with its spherical floats — 15 cm across — buoyant at the surface. From Alaska to California, *Nereocystis* forms offshore groves, being a fast-

Fig. 19.7 Wave-exposed and sheltered zonation at Execution Rock
(**top left**) (a) *Verrucaria* black lichen zone (b) littorine zone (c) *Balanus* zone (d) darker *Balanus* strip with *Pollicipes*
(e) coralline strip (f) brown algal zone
1 *Pelvetiopsis limitata* 2 *Mytilus californianus* 3 *Mytilus* with *Semibalanus cariosus* 4 *Hedophyllum sessile* (long-fronded)
5 *Eisenia arborea* 6 *Laminaria setchelii* 7 *Nereocystis luetkeana* 8 *Pterygophora californica* 9 *Fucus gardneri* 10 *Costaria
costata* 11 *Phyllospadix scouleri* 12 *Balanus glandula*.

growing annual reaching 25 m in a single season. The
slender, unbranched stipe attaches at a good depth terminating in a float carrying two clusters of blades, splitting dichotomously as in *Lessonia*. California sea-lions,
Zalophus californianus, disport among the floats and
foliage.

A still wider array of sublittoral laminarians may be
freshly gathered after an onshore blow. These commonly
include *Laminaria setchellii*, *L. saccharina*, the broad,
mid-ribbed blades of *Pleurophycus gardneri* and
Cymathere triplicata, and the five-ribbed *Costaria
costata*, as well as the two *Agarum* species. Both *A.
cribrosum* and *A. fimbriatum* have wide laminar blades,
and the former is perforated with small holes. Like most
of the kelps entering the sublittoral fringe, these are all
high boreal species ranging north to Alaska or the
Bering Straits.

Ranging south from Vancouver Island there is also a
distinctive group of low boreal species, including *Egregia menziesii*, *Dictyoneurum californicum*, *Pterygophora californica* and *Eisenia arborea*, all coming to
their maximum on the shores of north California.

Generally to be found in storm drift are the cold-
water brown algae of the genus *Desmarestia*. The most
frequent species, *D. ligulata*, with its narrow fronds
mid-ribbed and side-branched, is mostly sublittoral, but
occasionally reaches the lowest intertidal.

A wider spectrum

Brady Cove, as we have seen, gives the best introduction
to open shores, outside the Bamfield Inlet. A longer
stretch of open coast continues south towards Cape
Beale. It can be typified by Execution Rock, accessible
along the shore or by a taiga forest track through from
the top of the Bamfield Stream.

Differences of slope and orientation produce a wide
range of exposure and shelter. A high rock near Execution Point gives a contrast on its two sides. The open

face is more exposed than at Brady's. The extreme upper shore carries seasonally the membranous red alga *Porphyra perforata*. Below this is the olive green to tan *Pelvetiopsis limitata*, confined to high level exposed places. The mussel/*Pollicipes* zone follows as at Brady's, but with *Mytilus californianus* and patches of *Pollicipes* becoming more important.

The lower half of this exposed face carries laminarians in clear, straight order. At the top is *Hedophyllum sessile*, with its lobes far more pointed and attenuate than under shelter. Next comes *Eisenia arborea*, with woody upright stipes and a crown not unlike the fronds of *Ecklonia*. Below this and under continuous wave action is the strong but graceful *Laminaria setchelii*, with greater frond division and arched on a longer stipe than *L. groenlandica*. It forms the pure and at low tide fully visible fringe on surge-attacked shores. Beyond it and subtidal is *Nereocystis luetkeana*, well recognised as its large spherical floats break water. Its fronds are the favoured fodder of the California sea-lion, often to be watched browsing off these open coastlines.

The sheltered slope contrasts almost completely with the exposed face. Apart from seasonal *Porphyra* at the top, the only conspicuous alga long uncovered is the stretch of *Fucus gardneri*, represented only in a narrow strip at Brady's. *Fucus* mingles seasonally with smaller algae, including *Scytosiphon lomentaria*, and continues down to the large pools, diversely crowded with algae. The margin and sides are profusely clad with *Corallina vancouveriensis*. The eel-grass *Phyllospadix torreyi* often forms a pool border, and on its leaves should be noted a small membranous epiphyte, *Smithora naiadum*, related to *Porphyra*.

These pools have one of their largest browns in the shield-shaped but flexible *Costaria costata*. This is one of the most southerly reaching of a mainly cold-water group, prominent right to the Arctic. Also in the pool may be the smaller three-lobed *Laminaria groenlandica*, as well as short *Egregia* and very young *Nereocystis*, unlikely to reach full size. Ovate membranes of *Gigargtina exasperata* are common. On the fine, densely branching red, *Odonthalia floccosa*, grow brown epiphytes, such as young pads of *Colpomenia sinuosa* and *Leathesia difformis*. There are also small yellow brown nodules of the epiphyte *Soranthera ulvoidea*.

The tubular *Halosaccion glandiforme* lives beyond pools at mid-tide. On open rocks still moist and somewhat sheltered are banner-like reds, *Iridaea cordata* and *Iridaea lineata*, and on low tidal rocks deep wine-red *Constantinea simplex* with its circular saucer-shaped blades.

From near Execution Rock it is easy to push out further over the flat shores carrying immobile boulders that in many places stretch between or out to coastal islets

with surge coursing over the shallows between them at high tide. Such boulder flats are exceptionally rich. Fig. 19.7 shows algal zoning on one of these boulders, capped on top with *Hedophyllum* and at the sides carrying *Codium fragile* and disc-like *Codium setchelii*. The same boulder has a large fringe of brown algae: *Costaria costata* and small *Egregia* and *Alaria* with very young *Nereocystis*.

Further out over boulder flat the buoyant shores of *Nereocystis* are bigger. There are also upstanding axes somewhat like broomsticks, of the tall brown alga *Pterygophora californica*.

Dark habitats
Overhangs and caves
The boulders towards low water have a richness beneath, recalling the same habitats in New Zealand. On Vancouver Island dark communities become large and diverse under low tidal overhangs festooned with *Hedophyllum*, the blades of *Alaria* or the fringed streamers of *Egregia*. Entirely concealed until this curtain is drawn back, the rock faces present an abundance of colour, from ascidians, bryozoans, sponges and anemones, along with their gastropod and other predators.

Two sorts of bushy bryozoan are widespread under the algal curtain. The cyclostome *Filicrisia geniculata* forms a white, fleecy ground layer (Fig.19.9), easily distinguishable from the stronger tufts of the commonest cheilostome *Tricellaria occidentalis*. The ctenostome *Flustrellidra corniculata* grows in short fleshy straps, bristled all over. There are also low ground-spreads of a branching *Bowerbankia*, while an *Alcyonidium* species forms a smooth crust, at first sight mistakable for a compound ascidian.

Anemones living beneath overhangs include large, higher level *Anthopleura elegantissima*, and a smaller reddish *Diadumene*. Lower down near sand may be found the large greenish *Anthopleura xanthogrammica*.

A minor underworld of shade-loving red algae includes small *Rhodoglossum affine* and down at sand level the narrow, dichotomous *Gymnogongrus linearis*.

Sponge and ascidian caves
My first and unforgettable glimpse of the sciaphilic richness and colour of these shores was in penetrating such a cave at Execution Rock. No longer a narrow cleft, or a curtained underhang, the space expanded, inside a narrow-lit entrance, into a room of standing height. *Pollicipes* reached marginally within, up to our shoulders. Aside from wine-red *Lithothamnion* on the dimly lit floor, algae were exchanged for the red, orange and lighter pastel colours of sponges and compound ascidians. Next below the barnacles came heavy-ridged

Fig. 19.8 Zonation of steep, maximally exposed shore at Cape Beale, Vancouver Island

(**top right**) Landing place* and working site,** in relation to lighthouse point. a–b elevated supralittoral black zone with *Verrucaria*, and with *Potentilla* and other flowering plants; b–c littorines, limpets and barnacles, and *Pelvetiopsis*; c–d deep channel with black rhodophyte crust and *Pollicipes*; d–e *Balanus glandula* and *Pollicipes*, with *Postelsia* in profile; e–f *Postelsia* with *Mytilus californianus* and *Pollicipes*; f–g *Mytilus* and corallines; g–h brown algae *Alaria*, *Eisenia* and *Laminaria setchelli*

1 *Littorina sitkana* 2 *Lottia digitalis* 3 *Balanus glandula* 4 *Chthamalus dalli* 5 *Pelvetiopsis limitata* 6 *Postelsia palmaeformis* 7 *Alaria nana* 8 styelid ascidian 9 *Mytilus californianus* 10 *Pollicipes polymerus* 11 *Corallina vancouveriensis* 12 *Calliarthron scmittii* 13 *Bossiella californica*.

(**bottom right**) A. sea otter (*Enhydra lutris*) B. California Sea Lion (*Zalophus californianus*)

Fig. 19.9 Shaded overhang at low tide, Brady's Cove, Barkley Sound
1 *Alaria marginata* 2 *Egregia menziesii* 3 *Hedophyllum sessile* 4 *Anthopleura elegantissima* 5 *Distaplia* sp. 6 *Flustrellidra corniculata* 7 *Solaster dawsoni* 8 *Pisaster ochraceus* 9 *Filicrisia geniculata* 10 *Tricellaria occidentalis* 11 *Aplidium* sp. 12 *Didemnum* sp. 13 polyclinid heads 14 *Diadumene* sp. 15 halichondriid sponge 16 *Katharina tunicata* 17 *Archidoris montereyensis* 18 *Gymnogongrus linearis* 19 *Rhodoglossum affine* 20 *Anthopleura xanthogrammica*.

sponges.

As much as sponges, compound ascidians occupy the cave floor. The most pervasive are a pale yellow *Aplidium*, in thick, gelatinous sheets. A second polyclinid has pendent heads, translucent and studded with red zooids. A *Distaplia* species forms hemispherical or flat-topped colonies. There are also thin crusts of beige or white *Didemnum*. The translucent *Pycnoclavella stanleyi* can be found massed together in small sheets, being a 'social' ascidian with zooids along a common stolon rather than truly compound.

It is in the deepest shade or in near-dark caves the sponges come into their own. The widest occupants, coming up to waist-height on the walls, are a mauve to violet *Haliclona*, a buff *Halichondria*, a scarlet *Ophlitaspongia*, and the flesh-pink *Aplysilla glacialis*. On the most shaded walls grow pure white calcareous sponges, *Grantia* and *Leucosolenia*. In all these places, the grazing molluscs are in strong force, with the chitons *Katharina tunicata* and *Cryptochiton stelleri*, and the gastropods *Calliostoma*, *Megathura crenulata*, *Diadora aspera* and *Fissurellidea bimaculata* (Fig. 19.12). Some have become regular sponge-consumers, like the opisthobranch *Archidoris montereyensis*.

In high exposure

The Lighthouse promontory of Cape Beale is maximally exposed, and on nearly all days hazardous to landing. Out in front, there are sharp rock-stacks and dissected headlands, constantly swept by surge. Getting ashore on a quiet day, one can venture at a low tide along the

Fig. 19.10 A saline pond with zostera grass in Bamfield Inlet

A. zone with tawny grass B. salt meadow of *Triglochin maritimum* with *Lilaea scilloides* C. wide zone of *Salicornia virginica* D. stony zone with dwarfed *Fucus gardneri* E. middle shore with green *Enteromorpha* and *Cladophora* F. zone of purplish brown *Gracilaria* G. submerged sward of low water *Zostera marina*.

1 great blue heron, *Ardea herodias* 2 *Aeolidia papillosa* 3 *Epiactis prolifera* 4 *Idothea resecata* 5 *Melanochlamys diomedea* 6 *Phyllaplysia taylori* 7 *Zostera marina* var. *latifolia* 8 *Haliclystus auricula* 9 *Haminoea virescens* 10 *Leptasterias hexactis* 11 *Lacuna variegata* 12 *Melibe leonina*.

defiles and gullies, with mussels and *Pollicipes* reaching above head-level, high-elevated under the wave effects. Beyond these in turn rises a tawny expanse of balanoid barnacles. Still higher there is bare rock with littorines and finally — almost to the tree-line — a steep face blackened with the lichen *Verrucaria*.

The two periwinkles attain high levels under heavy spray, with *Littorina sitkana* far outreaching *L. scutulata*. Grey to purplish *Porphyra perforata* is seasonally conspicuous on the upper eulittoral faces. The highest reaching, most exposed fucoid is *Pelvetiopsis limitata*, with short fans of dichotomous branches, brown to olive-green. Its Pacific range is from British Columbia to California.

The barnacle zone begins with *Chthamalus dalli* at the top, soon to be superseded by *Balanus glandula*

spreading down to the beds of *Mytilus californianus*. *M. trossulus* is found here only in a few highly sheltered retreats. Among the lowest Californian mussels grows the large, sharply ridged barnacle *Semibalanus cariosus*. Pale *Pollicipes polymerus* seizes the high points where surge sweeps or swash runs back, imparting a chequered effect to the mussel zone. Around the mussels and barnacles, the rock is dark with black or brown *Petrocelis*, an uncalcified crust firmly fixed to the substrate.

A rough veneer of encrusting corallines and *Lithophyllum* reaches among the mussels, almost to the top of the eulittoral, and there is more jointed coralline turf than would be seen on shores of lower exposure. In addition to *Corallina vancouveriensis* and *C. pilulifera* as on more sheltered shores, there are pink or purple tufts of *Calliarthrion tuberculosum*, up to 15 cm tall, along with broader-fronded *Bossiella californica* and *B. plumosa*.

On the tops of stacks as far up as mid-tide, the sea-palm, *Postelsia palmaeformis*, may first be glimpsed, often in inaccessible places but standing out in distant silhouettes. This is a relatively small brown alga with its hollow trunk no more than 60 cm tall and holding aloft a head of light yellow streamers, dichotomous from the base. In its mid-eulittoral sites *Postelsia* is the foremost indicator of the highest wave-attack. Scarce on Vancouver Island, it will be seen coming to its maximum on the outer headlands of Oregon and north California.

On the same flat tops with *Postelsia* two fleshy rhodophytes are found, both also indicators of high exposure. *Constantinea simplex* is saucer-shaped, raised on a discoid holdfast, and *Iridaea cordata* forms clusters of thick, cartilaginous blades.

The low level laminarians found at Cape Beale are of strongly wave-resistant species. *Hedophyllum sessile*, *Laminaria groenlandicum* and *Egregia menziesii* have disappeared from the picture, along with the sea-grass *Phyllospadix*. The most prominent species here is *Laminaria setchelli*, with its erect half-metre stipe, arching into a divided blade of the same length. In some sites, too precarious to reach close-up, glimpses of *Lessoniopsis littoralis* have been reported, rising to 2 m from a woody holdfast, and branching dichotomously into mid-ribbed blades. In similar places grows the small, wave-exposed *Alaria nana*, only half a metre long, and with tattered margins when mature.

As each wave draws back, the tiered ranks of laminarians momentarily come into view, flexible and yielding under constant surf-wrenching. Lowest down can sometimes be glimpsed the broomstick stipes of *Pterygophora californica*, far more rigid than the rest. Furthest out, with its fronded bladders buoyant in the white surge, is the long-stiped *Nereocystis luetkeana*.

In shelter

Inside the Bamfield spit, the shore platform can become shaded down to the **littoral fringe** by low-spreading taiga vegetation, notably Sitka spruce. The black *Verrucaria* lichen strip is narrow or lacking, while *Littorina scutulata* and *L. sitkana* are both shortened in range. Right through the **eulittoral** runs *Balanus glandula*, with small brown algae, including *Fucus gardneri*, crusts of *Ralfsia*, and a weft of *Pylaiella* or *Elachista*. *Pollicipes polymerus* and *Mytilus californianus* are both entirely gone, with small *Mytilus trossulus* holding its ground alone.

The **lower eulittoral** is green with *Ulva fenestrata*, and an underlayer of *Cladopohora flexuosa*. A lower band is sometimes formed by purplish black *Cryptosiphonia woodii*. A silty turf of *Corallina vancouveriensis* occupies the **sublittoral fringe**, sometimes enriched into

a small algal garden. Fleshy vesicles of *Colpomenia sinuosa* and *Leathesia difformis* mix with the ovate bladders of *Halosaccion glandiforme*. The small red alga *Neorhodomela larix*, a notable cold-water species, is recognised by its small cylindrical branchlets. Bladders of the brown *Soranthera ulvoidea* grow upon it. Two *Gigartina* species are regularly to be seen, the branching *G. crispata*, and the papillose sheets of *G. exasperata*.

It is at low water on these enclosed shores of Vancouver Island that the wealth of Laminariales disappears. At sheltered places *Fucus* is now important in the middle intertidal in several poorly demarcated species, of which *gardneri* is held to be the most important. In warmer, protected enclaves of the Barkley Sound, at or beyond low water, may be seen thickets of Pacific North America's two sublittoral fucoids. The narrow-leafed *Cystoseira geminata* grows like a *Sargassum*, with terminal branchlets inflated into chains of vesicles. While *Cystoseira* runs north to the Bering Sea, the second fucoid, *Sargassum muticum*, is a low boreal species, cutting off to the north in British Columbia. The slender basal leaves are larger and mid-ribbed, and the ultimate twigs carry spherical vesicles in their axils.

The single laminarian left in shelter is *Macrocystis integrifolia* (Fig. 17.11) growing in small stands outside the Spit entrance. In its main occurrence it is a highly productive brown alga forming forests offshore, beyond the low tidal effects of wave action. This is a smaller-leaved species also with holdfast differences from *M. pyrifera*.

Backwaters

The south arm of the Bamfield inlet reaches up to a brackish tidal basin (Fig. 19.10). The shallow intertidal shows concentric zones of different hues around a rich green swathe of the sea-grass *Zostera marina* var. *latifolia* at low water.

Just before the tree-line, the supralittoral zone begins with a tawny surround (A) of saline grass. Next comes a salt meadow (B) of *Triglochin maritimum* with *Lilaea scilloides*, reaching out to a wider belt (C) of *Salicornia virginica*. Under small boulders or plant litter several gastropods take refuge: the ellobiid *Phytia myosotis*, the prosobranch *Assiminea californica* and the slug *Onchidella borealis*. A little nearer the centre, the stones carry a micro-zonation of the periwinkles *Littorina scutulata* and *L. sitkana*, *Balanus glandula* and dwarfed *Mytilus trossulus*. A dwarfed form of *Fucus gardneri*, loose-lying or attached to stones, forms a dull yellow zone further out. Under the same boulders lives the silt-tolerant crab *Hemigrapsus oregonensis*.

Inside this scatter of boulders, the middle shore (E) is green-tinted with bands of *Enteromorpha* and a paler *Cladophora*, succeeded next inwards by a felt of pur-

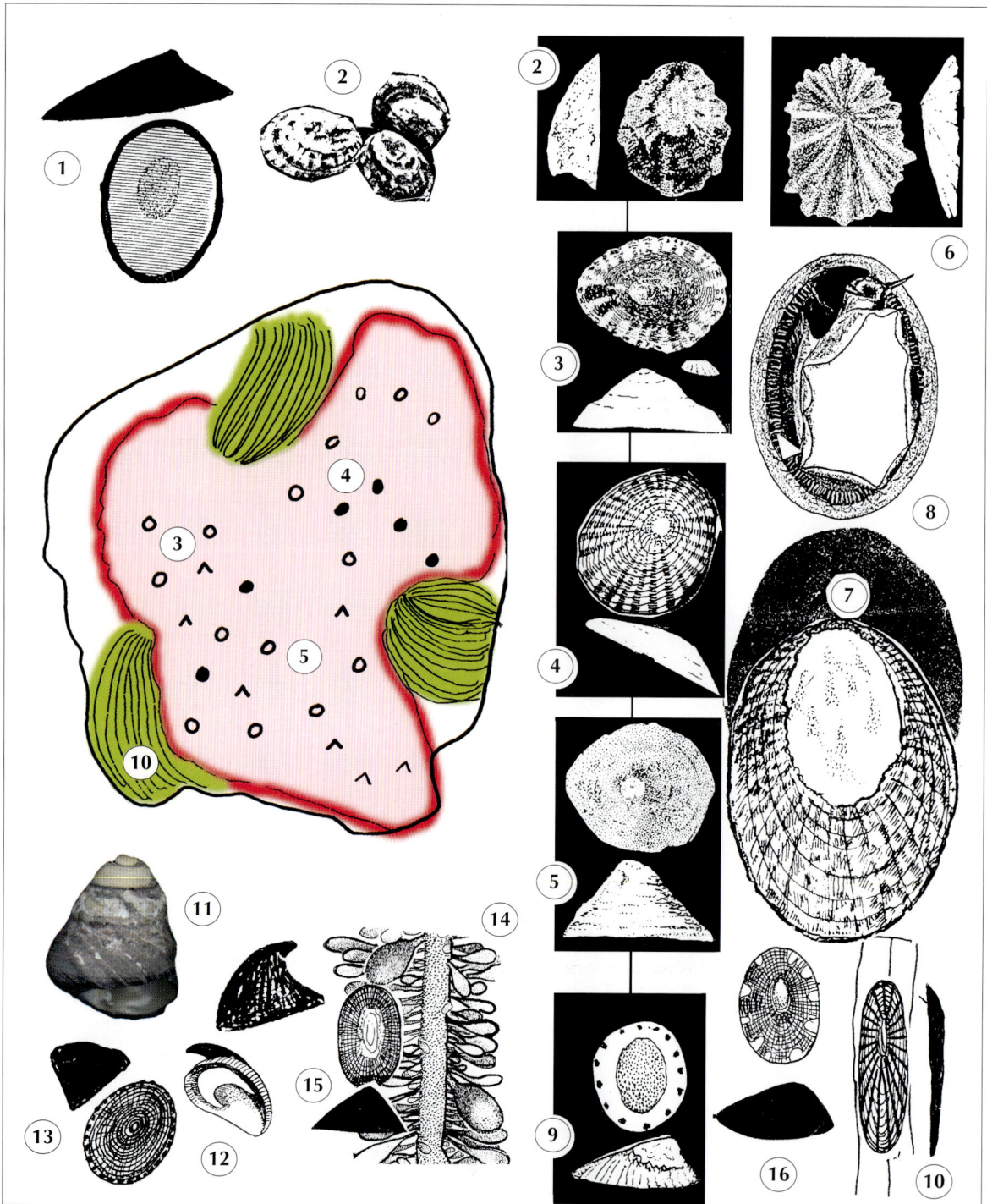

Fig. 19.11 The habitats of Pacific shore limpets
The species are shown according to shore location, and to range between Vancouver Island and California.
1 *Tectura persona* 2 *Lottia digitalis* 3 *Tectura scutum* 4 *Lottia pelta* 5 *Acmaea mitra* 6 *Collisella scabra* 7 *Lottia gigantea*, with homing scar part revealed 8 *Lottia gigantea*, animal from below 9 *Tectura rosacea* 10 *Tectura depicta*, on *Phyllospadix* leaf 11 *Tegula funebralis*, carrying 12 *Crepidula adunca* and 13 *Lottia asmi* 14 *Egregia menziesii* carrying 15 *Lottia incessa* 16 *Lottia instabilis*, from *Pterygophora californica*.
(left) sun-lighted eulittoral pool carrying *Phyllospadix* tufts, and on the pink-painted algal floor, three limpet species, 3, 4, 5.

plish brown *Gracilaria* (F).

The sward of sea-grass (G) stays just submerged at low water. *Zostera marina* var. *latifolia* is a large grass, with leaves a metre long and 2–3 cm wide. It harbours a

notably rich fauna, both in the sediments round its rhizomes and on the leaves. Scaled down to the width of a *Zostera* leaf, the miniature sea-hare *Phyllaplysia taylori* is concealed by its green venation as bright as the plant and reddish *Aeolidia papillosa* on the anemone.

Several crustaceans are fashioned for life on eel-grass: the narrow, close-clinging isopods *Idothea rese-cata* and *Idothea aculeata*, and green aesop prawns, *Hippolyte*, found swimming among the leaves.

Shore invertebrates
Molluscan grazers
The pre-eminence of the Pacific boreal in brown algae extends also to grazers and browsers among molluscs, most notably chitons and acmaeid limpets. Where eastern Canada can show just one intertidal limpet, the

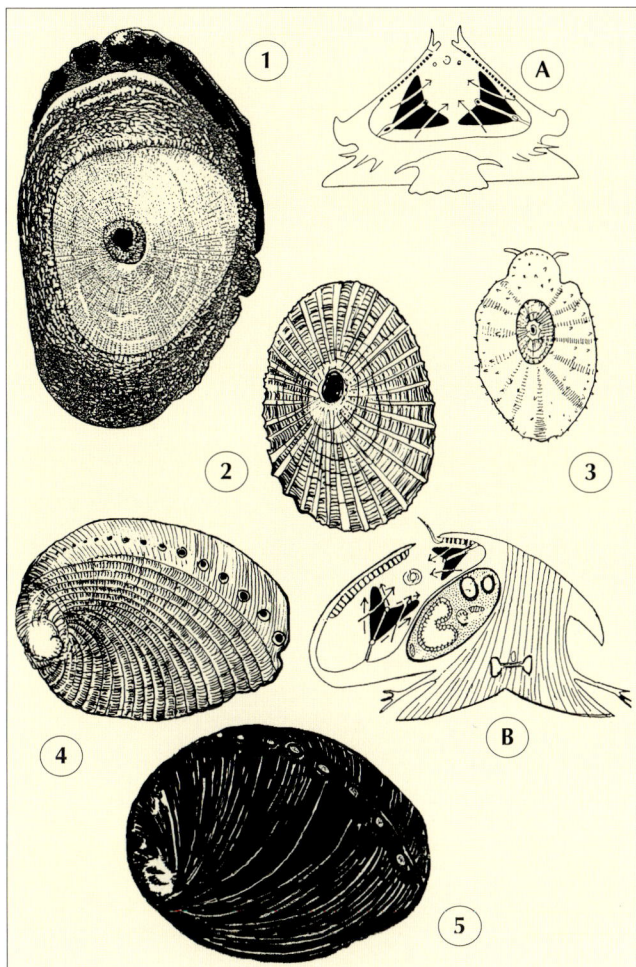

Fig. 19.12 The shore ranges of keyhole limpets and abalone

1 *Megathura crenulata* 2 *Diadora aspera* 3 *Fissurellidea bimaculata* 4 *Haliotis fulgens* 5 *Haliotis cracherodi*.
The cross-sections of a **fissurellid** (A) and a **haliotid** (B) show the mantle cavity with paired, bipectinate ctenidia.

it grazes on, though its pale spawn coils are much more easily detectable. The prosobranch snail *Lacuna variegata* also grazes on *Zostera*, being itself predated by the small five-rayed star *Leptasterias hexactis*.

Several more opisthobranchs are common among *Zostera*, including two cephalaspids, the bubble snail, *Haminoea virescens*, and the slug-like carnivore *Melanochlamys diomedea*. The largest and most outstanding is *Melibe leonina*, living in quiet water above and around the sea-grass. This is a soft, translucent slug, drifting in quiet water or breaking into leisured swimming with its side rows of oval paddles. Over the head is a spacious cowl fringed with sensitive tentacles, in which small crustaceans are snared.

The red to greenish 'dividing anemone', *Epiactis prolifera*, attaches to *Zostera* leaves, as does also the small rhizostome jellyfish *Haliclystus auricula*. The hydroid medusa *Gonionemus vertens* alights on the plant more briefly. Two aeoliid slugs graze upon cnidarians, yellowish green *Hermissenda crassicornis* on hydroid colonies,

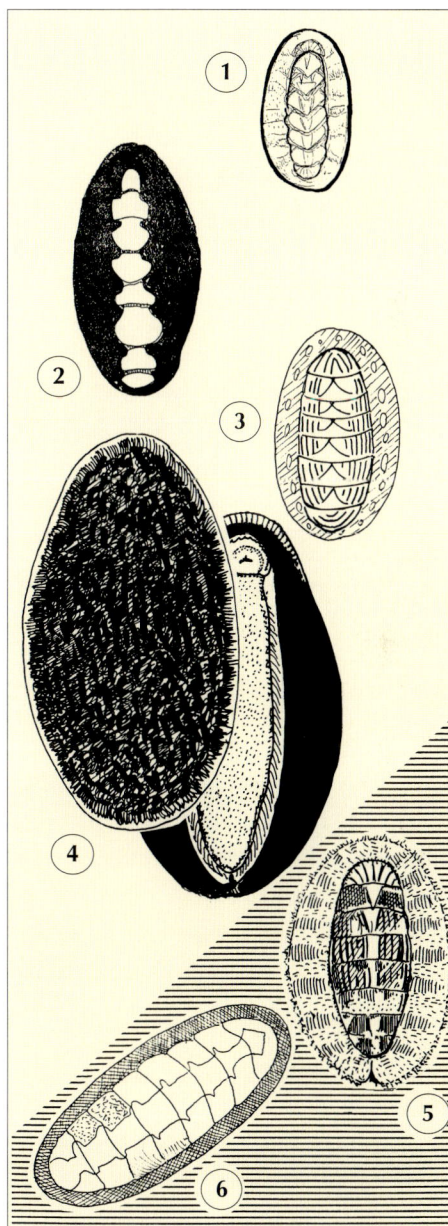

Fig. 19.13 Some chitons of the North American Pacific
(On the open surface) 1 *Nuttallina californica* (from Puget Sound south) 2 *Katharina tunicata* 3 *Tonicella lineata* and 4 *Cryptochiton stelleri* (in shade and in caves) 5 *Mopalia ciliata* (under boulders) 6 *Ischnochiton magdalenensis*.

Pacific has 17. The whole American Atlantic coast has half a dozen chitons, while the Pacific coast has more than a hundred. In Atlantic Canada's high boreal to sub-arctic Bay of Fundy, four littorines occur on the upper shore; two of these — with the lack of trochids, turbinids or patellids — extend down to pre-empt the whole eulittoral. The rest of the Fundy gastropods between tides comprise one thaid and one buccinid whelk, and several small herbivores, making up a species total of only 10.[1]

With the acmaeid limpets so rich, the American Pacific coast is devoid of Patellidae, while *Siphonaria thersites* is the single pulmonate limpet north of Point Conception. Highest-reaching in the littoral fringe (recalling *Notoacmea pileopsis* in New Zealand) is *Tectura persona*, ranging from Alaska to Vancouver Island. Next down, abundant in the fringe and entering the barnacle zone, comes *Lottia digitalis*. These are succeeded in the eulittoral by *Lottia pelta*, *Tectura scutum* and the coralline-painted *Acmaea mitra*. In California, the speckled owl limpet, *Lottia gigantea*, as big as a patellid, enters the picture, clustering upon mid-eulittoral rock or attaching to mussels. A little lower down comes *L. limatula*, and then lowest the small, pink-encrusted *Tectura rosacea*, moving about freely on coralline surfaces. *Tectura fenestrata* is restricted to smooth boulders partly sand-embedded.

Several smaller acmaeids have taken on special habitats. The boat-shaped *Lottia instabilis* rocks back and forward in its scar on an algal stipe, usually *Pterygophora*. *Discurria incessa* forms a scar in *Egregia*, while the narrow linear *Tectura paleacea* (in California) lives upon *Phyllospadix* leaves. *Lottia asmi* is a lustreless black limpet, often found on the topshell *Tegula funebralis*, while the small slipper limpet, *Crepidula adunca*, with hooked spire, sits upon the same snail.

Several keyhole limpets (**Fissurellidae**) are prominent between tides. The ribbed and high-built *Diadora aspera* lives under stones, and *Fissurellidea bimaculata* beneath overhangs. From Monterey south will be added the 2.5 cm long *Fissurella volcano*, and the giant *Megathura crenulata*, growing to 115 mm, with a broad yellow foot and a black mantle enveloping the shell.

Large ormers (abalones) — though over-exploited — were traditionally present in force in the Californian sublittoral fringe. The black abalone *Haliotis cracherodi*; the red *H. rufescens*; the green *H. fulgens;* and the corrugated *H. corrugata*. *Haliotis kamchatchana* reaches to Vancouver Island and the cold north.

There are Pacific trochids typical of both warm and cold waters. The *Calliostoma* snails are northern-centred browsers in the coralline zone. The chestnut and blue-tinted *C. ligatum*, the light brown, channelled *C. canaliculatum* and the elegant *C. annulatum*, with the

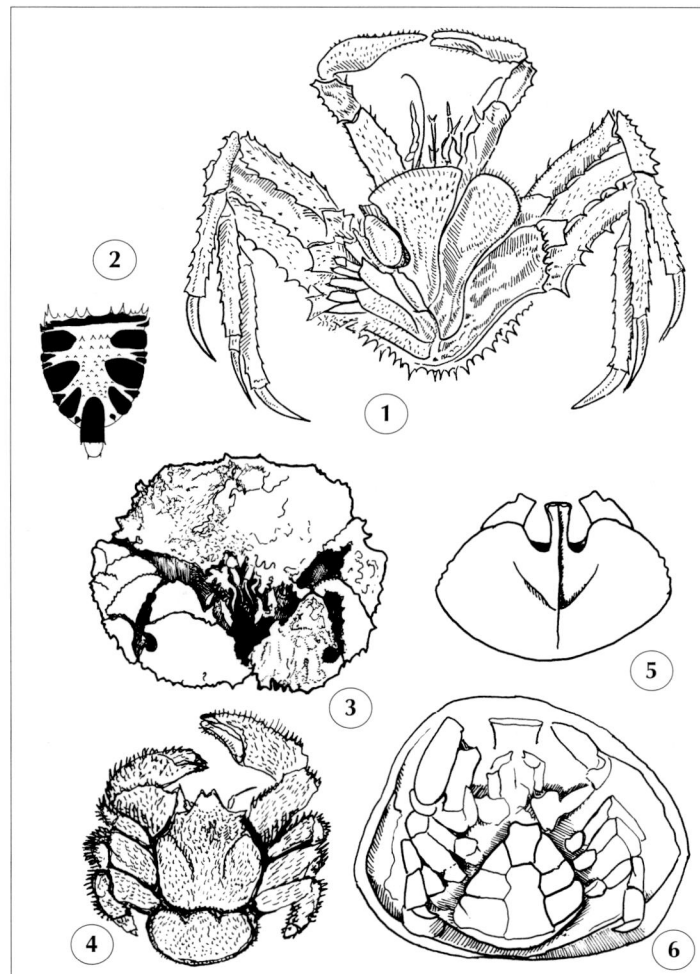

Fig. 19.14 Stone crabs: Lithodidae
1 *Lithodes maia*, ventral view 2 *Lithodes*, undersurface of abdomen 3 *Lopholithodes foraminatus* 4 *Hapalogaster cavicauda* 5 *Cryptolithodes sitchensis* 6 the same, ventral view.

animal salmon pink, all range from Alaska to San Diego. *Margarites* and its relatives are small boreal trochids browsing on laminarians. *Margarites pupillus* ranges from the Bering Sea to San Pedro, and *Lirularia lirulata* from Alaska to San Diego.

The common littoral *Tegula* species are more southern-centred. *Tegula pulligo* alone reaches Alaska. The dull black *T. funebralis* is common north to Vancouver Island, with *T. brunnea*, *T. montereyi*, *T. eiseni*, *T. aureotincta* and *T. ligulata* all Californian.

In the Pacific boreal, chitons vie in prominence with the limpets. The highest-reaching on Vancouver Island is *Katharina tunicata*, living among barnacles and mussels in the eulittoral, and known by its black, leathery girdle. From Puget Sound south, the top sites are held by *Nuttallina californica*, brown with white streaks, and the girdle spiny or scaly. Resting scars are excavated in the surf-swept rock. On the open surface lower down lives the attractively marked *Tonicella lineata*. At low water and just beyond is found the 'gum-boot' *Cryptochiton stelleri*, the largest of the world's chitons, up to 25 cm long, with the valves buried in the black to dark red girdle.

Shade-loving on low tidal faces or the ceilings of caves are the chitons of the Family **Mopaliidae**, with a wide bristled girdle. In *Mopalia muscosa* this carries stiff, curled hairs and the valves are often coralline encrusted. *M. ciliata* is lightly haired and may be partly sand-buried, while *M. lignosa* lives under boulders in sheltered bays. To the same family belongs *Placiphorella velata*, uniquely — for a chiton — carnivorous, and found on boulders near sand. The almost circular girdle is raised in front like the peak of a cap.

Confined under boulders often lying on sand are the long, narrow *Ischnochiton* species, highly mobile and fast retreating when exposed to the light. The largest species are Californian, including the olive-mottled *I. magdalenensis* (75 mm) and *I. conspicuus* (100 mm). *I. mertensii* (up to 35 mm), orange to brown and white-blotched, ranges from Alaska to California.

Crabs

The cold-water Family **Cancridae** is strongly represented on Vancouver Island, on low or subtidal shores. Ranging right to the Aleutians are *Cancer oregonensis* (with its southern limit in California) and *C. magister*, and as far as Alaska, *C. productus* and *C. gibberulus*. The deeper water *C. antennarius* is north-limited to British Columbia.

The swimming crabs (**Portunidae**) are thinly represented in Pacific America, with only *Portunus xantusii* ranging from southern California to Chile.

The **Grapsidae** are, in contrast with the cancrids, warm temperate and tropical crabs. On the enclosed shores of Vancouver Island there is the silt-tolerant *Hemigrapsus oregonensis*. The cleaner shore *Hemigrapsus nudus*, unusually for a grapsid, has a cold range north to Sitka, in Alaska. It is as belligerent as the rock crab *Pachygrapsus crassipes*, that dominates in warmer California, as the ecologic equivalent of the Australasian *Leptograpsus variegatus*.

The Family **Xanthidae** are also warm-centred, with *Lophopanopeus bellus* alone reaching north to Puget Sound. *Xanthias taylori* begins at Monterey and *Pilumnus spinohirsutus* ranges from Santa Monica to Ecuador.

The spider crabs (Family **Inachidae**) have important cold-water species. The slender-limbed *Oregonia gracilis* and the lyre crab, *Hyas lyratus*, reach north to the Bering Sea, with *Mimulus foliaceus* and *Scyra acutifrons* in Alaska. The large kelp crab *Pugettia producta* ranges from Alaska to lower California with *P. nutalli* found south of Point Conception. The smaller *P. gracilis*, living on seaweeds or *Zostera*, reaches north to Alaska.

The peculiar stone crabs, **Lithodidae**, are highly typical of the cold Pacific. They extend round the high boreal to Siberia, with some offshore species reaching southern New Zealand. Classed with the 'mixed bag' of the Anomura, the lithodids are superficially like true crabs, but details betray a hermit crab origin, probably from paguroids. The abdomen, though applied closely to the thorax, is fleshy and markedly asymmetrical, especially in the female with pleopods only on the left. With its original sclerites lost, new plates have been developed by the fusion of small spines and warts into central, lateral and marginal pieces. The Pacific coast stone crabs, ranging from Alaska to Monterey, include the spinose *Hapalogaster mertensi* and *H. cavicauda*, the smooth, convex *Oedignathus inermis* and *Cryptolithodes typicus* and *C. sitchensis*, both with the carapace winged at the sides to cover the walking legs. In the box crab, *Lopholithodes foraminatus*, the legs and chelae fold tightly against the carapace, leaving two large notches for the respiratory current.

Seastars: Asteroidea

The Pacific coast has an outstanding diversity of seastars. Of the three orders of the Class **Asteroidea**, the northern boreal could be seen as geographic centre for the **Forcipulata** and **Spinulosida**, while the **Phanerozonia** are tropical with their main American centre in southern California.

The seastar's calcified plates strictly form a sub-epidermal endoskeleton. In some stars the plates bear jointed spines moved by their own muscles, while defence is generally afforded by the minute stalked pincers called *pedicellariae*.

In the Order **Forcipulata**, the pedicellariae are principally clustered around the spines in rings. With the stimulus of an alighting organism these rings move up to ensleeve the spines, and the pedicellariae bend over to the site of disturbance. Any small animal that ventures over the surface finds itself seized, and held for days, if necessary, until death and decay take place.

Stalked pedicellariae with toothed jaws like pliers are found only in the Forcipulata. The rest of the asteroids have larger sessile pedicellariae scattered singly between the spines, with long jaws like alligator clamps.

The Asteroidea show a broad range of feeding styles. Most forcipulates live as predators on bivalves, barnacles or other shelled prey. Some of the Spinulosida are herbivores, like *Asterina*, scouring the surface algal film. The Phanerozonia include a number of forms, such as the comb-stars *Astropecten*, that swallow small bivalves and gastropods alive. Lacking the rasping jaws found in ophiuroids, comb-stars take in food by cilia. The anus is absent, so the mouth also extrudes waste.

The Pacific boreal is richest in **Forcipulata**. By far the

Fig. 19.15 Asteroids of the American Pacific Coast
Forcipulata 1 *Pisaster ochraceus* 2 *Evasterias troschelii*
3 *Leptasterias hexactis* 4 *Pycnopodia helianthoides*
Spinulosida 5 *Henricia leviuscula* 6 *Solaster stimpsoni*
7 *Solaster dawsoni* 8 *Crossaster papposus* 9 *Asterina
miniata* **Phanerozonia** 10 *Mediaster aequalis*
11 *Dermasterias imbricata* 12 *Linckia columbiae*
13 *Luidia foliolata*.

best known seastar onshore is *Pisaster ochraceus*, a northern species ranging from San Diego to Alaska, while *P. giganteus* runs from Vancouver to south of Point Conception. In Puget Sound and Vancouver Island, *Evasterias troschelii* is found on wharf piles around low tide, feeding on barnacles, mussels and other bivalves. On intertidal shores from Monterey to Sitka *Leptasterias hexactis* takes small gastropods. *Orthasterias koehleri*, from Alaska to Puget Sound, feeds on small crustaceans and bivalves.

The giant sunflower star, *Pycnopodia helianthoides*, grows to a metre across, and is found subtidally from California to Alaska feeding on small bivalves, urchins

and other asteroids. Common in California is the small, red-brown mottled 'soft star', *Astrometis sertulifera*. Subtidal stars on Vancouver I. include *Stylasterias forreri* feeding on clams, mussels and sand dollars, and *Hippasterias spinosa*, a predator on sea-pens (*Ptilosarchus*) and the bearded anemone (*Metridium senile*).

The stars of the Order **Spinulosida** are not sharp-spined but covered with a mosaic of rounded pin-heads. The Solasteridae or sunstars are many-rayed, with long arms like the forciculate *Pyncnopodia*, but distinguished by their shark-skin of flat-topped (*paxilliform*) spinelets. *Solaster stimpsoni* feeds beyond low water upon *Stichopus* cucumbers, and is itself predated by *S. dawsoni*. Both reach north to the Aleutian Islands.

Contrasting with these long-armed species are the cushion stars (Asterinidae), spinulosids with a thick disc and short triangular rays. The reddish or purple *Asterina miniata* is one of the commonest stars on rocks and sand from California to Alaska. *Dermasterias imbricata* (Family Asteropidae) is covered by a thick, soft membrane and feeds on the star *Mediaster*, crustaceans, holothurians and bivalves. The long-armed Echinasteriidae include the reddish five-rayed star *Henricia leviuscula*, found among stones around low tide and collecting particles by ciliary means.

The third order of asteroids, the **Phanerozonia**, consists mostly of stiff, inflexible stars, kept rigid by the aboral skeleton of flat, close-set paxillae. The arms carry large marginal plates. In the first suborder, **Paxillosa**, the tube-feet are suckerless and pointed, virtually rowing the animal in semi-fluid sand, as with the five-armed *Astropecten* and the 11-armed *Luidia*. In the second suborder, **Valvata**, the tube-feet bear suckers. The Goniasteridae includes the flat biscuit stars, such as *Ceramaster arcticus*, continuing to the cold north and *C. leptoceramus*, from San Diego to Point Conception. The vermilion *Mediaster aequalis* (California to Alaska) has the arms more pointed, and feeds on sponges, bryozoans and sea-pens. The Family Ophidiasteridae gives a foretaste of the tropics with the dull red or grey *Linckia columbiae*, found from La Jolla south.

Sea cucumbers: Holothuroidea

As well as the common cucumber *Stichopus californianus*, of the Order **Aspidochirota**, the Pacific coast has numerous **Dendrochirota**. These possess branching oral podia, and they generally become more closely affixed by the undersurface or 'sole'. *Cucumaria miniata*, ranging north to Sitka, fastens under rocks or in crevices. Rather than deposit-feeding like *Stichopus*, they keep the tentacles in motion to entrap suspended particles. A whole colony has a delicate beauty, with its underwater field of coral red tentacles.

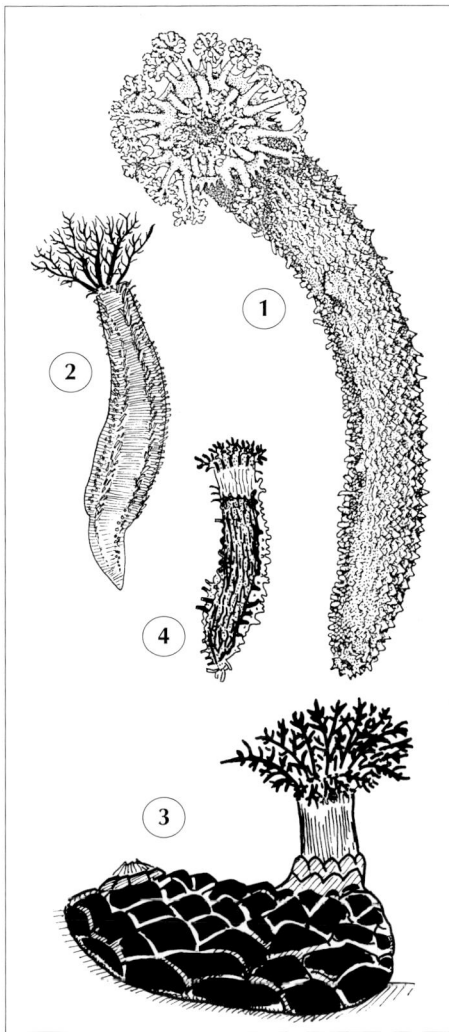

Fig. 19.16 Holothurians
1 *Stichopus californianus* 2 *Cucumaria miniata*
3 *Cucumaria vegae* 4 *Psolus chitonoides.*

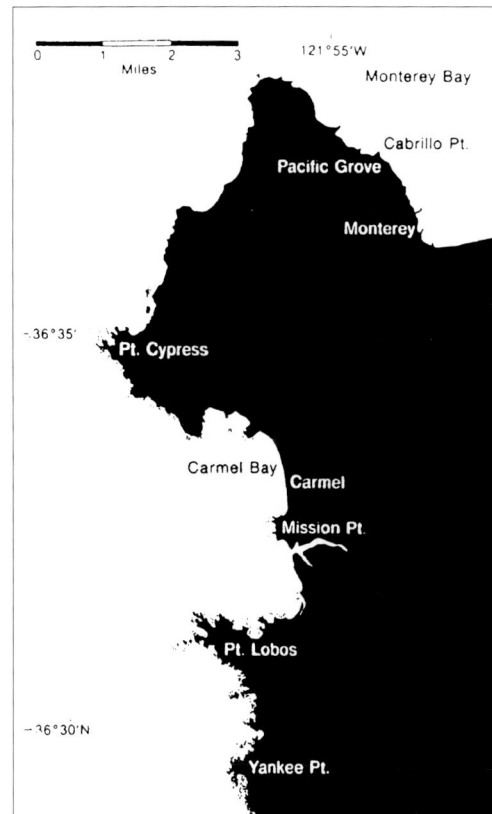

Fig. 19.17 A map of the Monterey Peninsula, California, and the coast to the south.

Smaller cucumarians include the black *Cucumaria vegae*, fixing down in crevices, and the 'pepper and salt' *C. piperata* with *Pseudocnus lubricus*, in kelp holdfasts. *Corallina* turf can sometimes be found speckled by centimetre-long *C. curata*, looking like bits of tar. *Psolus chitonoides* has become attached like a limpet, using the lower surface as a sole, with the convex upper side shelled with a mosaic of plates. The mouth faces up, with its tentacle circlet extended for suspension feeding.

California: Pacific Grove
For their last study of a Pacific shore, in 1947, Alan and Ann Stephenson chose two parts of the Californian coast. **Pacific Grove**, within the cool temperate Oregonian Province, was worked upon from Stanford University's Hopkins Marine Station. They then visited warm temperate **La Jolla**, in southern California, the site of the Scripps Institution of Oceanography.

Pacific Grove was found to belong to a different

world from the Vancouver shores around Nanaimo.[2]

Instead of quiet borders of inlets and straits, many below wooded slopes and far from the ocean, we now have a fully exposed coast of bays and headlands. On the most prominent of these headlands the wave action is probably as powerful as anywhere in temperate latitudes. Mountainous waves surge in from the Pacific, often terrifying in aspect when seen at close quarters, and especially impressive when they come tearing in from a fog, as they so often do on this coast. When they explode on the rocks and cliffs in cascades of foam, sending solid spouts of spray into the air, they create a turmoil of broken water, sometimes white throughout the whole of a small bay.

With the major headlands separated by small bays, the shores grade from steep cliffs to gentler slopes in some parts protected from waves by outlying reefs. The complex of promontories and islets is the domain of pelicans, cormorants, gulls and sea-lions. To landward, the tree-line recedes skyward where spray drenches the rock below. Curtains of laminarians drape from the lowest faces, with a close mosaic of algae up to mid-tide.

Monterey: shore zonation
Though the whole Pacific Grove region has an essentially open coast, there are differences in exposure

Fig. 19.18 High exposure zonation on steep slopes at Pacific Grove

(**left**) Zonation column in extreme exposure at Devil's Cauldron, Point Lobos.

a–b bare rock with littorines; b–c *Balanus*, joined by (**centre**) *Endocladia* and (**bottom**) *Tetraclita;* c–d corallines and *Lithothamnia;* d–e *Lessoniopsis littoralis*

(**centre**) Zonation in sub-maximal exposure at Yankee Point

a–b *Balanus* and *Pelvetiopsis;* b–c *Mytilus* and *Pollicipes;* c–d *Postelsia;* d–e corallines and *Lithothamnia;* e–f *Alaria marginata;* f–g rhodophycean undercover

1 *Pelvetiopsis limitata* 2 *Endocladia muricata* 3 *Lessoniopsis littoralis* 4 *Balanus glandula* 5 *Pollicipes polymerus* 6 *Mytilus californianus* 7 *Postelsia palmaeformis* 8 *Tetraclita squamosa rubescens* 9 *Pelvetia fastigiata* 10 *Megabalanus tintinnabulum* 11 *Pisaster ochraceus.*

to be easily picked out. Thus we see that the population of a wave-swept promontory at **Point Lobos** is very unlike that of the relatively sheltered inlet at **Cabrillo Point**. Great variety of intermediate conditions is to be seen, but it is Logos and Cabrillo that were chiefly studied by the Stephensons and have provided the data for our figures of each shore. Each affords differences from the exposed seaboard of Vancouver Island, and still more from the warm temperate south Californian shore to be seen at La Jolla.

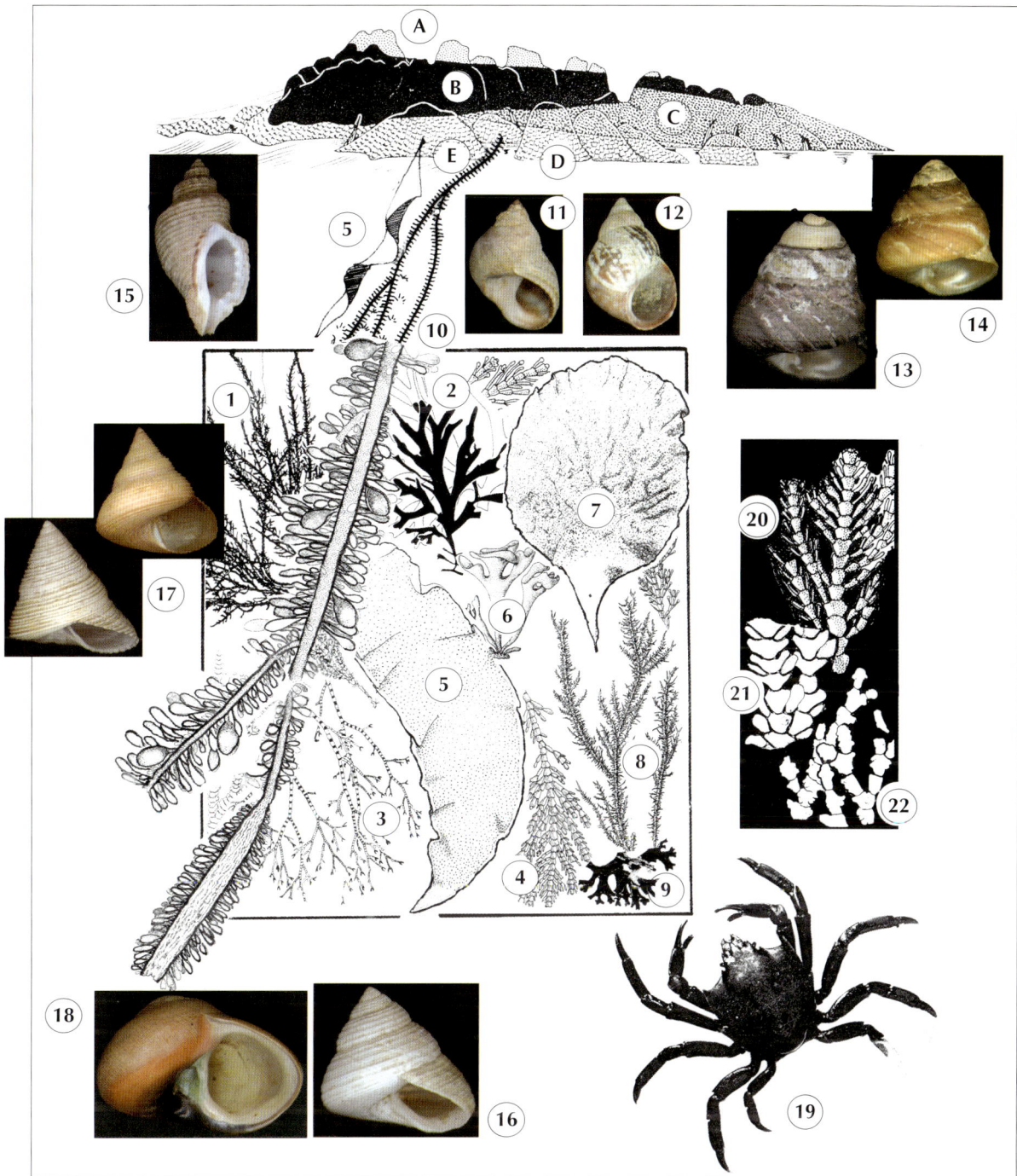

Fig. 19.19 Moderate exposure to shelter at Pacific Grove
Low-sloped rocky reef at Cabrillo Point

(**top**) Sequence from strong wave action (**left**) to shelter (**right**); (from the Stephensons)

A. upper eulittoral, with barnacles B. mussels and *Mitella* C. brown turf D. red turf E. *Egregia* and other brown algae.

(**centre**) Characteristic algal mixture of red turf

1 *Gelidium coulteri* 2 *Rhodymenia pacifica* 3 *Ceramium eatonianum* 4 *Corallina officinalis* var. *chilensis* 5 *Iridaea flaccida* 6 *Iridaea cornucopiae* 7 *Gigartina corymbosa* 8 *Gigartina leptorhynchos* 9 *Gigartina papillata* 10 *Egregia menziesii* 11 *Littorina keenae* 12 *Littorina scutulata* 13 *Tegula funebralis* 14 *Tegula brunnea* 15 *Acanthina spirata* 16 *Margarites pupillus* 17 *Calliostoma* sp. 18 *Norrisia norrisi* 19 *Pugettia producta*

(**right**) important corallines 20 *Corallina officinalis* var. *chilensis* 21 *Bossiella orbigniana* 22 *Calliarthron tuberculosum*.

From the **littoral fringe** the periwinkle *Littorina scutulata* extends down to the upper eulittoral. The cold-water *L. sitkana* has been replaced as the top level species with *L. keenae*, ranging south from Puget Sound. The two high level limpets *Lottia digitalis* and *Collisella* *scabra* (Fig. 19.11) reach up into the littoral fringe.

The **eulittoral zone** is primarily divided into an upper half with barnacles and mussels and a lower stretch clad by a dense algal turf. Under maximal exposure, this turf is replaced by a veneer of *Lithophyllum* with tufts of

371

jointed corallines. Standing out as a rich brown band at a distance, the turf has a predominance of small red algae. On the most wave-beaten promontories, around mean sea level, groves of the palm-weed, *Postelsia palmaeformis*, stand up in high profile.

In the **upper eulittoral** — as in British Columbia — the dominant barnacle is *Balanus glandula*, but two further species give a foretaste of the warm temperate shores we shall meet further south. The small *Chthamalus fissus* lives among *Balanus* but does not extend so high, while further down there is a sub-zone of the much larger, pinkish *Tetraclita squamosa rubescens*. The large, cold-water *Semibalanus cariosus* is occasional but never abundant. The common trochid is *Tegula funebralis*, and the open rock chiton at this level is *Lepidochitona hartwegi*. The upper eulittoral thaid whelks are *Acanthinucella punctulata* and *Nucella emarginata*.

Under moderate to high exposure the upper half of the eulittoral has close-packed beds of mussels and the stalked barnacle *Pollicipes polymerus* below its topmost band of *Balanus* and *Chthamulus*. *Mytilus californianus*, with its liking for high but not maximal exposure, is the favoured mussel, and *Mytilus edulis* will generally be looked for in vain. In strong exposure, *Pollicipes* forms white enclaves among the mussels, but — while overlapping — these two have distinct preferences. The roughest sites have no really large mussel beds, only stragglers. *Pollicipes* reaches higher than the mussel in strong surge and along narrow runnels, but is reduced under even moderate shelter. Far down in the sublittoral fringe we shall find the large barnacle *Balanus nubilus*, singly or in clusters. and reaching 5 cm across the base.

The small fucoid alga *Pelvetiopsis limitata* achieves a scatter among the topmost barnacles. A much larger fucoid, *Pelvetia fastigiata*, occurs in California in the lower eulittoral among the rhodophytan turf. With its slender dichotomous branches reaching a metre long, it attains an important status. Small *Fucus evanescens* sometimes grows alongside it.

Rocky shores of only moderate exposure as at **Cabrillo Point** were illustrated by the Stephensons with a diagram showing the relation on rather gently sloped shores between plant turf and animals and the amount of wave action (Fig. 19.19).

Below the cut-off of the mussels, and in less than maximal exposure, the turf displays a rich assortment of small to middle-sized red algae, generally with some green highlights of *Cladophora* and *Ulva*. *Gigartina canaliculata*, *G. leptorhynchos*, *Mastocarpus jardinii*, *M. papillatus*, *Endocladia muricata*, *Rhodoglossum affine* and *Porphyra purpurea* are common. The Stephensons provided a further list of important rhodophytes: *Ceramium eatonianum*, *Chondria decipiens*, *Cryptopleura lobulifera*, *Gastroclonium subarticu-latum*, *Gelidium coulteri*, *G. sinicola*, *Microcladia borealis*, *Pikea californica*, *Neorhodomela larix*, and short corallines, including *Haliptilon roseum*.

As exposure increases, some larger rhodophytes ultimately overshade the turf. The cold-water *Iridaea* species (*I. cordata*, *I. cordata* var. *splendens* and smaller numbers of *I. coriacea*) stand out as dull red to olive-green blades up to half a metre long. As *Iridaea* comes to command the lower eulittoral, *Endocladia muricata*, with its preference for shelter, is displaced upwards to abut with *Pelvetiopsis limitata* among the barnacles. Under the highest wave action the turf disappears altogether, to give place to a coralline veneer, or the brown algal curtain may reach right up into the middle of the eulittoral.

The **lower eulittoral** is a rich haunt of gastropods and chitons. *Tegula funebralis* lives here in vast numbers, often with the slipper limpet, *Crepidula adunca*, and the small *Lottia asmi* attached. A second topshell, the brown *Tegula brunnea*, is usually present too. The principal chitons are *Tonicella lineata*, *Lepidochitona hartwegi*, *Mopalia muscosa* and *Katharina tunicata*. *Nucella emarginata* descends to this level, and *Nucella lamellosa* may also be found.

The small sea-palm, *Postelsia palmaeformis*, found sparingly at outlying points on Vancouver Island (Fig. 19.8), comes to its maximum on the wave-beaten headlands of Oregon and north California. Confronting the breaking waves, *Postelsia* reaches higher on the shore than any other laminarian. In its main footholds on level outcrops or shelves, it grows like a small tree up to 60 cm high, sturdy but flexible against constant wave-lashing. *Postelsia* is a summer annual, destined to darken and disappear in autumn and return the next spring.

In the high-exposed sites of extreme exposure at Devil's Cauldron and the most exposed parts of Point Lobos and Yankee Point, the laminarian algae of the **sublittoral fringe** are far less diverse than on Vancouver Island. Several species, however, contribute to a strong curtain, tough and wave resistant as it is wrenched and lashed against the pink underlying rock face. In ultimate exposure, *Lessoniopsis littoralis* may hold the outermost promontories almost alone. The holdfast is woody and deeply gnarled, and carries dichotomous dark brown branches up to 2 m long. These are narrowly linear and like whip flails, up to 500 to a large plant, with wavy margin and flat mid-rib. Sporophyll branches are shorter and smooth, arising singly just below each dichotomy.

Receding from maximal exposure to places still rough enough for *Postelsia*, the sublittoral zone and its fringe are dominated by long-attenuate laminarians momentarily slack, then lashing and swirling in the turmoil of the surge. The fringed ribbons of *Egregia menziesi* form branching straps up to 4–5 m long. The smooth, mid-ribbed blades of *Alaria marginata* can reach almost the same length. Towards shelter *Alaria* cuts out before *Egregia* which continues into the mixed algal turfs that become enriched with reduced exposure.

Giant blades of *Costaria costata*, up to 3 m long and with five longitudinal ribs, can be found in comparative shelter, though scarcer here than in the cold north. The common *Laminaria* at Pacific Grove is *L. setchelli*, erect on its long stipe, with its blade dividing into straps a metre and a half long. The long, undulant blades of *Laminaria farlowii* can accompany it.

From the high cliff-tops both *Nereocystis luetkeana* and *Macrocystis pyrifera* can be recognised where their fringes break the water beyond the shore, even in such high-exposed sites as wave-beaten Point Lobos. The second of the two bladder-kelps, *Macrocystis integrifolia*, grows further into shelter, as we found it near the quiet Bamfield Inlet, joined at Pacific Grove by the longer *M. pyrifera*. Familiar off the cold westerly shores in the southern hemisphere, this species has here found its way north into the comparable and hospitable coastline of California.

The bright green shocks of the sea-grass *Phyllospadix scouleri* are common in the sublittoral fringe under reduced exposure. The sargassoid *Cystoseira osmundacea* is also found here, with its basal fronds simple or dissected like oak-leaves and the terminal branches constricted into rows of vesicles.

The vertical ranges of the limpets at Pacific Grove are shown in Fig. 19.11. As the Stephensons noted, the Family Acmeidae is here in full force. Most of its seventeen Californian species are now placed in *Lottia*. Nowhere on this coast are there any Patellidae, but the large speckled owl limpet, *Lottia gigantea*, reaches full patellid size at up to 90 mm.

The **Haliotidae**, with the ormers or abalones, traditionally reached high concentrations in California. Commonly visible intertidally in the 1950s, they were even then succumbing to over-exploiting and had been banned for out-of-state export. In 1947 *Haliotis cracherodi* was still reported as astonishingly abundant inshore. Singly on open rock or clustering in wave-washed crevices, it could reach even into the upper eulittoral.

Pacific Grove is memorable for its prodigality of urchins onshore. The smaller and more numerous species is *Strongylocentrotus purpuratus*, abrading its scars in the pink coralline crust or massing in pavements of thousands of individuals in pools. This urchin grows up to 50 mm across, and has short, fluted spines. Young specimens are greenish. More than twice this size, with the test up to 125 mm, the longer-spined *S. franciscanus*

generally lives in deep pools, in its two colour phases, red-brown and dark purple. *S. purpuratus* — along with the anemone *Anthopleura elegantissima* — may in places carpet the rock right down into the shallow subtidal. The larger *S. franciscanus* is deep purple or scarlet-spined, sometimes up to 20 cm across.

The smooth-backed, olive-green spider crab *Pugettia producta* clings and moves about among *Egregia* and other kelps. Both its carapace spines and the sharp, prehensile tarsi can puncture the skin. As Ricketts and Calvin advise, 'a kelp crab large enough to wrap itself around a bare forearm had best be left to follow the normal course of its life'.

From the mussel belt and reaching downshore, the seastar predator *Pisaster ochraceus* is also immensely common at Pacific Grove. Often urchin-mingled, its colours show up a patterned chintz, gold, orange, brown, lilac, dull purple and reddish.

Warm temperate California: San Diegan Province

The biogeographical divide between the cool temperate *Pacific Boreal Region* and the warm temperate *Californian Region*, has been placed at Point Conception, in southern California, at latitude 34 degrees 40 minutes north. Beyond this cape, the coastline swings east, with the benefit of a relatively warm current and virtual protection from the onshore north-westerly winds. North of the Point, the low boreal *Oregonian Province* thus takes in most of California, including Monterey, as well as Oregon, Washington and British Columbia.

Within the warm *Californian Region* we can recognise a clear division into the *San Diegan Province*, taking in the outer Pacific coastline, and the *Cortes Province*, comprising the land-locked Gulf of California which was the setting for John Steinbeck and Ed Ricketts' classic of travel and research, *The Sea of Cortes* (1941).

With that long, enclosed Sea we shall not deal here, observing only that it opens far south, to become confluent with the *Eastern Pacific Tropical Region* to which its fauna most closely relates. The Sea has thus rather few species in common with the *San Diegan Province*, the Cortes endemics being derived from tropical forms to the south, and most of its fauna composed of wide-ranging eurythermal tropical species.

The northern boundary for a warm province at Point Conception accords with the evidence from fish distribution and the known ranges of brachyuran crabs, littoral balanoids, molluscs and bryozoans. Briggs (1974) would regard the *San Diegan Province* as a zone of mixing. As well as its possessing some 300 endemic species,

Fig. 19.20 Moderately exposed shore at La Jolla, South California

(**top**) prospect of the zoned shore, after Stephenson

A. *Chthamalus* with limpets B. turf C. *Lithothamnion* D. *Phyllospadix* E. *Halidrys*

a–b *Lottia digitalis*; b–c *Chthamalus* and *Lottia digitalis*; c–d *Mytilus*; d–e *Tetraclita*; e–f algal turf and *Chama;* f–g *Halidrys, Sargassum, Cystoseira* and *Phyllospadix*

1 *Lottia digitalis* 2 *Littorina scutulata* 3 *Chthamalus fissus* 4 *Tetraclita squamosa rubescens* 5 *Chama arcana* 6 *Pseudochama exogyrata* 7 *Pachygrapsus crassipes* 8 *Mytilus californianus* 9 *Egregia laevigata* 10 *Halidrys dioica* 11 *Cystoseira osmundacea* 12 *Zonaria farlowii* 13 *Sargassum agardhianum.*

themselves from either tropical or northern genera, there is a commingling of eurythermal species derived in about equal proportions from north and south, with the northern ones living mostly offshore towards the edge of the shelf.

There has been a notable tradition of pioneering books on Pacific shore biology. In 1939 came Ricketts and Calvin's classic guide: *Between Pacific Tides,* the first to move confidently from a taxonomic to an ecological frame. An earlier treasure, during my childhood

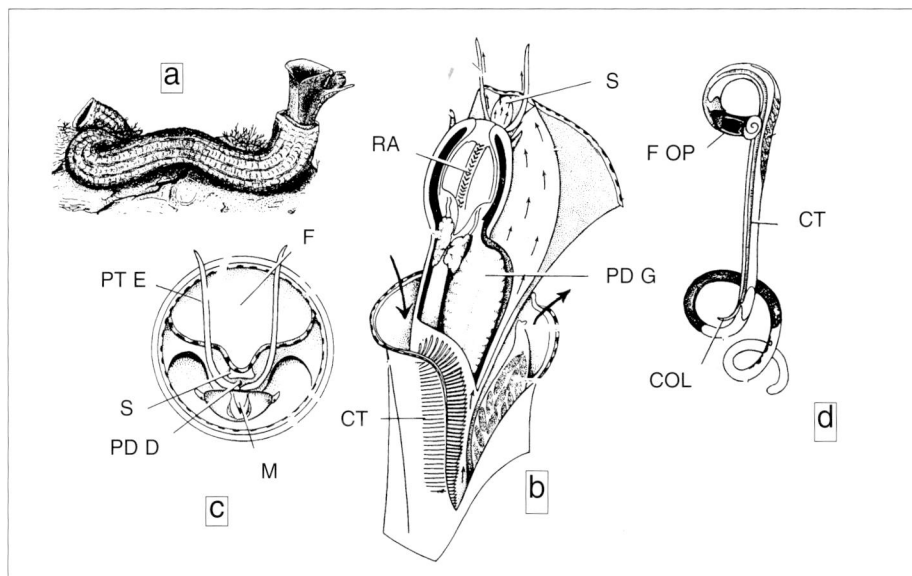

Fig. 19.21 Vermetid gastropods of California

Serpulorbis squamigerus (a) in shell (b) head, foot and mantle cavity dissected (c) animal at the tube-mouth (d) *Petaloconchus* sp. removed from shell.
(CT) ctenidium (COL) columellar muscle (F) foot (F OP) foot with operculum (M) mouth (PD D) duct of pedal gland (PD G) pedal gland (PT E) pedal tentacle (RA) radula (S) vestigial sole.

on New Zealand's shores, had been Johnson and Snook: *Sea Shore Animals of the Pacific Coast*. In advance of the time, a wealth of taxonomic and biogeographic information, with good species illustrations, were combined with unusually full treatment of invertebrate structure and function. More recent and definitive works have been *Intertidal Invertebrates of California*, and *Marine Algae of California*.

La Jolla

The rocky shores near **La Jolla**, below the site of the famous Scripps Institution of Oceanography, were studied in detail by the Stephensons in 1947–48, though publication was to be delayed until *Life Between Tidemarks on Rocky Shores* in 1973.

La Jolla lies in latitude 32 degrees north, with a sea temperature ranging between 14°C and 22°C, as compared with 11–16°C for Pacific Grove. Just north of the Scripps Institution pier, the intertidal shoreline is carved out of low sandstone and mudstone cliffs, with two broken platforms, an upper one at the cliff-base often eroded into ridges and stacks, and a lower one regularly uncovered even at the shortest tides. Though there are large waves with high impact of spray, La Jolla receives none of the colossal pounding experienced by the exposed shores at Pacific Grove.

As well as the provision of loose boulders, the rock structure has important effects on the siting of organisms. The softer sandstone has a texture friable with a knife or fingernail, and porous enough to retain water. This — in shaded situations — permits a high elevation of the pink algal veneer. By contrast, the outcrops of harder sandstone acquire a strong barnacle settlement. There are also beds of mudstone of varying hardness, riddled by the penetrant bivalves *Petricola californiensis* and *Penitella penita* and supporting a rich crevice fauna.

The periwinkles of the **littoral fringe** are the warm-water *Littorina keenae* and *L. scutulata* — the second being scarcer than at Pacific Grove, and confined to a lower level. In the **upper eulittoral**, barnacles are wanting on the softer sandstone, but abundant on harder rock. Though the cool temperate *Balanus glandula* is still present, the lead has now passed to *Chthamalus fissus*. In the **mid-eulittoral**, just above the algal turf, grows a larger warm-water barnacle, the pink *Tetraclita squamosa rubescens*. The giant *Megabalanus californicus* is found singly or in clusters at its accustomed level in the sublittoral fringe.

More conspicuous than barnacles on the soft rock are the limpets. *Lottia digitalis* abounds in the upper eulittoral, with *L. conus*, *Collisella scabra*, *L. pelta*, in

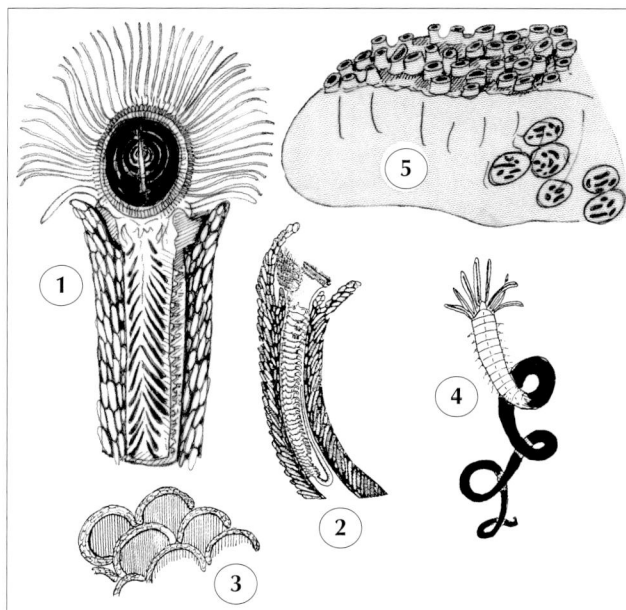

Fig. 19.22 Tubeworms at La Jolla

Phragmatopoma californicum: 1 head and anterior part of tube 2 tube in longitudinal section, 3 cluster of tubes, from above. *Dodecaceria fistulicola* 4 worm removed from tube 5 massed tubes.

the mid-eulittoral and the large *L. gigantea* a little lower down. The small chiton on the open surface is *Lepidochitona hartwegi*, while *Nuttallina fluxa* forms its homing scars a little further down. On the harder rock, with the operculate barnacles, there are also patches of the stalked *Pollicipes polymerus*. The leading mussel of the mid-eulittoral is *Mytilus californianus*, with *Mytilus trossulus* scarce or wanting. Mussels no longer form wide sheets as in the colder north, only scattered clumps or caps upon boulders. The fast crab *Pachygrapsus crassipes* runs about freely in this belt, having come into its own on the warm shores south of Point Conception. The small algae of shaded surfaces are the ridged and fleshy *Petrospongium rugosum*, with crusting *Ralfsia verrucosa*, seasonal *Scytosiphon lomentaria*, and — lower down — clumps of *Haliptilon roseum*.

In the **lower eulittoral**, a dull coralline veneer takes over from the barnacles and mussels. Here, the gregarious anemone *Anthopleura elegantissima* may become moderately dense. Still more characteristic of the zone is the small vermetid gastropod *Dendropoma lituella*, part embedded in the algal crust. Highly typical of the eulittoral of warm temperate and tropical as distinct from boreal shores, the vermetids at La Jolla include also clusters of much larger *Serpulorbis squamigerus* living in the sublittoral fringe.

Besides the calcareous crust, the lower eulittoral has patches of a strong algal turf, in general like that of Pacific Grove. In its upper part, the Stephensons found *Corallina chilensis*, *Haliptilon roseum*, *Jania* species and *Bossiella orbigniana*; with non-calcareous *Centroceras clavulatum*, *Gigartina armata*, *Rhodymenia californica*, *R. palmatiformis*, *Nienburgia andersoniana*, *Laurencia diegoensis*, *Gelidium cartilagineum* and *G. pulchrum*. Lower down came *Gigartina leptorhynchos* and *G. eatoniana*.

In these warmer waters the fucoid *Pelvetia fastigiata* is still locally to be found, as at Pacific Grove. Where large and dense these plants shelter an understorey of *Pterocladia pyramidale*, *Laurencia pacifica* and *Gigartina canaliculata*. The smooth spider crab *Pugettia producta*, deep olive and sharp-clawed, moves about actively among *Pelvetia*. On the rock surface are *Fissurella volcano*, and the barnacle *Tetraclita squamosa rubescens*.

Grazing gastropods at this level include the turbinid *Astrea undosa* and the handsome trochid *Norrisia norrisi*, with the shell dark chestnut, with vivid green on the base, and the animal black and scarlet. The trochids *Tegula funebralis* and *T. ligulata* are both common.

Fixed to the same surfaces, often as a continuous pavement beneath the algae, are two species of the sessile bivalve *Chama*, superficially resembling oysters (Fig. 19.20). The **Chamidae** are mainly tropical, but in Amer-

ica range as far north as Oregon, being found at Pacific Grove thinly scattered beneath boulders. Almost as high as broad, they are cup-shaped, with one valve deep and fixed and the other forming a flat lid. Unrelated to oysters, they are derived from higher bivalves (Anomalodesmata), with short but well developed siphons. *Chama arcana* is attached by its cupped left valve and has a frilly sculpture of translucent lamellae, while *Pseudochama exogyrata* is right side-attached and lightly sculpted.

Alongside the chamids, or reaching up to abut with the mussels, are not only clusters of the vermetid *Serpulorbis squamigerus* but also the crusts of two sorts of polychaete tubes. Large colonies are built up by the honeycomb spread of the sabellariid *Phragmatopoma californicum*, forming its wide-mouthed tubes of cemented sand grains. The operculum is a smooth dome, opening back to allow the feeding filaments to be deployed as in *Neosabellaria* (Fig. 2.8). A second tubeworm, *Dodecaceria fistulicola* (Family **Cirratulidae**), is most notable on limestone shores, but common also at La Jolla. The crusts of its massed tubes are washed with fast currents and the first few segments are extended to expose the branchiae and tentacular cirri. This worm produces a bright yellowish green fluorescence.

The laminarian algae of the **sublittoral fringe** and **upper sublittoral** point up the contrast of the warm temperate shores with those of the Oregonian Province. While *Macrocystis* and *Pelagophycus* are abundant offshore, the inshore laminarians are comparatively thin. *Eisenia arborea*, a small warm-water kelp recalling *E. bicyclis* of warm temperate Japan, is to be found only south of Point Conception. Almost as common as further north, around low water and sublittorally, is *Egregia laevigata*, a southern counterpart of *Egregia menziesi*, with fringed straps reaching up to 4.5 m long.

For the rest, the sublittoral fringe is covered with large swathes of the marine grass *Phyllospadix scouleri* (Fig. 19.7), as well as two low level fucoid algae, *Cystoseira osmundacea* and *Halidrys dioica* (Fig. 19.20). The stems of both may in some places be torn away to leave a stubbly cropped turf. *Sargassum agardhianum* covers the tops of low boulders among the turf. In channels and pools the fan-shaped *Zonaria farlowii* (Dictyotales) is common.[3]

South America

Nowhere else has a temperate shore zonation been brought to such low latitudes as in Pacific South America. The cold *Peru (Humboldt) Current* first becomes apparent at about 38 degrees south and flows northwards past Chile and Peru to about 4 degrees south, then setting west to wash the Galapagos and finally to

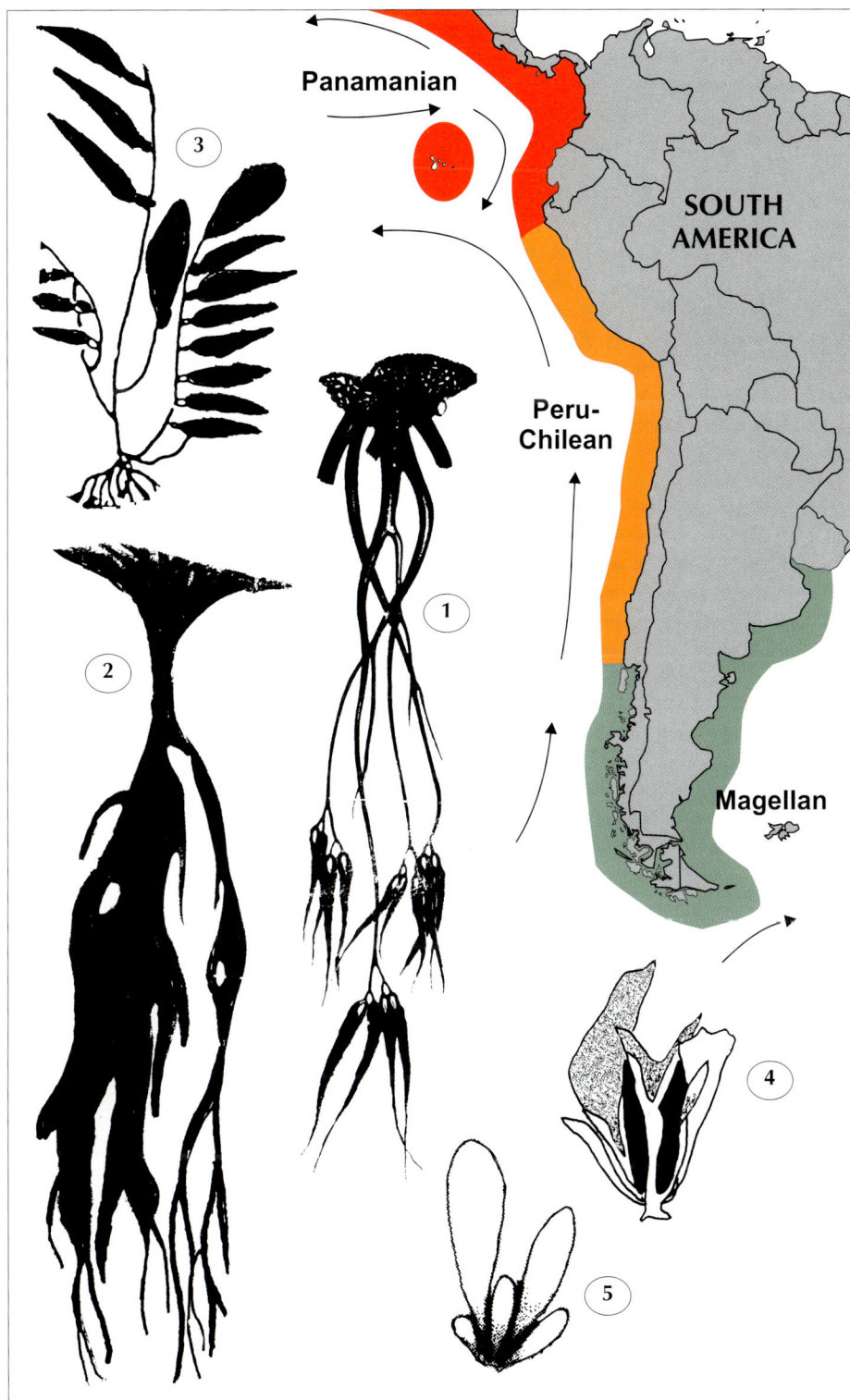

Fig. 19.23 Marine provinces and leading algae of the American South Pacific
1 *Lessonia nigrescens* 2 *Durvillaea antarctica* 3 *Macrocystis pyrifera*
4 *Iridaea laminarioides*
5 *Adenocystis utricularis.*

merge with the South Equatorial Drift. The *Western South America Warm-Temperate Region* thus gives place to the *East Pacific Tropical Region* only at the Gulf of Guayaquil in latitude 3 degrees south.

The prevailing south-westerly winds off Peru set in motion coastal currents that cause the upwelling of cooler water. The rich nutrient supply this brings to the surface accounts for the high oceanic productivity, culminating in part with the rich Peruvian anchovy crop. A warm temperate regime has here been extended far north. Sometimes the Trade Winds, normally south-easterly along the coast of Peru and North Chile, may fail,

being replaced with a north wind and with it a flow of equatorial water (El Nino) that can kill fish and plankton, especially in Peru from January to March.

Peru and most of Chile form a distinct warm temperate *Peru-Chilean Province*, with its southern boundary at the north end of Chiloe Island established on the evidence of molluscs, decapod Crustacea and Asteroidea. Some support exists for additional faunal breaks at Valparaiso (30 degrees south) and near Concepcion (36 degrees south), with certain eurythermic cool temperate species extending north to these points.

377

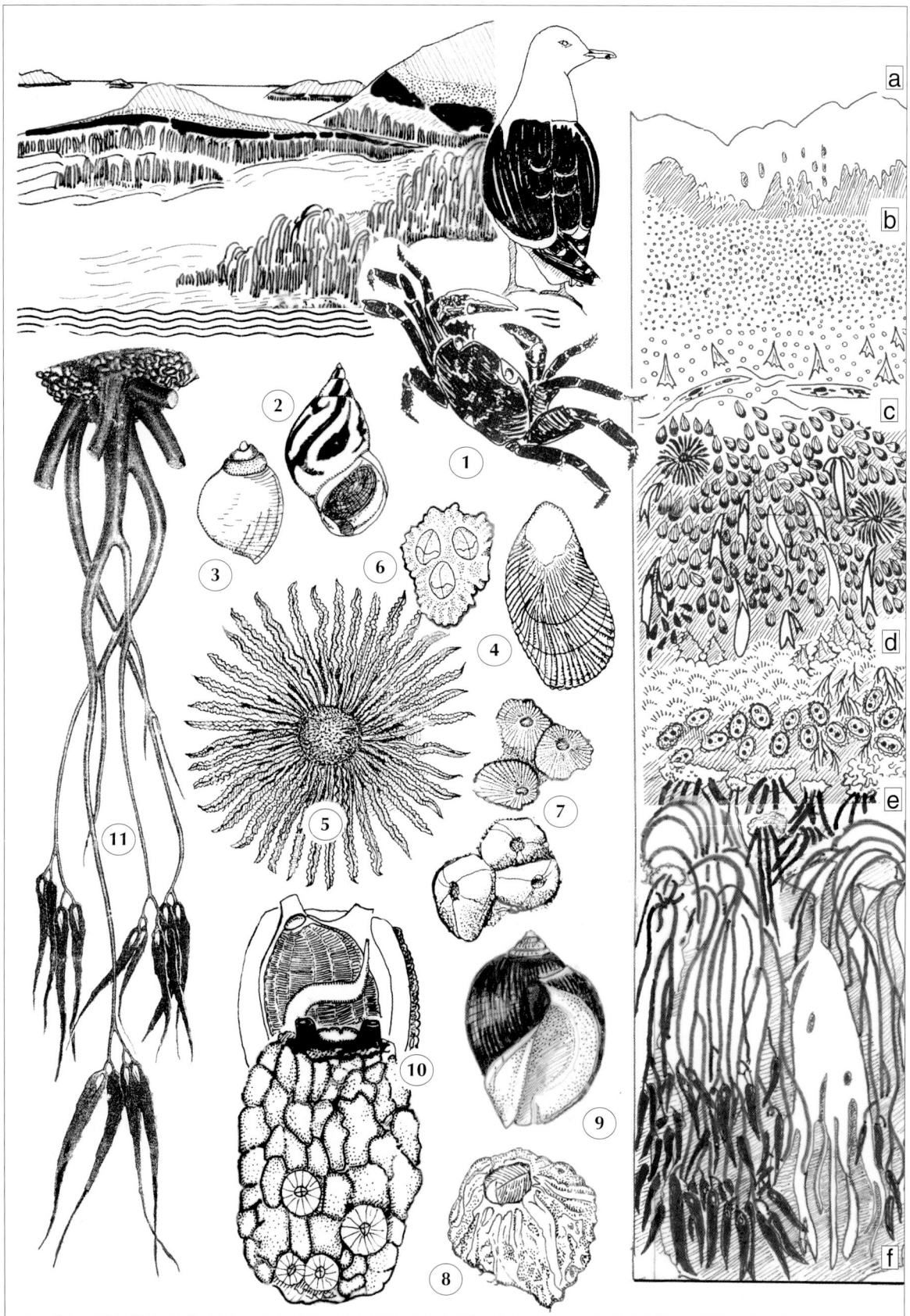

Fig. 19.24 Exposed zoned shore at Montemar, near Valparaiso, Chile
(**inset, top**) prospect showing bands of barnacles, mussels and *Lessonia*.
a–b white with guano; b–c littorines, *Chthamalus cirratus*, and *Porphyra*; c–d *Perumytilus*, *Ulva* and *Iridaea*; d–e coralline turf and *Pyura*; e–f *Lessonia nigrescens* and *Durvillaea antarctica*
(**top**) black-backed gull, *Larus dominicanus*
1 *Leptograpsus variegatus* 2 *Littorina peruviana* 3 *Littorina araucana* 4 *Brachidontes purpuratus* 5 *Stichaster striatus* 6 *Chthamalus cirratus* 7 (**above**) *Balanus laevis* (**below**) *Notobalanus flosculus* 8 *Austromegabalanus psittacus* 9 *Thais chocolata* 10 *Pyura chilensis* 11 *Lessonia nigrescens*.

Peru

The coast of Peru is entirely non-tropical. Corals are lacking, and the whole fauna and flora have an essentially temperate character. Winds on the coast are constant, with little temperature variation from day to day or month to month.

The largest of the Peruvian brown algae, *Macrocystis integrifolia*, is present in two distinct ecological forms. On surf-swept rocks just beyond low water, the holdfasts are deeply entangled, with the leaves small and narrow and the vesicles elongate. The larger form growing further offshore reaches more than 6 m long, with isolated holdfasts, broad leaves and the vesicles almost globular. The most important inshore brown algae are *Sargassum pacificum*, *Eisenia cokeri* (mainly subtidal and collected in drift), and *Lessonia nigrescens* which continues south through Chile. The smaller browns, belonging to the Dictyotales, include *Spatoglossum crispatum*, *Neurocarpus cokeri*, *Dictyota dichotoma* and *Glossophora kunthii*.

Peru has but few warm-water green algae, though *Caulerpa flagelliformis*, *Codium tomentosum peruvianum* and *Codium foveolatum* are common. The red algae include a goodly number of *Gigartina* (*G. lessonii*, *G. chauvinii*, *G. glomerata*, and *G. tuberculosa*), as well as *Chondrocanthus chamissoi*, *Ahnfeltiopsis furcellatus*, *Gymnogongrus disciplinalis*, and *Ahnfeltia durvillaei*.

Chile

The Chilean coast, from latitude 18 degrees to 42 degrees south is dominated by onshore south-westerlies in the zone of the 'roaring forties'. These winds extend north to Peru where they merge with the south-east trades blowing offshore.

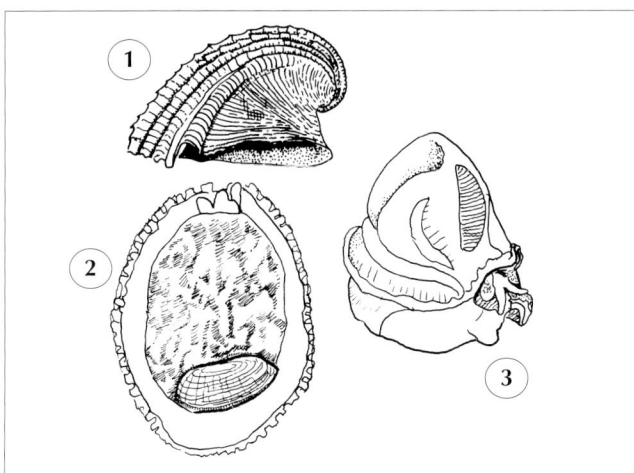

Fig. 19.25 Loco, *Concholepas concholepas*
1 shell 2 under-view, showing foot and operculum
3 animal removed from shell.

Wave action is much influenced by onshore sea breezes. In summer these arise in the late morning to blow for most of the afternoon, whipping the sea into short, sharp waves superimposed on the broad ocean swell, which impede shore-study for most of the day. Cold upwellings caused by winds blowing along the line of the coast are rich in phosphate. The uniformly low surface temperatures are comparable with those at 150 m depth, and offshore at the 100 degrees west meridian.

Under the full swell raised by the south-westerlies, the Chilean coast is exposed to some of the strongest, most continuous wave action in the world. North of Chiloe Island there are no bays and virtually no sheltered coast. Right through the 'warm temperate' province, bathed by cold upwellings, with prevalent mists, wave action takes full command, virtually overruling the regimen of the tides.

Down this long coast the zoning pattern is notably uniform, with little change in the sequence of species. **Montemar**, just north of Valparaiso, near the most northerly reach of *Durvillaea*, could in most respects stand for the Chilean seaboard at large. In the terrestrial climate at the coast there are, however, real differences. In the north, with rains sparse or rare, high insolation results in hot, dry desert; under nearly continuous spray and mists, most of the supra-tidal is permanently hypersaline. In Central Chile, the rains vary sharply with the seasons, producing a Mediterranean climate, with warm dry summers and mild wet winters.

Such differences are hardly reflected in the marine biota. Between north and central Chile there is only 3°C difference in mean winter sea temperatures. Owing to the current direction, tropical species are almost wanting. To the north, Arica, in latitude 13 degrees north, has no corals, while mangroves are insignificant south of Tumbes (latitude 3.3 degrees) in north Peru.

The whole Chilean coast divides naturally into two Provinces: the continuation of the *Peru-Chilean* and a cool temperate *Magellan*, with its northern limit at Chiloe Island. Briggs notes that Chile has only three exclusively warm-water echinoderms, the rest belonging to a cool temperate element from south of Chiloe Island.

The *Magellan Province* has shores with *Durvillaea* and mussels, clear characters of the Pacific cold south. The warm temperate *Peru-Chilean* coast is dominated by *Lessonia nigrescens*. There is a notable penetration of Magellanic elements into the *Peru-Chilean Province*, particularly Crustacea and the very prominent fissurellids. Of a total of 37 of these gastropods in Chile, 15 species are Magellanic, with five extending north of Chiloe Island.

Warm temperate Chile: Montemar

Almost all our understanding of the shore zonation of

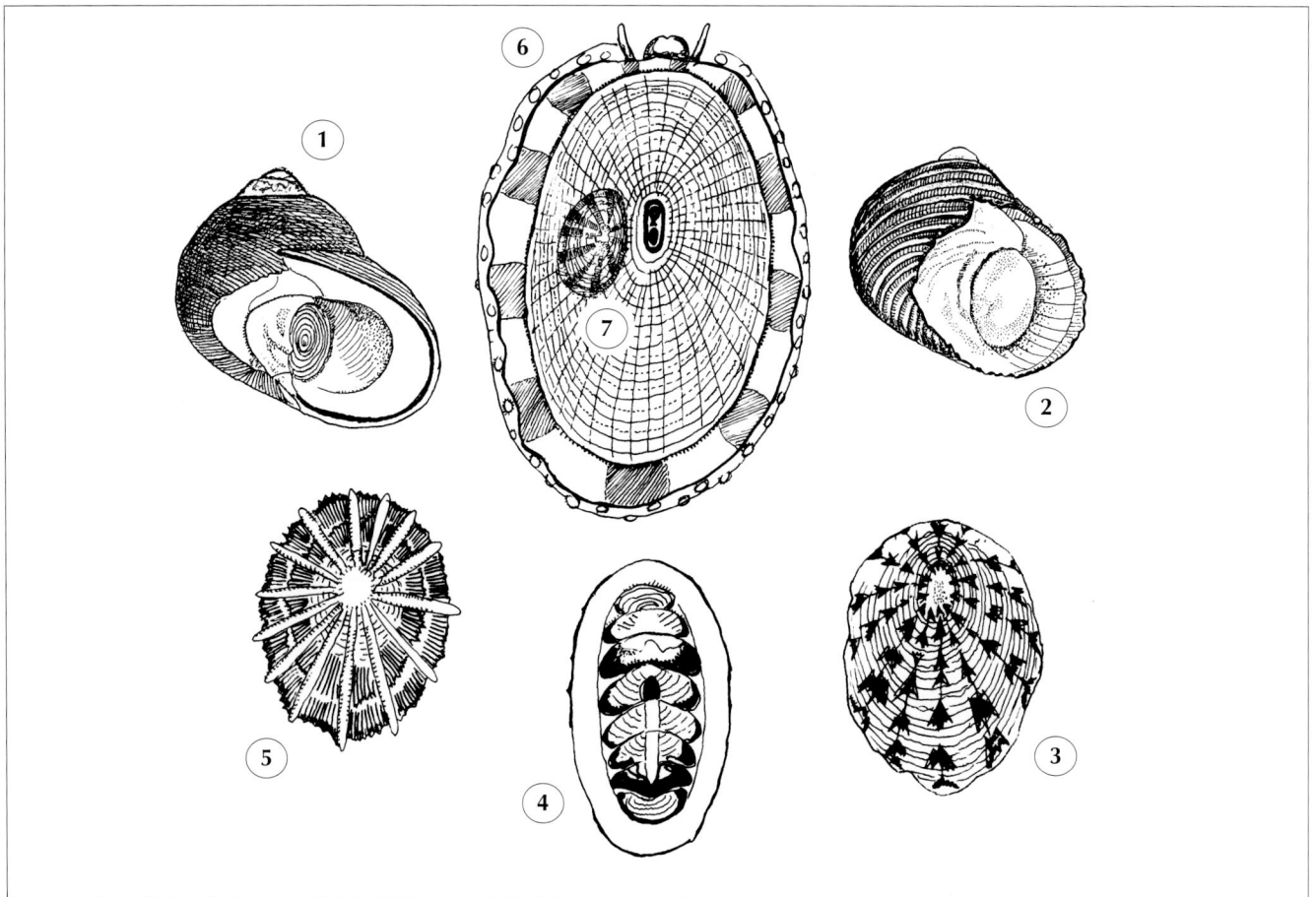

Fig. 19.26 Some grazing gastropods of the Peru-Chilean Province

1 *Tegula atra* 2 *Prisogaster niger* 3 *Lottia zebrina* 4 *Tonicia* sp. 5 *Lottia araucana* 6 *Fissurella crassa*, with 7 *Scurria parasitica*.

Chile's warmer province we owe to the papers of the Australian E.R. Guiler (1959).[4] Though sparing in illustrations, these give a wealth of information on which the present account has drawn freely. A debt is owing also to the first-hand acquaintance with these coasts of George Knox, and my late colleague Brian Foster, and not least to a short, well illustrated coastal guide by Juan Carlos Castilla.[5]

Montemar lies 16 km north of Valparaiso, in latitude 33 degrees south, being the location of the Marine Station of the University of Chile. Although the littoral fringe is clean and spray-swept the expected lichen colours are lacking, superseded by the smearing of white guano from the Dominican gull, shags and the night heron (*Nycticorax obscurus*).

The barnacle zone, drenched with spray, is the widest belt of the shore, with its top band consisting of *Chthamalus cirratus*, especially close-massed in cracks and rock junctions. The two periwinkle species are the northern-derived *Littorina peruviana*, recognised by its broad pale streaks, and the uniform coloured *Littorina araucana*. Both have receded from the **littoral fringe** to form a band with the barnacle in the **upper eulittoral**.

Next below appears a seasonal subzone of *Porphyra columbina*, a widespread southern Pacific species familiar also in New Zealand. Purplish brown in spring and early summer, this is gathered for food, or by full summer dies off to a yellow-brown. The purple crab, *Leptograpsus variegatus*, another New Zealander, runs over the rock face or lines up backed into secure crevices.

The lower reach of the barnacle zone is enriched with the pale, smooth *Balanus laevis* and smaller numbers of the mauve, radially ribbed *Notobalanus flosculus*. In crevices at this level live *Chiton cumingsi* and *Chaetopleura peruviana*.

The high-level limpets in Chile belong to the acmaeid genus *Lottia*. At Montemar, the first is *L. orbignyi*, extending above the barnacle level. Further down live *L. zebrina* and *L. ceciliana*. A fourth species, *L. araucana*, continues south only to Valparaiso.

The belt of the **mid-eulittoral**, coming constantly under wave attack, is conspicuous from a distance by its dark band of mussels. The principal species is the ribbed mussel, *Brachidontes purpuratus*, with the smaller, smooth mussel *Chloromytilus chorus* slightly lower down. The mussels are spangled in golden brown with the long-oval sheets of the red alga *Iridaea laminarioides*. Bright green *Ulva lactuca* is conspicuous through most of this belt, especially over the mussel shells, and reaches down into the next zone. In the **lower eulittoral**, the algae become enriched, the dominant species being *Centroceras clavulatum*, *Gelidium filicinum*, often collected for agar, and *Corallina chilensis*.

The mussel zone supports its asteroid predator in the many-armed seastar *Heliaster helianthus*. Among the mussels and upon the barnacles upshore, the common thaid whelk is the smooth, brown-banded *Acanthina calcar*.

The barnacle *Notobalanus flosculus* grows upon mussel shells, together with *Porphyra columbina*, and several species of limpets. Common underneath mussels are the crabs *Pilumnoides perlatus* and *Acanthocyclus gayi*, and the polychaete *Perinereis galapagensis*. Algae from the barnacle zone generally continue down to the next zone, to grow on or among *Pyura*, with golden saccate *Colpomenia sinuosa* seasonally very common. In sheltered places the bladder clusters of *Adenocystis utricularis* will be normally found, here as familiar as in cold southern New Zealand.

Large ascidians, *Pyura chilensis*, densely aggregate in wide sheets in the *lower eulittoral*, recalling the still larger cunjevoi zone on southern Australian coasts. The heavy tests are covered with large flat mamillae, and carry small barnacles. At Antofagasta Guiler found *Pyura* enhanced to a wide belt, replacing both the lower mussels and the *Lessonia* normally forming the sublittoral fringe. In the ascidian belt several limpet species are common, though sometimes difficult to reach. They may include large numbers of the large, colder water *Nacella magellanica*.

The rock clefts of the lower eulittoral are the haunts of large keyhole limpets, becoming increasingly abundant further south, but including at Montemar *Fissurella maxima*, *F. crassa*, and *F. latemarginata*. The common chitons at this level are *Chiton cumingsi*, *C. granosus* and *C. latus*. We may find as well the large *Enoplochiton niger*, up to 8 cm long, with the valves generally eroded and the girdle slightly scaly, and the scarcer *Acanthopleura echinata*, with black spinose girdle. The last is common also among the holdfasts of *Lessonia*.

The great brown algae of the Chilean coast are *Lessonia nigrescens* and — increasingly towards the south — the bull-kelp, *Durvillaea antarctica*. As compared with North America, and Australasia, the lack of fucoids is remarkable, as indeed the absence of any large browns save these two. The dark *Lessonia*, set off by its straight upper limit, is dominant everywhere in the **sublittoral fringe**. The holdfasts are strong and many-branched, and the thin, tough stipes repeatedly subdivide into whips ending in pointed flails about 0.3 m long. The *Lessonia* branches form impenetrable

thickets, standing out in rich brown and interspersed — towards the south — with the dull yellow thongs of *Durvillaea antarctica*. Second only in prominence to *Lessonia*, this bull-kelp has its northern limit near Montemar. Both these large algae are heavily cut for harvest.

The complex holdfasts of *Lessonia* are favoured habitats for the low tidal barnacle *Austromegabalanus psittacus*, with strong, gnarled ridges, and sometimes growing to 16 cm high. The chitons *Acanthopleura echinata* and *Enoplochiton niger* are common in holdfasts as is the large *Fissurella crassa*, usually with a small siphonariid limpet *Scurria parasitica* attached. The cousin species, *Scurria scurria*, lives in permanent scars in the *Lessonia* fronds.

The most distinctive predatory gastropod of Chile and Peru is the modified thaid whelk *Concholepas concholepas* that has taken on a virtual limpet form. Under the popular name *loco*, it is much gathered for food, and could stand as the molluscan 'icon' of the Peru-Chilean province. Common in clefts in the *Lessonia* zone, it shows its whelk affinities by the inhalant canal, and the large, non-functional operculum, concealed under the shell on the back of the foot. With typical thaid features, including purple gland and radula, *Concholepas* feeds upon *Austromegabalanus psittacus* and probably other shelled animals.

Common gastropods at the same level include the whelk *Thais chocolata* and the turban shell *Prisogaster niger*. In semi-sheltered places throughout this zone, the fleshy green alga *Codium dimorphum* can cover large areas of rock.

Four important species of crab shelter in *Lessonia* holdfasts: *Petrolisthes spinifrons*, *Pachycheles crassimanus*, *Pilmnoides perlatus* and *Taliepus dentatus*, along with the snapping shrimp, *Synalpheus spinifrons*. The maiid crab *Pisiodes edwardsi* crawls among the *Lessonia* fronds, and the strong-chelate *Acanthocyclus hassleri* and *A. gayi* live within the branching stipes.

On steep rocks and in gullies lives the remarkable cling-fish *Sicyases sanguineus* (Family **Gobiesocidae**), reaching 40 cm long. Attaching by its ventral sucker, it grazes with fine incisor teeth on minute algae. Effectively camouflaged, it moves spasmodically, fixing fast under wave assault.

In the sublittoral fringe grow the large anemones, *Phymactis clematis*, with three colour phases, red, green and bright blue. Frequently associated with them is the porcellanid crab *Petrolisthes angulosus*. The urchin *Loxechinus* often abounds in rock pools, at times eating out most of the small to middle-sized algae.

Beyond the limits of *Lessonia* the rock surface exposed at low tide is pink-veneered with a coralline alga left almost bare except for the chitons *Acanthopleura* and *Enoplochiton* and the gastropod *Prisogaster*

Fig. 19.27 Intertidal species at Tierra del Fuego

A rock slope as shown (right) supports *Durvillaea*, and to the left are *Macrocystis* and *Lessonia*, with holdfasts and branches giving some protection from offshore wave-aatack. (based on field notes and literature).

1 *Kerguelenella lateralis* 2 *Nacella magellanica* 3 *Lottia ceciliana* 4 *Laevilittorina caliginosa* 5 *Chthamalus cirratus*
6 *Brachidontes purpuratus* 7 *Mytilis edulis chilensis* 8 *Aulacomya atra* 9 *Nacella mytilina* 10 *Margarella violacea*
11 *Austromegabalanus psittacus* 12 *Trophon geversianus* 13 *Pareuthria plumbea* 14 *Lessonia nigrescens*, holdfast
15 *Gaimardia trapezina* 16 *Macrocystis pyrifera*, holdfast 17 *Macrocystis*, frond and single enlarged leaf-blade 18 *Lessonia*,
branch tips and laminae 19 *Durvillaea antarctica* 20 *Adenocystis utricularis* 21 *Iridaea laminarioides*.

niger. Two seastars are typically found at this level: *Stichaster striatus*, also living higher upshore among mussels, and *Meyenaster gelatinosus*.

The topshells *Tegula atra* and *T. tridentata* normally live in greater shelter. Under decaying bull-kelp, just as in southern New Zealand, live huge numbers of the graphite-black trochid *Diloma nigerrima*.

Close to Montemar, from Las Ventanas and Quintero, Guiler has recorded the bladder-kelp *Macrocystis integrifolia* in small beds offshore. Increasing further south, it forms dense offshore barriers on the cool temperate coasts that we must now describe.

Cold shores

South of the island of Chiloe, at latitude 38 degrees south, we pass into the cool temperate *Magellan Province*, extending around Cape Horn and up the Patagonian coast of Argentina. The shore zonation has been described all too briefly by Anne Stephenson, in *Life Between Tidemarks on Rocky Shores* (1972), based on the ecological notes of George Knox.

For the shores of Tierra del Fuego, I found a treasury of information, with clearly sketched animals and algae, in the admirable hand-book by Rae Natalie Prosser Goodall (1979) on the *Mountains and Shores of the Beagle Channel*.*

South from Chiloe the dominant large alga offshore is *Macrocystis pyrifera*, affording shelter to an inside channel where the wave force is broken. Here boats can be moored to the strong branches of low or subtidal *Lessonia nigrescens*. Right through the *Magellan Province Durvillaea antarctica* forms an unbroken belt onshore, especially notable in the Beagle Channel and at Staten Island.

The 80 listed species of brown algae south of 42 degrees include the cosmopolitan *Colpomenia sinuosa*, *Scytosiphon lomentaria* and the cold southern *Adenocystis utricularis*, all familiar from New Zealand. Onshore, there are zones of *Porphyra columbina* and *Corallina officinalis*. Prominent offshore are no fewer than eight *Desmarestia* species, some growing to bushes up to 2 m tall. Among the leading offshore reds are the decorative Delesseriaceae, with fine venation and secondary leaflets.

Going down-shore from the **supralittoral zone**, luxuriant green-grey lichens (*Usnea*) hang like torn curtains from the tree-line. Below these the orange lichens encrust the rocks. The single one of its family is the small and abundant *Laevilittorina caliginosa*. The bar-

* I am indebted also for vistas of shore structure and scenery to Geoffrey Payne and Margaret Hough, who in 1990 sailed and photographed the passage from Chile round the Horn and thence to the Subantarctic Peninsula.

Fig. 19.28 Tierra del Fuego: *Drimys winteri*
Flowers and foliage, with a vertical section of flower (A) and (top left) shore shrub showing the effect of wind.

nacle of the **upper eulittoral** is *Chthamalus scabrosus*, with *Balanus improvisus* and the giant *Austromegabalanus psittacus* towards low tide. The **middle eulittoral** is mussel-dominated with the small, ribbed *Brachidontes purpuratus* (mejillin) above and the larger, smooth *Mytilis edulis chilensis* (mejillon) below. Subtidally there is the ribbed *Aulacomya atra* (cholga) much esteemed as food. The tiny bivalve *Gaimardia trapezina* nestles in *Macrocystis* holdfasts and a pink *Lasaea* abounds among the byssus threads of mussels. Four species of scallops (Pectinidae) are a characteristic feature of these cold southern shores.

The Magellan Province has as many as 21 species of chiton, the commonest between tides being *Tonicia lebruni* and *Plaxiphora aurata*. The gastropod list (157 in all) has a heavy limpet bias, including *Nacella deaurata*, *N. magellanica*, *N. delicatissima*, *N. fuegensis*, *N. mytilina* and *Lottia ceciliana*. As well there are the keyhole limpet, *Fissurella maxima*, and the pulmonates *Kerguelenella lateralis* and *Siphonaria lessoni*, along with a calyptraeid limpet *Trochita trochiformis*. The most prominent of the cold-water trochids is *Margarella violacea* on kelp. Common whelks are *Pareuthria plumbea*, *Trophon geversianus* and *T. laciniatus*.

Tierra del Fuego, or 'Land of the Fires', has the most southerly human occupation. Across the Straits of Magellan, bounded by three oceans — the Pacific, Atlantic and Antarctic — lies an archipelago of more than a thousand islands, only half a dozen inhabited. The largest is **Isla Grande**, rugged and windswept, with daily rain, but with sheep-farming on the Argentine side.

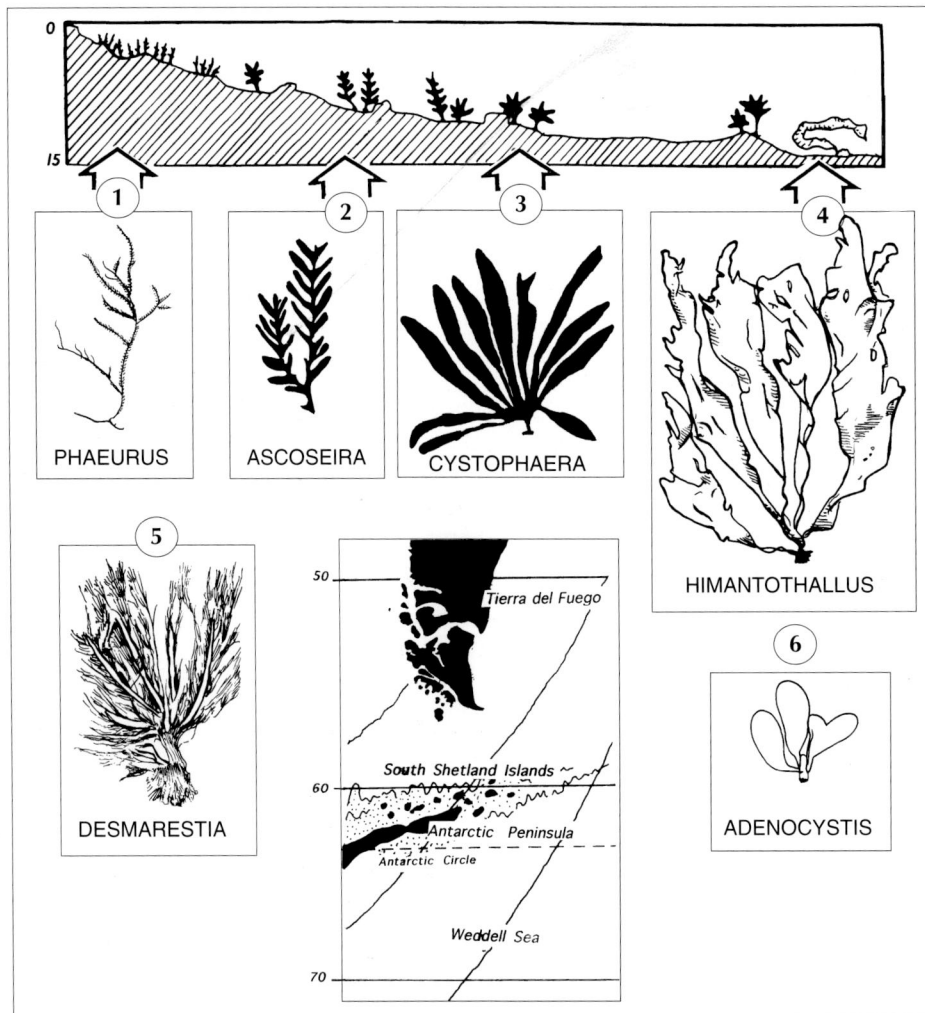

Fig. 19.29 A subtidal transect at the South Shetlands, with the principal brown algae
(After Lüning.)

The Chilean maritime stretch, with less snow and ice but permanent mist and cloud, forms a rainforest zone. The three trees are the evergreen beech *Nothofagus betuloides*, *Maytenus magellanica* and *Drimys winteri* of the Family Winteraceae, possibly the most primitive of living flowering plants. In wave-sheltered spots these reach in right to the water's edge, generally wind-flattened to form a ground layer of horizontal trunks. Further to the west the islands directly fronting the Pacific have a subantarctic climate, with fierce winds, frequent rain and heavy cloud. The vegetation is of heavy bogs with short trees only in protected places.

Of the marine fauna of the Tierra del Fuegian Province, the Crustacea and Mollusca all belong to distinctive cold-water taxa. Under the wide denomination of 'krill' come both the euphausiid shrimps *Thysanoessa*, and the lobster krill *Munida*. The largest decapods are the king crab, *Lithodes antarcticus* (centolla) and *Paralomis granulata* (falsa centolla). There is a slender-legged spider crab, *Eurypodius*

latreillei, and also the reddish tractor crab, *Peltarion spinulosum*, convex and oval, with heavy spanner-like chelae. *Halicarcinus planatus*, reminiscent of New Zealand, lives in *Macrocystis* holdfasts. The flat isopods *Serolis* (see Fig. 10.19) show a strong Antarctic affinity, appearing on the flats like large coins when the tide goes out.

The commonest urchin is a *Pseudechinus* species, with green spines and deep red podia, fished up on long poles from its abundant populations in bladder-kelp.

Tierra del Fuego's most representative fishes comprise 12 species of the Antarctic Family **Nototheniidae**. Two of these, *Patagonotothen sima* and *P. cornucola*, live in the intertidal, while the 'pike', *Champsocephalus esox*, also found in the Falklands, is closely related to the 'bloodless' fishes of Antarctica. A second Antarctic littoral group, the **Zoarcidae**, has two eel-like species of *Austrolycus* in tide pools and *Phucocoetes latitans*, the kelp eel, among the holdfasts of *Macrocystis*.

Antarctica

Our knowledge of Antarctic seaweeds, deriving mostly from collections made on the Antarctic Peninsula, has been well summarised by Lüning.[8] Fig.19.29 shows a subtidal transect at the South Shetlands as seen by divers.

The laminarian kelps are lacking in Antarctica. Instead the canopy role is taken over by the Order Desmarestiales, occurring all round the Antarctic and having several species down to 40 m depth. The giant of the desmarestians is *Himantothallus grandifolius*, growing fronds up to 10 m long. Before its life history became known this plant was regarded as the only Antarctic laminarian. The holdfast, stipe and wide blade have all been modified in *Durvillaea* fashion. A much smaller Antarctic desmarestian is *Phaeurus antarcticus*.

The Fucales are represented on the Antarctic Peninsula by *Cystosphaera jacquinotii*, up to 5–10 m, with air-bladders and found mainly in drift. *Ascoseira mirabilis* — placed in a separate Order Ascoseirales — has a massive holdfast and divided blades, and shows convergences with both the Laminariales and the Desmarestiales.

Zoned algae in the intertidal stretch are very sparse, owing to the abrading action of drift-ice. At King George Island, there are a number of green algae: *Prasiola* in the littoral fringe and *Urospora*, *Ulothrix*, *Enteromorpha*, *Spongomorpha* and *Monostroma* in the eulittoral. In the lower eulittoral grows the cold-water brown alga *Adenocystis utricularis*. The reds are represented by *Porphyra* and — in the lower eulittoral — large *Iridaea obovata* and *Palmeria decipiens*.

At Signy Island there is a luxuriant algal flora between 2 m and 20 m. Forests develop below 5 m, with the *Durvillaea*-like *Ascoseira mirabilis*; while thickets of *Desmarestia menziesi* and *D. anceps* grow up to 1 m high. The perennial brown alga *Himantothallus* flourishes below 9 m. Higher up will be found some species and genera already familiar from the cool temperate: *Gigartina apoda*, *Myriogramme* and *Porphyra*. On *Desmarestia* and *Ascoseira* hang the orange isopods *Antarcturus signiensis* and the byssus-held bivalve *Lissarca miliaris*.

Vertical faces not dominated by algae include immense numbers of holothurians (*Cucumaria*) wedged into crevices, among brachiopods and tunicates. In larger embayments pale yellow gorgonians (*Primnoella*) flourish, while yellow sheets of *Haliclona* sponges stand out against purple *Lithothamnion*.

PART THREE
PACIFIC CORAL SHORES

Fig. 20.1 The atoll of Aitutaki, Cook Islands
Seen from the east, the islets of Motukitiu to Akaiami lie in the foreground, with the higher island of Maungapu to the rear.

CHAPTER TWENTY
Reefs and Their Building

As long ago as 1856 Edward Forbes, the founding father of marine biogeography, wrote of an Indo-Pacific Province as the realm of reef-building corals, and of the wondrously beautiful assemblage of animals, vertebrates and invertebrates that live among them or prey upon them.

Beyond comparison, these are the richest marine communities the world has to show. Their high point of diversity has long been traced to the concentrated triangle bounded by Malaysia, the Philippines and New Guinea. Here — all through the Tertiary — there seems to have existed the continued climatic optima needed to generate and maintain such an unrivalled species flow. It is this centre that has contributed — albeit in diminishing numbers as we travel east through the Pacific or west into the Indian Ocean — the shallow water biota of the Indo-Pacific tropics, and ultimately of much of the world beyond.

In nearly all the major groups, many species tend to extend far outwards from the centre of origin, sometimes right to the wide tropical Indo-Pacific periphery. Hence it is that biogeographers have been quite unable to designate this central triangle as a discrete Region or even — by itself — a Province. Though in almost every taxon the species thin out as we move from the hub, we can reach east to the Tuamotus, north to Taiwan, south to Australia's Great Barrier Reef or west into Indonesia and Sri Lanka and still find few endemic species not present at the centre. So the triangle itself has far less than the 10% exclusive endemism that would qualify it for a province.

The triangle and its receiving areas through two oceans are together defined by John Briggs as an *Indo-Polynesian Province*, extending all the way from the entrance to the Persian Gulf to the Tuamotu Archipelago, and from Sandy Cape on the coast of eastern Australia to the Amami Islands of south Japan. Such a vast single Province has not only an unprecedented number of taxa, mostly large in their number of species, but also its own high endemism. Thus — taking some of Briggs' figures — the molluscan endemics include, for the Strombidae 22 out of 43 species and subspecies; for the

cowries 47 out of 127, and for the Volutidae 30 out of 39, while three species of the six giant clams (Tridacnidae) are also endemic to the Indo-Polynesian Province. The echinoids have a 50% Indo-Polynesian endemism, while the crinoids (sea-lilies) reach 80%.

From the Philippines to Japan, the north-western Pacific seaboard is washed by the *North Equatorial Current*, that turns northward as the *Kuroshio Current*. The Polynesian islands north of the equator come under the influence of the *North Equatorial Current;* while the Indo-Australian archipelago and north-east coast of Australia are supplied by the *South Equatorial Current* flowing south as the *East Australian Current*. The *South Equatorial Countercurrent* washes such island groups as Fiji, Tonga, Samoa and the Marquesas.

The *Indo-Polynesian Province*, easily the world's largest, itself forms the major part of an *Indo-West Pacific Region*, that extends two thirds of the way round the globe, and through as much as 60 degrees of latitude. In comparing it with the tropical West Atlantic, Briggs showed that the Indo-Pacific hermatypic (i.e. reef-building) corals, with 500 species, are 10 times as numerous. The 1000 bivalve molluscs total twice those of the Atlantic, while the Indo-Pacific has four times as many cowries and five times as many strombids. The 3000 species of Indo-Pacific shore fishes compare with under 1000 for the Atlantic.

In nearly every big taxon the diversity tails off towards the periphery. Thus, the 2000 species of shore fish at the Philippines reduce to 400 at Raroia in the Tuamotus, and only 384 in Hawaii. Of 2000 molluscan species in Vanuatu, only about a quarter remain at the Tuamotus and Gambia. Fig. 20.2 follows the diminution in coral species, west and east of the triangle, and records also for each main island group the numbers of Strombidae. The species that reach the periphery are, however, almost all widespread Indo-Malaysian entities, with little marginal endemism.

In areas that reach the required 10% of endemism, several distinct Provinces can be demarcated at the edge of the Indo-West Pacific Region. Thus, in the Indian Ocean, the African coasts justify recognition as *West*

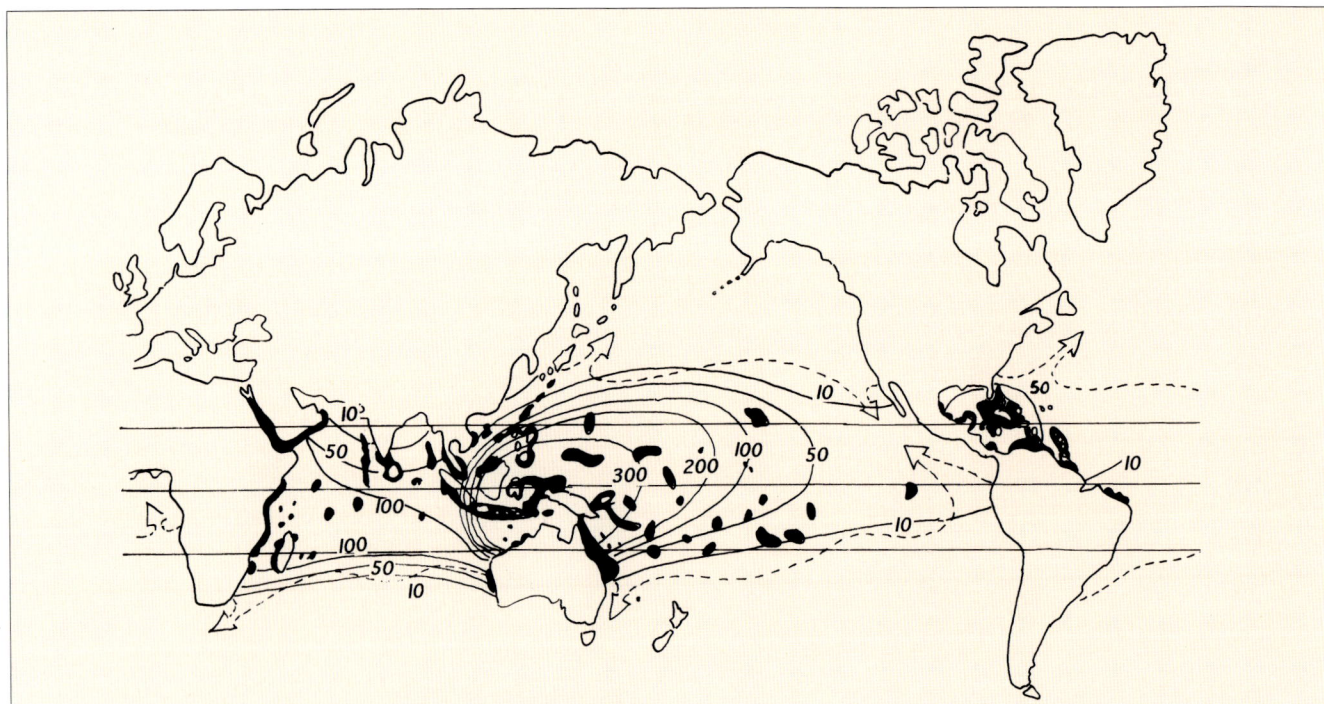

Fig. 20.2 The coral coasts of the Indo-Pacific and Atlantic
Coral regions shown black, with the Indo-Malaysian peak and the outward diminution illustrated by the number of species of Scleractinian corals.

Indian Ocean and *Red Sea Provinces*. A well enough marked *North-west Australian Province* runs west from Cape York to Shark Bay in Western Australia, leaving the east coast of Queensland, including the Great Barrier Reef, as part of the undivided *Indo-Polynesian Province*. At this Province's south-west periphery, there are two small separate Provinces, *Lord Howe/Norfolk Island* and *Kermadecian*, the second falling within the warm temperate *North New Zealand Region*. At the North Pacific edge of the tropics, the *Hawaiian Islands* form a separate Province. So also, in the far east Pacific, do the *Marquesas*, the *Galapagos* and the far remote *Easter Island*.

After our perspective of the Pacific's temperate rim, we can offer nothing like exhaustive treatment of its coral reefs. With its detailed exposition of a part of the Indo-Pacific Province, the *Readers Digest Book of the Great Barrier Reef* must stand as the finest illustrated account of any tropical shore yet produced.

Our prime purpose here will be to compare the rich coral shore — in its basic patterns — with the temperate zonation we have already seen. To begin, we have chosen to focus upon one atoll system, Aitutaki, 230 km north of Rarotonga in the Cook Islands, offering the full spectrum of reef topography, with windward, leeward and lagoon shores. The shores of Aitutaki could be accounted rich, but not maximally diverse. In this they lose nothing of their novelty and wonder for the beginner, and they will offer in some ways a clearer introduction than the highest levels of diversity to the west.

The characters of coral shores

In all their essentials tropical shores are faithful to the 'universal pattern' of zoning we have found in the temperate regions. They have also, however, some notable differences. Unlike any other hard shores coral reefs are not on balance receding, but are being *prograded* or built forwards. But such slow advance is being itself held in check, since the growing margins of the reefs are in turn liable to constant cutting back. Such erosion — like the reef's formation — is the work of living organisms. Growth is thus kept in equilibrium with erosion. In either process the main force is biological.

The first and salient fact about a coral reef is its wide *horizontal* extent. With a tropical forest — the reef's complex counterpart upon land — all the growing thrust is *vertical*. The multi-storeyed canopy soars to more than 50 m high, with every bit of it living or carrying other dependent life. In a coral reef, the whole substrate is likewise of living origin. But its scope for elevation is abruptly cut short, at a little above the low water mark of spring tides. In only a few places can living corals grow further up than this, as where pools or the seaward margin are replenished or swept by surge. Normally the microclimate between tides proves altogether too harsh for live coral, and on low islands and sand-cays few elevated substrates would be on offer. Instead there is the vertical extension offered by depth, with communities

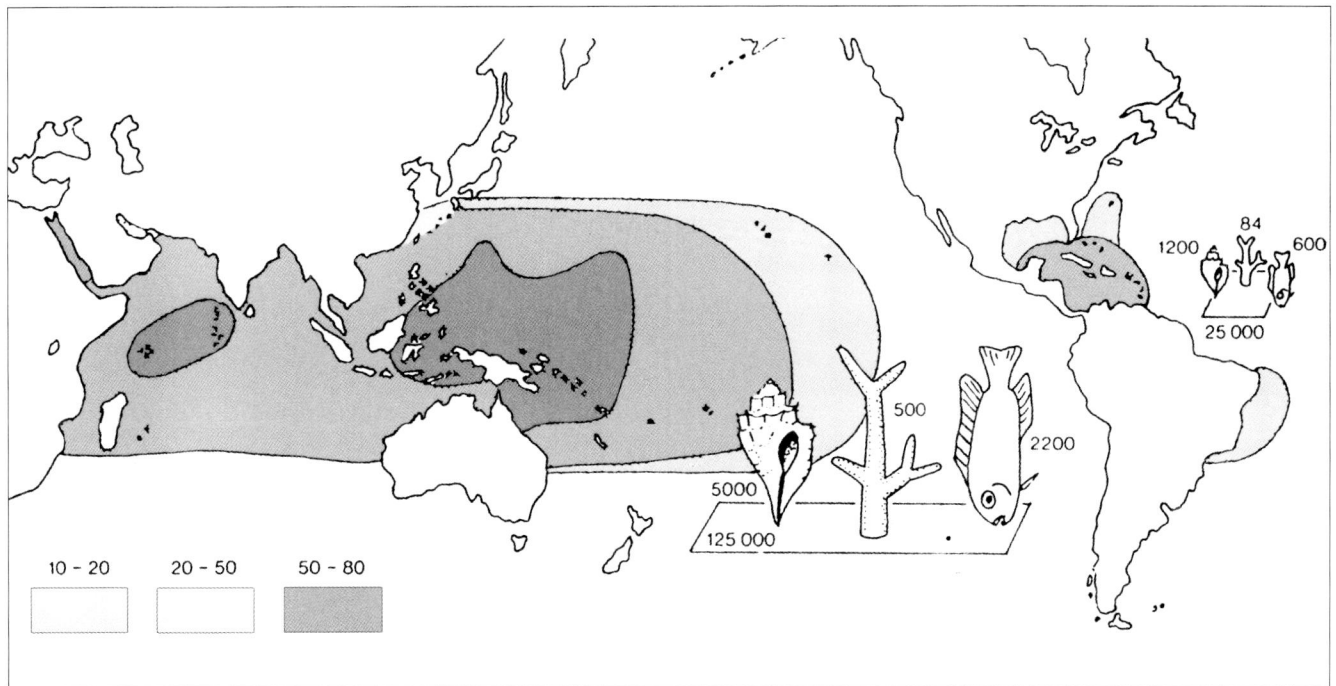

Fig. 20.3 The Indo-Pacific and Atlantic coral regions
In a comparison of the surface extent and biotic richness, the reef area of the Indo-Pacific totals 125,000 sq km and for the Atlantic 25,000 sq km. The number of coral genera in each zone is indicated in shading. Relative numbers of shelled molluscs, acroporid corals and fishes are also shown.

proliferating within the dark interspaces or boring into the coral itself. Spatial diversity is enhanced by positive feedback, leading to the establishment of still more niche space. Each of these opportunities, and the kinds of community they have given rise to, will be traversed in our detailed dissection of the reef habitat.

A coral reef's nearly level outward growth happens at the threshold of the **sublittoral fringe**, just where — on temperate shores — the biota so much begins to diversify, with its release from the climatic constraints experienced upshore. In the temperate region we have been regarding the sublittoral fringe as a narrow, high-enriched tail-piece to the wider eulittoral shore. On a coral reef, the sublittoral fringe is so horizontally pro-longed — sometimes to as much as 1000 m — that the zoned eulittoral is left as a near-forgotten prelude to this new, enriched substrate, with all its surface alive or the direct product of living things.

The horizontal extension of reef-building is the work of two kinds of lime-secreting organism — animal and plant. Above its thickness of accumulated limestone, the reef's still living skin is surprisingly thin. The first and — as commonly assumed — the prime reef-builders are the stony corals (Scleractinia), along with a few other kinds of calcified Cnidaria. Second are the calcareous red algae (Rhodophyta), found right through the temperate, but making their most prodigious contribution in the tropics.

Corals essentially like today's have existed since the

Triassic age, and reefs of more ancient corals are as old as the Cambrian. Modern reef corals are geographically limited by sea temperature, as well as by illumination, water movement, salinity and turbidity. Thus, the water temperature must not fall below 18°C and should preferably average several degrees higher. Most coral grows in waters shallower than 30 m; beyond this corals are smaller or solitary, never *hermatypic* (i.e. with growth forms massive enough to construct a reef). The upper limit of coral growth is held at around low water. Above this level, most polyps are killed by more than a brief exposure to desiccation, being at once vulnerable to temperatures higher than 36°C.

All hermatypic corals live in symbiosis with unicellular algae called *zooxanthellae*, said to be 'farmed' in their tissues. Their prime requirement is thus shallow horizontal spread under high illumination. Though the polyps of most species are able to capture zooplankton, chiefly at night, daytime nutrition switches to reliance on symbiotic algae, that in turn are thought to receive their nitrogen requirements from the coral's excretion.

Coral growth is favoured by moderate water movement, bringing a constant supply of nutrients and oxygen. On weather coasts, corals are in danger from storm damage by breakers, most of all in hurricanes. Under extreme shelter, high turbidity is inimical to most corals. Clouding with fine suspended matter reduces light for photosynthesis, and the feeding systems of the

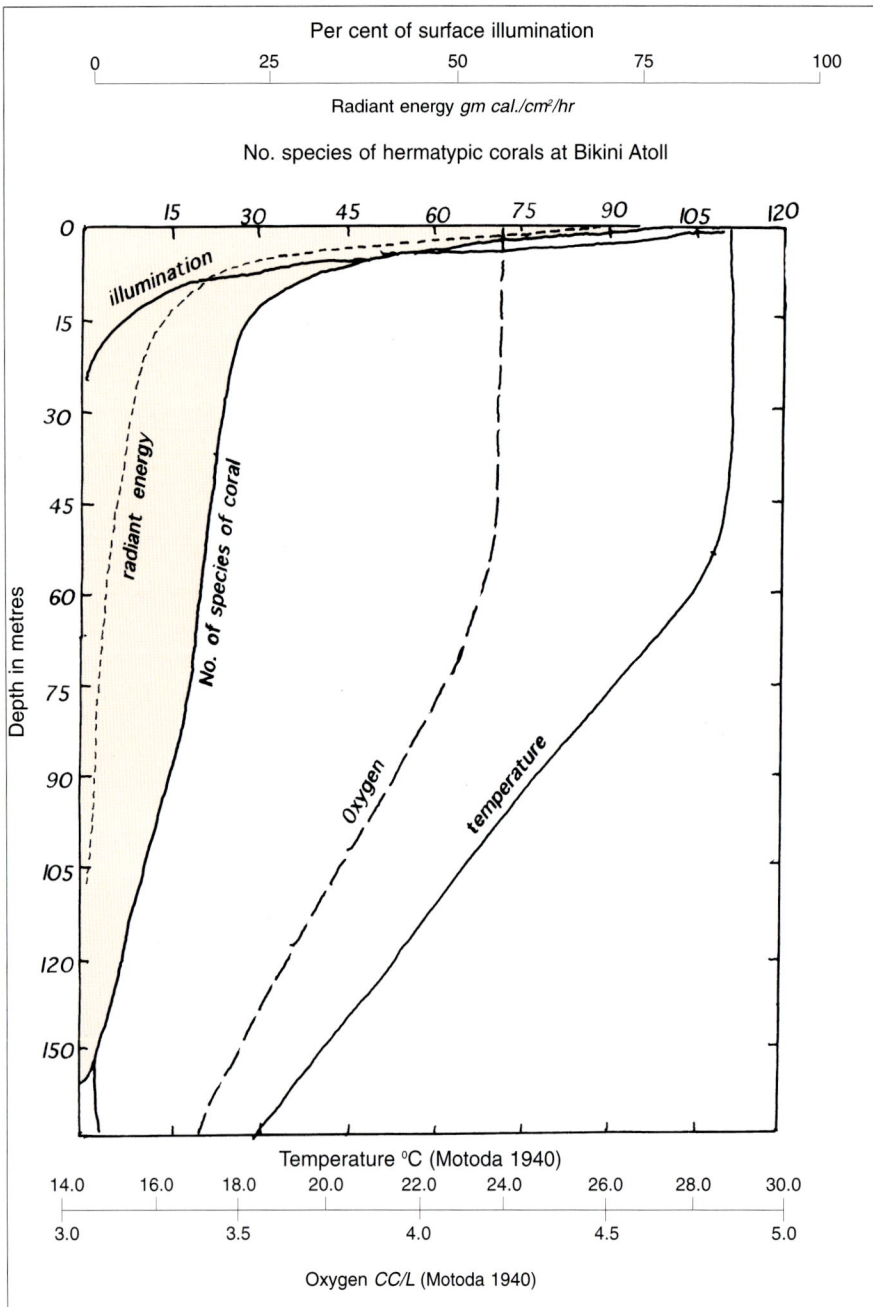

Fig. 20.4 Bikini atoll: coral distribution
Graphical analysis of number of hermatypic coral species according to depth, oxygen, temperature and radiant energy. (After J.W. Wells.)

polyps become clogged with silt. Certain corals, however, can remove sediment by efficient cleansing cilia. Some can also tolerate lowered salinity, such as the *Porites* species that can occupy sheltered, land-locked reaches, but the rigorous limits of salinity lie between 27 and 40 parts per thousand.

Zooxanthellate corals live close to their upper limit of temperature tolerance and become stressed with even a slight maximum temperature increase of 1–2°C. They will then expel their algae and turn white. Such 'bleached' corals may recover in part or die completely. In the 1990s bleaching throughout the tropical world has increased dramatically during summer, with large-scale loss of many corals at all depths. It is now clear that with continued climate-warming reef corals around the Pacific are likely to suffer even greater bleaching, with increasing coral die-off in the coming decades.

Of the geographical distribution of corals, John Wells has written:

Coral reefs are scattered over an area of 190,000,000 square kilometres, wherever a suitable substratum lies within the lighted waters of the tropics beyond the influence of continental sediments, and away from the cool upwellings of the sea in the eastern part of the oceans' basins.[1]

The confined deltaic coasts of south-east Asia, with their fluctuations and high extremes of temperature, turbidity and salinity, are in contrast largely avoided by corals. In the Indian Ocean corals less often form emergent reefs. The Caribbean and parts of the Gulf of Mexico are the only coral coasts that faintly rival the Pacific, though corals flourish southwards through the

Fig. 20.5 Zonation above the coral-line: exposed south coast of Upolu, Western Samoa

a–b littorines, *Grapsus* and *Siphonaria*; b–c *Chthamalus* and *Capitulum*; c–d *Planaxis* and *Isognomon*; d–e dappled coralline, *Clypeomorus* and *Modiolus*

1 *Scaevola taccada* 2 *Pythia scarabaeus* 3 *Melampus fasciatus* and *M. lutueus* 4 *Nodilittorina feejeensis* 5 *Nerita plicata* 6 *Littorina coccinea* 7 *Siphonaria normalis* 8 *Morula granulata* 9 *Planaxis sulcatus* 10 *Clypeomorus batillariaeformis* 11 *Conus ebraeus* 12 *Coenobita rugosus* 13 *Grapsus albolineatus* 14 *Isognomon perna* 15 *Capitulum mitella* 16 *Chthamalus intertextus* 17 *Modiolus auriculatus* 18 *Periophthalmus.*

Antilles to east Brazil. Corals are sparse in the eastern parts of oceans, where the west coasts of the great continents experience cool currents and cold upwelling. In western America, coral reefs are restricted to two relatively short stretches from the Gulf of California to around latitude 2 degrees south. The eastern Indian Ocean has better reefs, with the shorter reach of West Australia deflecting less cold water north than along the west coasts of South America and Africa.

Remote from the great continents, round the small islands of the tropical Pacific, the coral sublittoral raises its skin of polyps-with-symbionts up to the light. The white line of surge marks off the living reef from the deep blue ocean beyond, while in the clouded green lagoon, the diversity of life can outrun even that of the reef edge.

Above the coral-line

Before we reach the living coral, the upper and middle tropical shore will show us a zonation essentially comparable with the temperate. We shall thus begin with a representative description of a tropical Pacific supralittoral zone, littoral fringe and eulittoral zone. Fig. 20.5 depicts a zonation on basalt volcanic bedrock, on Upolu Island, **Western Samoa**. In the far west Pacific, with terraced shorelines becoming progressively elevated, the comparable zones are bedded upon upraised coral limestones.

As most of the world, the prime denizens of the **littoral fringe** are periwinkles (**Littorinidae**). In the tropics the larger snails of the **Neritidae** rival or even eclipse the periwinkles at high levels. The familiar limpets of the temperate, the patellids and acmaeids, are outnumbered, if not quite ousted, by the air-breathing **Siphonariidae**. Aberrant pulmonates common on sun-warmed upper shores are the naked, rubbery skinned slugs, *Oncidium*, attaining to the large size of 10 cm and lying cool and clammy in their home scars. A scar is formed also by the large chiton *Acanthopleura gemmata*, up to 50 mm long, with its wide girdle set with stiff papillae. This is one of the rather few tropical chitons, and the only one common on the open surface.

Above the littoral fringe, not always sharply delimited, is the **supralittoral zone**. In the tropics, maritime lichens are almost lacking; the splash and spray they most need is held back by the far-out reef edge breaking the main force of the waves. Often too the dense land vegetation of the adlittoral zone will press right down to the littoral fringe. The typical gastropods of the supralittoral zone are the primitive pulmonates of the **Ellobiidae**, especially the *Melampus* species. The larger *Pythia* species live in the adlittoral zone, being thus entirely terrestrial.

Important too in the supralittoral zone are the semiterrestrial hermit crabs, **Coenobitidae**. A little lower down, grapsid crabs appear, with small, fast *Cyclograpsus* under high level litter, and a larger *Grapsus* species scuttling about on the open faces.

On the basalt slopes of southern Upolu, Western Samoa, the first and widest stretch below the adlittoral vegetation is a **littoral fringe**, bare or covered with the brown tomentum of a *Bostrychia* species, or the dark film of a cyanobacteria (blue-green 'alga'). The two periwinkle species are *Nodilittorina feejeensis* and, in open situations, the pale, apricot-mouthed *Littorina coccinea*. The cream *Nerita plicata* is common at high levels. Here and below, *Siphonaria normalis* is the widespread limpet.

The top of the **eulittoral zone** on temperate shores — we will recall — was marked with a strong barnacle zone. In the tropics, especially on hot, inhospitable basalt, the barnacle strip is far more fragmentary, being well developed only in shade. Many tropical shores have also a middle eulittoral bivalve zone, below the barnacles, made up of oysters (*Crassostrea*) with small mussels (*Septifer* or *Brachidontes*). In Samoa, eulittoral bivalves are represented only by the small wafer-flat *Isognomon*, scattered in crevices up to the very top of the eulittoral.

The highest-reaching barnacle species is the pedunculate *Capitulum mitella*, wedged singly or in clusters into crevices. Like its temperate relatives, this is an ancient splash-feeding form, adapted for a high relict habitat. The smaller operculate barnacles are *Chthamalus malayensis* and *Chthamalus intertextus*.

In the eulittoral, the littorines are replaced by their larger relative, *Planaxis sulcatus*, black and white chequered and often densely congregated. Lower down, *Isognomon perna* forms continuous rows in crevices. Here or still lower lives the small mussel *Modiolus auriculatus*.

On the pink dapple of basal coralline below the barnacles, the cerithioid snail *Clypeomorus batillariaeformis* clusters in dozens. Common snails of the mid-eulittoral surface are the thaid predator *Morula granulata*, with the mitrid *Strigatella scutulata* a little lower, and the turban shell *Turbo cinereus* on coralline. Along the lowest fringe, there is the small, black spotted *Conus ebraeus*. Large numbers of mudskippers, *Periophthalmus*, sit up vigilantly at this level. This is a small, alert fish, with spherical periscopic eyes. Raised on its arm-like pectoral fins it makes parabolic leaps with thrusts of the tail.

Under strewn boulders at eulittoral level further gastropods will be found, including the smooth shelled *Nerita polita*, the flat-based *Nerita atramentosa*, a small pointed nerite, *Clithon* species, and the spiral banded *Planaxis lineata*.

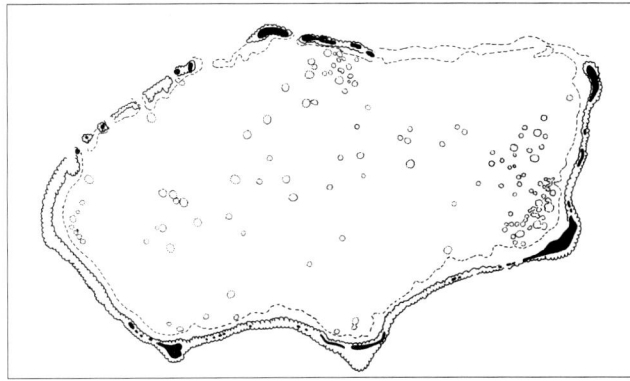

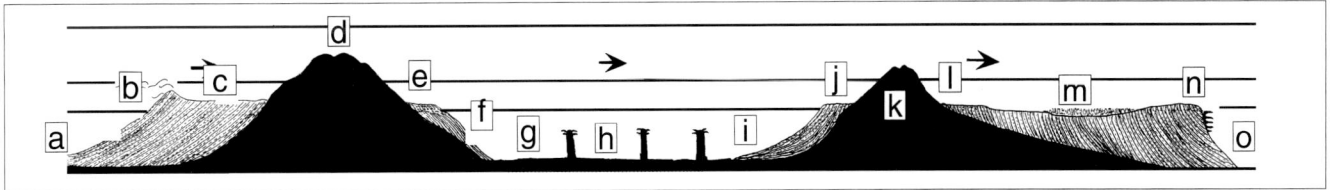

Fig. 20.6 An atoll system in section showing reef structures from windward to leeward

(a) seaward slope to windward rising to an algal ridge (b) behind which is the windward reef flat (c) fringing the islet (d). The islet fronts the lagoon with a short lagoonal reef-flat (e) falling by a slope (f) to the lagoon floor (g). The lagoon has pinnacles and coral knolls (h) and at its other side, facing into the prevailing easterly wind is a windward lagoon slope (i) topped by a reef-flat (j). Leeward of the island (k) begins the leeward reef with its moat (m) and raised edge (n) with a steep slope to seaward (o).

The effect of the east-west trade wind is shown, with a surface current windward to leeward, and return of water as a deep current, upwelling to the leeward side of the islet (d).

(**top right**) Map of an atoll, showing islets, and submerged lagoonal knolls and patch-reefs.

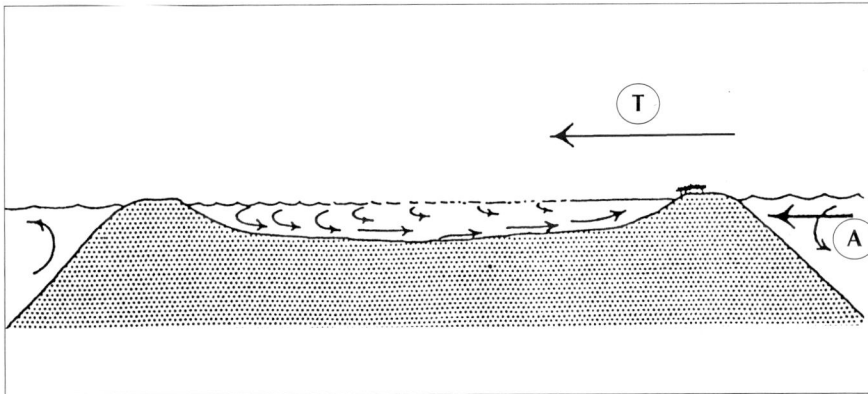

20.7 Circulation in an atoll lagoon
The surface current is caused by the trade wind (T), but most of the water returns as a deep current that upwells on the sheltered side of the lagoon. Energy is thus conserved within the lagoon, but the atoll's maintenance and growth depends on the constant entry of a supply of energy (zooplankton) from the open ocean (A).

Reefs and reef building

The classical oceanic reef form is the *atoll* — a rim of emergent coral breached by channels and enclosing a central lagoon. Here alone can we find a reef system complete in all of its possible transitions, from leeward to windward, from reef-front to backshore, and from oceanic exposure to close lagoonal shelter.

The atoll lagoon may be as much as 60 km across, and is ordinarily more than 50 m deep. At one or more points the broken reef circle may rise to a fully emergent islet. Built of sand from the weathering of corals, foraminifera and calcareous algae, such islands tend to

shift progressively in the direction of the predominant wind. The cross-section (Fig. 20.6) shows the morphology of a whole atoll system.

Under the force of south-easterly trades, the *windward* side of the atoll develops notable differences from the leeward. To the windward we can recognise a surf-exposed seaward slope (a), rising to a calcareous *algal ridge* (b) that forms the site of the greatest wave action. Behind this raised strip, and protected from direct wave impact, a *reef flat* (c) of varying width runs back to landward. Though still carrying live coral, this is built chiefly upon dead and bio-eroded coral limestone and detritus

Fig. 20.8 Darwin's theory of atoll formation
Three stages of the submersion of a volcanic island are shown, with rising sea level at the stages: A. fringing reef B. barrier reef and C. atoll. (Modified from Davis 1928.)

re-cemented and encrusted by calcareous algae. There can be also large storm blocks of coral broken off from the seaward face and cast up on the reef-flat by hurricanes and typhoons.

The reef flat approaches the islet or low rubble bank or by an outer beach slope. A *lagoon beach* (d) of white sand drops to a *lagoon reef* (e) falling by a slope (f) to the lagoon. Here the floor (g) is strewn with detritus washed in from the reef, or the calcareous remains of lagoonal organisms. In the quiet lagoon waters, living corals are able to build up fragile *pinnacles* and *knolls*. (h) Sometimes more extensive *patch reefs* develop, remaining submerged, or even producing a second order of emergent '*lagoon atolls*', each enclosing a little lagoon of its own.

Though securely closed off from ocean waves, the lagoon waters can be stirred by tidal currents through the passages, or become turbid from waves raised by storm winds. But they are normally quiet, and their colour over the sand shoals is a cloudy turquoise green, in pallid contrast with the intense blue of the waters outside.

The further shore of the lagoon, as it faces the prevailing wind from across the enclosed waters, often has a *windward lagoon reef* (j) with its own slope (i) and reef flat fronting the lagoonal side of the principal island. (k). From the seaward front of the island, we can finally look out upon the wide flat of the *leeward reef* (l–n), far more extended than the reef lying to windward. The leeward reef has no distinct algal ridge, and

its margin, deeply indented and crevassed falls to the open sea, by a steep slope (o) typically overhung by a parapet of coral tables.

Fringing reefs

The familiar reef shores, easiest and most often visited, are the *fringing reefs* close-attached to pre-existing islands or continental shores. These may reach out from a few metres to half a kilometre across, and are usually submerged at high tide. They are often embayed or interrupted by stream mouths. Like other reefs, these fringes consist mostly of dead coral rock, with live corals on the wave-washed surface and the submarine outer face. During storms large coral blocks may be torn up from the subtidal reaches and thrown over to the reef flat; while strong waves are constantly removing detritus from the reef front and carrying it over into the lagoon.

Some fringing reefs may be protected by a barrier reef to seaward, with a deep boat channel in between. Or the fringing reef may itself stand off sufficiently from the shore to have its own boat channel, generally wadable at low tide and carpeted with a field of sea-grasses.

Barrier reefs

Much wider than fringing reefs, being sometimes up to a kilometre across, barrier reefs stand off further from the shore, with a lagoon perhaps some 40–80 m deep

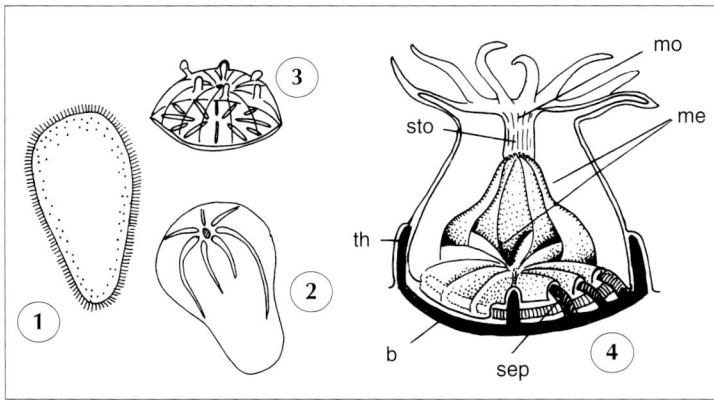

Fig. 20.9 Larva, corallite and polyp in Scleractinia
1 ciliated planula 2 post-settlement stage 3 rudiments of
tentacles and septa 4 simplified polyp structure; (b) base,
(th) theca, (sto) stomodaeum, (mo) mouth, (me) mesentery,
(sep) septum.

and from one to as many as 15 km across, sometimes
with its own inshore fringing reef.

The block diagram (Fig. 20.6) shows a submarine
section of a terraced mass of reef deposits crowned by
an emergent rim. Where a fringing reef forms a small
terrace, rising from its deep foundations, the barrier reef
builds a massively larger terrace. In either case the total
structure, not just its exposed rim, must be referred to as
the 'reef', with its whole mass derived from the long-
continued activities of reef builders.

The barrier reef — like the fringing reef — is strewn
with sediment and rubble, including dead coral blocks
of all sizes. Living coral is generally in much smaller proportion. Active cementation takes place below the sur-

face, with the solution and redeposition of calcite. There
can be also chemical alteration from calcite to the alternative molecular form of dolomite, giving rise to the special formation called 'beach-rock'. The seaward edge of
the barrier reef may be either precipitous or show a gentle outward slope, mainly of detritus, to about 100 m,
followed by a steep drop at a near 45 degree slope.

Colonial polyps

The *hermatypic* or reef-building corals — along with the
sea anemones, jellyfish and hydroids (including the
freshwater hydras) — belong to the large Phylum
Cnidaria. The older name, Coelenterata, still to be
found in some books, refers to the single interior space
or *coelenteron*, that serves both as body cavity and gut.
A thin *mesogloea* lies between the external body wall
(*ectodermis*) and the inner layer (*gastrodermis*) that lines
the gastro-vascular cavity. A single opening, the mouth,
is usually encircled by tentacles beset with stinging cells,
each equipped with the explosive device known as a
nematocyst. These kill or numb small zooplanktonic
prey that alight against the tentacles.

In the stony corals the polyps are like skeleton-form-
ing anemones. Some corals are solitary, such as *Fungia*
with polyps as much as 25 cm across, but most polyps
are small and colonial, mostly between 1 mm and a few
millimetres in diameter. The entire colony (*corallum*)
results from the continued asexual budding of a polyp
system from an initial sexually produced parent polyp.
There are two methods of reproduction. Either a sexually ripe male polyp releases sperm into the surrounding
water, which swim through the mouth of a female polyp
to fertilise the eggs internally, or both the males and
females release their gametes into the water column for
external fertilisation. The eggs develop into ovoid larvae
called *planulae*, up to 1.6 mm long and covered with
cilia, destined for a spell of free-swimming.

Planulae of some species settle within a few hours,
others drift for days to months before fixing to a hard
substrate to grow into tiny polyps. Each young polyp
secretes a supporting cup (*calyx, calice*) of calcium carbonate (*aragonite*) into which it can withdraw for protection. From the bottom of the cup, thin radial
partitions (*septa*) grow upwards into the polyp base.
Each septum lies between a pair of *mesenteries*, which
are the radial folds dividing up the coelenteric cavity. In
most corals, the calyx has a vertical central pillar or *columella*. The calices with their skeletal parts grow
together to form the whole *corallum*. It is the details of
hard skeletal structure that provide the basis for the classification into families, genera and species.

The living polyps remain interconnected where a
double-layer of soft tissue spreads over the calyx rim to

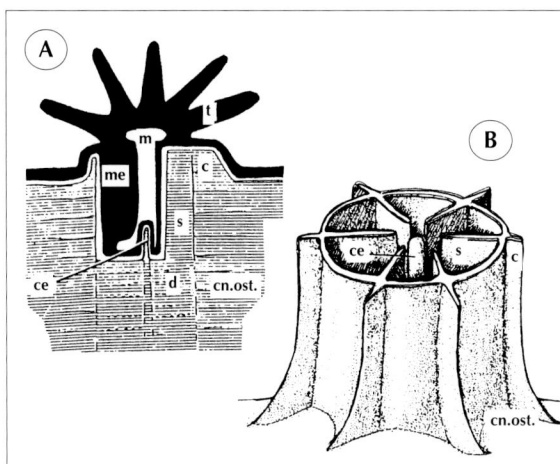

Fig. 20.10 Polyp and theca
A. Polyp overlying calcareous corallum B. Structures of
theca.
(cn.ost.) coenosteum, (col) columella, (c) calyx,
(d) dissepiment, (m) mouth, (me) mesenetery, (s) septum,
(ce) columella, (t) tentacle. (From Schuhmacher.)

Fig. 20.11 The growth forms of Acropora

1 *Acropora* cf. *formosa*, Stag's-horn arborescent 2 *A. corymbosa*, narrow-based and short-branched, wide-spreading 3 *A. hyacinthus*, central stalked, wide plates 4 *A. reticulata*, side-stalked, horizontal brackets 5 *A. humilis*, low bushes with short, strong branches 6 *A. palmeri*, encrusting, with branches reduced or lost 7 *A. palifera*, massive, club-branched, sometimes in micro-atolls 8 *A. echinata*, in dense bottle-brushes 9 *A. longicyathus*, long, fragile calices 10 *A. granulosa*, dense bushes of small calices. Branch-tips with terminal calices shown with (1) and (9).

fuse with a neighbouring fold. These tissues are referred to as *coenenchyme*, and contain an extension of the gastrovascular cavity. The lower (epidermal) layer secretes the *coenosteum*, which is that part of the skeleton covered in life with soft tissue, that spreads between the individual polyps.

In each polyp the oral disc carries a slit-like mouth. Tentacles are arranged in cycles of six (hence the alternative name for the scleractinian corals, **Hexacorallia**). There may be a single marginal cycle of 6–12 tentacles or several alternating cycles, in sixes or multiples (6, 12, 24, etc).

From the stinging cells the nematocysts shoot out barbed threads to paralyse microscopic prey which is then passed by cilia up the tentacles to the mouth. Food is digested extracellularly by enzymes from cells in the glandular edges (filaments) of the mesenteries. These folds provide also a wide internal surface for absorption and for excreting waste products. After digestion, all waste and irreducible remains are ejected by the mouth.

One important characteristic of reef corals is that their polyps contain minute, single celled algae called zooxanthellae. The presence of these symbiotic algae

enables reef corals to grow much more rapidly than other corals lacking such algae, but because zooxanthellae require light to photosynthesize, reef corals are restricted to shallow sunlit waters.

Classification of corals

The first and only enlightened advice must be not to collect living specimens at all. Field notes and colour photographs, with details of growth habit and branching, can generally give important help with species identification. Dead and dried portions of a corallum can be examined with a hand lens in the field to verify detailed structure. We have come to an age where both conscience and international regulation (CITES) firmly rule out over-collecting.

For the beginner, John Veron's *Corals of Australia and the World* and the recently published *Corals of the World* provide excellent colour photos and diagrams for identifying species. In more advanced study, especially with the prolific and difficult *Acropora*, *Montipora* and the Faviidae, monographs published by the Australian Institute of Marine Sciences can be consulted.[2] There are

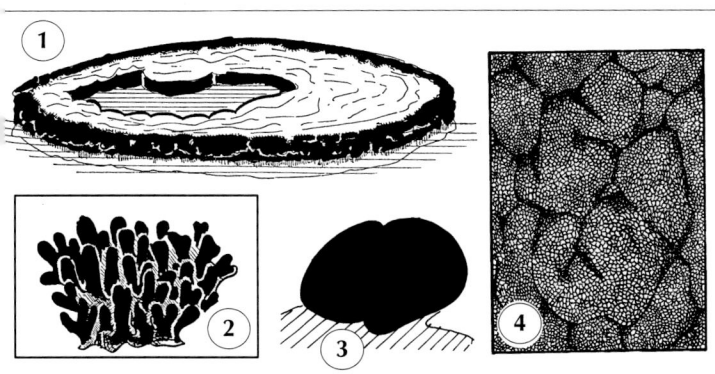

Fig. 20.12 The features of *Porites*
1 micro-atoll of *Porites lutea* 2 blunt finger structure of
P. nigrescens 3 convex mound or helmet of *P. lobata*
4 surface detail showing corallites.

few western Pacific species that these works do not include.

Order Scleractinia

By way of introduction we shall look at just four scleractinian families likely to be encountered on the first crossing of a reef. However hesitant as to the naming of species, the beginner need find no difficulties with families, and there are few pitfalls identifying down to genera. At this level, distinctive colours, growth forms, and habit will all be helpful, and can be picked up at once from field observation.

Family Acroporidae

Contains two species-rich genera: *Acropora* and *Montipora*. At present, around 170 species of *Acropora* are recognised, accounting for something like 25% of all living corals. *Acropora* contain a wide variety of growth forms, even within the same species in different habitats. Growth forms include the following:

A. **Arborescent:** colonies with tree-like branches, interlaced and usually fragile, e.g. *grandis*, *formosa*.
B. **Bottle-brush:** fragile, attenuate, densely branching, with long calices, e.g. *echinata*, *longicyathus*, *carduus*, *subglabra*.
C. **Corymbose:** heads wide-spreading, with branches short, upright and closely massed, e.g. *millepora*, *aculeus*, *tenuis*.
D. **Digitate:** colonies with short, non-dividing branches, e.g. *humilis*, *digitifera*.
E. **Table or plate-like:** forming flat colonies, either with basal stalk or encrusting on substrate, e.g. *hyacinthus*, *clathrata*, *cytherea*.
F. **Massive or club-branched:** forming ridges, flanges and low crusts, even micro-atolls, e.g. *palifera*, *cuneata*.

A second large genus, *Montipora*, is almost as variable in form as *Acropora*, though few of its species are branched. They lack the large terminal calices, and the corallites are small and somewhat featureless, with the surrounding coenosteum spiny, granular or verrucose. Common colours are grey, khaki, dull bronze or green.

Family Pocilloporidae

This family includes three genera having species that typically form branched colonies. The brittle thickets of *Seriatopora hystrix* can be recognised by their slender branches, tapered to sharp points, and — as the name picks up — by their linear rows of small, dark-rimmed corallites.

The *Pocillopora* species are of more robust build, mostly familiar throughout the west Pacific. The short, blunt branches bear small tubercles, each studded in turn with several corallites having only rudimentary septa and column. Their common colours are magenta, or sometimes greenish, with their branches chocolate-stained around the base.

Pocillopora verrucosa and *P. damicornis* grow in both exposed and sheltered habitats, with denser and more robust branching in exposed reef-fronts, and more delicate and open branching on sheltered reef flats. *P. verrucosa* has uniform upright branches, whereas *P. damicornis* is more irregularly branched.

The *Stylophora* species, typified by *S. mordax*, have short, club-like branches, with the surface made rough and scaly by the lamellae hooding each calice above. The columella is prominent.

Family Poritidae

The *Porites* species include some of the most widespread and successful Scleractinia. These are usually the first corals to be encountered in shallow moats and on the landward stretch of the reef. Their small calices are closely set, with little coenenchyme in between, and have porous walls which make the whole corallum very light. The fully extended polyps reach out like long villi. The light corallum in *P. lobata* or *P. lutea* may grow rapidly, to form a large hemispherical or helmet-shaped colony in deep water or a micro-atoll a metre or more across in intertidal areas. The growth of a micro-atoll is planed off flat just beneath water level at low tide. Incremental growth continues peripherally, as the centre dies, or hollows out to form its own 'lagoon', lined by living coral. In addition to massive coralla, *Porites* has bluntly branched species: (*P. cylindrica*, and *P. nigrescens*) and these too may form micro-atolls intertidally.

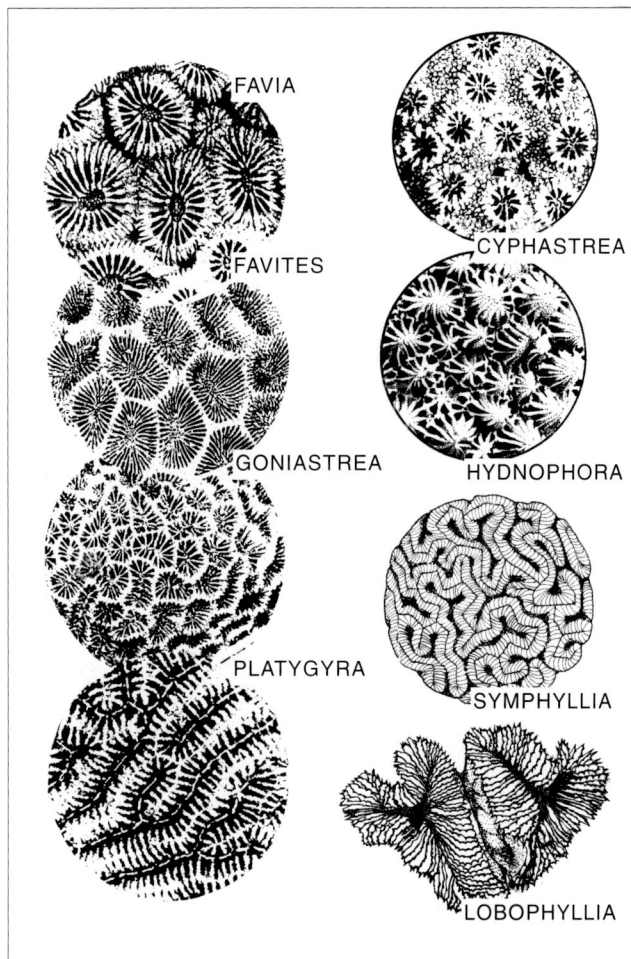

Fig. 20.13 The morphology of faviid and faviid-like corals
Faviidae: *Favia* (= plocoid) *Favites* (= cerioid) *Goniastrea* (= cerioid towards meandroid) *Platygyra* (= meandroid) *Cyphastrea* (= plocoid)
Merulinidae: *Hydnophora microconos* (= cerioid to meandroid)
Mussidae: *Lobophyllia* (= phaceloid to meandroid) *Symphyllia* (= meandroid)

The massive or branched *Goniopora* species have their calices more separated than in *Porites*, with polyps having 24 tentacles rather than 12. *Alveopora* polyps have 12 tentacles, the septa are reduced to spinose rudiments and the thecal wall forms a light lattice. Polyps of *Goniopora* and *Alveopora* normally extend during the day.

Family Faviidae

Corals in this family generally form convex heads or crusts that — like poritids — may ultimately grow into micro-atolls. The faviids commonly have rather large corallites and massive growth form, though some species are explanate or branched. There are several different kinds of corallite arrangement within the

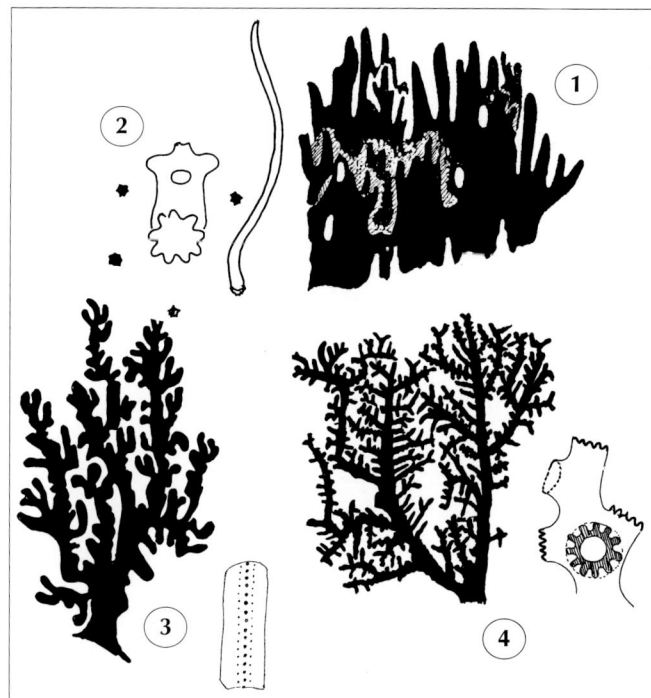

Fig. 20.14 The hydrocorals
1 The common growth form of *Millepora platyphyllia* 2 reticulate branching of *Millepora*, showing tentaculate zooids 3 *Millepora*; arrangement of gastrozooids and dactylozooids 4 *Distichopora* 5 *Stylaster*, each with detail of zooid arrangement.

Faviidae, as typified by the genera *Caulastrea*, *Favia*, *Favites*, *Goniastrea* and *Platygyra* (Fig. 20.13).

Caulastrea has *phaceloid* form with separate monocentric corallites in free-standing, parallel branches. The widely familiar *Favia* and *Favites* both have large calices with bright polyps, green or brown-centred. *Favia* species are *plocoid* with circular or sub-hexagonal cups, separately rimmed. In *Favites* the cups are cerioid, forming close-packed hexagons with sharp common divides. Species of *Goniastrea* are cerioid or rarely *sub-plocoid*, while some become *sub-meandroid*, with branching or polystomodaeal calices.

The *Platygyra* species (together with *Leptoria phrygia*) constitute the 'brain corals' of the Indo-Pacific, having *meandroid* corallites with multiple centres, arranged in linear, convoluted and branching series.

The faviids described above all grow by intra-tentacular budding off of corallites, but some other faviids grow by extra-tentacular budding. Among the latter, *Cyphastrea* species possess small, upstanding corallites, with the coenosteum between them wide and spinose. In *Montastrea* the corallites are larger, with strong, beaded costae (longitudinal ridges on the outer wall of corallites). *Plesiastrea* has smaller corallites, sub-cerioid or plocoid, with a well developed crown of pali round the columella. *Leptastrea* is similar with a columella composed of vertical pinnacles. *Diploastrea* is plocoid, with well developed columella, and characteristically forms large dome-shaped colonies.

Finally, the **Echinopora** species are unusually built fragile faviids, living remote from strong wave action. They are plocoid, generally with a spinose coenosoteum. They are foliaceous or branched: *E. lamellosa* is typically foliaceous, built of flat sheets, whereas *E. horrida* is arborescent with contorted branches.

Hydrozoan corals

In two families — not closely related — of the cnidarian Class Hydrozoa, the hydrorhiza has become calcified to give a hard branched axis, in which the polypides rest in small pits. The bright-coloured **Stylasteridae** are confined to deep-shaded retreats. The **Milleporidae** or 'fire corals' live in the open and will be recognised on the first visit to a reef edge.

The fire corals can at the first encounter be distinguished from Scleractinia by the painful sensation from their stinging cells on the more sensitive skin, as beneath the wrist. They have no projecting corallites, but a rough shark-skin surface with tiny polyps sunk in pits. Larger pits contain the *gastrozooids*, surrounded by a circlet of smaller pits for the more numerous *dactylozooids*. The *Millepora* species are variable in growth habit, forming thickets of branches, or — under high water movement — strong upstanding lamellae.

Darwin and coral reefs

In the few brief island stops that the *Beagle* made in the Pacific and Indian Oceans, Darwin saw only three coral reefs close up. His comprehensive but essentially simple theory of reef formation had been worked out in its essentials in South America before he reached the coral seas. Like his grand doctrine of natural selection, it was based on a discerning synthesis of the literature with the penetrating observations Darwin was to make for himself.

Setting out west from the Galapagos in late 1835, Darwin first observed from a distance the atoll reefs of the Low Archipelago. Later, from the mountain tops of Tahiti he was to look across to the volcanic island of Eimeo. The surrounding barrier reef with its circle of white breakers he compared to the frame of a picture. Its marginal mounting was the glassy lagoon and the drawing itself was the island.

Darwin first landed on a coral shore when the *Beagle* anchored within the lagoon of Cocos Keeling Atoll, 900 km south-west of Sumatra. Viewing the atoll first from within, he wrote that 'the shallow clear and still water of the lagoon, resting in its greater part on white sand, is, when illuminated by a vertical sun, of the most vivid green'.

This expanse of lagoon, within its atoll ring, several kilometres across, was 'divided by a line of snow-white breakers from the dark heaving waters of the ocean'. Landing on the lagoonal side, with its rookery of gannets, frigate birds and terns at rest on the trees, he then walked over the outer reef flat to gaze down upon the gullies and hollows where the surf pounded in. Finally, winding through the delicately branched stags-horns, of *Madrepora* (= *Acropora*), he reached the island at the head of the lagoon, withstanding at its outer edge the full force of the south-east trades. The ocean 'throwing its waters over the broad reef appears an invincible, all-powerful enemy, yet we see it resisted and conquered by means which at first appear most weak and inefficient'.[3]

Having thus visited briefly a barrier reef and an atoll, the *Beagle* stopped from 29 April to 6 May 1836 at a fringing reef in Mauritius. Though he could not visit the eastern side exposed to heavy surf, Darwin inspected the gently sloping leeward reef and made some soundings.

His classic book *The Structure and Distribution of Coral Reefs* appeared in 1852. A comprehensive summary of nearly three-quarters of a century's subsequent debate is to be found in W.M. Davis' volume *The Coral Reef Problem* (1928).

By their continued growth, coral reefs resting on slowly subsiding foundations would normally maintain their crests at sea level. Fringing reefs could be established on island or continental coasts and Darwin showed such a reef would if submerged become the precursor of a barrier reef. With the sinking of its foundations, the reef lagoon would deepen, with its ultimate depth determined by the rate of sinking and infilling with sediment. Though some sediment may be lost by flowing out through the narrow passes, the lagoon floor is always smooth with greater deposition than removal. With a long stationary period, a lagoon may almost be filled up, but no large lagoon has ever been found completely occluded.

Fringing reefs were held by Darwin to be the foundation of all others. On subsiding coasts, they could first arise as new reefs attached to the land. Subsidence of the foundation would lead to continued upgrowth of a fringing reef, transforming it — as Darwin postulated — into a barrier reef with a lagoon. New fringing reefs of a second generation may grow along the receding shoreline of the subsiding island, as the lagoon widens. These do not in their turn become barriers but, as Davis put it, 'sidle progressively up the slope and hold to the shore'. This could be due to the enfeebled growth of corals within the lagoon, though a few good examples are known of double barriers, the best perhaps at the

Marovo Lagoon at New Georgia in the Solomon Islands.[4]

Building up during subsidence, barrier reefs become very thick. Reefs 70–100 m through preclude any idea of a barrier reef growing out upon a stationary platform at small depth. By extending the land profile below low water, Darwin could estimate depths of up to 400–1000 m for the barrier reef's foundation.

From barrier reefs Darwin went on to account for atolls that would come into being by the continued upgrowth of a barrier reef after the central island had entirely disappeared by submergence. The penultimate stage was an 'almost atoll' with an island summit at the middle of a wide lagoon.

The alternative theory of atoll formation — held by the great geologist Sir Charles Lyell for many years until he was convinced by Darwin — involved upgrowth from shallow submerged crater rims. Darwin showed the unlikelihood of so many craters forming just a few fathoms below the surface with none appearing above it.

Surveying the distribution of all the coral reefs and the volcanoes active in historic or prehistoric time, Darwin constructed an early map, colouring atolls dark blue, barrier reefs light blue and elevated reefs — as well as active volcanoes — red. He wrote:

It is impossible not to be struck, first by the absence of volcanoes in the great areas of subsidence, tinted pale and dark blue, and secondly with the coincidence of the principal volcanic chains with the parts coloured red. Reefs coloured red and blue, being produced under widely different conditions, are not indiscriminately mixed together. No large group of reefs and islands supposed to have been produced by long continued subsidence lies near extensive areas of coast coloured red, which are supposed to have remained stationary or to have been upraised.[5]

As a comprehensive explanation of reef origins, Darwin's theory of upgrowth on subsiding foundations won broad acceptance. In its simplicity and grandeur of conception, wrote the geologist Sir Archibald Geikie, in 1882, 'no more admirable example of scientific method was ever given to the world'.

It is only when it can account for other things it was not invented to explain, even that were not known to exist when it was invented, that a theory concerning the facts of natural science can be said to be abundantly demonstrated.

Certain implications of his theory Darwin was never able to verify, such as the internal structure as disclosed by a sectional profile of any reef afterwards elevated.

One independent verification would have been the embayed shoreline an island could be expected to show after subsidence. Four years later Darwin was to observe in the Fiji group the dissected contour of drowned islands encircled with barrier reefs. As well as embayed shorelines Darwin's theory can be seen to imply unconformable shorelines between a reef and its foundation rock already weathered before its submergence. Though not envisaged by Darwin the theory's ability to explain such unconformities observed in elevated reefs is one of its strongest verifications.

Coral reefs and plate tectonics

Darwin's theory of coral reef formation is compatible with the modern theory of plate tectonics, which was formulated nearly 120 years later (in the 1970s). The areas coloured red on Darwin's world map, showing fringing and elevated reefs and active volcanoes, correspond well with the collision boundaries between the crustal plates (e.g. the Pacific Ring of Fire) and with active hot spot volcanoes. The collision forces between crustal plates result in rapid vertical displacements often pushing up the areas where volcanic arcs are erupting. Rising magma beneath active hot spot volcanoes heats up the surrounding crust and it becomes buoyant and bulges up.

The large blue areas showing the occurrence of coral atolls and barrier reefs, that he proposed had grown on subsiding islands, largely occur within crustal plates away from their margins. Many of these localities are where hot spot volcanoes have been active in the distant past, heating the crust and resulting in elevated volcanoes. These volcanoes are now extinct, as the crustal plate has now been moved past the hot spot by the plate tectonic conveyor belt. As a result the crust beneath the extinct volcano has slowly cooled with consequent slow submergence of the volcanic island and upwards Darwinian growth of barrier reefs, finally developing into coral atolls.

Another plate tectonic setting for these submerging land areas is mid-plate passive continental margins, such as Australia's Great Barrier Reef. Here again the crust is slowly cooling and the coast very slowly submerging with the development of the world's largest barrier reef, of Miocene age in the north.

Coral reefs and ice age sea level fluctuations

Our modern understanding of the astronomically induced Ice Age climate cycles during the last 2.5 million years, with dramatically fluctuating sea level every 40,000 to 100,000 years (Fig. 9.26) is a product of research on deep-sea cores in the latter part of the twentieth century. Thus Darwin never imagined that sea level had alternated 10 times in the last million years, between its present level and 100 m or so lower. These wild and often rapid, sea level fluctuations would have had a major impact on the coral reefs and their associated biota. Since most coral reef growth occurs in water shallower than about 50 m, there would have been a complete die-off of existing reefs with a sea level fall in excess of 50 m and similarly with a sea level rise of the same amount. With gradual sea level rise, upwards growth of the coral reef may have kept pace, but in many instances sea level rise was rapid (around 20 m per 1000 years) and corals could not keep up. Studies of modern barrier reefs and coral atolls[6] generally show that they have become established and grown up on top of the rubbly remains of the coral reef that lived on the site last time sea level was up this high (at least 70,000 years ago).

During the peak of each Ice Age period, when sea level was over 100 m lower, the tropics were not much cooler than present and coral reefs still grew in the tropical Pacific. In some places the remnants of coral reefs that lived 18,000 years ago during the peak of the Last Ice Age are present 130–150 m beneath the waves on the lower slopes of the volcanic islands or coral atolls, no matter whether the region is in the long term submerging or emerging. In many places in the tropical Pacific, the submarine topography at these depths is cliffed or steeply sloping and the amount of suitable substrate for coral reefs was limited.

Further up the slope there are often remnants of additional small reefs that became established and grew for a while at intermediate levels as sea level was rising. As a consequence our modern coral reefs are extremely young, largely having grown up in just the last 8000 years or less, since sea level reached within 20 m of its present level. While reef growth could not keep pace with the rate of sea level rise, the corals themselves and other reef organisms would have had little problem migrating upslope or in planktonic form drifting in by ocean currents from more distant refuges to colonise new submerging substrates. Modern coral reefs are generally 5–30 m thick (growth rates of 1–8 m per 1000 years). If they are barrier reefs or atolls they may cap a sequence of successively older reefs that grew during each higher sea level stand. Modern fringing reefs have usually become established directly on the bedrock of the island or mainland.

CHAPTER TWENTY-ONE

Crossing the Reef

Every stretch of coral reef has its own individuality. This may turn upon many features: its frontal exposure to wave attack, windward or leeward; the strength of the currents that sweep its lagoon, and the sediment or nutrient content in the runoff from the land behind.

The full extent of a reef can still be gauged at high tide from the constant line of white surge, breaking over a crest only just submerged when the tide is in. Zone by zone, the whole reef can be appreciated by a walk out to its margin at low tide. The tidal range is shorter in the Pacific tropics than in the temperate, so the spring-neap difference is less significant. More important for a day

on the reef will be calm weather, with no more than a light onshore wind.

The reef's character is influenced most of all by the winds. Every atoll ring or island in the south Pacific will have a windward exposure to south-easterly trades, and a leeward side with the winds offshore. A gradient from weather coasts to sheltered is thus set up.

As the map shows (Fig. 21.1), the reefs of the lee-shore are much wider than those to windward. On weather coasts, with greater wave attack and stronger currents behind the crest, reefs will be cleaner of fine sediment with their zones shorter and more clearly demarcated.

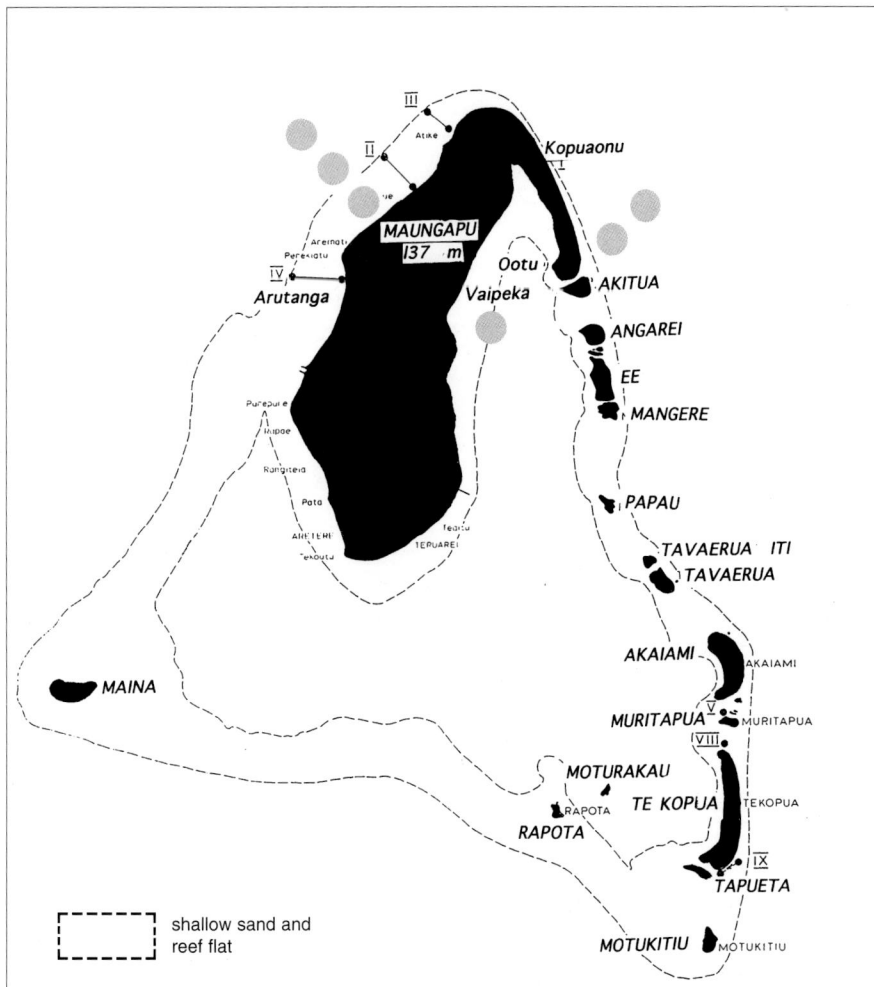

Fig. 21.1 Aitutaki, Cook Islands, showing location of shore transects (●, ●●, ●●●)

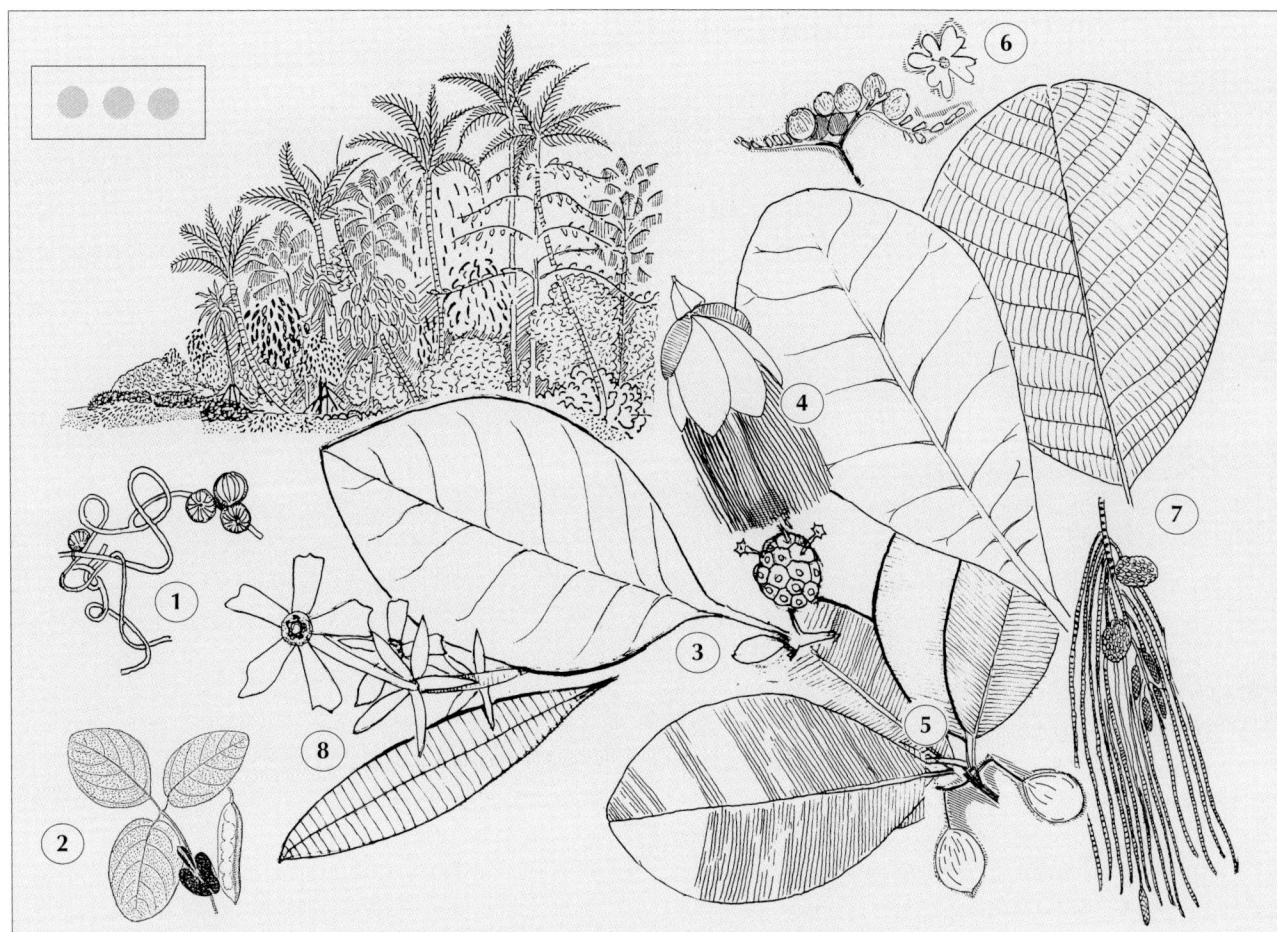

Fig. 21.2 Adlittoral vegetation behind leeward reef, Aitutaki
1 *Cassytha filiformis* (trailing on sand) 2 *Canavalia sericea* 3 *Morinda citrifolia* 4 *Barringtonia asiatica,* with flower
5 *Calophyllum inophyllum* 6 *Guettarda speciosa* 7 *Casuarina equisetifolia* 8 *Cerbera manghas.*

The reefs of Aitutaki

Aitutaki — one of the Cook Islands — lies 225 km north of Rarotonga in eastward Polynesia. Its reef system has the technical status of an 'almost atoll', and as well as its largest, elevated and inhabited island, the reef rim carries a half circle of small islets (*motu*). The whole atoll complex occupies a total of 106 sq km.

The wide-dispersed entities under Cook Islands administration include all types of oceanic island, some practically devoid of soil, others richly productive. Some are just low specks with the inhabitants dependent on fish, coconuts and the pandanus palm. Others are lofty volcanic islands, with a richly productive soil and vegetation. Of intermediate status there are raised coral platforms with thin soil and dry-scrub vegetation.

Though rich by any temperate comparison, the Cook Islands reefs are somewhat less diverse than in the west Pacific. The reefs of Aitutaki were surveyed by the *Cook Islands Bicentennary Expedition* of 1969[1] and were visited twice by the writer in 1985. The 46 genera of hermatypic corals recorded from Fiji, and 41 from Samoa, contrast with 26 from Tahiti and 18 from the Tuamotus. For the Cooks, Stoddart listed 24 coral genera, mostly from collections at Aitutaki.

Aitutaki's 'almost atoll' nicely combines almost all types of reef habitat. There are good outside reefs, both windward and leeward, also motu shores, and patch-reefs and pinnacles as well as soft flats within the lagoon. The main island is a volcanic residual of 16.8 sq km, rising to 119 m. On the atoll's southern rim there are two basalt and agglomerate outcrops, Moturakau and Rapota, with tails of calcareous sand. To the east is a line of motu built of coral debris, 13 if we count the largest, which is the long Ootu Peninsula, joined at the north to the main island. Maina, an isolated motu at the south-west, is a sand cay formed by wave refraction with a margin of beach-rock.

Landward vegetation

Each type of shore has its own backdrop in the vegetation of its adlittoral zone. The coastal bush backing the **leeward reef** is the richest and most luxuriant, with its skyline of *Cocos* palms and a

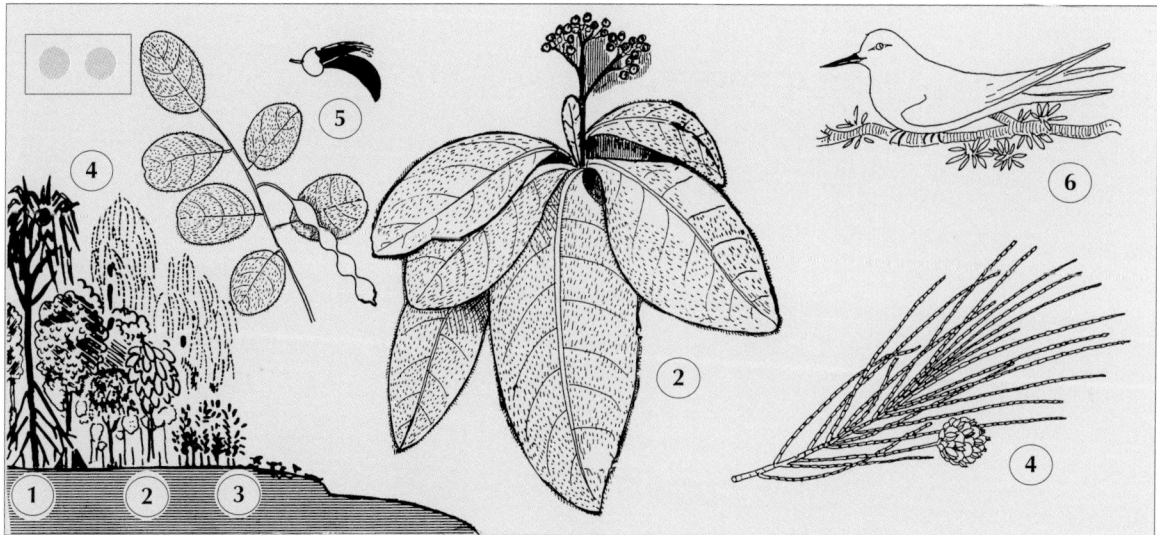

Fig. 21.3 Adlittoral vegetation behind windward reef, Aitutaki
1 *Pandanus tectorius* 2 *Tournefortia argentea* 3 *Scaevola taccada* 4 *Casuarina equisetifolia* 5 *Sophora tomentosa*, with flower (**inset**) 6 white tern, *Gygis alba*.

diversity of broadleaf species, including *Calophyllum inophyllum*, *Morinda citrifolia*, *Guettarda speciosa*, *Cerbera manghas*, and *Barringtonia* species. In the front rank are generally needle-branched *Casuarina*, with a shrub-line of the Pacific-wide *Scaevola taccada*, and the leguminous vines *Canavalia* and *Vigna*. The white beach has trailing, purple-flowered *Ipomoea pescaprae* and tangled yellow stems of the dodder-like *Cassytha filiformis* (Lauraceae).

The **windward reef** is backed with a tighter scrub, compacted and shaped by constant wind. The beach-front is pre-empted by shrubby *Pemphis acidula*, with its older trees on the foremost rock stacks developing root buttresses, reminiscent of the mangroves that are absent from the Cooks. Behind the sand strip, with *Scaevola* and *Ipomea*, the rising profile is made up of large-leafed *Tournefortia argentea* (Boraginaceae) with the legumi-nosan *Sophora tomentosa* (cousin of the New Zealand kowhai). The leaves of both have a silver-grey tomentum giving resistance to evaporation. The near-impenetrable main thicket is made up of the screw-pine, *Pandanus tectorius*, with its saw-edged foliage, spiny trunk and heavy strut roots. To landward there is the small-leafed leguminosan tree *Leucaena leucocephala*.

Each small motu faces the lagoon with a lee-shore veg-etation, rich in broadleafed trees like the open coastline to the west. The windward lagoon shore off **Vaipeka** has instead of coral a long expanse of silty sand. Lofty *Cocos* palms are mixed with *Pandanus tectorius* and *Inocarpus fagifer*, and are fringed with small trees of two common Pacific malvaceans, *Hibiscus tiliaceus*, of bluish green foliage and the more golden green *Thespesia populnea*. In front, where the western Pacific would have mangroves, there are windward outliers of *Pemphis acidula*.

The shore transects

Two sorts of reef system will be surveyed at Aitutaki: the east-facing **windward reef** south of Ooka, and the much wider **leeward reef** girding the whole west coast of the main island, from Atike in the north to the settlement of Arutanga in the south. The lagoonal shore at Vaipeka will also be briefly visited.

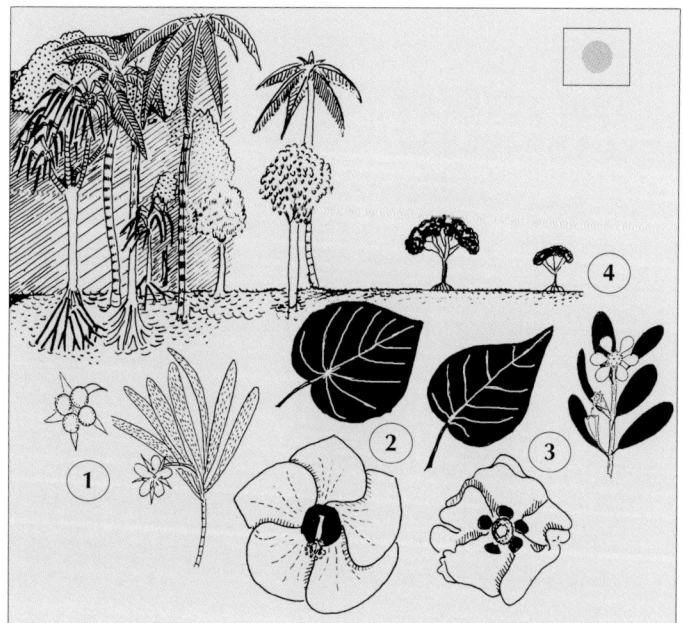

Fig. 21.4 Vegetation of the lagoonal shore at Vaipeka
Bush fringe with *Cocos nucifera*, *Pandanus tectorius* and *Inocarpus fagifer*, with edge zone of *Hibiscus tiliaceus* and *Thespesia populnea* and out in front *Pemphis acidula*.
1 *Suriana maritima* 2 *Hibiscus tiliaceus* 3 *Thespesia populnea* 4 *Pemphis acidula*.

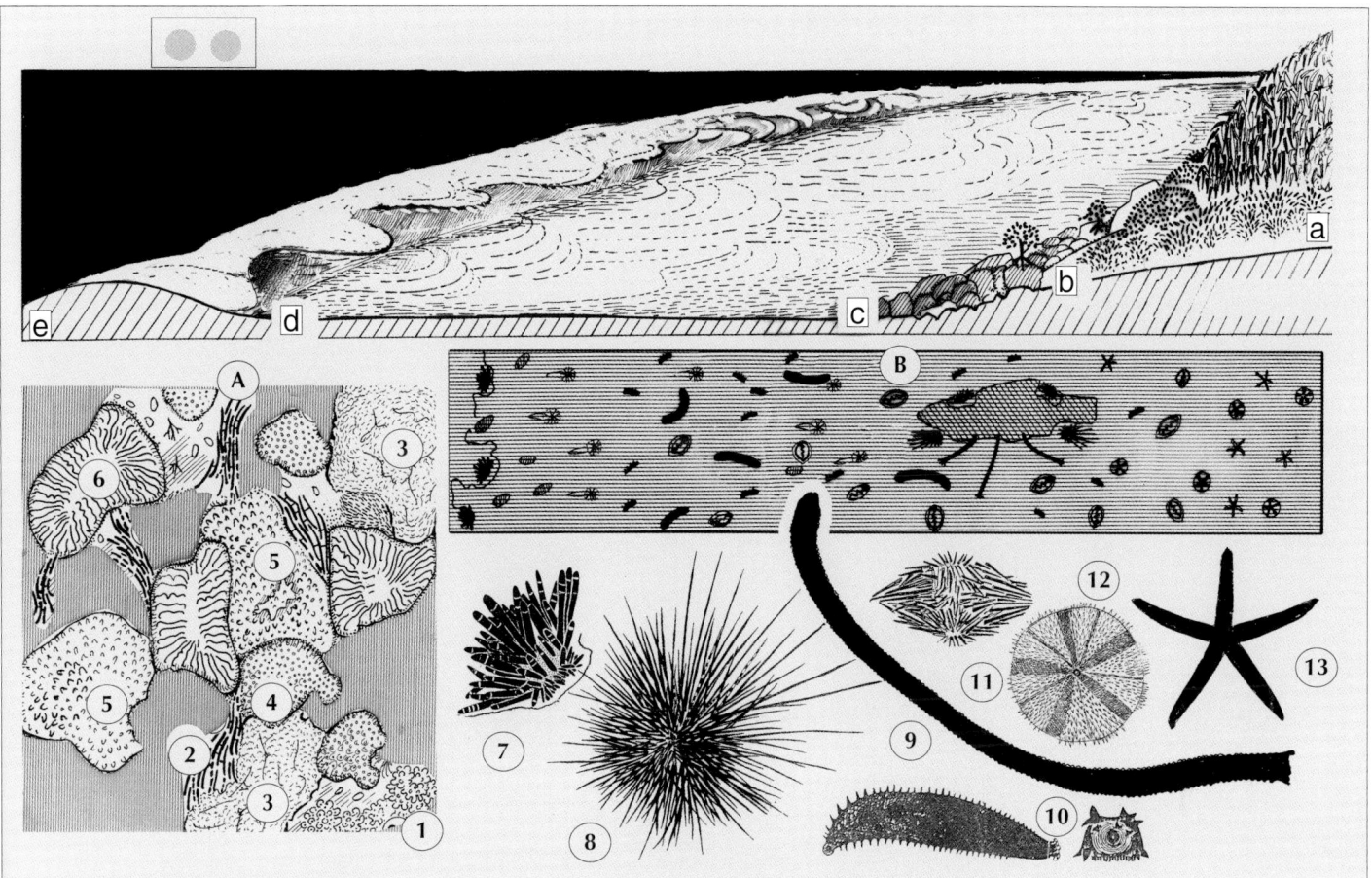

Fig. 21.5 Aitutaki, Cook Islands: windward reef off Ootu Peninsula

a–b coastal scrub with *Pandanus*, *Tournefortia* and *Sophora*; b–c edged by *Scaevola*; c–d reef moat; d–e reef crest and seaward margin. A. Surface mosaic under surge at reef edge B. Distribution of echinoderms across the moat.

1 *Pocillopora damicornis* 2 *Millepora platyphyllia* 3 crustose *Montipora* 4 *Palythoa caesia* 5 encrusting *Acropora* 6 *Lobophytum expansum* 7 *Heterocentrotus mammillatus* 8 *Diadema setosum* 9 *Holothuria leucospilota* 10 *Stichopus chloronotus*, with anterior end 11 *Echinometra mathaei* 12 *Tripneustes gratilla* 13 *Linckia laevigata*.

The windward reef

Opposite the airfield, on the **Ootu Peninsula**, we may walk out across a windward reef, extending some 140 m from high water to its seaward margin. From a terraced backshore, there is first a short drop to a shallow moat, which continues out to the visible rim. The reef edge then rises as a low convex algal ridge, well emergent at any low tide. From its summit, this ridge slopes out at a gentle incline into the white line of surge. Beyond wave-break there is the almost cobalt blue of the open sea. The moat is in contrast pallid bluish green. Hard-floored or thinly strewn with sand, it is knee-deep at low tide, and about 60 m in full width.

1 Littoral fringe and upper eulittoral

A conglomerate platform, built of shell and coral debris, rises a metre above the low water level of the lagoon. Its margins and off-lying stacks have a zoning comparable with temperate shores. The sun-warmed **littoral fringe** has the open-shore periwinkle *Littorina coccinea*. Among the **adlittoral** vegetation, along with ellobiids

(*Melampus*), there is a higher level littorine, the top-shaped and spinose *Echininus cumingi*. Both littorines are denizens of windward Pacific shores, as are the two upper shore nerites, *Nerita undata* and *N. plicata*.

The **upper eulittoral** lacks a proper barnacle zone, showing only a sparse scatter of *Chthamalus intertextus* and pits of the pedunculate barnacle *Lithotrya* (Fig. 22.17). Above or among the barnacles is the limpet *Siphonaria normalis*. The thaid predator is *Morula granulata*. Large *Grapsus* crabs run about freely; and small watchful mud-skippers *Periophthalmus* are typical of high eulittoral sites.

2 The reef moat

This whole stretch of the moat or shallow lagoon can best be likened to a prolonged eulittoral pool. The hard, sediment-strewn floor has a profusion of echinoderms, most notably holothurians. These include *Holothuria atra*, like black sausages white dappled with sand, and the smaller, stiffly papillose *Stichopus chloronotus*. *H. leucospilota*, also black, is longer and perfectly

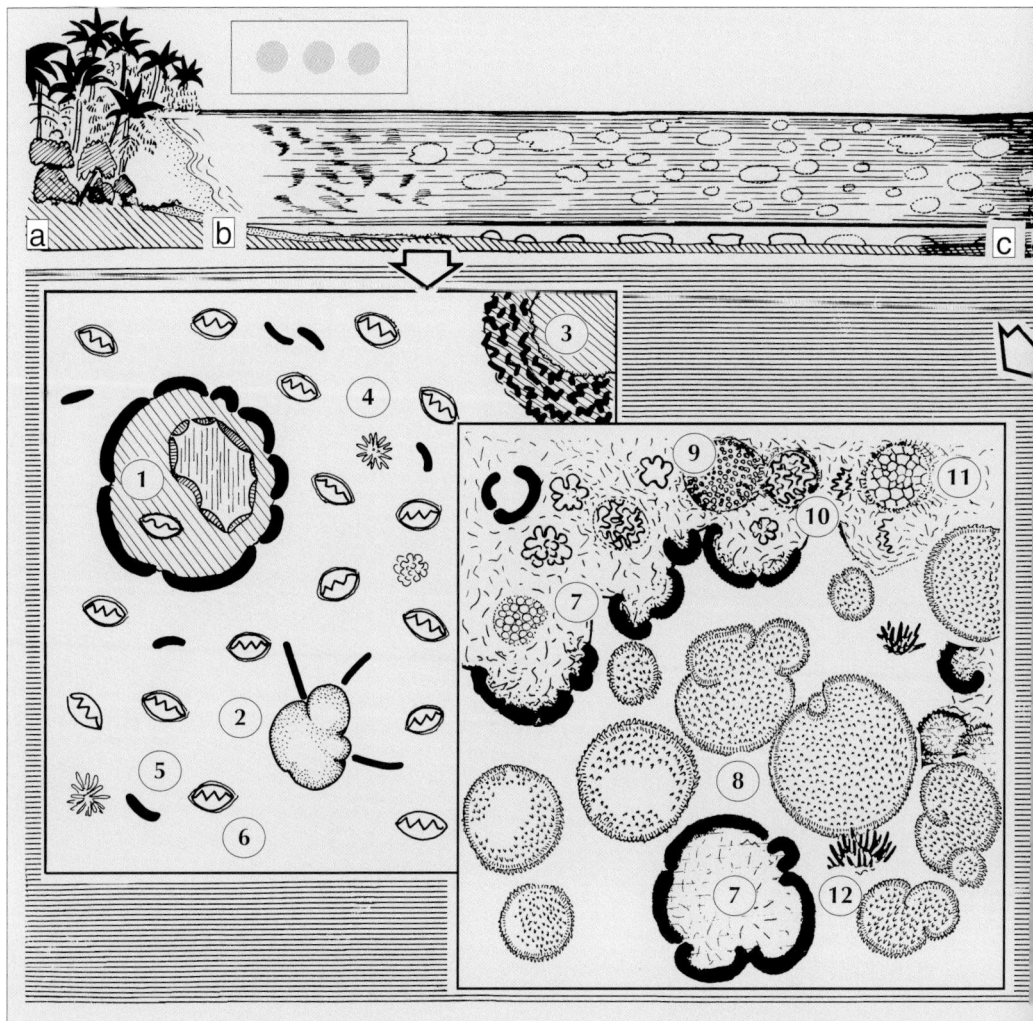

cylindrical, stretching forth from its attachment by the anal end beneath a rock.

Three sorts of echinoid especially abound in the moat. The commonest is everywhere the semi-ovoid *Echinometra mathaei*, with short, strong spines scouring into the coral rock. *Tripneustes gratilla*, with rusty spines and dark ambulacra, lies free on the sandy floor, often bedecked with attached rubble. The toxic and fearsomely black-spined *Diadema setosum* hides itself under boulder overhangs. As familiar as the echinoids is the vivid blue seastar, *Linckia laevigata*, sometimes accompanied with a related brown *Ophidiaster* species.

Fig. 21.6 *Holothuria cinerascens*, buried in sand, with circlet of branched podia

Giant clams, *Tridacna maxima* (Fig. 21.7), lie deep-entrenched in their eroded slots, displaying their attractive-coloured mantle edges. They can reach densities of 3–4 sq m.

Corals growth within the moat is mostly subdued. In the middle depths, massive *Porites lutea* form micro-atolls just beneath the water level. Strongly crusting faviid species include *Plesiastrea versipora*, *Leptoria phrygia*, *Favia favus* and *Favites acuticollis*. *Acropora humilis* with its blunt, abbreviated branching appears towards the moat's outer edge. Tables of *Acropora hyacinthus*, fragile but well adapted to currents, are not uncommon. Freely branching *Acropora*, as well as finger *Porites*, *Montipora* and also the fungiids, all appear to be lacking on the windward reef. Small colonies of *Pocillopora damicornis*, *Hydnophora microconos* and *Lobophyllia corymbosa* are, however, common in the moat.

3 The reef crest

In contrast with the pale pastels of the moat, the reef crest offers bright splashes of colour. First and most vivid may be the salad-like green algae in shallow pans. But a little further out, the whole surface is a coloured mosaic, where encrusting corallines interlock with

Fig. 21.7 The leeward reef from in front of Arutanga, Aitutaki, Cook Islands

a–b adlittoral bush fringe and beach; b–c the inshore shallows; c–d the middle stretch; d–e the reef edge

The quadrats (2 sq m) show details of shallows, middle stretch and reef edge.

1 micro-atoll of *Porites lutea* 2 helmet-shaped mound of *Porites* with *Holothuria leucosticta* 3 micro-atoll of *Pavona decussata* 4 small *Acropora gratilla* 5 *Holothuria atra* 6 *Tridacna maxima* 7 massive *Porites* stack edged with living corals 8 *Acropora hyacinthus* table 9–11 faviid heads 12 corymbose *Acropora* 13 short-branched *Acropora* 14 *Lobophytum expansum* 15, 16 crustose *Acropora* with reduced branching 17 *Porites* heads 18 *Millepora platyphyllia* 19 *Favia* sp. 20 *Pocillopora meandrina* 21, 22 meandroid faviid heads.

sponges, ascidians and alcyonaceans and frequent stony corals. From the first intimations of the living reef edge, we can walk out through surge that even on a calm day rolls in continuously. At minute-or-two intervals a blue wall rises beyond the reef, then — topped with white — rolls in to submerge the reef crest waist-deep.

To everything happening to seaward we should be constantly alert. If an unexpected wave should knock the bystander off his feet, he must let it take him shoreward, never trying to cling to corals, at risk of injury or abrasion. Such danger can be minimised with long-sleeves, a protection also against sun-exposure. It is sensible not to approach the reef front alone. One or more extra make not just for safety, but help share in the observation of many things only momentarily glimpsed. At each wave drawback a wide stretch will be exposed for just a few moments of safe, sure-footed access.

Up the slope from the moat to the reef crest, we may first notice a band of black circlets like tiny lace doilies. These are the branched oral podia of the cucumber *Holothuria cinerascens*. Buried with sediment in eroded scars, the animal puts out its tentacles at the front, with the anus brought to the surface at the rear (Fig. 21.6).

The reef crest is topped with a veneer of calcareous algae. The chief species is a *Porolithon*, much more friable than coral and forming a creamy cement, mauve to

pink on its living surface. Most of the wave-swept corals assume the form of low plaques. The prominent *Acropora* are either low-corymbose, with a broad-base, or even crustose, welded to the rock by their entire surface. Some of the crusts have short, upright fingers, as in the surge-form of *Acropora humilis*. In others branching is altogether suppressed.

A little further out, exposed to full surge, the low corals are chiefly *Pocillopora*, with heads of clubbed branchlets between which the surge is constantly driven. Each branchlet is studded in turn with smaller tubercles bearing clusters of corallites with bright magenta polyps finely ringed in black. The two important species at the reef edge are *Pocillopora meandrina*, with compact, rounded heads and *P. eydouxi*, branching in broad, spatulate flanges.

Just as constant as *Pocillopora* at the reef edge is the fire coral, *Millepora platyphyllia*. Greyish or cream, sometimes mustard tinted, this species puts up strong lamellae, broken into spiky crests in surf runnels or high-exposed sites. Not a stony coral at all (Fig. 21.7) *Millepora* has a matt shark-skin surface, and unusually virulent nematocysts. Brushing against it with bare skin should be avoided.

Next after the corals, algae are prominent at the reef edge. Small ribbon-leafed *Sargassum* abound in fast runnels. Under direct surge, there are expanses of short *Sargassum cristaefolium* and *Turbinaria peltata*. Of smaller extent but still prominent are a semi-cartilaginous *Gigartina*, crisp turfs of jointed *Amphiroa*, and other small reds, both calcareous and succulent. Shallow splash pans carry green algae: *Caulerpa* species with their diversity of branching form, vivid tufts of *Chlorodesmis fastigiata*, and the more pallid clusters of vesicles of *Dictyosphaeria*.

Observing the reef

With first enthusiasm, many will want to swim straight to the reef edge with mask and snorkel. For all its novelty of encounter, this could prove less informative than a methodical walk.

If the prime object is to store up observations, the basic equipment should be a clip-board, with sheets of thick A4 cartridge paper, or the plastic sheets that can be marked in the wet with lead pencil. Otherwise ballpoint pens, ideally in several colours, are for speed of execution far better than pencil. Pens and paper must at all cost be kept dry, carried in a plastic envelope held clear of the water. A plastic bucket will carry small pots and tubes, capped or stoppered. Polythene is lighter and better than glass. It is a folly to take a book already full of previous notes back to the reef. One false step could see its precious store soaked and lost. But on even the shortest reef walk, a few blank pages should be taken for the unexpected observation that is certain to want recording.

At first only two quite simple measurements need be attempted. Horizontal distance can be measured by pacing, knowing the length of your stride and keeping as far as possible in a straight path. Vertical profile can be recorded against distance by noting water depth on a graduated staff, of just up to waist, knee or ankle, and by recording the height of any emergent feature. The

type of terrain — micro-atolls, coral thickets or tables, rubble beds or sea-grass — should also be noted.

Every shore picture in this book is derived from hand-sketches, producing a field record suitable for easy retrieval. The camera has lots of advantages, but on a reef shore it has limitations. Even today's sophisticated cameras, secured against water damage, will not produce pictures instantly accessible. Quick photography is also a substitute for conscious appraisal. Sketching — in the mode of the master naturalists — is most of all valuable for the recognition and analysis it calls for on the spot. Eye and brain are matched together in a way that routine exposed rolls of film or megabytes of disc space can never ensure.

A viewing box (Fig. 21.8) will be found invaluable when wading to waist-deep. Through a rippled surface it can offer a clear field of vision in places too shallow for mask and snorkel. It can thus be pushed ahead as a continuous viewing screen, or serve too as a carrier for notebook and pots, while giving the foot-traveller a third point of stability. Made from 5 mm thick plastic sheeting, the box can have a 4 cm rim to exclude splash, and a transverse bar as a handle for carrying.

The leeward reef ● ●

The **leeward** reef fringing the main island of Aitutaki to the west is notably wider than the reef to windward. At its easiest crossing, in front of the village of Arutanga, it is some 800 m out to the edge. Most of this width comprises a reef flat with a shallow lagoon, in its outer part reaching waist-deep at low water. There is no algal ridge, but the final stretch, emersed at low water, becomes level-topped by the ultimate closing together of the coral stacks in the outer lagoon.

The seaward face of the reef is indentured with narrowly incised channels and its fall-off to the deep is steeper than on the windward reef, being almost vertical and overhung with tiers of coral tables. The open sea is notably calmer than to windward, with its clear blue contrasting with the lagoon's pallid green. In calm weather the reef can be safely traversed right to the edge, with little hazard from incoming surge.

From the line of villages on the coastal road from Arutanga to Amuri, runoff enters the lagoon by a stream that becomes seasonally dry. North from Amuri there is vehicle access along the sand beach above high water, to where the west-east arm of the airstrip cuts across the island. The reef flat is constantly at risk from sediment and eutrophicants originating on the land. With the threatened increase of tourist accommodation, it will be vital to prevent the entry of nutrients or toxic substances to the lagoon, which is not only a traditional fishing ground, but is itself the island's most precious tourism

Fig. 21.8 Using a viewing box

resource. With pollution or eutrophication the reef and lagoon could become a virtual desert like the fringing reef of Rarotonga under the cumulative runoff from a century's intense cultivation and settlement.

The reef flat and lagoon
The shallows inshore

Approached from the roadside, the leeward reef can in a few places at low tide be traversed dry out to the edge, along a natural causeway of shell-sand and rubble. From this path detours can be made at any point into the waist-deep moat. At Arutanga the reef edge has been breached by a narrow constructed passage to the island's wharf and slipway.

The white beach drops off straight into the sandy lagoon shallows, initially less than knee-deep and often tepid on a warm day at low tide. This first stretch is rich in inshore green algae, evidently promoted by nutrient runoff. On a bright day, flossy clumps of pale green *Boodlea* become buoyant with oxygen bubbles released from super-saturation after high photosynthesis. Salad green *Caulerpa racemosa* stretches as rhizomes across the white sand. Common also is a close-branching tubular *Enteromorpha*.

Brown algae are already numerous too. The warmest shallows have clusters of the small fans of *Padina boryana*, and the branched ribbons of *Dictyota*. The fleshy lattices of *Hydroclathrus clathratus* adhere loosely to the bottom. The principal red alga is a slender-branched *Hypnea*, in actual hue a dull greenish brown.

With slight deepening, black cucumbers begin to appear, both *Holothuria atra* and *H. leucospilota*. As on the windward reef, where sand gives place to bare coral rock, we can also find the black-laced tentacle spreads of *Holothuria cinerascens*. Equally abundant here is the smaller giant clam, *Tridacna maxima*, rock-embedded with several to a square metre, located by their vivid green and blue siphon edges. Just as bright-coloured is the widely strewn blue seastar *Linckia laevigata*.

Such warm shallows will also reveal the first live corals. *Porites lutea* scatters the bottom, with tawny brown or mauve nodules, growing to larger heads and — a little further out — forming flat micro-atolls a metre or more across. A branching *Montipora* species is also common. Greenish *Galaxea fascicularis* occurs in small patches or grows up into fragile mounds. But generally commonest after *Porites* is the light pink *Pocillopora damicornis*, distinguished by its slender, fragile branches from the massive *Pocillopora* typical of the reef-front.

The middle stretch

Though a wadable path can be picked out right across, there are parts of the lagoon's middle stretch that deepen to 2 m at low tide. With care, the crossing can be completed either by snorkelling through narrowed passages or by using the solid closing-in stacks of *Porites* as stepping stones. Here the sides of the channels become increasingly shaded, topped in places with fire-coral *Millepora platyphyllia*. Further down the sides, a wide offering of the scleractinian corals grow in sites protected but not too shaded.

In the middle stretch *Pocillopora ligulata* and *Montipora danae*, *M. capricornis* and *M. aequituberculata* are common almost everywhere. The same coral gardens also have *Psammocora haimeana*, and *Galaxea fascicularis*, as well as a number of finger or corymbose *Acropora valida*. These include short-branched *A. secale*, digitate brackets of *A. valida*, and fragile brushes of *A. rosaria* and *A. paniculata*. Among the faviids contributing to the solid stacks are the meandrine 'brain corals', a *Leptoria* species and *Platygyra lamellina*. *Hydnophora microconos* often grows massively, and loose *Fungia* lie freely on the sanded bottom.

Open stretches between the coral stacks are strewn with clean rubble, prettily veneered with pink lithophylla. In well-lit expanses there are panoramic spreads of the flat *Acropora hyacinthus*, with their pale-grey tables reaching to just beneath the low tide water level. Each forms a close mesh of horizontal branches, covered all over with short vertical fingers. An upturned margin is formed by slightly taller fingers, mauve or blue at the tips. Unlike firm *Porites*, the closely crowded tables of *Acropora* call for careful picking of the way to avoid damage to their fragile margins.

On a table's underside, in low light reflected from the white sand, grow shade-tolerant algae: bottle green vesicles of *Valonia*, tender rhizomes of *Caulerpa*, and the reddish crusts of *Peyssonnelia*.

Standing between the tall stacks or among *Acropora* tables, we can now look seaward over the reef edge, wavy and indentured towards the lagoon. As the coral massif finally closes right together its channels narrow ultimately to covered passageways. All over the exposed top, the shallow pans are replenished by surge, and the superficial coral plaques are never allowed to become dry. At the channel edges cream or white fire-coral, *Millepora* usually stands out, while the branching *Acropora* species crowd the deeper wave-runs further down.

The reef edge

We can finally step up from the lagoon on to the seaward reef's level top. Exposed to the air at a normal low tide, this surface is swept intermittently by low swell.

But wave movement is usually slight as compared with the windward reef; the seaward drop can be safely approached, to gaze — with an ultimate wonder — into the clear blue depths.

As far in as it receives intermittent light surge, the reef edge is enriched with the gold, tawny pink or mauve of living corals. Most take the form of low, convex heads or encrusting plaques, all with their branching reduced. There are no brittle-fingered *Acropora*, nor many with corymbose heads. A branchless *Acropora* encrusts much of the surface, putting up in some places a few blunt marlin spikes, and with larger blue calices showing up where branches would normally be.

At the furthest reef edge, heads of magenta-coloured *Pocillopora meandrina* stand out, with surge coursing through their labyrinths of strong tubercles. The vertical lamellae of *Millepora*, strongly serrated along their edges, share space with the low *Acropora* crusts. These two corals give place a little further back to pink *Lithophyllum*, with the short sargassoid alga *Turbinaria* in slightly elevated sites.

The soft corals (**Alcyonacea**) are far less evident at the Cook Islands than further west in the Pacific. But under flowing surge, parts of the reef edge may carry the corrugated, rubbery folds of *Lobophytum expansum*. There are also patches of tan-coloured *Palythoa caesia* kept awash in breaking surge.

For the last 25 m the reef edge, with a hardly perceptible slope, is deeply incised. On a calm day, we can look down directly into the drop-off beyond the shelving tables of *Acropora hyacinthus*.

Snorkelling off the reef front is generally safe and rewarding. The deep corridors open up to view as the surge sucks back, revealing corals tier upon tier, with their squadrons of small reef fish: blue chromids and black and white dascyllids, along with larger, brightly coloured butterfly fish. Then, in a moment, all such visibility will be lost, as the mounting surge pours in. Unlike the gradual sublittoral incline to windward the leeward reef has a vertical drop-off, with seemingly fragile *Acropora* tables just strong enough to resist the waves.

At the leeward edge of the summit, just as to windward, all the animals not fastened down are streamlined and broad-based, able to take firm hold as surge passes, or lodge securely in crevices or excavations. The wide, flat holothurians, *Bohadschia*, need leverage to dislodge. *Holothuria cinerascens*, deep-embedded in furrows, puts up its black tentacle circlets.

In eroded potholes, the slate pencil urchin, *Heterocentrotus mammillatus*, is commonly seen, but difficult to prise out intact. The far more abundant urchin *Echinometra mathaei* everywhere studs the eroded surface.

The Vaipeka sandflat ● ● ●

In a larger atoll, the windward lagoon shore would have its own fringing reef. But in the absence of coral, the corresponding shore at **Vaipeka** deserves mention for its vegetation and rich sequence of crabs. Among old stands of *Pemphis* and the open fringe of *Hibiscus tiliaceus* and *Thespesia populnea*, there are fiddler crabs: *Uca dussumieri* with cherry-red chelae. Further forward, in cleaner sand with regular tidal reach there are pallid ghost crabs *Ocypode cordimana*, flitting across the dry beach on two pairs of claw tips, as fast as a blown leaf. Further out — in silty sand — is the shallow-buried *Macrophthalmus* with long emergent eyestalks. Under rock rubble over the upper part of the immersed beach there are fast and aggressive *Thalamita* (Portunidae).

On the high beach among litter, small fast-running grapsids *Cyclograpsus* abound. There is also a swift-footed *Metopograpsus* with violet and white-tipped chelae, and under logs a larger *Geograpsus* with yellowish-brown chelae. At the highest level, amid the coastal bush live the virtually terrestrial hermits, *Coenobita rugosus*. The big *Cardisoma carnifex* is smoothly convex and flesh-coloured, common in holes under logs from which it runs out to brandish its chelae.

Lagoon corals
Patch reefs

The coral communities of the lagoon bottom never rise above low water. But snorkelling will reveal scatters of small patch reefs, often luxuriant in growth. In contrast to the surge-swept corals of the reef front, these secluded thickets may rise 2 m above the bottom, so that the tallest corals stand free of the fine sediments to which they would otherwise be vulnerable.

Mounds and helmets of *Porites*, stable and broad-based at the bottom, are good at withstanding sediment. The surrounds of the patch reef may also be strewn with fungiids (Fig. 21.10) with their big corallites facing upward, with their cleansing cilia efficient in dealing with fine particles.

Above the *Porites* arise long-branched corals, wholly different from those on the wave-front or emersed at low tides. Stagshorn *Acropora* easily predominate. They may include *Acropora grandis*, *A. formosa*, *A. horrida* and *A. florida*, in fragile forests, with the terminal calices distinctive in violet, pink or lemon. These formations may cover wide areas, and call for the utmost care by divers, since they could be damaged by even glancing contact. The scene is enhanced by the shoals of little fish, such as sapphire-blue *Chromis* or black and white banded *Dascyllus*.

Below the trees is usually a shrub layer of bottle-

Fig. 21.9 The coral assemblage of a submerged patch reef
1 stagshorn *Acropora* cf. *grandis* 2 stagshorn *Acropora* cf. *formosa* 3 'shrub layer' of *Acropora granulosa*
4 *Acropora echinata* 5, 6 short-branched *Acropora* 7 erect candelabra sponge with symbiotic green
algae 8 *Porites lutea* 9 *Turbinaria peltata* 10 *Montipora* sp. 11 free-lying *Fungia* 12 free-lying *Polyphyllia*
13 *Pavona* sp. 14, upright branching *Montipora* 15 *Merulina ampliata*.

brush *Acropora*. The largest is the lightly built *Acropora echinata*, with tubular calices, dense on each branch-like spruce or fir twigs. A second bottle-brush is the smaller *A. longicyathus*, with slender calices curved and wider-spaced. In *A. granulosa*, growing in festoons like a cake-icing, the pointed calices are ornamented with mamillae.

Brittle meshworks of *Seriatopora hystrix* are also a common feature of low energy patch reefs.

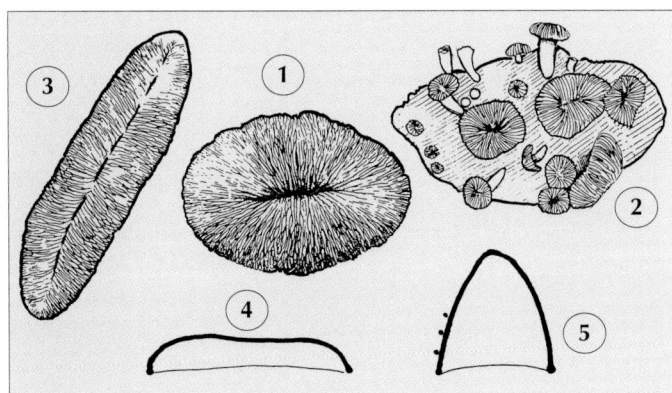

Fig. 21.10 The growth forms of Fungiida
1 Mature, free-lying *Fungia* 2 juvenile, stalked *Fungia* 3 a corallum of *Herpolitha* 4 cross-section of *Polyphyllia* 5 cross-section of *Halomitra*.

Classed next door to *Acropora* (Fig. 21.9), the *Montipora* species grow profusely beyond low water. They include cornet-like forms (*M. foliosa*), slender fingers (*M. digitata*), and surface crusts of varying relief, some with thin, fragile exfoliations. *Montipora* has fewer good recognition characters than *Acropora*, with no large apical calices. Their corallites are small, with the peritheca rough or verrucose, coloured grey, khaki or dull green.

Many Scleractinia in these quiet conditions grow into fragile sheets and scrolls. *Merulina ampliata* (Family **Merulinidae**), common in patch reefs and lagoonal slopes, is perhaps the most widespread of these 'explanate' corals. The separate calice walls are lost, with the lamellae covered instead with a meandroid pattern of branched corallites.

Massive faviid heads will be looked for in vain, but the fragile faviid *Echinopora lamellosa* is often abundant, as curling scrolls with corallites on only one face. In the patch reefs of the west Pacific — as in Fiji — may be found other scrolled and branched growth forms unusual for their families. The branching faviid *Echinopora horrida* has a sharp spinulose peritheca between the corallites. Equally untypical of their families are a

413

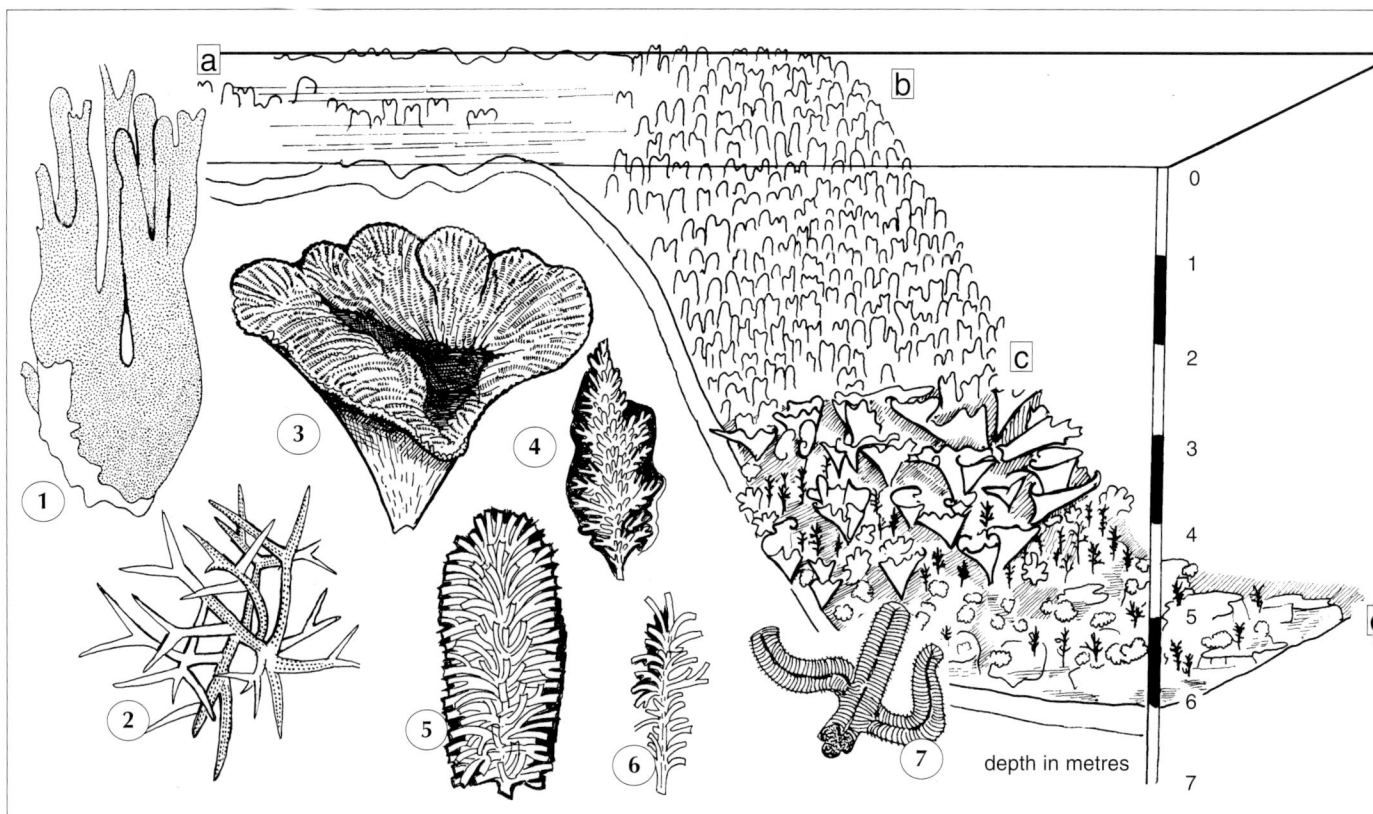

Fig. 21.11 Sheltered offshore slope of a lagoon reef: Marau Sound, Guadalcanal, Solomon Islands
a–b intertidal stretch of sparse *Porites*; b–c upper slope with *Porites* cf. *columnaris*; c–d lower slope with *Merulina ampliata*, bottle-brush *Acropora* and other fragile corals
1 *Porites* cf. *columnaris* 2 *Merulina ampliata* 3 *Seriatopora hystrix* 4 *Acropora granulosa* 5 *Acropora echinata* 6 *Acropora longicyathus* 7 *Clavarina* sp. (Merulinidae).

branched *Alveopora* with perforate calices like lace, and the branched *Hydnophora exesa* (see *H. microconos*, Fig. 20.13).

The corals of the Suborder **Fungiida**, are important subtidally, mostly lying free on sand, as around patch reefs. Their simplest forms are the *Fungia* species, with corallites up to 15 cm across, being the largest in the Scleractinia. Blade-like septa radiate from a long mouth-slit, and the polyp colours range from pale green to pink or maroon, with tentacles over the whole oral face. The *Fungia* corallite is at first stalked, until its disc cuts off like an overturned mushroom, leaving the pedicel to produce a new one.

Of the colonial fungiids, *Herpolitha* is larger than *Fungia*, and elongate, with an axial row of separate mouths along the summit. *Polyphyllia* is like an everted bowl, with its axial mouths augmented with numerous lateral mouth centres. *Halomitra* is elevated like a helmet, lacking the axial mouths, but with scattered mouth centres, each having a small radiate septal system.

Finally we must mention two families of advanced structure that are largely or wholly subtidal, in the lagoon or below wave-break on the reef-front. The first, the **Agariciidae**, however, contributes several *Pavona*

species to the reef's inner shallows. Coloured reddish, yellow, brown or purple, the calices are highly modified. The agariciid polyps are shallow and without boundaries, the tentacles being rudimentary. The lines of mesenteries run continuously between them.

Intertidal *Pavona* species may form ground-crusts or bifacial lamellae. The coralla of *Leptoseris* and *Pachyseris* are light and foliaceous, only to be seen by diving. *Pachyseris* is meandroid, with calices in concentric rows. In *Leptoseris* the corallites are crateriform, scattered over only one side of a thin unifacial lamina.

The offshore Family **Pectiniidae** also has highly modified explanate or foliaceous coralla. As in agariciids, the corallites have no boundary walls and the septa are spinose, often reduced. The fragile corallum has calices only on one side, horizontal in *Echinopora*, and in *Mycedium elephantatus* nariform (nostril-shaped). In the foliaceous *Pectinia*, the 'sea lettuce coral' submeandrine series are formed by budding, separated by high sharp collines.

Lagoon shore reefs

The sloping offshore face of a lagoon reef has a very different character from an outside reef margin. The example illustrated in Fig. 21.11, on east Guadalcanal, **Solomon Islands**, forms the leeward back-slope of a spit cutting off a lagoonal backwater from the open sea. Dropping from a rubble-strewn shore, just awash at low water neap tide, is an upper slope built chiefly of the columns of greenish brown *Porites* (cf. *columnaris*). Beyond 3 m come the brackets and cornets of *Merulina ampliata*, mingled with a delicate brush-like *Acropora*. There is also a brittle foliose agariciid *Pavona cactus*. The ground at the bottom is littered with coral-boulders, with live *Seriatopora hystrix*, *Pocillopora damicornis*, and cords of a peculiar ramose merulinid *Clavarina* species.

Coral pinnacles

Pillars of living corals are a spectacular growth form not well represented at Aitutaki, but important in lagoons in the westward Pacific. The example in Fig. 21.12 is typical of Fiji and the Solomons.

Far more than the patch reefs, these tall columns abound in the stronger, short-branched acroporids and faviids. Their bases — like the lagoon floor at large — are also rich sites for the soft corals **Alcyonacea**. The summit of a column, affected by currents but free from wave action, is ideal for bracket and shelving *Acropora hyacinthus* and *Anacropora reticulata*. Next downward come corymbose *Acropora* heads with short upright branches, generally with secondary branching. The most massive branched *Acropora* is the heavy-clubbed *A. palifera*, with its large terminal calices blue.

Towards the bottom, the pinnacles become dominated by faviid heads, with a range of species from *Favia*, *Favites*, *Goniastrea* and the meandroid genera *Platygyra* and *Leptoria*. Related to the Faviidae is the small Family **Mussidae**, with its big polyps massed in coralla that may grow into micro-atolls. The large corallites have strong, marginally dentate septa. The colonies are cerioid, with the corallites monocentric, in *Acanthastrea*. In *Lobophyllia* they have become phaceloid, with the cups branching or pinched off separately; *Symphyllia* is meandroid, with corallites united back to back, separated by sharp summits.

Classed traditionally with the faviids, the genus *Hydnophora* is today placed in the **Thamnasteridae**. *Hydnophora microconos*, on the wave-exposed fringe, forms a convex head like a faviid, beset with conical 'monticles', in effect everted calices.

Towards the column base (Fig. 21.12) are shown two *crateriform* or vase-shaped corals: *Podabacia* is a fixed, bowl-shaped fungiid with polyp mouths deployed all over its interior. *Turbinaria* (confusingly with the same name as an alga!) belongs to the **Dendrophylliidae**, chiefly branching corals in darkness or deep shade and ahermatypic (lacking algal symbionts). To all these features *Turbinaria* is the exception. *T. peltata* is dull yellow, folded into deep bowls, studded with polyps on the inside face. On windward reefs it can reach up to low water mark.

The soft corals

The **Alcyonacea** — the first and largest order of the Subclass **Octocorallia** — consists of the soft corals, abundant in moats and atoll lagoons as well as on leeward reefs. Their fleshy colonies are branched or lobate, studded with star-like polyps. The stiff mesogloea is permeated with calcareous spicules. The polyps invariably have eight tentacles fringed with side pinnacles. Only their *anthocodia* or distal parts protrude from the surface, and their gastrodermal tubes are prolonged into a coenenchyme, where they are linked by cross-connections (*solenia*).

In the Family **Alcyoniidae** the anthocodia are completely retractile. *Lobophytum expansum* forms low-contoured rubbery sheets, with transverse folds, and — unlike most of the family — lives in strong surge at the reef edge. The *Sarcophyton* species (such as *S. latum* and *S. trocheliophorum*) form smooth or undulant cushions. Both these genera are dimorphic, with two sorts of polyp, large autozooids and small siphonozooids which take in water to supply the tube system of the colony.

The rest of this family are monomorphic. The *Alcyonium* species are bluntly lobed, with characteristic diamond-shaped spicules. *Sinularia* species are more diverse in build, with the branches whip-like in *S. variabilis* and shorter in *S. polydactyla*. *Sinularia* are always distinguishable from *Lobophytum*, or *Sarcophyton*, by the coarse spicules at the colony base. *Cladiella* is built of close-set fleshy lobes, with the polyps standing out in dark brown.

The soft corals of the Family **Nephthyidae** form bushy colonies, with the branches heavily spiculated. In *Dendronephthya* the anthocodia are clustered in tufts at the branch ends, while in *Nephthya* and *Stereonephthya* they are more evenly distributed. The very large spicules are clearly visible through the translucent stems with the mesogloea much reduced. In several species a protective armature of giant spicules projects from the surface.

The final family of Alcyonacea is the **Xeniidae**, with the spicules minute or entirely lost. In *Xenia*, clusters of soft polyps spring like a medusa's head from a common fleshy base. Respiratory water is taken in through the mouth by pulsing contractions. Nutrients are thus brought to the internal zooxanthellae — lighted through the body wall — on which *Xenia* relies for its food supply. At least partial symbiosis seems to be a feature of most tropical Alcyonacea.

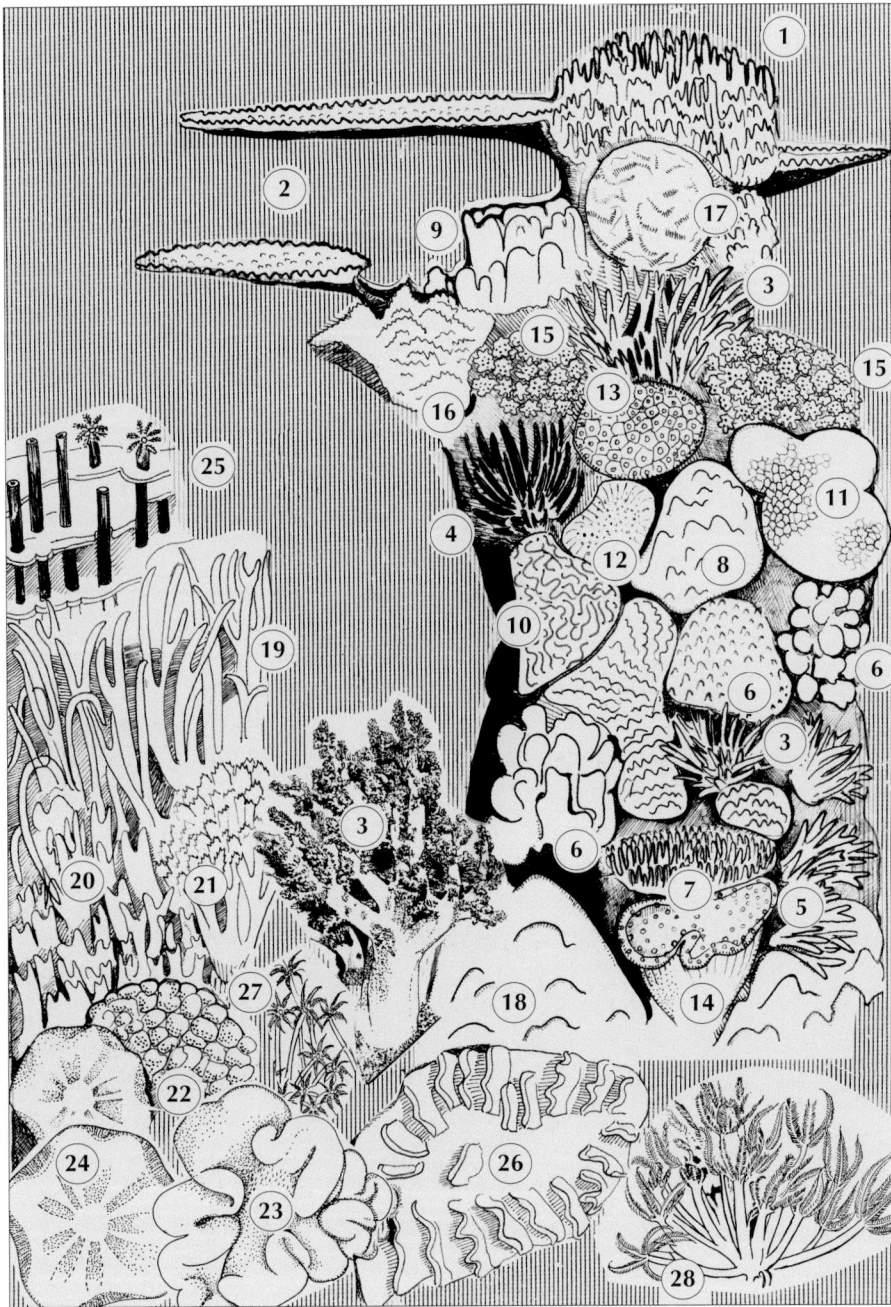

Fig. 21.12 A coral pinnacle, and lagoonal soft corals (right)

1 a columnar *Porites* 2 *Acropora hyacinthus* tables 3, 4, 5 short-branched *Acropora* species 6 *Acropora palifera* 7 corymbose *Acropora* 8 encrusting *Acropora* 9 *Goniopora* sp. 10 *Platygyra lamellina* 11 *Goniastrea* sp. 12 *Cyphastrea* sp. 13 *Favites* 14 *Turbinaria peltata* 15 *Pocillopora meandrina* 16 crustose *Montipora* 17 *Symphyllia recta* 18 *Porites lutea* 19 *Sinularia flexibilis* 20 *Sinularia polydactyla* 21 *Dendronephthya* sp. 22 *Cladiella* sp. 23 *Sarcophyton latum* 24 *Sarcophyton trocheliophorum* 25 *Tubipora musica* with (**inset, right**) polyp 26 *Lobophytum expansum* 27 *Xenia*, with (28) detail.

The dull red organ-pipe coral, *Tubipora musica*, belongs to the Order **Stolonifera**, having an essentially simple, non-fleshy structure. The polyps stand up from a hard creeping coenenchyme, enclosed in long, calcareous tubes interconnected at intervals by horizontal platforms.

Reef building and succession

The whole extent of living coral, emersed or accessible at low tide — in some places hundreds of metres across — has been treated here as a **sublittoral fringe**. In its vastly enhanced scale and diversity in the tropics, this 'fringe' calls for further subdivision under its own descriptive terms.

The section of a fringing reef (Fig. 21.13) shows the percentage of full daylight from the subtidal slope and reef margin back across the lagoon to the high-illuminated backshore. Water movements are also indicated, with the *along-reef currents* offshore, and the inner surf zone with *rip currents* and *wave orbital movements*. Two lines of wave break are thus generally set up, the strong one at the reef margin, and a weaker one, effective only around high tide, involving the eulittoral and the littoral fringe.

First, as in Fig. 21.14, we may show the order in which the primary coral types appear across the reef, in relation to waves and currents, sediments, salinity and illumination. The distribution of corals on the reef flat and reef front follows a concept of Michel Pichon, based on reefs at the Seychelles, in the Indian Ocean, and found generally applicable through the Indo-Pacific.

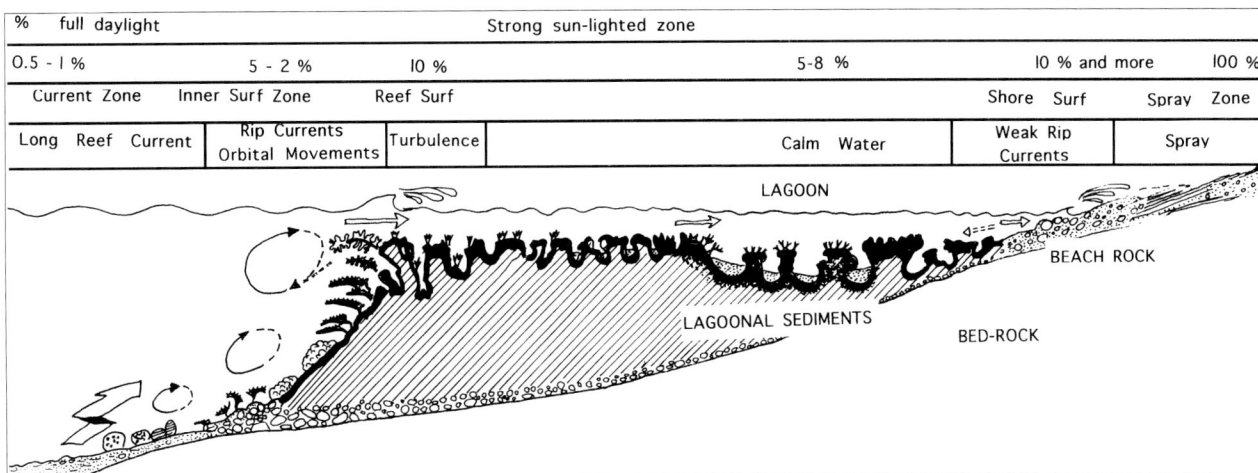

%	full daylight		Strong sun-lighted zone				
0.5 - 1 %	5 - 2 %	10 %		5-8 %		10 % and more	100 %
Current Zone	Inner Surf Zone	Reef Surf				Shore Surf	Spray Zone
Long Reef Current	Rip Currents Orbital Movements	Turbulence		Calm Water		Weak Rip Currents	Spray

Fig. 21.13 Section of a fringing reef, with water-flow and illumination from seaward to beach margin

1 In high-energy environments, **Pocillopora** assemblages (especially *P. eydouxi*, *P. meandrina* and *P. damicornis*) are dominant. Their regular associate is the fire coral **Millepora** and these two appear together wherever surge breaks over a reef front, to windward or leeward.

2 Where water movement is appreciably less, the **Acropora** species come into their own, highly plastic in growth form according to depth, current speed and wave attack. On a leeward reef, table *Acropora* may intermix with *Pocillopora-Millepora*, right to the reef edge. To windward, the *Acropora* species out front are normally reduced to low plaques with branches ultimately suppressed.

Acropora thus becomes dominant, in place of *Pocillopora-Millepora*, in reduced water movement, both behind the reef edge, and at depths beyond wave-break on the reef front.

3 Furthest back on the reef, under minimal wave action, an assemblage of *Porites* comes to prominence, generally as convex heads or micro-atolls. With them are sometimes associated inshore *Pavona* species (*P. varians* and *P. decussata*). Likewise, on the reef front, with water movement diminished below wave-base, *Porites* will become predominant as large, rounded heads beyond *Acropora*. The agaricids *Pachyseris* and *Leptoseris* may grow here too.

Such a sequence: **Pocillopora (+ Millepora) > Acropora > Porites (+ Pavona)** is thus associated with a gradient of diminishing water movement in the three directions:
1 shorewards across a reef flat
2 along a reef front with a transition from windward to leeward.
3 vertically, from the high energy zone of a reef front, to the reduced water movement below wave-base.

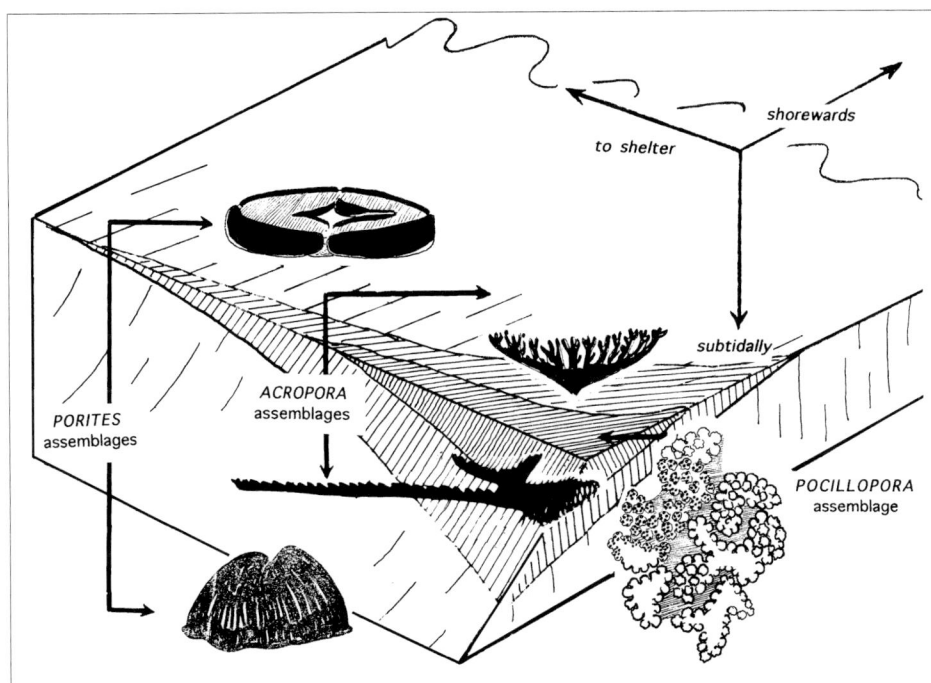

Fig. 21.14 Relation of *Porites, Acropora* and *Pocillopora* to water movement

Wave action decreases in the three progressions: (1) *upshore*, (2) *longshore towards shelter* and (3) with *increasing depth sublittorally*.

(Based on a diagram by Pichon, at Mahe, Indian Ocean.)

417

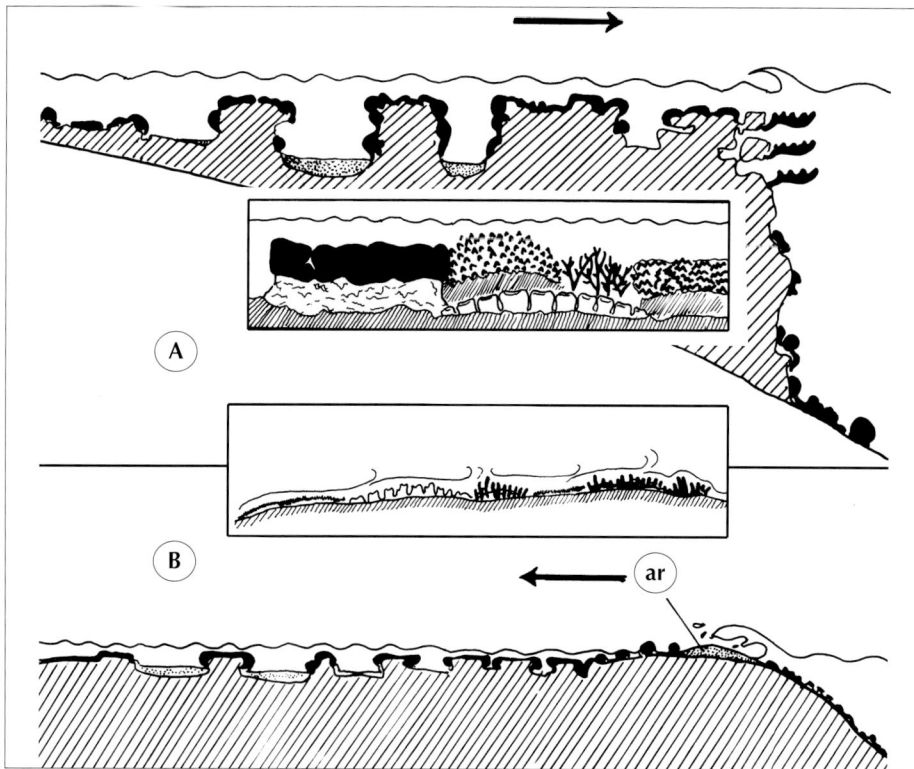

Fig. 21.15 Comparison of the outer front in A. a leeward and B. a windward reef (**Insets**) detail of coral diversity and growth profile. (ar) algal ridge.

Reef construction

The **reef front** is generally well enough lit for the growth of hermatypic corals down to approximately 50 m. On weather coasts, however, two factors may operate to reduce the diversity of frontal corals: *wave action* will restrict the available options in growth form, while the quantity of *illumination* is reduced by back-reflection of light at the white line of surge.

On the front of leeward reefs, tiers of *Acropora* brackets commonly shelve out just beneath low water. At the windward reef edge, *Acropora* are forced to withstand not just fast current flow but strong wave battery. Corymbose growth forms thus tend to predominate, with low spreads of branched fingers on a base with a short pedicel. Under the highest wave energy, the *Acropora* fuse by a wide base to the surface, with the fingers short and blunt. Branching may finally be suppressed altogether, with their old position of fingers still marked by the coloured terminal calices, scattered over the crust.

Behind the crest

Most of the corals inward of the reef crest will remain shallow-immersed even during low water. In quieter spots with enough permanent depth of water, the earliest to appear are likely to be branched *Acropora*. These can first establish upon rubble, and once initiated, on a pool or at the edge of a moat, a dense thicket will strike upwards. Ultimately there may be a well-lit top growth

supported on the stilts of dead branches. These provide under-space for small solitary corals (ahermatypic), and also *Pavona*, *Montipora* and young (stalked) *Fungia*.

Filling in

Consolidation of a thicket of branched coral follows in steady course. Cementation by calcareous red algae can progressively create a fragile canopy from which the still living tips of largely dead branches may project. As this roof strengthens, towards the lagoon edge low corymbose *Acropora* can sometimes establish on top of it.

To seaward of the moat system the reef rises to a low summit. This may take its shape in variant ways. In stronger exposure, calcareous red algae can weld together into a mass softer and more friable than coral rock, yet strong enough to constitute a standing 'algal ridge'. The algal contribution has been so great that some authorities prefer to speak of 'coral-algal' rather than simple 'coral reefs'.

The algal ridge hardly ever appears on leeward reefs, nor is it always complete on the windward front. Some reefs rise to a summit made of finger rubble, dead corymbs and tables or fusing coral stacks and micro-atolls. The most massive bits of the profile — with their tops permanently emersed — are often the dead storm blocks, up to 2–3 m high, that have been detached from the front and carried over its summit by heavy swells.

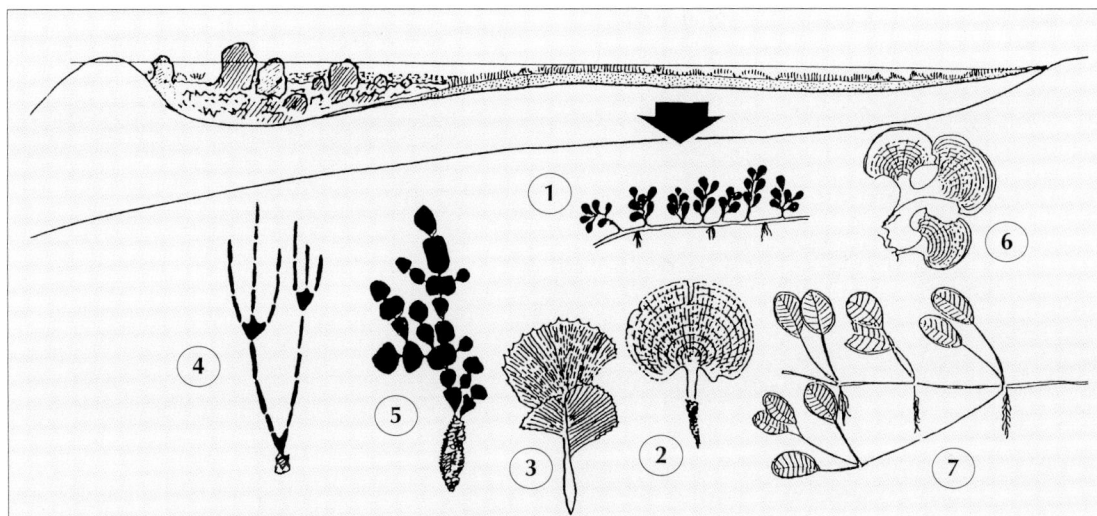

Fig. 21.16 Profile of a post-mature reef, with accumulated storm-blocks and sediments carrying sea-grasses and algae
1 *Caulerpa racemosa* 2 *Udotea* 3 *Avrainvillea* sp. 4 *Halimeda cylindracea* 5 *H. macroloba* 6 *Padina fraseri* 7 *Halophila minor.*

Reef succession

Meanwhile the inner shallows of the lagoon will have been establishing their own coral communities. *Porites*, the hardiest of all the inshore corals, is fast-growing, forming massive heads or ultimately flat-topped micro-atolls, sometimes a couple of metres across. In these quiet waters the dominant coral is *Porites*, alone or with an inshore *Pavona* and usually small *Acropora* or the sheltered *Pocillipora damicornis*.

The shores described in this chapter have shown varying stages of development of a mature reef, through a life cycle over hundreds, sometimes thousands, of years. The diagram (Fig. 20.5) is based on a reef on Upolu, Western Samoa, on top of sub-recent basaltic lavas of an age we can reliably estimate. There are today the beginnings of a canopy, with fine living Acropora tables in the outer reach of the moat. Sheltered embayments offer representative examples of an interlaced forest of long-branched and fragile Acropora.

Regression

The inward shallows of the post-mature reef have generally been colonised by *Porites* mingling with groves of soft corals, notably *Sarcophyton*. Transported boulders or sediments contribute to the regressive stage of reef evolution. The final state will turn upon the balance between deposition and current flow across the reef flat. Stream discharge of fresh water may inhibit reef growth altogether. A permanent flow across the flat will carry sediments further seawards. Where flow is stronger, terrigenous material will be swept right out to sea, with only rubble or boulders allowed to remain. Clumps of calcareous green algae, notably *Halimeda opuntia*, or red algae such as *Amphiroa*, may entrap sediment. With higher salinity, zoanthid polyps (*Zoanthus, Isaurus and Palythoa*) come to cover wide expanses.

On some reef flats regressive evolution is marked by accumulation of sediment that the tidal flow is too weak to remove. Coral communities have generally disappeared, save for small *Porites* nodules at a few spots. The silty deposits carry instead sea-grass beds, as of *Halophila minor* (Fig. 21.16). Green algae well adapted to sediments include *Caulerpa racemosa*, *Halimeda macroloba* and *H. cylindracea*, also *Cladophora*, *Boergesenia forbesii*, and species of *Avrainvillea* and *Udotea*.

Reef metabolism

Coral reefs are sites of high productivity, probably unmatched by any other naturally occurring community. Their annual primary production has been estimated at 1500–3600 g calories per square metre, higher by a hundred-fold than in the surrounding ocean, and of a similar order to the most intensive tropical agriculture. A coral atoll has been likened to a highly productive lake in mid-ocean, 'cupped in a shallow saucer on an emergent mountain peak', with the bottom raised from the aphotic zone into one of strong illumination. Its current system is essentially isolated and self-maintained.

Fig. 20.7 has shown the lagoonal circulation in relation to trade wind direction. On the windward side a steady influx of water is driven by trade winds through channels in the reef. The leeward side receives no wind-driven water, only a tidal inflow. Surface waters are impelled by trade winds across the lagoon surface from windward to leeward. Unable to escape, this water is

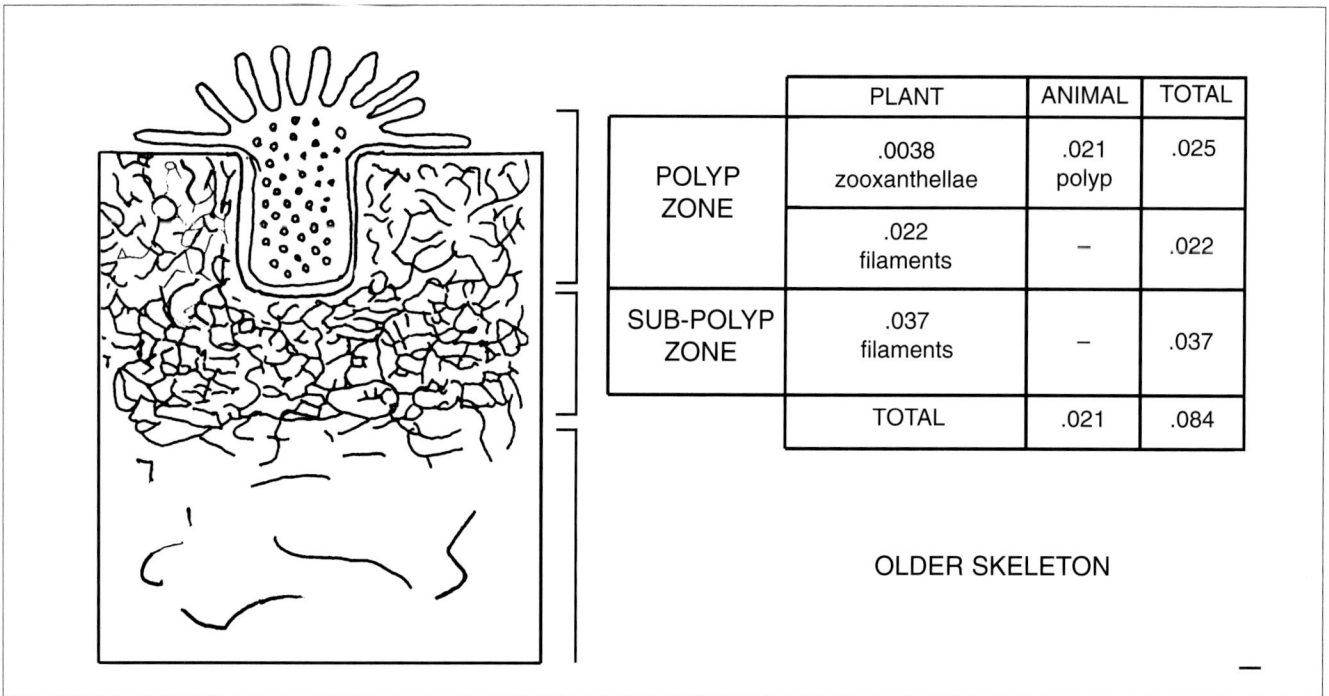

	PLANT	ANIMAL	TOTAL
POLYP ZONE	.0038 zooxanthellae	.021 polyp	.025
	.022 filaments	–	.022
SUB-POLYP ZONE	.037 filaments	–	.037
TOTAL		.021	.084

OLDER SKELETON

Fig. 21.17 Data from the Odum Reef Survey
(**above**) A polyp in section showing the comparative distribution of plant (algal) and animal (coral polyp) tissue as claimed by the Odums, in the polyp and sub-polyp zones and in the older skeleton
(**right**) Reef section showing the changes in content of the water as it passes over the crest and across the reef flat.

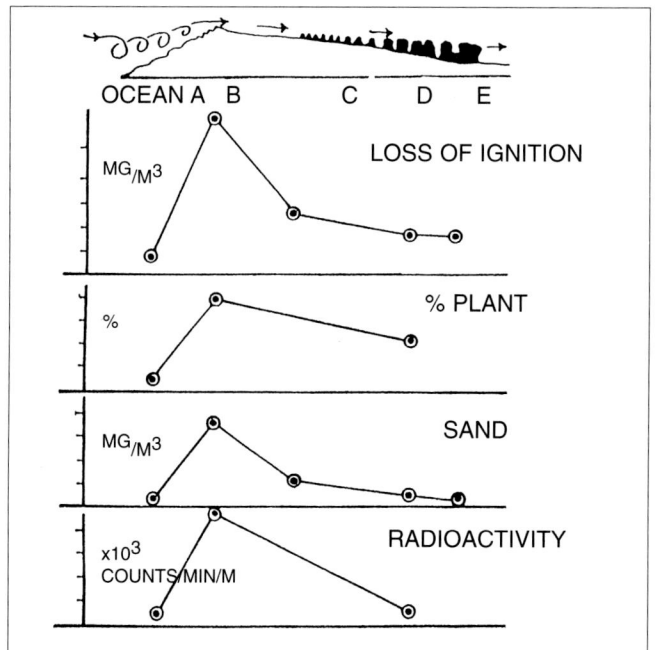

driven below to return upwind as a bottom current, and upwell on the windward side. At Bikini Lagoon the replacement of the total water volume took an estimated 30 days during trade winds, but double that time in summer.

Clearly the growth and maintenance of an atoll system requires replenishment by intercepting materials from outside and retaining them within the system. The normal plankton within or overlying the reef system has in several studies been found insufficient for the total growth and maintenance of a reef.[2]

For the huge primary productivity of coral reefs we must clearly resort to their algae. These unicellular plants, known as *zooxanthellae*, are deployed in the superficial layers of the polyp tissues so as to capture optimum light for photosynthesis. They are to be regarded as an imprisoned and farmed phytoplankton.

Dependence of corals on algal productivity was one of the assumptions in the Odums' classic study of reef metabolism on Eniwetok atoll.[3]

Corals were seen as highly integrated ecological units, dependent not only on plankton-catching by the polyp but on the photosynthesis of algae. The plant biomass involved was held to be contributed only in small part (as little as 6%) by the zooxanthellae (as little as

6%) and to include filamentous green algae penetrating the coral skeleton, as well as the entire algal mass, low in standing crop but fast in turnover, of the whole reef.

The Odums made a field estimate of the overall production and respiration of the entire reef system. From the water flowing tidally over the reef flat they made simultaneous hourly measurements day and night, at points upstream and downstream. The increase in oxygen content between these two points during the day gave the net photosynthetic production of the community. Conversely the decrease in oxygen by night yielded a value for the total respiration. The respiration by night and the net production by day would then equate with the total production.

This was found to be so high that the reef was

envisaged living largely on its own primary production. Plankton passing over the reef might still, however, make some contribution. Even if its own productivity were low during passage of the reef, the large volumes of water filtered by reef animals might secure appreciable amounts of organic matter previously accumulated by the planktonic organisms.

The content of suspended matter was measured (a) in outside water to windward of the reef, (b) in water crossing the reef front over the breaker zone, and (c) in water leaving the reef over the back-reef zone. Water on the reef front was found to have a high organic content — picking up from the buttress zone a lot of coarse plankton and algal fragments. Then — as the water passed over the reef shallows — much of this was removed. What remained could settle in backwaters or could be filtered by small anchovy-like fish. The whole reef is a highly efficient filter even though fluorescine dye shows water may cross it in as little as 10–15 minutes.

The volume and composition of reef plankton vary with the tide. At low tide, breakers at the reef front continue to pull off fragments of algae, but because little water is thrown over the reef, these accumulate in the eddies until the incoming tide. Plankton and other suspended material is then heaviest in the first water to cross the reef. The graph (Fig. 21.17) shows with passage over the reef (1) loss on ignition of combined suspended matter, (2) richness of plankton estimated by chlorophyll extraction (3) labelled carbohydrates, and (4) suspended sand.

The flux of large suspended plankton on to the reef was measured by determining the volume in unit time passing through a net fastened to a stake. This contained copepods, radiolarians, pteropods, and tiny filamentous algae. The large increase found as water crossed the breaker zone was removed in passage over the back portion of the reef — with most of the 'pseudoplankton' consumed.

So the amount of suspended material in water leaving the back of the reef is the same or slightly greater than in the water approaching from outside. The reef thus would appear to be energetically self-sustaining, and derive no net gain from the incoming water.

The balance sheet for the whole reef community thus suggests that respiration is not far out of balance with production of new biomass. Energy gain comes thus from photosynthetic production, plus influx of organic material in the water. Losses are by respiration or removal of inorganic material on the ebbing tide. The Odums found the two totals only 4% apart, close enough to equilibrium. Coral reefs could thus represent true 'climax' communities in that they are making little net increase to their living biomass. Such a highly productive ecosystem may not be far from a steady state. Through its long evolution as an open system, the construction of self-regulating interactions has led to the survival of the stable, a stability based upon high diversity.

CHAPTER TWENTY-TWO
The Richly Inhabited Shore

Our skimming of the reef's visible surface has inevitably been superficial, somewhat like a flight over the top canopy of a tropical forest. Such forests are indeed the only communities on the planet that in their intricate complexity can be compared with coral reefs.

Yet the reef system spread at our feet or viewed through a snorkel mask is far more accessible than the high rainforest. Where the plant kingdom reaches its peak of achievement in the forests, it is the tidal shore and — more than any other part it — the coral reef that has been the lavish provider of invertebrate animals, from their Indo-Malayan-Australian highpoint of diversity, out to the less-endowed but still rich Atlantic tropics and to the far more restrained diversity of the temperate zone.

Why are the tropics so rich?

Far from their habitat-space becoming saturated, tropical habitats give the impression of a biodiversity that is self-enhancing. Each niche carved out and occupied would seem to provide the feedbacks in favour of still more innovation. Some of the explanation must lie in the living substrate itself. Nowhere else are the dependent life-forms so adaptively stamped to the ground on which they reside. The coral milieu promotes biotic associations of every degree of intimacy. Thus, we find not only special dietary dependence, but camouflage, commensalism and inquilinism. Animals live around and upon and even inside others.

The physical topography of the reef in itself provides far more spatial diversity than a normal bedrock. Though capped in vertical ascent — for corals at approximate low water — the reef biospace reaches to depths finally cut off from light. As well as in the dark but still open interspaces, there is a lavish infauna boring and channelling into live and dead corals. Green and blue-green algal symbionts are imprisoned and farmed within soft and stony corals and in the tissues of clams and ascidians. Instead of bare rock, the visible reef surface becomes a living mosaic, exploited by a grazing

force of both primary herbivores and dependent carnivores. Even though mobile, the benthic fishes stay close-tied to their favoured substrate, coral or other. Some bite off individual corallites, others scour off the wide surface film. Numerous, colourful and easily studied, the fish are as integral to the reef as birds to the tropical forest.

The coral reef and the equatorial forest, then, are the most complex biosystems the planet has ever produced. Each bears witness to some biological principles we could call the 'rules of tropical enrichment'.

First, for both plants and animals, the number of entities is spectacularly increased. Not only are there more species, but new higher taxa can emerge, that the temperate has given no hint of.

Second, even where some of these taxa have a small overspill into the warm temperate, the zenith of each is tropical, with species multiplied by one, even two orders of magnitude.

Third, in almost all of the great invertebrate taxa, the biomass of the largest individuals can increase by the same order.

The genesis of diversity

Questions so arise not just about growth and size, but about the generating of genetic diversity: why this happened so richly in the past time, and how — in terms of niche-space — it is maintained today.

There is first the truism that the tropics are hotter. We may presume that the generation of new diversity is, like other metabolic processes, temperature-dependent. Some tropical species are known to reach their upper thermal limits not far above their ambient summer temperatures. Maximum growth and production of biomass thus take place rather close to their lethal or sublethal temperature points. Do marine species at such temperatures produce more genetic variation than in those at more central parts of their temperature range? Much of the chromosome recombination that results from breakage and crossover at meiotic division has been found to occur at sub-lethal temperatures. In the tropics, with

Fig. 22.1 Mosaic of algae, corals and encrusting animals at the surge-swept reef edge, with some small commensals
The quadrat from aspect A occupies 0.5 m².
1 encrusting branchless *Acropora* 2 a meandrine faviid 3 small *Favites* head 4 *Pocillopora meandrina* 5 *Zoanthus* 6 *Xenia* 7 *Amphiroa foliacea* 8 *Palythoa caesia* 9 a green didemnid ascidian with algal symbionts 10 *Millepora platyphyllia* 11 *Turbinaria peltata* 12 *Caulerpa peltata* 13 colourless didemnid colony.
(Detail) 14 *Pocillopora meandrina* with (15) commensal crab *Trapezia* 16 *Xenia* with (17) *Caphyra laevis* 18 *Caulerpa racemosa* with its sacoglossans 19 *Placida* sp. 20 *Oxynoe viridis* 21 a large *Elysia* sp. and (22) *Julia* 23 sacoglossan *Cyerce nigra* and 24 crab *Caphyra rotundifrons* on (25) green alga *Chlorodesmis comosa* 26 commensal pinnotherid crab *Xanthasia* from clam *Tridacna* 27 xanthid crab *Lybia tesselata* with anemones borne on chelae 28 *Huenia heraldica*, male (above) and female.

often year-long sexual reproduction, such opportunities must be enhanced.

During their long history the tropics have experienced a relative climatic stability (notwithstanding the Pacific's high incidence of volcanism). There is the further fact that a great proportion of tropical shore space happens to exist in archipelagoes. The small islands, if not bathymetrically permanent, could have been separated long enough to offer genetic isolation.

New entities — it is traditionally accepted — have evolved with such geographic isolation by *allopatric* speciation. Spatial isolation must be available to allow a spell of genetic separation of a daughter population from its original parents. Only after genetic re-making in such temporary isolation can a new entity overlap and co-exist without admixture by hybridisation with the old. The alternative hypothesis of *sympatric* speciation on the home ground would assume that daughter species emerge and acquire an apartheid while still within the parental territory. The tropics also can offer plenty of mosaics of segregated space (including restriction to particular host species) so that here — if anywhere — sympatric speciation might have its best chance to operate.

For *allopatric* speciation the tropics — as we know them — could thus offer far more opportunities of medium-scale geographical isolation than the temperate. The archipelagoes of Malaysia, the Philippines and the south-west Pacific — like the Caribbean and the Antilles — all present clusters of small islands. The bathymetric bands around a small island would set up temporary enclaves of isolation, all within a biogeographic realm that covers a quarter of the world's oceans. Such isolation by distance and depth would be intensified through time as volcanism, subsidence and wildly fluctuating sea levels changed the island contours. Multiple invasions would continue, with new species, re-introduced at successive times, able to co-exist side by side.

Yet geographic isolation must be only the initial — and provisional — part of the story. Indo-Pacific shallow water species tend to have wide geographical ranges, many of them covering the entire Region. Gene migration and inter-availability must be favoured by the high incidence of long-lived free-swimming larvae. Any hypothesis depending on island isolation must thus be balanced against the relative ease of gene-spread in the tropics. For a high majority of marine taxa, the Indo-Malaysian-North Australian triangle has been shown to be the primary base of dispersion. With terrestrial plants, it is equally clear that the primaeval tropical forest — to which Humboldt gave the name *hyalea* — has been a bountiful provider, from which the deciduous forests, grasslands, taiga and tundra have been populated along axes of decrement. Today's temperate marine animals and land plants must then be regarded as reduced, low-diversity communities, having the special physiological capacity to survive at lower temperatures and in conditions more marginal than at their tropical centre of origin.

Habitat space

The atoll system, with its interrelated parts conceived as in Fig. 22.1, we shall now begin to dissect. For each of its light and dark communities, the description in the following pages is founded on reefs like those seen at Aitutaki. But the picture has in some places been widened to be broadly applicable to the reefs of the whole far-flung Indo-Polynesian Province.

The reef edge

A *windward* reef edge under moderate exposure, briefly uncovered on a calm day, will reveal a bright-coloured living mosaic, almost always coral-dominated. The foremost corals under the constant surge are *Pocillopora*, *Millepora* and low, plaque-forming *Acropora*. There are also the low, convex heads of faviids; and on a leeward margin, tiers of fragile *Acropora* tables can be glimpsed at each drawing back of a wave.

In only moderate water movement, there is such richness that all the principles of straight zoning come to look confused. It is on higher exposed windward coasts that the zones assume a stricter regime. The reef edge chosen for Fig. 22.1 can thus be resolved into three-tiers. The uppermost expanse is of rich brown *Sargassum*. Furthest out and constantly washed with surge are crusts of calcareous red algae, like heavy pink cake icing, with high points of wine red and mauve. Between these two strips runs a narrow pavement, with a diversity of encrusting animals, some of them soft and uncalcified.

The important brown algae at the reef edge are low-growing sargassoids. Species with vesicles and a long leafy axis are confined mainly to runnels or the outer reaches of the lagoon. But in sweeping surge, the short, crisp *Sargassum cristaefolium* presses against the rock, with a cartilaginous texture resistant to wave damage. The golden brown leaves have a duplicated spiny margin. *Sargassum oligocystum* is darker brown, with the leaves longer and mid-ribbed, and also without bladders (Fig. 22.2).

The *Turbinaria* species are tropical sargassoids with their leaflets modified to stalked discs or pedestals. The familiar *Turbinaria peltata*, forms a low cover under surge. In more sheltered spots its ball of clustered leaflets is attached by a stalk. In the west Pacific there is a second common species, the triangular topped *T. murrayana*.

Beyond *Sargassum* and *Turbinaria* but before the rough-cast of lithophylla, the mosaic of encrusting animals begins, competing with low algae for the available space. Small patches interlock like a jigsaw, with every bit of terrain pre-empted. The diversity can be realised from the typical 0.5 sq m patch (Fig. 22.1), with more than a dozen contributing species. The small stony corals include a *Favites*, a branchless *Acropora*, *Pocillopora meandrina* and low flanges of *Millepora*.

After corals, the zoanthids are the most important cnidarians. The *Zoanthus* species are massed together to form leaden grey carpets where emersed, but their bright

Fig. 22.3 Zoanthidea

1 cluster of *Zoanthus* polyps around a single *Protopalythoa* 2 *Zoanthus* polyps
3 *Protopalythoa*, polyps seen under water
4 *Palythoa caesia*, with (5) detail of a polyp.

Fig. 22.2 Brown and red algae of the reef front

1 *Sargassum cristaefolium* 2 *S. oligocystum* 3 *Turbinaria peltata*
4 *T. conoides* 5 *Jania* sp. 6 *Amphiroa foliacea* 7 *A. crassa* 8 *Galaxaura*,
with enlarged frond detail (below) 9 *Peyssonnelia rubra* 10 *Martensia
flabelliformis* 11 *Neorhodomela* sp. 12 *Gracilaria eucheumatoides*
13 *Lithophyllum pygmaeum* 14 *Hypnea* sp. 15 *Laurencia* sp.
16 *Rhodymenia* sp. 17 *Jania* sp. 18 *Lithophyllum* sp.

green oral discs open out under the flow of surge. Their patches are studded with the larger fawn zoanthid, *Protopalythoa*, having the tentacles vestigial. The shiny buff colonies of *Palythoa caesia* are present almost everywhere with breaking surge, either in small convex heads or merged into crusts, with the polyps embedded in a stiff coenenchyme.

The two most frequent alcyonaceans in the surge mosaic are the dull red organ-pipe coral *Tubipora musica*, and the soft grey polyps of *Xenia*, with naked polyps arising snake-like from a fleshy base.

A common sponge under surge is the blue-grey *Adocia caerulea*, elastic and spicule-free, but with sharp-pointed papillae. Next in importance to the cnidarians may often be the compound ascidians. The commonest sorts visible to the light, as shown at the tops of rubble-

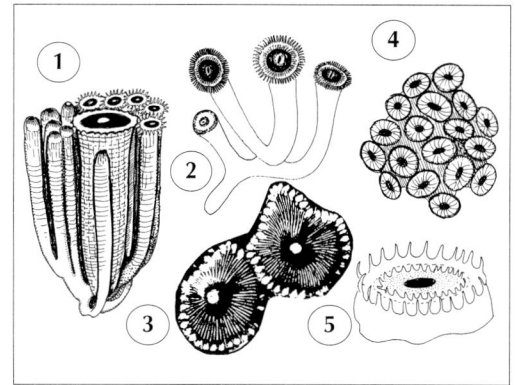

beds, are Didemnidae that have become green-tinted by incorporation of algal symbionts in the atrial wall (Fig. 22.6).

Algae

Even ahead of the sargassoids, the most pervasive algae are generally the calcified reds. These fall into two series. First are the jointed forms, including the common *Jania rubens*, but led by the larger species of *Amphiroa*: heavy-jointed *A. crassa*, the spear-branched *A. foliacea* as well as the finer textured *A. anastomosans* and *A. fragilissima*. Second, the reef has its overlay of crustose reds, giving a dull sound with a hammer, in contrast to the sharp ring of coral limestone. Some forms retain their branching but with the joints welded stiff, as for example the fist-sized clumps of *Lithophyllum pygmaeum*, like hedgehogs wearing blunt spatulate spines.

The pink or mauve *Galaxaura* species have annulated, dichotomous branching, and though still soft and flexible, their tissues are permeated with lime grains. Non-calcified rhodophytes include the dull olive brown *Laurencia* species and *Hypnea pannosa*. The fan-shaped laminae of *Martensia flabelliformis* frequently appear in shallow pools.

Almost rivalling the reds, the small green Chlorophyta stand out most vividly. Simple in build but often exquisite in detail, these are mostly soft and succulent, with *Halimeda opuntia* as the only calcified green at the reef edge. Stiffer than the rest is the glaucous green *Dictyosphaeria versluysii*, scattered everywhere near the edge, as small aggregates of vesicles like plastic spittle. Contrasting with these are the *Caulerpa* species, of bright salad green and soft texture. Each has its distinctive form. Most widespread are the bunches of grapes of *C. racemosa*. In *C. peltata* the branchlets form stalked discs. In *C. serrulata* they are twisted and saw-edged, while *C. sertularioides* carries stiff branchlets in two (sometimes three) rows, and *C. cupressoides* has broad pinnae like cypress leaves.

Of even more delicate construction are the green feathers of *Bryopsis* and the elegant mushrooms of *Acetabularia*, long-stalked with a single nucleus and a crown of whorled branchlets.

The largest of the green algae, common on all shores except those in extreme shelter, is *Chlorodesmis fastigiata*, forming soft tresses of branching filaments up to 25 cm long.

Grazers and browsers

In the grazing force at the reef edge it is fishes, echinoids and gastropods that predominate. As in the temperate, the alga are cropped by snails belonging to the primitive Archaeogastropoda. True limpets are fewer than on temperate shores, but the streamlined *Patelloida saccharina* (an acmaeid) and *Scutellastra stellaeformis* (a patellid) are common at the surf edge. The most widespread Pacific ormer is the small *Haliotis varia*. In the west there is also the 'donkey's ear', *Haliotis asinina*, with a light, reduced shell and the exposed skin handsomely marbled in green and cream.

The same reduced shell, though lacking the former's row of perforations, is adopted by the small trochid relatives, *Stomatia* and *Stomatella*, moving about the algal-filmed surface, even under breaking surge. The back of the fleshy foot is able be autotomised or cut off defensively.

The more normal trochids, conical with a flat base, have also a secure foothold for prehension in moving water, though their largest, *Trochus niloticus*, is typically found in moats behind the reef edge, along with the large turbinids. The smoothest of the turbans, *Turbo petholatus*, polished and marbled in emerald and rich brown, is better contoured for life at the wave front.

The cowries (**Cypraeidae**) are chiefly domiciled in the moats or — with their wealth of small species — under boulders, but there are two important species, *Cypraea mauritiana* and the smaller *Cypraea caputserpentis*, found in crevices at the reef edge. Both are broad and streamlined, with the shell base flat and heavy.

The aplysioid slug *Dolabrifera brazieri* is a common algal browser at the reef edge. Mossy green, mottled with pink and white, it is low-built and broad-footed, with the parapodia almost completely fused. Locomotion is essentially leech-like with head-extension and front-attachment, followed by shifting the broad sole forward.

Predators

On the reef edge — as in nearly every temperate or tropical shore — leading gastropod predators are thaids. First, the *Drupa* species are found lying low to the surface near the wave front. With shortened spire and wide aperture, they have taken on a virtual limpet form. Lifted up, each species can be recognised by the colour of the mouth that has prompted its specific names: *Drupa morum* (mulberry), *D. grossularia* (gooseberry), *D. ricinus* (castor oil), *D. rubusidaeus* (cut strawberry). The drupids typically prey upon coral-piercing mytilids or sipunculans, by extending the long proboscis into their borings. The more normally shaped *Drupella cornus* and related thaids live among branched corals. In crevices and surge runnels we shall find the smooth, ovoid *Nassa francolina* and the larger *Thais persica*.

The coral predators of the Family **Magilidae** appear to have been derived from near the thaids. All have lost the radula and feed suctorially by extruding salivary enzymes and pumping up fluid food with the muscular pharynx. The small, very common *Coralliophila neritoidea* is thick-shelled and purple mouthed, sitting in a permanent scar on top of *Porites*. *Quoyula monodonta* has the same relation with *Montipora*. The thinner *Leptoconchus* has evolved further, to live permanently in a flask-shaped cavity within a faviid head. The most advanced of the family is *Magilus antiquus*, detected from its navel-like pit in the head of a meandrine faviid. Split open such a coral and revealed will be a straight tube like a sleeve, with the spiral tip filled solid. Lacking a radula, *Magilus* appears to explore the surrounding coral surface with a capilliform proboscis that can be inserted into the mouths of the communal polyp systems. The largest of the Magilidae is the thin, fragile

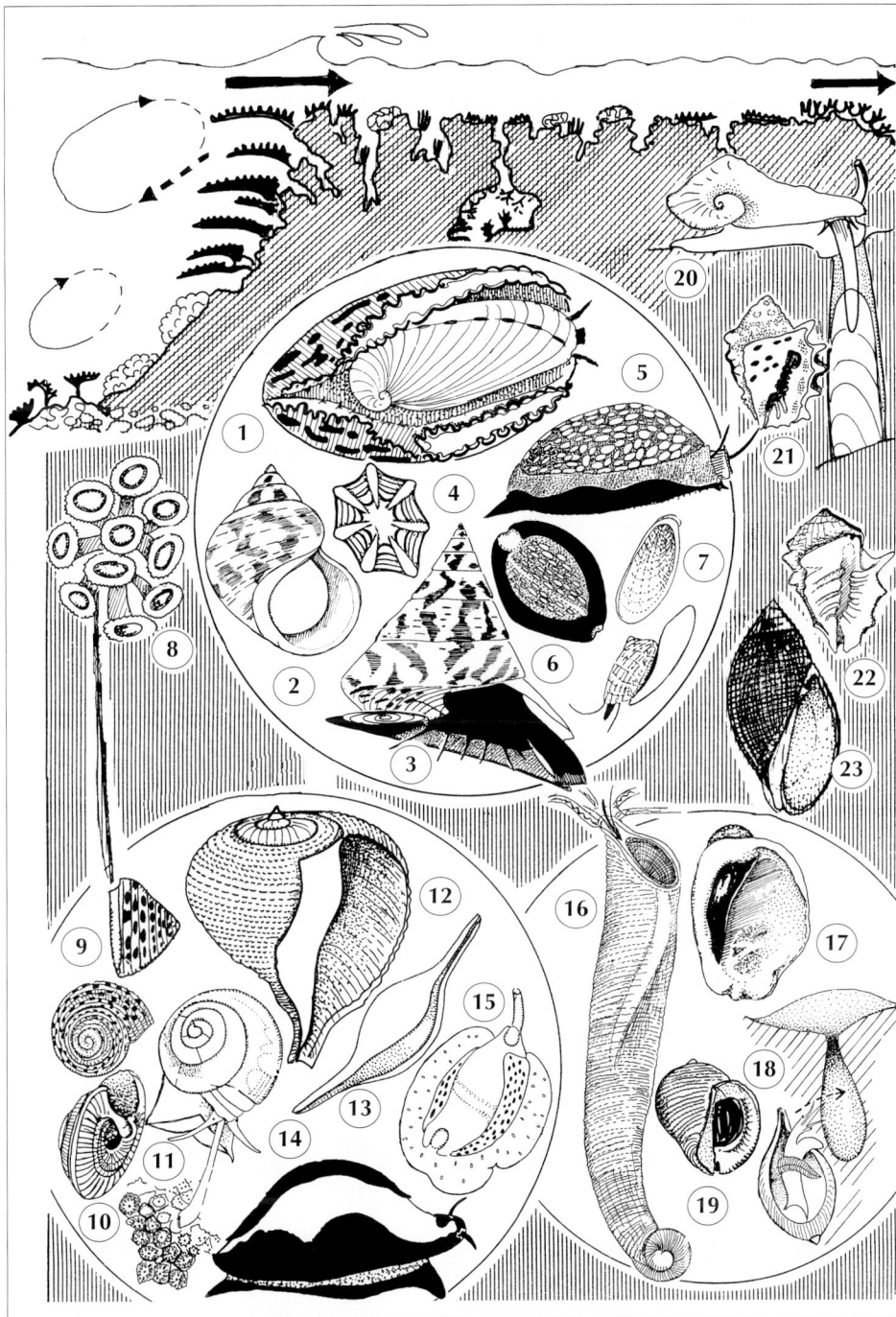

Fig. 22.4 Some gastropods of the reef crest
1–7 **herbivores** 1 *Haliotis asinina* 2 *Turbo petholatus* 3 *Trochus niloticus* 4 *Patelloida saccharina* 5 *Cypraea mauritiana*
6 *Cypraea caputserpentis* 7 *Stomatia* sp. 8–15 **predators on Cnidaria** 8 Zoanthid polyps with 9 *Heliacus variegatus*,
suspended by mucus thread 10 *Architectonica* 11 *Philippia lutea* feeding on *Pocillopora* 12 *Rapa rapa* 13 *Volva volva*
14 *Ovula ovum* 15 *Calpurnus verrucosus* 16–19 **coral penetrants** 16 *Magilus antiquus* 17 *Quoyula monodonta*
18 *Leptoconchus* with chamber in faviid head 19 *Coralliophila neritoidea* 20–23 **thaids** 20 *Drupa* inserting proboscis in
Lithophaga burrow 21 *Drupa ricinus* 22 *D. morum* 23 *Nassa francolina*.

Rapa rapa, bulbous like an onion and burying in tissues of the soft coral *Sarcophyton*.

In the Family **Ovulidae** the cowry tribe has produced its specialist predators, chiefly interested in alcyonaceans. *Ovula ovum*, the egg cowry, is pure white but covered with a purplish black integument, and preys on

fleshy soft corals. The small and thicker *Calpurnus verrucosus* is found on the soft coral *Sarcophyton*, while the narrow-shelled *Volva* and *Primovula* live on gorgonian colonies.

The snails of the Family **Architectonicidae** are at home on corals, having elegant, low-conical shells with

427

a wide, stair-cased umbilicus. *Philippia radiata* feeds on *Pocillopora* corallites, while among coral rubble large numbers of the black and white chequered *Heliacus variegatus* crawl over or hang by mucus strings from the zoanthid polyps on which they prey.

Camouflage and commensals

More intensively than any other habitat, the reef edge offers a wealth of camouflage-matching. Many species have gone on to evolve an exclusive commensalistic relation with a host plant or invertebrate. Foremost among these are the many species of commensalistic crabs. The pallid white *Caphyra laevis* (Family **Portunidae**) lives among *Xenia* heads. A second species, the green-lined *Caphyra rotundifrons*, can be detected by feeling among the tresses of almost any bunch of the green alga *Chlorodesmis fastigiata*. To the same family belong the harlequin crabs, *Lissocarcinus orbicularis*, found in sexually dimorphic pairs in the rectum of large holothurians.

The polished, wide-armed *Trapezia* crabs, belonging to the **Xanthidae**, are to be found throughout the tropical Pacific, pressing between the branches of *Pocillopora*. Food is scraped from the coral by the strong chelae, chiselled like rodent incisors. *Cymo* and *Tetralia* have a similar relation with *Acropora*.

The sargassoid algae generally shelter a distinct faunule of their own. *Sargassum cristaefolium* has its small fast-clinging maiid crab *Huenia heraldica*, well camouflaged and sexually dimorphic. Also present may be a small ophiuroid, a gold-tinted anemone, and a small predatory gastropod *Pyrene* hung by a pedal mucus string. Yellow and brown amphipods (*Hyale* species) abound.

The green algae, especially *Caulerpa*, are the classic haunt of the small sap-taking opisthobranchs of the Sacoglossa (Fig. 22.1), virtually the aphids of the molluscan world. Fronds should be carefully washed in sea water with a drop of formalin, in a white dish where the sacoglossans can be spotted upon release. All these slugs lance the green algal cells one by one with the radula followed by sucking with the pharynx. The handsome *Cyerce nigra*, with its ornament of black and gold-embossed leaflets, feeds on *Chlorodesmis fastigiata*. With *Caulerpa racemosa* in Fiji, three sacoglossans were found: an *Elysia*, a *Placida* and the primitively shelled species *Oxynoe viridis*.

Thinly shelled sacoglossans are found also in the early genera *Cylindrobulla* and *Lobiger*. But the most extraordinary dwellers with tropical *Caulerpa* are the 'bivalved sacoglossans' of the genus *Julia*. The two-pieced shells, only collected as dead valves, had been previously classed as lamellibranchs, though the left valve was observed to carry a spiral protoconch. When a living *Julia* — then described as *Tamanovalva* — was first discovered in Japan in 1959, it was recognised as a gastropod, with a right-hand piece hinged on to the left (spiral) shell and derived from a separate calcification centre in the larval mantle. Green and with lancing radula teeth and ascus-sac (Fig. 22.1), this 'bivalve' turned out to be an unmistakable sacoglossan.

The 'dead reef'

Back from the crest, but still out beyond the lagoon, large stretches of the reef are covered with moveable rubble. Strewn with old coral, especially broken and overturned *Acropora* tables, *Porites* heads and storm blocks, this is the wide expanse of the reef often dismissed as the *dead reef*. It is in fact one of the most diversely teemingly alive parts of the shore. Echinoderms, crustaceans, molluscs and polychaetes abound. Rich and colourful sciaphilic communities are revealed in its dark spaces. Plenty of time then must be given to this dingy coloured stretch and all that lives beneath it, on the way back — from the more instantly alluring reef edge.

Under boulders

With experience of temperate shores, our first instinct will be to look under coral blocks and boulders. Here — as on the underhangs of *Acropora* tables, or in the narrow spaces under faviid or *Porites* heads — rich assemblages await us.

The primary elements of this *sciaphilic* or dark-loving fauna are encrusting invertebrates. Three leading taxa — the sponges, compound ascidians, and bryozoans account for most of the surface cover. Any of these may locally predominate. Furthest away from light, serpulid tubeworms generally have the lead, accompanied — in the tropics — with a new prominence of sessile Foraminifera.

For some visitors, the prized items under boulders — to the exclusion of all else — will be the small and exquisite gastropods. On the more accessible coral reefs, gastropods are being imperiled by thoughtless ransacking. The time is come when live reef shells need protection against commercial depredation. While frugal specimen-taking can be legitimate, this is an invasion that can be justified in conscience only by the intention of serious study. It is even more important that no coral boulder should be kept long over-turned or exposed to the sun's heat and light. After careful scrutiny and perhaps photography each boulder should be restored exactly to its former position.

Lifting up first a small round *Porites* head, we should be able to recognise a concentric pattern of zoning. Fig.

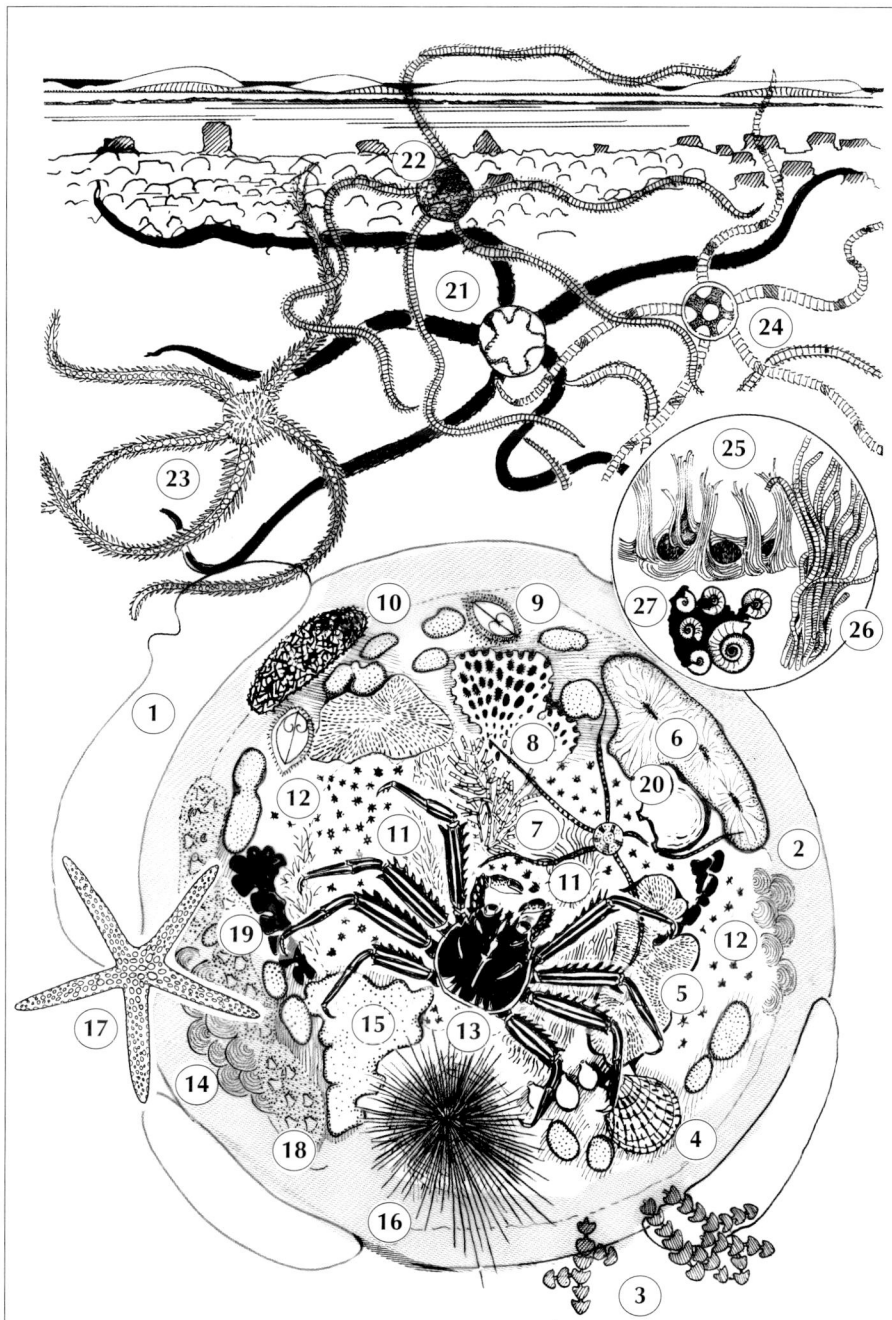

Fig. 22.5 The sciaphilic community of the 'dead reef'
The inhabited surface beneath a small Porites head, showing: 1 Periphery of living coral 2 mauve *Goniolithon* edge zone
3 *Halimeda opuntia* 4 *Chlamys* sp. 5 cheilostome bryozoan 6 sponge, *Spirastrella coccinea* 7 cheilostome bryozoan
Margaretta 8 bryozoan *Reteporellina* 9 penetrant *Arca* 10 sponge *Hyrtios erecta* 11 foraminifer *Carpenteria* 12 foraminifer
Homotrema 13 *Percnon planissimum* 14 *Peyssonnelia rubra* 15 compound ascidian *Didemnum* 16 *Diadema setosum*
17 *Fromia* 18 halichondriid sponge 19 young colony of *Distichopora* 20 *Corella japonica*
(**top**) ophiuroids of four genera: 21 *Ophiolepis* 22 *Macrophiothrix* 23 *Ophiomastix* 24 *Ophiarachnella*
(**inset**) serpulids 25 *Filograna*, fine structure of aggregated tubes 26 the same enlarged 27 *Spirorbis* tubes.

22.5 shows first a lighted subperiphery with its rim of live coral. Inside this, three zonules are usually to be made out:

1 a shade veneer of a calcareous alga, perhaps *Lithophyllum* or *Neogoniolithon*, in low reflected light
2 a zone dominated by bryozoans, sponges and both simple and compound ascidians
3 in total dark, towards the centre, a zone with serpulid tubeworms, especially *Filograna*, and sessile Foraminifera.

Ascidians

Simple ascidians are larger but less numerous than compound. They may include the spherical tests of *Corella*

Fig. 22.6 Zooid structure of colonial ascidians
a Clavelinidae b Polycitoridae
c Polyclinidae d Didemnidae
the common test is shown
diagonally hatched;
(c c c) common cloacal cavity.

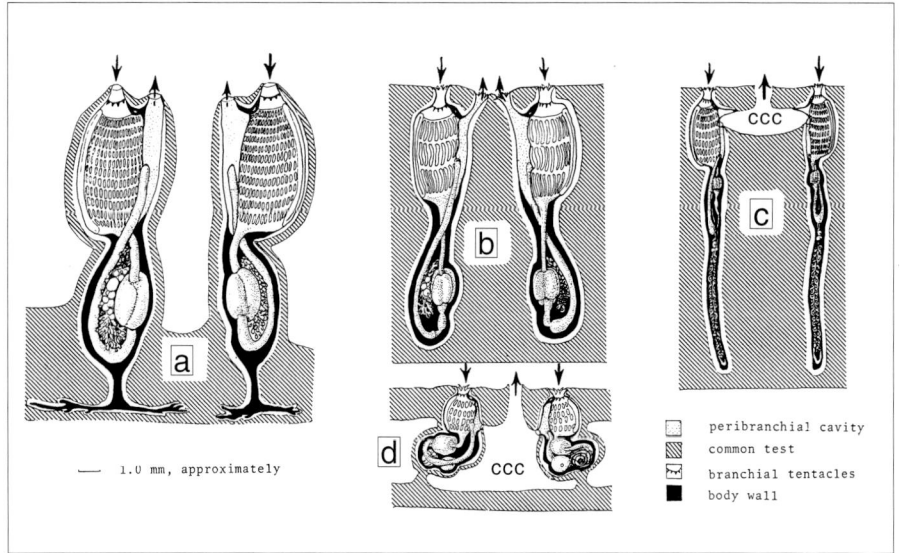

— 1.0 mm, approximately

peribranchial cavity
common test
branchial tentacles
body wall

japonica and *Herdmania momus* and towards the boulder edge the yellow and magenta *Polycarpa aurea*. The compound habit has been evolved in separate groups, and taken to different stages. The small interlinked tests of *Polyzoa depressa* (**Styelidae**) are still separate, merely disposed upon a common basal crust. The **Clavelinidae** (including *Clavelina* and *Pycnoclavella*) have miniature stalked tests, mostly transparent, linked together by their common stolon. In the **Polycitoridae** (*Polycitor* and *Cystodytes*) the zooids have become embedded in a common test, but each opening by a separate cloaca. The numerous species of the **Polyclinidae** form knob-shaped heads, sheets or cushions, studded with long, attenuate zooids using a common cloaca. Polyclinid heads may be yellow, apricot, scarlet or jet-black and some are sand-impregnated.

The most advanced compound ascidians are those of the large Family **Didemnidae**, where low-encrusting colonies have zooids sharing a common cloacal space. The test is rendered putty-stiff with microscopic stellate spicules. The *Didemnum* and *Trididemnum* species have a wide colour range, red orange and yellow, as well as white. On top of lighted rubble some of their species are blue-green with their contained symbiotic algae.

Bryozoans

The under-boulder Bryozoa in the tropics are noted for their reticulate (or net-like) growth. These include the fenestrate fans of some cyclostomes (*Mesonea* and *Nevianopora*), but most of the reticulate forms are anascophoran cheilostomes.

First among these are the species of *Margaretta*, with cylindrical branches forming long internodes with chitinous joints, and having the zoooids in alternating whorls of two to six. Next are the fully fenestrated colonies known as 'lace-corals'. The *Reteporella* species have almost parallel rami, with slender cross-connections leaving long, narrow fenestrae. The other reticulate forms have trabeculae of even width, and round-ovate

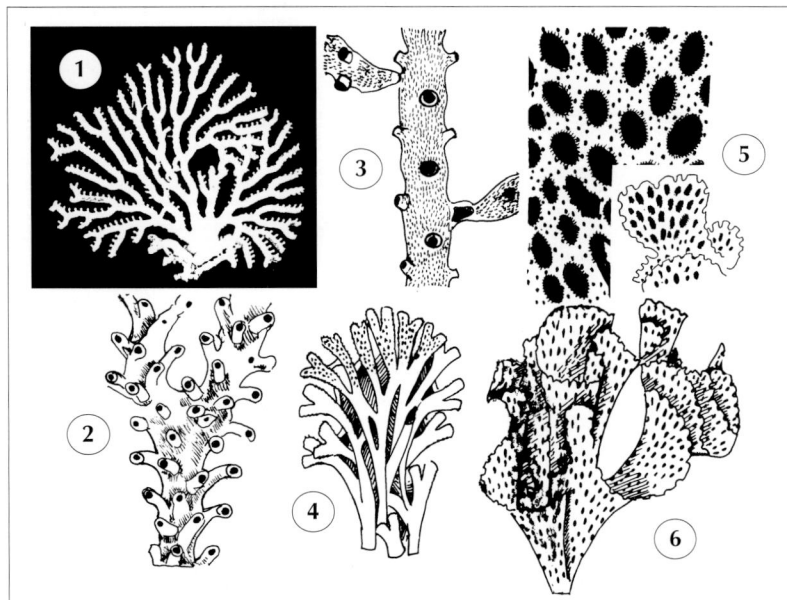

Fig. 22.7 Reef Bryozoa
Cyclostomata 1 *Mesonea radians*
2 *Nevianipora pulcherrima*
Cheilostomata 3 *Margaretta gracilior*
4 *Reteporellina denticulata* 5 *Triphyllozoon*,
colony and enlarged detail 6 *Reteporella* sp.

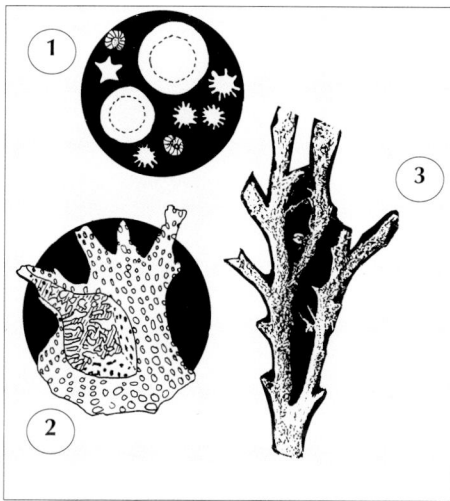

Fig. 22.8 Reef Foraminifera
1 free-living Foraminifera from sandy rubble,
Baculogypsina sphaerulata (star-shaped), *Calcarina* spp.
(large spines), *Marginopora vertebralis* (large disks, used in
necklaces) 2 *Homotrema rubrum* 3 *Dendrophrya* sp.

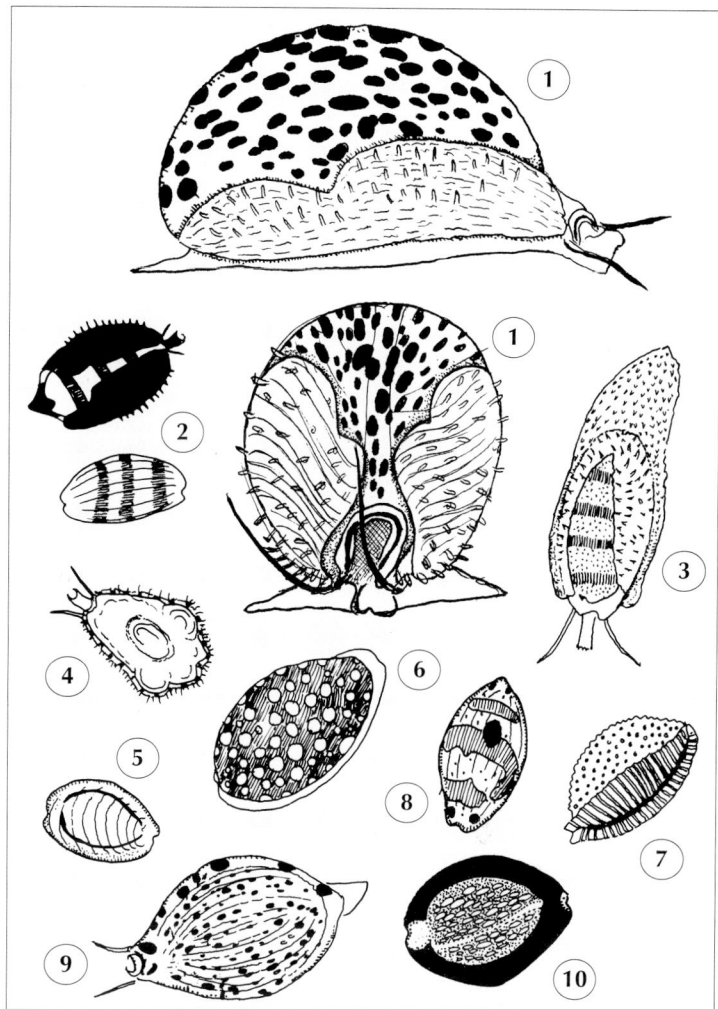

Fig. 22.9 Cowries of boulder flat and lower reef
1 *Cypraea tigris*, lateral and (**beneath it**) from in front
2 *Cypraea asellus* 3 *Cypraea lynx*, 4 *Cypraea moneta*
5 *Cypraea annulus* 6 *Cypraea cribraria* 7 *Cypraea nucleus*
8 *Cypraea ursellus* 9 *Cypraea lynx* 10 *Cypraea
caputserpentis*.

windows, as in the stony pink or violet nets of *Iodic-tyon*. The most delicate of the lace corals, *Triphyllozoon* are rolled up like brandy-snaps with the zooids opening inwards.

The siliceous sponges (**Demospongiae**) take numerous forms and textures: firm or compressible, massive and spiculose, or as tough elastic sheets. Distinct from all these are the small white tubes and vases of calcareous sponges.

The common serpulid worms, in the central dark-space of almost every boulder, are *Filograna* species, forming a filigree of cross-linked tubes, sometimes building up patterns of high relief to give their crowns the most effective arrangement for filtering.

In the same deep shade, the sessile colonial Foraminifera are prolific. The *Dendrophrya* species have excessively delicate calcified tubes teased out like finely spun glass. From these they put out the microscopic pseudopodia used for food-catching. The colonies of *Homotrema rubrum*, with shorter clumps of tubes like daubs of sealing wax, are more robust than *Dendrophrya*. They are drawn into bundles of tubes, perforated all over with microscopic foramina.

Gastropoda
The shelled gastropods under coral blocks are profuse and often colourful. The most frequent species in the boulder field are probably the smaller cowries (**Cypraei-dae**), headed by the pale, orange-ringed money shell, *Cypraea annulus*, followed by the polished, heavily cal-lused *C. moneta*. The wide, flat-based, *Cypraea mauri-tiana* and *C. caputserpentis* we have seen (Fig. 22.4) to be denizens of the wave zone, while the largest cowries

such as the tiger (*C. tigris*) and the tortoise (*C. testudi-naria*) live in quiet lagoons. Typical under boulders are the medium to small *C. talpa*, *C. lynx*, *C. isabella* and *C. errones*, with the tiny and exquisite *C. asellus*, *C. ursel-lus*, *C. cribraria* and *C. nucleus*.

To call the cowries omnivorous grazers is a cloak for partial ignorance, but the related families **Ovulidae** and **Triviidae** have better defined specialist dietaries. More distant from the cowries, the **Lamellariidae** have a predilection for compound ascidians, for feeding and implanting their egg capsules. Their most conspicuous tropical species is the fleshy jet-black *Coriocella nigra* with clubbed appendages and glassy internal shell.

Several lines of sedentary mesogastropods have evolved towards microphagous feeding. The **Hipponici-dae** are cup-shaped and conical, quite immobile with an accessory shell-plate secreted beneath the foot. Some species such as *Hipponyx conicus* attach to the shells of

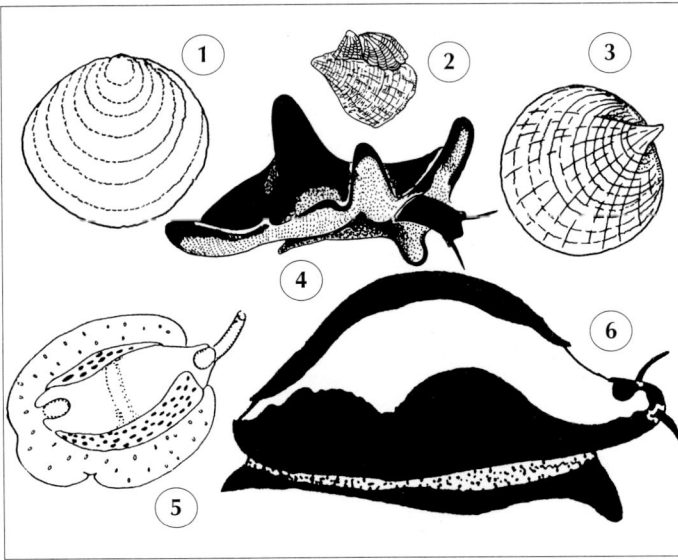

Fig. 22.10 Mesogastropods with special habits
1 *Cheilea equestris* 2 *Sabia conica* 3 *Capulus* sp.
4 *Coriocella nigra* 5 *Calpurnus verrucosus*, a predator of
Montipora 6 *Ovula ovum*, a predator of soft corals.

Turbo, to feed on their faecal pellets. The **Capulidae** sit strategically at the edge of bivalve shells, and intercept food brought in by the currents of the gill. The **Calyptraeidae** are finally established as ciliary feeders, commonly represented on reefs by the cup and saucer limpet *Cheilea equestris*.

Ophiuroidea

The snake-stars or brittle stars are the most mobile of echinoderms, retreating into all kinds of cover, or twining upon other organisms. They attain their greatest range of form and colour under the boulders of the *dead reef*. Despite their basic uniformity they have various methods of food-catching. First, *Ophiocoma scolopendrina* — probably the commonest Indo-Pacific brittle star — is a highly versatile feeder. Abounding in loose rubble, it responds to the rising tide by unfolding and deploying the arms, with their flexible tips sweeping for particles at the water surface. Mucus nets between the arm-spines catch material to be carried to the mouth by the pointed tube-feet. This same star also grazes hard surfaces with the jaws.

To the same family, **Ophiocomidae**, belong *Ophiomastix*, with arms and disc carrying long, slender spines, and *Ophiarthrum* with a naked but attractively patterned disc. The Family **Ophiotrichidae** is made up of fragile stars, with slender spines and tubercles covering the disc scales. The arms bear long, glassy spines, as in the *Macrophiothrix* species, commonly found on sponges and gorgonians.

The **Ophiuridae** are by contrast robust, strong-armed carnivores taking large fragments by arm-looping. They

are typified under boulders by the handsome *Ophiolepis superba*. Similar habits are shown in the **Ophiodermatidae**, with the disc scales of both sides covered by fine granules. *Ophiarachnella gorgonia* has short arm-spines, while in the *Ophiarachna* species they are much longer than the segment that bears them.

The **Amphiuridae** have small discs bearing long arms, often convoluted. They burrow in pockets of sediment, sampling the surface deposits with the emerging arm-tips. The **Ophiactidae**, with *Ophiactis savignyi*, widely common in the tropics, similarly extend long arms from small concavities.

Crinoidea

While the five-armed ophiuroids are mostly secretive and hidden, the crinoids or feather stars openly display their 10 or more arms. The most familiar crinoids of coral reefs are the comatulids, belonging to the Order **Articulata**. Early in life these break free from the attaching stalk possessed by the classical crinoids. The body forms a cup (*calyx*) with a roof (*tegmen*) of small plates covering the oral surface. From its margins the basic five arms arise, each dividing first into two, though the final arm number often exceeds 10. The oral surface carries chan-

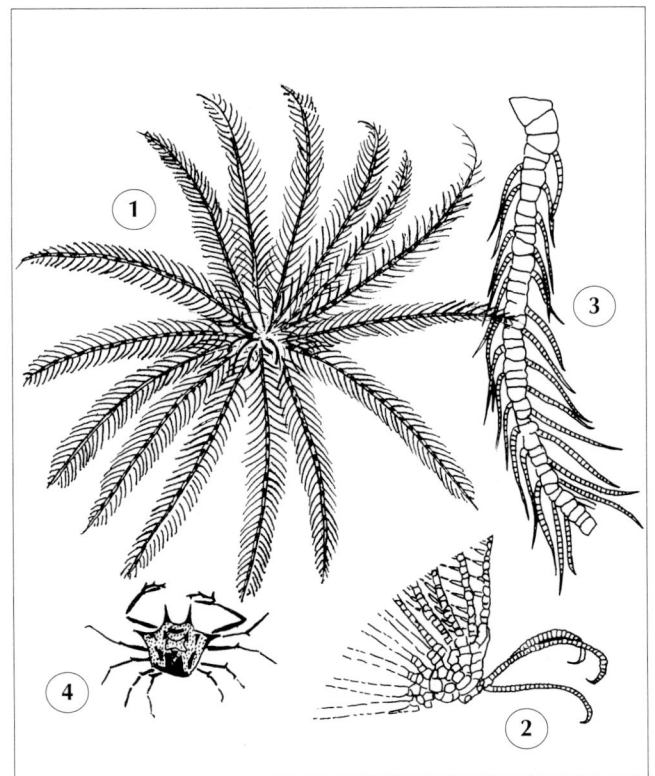

Fig. 22.11 Crinoidea
1 a comatulid crinoid, viewed from below 2 detail of calyx, attachment of pinnae and cirri 3 portion of a pinna with details of cirri 4 *Ceratocarcinus trilobatus*, commensal with *Comanthus*.

nels to the mouth from their converging ambulacral grooves. The anus as well as the mouth lies on the oral face.

Diatoms, dinoflagellates and small zooplankton alight upon the arm pinnules, and — so stimulated — the erect podia propel this food towards the ambulacral groove, where cilia carry it bound in mucus to the mouth.

The largest Indo-Pacific crinoid family, the **Comasteridae**, has 19 genera, with some strikingly coloured species. Among the commonest are the fluorescent yellow *Comaster gracilis* and the light green *C. multifida*. There are other species magenta-coloured, or black and white-striped. The comasterids have lost the graceful arm-swimming ability of other comatulids. Hiding by day with the arms curled, they creep about at night by using the prehensile combs near the pinnule tips. The calyx is flattened, and lacks the claw-like basal cirri found in other families.

The larger crinoids of the **Tropiometridae** have a deep calyx with large attachment cirri, and 10 pinnae not further branched. The very common black *Tropiometra atra* lives not under boulders like comasterids but attaches to large overhangs.

Crinoids are notably hospitable to commensal species of many sorts: small galetheid crabs often sharing the host's banded colouring, alpheid shrimps, small brittle stars and minute gastropods, mostly subsisting on food picked up from the crinoid's arm grooves.

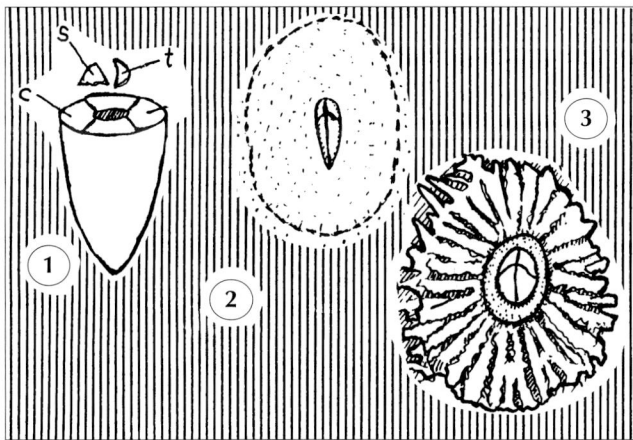

Fig. 22.12 Coral Cirripedia
Low tidal operculate barnacles are not at first noticeable on reefs. But boring into living corals is the special habit of operculates of the Pyrgomatidae, with fused column plates almost buried in the coral, and only the operculum visible.
1 *Cantellius pallidus* with column plates and operculum found in *Acropora, Pocillopora, Cyphastrea* and on the back of *Fungia* 2 *Savignium milleporum* 3 *S. crenatum*, found in faviids.

Thickets and rubble

The first corals likely to make their appearance in quiet places are long-branched *Acropora*. In light water movement these establish easily upon hard debris, and — once initiated — a dense thicket can grow upwards, ultimately supported on the stilts of dead branches. The interspaces will accommodate solitary (*ahermatypic*) corals, and also give space for colonies of *Pavona* and *Montipora*, and stalked juvenile *Fungia*.

Consolidation of this loose forest will follow in slow course. Long-branched stags-horn corals continue to stand vertically long after they are for the most part dead. Their interspaces eventually become cut off from light by the spread of a ceiling of calcareous lithophylla, topped with a wealth of small rubble algae. The branches, often with the tips still living, become cemented into a fragile pavilion, remaining upright and intact until broken by a tornado or crashed into by incautious treading.

The bio-space under the canopy may reach down for as much as a metre, with illumination — from the sides or through the few openings left above — progressively reduced. The surfaces of the branches carry a descending set of sciaphilic communities, essentially like those under coral tables and boulders, but more amply spread.

Over the visible roof, a pervasive green tinge is given by small rubble algae, numerous if unspectacular: *Laurencia, Hypnea, Gelidiopsis* and *Gigartina* are well represented among the smaller reds, in fact mostly dull olive-green. Golden green *Boergesenia forbesii* masses its cylindrical vesicles side by side. Wedged into rubble are the much larger bottle-green *Ventricaria ventricosa*, the size of ping-pong balls. Fine-branched networks of green *Boodlea* and *Struvea* grow up in tufts. Small pink *Rhodymenia* are found in local spots of shade. A ground cover can be formed by the ubiquitous *Halimeda opuntia*, or the sharp-speared *Amphiroa foliacea*, often with the finer branching *Jania* species.

Large Foraminifera live freely among these rubble algae, like tiny cog-wheels or biconcave discs. Vermetid gastropods, part-uncoiled, attach to the finger rubble. Dull black snake-stars, *Ophiocoma scolopendrina* and urchins, *Echinometra mathaei*, take refuge in these upper levels.

Spaces among the surface rubble may be filled up with fleshy or cartilaginous red algae such as the thick *Gracilaria eucheumatoides*. Colonies of branching zoanthids are also, sometimes swarming with their gastropod predator, the chequered black and white *Heliacus variegatus*.

The rubble algae also support numerous grazing and browsing opisthobranchs. The largest is an ungainly aplysioid, the shaggy, dull green *Dolabrifera dolabrifera*, broad and planed off behind, where the fused parapodia

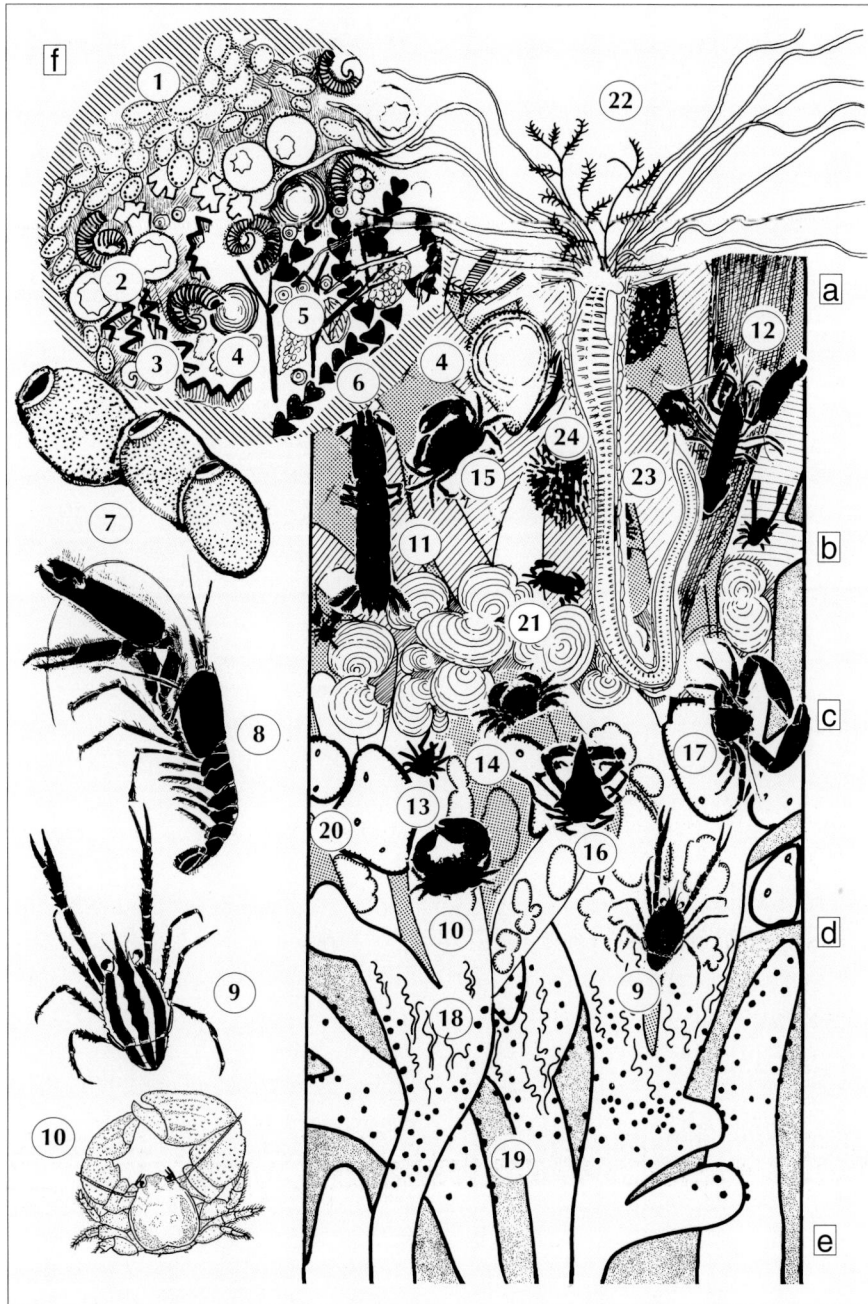

Fig. 22.13 A bed of Acropora finger rubble, showing the zonation with depth
a–b surface rubble algae with *Goniolithon* veneer below; b–c low algal zone of *Peyssonnelia*; c–d zone of compound
ascidians and sponges; d–e deep zone of serpulid worms and sessile Foraminifera
(**top left**) f detail of rubble surface
1 didemnid ascidian with zooxanthellae 2 *Dendropoma* sp. 3 *Pavona* sp. 4 *Valonia* sp. 5 *Amphiroa foliacea* 6 *Halimeda
opuntia* 7 *Didemnum molle* 8 *Alpheus* sp. 9 *Galathea* sp. 10 *Pachycheles* 11 stomatopod *Gonodactylus* 12 *Lyngbya*
'stocking' with in-dwelling alpheid shrimp 13 maiid crab 14 *Pilumnus* 15 xanthid crab, *Liomera*, 16 *Menaethiops*
17 *Porcellana* 18 *Filograna* 19 *Homotrema rubrum* 20 demospongian 21 *Peyssonnelia rubra* 22 *Loimia medusa*, tentacles
23 *Loimia medusa* 24 *Echinometra mathaei*.

have their narrow respiratory opening. A smaller, more
graceful relative, *Stylocheilus longicauda* may be found
nearby in patches of sea-grass *Thalassia*. The sacoglossan slugs found with green algae among the rubble
include several large elysioids. *Plakobranchus ocellatus*
has cream parapodia lined with green striping and feeds

suctorially on *Enteromorpha*. A second elysioid is lilac
and brown-spotted, another yellowish-flecked on pale
green, with dark border.

Compound ascidians can have an important role at
the lighted surface, with rubble patches covered by
colonies of didemnids, carrying symbiotic blue-green

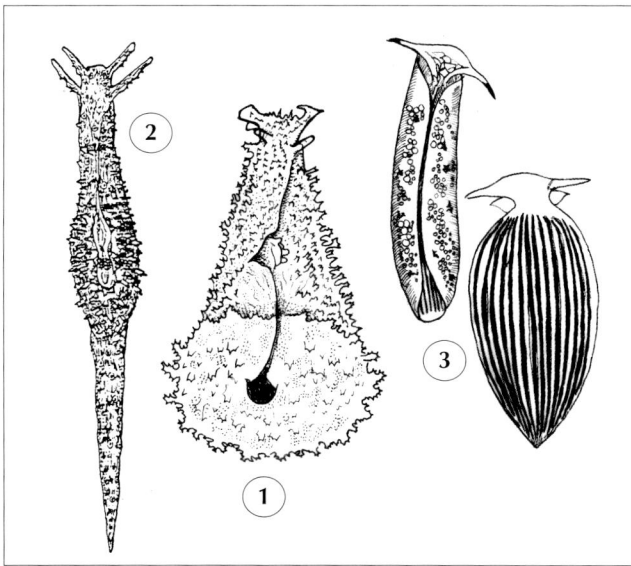

Fig. 22.14 Algal-feeding opisthobranchs of the rubble-field

1 *Dolabella auricularia* 2 *Stylocheilus longicauda*
3 *Plakobranchus ocellatus,* with side lobes closed and spread.

reach their highpoint of diversity. The same dark places also abound in small Crustacea, especially the miniaturised species of three prolific groups, the **Xanthidae, Galatheidea** and **Alpheidae.** The xanthids or black-finger crabs are often brightly coloured, with species of the widespread genera *Liomera, Chlorodiella, Pilumnus, Actaea* and *Daira.*

Both the families of **Galatheidea** contribute to the rubble fauna. First are the small squat lobsters (Galatheidae), attractive in colour and with some species camouflaged. The abdomen is folded on itself, but still visible behind, and the rostrum is prominent and triangular. From the galatheids have evolved the **Porcellanidae,** virtually finished crabs with the abdomen folded under the thorax but with a still mobile tail-fan. Porcellanids feed on fine particles strained with the third maxillipeds. As with galatheids, several have become commensals. Slow-moving *Pachycheles* differs from the agile *Porcellana* and *Petrolisthes* in its heavy chelipeds with shaggy hairs.

The snapping shrimps, **Alpheidae,** are some of the most numerous tropical decapods. The first thoracic legs are heavily chelate, with strong asymmetry in the male. Alpheids are never good swimmers, and most stay con-

algae in their atrial lining. From Viti Levu, Fiji, Patricia Mather has reported *Diplosoma virens,* pale grey-green to rich green, as the most conspicuous rubble species, along with *Trididemnum cyclops* and *Lissoclinum bistratum.* Flimsily attached to dead coral antlers and remaining beneath water at low tide are the very distinctive colonies of *Didemnum molle,* soft and spherical and the size of a grape. The wide exhalant opening, like the rim of a bottle, reveals the green algal lining inside.

Beneath the shading canopy, the dead coral is encrusted by the calcareous algae, *Neogoniolithon, Lithothamnion* and *Dermatolithon,* still receiving low light. A little further down, near the limit of photosynthesis, the rubble carries lightly attached sheets of *Peyssonnelia,* thinly calcified and burnt sienna coloured. Here as well can be found small specimens of the hydrocorals, orange *Distichopora* and pink *Stylaster* (Fig. 22.16).

Beyond the shade algae, there follow small sponges, both Demospongiae and Calcarea, and compound ascidians. Of the last, pink, white or apricot *Didemnum,* particularly *D. moseleyi,* are typical. The deepest level, virtually unlit, carries encrusting tubeworms, especially white *Filograna* and small coiled spirorbids. Bryozoans may include the white *Disporella* and jointed *Margaretta.*

In the ultimate secluded reaches, there are sessile Foraminifera — spots of red *Homotrema* and the brittle spun-glass of *Dendrophrya* — scattered over the otherwise bare surface.

With the rubble-bed's greater space, the mobile fauna is even richer than under boulders. Ophiuroidea here

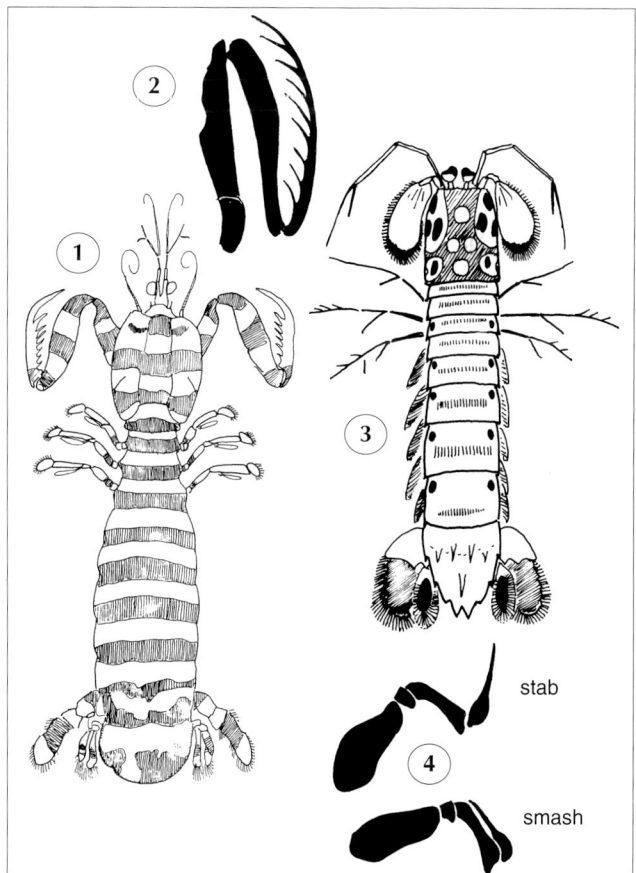

Fig. 22.15 Mantis shrimps (Stomatopoda)

1 *Lysiosquilla maculata,* with (2) detail of chela
3 *Odontodactylus japonicus* with (4) detail of chela, as used to 'stab' and 'smash'.

cealed under rocks, in corals or in their own burrows. In rubble beds certain alpheids live in sex-pairs in long stockings spun from a filament mesh of the blue-green alga *Lyngbya*.

Stomatopoda

Finally, the mantis shrimps are highly distinctive — it could be said 'off-beat' — decapods active in rubble-beds. Long and narrow and agile, they swim, curling up and somersaulting by use of the large abdominal pleopods. All the mantis shrimps are raptorial carni-vores with the second thoracic limbs forming powerful chelipeds folding beneath the carapace like jack-knives. In the Family **Squillidae** the terminal dactyl of the chela is armed with sharp, curving teeth. Moving prey is impaled by its sudden extension and strike. Common in rubble and under boulders are species of *Squilla*, *Oratosquilla* and *Harpiosquilla*. Burrowing deep in sand, there is the cross-striped *Lysiosquilla maculata*, as large as a man's forearm.

In the narrow, semi-cylindrical **Goniodactylidae** the chelipeds have — instead of prehensive teeth — a sharp stabbing point and a heavy calcified 'heel'. Small gas-tropods, hermits and bivalves are smashed by the action of the claw, wielded like a loaded boxing-glove. Perhaps the most beautiful of all rubble stomatopods is *Odonto-dactylus japonicus*, cream-enamelled, with high points of blue and carnation. This genus is distinguished by the blunt tooth behind the dactyl spine. In *Gonodactylus* this is lacking.

Polychaetes

Coral debris offers space for a wealth of polychaete worms. Common here is the widespread *Eurythoe com-planata* (**Amphinomiidae**) with its bunches of fine, glass-like setae, painful to the touch and indeed toxic. There are above all numerous euniciid species, along with cir-ratulids such as *Cirriformia filigera*, and the opheliid *Armandia*. By far the largest tubeworm in rubble-beds is the terebellid, *Loimia medusa*, constructing tubes of coral, shell and *Halimeda* fragments. The tube descends vertically to as much as 30 cm and the feeding tentacles extend around the mouth like pale strands of wool up to half a metre long. They withdraw at once if touched.

Stone-fish

The stone-fish *Synanceia verrucosa*, one of the Scor-paenidae well respected for its venomous powers, lurks among rock and rubble. Perfectly camouflaged and nor-mally immobile, it makes a lightning upward lunge, opening a cavernous mouth to seize small fish. Poison from glands associated with the dorsal fin spines pro-duce excruciating pain and serious secondary effects.

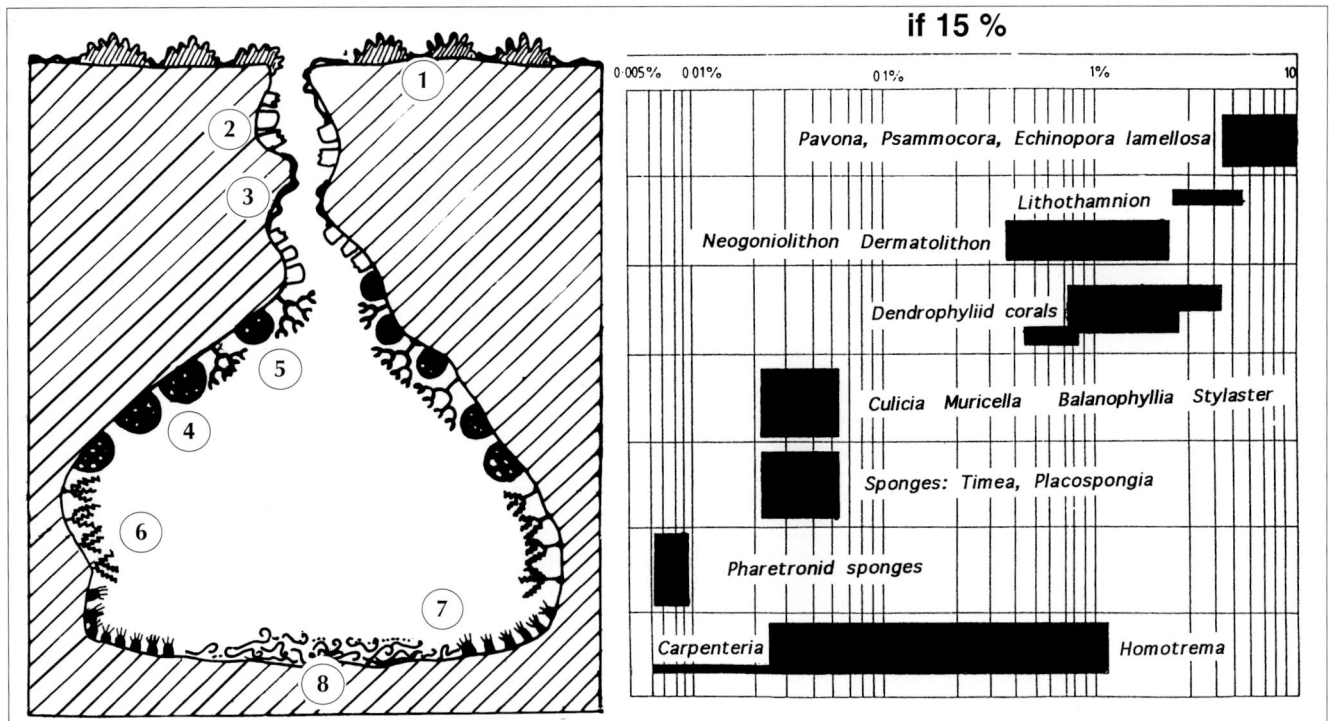

Fig. 22.16 A blowhole passage in vertical section at the reef edge, with its sciaphilic community
(**left**) 1 lighted Scleractinia 2 *Tubastrea* 3 *Goniolithon* crust 4 tethyid sponge 5 *Distichopora* 6 *Stylaster* 7 *Homotrema* 8 serpulid tubeworms. (right) Distribution of sciaphilic organisms in relation to light. Illumination (if 15%) is expressed as a percentage of the light value on the open surface. (Data from Jaubert & Vasseur's study of Madagascar reefs.)

436

Dark passages

The deepest habitats of the coral system, hardly represented on a temperate shore, are the crevasses and tunnels that intersect the reef front. Surge is all the time driven in forcibly, while the low illumination falls ultimately to total darkness.

Where surge breaks on the open front, the typical algae are calcified and jointed reds such as the strongly built *Amphiroa* and *Cheilosporum*, the flexible but lime-impregnated *Galaxaura*, along with some pink, non-calcified *Rhodymenia* species. As the crusts coalesce to roof over channels, shade-loving Melobesiae (*Neogoniolithon* and *Dermatolithon*) appear in profusion, as well as Squamariaceae (*Peyssonnelia*). Some scleractinian corals (*Pavona*, *Montipora*, *Echinopora* and *Leptoseris*) flourish in the same low lighting.

Finally, on the ceilings of overhangs and in caves, hermatypic corals give way to the apricot cups of *Tubastrea coccinea*, and the paler spreads of *Culicia*. The bright-coloured, branching cup corals *Dendrophyllia* and the solitary cups of *Balanophyllia* and *Caryophyllia* may also occur here, though usually reserved for greater depths.

The same dark reaches are the haunt of the bright-coloured **Stylasteridae**, 'hydrocorals' forming brittle, single-planed fans. Orange to mauve *Distichopora violacea* is the more strongly built, and reaches up to 15 cm across. The smaller pink fans of *Stylaster* species are light and fragile, and exquisitely branched. Along with the hydrocorals may grow a small, scarlet gorgonian *Muricella*, one of the few of its group to have become intertidal.

In the ultimate narrow tunnels beneath the outer reef crest, the surge reverberates, sometimes spouting up through blowholes in the ceiling. Such blind galleries, with illumination less than 0.1% of daylight, were investigated by Jaubert and Vasseur, in the Malagasy reefs. The surface has under 10% of living cover, made up chiefly of sessile foraminifera (*Homotrema* and *Carpenteria*) as well as some special sponges. At the darkest sites were reported species of lithistids, pharetronids and species of the silico-calcareous sponge genera such as *Aciculites*, *Callipelta*, *Plectronimia* and *Astrosclera*.

Bio-erosion

Not only do boulders and rubble provide shade. The dead and sometimes living coralla are deeply eroded by boring and abrading organisms. First, their exposed tops are scarred and channelled, principally by bivalves and urchins. Second, the same surfaces are penetrated by long, narrow shafts striking deep into the dead coral. The extent of boring, with the density of burrows, all contrived to avoid intersecting, is shown from the split section of coral rock in Fig. 22.18.

The primary bio-eroders are those species that actively bore their substrate. There is in addition a succession of later species that nestle and enlarge the cavities that the firstcomers have prepared.

The tops of coral blocks, or of light-structured *Porites* micro-atolls, are densely pierced by sipunculan worms and — lower on the shore — with the borings of polychaetes. Wider-spaced among these are the larger slits made by boring bivalves, chiefly the date mussels, *Lithophaga*.

Pholad bivalves that bore mechanically are scarce but not lacking in the tropical Pacific. The globular-conic *Jouannetia globosa* is abundant where soft sedimentary rocks occur. The small bivalve *Gastrochaena cuneiformis*, superficially like a pholad, bores in dead coral in the same way by rotation of the shell. The file-like sculpture is only slightly developed and the shell abrades with a sharp rostrum or prong as seen in Fig. 22.18.

The narrow *Lithophaga* mussels are by contrast fragile and ill-adapted for mechanical boring. Instead of rotating like pholads, they fasten to the burrow wall by byssus threads. The boring is heart-shaped in section, resulting not from shell-rotation but from secretion at the mantle edge of mucus with an acid phosphatase.

The eroding ark shells, such as *Arca avellana*, lie horizontal and remain well visible from outside. Their heart-shaped cavities expose the shell's spiral umbones, flush with the surface. The largest bivalve excavations — usually in *Porites* micro-atolls — are carved out by the lesser species of the giant clams. *Tridacna squamosa* attaches in coral rubble by weak byssus strands. *Tridacna maxima* excavates shallow depressions in coral, while the smaller *T. crocea* is deep-embedded and anchored by a dense byssus.

Both the ark shells and the tridacnids erode by the rocking motion of the shell. The latter are, however, far more modified, as will be seen from Fig. 22.18. The embedded *Tridacna* sits in the rock hinge-downward, while the strong-toothed ventral gape has been turned round to face upwards. The foot and byssus have been re-oriented to end up close to the dorsal but now deep-lying hinge. The upturned gape reveals the expanded lips of the exhalant siphon, attractively patterned in green, blue, brown or yellow.

From the mid-eulittoral down, a prominent abrader is the tropic-wide urchin *Echinometra mathaei*, possibly the world's most numerous echinoid. Smooth galleries are excavated in coral rock by its short, strong spines, like blunt gramophone needles. Ovoid-rectangular in shape, *Echinometra* is adapted to move along its eroded

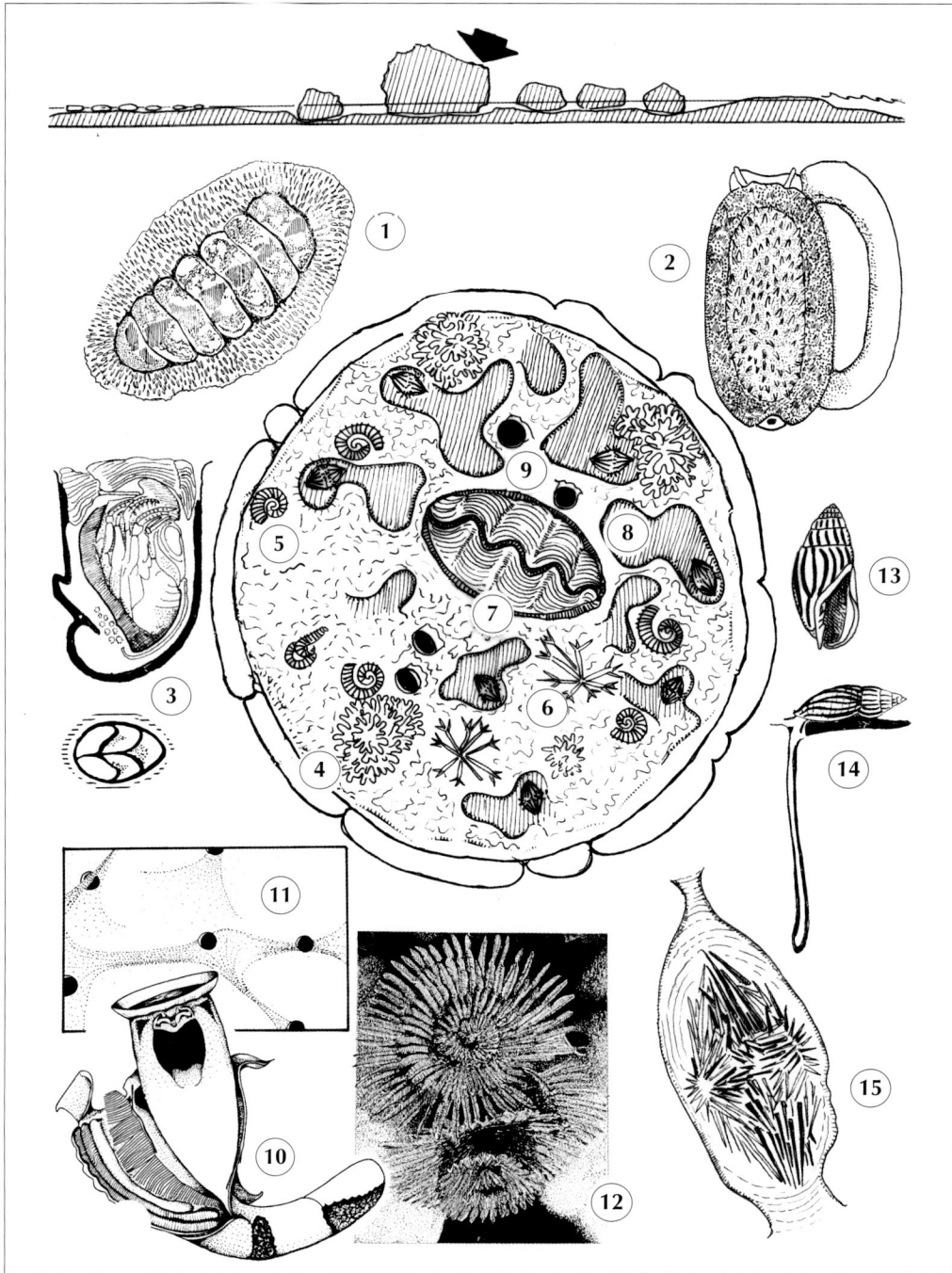

Fig. 22.17 Nestling and eroding species from upper surfaces
(1–3) from high level on storm blocks 1 *Acanthopleura gemmata* 2 *Onchidium* sp. 3 barnacle *Lithotrya, in situ* with (**beneath it**) opercular plates. (4–12) in well eroded small *Porites* heads 4 cluster-branching *Lithophyllum* 5 vermetid *Dendropoma* sp. 6 *Amphiroa foliacea* 7 *Tridacna maxima* 8 *Echinometra mathaei* in its eroded gallery 9 opening of *Dendropoma maxima* tube 10 *D. maxima* with mantle cavity opened to show gill 11 secreted mucus traps of *Dendropoma*. 12 *Spirobranchus giganteus* 13 *Strigatella* sp. 14 *Strigatella* sp., with proboscis extended into a sipunculid tube 15 *Echinometra mathaei* in its eroded gallery.

passage and make feeding forays out of the open end. The far reef edge is scoured by the much larger slate-pencil urchin, *Heterocentrotus mammillatus*, also one of the Family **Echinometridae**. Armed with blunt spines, like heavy pendants, it bores the softish lithophylla of the algal ridge, an easier substrate usually already bio-eroded.

Near the tops of coral storm blocks only briefly immersed at full tides, sipunculans and bivalves are scarce. The principal borer is here the specialised pedunculate barnacle *Lithotrya*. Basically like a small *Capitulum* straining food with passive cirri, this barnacle sits in a narrow slot like a foot in a sock. Tight-closing terga and scuta form an operculum, inset a little below the surface.

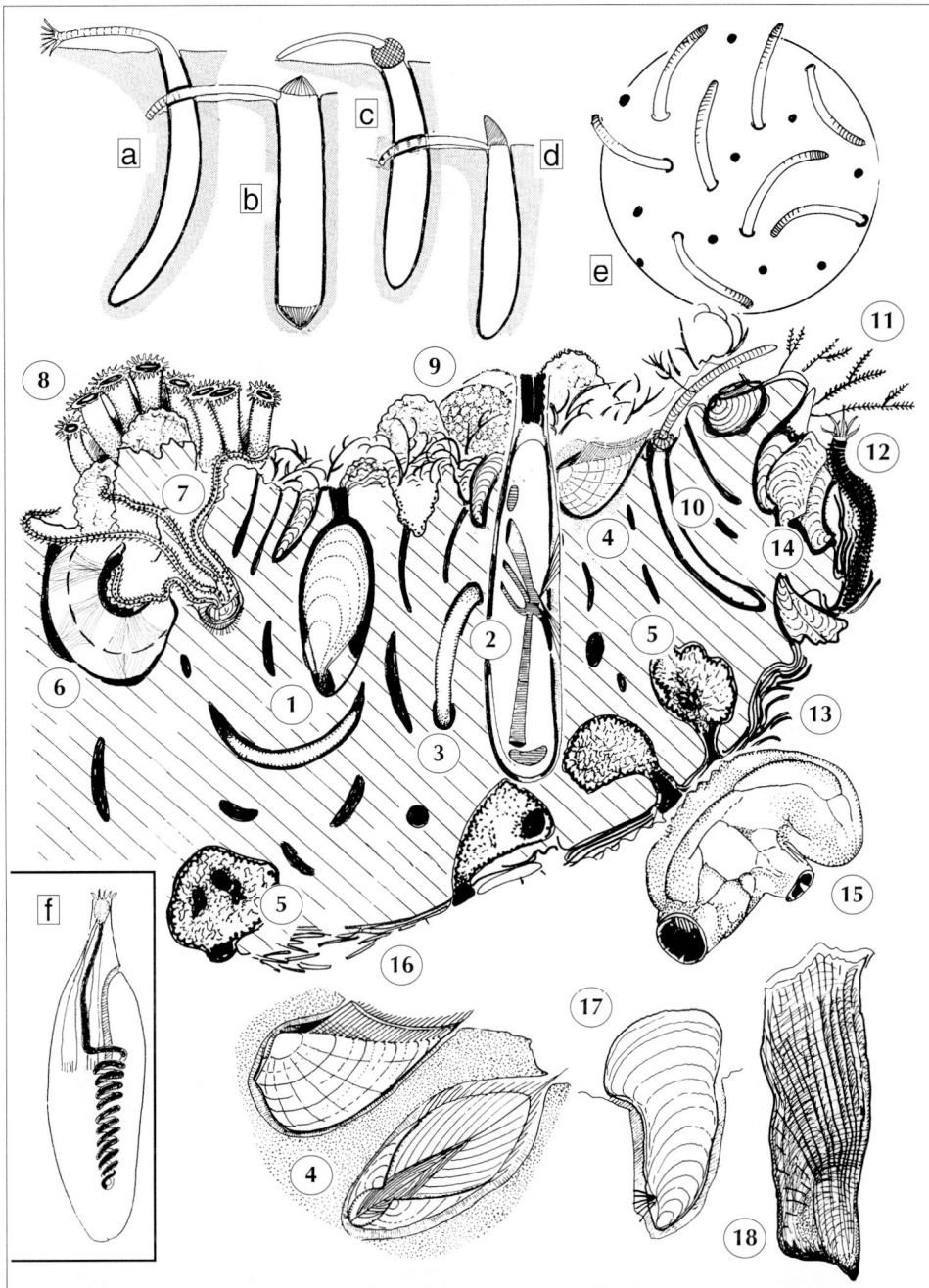

Fig. 22.18 Vertical section of a dead coral showing bio-eroders and other colonisers
(**top left**) (a) *Phascolosoma* (b) *Aspidosiphon* (c) *Cloeosiphon* (d) *Lithacrosiphon* (e) surface view with sipunculid holes and extroverts extended (f) (**bottom left**) sipunculan showing course of gut.
1 *Gastrochaena cuneiformis* 2 *Lithophaga zitteliana* 3 portion of *Phascolosoma* in boring 4 eroding ark shell, *Arca avellana* 5 eroding sponge entering from undersurface 6 *Cryptoplax japonicus* 7 ophiuroid 8 zoanthid colony 9 *Dictyosphaeria* 10 *Cloeosiphon* in boring 11 sertularian hydroid 12 euniciid worm 13 shade alga, *Peyssonnelia rubra* 14 nestling *Isognomon* 15 *Polycarpa aurata* 16 cheilostome *Margaretta* 17 *Isognomon* in crevice 18 *Streptopinna saccata*.

At the same level, homing into a scar, lives the large *Acanthopleura gemmata*, one of the few tropical chitons encountered on the open surface. The girdle is beset with horny but flexible spinules. Reaching even higher levels and keeping cool and clammy in shade may be found large pulmonate slugs *Onchidium*, grey-brown mottled and reaching 8 cm in length.

Towards low water, *Porites* micro-atolls may have holes large enough to insert a finger. These are the straight tubes of the largest of the vermetid gastropods, *Dendropoma maxima*. They are immured right up to the aperture in coral rock, not closely aggregated but usually with several in a neighbourhood. The bowl-shaped operculum sits on the yellow and jet-black foot. Like all its family *Dendropoma* feeds by mucus traps. Wide strands of secretion are deployed from the pedal

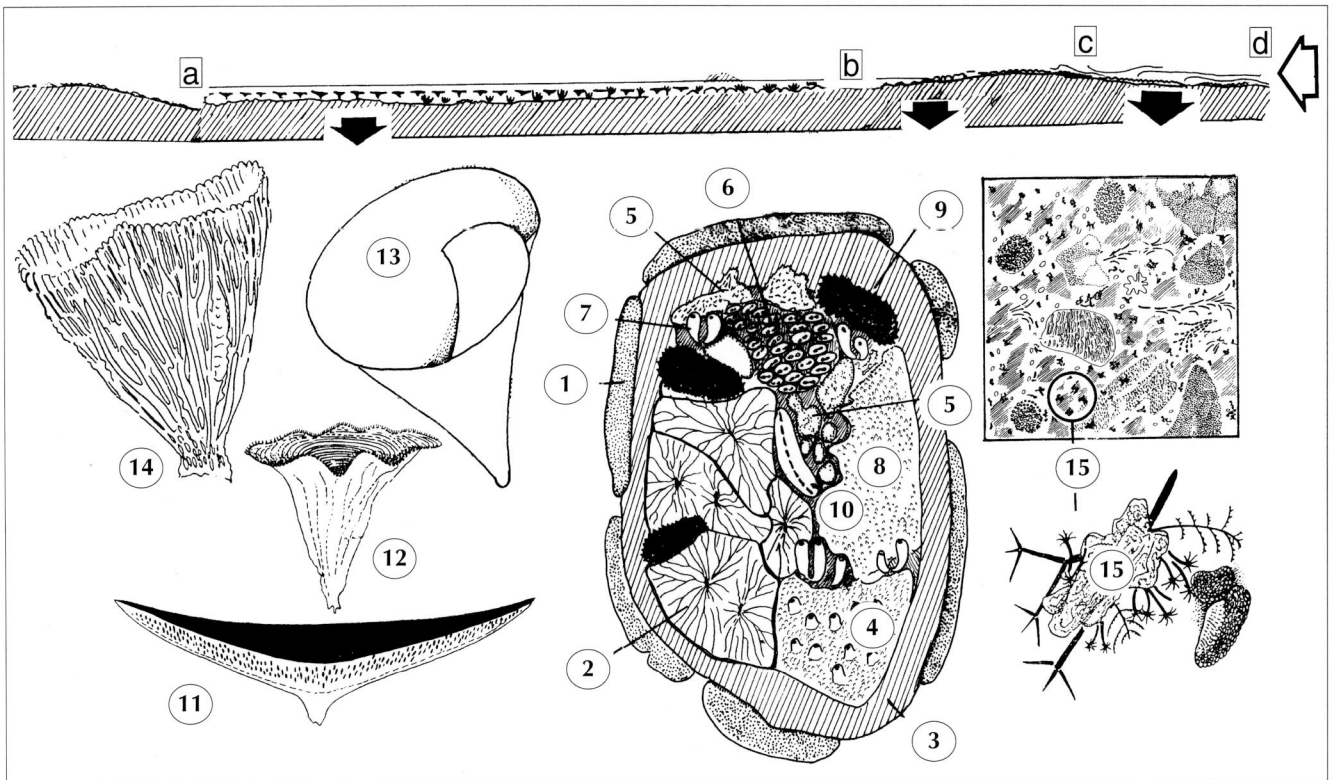

Fig. 22.19 Sponges of lagoon, boulders and reef face
a–b lagoon with dead coral tables; b–c higher outer stretch with emersed boulders; c–d surf-swept slope, with low-relief encrusting corals and blue sponge *Adocia*.
(1–10) beneath a *Porites* micro-atoll 1 living coral edge 2 *Spirastrella coccinea* 3 *Goniolithon* 4 *Halichondria* sp.
5 *Didemnum* sp. 6 ascidian cluster 7 *Sycon* sp. 8 crustose *Haliclona* 9 *Hyrtios erecta* 10 chiton, *Cryptoplax*;
(11–14) sponges of the lagoon floor 11 *Hippospongia metachromia* 12 *Callyspongia* sp. 13 *Dactylia infundibuliformis* 14 *Phyllospongia papyracea*; 15 *Adocia* cf. *caerulea* under surge at reef front in small black clumps with (below) a clump associated with *Xenia* polyps, a pennarian hydroid, *Amphiroa* and *Dictyosphaeria versluysii*.

gland in front of the mouth forming a spider-web mesh to intercept particles from the gill-drawn current.

A second tube-dweller, the polychaete *Spirobranchus giganteus*, is the largest, most conspicuous serpulid of the reef. It embeds, like *Dendropoma*, in the tops of coral rock and its twin-spiralled tentacle crown varies from orange brown to pink.

The rich bio-eroding community is represented at an advanced stage in Fig. 22.18. Bivalves *Lithophaga* and *Arca* have penetrated, along with the sipunculans, from the top. The upper surface is shown covered with the small algae *Dictyosphaeria* and *Gelidiopsis*, sertularian hydroids and zoanthid polyps. Shallow penetrants include the wafer-thin bivalves *Isognomon*, that do not bore but secure themselves with byssus threads. Their fragile valves are distorted to the shape of the cavity. The shaded undersurface is laminated with the calcareous red alga, *Peyssonnelia*. Small nests of penetrant sponges are eroding upward from the dark face, eventually to meet the organisms moving down. A large shade ascidian, the gold and magenta *Polycarpa aurata*, is shown attached below. In narrow crevices and between living corals near low water, the flat scallop-relative *Pedum spondyloideum* may be found wedged. *Streptopinna saccata*, the smallest and most modified of the fan mussels, is compressed to a narrow rectangle. Disused ark shell or urchin galleries can harbour the chiton *Cryptoplax japonicus*, with shell valves buried in the stiff shark-skin girdle.

Rock-boring worms

The non-segmented worms of the small Phylum **Sipuncula** are far and away the most numerous bio-eroders of the coral shore. They increase from the upper eulittoral downwards, to be joined on the lower shore by the destructive polychaetes of the **Euniciidae** and **Nereidae**. All these worms freely penetrate the coral rock and finally riddle it to destruction.

The polychaetes, as carnivores or scavengers, are able to part-emerge from their burrows. The sipunculans stay tightly enclosed, only everting the proboscis like the finger of a glove, to scour the surrounds for fine particles. In all the sipunculans the spiral intestine is reverse-coiled,

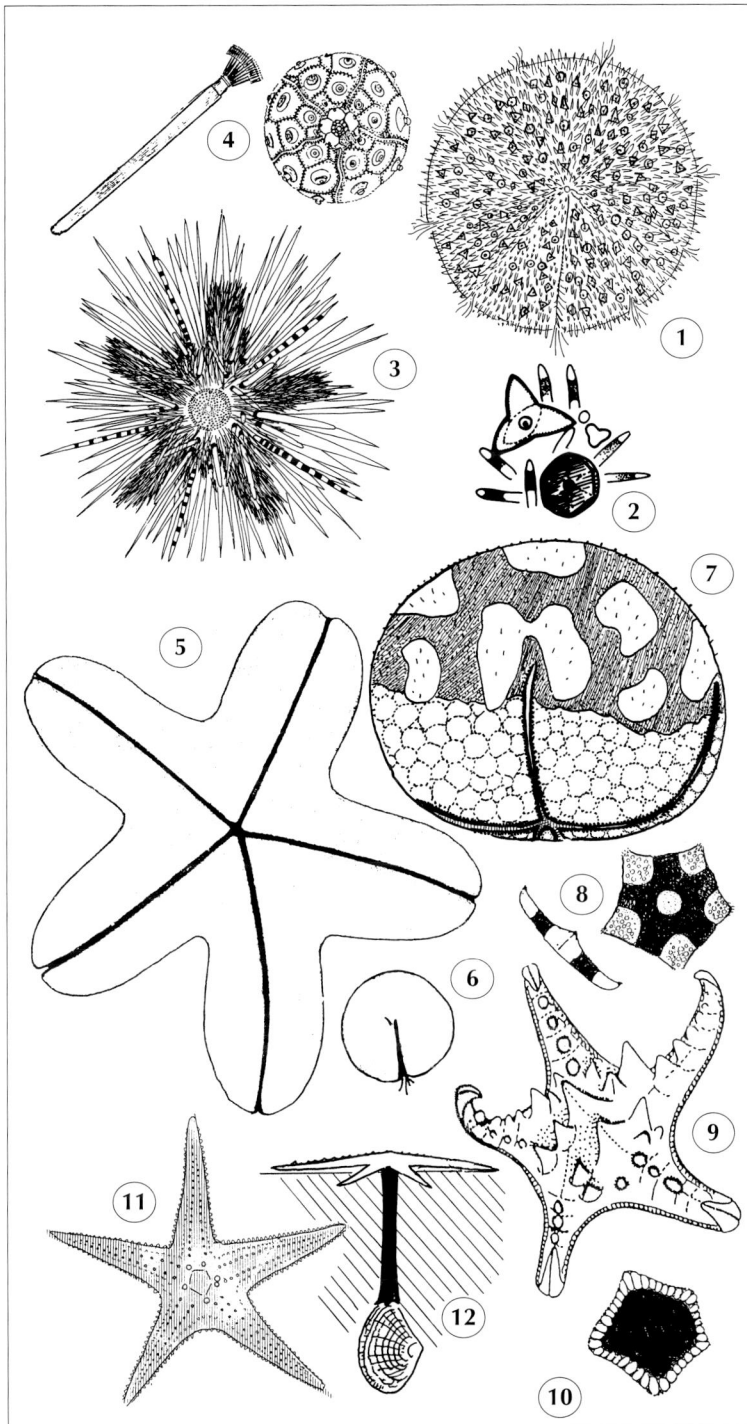

Fig. 22.20 Lagoonal echinoids and asteroids
1 *Toxopneustes pileolus* with (2) detail of pedicellariae 3 *Echinothrix calamaris* 4 *Phyllacanthus imperialis*, with detail of a single spine 5 *Choriaster granulatus*, with (6) end view of arm 7 *Culcita novaeguineae* with (8) juvenile, aboral and side view 9 *Protoreaster nodosus*, with (10) juvenile, aboral view 11 *Pentaceraster regulus*, with (12) stomach everted to prey on a burrowing bivalve.

bringing the anus up to the base of the head-shield, so the faeces can be voided outside the burrow.

Aspidosiphon carries a horny shield at either end, while *Lithacrosiphon* has a single opercular shield built up to a calcified cone covered with cuticle. In *Cloeosiphon aspergillus*, the head-shield is built of spiral platelets to form a spherical knob, with the proboscis passing through its middle. The *Phascolosoma* species lack shields. In *Aspidosiphon*, *Lithacrosiphon* and *Cloeosiphon* the proboscis carries rings of microscopic hooks, in some sense comparable with molluscan radula teeth. Particles are ingested as the proboscis is rolled back by its retractor muscles. In *Phascolosoma* by contrast the particulate food is directed to the mouth by a

circlet of ciliated tentacles.

It is not yet clear how the close-fitting burrows of Sipuncula become eroded. Those of *Phascolosoma*, *Cloeosiphon* and most *Aspidosiphon* are long and sinuous, but *Lithacrosiphon* borings are shorter and straight. Even at their densest they never intersect. The anterior shield in Sipuncula forms an operculum, but is evidently not used as an abrading tool, since it can carry a green algal film. *Aspidosiphon* performs strong contractions when removed, from the rock, suggesting that its posterior shield could be used to carve out a burrow. In shieldless *Phascolosoma*, the strong papillae of the cuticle at the posterior end could do the same. The Sipuncula have in addition epidermal glands whose secretion might dissolve calcium carbonate. The scanning electron microscope has shown burrow walls mechanically scratched, with calcite crystals etched out and displaced.

The annelid boring worms come to their peak at the reef edge. The **Euniciidae** have many colourful species, notably the famous palolo, *Palola viridis*, near the wave margin. The polychaete burrows are mostly serpentine in course, formed along cleavages or planes of weakness in the rock. How they are bored, whether with parapodial setae or by the hard jaw-parts, is still uncertain. Secondarily important as crevice-nestlers, entering rock previously bored or eroded, are the freely errant families **Syllidae** and **Phyllodocidae** and the sediment-buried **Terebellidae** and **Cirratulidae**.

Moats and lagoons

The permanently submerged stretches are generally styled *moats* if they are wadable, with the term *lagoons* reserved for deeper waters needing boat-crossing. Well suited for snorkelling, the lagoons are the special domain of soft corals, sponges and echinoderms that grow here with a freedom never possible under strong wave force. Macroalgae, browns in particular, may become abundant.

Some of the largest lagoonal sponges are funnel-shaped, with a lamellate or foliose structure. These have a marked association with symbiotic cyanobacteria and have so become thin and flat to capture sunlight. Sponges live typically on moat bottoms strewn with sediment. In the west Pacific, the following were found common: *Phyllospongia papyracea* forms brown perforate funnels attached by a narrow stalk to some hard base and sometimes rolled into an incomplete rubbery funnel; *Carteriospongia foliascens* has the funnel deeply pleated; a common upright sponge is *Dactylia infundibuliformis*,[1] *Dysidea herbacea* has branches arranged in a digitate lamella. One of the largest infundibular sponges, probably a *Pseudaxinyssa*, forms a wide funnel with a yellow matrix like cheese with holes inhabited by alpheid commensal shrimps.

The sponge *Xestospongia exigua*, common in moats, is dark brown externally and paler within, forming upright branches or thick crusts investing the hard substrate. In rather heavier silt lives the massive digitate *Myrmekioderma granulata*, with an orange surface hexagonally sculpted and yellow inside. With it is often associated the lumpy *Monanchora dianchora*, exuding scarlet mucus when pressed. In similar places grow the small globular *Tethya*.

Porites corals lying in shallow sedimented lagoons, with quiet waters, have a dominance of sponges on their dark undersurface. Fig. 22.19 shows the under-space of

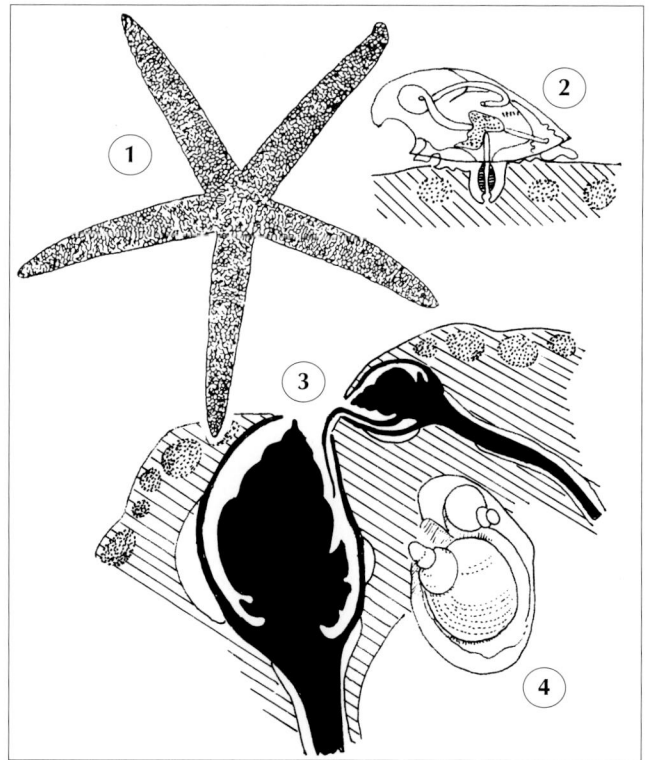

Fig. 22.21 The blue starfish and its gastropod parasites
1 *Linckia laevigata* 2 *Thyca callista* penetrating the integument 3 *Stilifer*, immersed in tissues, and showing (4) spired shell.

a *Porites* micro-atoll, with tomato-red *Spirastrella coccinea* and tough elastic cylinders of black *Hyrtios erecta*.

Echinoderms

Sea cucumbers of the families **Holothuriidae** and **Stichopodidae** (Order **Aspidochirota**) are the most prolific echinoderms of soft flats and sea-grass. All ingest fine particles by the circlet of 20 oral podia, with branchlets tipped by circular discs.

The species of *Holothuria* show two extreme feeding styles with numerous intergrades. At one extreme, the thick black *Holothuria nigra* is a deposit swallower, largely non-selective. By contrast *H. leucospilota*, thin, attenuate and attached at its tail-end to rubble, uses its tentacles to sweep up the fine suspended particles, including diatoms. *Holothuria scabra*, ashen-coloured with dark patches, burrows in coral sand, particularly near estuaries.

The *Bohadschia* cucumbers include the handsomely marbled leopard fish *B. argus*, and the prickly, chalk-coloured *B. marmorata*, common on lagoon bottoms. *B. vitiensis* is particularly found among lagoon grasses.

To a third genus of Holothuriidae belongs the slender, cylindrical *Actinopyga miliaris*, with its similar dorsal and ventral surfaces and small papillae scattering the whole skin. A notable feature is the presence of five calcareous anal teeth. *A. mauritiana*, tan above, and lighter below, clings firmly in the moving surge.

The cucumbers of the **Stichopodidae** are recognised by the large pointed papillae of their convex upper surface. The commonest species is the moderate-sized *Stichopus chloronotus*, shiny black with a green tinge and carrying two rows of long papillae. The beige *S. variegatus* has shorter and more crowded, prickly papillae. The largest of this family, the rubbery-skinned *Thelenota ananas* (the 'pineapple'), is clad with flat, branched papillae like flexible scales, ashen grey and tipped with pink.

To the Order **Apodida** belongs the peculiar holothurian *Synapta maculata*, a regular denizen of *Enhalus* and other sea-grass beds. The tube-feet have all been lost except for the oral podia, and there is no longer any 'dorso-ventral' differentiation. The translucent integument is roughened with skeletal spicules, by which the synaptid is able cling vertically to plants. When the water imbibed through the anus is discharged, the whole body collapses to a flaccid state, or contracts to a short sausage. Renewed intake of water will extend it to a metre long, when the feathery oral podia are again put out to feed.

Of the sea urchins (**Echinoidea**) met with in moats *Tripneustes gratilla* is generally the most numerous, distinguished by the dark blue or brown ambulacra, and its small white or rusty spines. Camouflaging algae and rubble can be held attached by the tube-feet. To the

same family (**Toxopneustidae**) belongs the beige-coloured *Toxopneustes pileolus*, also short-spined and with greatly enlarged pink pedicellariae that can deliver an irritant nip to the human skin.

The most conspicuous asteroid of the reef moat is the bright blue star *Linckia laevigata*. The stiff, blunt-pointed arms are cylindrical in section, and may carry a set of commensals in matching blue: a shield-shaped copepod, a polynoid worm and a hippolytid shrimp. The ectoparasitic gastropod *Thyca callista*, with its transparent shell-cap, sits in a scar on the skin. The further evolved *Stilifer* can sometimes be found buried deeper in the tissues, on the way to becoming virtually an endoparasite.

The Order **Phanerozonia**, to which *Linckia* belongs, in the Family **Oreasteridae** present some important reef asteroids. The hard skeletal frame and marginal plates are concealed in the adult under the stiff, naked skin. *Protoreaster nodosus* is brilliant red, studded above with blunt spikes. The stiff, five-armed star *Pentaceraster regulus*, grey with orange spinules, thrusts a tubular stomach into the sand to feed on bivalves. The adult *Choriaster granulatus* is a large flesh-coloured star, with blunt, rubbery arms circular in cross-section. *Culcita novaeguineae* has in the adult lost all sign of normal asteroid contours, being elevated into a massive, five-sided loaf shape.

Cnidaria

Stretches of coral sand and sea-grass beds can harbour some special sorts of cnidarian. The large benthic jellyfish *Cassiopeia* belongs to the Order **Rhizostomeae**, with the mouth subdivided into hundreds of pores, by partial fusions of the four much-divided oral lips. The oral aspect faces upwards, so the medusa lies permanently on its back in the well-lit shallows. Undulation of the umbrellar margin produces currents amid the oral tentacles, or shifts the whole disc in a sluggish swimming. In some parts of the tropics *Cassiopeia* can grow to a metre across. The tissues around the mouth are loaded with symbiotic algae (zooxanthellae), and like certain of the soft corals, these medusa derive their food wholly from algal productivity.

The sea-pens, *Pennatula*, are octocorals anchored firmly into sand by a fleshy peduncle. The large axial polyp has side-branches bearing the normal feeding polyps (autozooids) while other polyps (siphonozooids) on the back of the axis maintain a water current over the pen.

Like molluscs and echinoderms, cnidarians are lavish in their support of commensals and attendant species.[2] Richard Willan has cited *Stichodactyla hadoni*, an anemone in rubble and lagoons. Two gastropod preda-

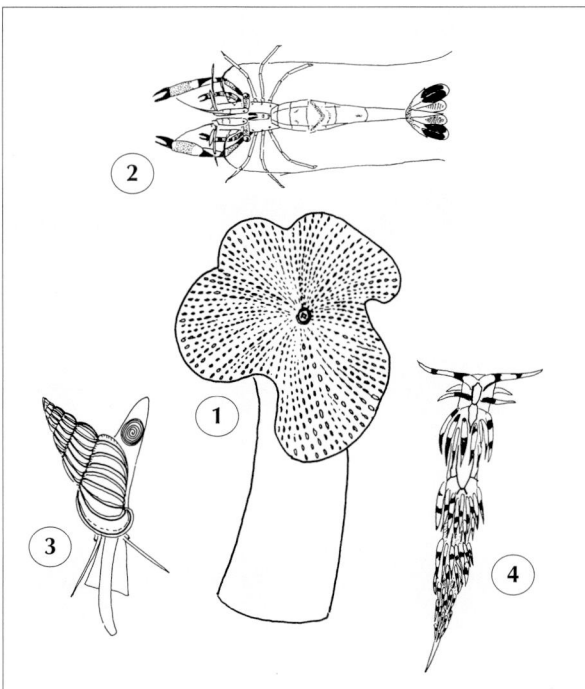

Fig. 22.22 The anemone *Stichodactyla hadoni*, with some associates
1 host *Stoichodactyla* 2 *Periclimenes holthuisi*
3 *Epitonium lamellosum* 4 *Cerberilla* sp.

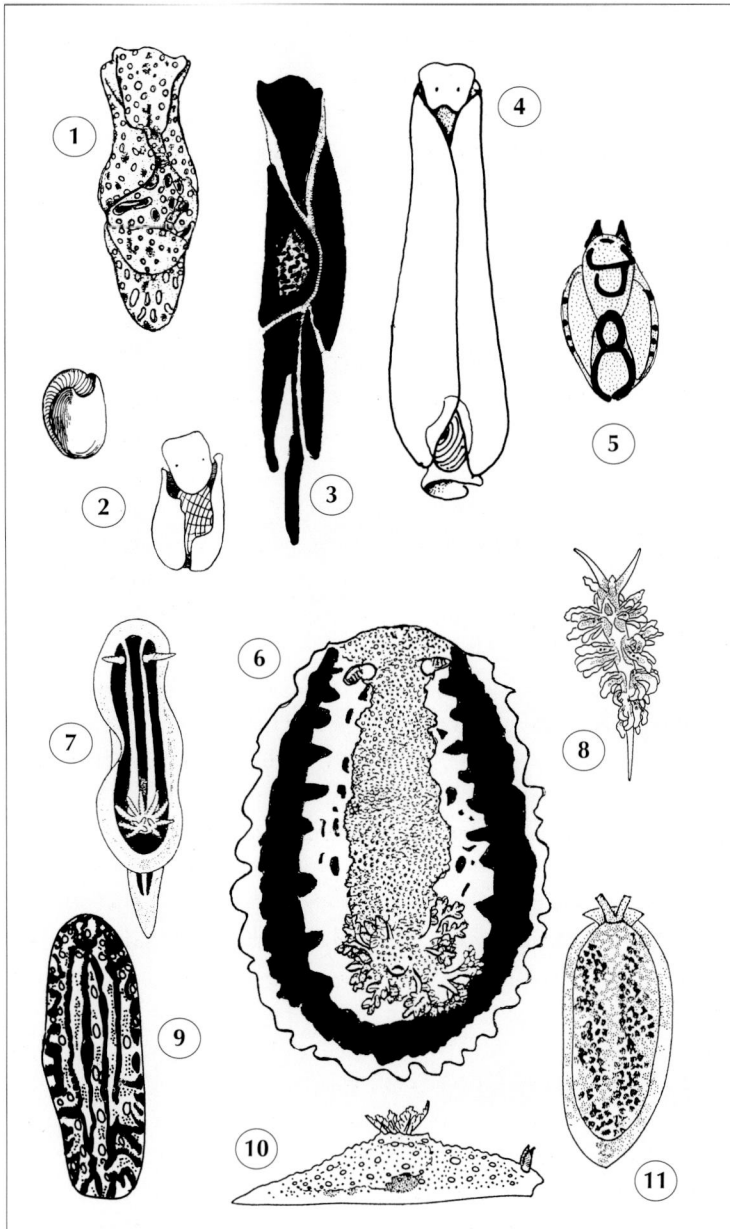

Fig. 22.23 Opisthobranchs
1 *Haminoea cymbalum* 2 *Smaragdinella viridis*, with shell enlarged 3 *Philinopsis pilsbryi* 4 *Phaneropthalmus luteus* 5 *Chelidonura* sp. 6 *Hexabranchus sanguineus* 7 *Chromodoris quadricolor* 8 *Favorinus* sp. 9 *Phyllidia varicosa* 10 *Gymnodoris* sp. 11 *Berthellina citrina*.

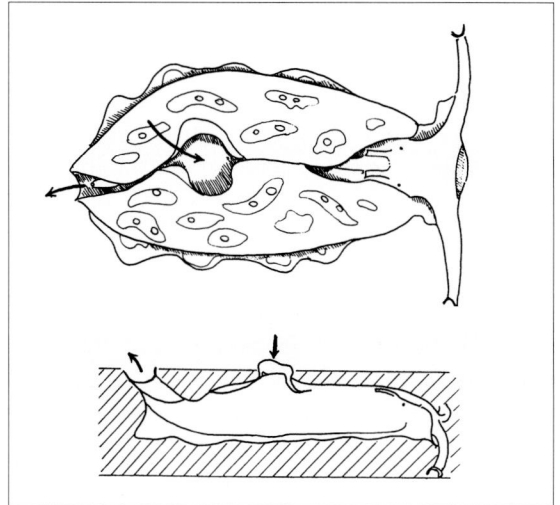

Fig. 22.24 *Syphonota geographica*
(**top**) upper surface (**bottom**) buried in sand, with respiratory currents.

tors, *Epitonium imperialis* and *Gyroscala lamellosa*, pierce its column with the proboscis. A new aeoliid slug, *Cerberilla* species, has been sieved from the surrounding sand. Schools of planktonic crustacea swim close around the anemone: a cloud of a commensal mysid above it, and over the top of the column the cyclopoid copepod *Lichomolgus myorae*. Dancing over the disc is the shrimp *Periclimenes holthuisi*, with translucent orange and purple joints, in company with the shrimp *Thor amboinensis*.

The best documented and illustrated account of commensalism, inquinilism, mutualism and parasitism in the Pacific has been given by Brian Morton in a book based on Hong Kong.

Opisthobranchia

The reef flat, especially its litter of rubble bordering the boulder zone, is the favoured haunt of tropical sea-slugs.

No single species is generally numerous, but their range and diversity are shown in studies made by Michael Miller in Fiji and the Solomon Islands.[3] The leading species are generally sponge-eating dorids, including *Platydoris flammulata*, *P. cruenta* and *Glossodoris atromarginata*. Most often to be found on sponges under boulders are the small, bright-patterned *Chromodoris* species (*C. adspersa*, *C. quadricolor*, *C. reticulata* and *C. geometrica*). All these rasp the sponges with the radular teeth. Suctorial feeders that have lost the radula include *Dendrodoris fumata* and *D. nigra*. *Gymnodoris citrina* is a predator of other nudibranchs.

The largest and most often photographed of the dorids is the superb rose-coloured *Hexabranchus sanguineus*, growing to 40 cm long. Taking off from the bottom it swims by alternate dorsal and ventral flexing of the body, synchronised with waves of contraction passing along the skirt of the mantle. In these movements the vivid red colouring, pointed up with blue and white, is prominently shown off.

Pleurobranch slugs, feeding on sponges and ascidians, include *Pleurobranchus forskali* (reddish brown), *P. albiguttatus* (reddish mottled in white), and the bright orange *Berthellina citrina*.

Widespread in the tropical Pacific on grasses and soft flats is the sand-buried aplysioid *Syphonota geographica* (Fig. 22.24), shown also *in situ* with inhalant and exhalant channels.

Finally the **Cephalaspidea**, beginning with fully shelled species, are well represented in reef moats and shallows. They show all stages of advance towards a cylindrical slug form with the shell lost or vestigial. What may be seen as the herbivorous line has proceeded from *Atys*, still fully shell-contained. A thinner shell, enclosed by parapodia, is shown in the small and attractive *Haminoea cymbalum*, green and orange-marbled and found among algae on the tops of boulders. In *Smaragdinella viridis* (Fig. 22.23) and *S. calyculata*, the shell is a concave plate invested by integument. *Phanerophthalmus* has the body almost bilateral with a vestigial shell relegated to the posterior end; these species are grazing herbivores, creeping through eelgrass or over the reef surface.

A carnivorous line of cephalaspids begins primitively with the cylindrical-shelled *Cylichna*, with shell reduction in *Scaphander*, to a thin internal shell in the burrowing *Philine*.

The shell-less carnivores include *Chelidonura*, with the long, cylindrical parapodia ensheathing the body and the foot forked and prolonged behind. Two common west Pacific species, one dark blue and orange, *Chelidonura hirundinia*, the other milk-white with a yellow margin, *Chelidonura electra* feed on polychaete worms. The related *Philinopsis pilsbryi*, a muscular yellow and black slug, takes worms and even shelled cephalaspids.

The sea-grasses

Marine monocotyledons, mostly grass-like, stretch as wide meadows on soft flats between and beyond the tides. In the tropics, several families are represented, on sites where few algae, save the green *Caulerpa* and sometimes *Avrainvillea*, can out-compete them.

Some botanists would regard the sea-grasses as evolved from freshwater groups by a special tolerance of salinity. Alternatively they can be seen as deriving from salt-tolerant terrestrial plants near the strand-line. Their tolerance is finely balanced. Even brief exposure to less than 35% salinity can be lethal.

Marine grasses are not always easy to classify without recourse to their flowers. These are seldom found in herbaria, and over long periods the plants may not produce them. To the first family, **Hydrocharitaceae** (classed

in the primitive Order Butomales) belong the two grasses *Thalassia* and *Enhalus*, forming extensive intertidal and submarine swards. They disperse only slowly, since the seeds ripen beneath the water and lack buoyancy.

Much the largest sea-grass in leaves and fruit is the west Pacific *Enhalus acoroides*, living subtidally in sheltered areas. The leaves are up to 2 cm wide and more than a metre long. The rhizomes are as thick as a finger, with a sheath of black fibres persisting from decayed leaves. Small male flowers are liberated from opening spathes only at spring tides, to float on the surface with the anthers pointing down. Meanwhile the female flower is raised on its peduncle to be fertilised as floating male flowers make contact with it. Surface tension pulls together the female petals, sucking in the male flowers sticking to their margins. Pollen is transferred in an air bubble, staying entirely dry during fertilisation.

The smaller grass *Thalassia hemprichii*, widespread in the western Pacific and Indo-Malaysia, forms shallower meadows than *Enhalus*, often intertidally. The short leaves, 4–16 mm wide, are slippery to the touch. The flowers, on separate male and female plants, are relatively large. Pollination takes place entirely under water, with the grains adhering in long strings, which helps their flotation. The fruits dehisce to nut-like seeds.

The least grass-like of the **Hydrocharitaceae** are the *Halophila* species, with the rhizomes putting up pairs of ovate leaves. Common in soft patches on coral reefs, *Halophila ovalis* has leaves 2–4 cm long. *H. minor* is smaller, with only slight tolerance of wave action.

The slender marine grasses of the Family **Zosteraceae** are mostly temperate plants, represented by the familiar *Zostera* (Fig. 19.10), and on rocky shores by *Phyllospadix*, as on the west American seaboard (Fig. 19.11).

The Family **Zannichelliaceae** consists of narrow-leafed sea-grasses related to the monocotyledonous weeds of fresh water. The largest species is *Syringodium isoetifolium*, forming wide low tidal and subtidal swards, as on the sheltered flats of Fiji. Small flowers, male and female, are borne on different plants. *Halodule uninervis*, also linear-leaved and no more than 1 mm wide, is common in lagoons, flowering only rarely.

Coral reef fishes

The fishes of the Pacific reefs are rich in species and prolific in individuals. The small or middling sized reef fishes can be regarded as a high plateau of 'teleost' or bony fish evolution. John Briggs has noted that of the 20,000 species in the 30 orders of teleost fish — by far the largest grouping of vertebrates — it is the three most advanced orders (Perciformes, Pleuronectiformes and Tetraodontiformes) that dominate those shelf waters of

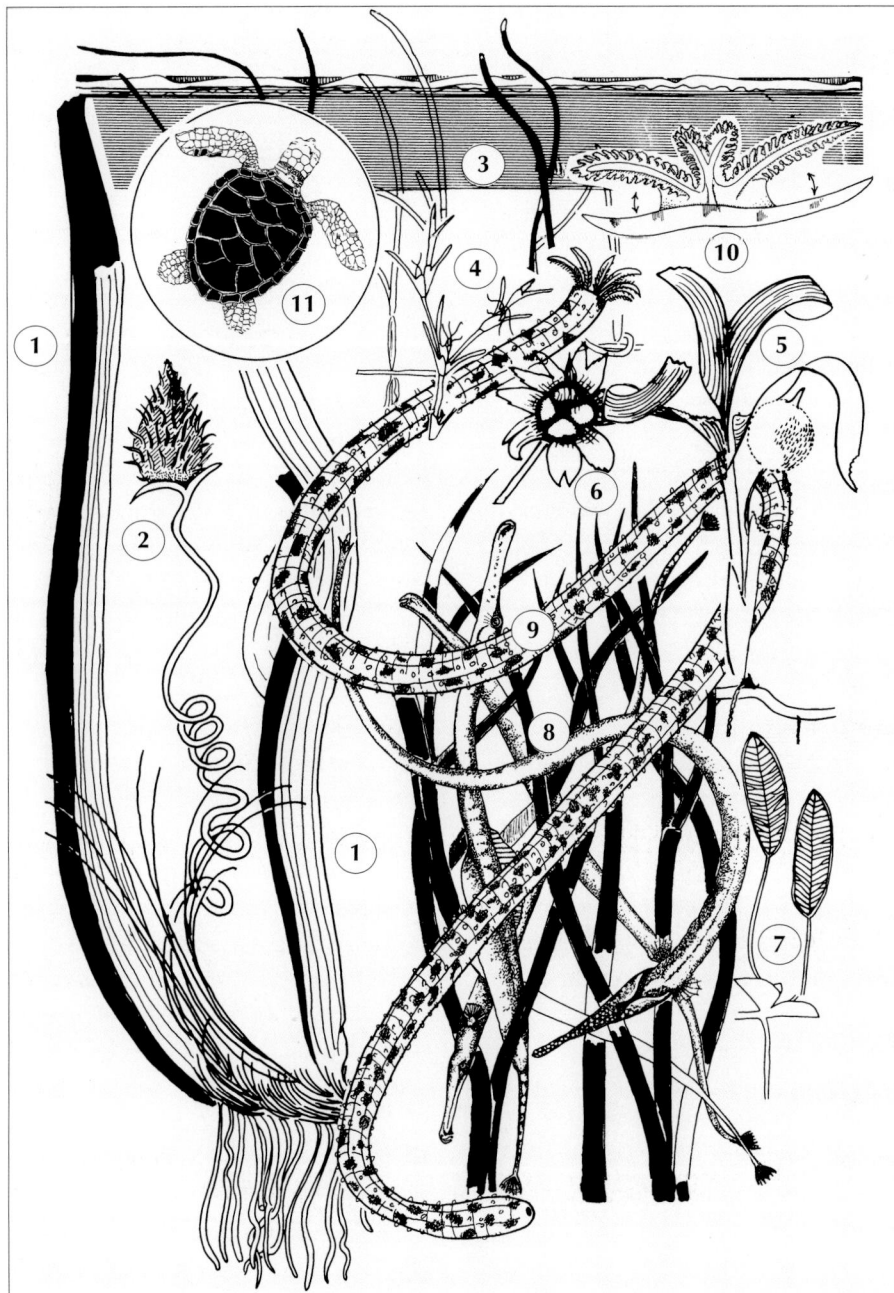

Fig. 22.25 Sea-grasses of the soft flats and lagoon
1 *Enhalus acoroides* leaf 2 *Enhalus acoroides* fruit 3 *Syringodium isoetifolium* 4 *Syringodium*, female plant with inflorescence 5 *Thalassia hemprichii*, with fruit 6 *Thalassia* flower 7 *Halophila ovalis* 8 pipefish *Syngnathus* 9 holothurian *Synapta maculata* 10 *Cassiopeia* 11 newly hatched turtle.

the tropics where speciation is greatest. In their flocking and movements, as in their wealth of colour, the reef teleosts have been likened to the birds of tropical forests. Among the largest denizens of the reefs are the sharks and rays, primitive cartilaginous fishes much older than the teleosts. Some of the smallest of fishes (and indeed of all vertebrates) are the species of goby only 10 mm long that fix by a sucker round a coral branch.

The Malaysia-Philippines-North Australian triangle, with upwards of 2000 species, is itself the mother source of all the radiations of coral reef fish.

The admirably detailed account of the Marshall Islands fishes, by Hiatt and Strasburg[4] and much writing that has followed, have stressed the huge variety of fish habitat space in the Pacific.

Out in front the long surge channels cut back into the windward ramparts of the reef. A view down one of these is awesome and spectacular, with corals of many shapes and sizes, and the fish diversity even greater. Here were noted a surgeon fish, *Acanthurus guttatus*, in fast-swimming schools in the breaking waves, with sweepers *Pempheris oualensis* holding positions just beneath the white water. Other surgeons, *Acanthurus achilles*, *Naso unicornis* and *Zebrasoma veliferum*,

Fig. 22.26 The habitats of reef fishes
1 *Pomacentrus* sp. 2 *Chromis viridis* 3 *Abudefduf* sp. 4. *Dascyllus* sp. 5 *Oxymonacanthus longirostris* 6 *Chaetodon lunula*
7 *Chaetodon auriga* 8 *Forcipiger* sp. 9 *Epinephelus* sp. 10 *Scarus* sp. 11 *Balistapus* sp. 12 *Ostracion* sp. 13 *Arothron* sp.
14 *Diodon* sp. 15 *Pterois volitans* 16 *Mugil* sp. 17 *Mullus* sp. 18 *Muraena* sp. 19 *Acanthurus* sp.

haunted the channels along with sharks, *Carcharhinus melanopterus* and *C. amblyrhynchos*, smooth dogfish (*Triakis*), moray eels, groupers, parrot fishes, blennies and trigger fishes.

Inside the breaker zone the reef flats have their own numerous fish: sharks, moray eels, red mullet, goatfishes, butterfly fishes, surgeon fishes, damsel fishes, wrasses, parrotfishes. Small gobies and blennies are denizens of tide pools, well able to resist heating up by the tropic sun.

The fishes of the reef show many different degrees of dependence on the substrate. The most free-ranging and least dependent are the large roving carnivores: sharks, tunnids, serranids, carangids. It is these that constitute the tertiary and quaternary consumers, occupying the highest 'trophic levels' in their feeding chains. Their food-taking is neither specialised nor diet-specific, but encompasses any of the smaller fast-swimming fishes that can be captured on the move.

Far more intimately adapted to their substrate are the smaller fishes living among the corals, and as typical of the reef as the corals themselves. These include the small schooling **Pomacentridae** (damsel fishes), the nar-

row, high-built **Chaetodontidae** (butterfly fishes), angel fishes (**Pomacanthidae**) and the **Balistidae** (trigger fishes). As well, there is a cluster of more spherical fishes, capable of inflating their bodies, in the Families **Tetraodontidae** and **Diodontidae**. There are also the **Canthigasteridae** and in addition the hard-cased boxfishes, **Ostracionidae**.

The closely territorial coral fish were well introduced in Norman Marshall's *The Life of Fishes*.[5]

Whether high-built, or globular or box-like, all these fish are nimble and agile. They typically hang head-down, not tail-swimming in the standard way, but held poised to fossick and nibble, with the pectoral fins employed in a kind of breast-stroke. Trigger fish undulate their opposed dorsal and anal fins, as do the boxes and puffers and file-fish and porcupines, with the tail-fin now used as a rudder.

Such swimming allows the fish to thread accurately through narrow defiles. With their pointed snouts directed to the precise spot, they can nip off a polyp by the tiny incisor teeth in the protrusible mouth. As well as by quick manoeuvre, they also defend themselves by spines or unpalatable taste. Their colours are generally protective, whether by way of warning, or disruptive camouflage.

Viewing distances in clear water are much greater than in forest foliage; so the visual parameters of focus, acuity, and discrimination need to be unfailingly good. Fishes focus not by reshaping the lens but by moving it in and out of the goggle eye. In neckless fishes, in potential danger from any direction, the eyes protrude at the sides with fields of 180 degrees, partly overlapping in front. Cones for colour vision are highly developed in bright coral fishes, where sharks and rays, themselves sombre-coloured, have only rods.

The first fishes to come under notice around the thickets of branched corals will be the schools of small **Pomacentridae**: black and white striped damsels (*Dascyllus*), and brilliant blue damsels (*Chromis*). Relying on the corals for their protection, these fishes swarm out to feed on zooplankton. The larger *Pomacentrus* species and the striped sergeant major fishes (*Abudefduf*)) use their small villiform or incisiform teeth to graze on algae or forage around coral-heads.

In the deep interstices of the coral heads small gobioids (*Gobiodon* and *Paragobiodon*) and caracanthids partake of the wealth of small crabs, shrimps, worms and molluscs. Other fish that cruise further out also come in to search among the coral branches. The wrasses *Gomphosus varius* and *Epibulus insidiator* thrust in with their tubular, protractile snouts. Some chaetodontids have probing snouts, as in *Forcipiger longirostris*, and also the file fish *Oxymonacanthus*, that bite off single polyps with fine incisor teeth.

The bright-coloured butterfly fishes (**Chaetodontidae**) and angel fishes (**Pomacanthidae**) are intimate coral-browsers. These are deep-bodied and compressed, gliding or sidling gracefully through the branches, singly or in pairs, and poking or thrusting with tubular snouts. *Chaetodon lunula* takes polyps alone. *C. vagabundus*, *C. ephippium* and *C. citrinellus* feed as well on the algae from living coral heads, while *C. reticulatus* and *Centropyge flavissimus* subsist chiefly as herbivores grazing round the coral bases.

In biting off polyps, some chaetodontids also scrape away the corallite tips. The parrotfishes (**Scaridae**) go further, biting into the corallum itself with strong fused teeth. The trigger fishes (**Balistidae**) and puffer fishes (**Tetraodontidae**) have strong dentition to break off corallites of *Acropora*, *Pocillopora* and *Stylophora*. For quite a small extraction of food, the gut is kept permanently crammed with coral fragments.

Carnivores

Farther out from the coral there are larger carnivorous fishes taking a wide variety of prey. They have been conveniently classed into the following *feeding guilds*:

1 **Carnivores digging up burrowing food** as they cruise over the bottom include the **Sparidae** (breams), **Lutjanidae** (snappers), **Labridae** (wrasses), **Holocentridae** (squirrel fish) and the **Mullidae** (goatfishes). The last are bottom-confined, probing the sediments with sensitive barbels under the jaw.

2 **Carnivores taking hard benthic fauna** comprise the **Congridae** (conger eels), **Orectolobidae** (wobbegongs), **Bothidae** (flounders), **Holocentridae**, (squirrel fish), **Lutjanidae** (snappers), **Pempheridae** (sweepers) and the **Eleotridae** (sleepers).

3 **Lurking carnivores.** Among the groupers (**Serranidae**), the largest species hover motionless near coral caverns, while smaller *Epinephelus* hide in crevices and under ledges. With large mouths and villiform teeth, as well as short canines, groupers pounce on small fish and mobile crustacea. The scorpion fishes (**Scorpaenidae**) are also voracious lurking carnivores, seizing fish and crustacea coming within their range. The venomous stone-fish (*Synanceia*) buries itself in rubble where its cavernous mouth snaps up unwary fish. The lion fish, *Pterois volitans*, with its bizarre fin expanse like a turkey's tail, lies in crevices and its thread-like fin-rays attract crustacea to swim in close. The moray eels (**Muraenidae**) have also perfected the lurking habit, as voracious night-feeders, pouncing at small fish with their fang-like canines.

Herbivores

Here we descend to the 'primary consumers', at the bottom of the food chain, grazing the reef's inconspicuous filamentous algae. Though the reef's standing crop at any one time seems meagre, there is continuous productivity and a high turnover. Much comes from the algal films upon rubble, or the microscopic algae that penetrate the coral skeleton or live in its soft tissues (zooxanthellae). A single coral head can thus present us with a virtually complete trophic community from primary producers upward. The primary consumers (herbivores) are in fact some of the most advanced fishes in their evolution, with special dentition and gut characters that evolved only when plant food rather late 'in the day' became abundant. Herbivorous fishes are diversified into the following guilds:

1 **Scrapers** such as the parrotfishes (**Scaridae**), that remove the algal film and leave their beak-marks on coral heads. The calcareous faeces reveal their major role as bio-eroders and sand-producers. With their strong mill of pharyngeal teeth the scarids crunch not only the coral but a wide range of hard invertebrates.

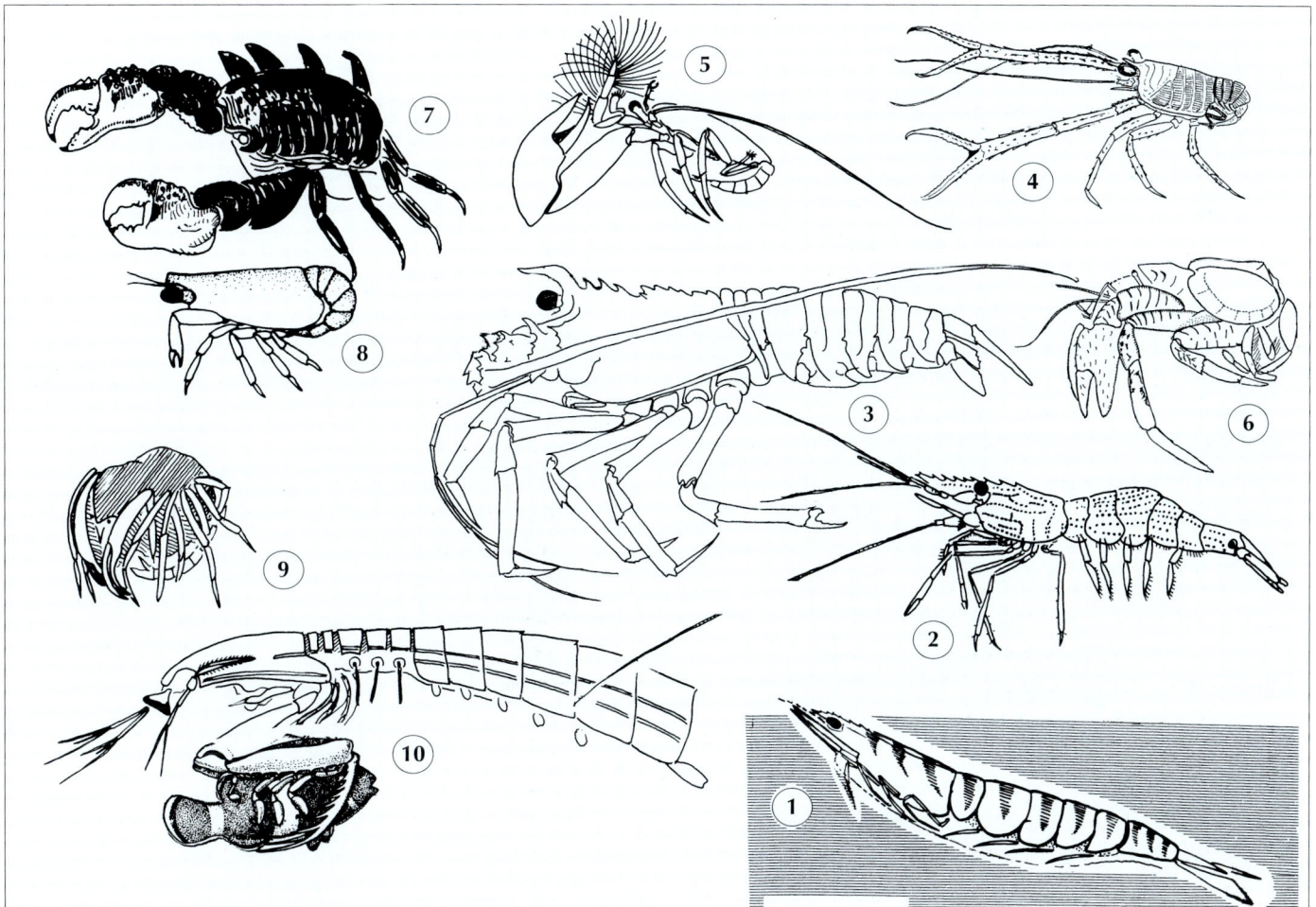

Fig. 22.27 The progression of decapod Crustacea from swimming to benthic (reptant) habit
The series ascends from lower right, with natant Decapoda (1 and 2) to reptant decapods (3) with the crab facies emerging
(4 to 9) and hermit offshoots parallel arising (6). 9 shows a reversion to burrowing and fast forward running, and
10 (*Hoplocarida*) is a parallel to reptant-natant lines of Eucarida.
1 *Penaeus* 2 *Palaemon* 3 *Panulirus* 4 *Galathea* 5 *Porcellana* 6 *Birgus* 7 *Grapsus* 8 megalopa juvenile stage of grapsid
9 *Dotilla* 10 *Lysiosquilla*.

2 **Bottom grazers** par excellence include the surgeon fishes (**Acanthuridae**). These fish are high-built and compressed, and are named from the retractable scalpel-spine at the tail-base. The dentition is adapted to grazing and browsing. Some species take up much sand with algal filaments, triturating it in a strong gizzard (*Acanthurus mata, A. olivaceus, A. xanthopterus, Acanthurys gahhm*). Others that lack a gizzard scrape over the rock for fine *Ectocarpus*, also *Dictyota* and *Padina* (*A. achilles, A. nigricans, Naso unicornis, Zebrasoma veliferum*).

3 **Sand-sifting** is the special expertise of the grey mullets (**Mugilidae**). With wide mouth but only minute teeth, sediment is scooped up, spun over to extract its uncellular algae, and the mouthful then blown out. The sweepers (**Eleotridae**) make permanent burrows in sandy rubble, pulling in balls of algae to close the burrow mouth. They also forage by mouth-sifting of sand, rejecting the residue not by the mouth but through the gills.

Reef Crustacea

In the plankton outside the reef, the tropical Crustacea are not rich by temperate standards. It is in the wealth of habitat space in the coral reef shallows that crustaceans have attained an unmatched diversity in species and higher taxa.

The Subphylum Crustacea has radiated in shallow seas in a way only comparable with the phylum of molluscs. Their large members are entirely benthic. Hard exoskeleton and the restrictive joint-planes of the limbs ensure that crustaceans can move fast only when they are small and light. Major success had thus been at the sacrifice of speed and agility. There are no swift pelagic crustaceans, to compare with mackerel or squid, nor any giants like shark or albacore.

The crustacean limb-set is complicated, and looks archaic. Its ultimate success has come with the crabs. As we have seen the abdomen is virtually lost and with scuttling gait sideways or backwards the antennae can

449

afford to be short, and the eyes to be turned to the sides, with diminished regard to the world straight ahead.

Swimming decapods

The higher Crustacea or Decapoda began with the *natant* forms, swimming competently but not fast. First come the shrimps and prawns built to the so-called *caridoid facies*. The earliest of these belong to the tribe **Penaeidea**, smooth-tapered prawns with a long, straight abdomen. The fertilised eggs are released into the water to hatch in the primitive crustacean way as *nauplius* larvae. Small exopodites are preserved at the bases of the walking legs. The basic penaeid habitat is in sand; we shall see a refuge in burrowing as a recurring option for many sorts of tropical decapods, including many crabs.

In the second and larger tribe, the **Caridea**, many species have taken up sites on the bodies of corals, anemones, gorgonians, sponges, echinoderms, or around fishes waiting to be cleaned. With this new lease of speciation, coral reefs have become extraordinarily prolific in shrimps, running into hundreds of species where six would be rich for a temperate shore. A.J. Bruce cites a reef at Malindi, Kenya, where 57% of 67 species collected were commensally associated with other animals.

Commensal crustaceans have to stay small, to enter tubes or burrows, or attach or hide upon the host. The exoskeleton must be thin enough for the colour cells involved in camouflage to show through. The major families containing commensals are the **Palaemonidae** and the **Alpheidae**, with a role also for the hump-back prawns, **Hippolytidae**. The **Hymenoceridae** (with the striped *Hymenocera* and the large-spotted *Gnathophyllum*) are a small, entirely commensal family. The *Stenopus* species, in their separate tribe, **Stenopidea** are cleaners of fish.

Walking decapods

Increase of size leads from the natant to the *reptant* or walking condition, epitomised in the spiny rock-lobsters of the **Panuliridae**. Common in the Pacific tropics are *Panulirus ornatus* (orange and green) and *P. versicolor* (black and white-banded and ornamented with cinnabar). The abdomen is of full-length, with the pleopods forming paddles for swimming. Its muscles produce a backward dart from a powerful tail-smack against the thorax. Poised on their thoracic legs (*pereiopods*), the *rock lobsters* also walk, still helped by the slow beat of the pleopods, as if treading water. Sweeping antennae and the smaller vibrant antennules serve in forward exploration.

With the nextdoor Family **Scyllaridae**, most of this mobility is lost. *Scyllarus* — and still more *Ibacus* — are like flat caricatures of rock-lobsters, with the side plates of the carapace and the curiously shield-like antennal bases held against the ground.

The crab achievement

The propitious transformation from the *caridoid* to the *carcinoid facies* — from crayfish to crab — has happened several times in the Decapoda. The classical division of the **Anomura**, suggesting many of the transitional stages, is probably not one group but involving a convergence of several evolutionary lines. Their most familiar members are the hermit crabs, themselves now considered polyphyletic.

Numerous anomurans have turned from a carnivorous life to particle-feeding or microphagy. The third and second maxillipeds are used in gathering food, as described with the half-crabs of the **Porcellanidae**. With flat, expanded chelae, these small crabs typically scuttle upside down against a boulder face or hide in crevices. They have also produced many commensals.

The Porcellanidae are almost finished crabs, evidently derived from the small squat lobsters of the Family **Galatheidae**. These last still keep the abdomen extended behind and the long, slender chelae held out in front. Often brightly coloured, they can be found under coral heads and in rubble, with a number of their species commensal. *Galathea* is a genus as typical of tropical reefs as is the related *Munida* or 'krill' in the macroplankton of cold seas.

Two groups of microphagous Anomura are *infaunal*, burrowing in sand or mud. The mole-crabs (**Hippidea**) are polished and egg-shaped, with the common species *Hippa pacifica* living gregariously in the middle beach. Mole-crabs push actively through clean, fluid sand by means of the pleopod beat. From Oregon down the American Pacific coast, the sand crab *Emerita analoga* lies just below the surface, to thrust its eye-stalks and plumed particle-catching antennae into the water above. The tropical *Albunea* sits stably at the bottom of a consolidated burrow, putting up a long inhalant tube formed by its closely apposed antennae.

The second group of burrowing Anomura, the **Thalassinidea**, are confined to deep, semi-permanent galleries in silty sand or mud. Dug out, they appear limp and disabled. Left *in situ*, the frail-bodied *Callianassa* and *Upogebia* actively dig with their pereiopods to maintain the burrow against subsidence. With the fringed maxillipeds they are able to intercept particles from a strong current driven by the pleopods. *Thalassina anomala*, the so-called 'mud lobster', forming its conspicuous volcano mounds amid mangroves, is the

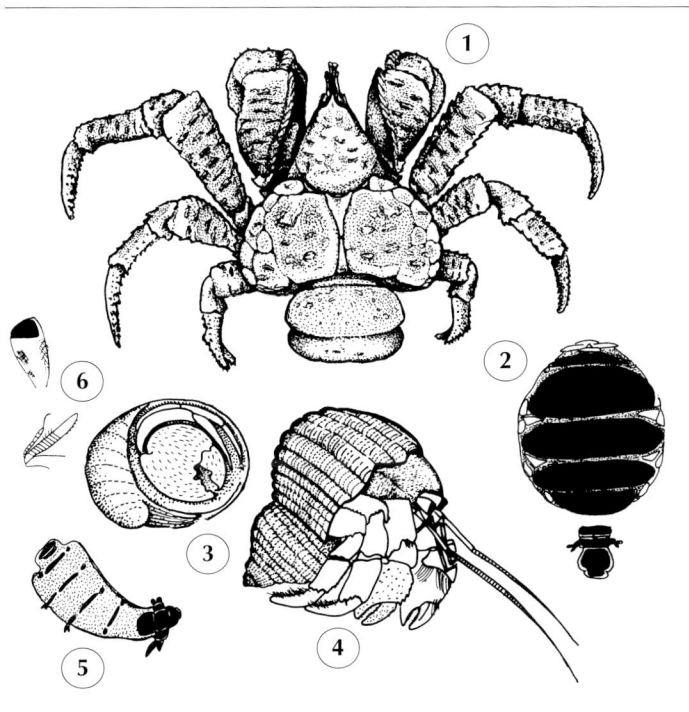

Fig. 22.28 Land hermits: Coenobitoidea
1 *Birgus latro* 2 *Birgus* abdominal sclerites and telson
3 *Coenobita perlatus* in shell 4 *C. rugosus*, in shell, from
the right side 5 *Coenobita*, abdominal sclerites and telson
6 *Coenobita* detail of first antenna and eye.

largest of this group. Evidently macrophagous, its feeding habits are still poorly understood.

Hermit crabs: the paguroid facies

A large number of Anomura have side-stepped the *carcinoid* path, opting for the hermit crab design, known as the *paguroid facies*. Their salient feature is the soft, spirally twisted abdomen typically protected within an old gastropod shell. The second and third pereiopods provide the scuttling gait, while the reduced fourth and fifth, with the uropods and telson, secure the body to the shell pillar. One or sometimes both of the chelipeds serves as an operculum, while the two can be used for aggression or food-getting. With paguroids, just as in galatheids, microphagy is common, with the third maxillipeds used for filtering.

Hermit crabs have evolved at least twice on a considerable scale. Separate stocks are believed to have originated from *panulirid* (rock-lobster) and *astacurid* (crayfish-lobster) forbears. First, the thoroughly marine hermits of the eulittoral and subtidal belong to the **Paguroidea**, familiar in temperate and cold seas as well as in the tropics. With several large genera, including *Dardanus*, *Pagurus*, *Clibanarius*, *Calcinus*, *Aniculus*, the paguroids are probably themselves of mixed origin.

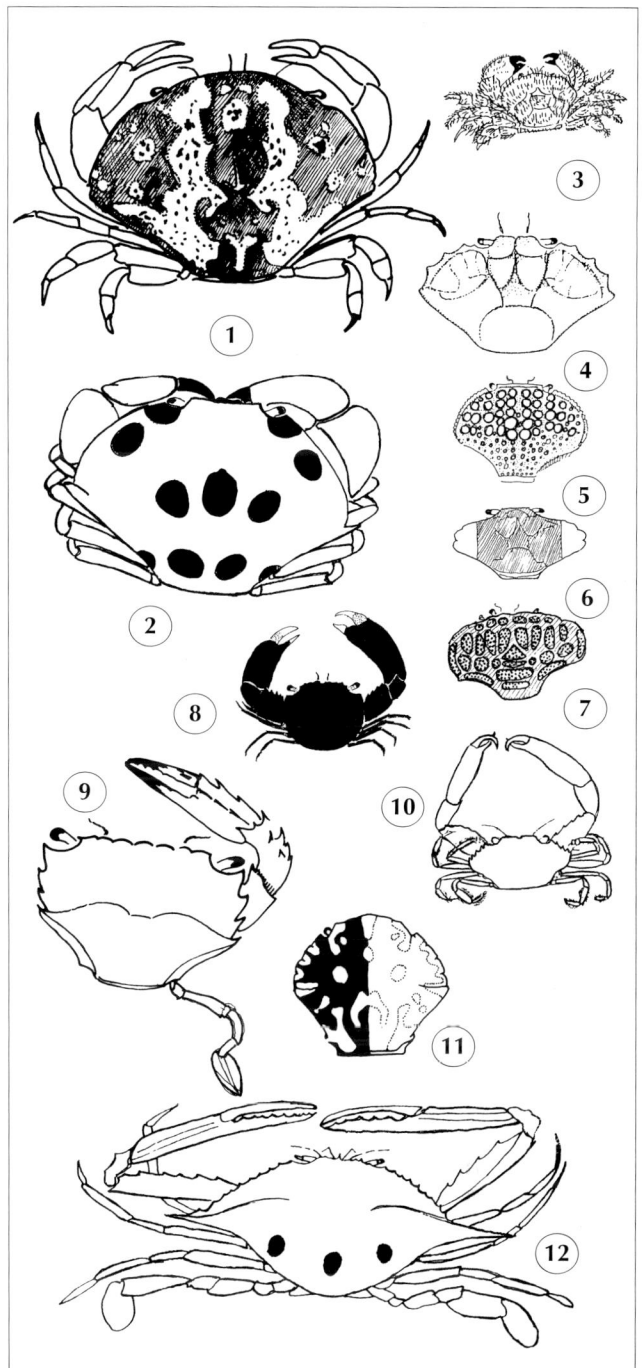

Fig. 22.29 Coral reef crabs
(1–8) **Xanthidae** 1 *Atergatis floridus* 2 *Carpilius maculatus*
3 *Pilumnus* 4 *Xantho* 5 *Daira* 6 *Liomera* 7 *Actaea*
8 *Trapezia* (9–12) **Portunidae** 9 *Thalamita* 10 *Caphyra laevis* 11 *Lissocarcinus* 12 *Portunus sanguinolentus*.

In contrast, the **Coenobitoidea** are tropical, semi-terrestrial hermits, active in the vegetation and litter of the supralittoral zone. *Coenobita* species (especially the small *C. rugosus*) are common everywhere in the Pacific tropics. They can be told at once from pagurids by their short antennules that terminate without a flagellum but in a laminate palp. These hermits are actively aggressive, and may escape from capture by slipping out of the shell and nimbly taking off.

Neither the pagurids nor the coenobitids have reached their end-point as normal hermits. In both groups some members have further evolved by withdrawing the abdomen from protection in a shell and reinvested it with hard plates. They have thus reverted almost to bilateral symmetry, though the new abdominal plates never reappear in their original form. In the cool temperate and arctic, the Paguroidea have undergone such remaking in the Family **Lithodidae**, with the stone crabs or king crabs.

In the Pacific tropics, the Coenobitoidea have produced the large, terrestrial 'coconut crab', *Birgus latro*. Nimbly climbing palms or pandans, *Birgus* can employ the uropods to anchor the globular abdomen in a hole or at a tree base. With the powerful chelae coconut shells can be husked and opened, and the crab feeds also on carrion, and even others of its own kind.

Both *Birgus* and *Coenobita* remain tied to the water, returning once or twice daily to moisten the gills, as well as seasonally to breed. The larva, known as a *glaucothoe*, is symmetrical, and swims or crawls in the splash zone. In *Birgus* it metamorphoses after a month, temporarily reverting to a hermit style in a tiny gastropod shell. This is cast off after six months, whereupon the spiral abdomen returns to its bilateral symmetry.

Brachyura: the carcinoid facies

These decapods with the final crab form are generally regarded as a unified group. Monophyletic or not, the division **Brachyura** comprises the most successful of all the Crustacea. With their wide evolution in the tropics, we have room to mention only the principal families, some of them already met with in the temperate.

The **Xanthidae** (or black-fingered crabs) belong typically to the lower and middle shore. Compact and sturdy, often bright coloured, they have short legs, with no great turn of speed. In the first set of genera (*Xantho, Leptodius, Phymodius, Actaea, Liomera, Actumnus, Daira*), common under boulders and coral heads, the carapace is cut up by grooves into *areoles*. In a second series, the sculpture is smooth, as in the large *Carpilius* species (including the red-spotted *C. maculatus*), *Atergatis* with the shawl crab (*A. floridus*), the large, red-eyed crab *Eriphia sebana*, and the bristly *Pilumnus* species.

Xanthids are not active chasers of prey. Some larger ones are evidently shell-crackers, with the short, strong fingers of the chelae. Others have chitin-tipped fingers like rodent chisels, as with the flat, wide-armed *Trapezia*, pressing between the branches of *Pocillopora* corals and scraping them with the chelae.

The **Portunidae** are distinguished by the flat paddles of their fifth legs and include many large crabs, coming to their zenith in the tropics as do the Cancridae in cold seas. With long, sharp-toothed chelae, portunids are

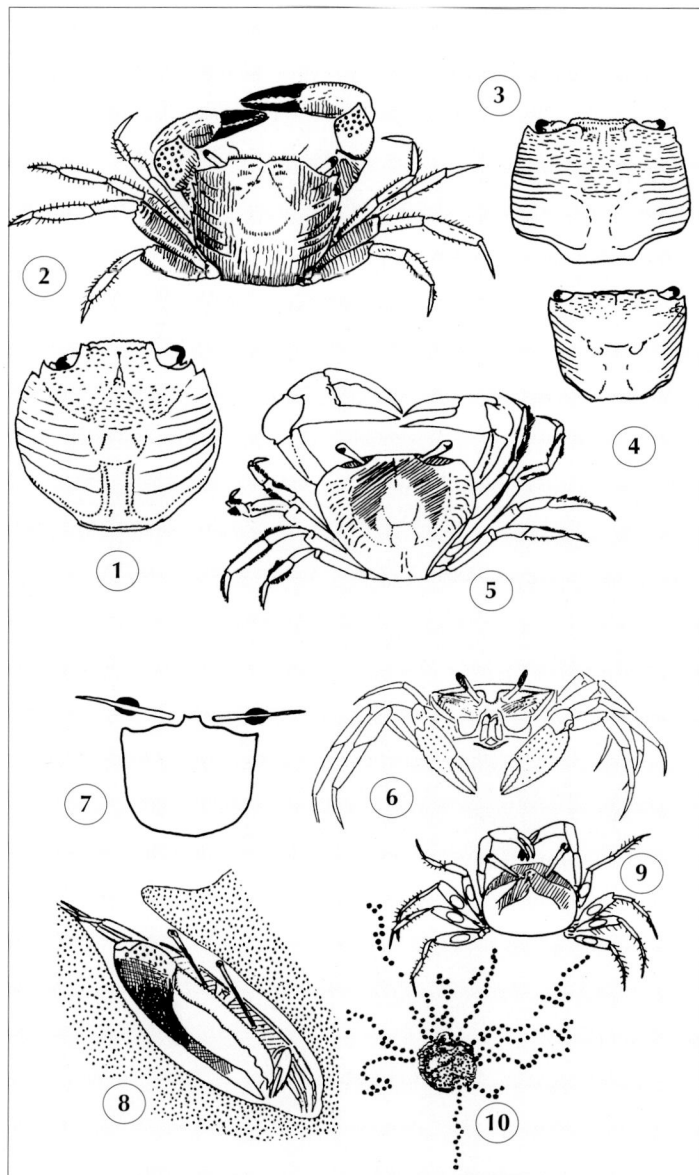

Fig. 22.30 Crabs of sheltered and soft shores
Grapsidae 1 *Grapsus* carapace 2 *Sesarma* sp.
3 *Geograpsus* carapace 4 *Metopograpsus* carapace
Gecarcinidae 5 *Cardisoma carnifex* **Ocypodidae**
6 *Ocypode cordimana* 7 *O. ceratophthalma*, carapace
with eyestalks 8 *Uca* in muddy sand burrow 9 *Scopimera*
sp. with (10) burrow and sand balls below.

active predators. Some forms such as the smooth *Ovalipes* bury themselves in low tidal sand. *Charybdis* live at mid-tide in coarser sand, while the fast and aggressive *Thalamita* species retreat under coral boulders. The largest of this family are some of the side-spiked *Portunus* species — such as the spotted *P. sanguinolentus* found in subtidal grass beds, and the heavy *Scylla serrata* living in mangrove swamps.

The land margin

Immensely successful in the tropics, the crabs of the **Grapsidae** almost all belong to the upper shore, having their great radiation in the supralittoral zone. Their high achievement in the tropics has been with the large Sub-family **Sesarminae**, with the migration to estuaries and mangroves. Here they have evolved special capacities to resist desiccation, and to cope with wide extremes of salinity and high ambient temperatures. For respiration, water can be pumped out of the exhalant branchial aperture through the orbit, being directed sideways onto a wide expanse of the cheek. Here it is re-oxygenated in a network of bristle-lined channels before its return to the branchial chamber.

The Sesarminae have the carapace almost square with the orbits at the extreme corners. They include fast-running species in mud-banks, marshes and among the roots of mangroves. The *Metopograpsus* species have a rhomboidal carapace, widest in front, and are active under cobbles or spiralling over the trunks of mangroves. *Varuna*, with the carapace near-circular, has species entering estuaries and rivers. Finally, in the fully terrestrial *Geograpsus*, the carapace is high-built, rather wider behind than in front.

Some of the best known terrestrial brachyurans of the Indo-Pacific are the large *Cardisoma* species of the grapsid-related Family **Gecarcinidae**. The carapace is smooth and roundly vaulted. Burrows are made in earthworks or at the edge of waterways.

Beaches and flats

On tropical intertidal flats, the Family **Ocypodidae** have taken the clear lead. Broadly, speaking, the ghost and soldier crabs (*Ocypode* and *Dotilla*) have occupied the cleaner sands, while the fiddler crabs (*Uca*) have possession of the fine sediments including deep muds. The latter we have already described on the enclosed shores of Hong Kong (Fig. 18.6). More than grapsids or xanthids, both the ocypodids and ucids may run to huge numbers of individuals. *Scopimera* or *Dotilla* species can dominate whole sandflats in armies of thousands. Colonies of fiddler crabs, segregated by their distinctive species colours, partition the territory of estuarine mudflats.

The *Ocypode* species are the largest and swiftest of their family, living as omnivores and scavengers on the slope of sand beaches. They move as fast and silently as a shadow, with their long eye-stalks held up for periscopic vision.

The small and far more numerous *Scopimera* species make deep vertical shafts. Around the opening the surface is littered with small sand balls like lead shot, forming patches over large tracts of a beach. These originate from the sand picked up with the chelae and worked round in the basket between the maxillipeds. After winnowing out the nutrients, the loose sand changes from a fluid sol to a firm gel, to be deposited as a ball. Larger than *Scopimera*, the soldier crabs (*Dotilla*) are the size of a cherry, powder-blue and with pink-banded legs. These are some of the few Brachyura adapted to walk forwards, and make vast concerted marches, coming out just after the ebb of the tide, to feed over the still moist sand surface.

Oxystomata

Set off from the mainline Brachyura, the **Oxystomata** are specialised crabs burrowing in low tidal sand. Their respiratory parts have the buccal area carried forward as an excavated gutter and the exhalant channels are enclosed by the endopods of the first maxillipeds. *Matuta* has the carapace marlin-spiked at the sides and has its fifth limb expanded, like a paddle crab's, as well as trowel-like tarsi on the others. It preys actively on other crustaceans.

The box crabs (*Calappa*) are slower-moving, with slender legs hidden under side-saddles of the carapace. The expanded chelipeds are drawn close to the carapace, and the knobbed finger of the right one serves as a nut-cracker for hermit shells.

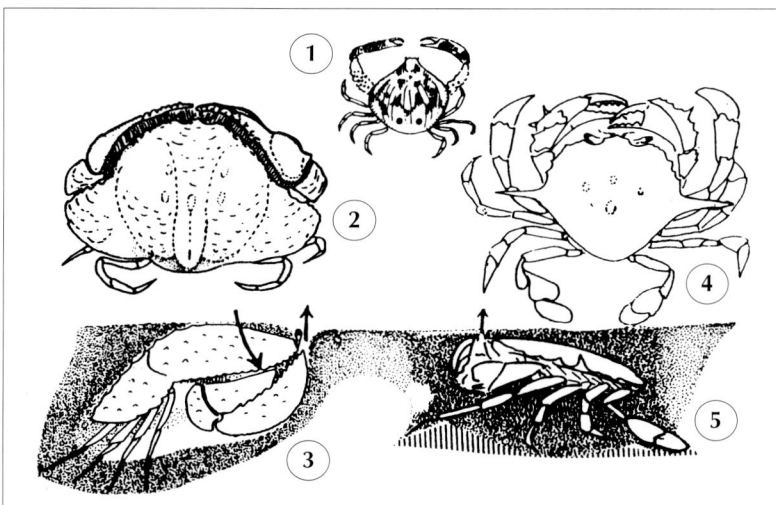

Fig. 22.31 Crabs of the Oxystomata
1 *Leucosia* 2 *Calappa hepatica*, dorsal view and (3) sand-burrowing *in situ* 4 *Ashtoret lunaris* dorsal view and (5) buried.

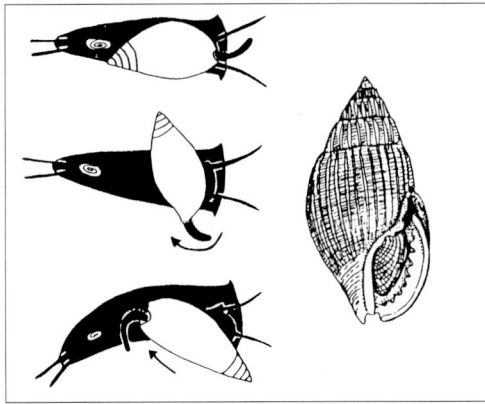

Fig. 22.32 *Nassarius olivaceus*
Shell, and animal in locomotion.

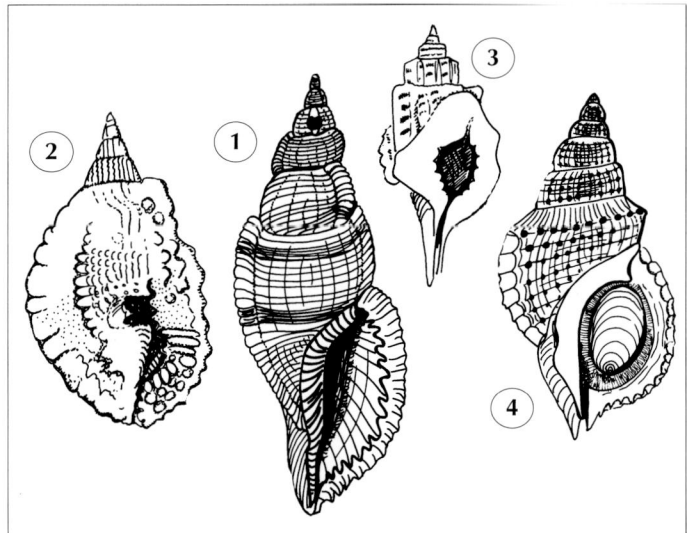

Fig. 22.33 Some sand-dwelling mesogastropods
1 *Cymatium pileare* 2 *Distortio anus* 3 *Cymatium muricinum* 4 *Bursa granularis*.

Sand Gastropoda

Associated with most sorts of coral reef there are sand formations of several kinds. These can range from the clean white beaches derived from finely comminuted coral or foraminifera, to the sand admixed with silt in the lagoons and the shallow subtidal. Only from these soft substrates can we realise the beauty of form and colour of the larger tropical gastropods, so many of which are normally hidden.

Much may be learned by night collecting with a tilley lamp. A shore walk at a low tide will reveal many sorts of gastropods emerged or lightly sand-covered or buried at the end of short, identifiable trails. But a truly representative sampling of all the molluscs that the sand supports can only be achieved by going out knee-deep from the water's edge and using a sieve with a coarse (1 cm) mesh. This can be filled in an approximately quantitative way, using a light, long-handled trowel with the edges turned up like a sugar trowel, to reduce spillage. When loaded the sieve can be shaken under water.

The small scavenging whelks of the Family **Nassariidae** are certain to come up in every sieving. The nassariids are active and agile, with their long, narrow foot, at or just below the surface. Their numerous species are short-spired and ovate, distinguished by details of sculpture.

The high success of the sand gastropods has been achieved with the evolution of specialist predation on bivalves and echinoderms. Through their diverse families there are several trends in common. They are more continuously mobile than hard-shore gastropods, with the shell either smoothly rounded or more often tapered for progression through or over fluid sand. The foot is generally broad and in several families its fore-part (*propodium*) has been shaped into a sand plough, while side-lobes (*parapodia*) enclose the shell. All these gastropods specialise in chemo-orientation, with the inhalant siphon extended and held aloft, or exploring as it ranges from side to side.

Mesogastropoda

In the large Indo-Pacific Superfamily **Tonnacea** the trumpet shells (**Cymatiidae**) have in numerous ways evolved parallel with the muricid whelks. Many of them are but slightly adapted for burrowing, having the whorls angled or turreted, and ornamented with axial ribs (varices), spiral cords and tubercles. *Cymatium pileare* is found under coral blocks, while *C. muricinum*, with its drawn-out canal, lives shallowly buried, as is also *Distortio anus* (from the Freudian detail of its shell aperture). The cymatiids feed upon molluscs and echinoderms. The largest of the tritons (*Charonia*) are notable predators of seastars, including the crown-of-thorns.

The helmet and bonnet shells (**Cassididae**) are all sand dwellers with a short, upturned inhalant canal. The largest and heaviest are the saffron-footed *Cassis cornuta* and the red-mouthed *Cypraecassis rufa*, both sublittoral predators on echinoids. The smaller *Casmaria erinacea* is common on intertidal lagoon sands.

The casques and tuns (**Tonnidae**) are by contrast thin-shelled and ovate, being chiefly holothurian-eaters. A familiar reef-flat shell is the partridge, *Tonna perdix*, while the small *Malea pomum* is to be found in or upon sand.

The necklace shells (Superfamily **Naticacea**) are of smaller size and smoothly polished. All are accomplished sand-burrowers preying chiefly on bivalves. In *Natica* the shell is spherical and brown or yellow banded, with the operculum shelly. *Polinices* are pure white and breast-shaped with a dark, chitinous operculum.

The *Sinum* species have the shell reduced to a small

Fig. 22.34 Sandflat and lagoonal gastropods and their prey
1 *Charonia tritonis*, with (2) crown-of-thorns, *Acanthaster* 3 *Cassis cornuta* with heart urchin 4 *Melo* sp. 5 *Oliva* sp.
6 *Tonna perdix* with holothurian 7 *Terebra subulata* 8 *Terebra maculata* 9 *Terebra dimidiata* 10 *Malea pomum* 11 *Polinices*
sp. 12 *Conus striatus* with blenny 13 *Harpa* sp.

plate, sitting on an animal as flat as the sole of a shoe. In naticids, the back of the head-shield and the parapodia enwrap the shell as the animal progresses. These extensible soft-parts are inflatable — as generally assumed by taking up water through aquiferous pores; they can be entirely deflated and withdrawn into the shell. The bivalve prey is fully enveloped by the foot and the proboscis applied to its shell. A hole is bored with the radula, assisted by solution with an acid saliva. Naticids deposit their spawn in circular *nidus*, a spiral ribbon cemented with sand and mucus.

Neogastropoda

Passing to the largest range of predatory gastropods, we find two great series represented on sandflats, the **Volutacea** and **Toxoglossa**. The first family of the former is the **Mitridae**. Their largest mitre shells include the smooth, vermilion-spotted *Mitra mitra* and the granular-sculpted *M. episcopalis*. The smaller *Vexillum* species have a strong axial sculpture and banded coloration. The lesser sand mitrids include the numerous *Cancilla*, *Pterygia* and the charming *Scabricola* and *Imbricaria* species. *Scabricola casta* is chestnut and white-banded and

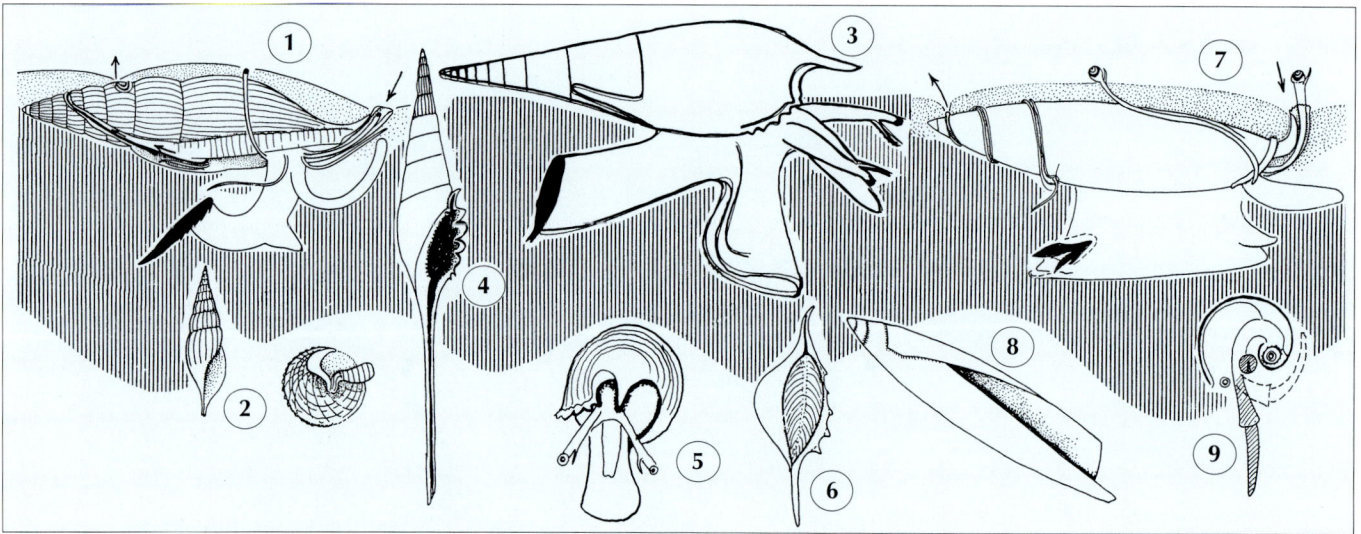

Fig. 22.35 The sand-burrowing 'classic' Strombidae

1 *Rimella* in sand-buried posture 2 *Rimella*, immature shell (**left**) and mature shell from anterior end (**right**) 3 *Tibia* in sand posture 4 *Tibia longirostra*, mature shell 5 *Tibia*, animal in shell from in front 6 *Tibia*, operculum closing aperture 7 *Terebellum*, submerged in sand 8 *Terebellum* shell 9 *Terebellum*, animal in shell, from in front.

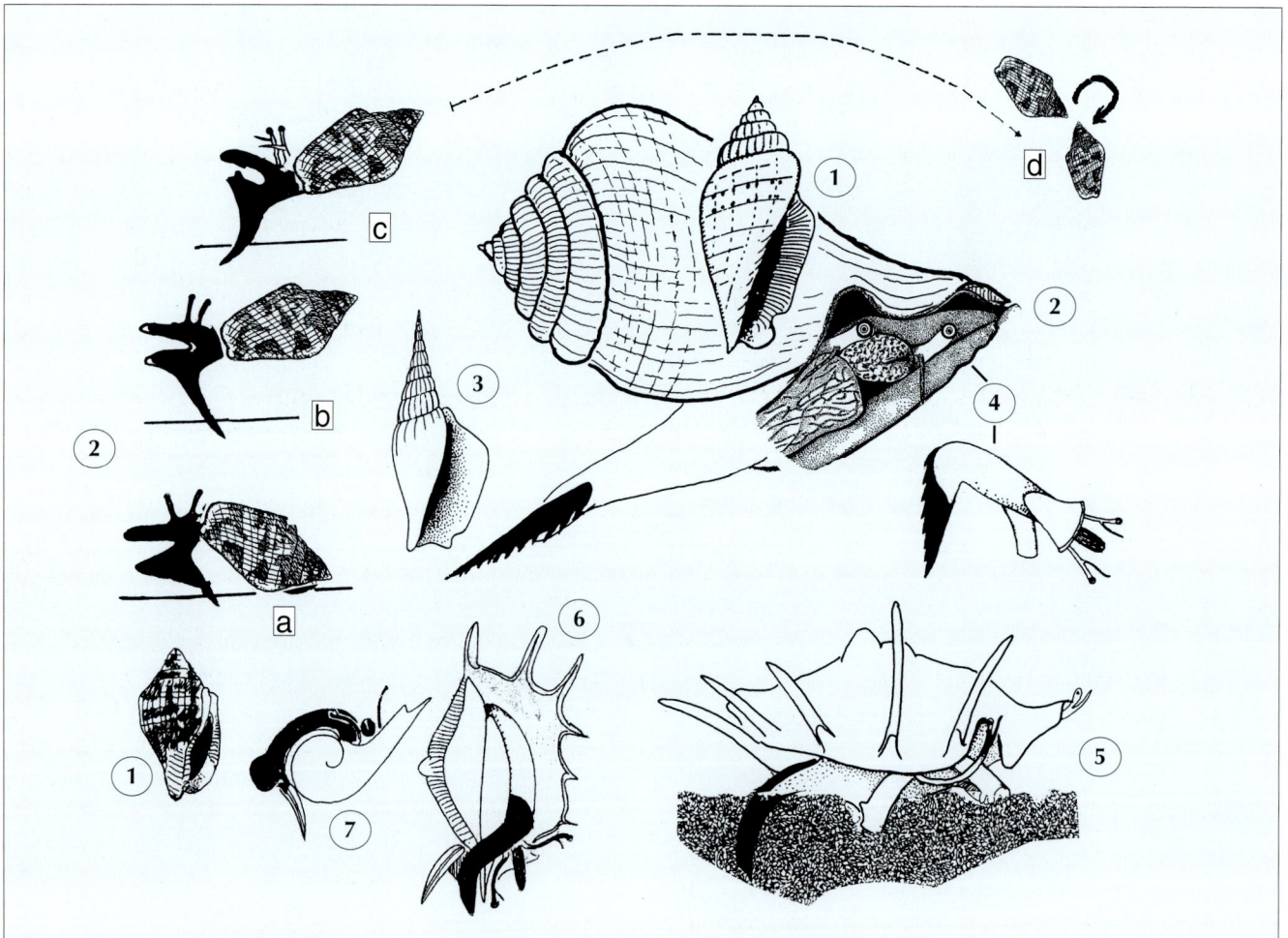

Fig 22.36 The later Strombidae

1 *Strombus maculatus* 2 *Strombus* in successive positions (a to d) with the backward flip 3 *Strombus vittatus* 4 *Oostrombus gibberulus* with shell and shallowly submerged in sand, showing foot with propodium and operculum, and head with proboscis and eye-stalks 5 *Lambis lambis* with foot extended, on a silty gravel surface 6 *Lambis* with aperture upwards to show animal 7 side-view at start of overturn, with operculum.

Imbricaria olivaeformis butter yellow with a black tip.

The olive shells (**Olividae**) have a smooth, long-ovate shell, highly polished and elegantly patterned. From the narrow rectangular foot, the parapodia fully enclose the shell, and behind the crescentic head-shield a short inhalant siphon stands erect. Olives are very common in sand-siftings: *Oliva oliva*, *O. mustelina*, *O. miniacea* and — smallest and most exquisite — the orange and white-banded *O. carneola*.

The largest of the Volutacea — and indeed of all the Gastropoda — are the smooth, melon-shaped baler shells (**Meloidae**) progressing or burrowing by the wide oval foot. The smaller harp shells (**Harpidae**) have in contrast a sinuous axial sculpture and are handsomely marbled in shades of brown. The harps are broad-soled, with a wide propodium, and can defensively autotomise the back of the foot, cutting it off with the sharp lip.

The arrow-tooths (**Toxoglossa**) take their name from the transformation of the radula into a sheath of barbed arrows. In the most specialised family, the **Conidae**, a single arrow can be shot out to impale the living prey, injecting it with a toxin from the salivary gland passed along a groove in the tooth. The smaller, narrow-mouthed cones, like the common *Conus ebraeus* (Fig. 20.5) and *C. chaldeus* on reef platforms, prey upon polychaete as well as sipunculan worms. Larger worm-catching cones on sandflats include the spotted *C. litteratus*, as well as the black-netted *C. marmoreus*. Those cones with a wider aperture and a somewhat convex, urn-shaped shell are mostly stalkers of live fish, such as small blennioids. The proboscis is poised over the fish for a successful strike, then dilated for a massive swallowing. The piscivores include *Conus textile* and *C. striatus*, both dangerously venomous, and *Conus geographus*, that has even proved to be fatal to humans.

The auger shells (**Terebridae**) are also toxoglossans, finer in colour and ornament than the cones, and more adept at burrowing. The slender shell is pulled down by the thrusting of a rather small foot on an extensile 'neck', with the inhalant siphon keeping surface contact. This family includes active carnivores and scavengers. In the mid-tidal sand beach live smaller terebrids, such as *Hastula*, with a pair of slender teeth in each row, and *Imrages* with clustered teeth like a cone. The *Terebra* species, largest and most colourful, have wholly lost the radula. They include some of the most prized and beautiful sandflat shells, *T. subulata*, *T. dimidiata*, *T. babylonia*, *T. areolata* and *T. maculata*.

In all these exquisite conids, olives and mitrids, the significance of their colours leaves us still mystified. Just as the cowries keep their attractive shells covered in life by an integument, most of these sand-dwellers pass their lives buried in sand and — at least by day — seldom seen.

The utility of ornamental sculpture in sand gastropods is better understood, serving as a defence against predatory seastars and crabs. In the trumpet and harp shells, the strong ribs and varices (the apertural lips from earlier stages) protect the shell against chipping away by sand crabs such as *Calappa* and *Matuta*. The naticids, which are unsculptured, employ another safeguard, by raising a fold of skin to cover the shell from behind, to prevent the tube-feet of a seastar from taking hold.

Another technique of defence is the 'escape behaviour', with shock movements of the foot released by the proximity of a predator. Most sand gastropods have such a faculty, but it is the conch shells and their relatives in the **Strombidae** that have adopted it to their standard locomotion. To this numerous and successful Indo-Pacific family we must briefly turn.

Strombidae

The strombids or conch shells have evolved on different lines from any of the gastropods yet mentioned. They are narrow-footed, with only the fore-part or propodium remaining of the sole or plantar surface for progression or burrowing in sand. The Strombidae are not predators but herbivores and deposit-grazers. One of the commonest tropical gastropods is *Oostrombus gibberulus*, turning up in almost every sieving of low-tidal sand. The animal advances by disconnected steps, with the upraised shell thrown forward and the foot then brought up from behind. If the shell gets overturned, the column of the foot is extended to thrust the blade-like operculum under the shell, which is then flipped back the right way up. In escape from a predator, the shell is thrown back by a forward kick of the metapodium, with the operculum used for purchase on the ground.

The postures and habits in the genera of Strombidae are shown in Figs. 22.35 and 22.36. The eyes are always bizarrely enlarged, set on long stalks to peep out with seeming vigilance, one from the anterior canal and the other from a special 'squint' in the shell lip. During strombid evolution the margin of the aperture has become variously elaborated. Beneath it the proboscis roves about to graze fine algae or glean detritus.

The Strombidae are a long-established family. The early *Rimella*, with a spindle-shaped shell having a long exhalant channel applied to it, reaches back to the Cretaceous. Almost as primitive is the spindle-shaped *Tibia* — with a shorter exhalant channel, but a long anterior spike — that first appeared in the Eocene. The thin-shelled *Terebellum* is notably simplified from this early pattern, being specialised for living deeper in fluid sand, where it moves along with the optical peduncles alternately put up like periscopes to the surface.

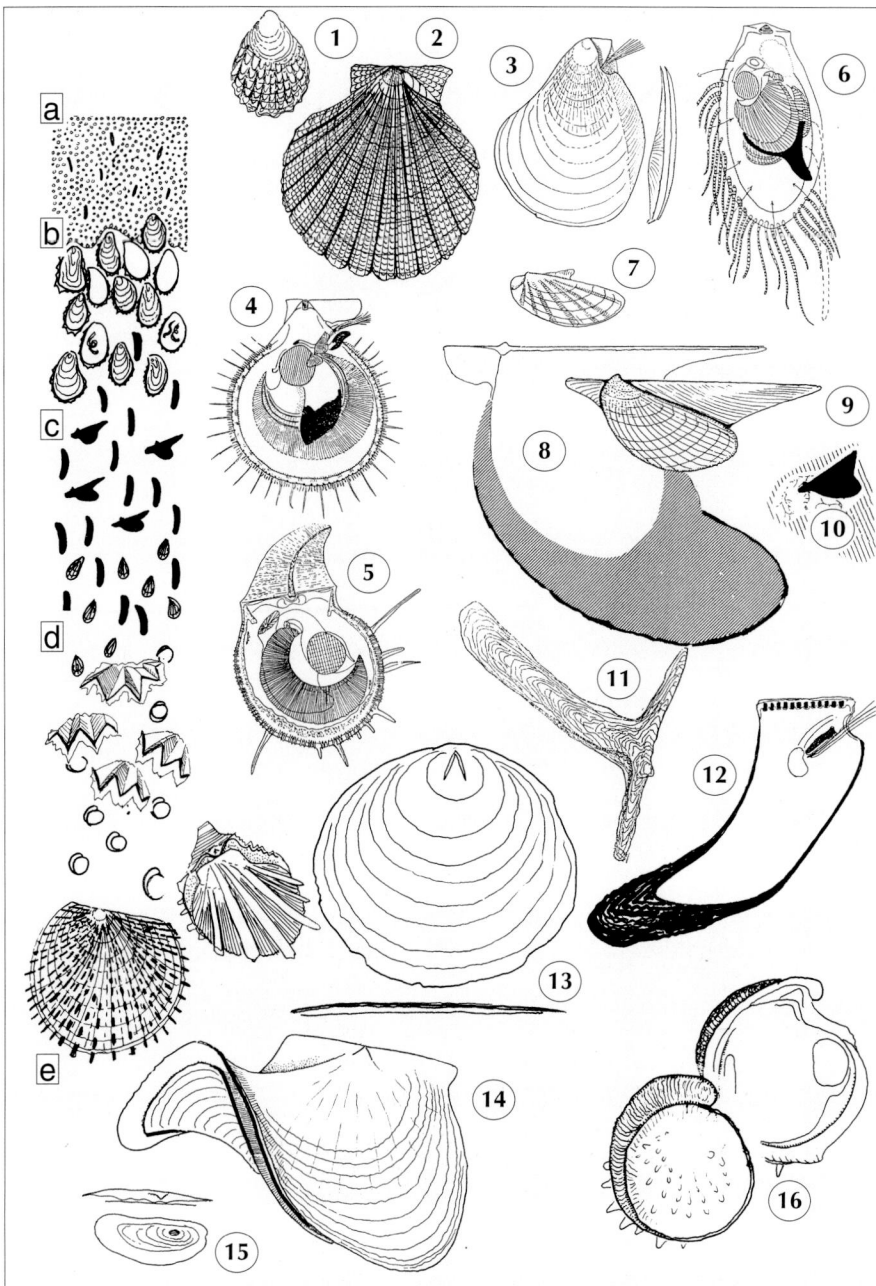

Fig. 22.37 Morphology of epibenthal bivalves

(**left**) Vertical distribution, based upon a wharf pile at Guadalcanal, Solomon Islands.

a–b *Isognomon perna* in zone of *Chthamalus* barnacles
b–c *Crassostrea*
c–d *Pteria*, mussels *Septifer*, *Malleus regulus*
d–e Cock's comb oysters, *Lopha cristagalli*, *Pseudochama exogyrata*, with *Pinctada* and *Spondylus*
1 *Plicatula* 2 *Gloripallium* 3 *Pedum spondyloideum* 4 *Chlamys* and
5 *Spondylus*, arrangement of pallial cavity 6 *Lima*, pallial organs and tentacles 7 *Electroma* 8 *Pteria penguin* 9 *Pteria* sp. 10 *Pteria* embedded in alga 11 *Malleus malleus*
12 *Isognomon* 13 *Placuna placenta*
14 *Placuna ephippium* 15 *Enigmonia aenigmatica* 16 *Pseudochama exogyrata*.

Strombus is the largest genus, abundant in species from the Pliocene onwards, and showing a diverse evolution of the lip, for life at or near the surface. In the large scorpion shells, *Lambis*, which evolved only in the Pleistocene, the lip carries long spines, and its species dwell on the surface amid the coral rubble of the dead reef.

Bivalves of the reef

The most numerous and commonplace bivalves, at least in temperate regions, are those plain, white-shelled species burrowing in soft shores. Relatively late in origin, these have the mantle edges extensively fused, but drawn out into siphons posteriorly, while the foot has generally remained large for active burrowing. The shell has retained its fore-and-aft symmetry, with anterior and posterior adductors.

Such sand-buried or *infaunal* bivalves — that came into fashion with the Mesozoic — are also well represented in the tropics. But on hard surfaces — no doubt including coral reefs — they were preceded by more ancient bivalves of an *epifaunal* habit, belonging to families already in being in the Paleozoic. Their early structure was to become far more radically specialised than in the successful stereotype of the sand-dwellers.

One of the oldest bivalve lines, well under way by the Silurian, is the **Arcoida** or ark shells, with the hinge distinctive in its *taxodont* or many-toothed dentition. Among coral and under boulders we find the relatively unmodified *Barbatia* species held down edge-on by a byssus fringe running along the foot. Others such as *Arca avellana* have begun to abrade dead coral by rocking movements of the shell (Fig. 22.18). In all the

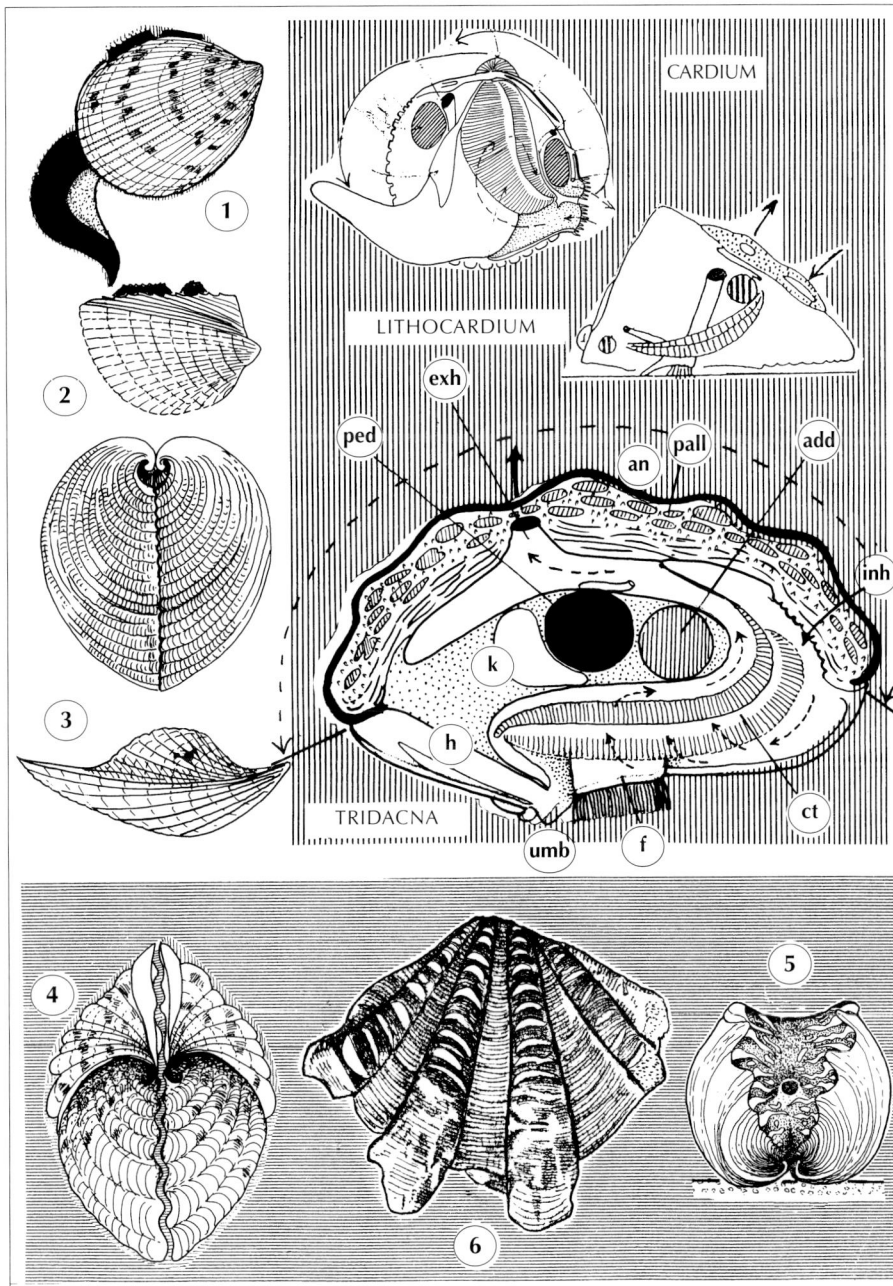

Fig. 22.38 Cardiacea: cockles and giant clams

1 *Acrosterigma* 2 *Fragum* 3 *Corculum cardissa* 4 *Hippopus*, with (5) open gape 6 *Tridacna*.
The rearrangement of symmetry in *Tridacna* is a derivation from *Cardium* and with the fossil *Lithocardium* as an
intermediate stage: (add) adductor muscle (an) anus (ct) gill (exh) exhalant siphon (f) foot with byssus (h) hinge (inh)
inhalant site (pall) coloured pallial lips (ped) pedal retractor muscle (umb) umbone.

arcoids the feather-like gill-plume, with filaments only loosely inter-attached, remains highly primitive.

The mussels (**Mytiloida**), form a lineage almost as old as the arks, dating back to the Devonian. We have already described their acquired asymmetry (*heteromyarian*) with the anterior end pointed and reduced and a compensating enlargement of the rounded posterior end. The tropics are notable for their coral-boring mytilids (*Lithophaga*) (Fig. 22.18). But the ultimate specialisation for boring is seen in the extraordinary *Fungicavia eilatensis*, living within the coelenteric cavity of fungiid polyps.

The thin shell valves, held open at 180 degrees, are enveloped by the mantle and the wide inhalant siphon makes contact with the polyp's internal mesenteries to draw in some of the ingested phytoplankton.

The later epifaunal bivalves have completely lost the anterior adductor muscle, to become '*monomyarian*'. First, the pearl shells (**Pterioida**) have given up the edge-on posture, to lie recumbent on the right valve, which has a deep notch for the byssus. The wing shells *Pteria* (= *Avicula*) have the hinge axis (antero-posterior) long and straight, offset by an oblique dorso-ventral axis of elongation.

The pearl oysters (*Pinctada*) are almost circular and the posterior adductor is now centrally placed. The inhalant currents enter nearly all round, except at the hinge-line. The small and light pearl shell *Pinctada martensii* attaches beneath boulders and the very flat *Pinctada fucata* lives in rock fissures. Blister pearls are formed by the larger, deeper water species, stabilised mainly by weight, the pale-lipped *Pinctada maxima* and the very large, black-edged *P. margaritifera* (*Avicula*).

The Family **Isognomonidae** have — like the wing shells — an extended hinge line, but it is distinguished by its long row of teeth. The *Isognomon* shell may be narrow and crevice-wedged, as in *I. legumen*. Some smaller species are wafer-thin, piled or clustered under boulders (*I. isognomon*). *Crenatula modiolaris* lives buried in *Haliclona symbiotica*, a green branching sponge. *Vulsella vulsella* is radially ribbed, embedded in the sponge *Suberites*. The hammer oysters, *Malleus*, are much elongated in the dorso-ventral axis, from hinge to growing margin. *Malleus regulus* is compressed like a flat pod, attached at the hinge end. *Malleus malleus* has the hinge-line drawn out as well, to give a hammer shape. Living in clean sand near rock, it attaches by a byssus with the long emergent 'handle' carrying both inhalant and exhalant apertures.

With a similar byssus-held burial in sand, the triangular **Pinnidae** are transformed into compressed cornets, opening and closing not by the action of the hinge but by elasticity at the posterior end of the shell. From the anterior end byssus threads radiate out to separate grains in the sand. The small pinnid *Streptopinna saccata* has become a crevice-nestler like an *Isognomon* (Fig. 22.37).

From a byssus-fixed habit, the scallops and their relatives (**Pectinacea**) have advanced in two directions. The earliest of the scallops are the small bright-coloured *Chlamys* and *Gloripallium*, lying at the surface with the right valve notched for the byssus. The hinge-line is shortened and the whole inhalant margin is fringed with tactile tentacles, with the addition of stalked eyes of a metallic sheen. Though relatively complex, these seem to function as no more than light and shade perceptors.

The scallop mantle edge has developed a wide fold or *velum* to preclude the entry of sediment. The pallial cavity is cleansed by clapping the shell valves with fast-repeated contractions of the adductor muscle. Accumulated particles are thus expelled as with a vigorous 'sneeze'. From the *Chlamys* level, the large *Pecten* scallops, mostly in colder seas, have gone on to a new agility, taking off from the ground with swimming spurts created by valve adduction. Much lighter than *Pecten*, with both valves almost flat, are the sun and moon shells, *Amusium japonicum* and *A. pleuronectes*,

tropical scallops with a red valve above and a white below. Both are accomplished swimmers close to a sand or gravelly bottom.

In contrast, the thorn-oysters (**Spondylidae**) are scallop descendants that have become fixed down like the true oysters (Fig. 22.37). The shell is heavy and strongly spined, and has evolved to resemble a *Chama*, cemented by a deep right valve, with the smaller left one forming a lid. With only the older part of the valve attached, spondylids form brackets as at the edge of coral boulders. *Plicatula plicata* is another pectinacean that has paralleled the spondylids in becoming cemented down.

Many of the file shells, **Limidae** — like the scallops — have become improvised swimmers, by use of the velum with fast adduction of the valves. Some species, like the thicker-shelled *Lima vulgaris*, with strong, spinose ribs, stay permanently attached. Others are light and thin-shelled, like *Limaria fragilis* in coral heads, that can take off to swim. *Lima* can also crawl upright in mussel fashion (but 'backwards' or hinge-first), using the narrow, extensible foot. The long, bright red pallial tentacles assist in swimming by their flexions that provide a 'rowing' action.

With the **Anomiidae** or saddle oysters we reach the most highly modified of the epifaunal bivalves. *Anomia* (Fig. 13.13) has the upper (left) valve clamped like a limpet, drawn down by the byssus retractor muscle. The byssus threads fuse into a calcified cable passing through a deep notch in the almost vestigial lower valve.

Just as with scallops and file shells, some anomiids have broken free of the byssus. On the leaves of *Rhizophora* mangroves in Malaysia and New Guinea, *Enigmonia aenigmatica* — while young — are like thin, oval limpets, creeping by extensions of the narrow foot. The window shell, *Placuna placenta*, is a light and circular anomiid, adapted for living in sand, with both valves flat and translucent. The very large *Placuna ephippium* is by contrast epifaunal and raised well above the surface, with both valves twisted to a saddle shape.

Giant clams

By comparison with the ancient bivalves just described, the giant clams are descendants of infaunal bivalves that came in with the Mesozoic. Beginning with sand-dwelling cockles, the Superfamily **Cardiacea** has culminated with the epifaunal clams of the Family **Tridacnidae**, including the largest shelled molluscs ever evolved.

The earliest tropical cardiaceans are normal cockles (*Fragum* and *Acrosterigma*), long-footed and active in sand. Some species already possess simple light-receptors in the inhalant siphon. Reliance on light is carried furthest with the farming in the siphon tissues of captive zooxanthellae. This habit is perfected in the giant clams. Its early stages can be seen retraced in the fashion of the burrowing *Corculum cardissa*, a recent cockle-derivative with a heart-shaped shell as if a *Fragum* cockle had been planed off behind the beaks (Fig. 22.38). Flush with the sand surface, the transparent *Corculum* shell admits light to symbiotic algae concentrated in the siphonal tissues.

In the **Tridacnidae** the normal pattern of symmetry has been radically altered. The beaks and hinge of the cockle have moved through 180 degrees to lie at the lower side, alongside the byssal gape. The interlocking margins of the valves have meantime been displaced to face upwards, and the enlarged lips of the exhalant siphon, patterned in green, blue or brown, occupy this whole gape. These pigments evidently serve as light fil-ters for the symbiotic algae (zooxanthellae) massively concentrated in the siphonal tissues. Algae find their way also to other parts of the clam, as in the amoebo-cyte cells where they are eventually digested. Their tis-sues have long been considered a major source of nutriment, making possible the giant size of the tridac-nids. From tracer studies, the clams would appear to gain the greatest benefit from the carbohydrates of pho-tosynthesis, rather than by digesting the algal them-selves.

Important Indo-Pacific species are the rock-eroding *Tridacna crocea* and *T. fossor*, and the still larger *T. squamosa* and *T. maxima*. The horseshoe clam, *Hippo-pus hippopus*, has lost the byssus and its heavy shell lies free at the surface.

CHAPTER TWENTY-THREE
Mangroves

The second great formation with focus on the tropical Pacific is that of the mangroves. Unlike the reef corals they are communities of soft shores, though they can border and overlap with the edges of rocky shores. They replace corals in their immense dominance where the shoreline withdraws into muddy and estuarine stretches. Though they are warm-water species, mangroves are not all insistently tropical. In their southern limits they extend down to northern New Zealand, where we see one lone species of *Avicennia*, and to New South Wales and Victoria, with the outposts of three mangrove genera. Something like 11 families of flowering plants (10 of them dicotyledons and one palm) share the convergent adaptations of the mangrove form for a tidal edge scrub and even — with increasing tropical complexity — a forest.

Just as for corals, molluscs, crustaceans and fishes, the rich Pacific shoreline becomes more luxuriant in mangroves as we travel west. To survey the mangrove communities at their most complete, we must shift from the main centre of this book to a concentration with the Indo-Malaysian Archipelago. Beside the floristic richness to be seen with the mangroves here, every other estuarine stand looks poor by comparison. We shall culminate our account of the edge of the tropical Pacific with the description of the mangroves at their high point of diversity in south-east Papua New Guinea, most easily accessible to the naturalist from Port Moresby.

At tropical temperatures, coastal scrublands — virtually forests — can reach down below high water. Where temperate shores would be colonised by salt-marsh and estuarine meadows, the tropical strand — on progressively enriching soils — can develop the tidal forests commonly styled *mangroves* or *mangals*. As well as being applied collectively to a whole plant formation, the term mangrove can be used also for any of its component species. The trees or shrubs referred to as mangroves thus belong to no single genus or family, but have separately evolved in a number of families with convergent or similar adaptations to a demanding mode of life.

Even on the coral reef itself, some mangrove species can grow in sandy patches but really complex mangals are achieved only in muddy estuaries or deltas. In total shelter from waves, they establish most diversely on land-derived sediments built up by organic breakdown. The surface layer is periodically shifted by the currents of the stream-course, with only slight disturbance by tidal flow. The mangrove muds are of low salinity and anaerobic below the surface, and would have been soft and unwalkable until stabilised by the trees.[1]

Mangroves colonise the shore from about mid-tide level up to stretches only exceptionally inundated at equinoctial spring tides. In the estuaries of Indo-Malaysia and northern Australia, with high enough humidity, complex mangals can occupy this whole extent. In drier places, away from the meanders of tidal rivers, mangroves are kept to the lowest reaches, while above them stretch the wide saltings of the *high marsh* where the surface desiccates and salt crystallises. Such saltings may drop by a short 'cliff' to the *low marsh* further out, carrying a pioneer mangal of *Avicennia* and *Sonneratia*. These — and especially the former — grow furthest to seaward and are thus regularly inundated. Neither genus has reached Polynesia or the oceanic Pacific.

Mangroves are able to reach far upstream, where a bottom wedge of heavier sea water raises the salinity of the surface mud. Secure from stress by waves or strong currents, mangroves have only superficial rooting systems. Rather than promote soil formation, they follow it. Land reclamation is helped by their roots and pneumatophores retarding soil removal, but sediments laid down in close shelter will not be colonised until they reach a suitable depth and consistency. Seaward of the mangrove front, the tidal muds remain bare and unconsolidated.

The mangrove community may then be thought of as a 'forest between tides', made up of trees and shrubs with a few non-woody plants and vines. The first and most widely familiar mangrove family is the **Rhizophoraceae**. Thickets of the arching strut-roots of *Rhizophora* are a near-universal feature of tropical estuaries. Next in prominence, though generally on a more sandy terrain, are likely to be the pioneer mangroves of the families **Sonneratiaceae** (classed near Myr-

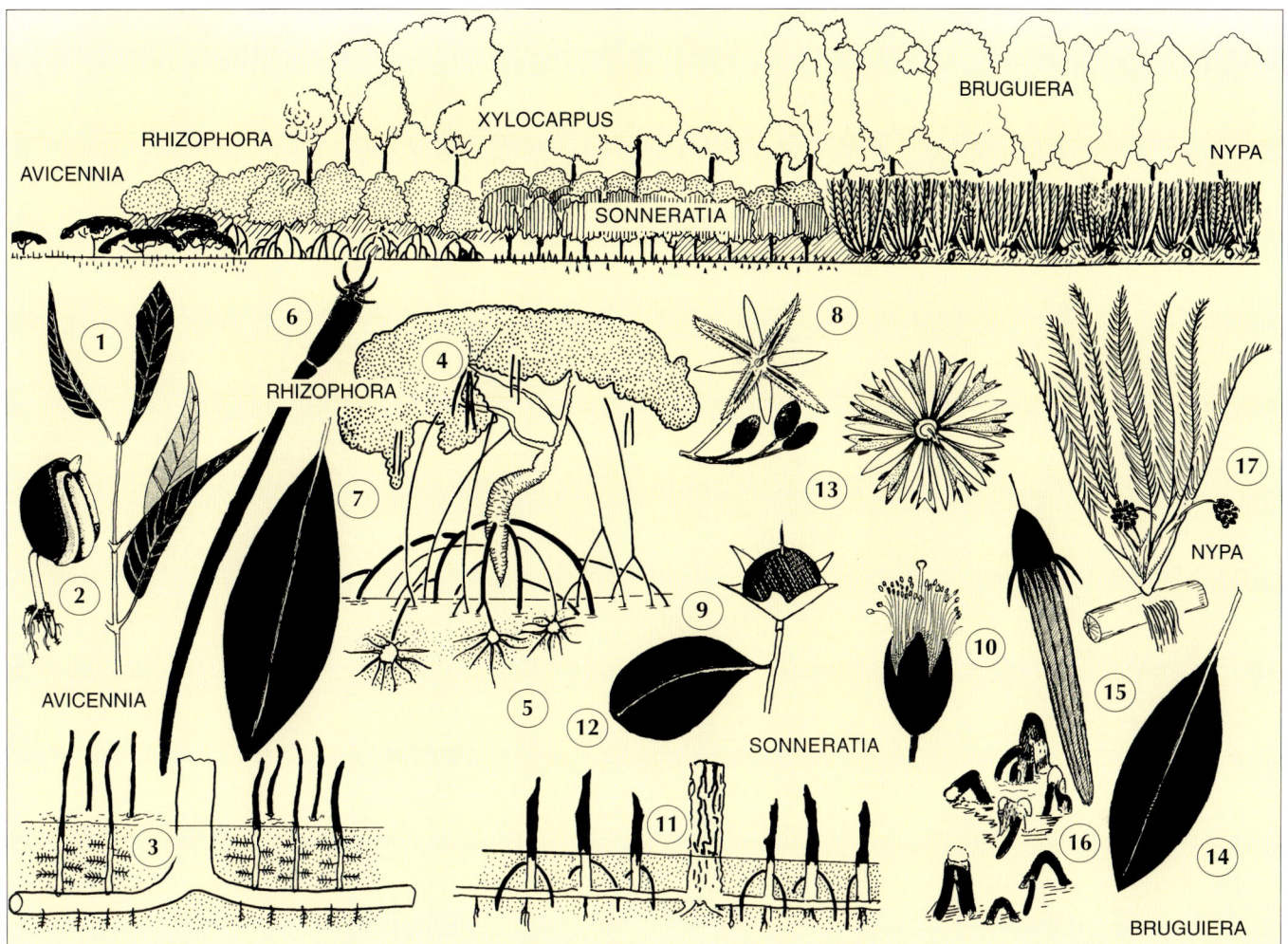

Fig. 23.1 A complex mangrove sequence near Port Moresby, Papua New Guinea

A succession is shown in the Galley Reach inlet from its opening to the sea (**left**) up to *Nypa* (**right**).

To seaward grow *Avicennia* pioneers, followed by *Rhizophora apiculata* and *R. stylosa* backed with *Bruguiera* and *Xylocarpus*. Then follows a forward rank of *Sonneratia alba*, with finally to landward *Bruguiera cylindrica* and *B. parviflora*, or with freshwater influence a pure margin of *Nypa fruticans*.

The characters of five mangrove genera are shown in detail. *Avicennia* 1 foliage, 2 viviparous embryo 3 radial roots, pneumatophores and nutritive roots; *Rhizophora* 4 mature tree with strut and aerial roots and 5 submerged roots 6 reproductive 'dropper' 7 leaf 8 flower; *Sonneratia* 9 advanced fruit 10 flower 11 submerged radial roots, with woody pneumatophores and nutritive roots 12 leaf; *Bruguiera* 13 flower 14 leaf 15 'dropper' 16 hooped pneumatophores; *Nypa* 17 palm with mature fruits.

taceae) and **Avicenniaceae** (close to and formerly forming part of the Verbenaceae).

Upstream, this simple association diversifies. The larger trees, now including *Heritiera* and *Excoecaria*, become intermixed with shrubs and lianes and the mangrove fern *Acrostichum aureum*. It is such a community that straddles the high water mark of normal spring tides.

To realise the full possible spectrum of the mangal, from backshore forest down to mean sea level, we must travel to the high point of the mangrove world in Malaysia, New Guinea or northern Australia. In the oceanic Pacific, the species diversity much attenuates, not only by geographic distance from the centre, but largely from the want of spacious estuarine terrain.

Fiji has achieved a moderately rich mangal in the river mouths of the two larger islands. Even better endowed, as we have seen, are the embayments of continental Hong Kong, lying near the northern limit of mangroves. Reaching furthest north in the Pacific is the small rhizophoracean *Kandelia candel*, touching the Ryukyu islands of Japan. Outposts of *Avicennia marina australasica* form the southern extreme for mangroves, in New Zealand and (with *Aegiceras*) in Victoria. Such extensions apart, mangroves are subtropical and tropical, with frost-point temperatures imposing final limits.

Along with soil type, salinity and shelter, an important determinant of mangrove distribution is the length of time of inundation. Five species groups are generally recognised in a full sequence across the shore, following Watson's classification:

Fig. 23.2 Some mangrove plants of Hong Kong
1 droppers of *Bruguiera conjugata* 2 droppers and flowers of *Aegiceras corniculatum*
3 droppers and flowers of *Kandelia candel* 4 flower of *Acanthus ilicifolius*
5 flower of *Lumnitzera racemosa*. (From Morton & Morton. Paintings by Karen Phillipps.)

1 **inundated at all high tides:** only *Rhizophora stylosa*, along stream courses with foliage still above water

2 **inundated at medium high tides:** *Avicennia* and *Sonneratia* species, forming an edge community along rivers, where *R. stylosa* predominates behind

3 **inundated by higher than normal tides:** the greater part of the mangal, with most of the *Rhizophora* species

4 **inundated by spring tides:** *Bruguiera* now replaces *Rhizophora*, on stiff soil

5 **occasionally inundated by exceptional or equinoctial spring tides,** *Bruguiera gymnorrhiza* often with the mangrove fern *Acrostichum*.

V.J. Chapman (1976) gave a schema of the succession of mangroves and associated plants in Micronesia:

Mud/silt	Coral sand
Rhizophora stylosa	Sonneratia alba
Bruguiera gymnorrhiza	
Lumnitzera	Xylocarpus
(swamp forest)	
Nypa fruticans	
Barringtonia racemosa	
Acrostichum	
Excoecaria agallocha	

The mangal diversity thins out northward to Guam, and east to the Marshall Islands. In the last, two other species, *Pemphis acidula* and *Intsia bijuga*, appear.

The mangroves of Tonga include *Rhizophora stylosa*, *R. mangle*, *Bruguiera gymnorrhiza*, *Lumnitzera littorea* and *Xylocarpus granatum*. Western Samoa has very limited mangals with *Rhizophora samoensis*, *Bruguiera gymnorrhiza* and *Xylocarpus granatum*. No mangroves have reached Tahiti naturally, and they are lacking also in the Cook Islands, where at Aitutaki the lagoon has well grown *Pemphis acidula* (Fig. 21.4).

Mangrove adaptations

Living in intense light and at high temperatures on saline, anaerobic soils has called for a special cluster of adaptations. First, paradoxically, the mangrove environment is physiologically 'dry'. The high surrounding salt content, by tending to draw water from tissues, has called forth some features elsewhere found only in xerophytes. Thus, mangrove leaves are semi-succulent, with a thick epidermis, waxy cuticle above, and (sometimes) hairy tomentum beneath. Somewhat reduced in area, the leaves have sunken stomata, small cells, and maintain a high internal osmotic pressure.

Living in soils not only saline but over-watery for normal angiosperms, mangroves exhibit also some of

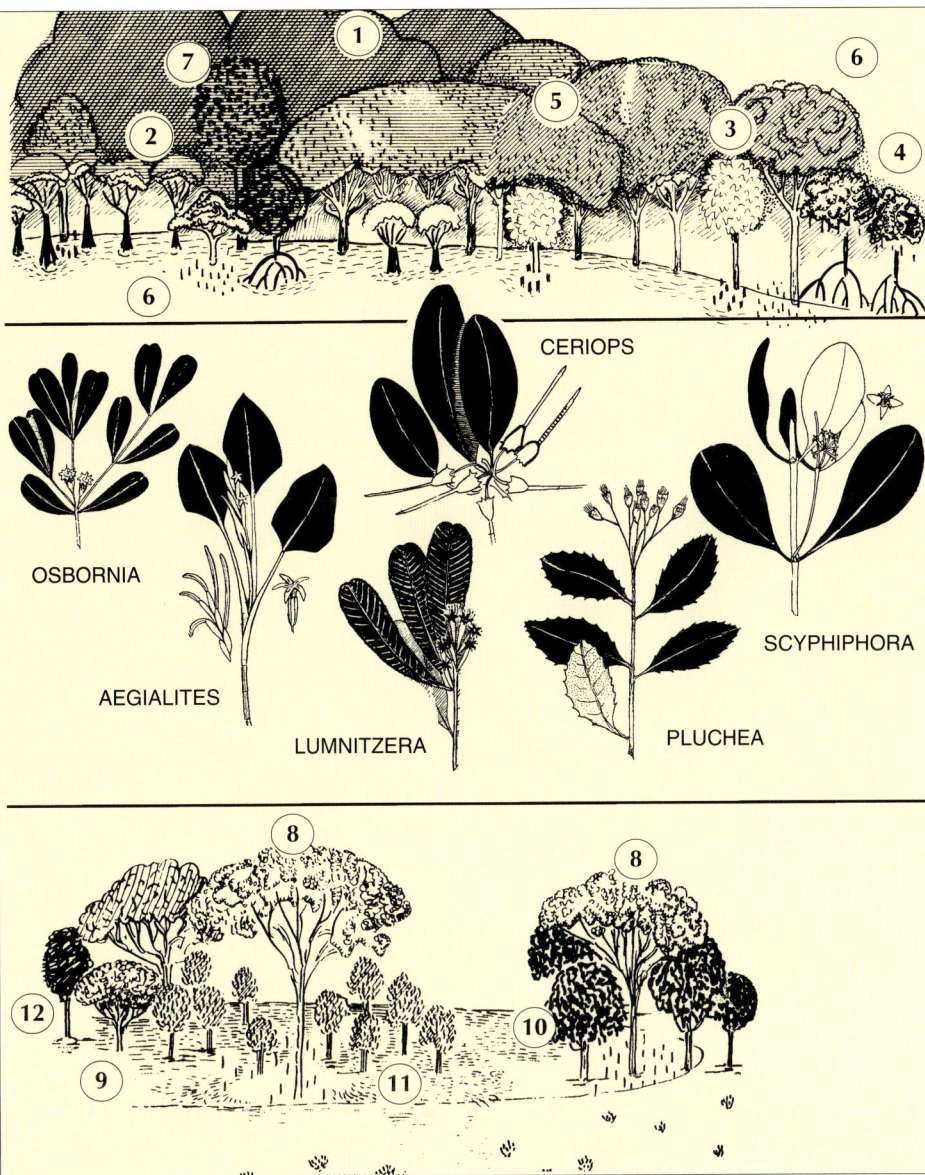

Fig. 23.3 Mangroves of the shoreline, near Port Moresby, Papua New Guinea

(**top**) Mangrove species at the edge of a rocky shoreline at Motupore.
1 *Hibiscus tiliaceus* (backdrop to the mangal) 2 *Aegialitis annulata* 3 *Sonneratia alba* 4 *Rhizophora stylosa* 5 *Osbornia octodonta* 6 *Avicennia eucalyptifolia* 7 *Lumnitzera littorea* (**middle**) Foliage and inflorescence for some New Guinean mangrove genera. (**bottom**) Mangroves on sandy stretches ar Fairfax Harbour, near Port Moresby.
8 *Avicennia marina* 9 *Osbornia octodonta* 10 *Ceriops tagal* 11 *Lumnitzera racemosa* 12 *Pluchea indica*.

the features of halophytes. They imbibe water easily, with high rates of flow and abundant 'guttation' or exudation from the leaf surfaces. The roots can achieve much lower chloride concentrations than in the outside sea water, involving an active return of electrolytes from the tissues to the exterior.

Mangroves have acquired various means of coping with excess salt. They can tolerate higher salt levels in the sap than most land plants, while some — like *Avicennia*, *Aegialitis* and *Aegiceras* — are active excreters, concentrating and removing salt through special glands on the leaves. These can produce a salty secretion 20 times as concentrated as the cell-sap. A second class of mangroves are salt-excluders (*Rhizophora*, *Bruguiera*, *Sonneratia* and *Lumntizera*), allowing less salt to enter and concentrating what passes in by retaining it in the leaves. Old leaves thicken up by developing salt and water storage tissue, especially in *Sonneratia*. *Rhizophora* reduces photosynthesis in old leaves as the storage layer thickens.

The mangrove genera have their different and char-

acteristic root systems, serving both for shallow anchorage, and for obtaining water and oxygen in low-aerobic soils. *Avicennia* and *Sonneratia* possess no taproot, being moored upon rafts of radiating cable roots, 25–50 cm deep. From the radials, smaller anchoring roots are given off, together with vertical breathing roots or pneumatophores that emerge above the surface. In *Avicennia* the pneumatophores are slender and flexible, while in *Sonneratia* they form strong woody spikes. Both the radial roots and the pneumatophore bases put out tufts of absorption roots into the superficial soil layer.

Lengthening as the mud accumulates, the pneumatophores bring the nutritive roots to just below the surface, while their lenticels admit atmospheric air to the internal aerenchyma tissue. When the lenticels are covered at high tide, root pressure begins to drop. As the tide goes down, air is sucked in again.

In place of a taproot *Rhizophora* has developed branching prop-roots from which long arched struts are put out radially. Just beneath the surface each prop-root divides into air-filled anchoring roots that give rise in

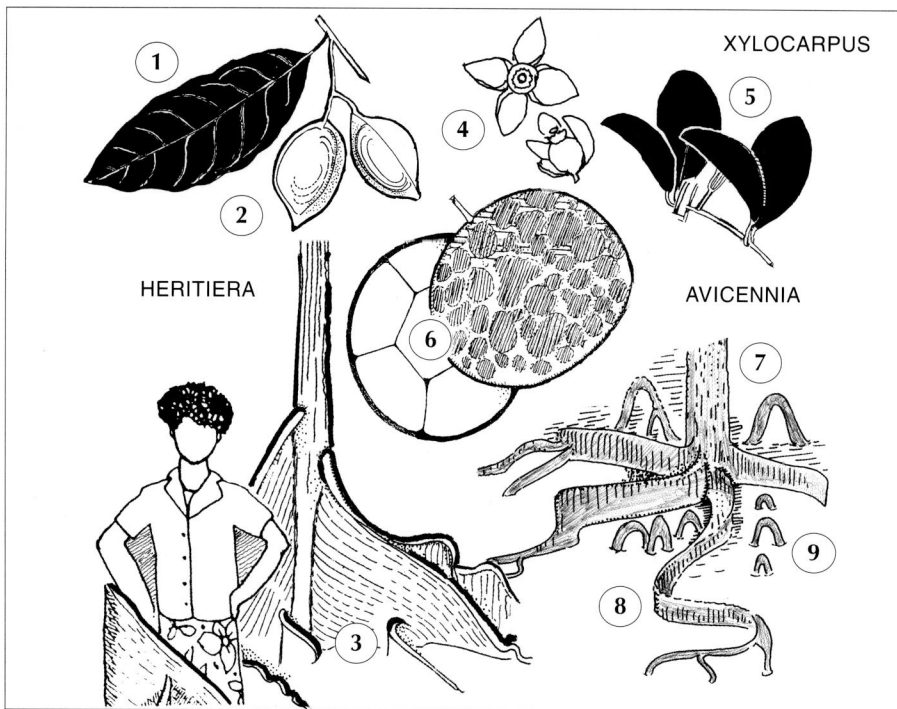

Fig. 23.4 *Heritiera* and *Xylocarpus*
1 *Heritiera* leaf and 2 fruit 3 root flange pneumatophores of *Heritiera* 4 *Xylocarpus granatus* flowers 5 *Xylocarpus* leaves 6 *Xylocarpus* fruit 7 *Xylocarpus* main trunk 8 radial horizontal root 9 emergent hoop pneumatophores.

turn to nutritive roots. Prop-roots, like stilts, up to 10 m long in a large tree, are put down from the upper branches, to take additional hold in the ground. Equipped with lenticels, these prop and strut-roots also serve as pneumatophores.

Other **Rhizophoraceae**, such as *Bruguiera* and *Ceriops*, have air-filled cable roots underground, from which thick loops arch up like knees to serve as pneumatophores. *Lumnitzera* produces clusters of thin, hoop-like pneumatophores, while in *Xylocarpus*, the radial roots have long vertical flanges running above ground (Fig. 23.4). Much larger flanged pneumatophores are formed by the branched buttresses of *Heritiera*.

Mangroves have in various ways become viviparous, allowing a water-borne seedling to settle at a relatively advanced stage. In *Aegiceras*, the small banana-shaped fruit germinates within the pericarp and its cotyledons form a long tube enclosing the plumule. *Avicennia* has a more advanced viviparity, with its bright green cotyledons folded double, and rupturing the pericarp while still on the tree (Fig. 23.4).

The Rhizophoraceae are characterised by their torpedo-shaped embryos, hanging in clusters from the mature inflorescence. These do not fall directly into the mud, but float horizontally, bending up to the vertical only when anchored. Successful colonisation requires a wide intertidal expanse, uncovered for some hours each day. In *Bruguiera*, the whole fruit falls from the plant with the hypocotyl still attached and piercing its apex, as the cigar-like embryo grows out. The cotyledons are fused only at the base, and their glandular epidermis takes nutrients from the pericarp to feed the hypocotyl.

Rhizophora has here evolved furthest. The hypocotyl (which in *R. stylosa* can reach a metre long) falls out of the fruit-base which stays attached to the plant. The cotyledons are completely fused, with the plumule entrenched in them, and suck in nutrients from the pericarp for the growth of the hypocotyl. The plumule often protrudes from the fruit and is demarcated by a groove from the hypocotyl.

The zenith of mangroves: Papua New Guinea

Illustrated in Fig. 23.1 is a full mangrove sequence lining the deep inlet of Galley Reach near Port Moresby, **Papua New Guinea**. The outermost fringes are young communities with *Avicennia marina* and *A. eucalyptifolia*. These seaward pioneers form low stands with profuse pneumatophores at the tide's edge. *Avicennia lanata* and *A. eucalyptifolia* are by contrast pioneers in places of lower salinity upstream.

Behind *Avicennia*, on slightly firmer ground, *Rhizophora apiculata* and *R. stylosa* grow to a high canopy into which *Bruguiera parviflora* and *Xylocarpus granatum* can enter. Upstream, the *Avicennia* front gives way to *Sonneratia*. The lanceolate-leafed *S. caseolaris* is virtually a freshwater species, though *S. alba* often grows to seaward of *Avicennia* at full salinity.

At its outermost edge the *Rhizophora* forest is dominated by *R. stylosa* inundated at each tide. Behind or upstream from this species grows the common estuarine *R. apiculata*. *R. stylosa*, never as tall as the other two, is found as isolated shrubs on sandy shores or reef flats.

Estuarine forests are rendered all but impenetrable

by the strut roots of *Rhizophora*, while nearest the water's edge *R. stylosa* also puts down long adventitious stilts from its branches. From further back, arched roots of *R. apiculata* leap forward in ongoing relays of three or even four orders. In front of *Rhizophora*, the bare mud slopes down to mean sea level, consolidating when it reaches an angle low enough to arrest its flow. In such forward sites grows *Avicennia eucalyptifolia*, confined mostly to creek edges.

The climax estuarine forest at Galley Reach consists of wide swamps, daily flooded from the creeks and gullies and reaching back far behind *Rhiziphora*. We find here the tall trunks of *Bruguiera*, with the main species *B. gymnorrhiza*. Under the closed canopy the air is humid and clammy. The consolidating floor is scattered with the conical chimneys of the mud-lobster *Thalassina anomala*. The mangrove fern, *Acrostichum aureum*, occupies much of the ground; where direct sunlight can get through, the holly-leaved shrub *Acanthus ilicifolius* becomes common. Climbing lianes include the heavy-scented *Hoya*, the leguminose *Derris* and the tendril-leafed monocotyledon *Flagellaria*.

Back beyond normal high tide, the tall forest enriches, leading ultimately to a dry vegetation. Here the canopy trees are *Bruguiera cylindrica* and *B. parviflora*, with the smaller rhizophoraceans *Ceriops tagal* and *C. decandra* beneath. *Bruguiera gymnorrhiza*, *Excoecaria agallocha* and *Xylocarpus granatum* are also present, along with narrow-buttressed *Heritiera littoralis*, that is here a smaller tree hardly reaching the canopy.

Further up the estuary, the transition at Galley Reach leads not to land but to freshwater swamps. The margin of *Rhizophora* here gives place to the small mangrove palm, *Nypa fruticans*, which can by itself dominate whole areas of swamp. *Nypa* becomes inundated at the highest spring tides and thrives when at least partly sub-merged. A horizontal trunk puts up tall fronds, and the massive root system is able to resist strong currents; the trunk cortex has air-spaces connected with large cavities in the buoyant leaf-bases. The leaves of *Nypa* reach 3–9 m long, with pinnae of up to a metre. The fruiting heads are 30 cm across, with clumps of woody carpels drying before shedding. The seed inside the carpel is white, like a pigeon's egg.[2]

Rocky shores

Sandy gravel in Papua New Guinea has its own distinct mangals. At Motupore Island, near **Port Moresby**, low *Rhizophora stylosa* stand in front of *Osbornia octodonta*, a small-leaved mangrove of the Family **Myrtaceae**. The seaward fringe is of *Avicennia marina* and *Sonneratia alba*. Typical also of stony shores is *Aegialites annulata*, belonging to the sea lavender family (**Plumbaginaceae**). *Aegiceras corniculatum*, never numerous, grows near the mouths of streams; *Lumnitzera littorea*, with small scarlet flowers, can generally be found in the same groupings.

Behind saltings

At Fairfax Harbour, Port Moresby, a mangrove shrubland is fringed by sand with the succulent chenopod *Tecticornia australasica*. The mangal is dominated towards the water's edge by the small rhizophoran *Ceriops tagal*, with *Avicennia marina*, *Lumntizera littorea* and in some spots *Sonneratia alba* and *Osbornia octodonta*. A mangrove of the Family **Rubiaceae**, *Scypiphora hydrophyllacea* grows here into good-sized trees. Further to landward there are small *Pluchea indica*, a shrubby mangrove of the Family **Compositae**.

CHAPTER TWENTY-FOUR
Coral Outposts

At the far extremes of the wide *Indo-West Pacific Province* (itself extending east to the Tuamotus) are some marginal islands, not in the full sense tropical, but each endowed with its complement of reef-forming stony corals. We shall describe here four such groups or single islands that have each been treated as separate provinces: *Norfolk-Lord Howe* to the south-west, *Hawaii* in the north, the biologically famous *Gala-pagos* off the Ecuador coast and crossed by the equator, and finally the isolated *Easter Island*, lying furthest east.

Norfolk-Lord Howe

Two small and isolated islands, together with Middleton Reef and Elizabeth Reef, are placed in their own Province of the *Indo-West Pacific Tropical Region*, lying in the path of the *East Australian Current*, a westerly derivative of the *South Equatorial Current*. Lord Howe Island — though more southerly — has good coral reefs and is rich in tropical fish. The shore coral at Norfolk Island is much slighter, and for both islands the winter surface temperatures are too low for a truly tropical marine biota, leaving some doubt that they belong truly within this Region.

Norfolk Island

Norfolk Island lies 660 km north of New Zealand and 900 km from Australia. Built of volcanic basalt, it is today a pasture-covered plateau with only small remnants of the original rich forest cover. More than half its perimeter is formed of precipitous cliffs under constant wave attack, so refracted as to bring the island's whole coastline into high exposure.

At **Point Ross**, where the shore strip of Fig. 24.1 was based, wave impact is reduced by submerged reefs. Under the fullest exposure, waves regularly move more than 3 m up and down, even on vertical slopes without breaking. On shores where the waves are able to break, the resultant white surge rises 5–10 m up a sloping face, fully inundating rock-stacks up to this height.

The zonation on steep basalt at **Point Ross** begins

with a **littoral fringe** of shiny black cyanobacteria with *Siphonaria normalis* as the sole limpet. The upper shore crab *Leptograpsus variegatus* runs freely over the rock, and shelters in crevices. *Nerita atramentosa* is common, with small numbers of the tropical *Nerita plicata*. The single littorinid is the conical *Bembicium flavescens*, familiar from the east Australian mainland. The **eulittoral** carries neither a barnacle zone, nor patelloid limpets, but has huge numbers of *Siphonaria normalis*. This limpet attains its greatest size in the **littoral fringe**, where smaller numbers of reproductive individuals reach 14 mm long.

In the **lower eulittoral**, at 3 mm length, its extraordinary densities can reach up to 16,000 sq m. Through the limpet's whole range, the total biomass (size x number) remains nearly constant.

Around the **mid-eulittoral**, crustose *Lithophyllum* increases and a crisp turf cover begins, with short *Laurencia*, *Gelidium* and *Gigartina*. Beyond this, in the **lower eulittoral** the *Lithophyllum* becomes corded and coral-like.

The **sublittoral fringe** has an almost total cover of wine-red *Pterocladiella pinnata*, up to 10 cm in length. Big stable blocks carry *Pterocladiella* on top, and large numbers of the urchin *Heliocidaris tuberculata* beneath.

In further shelter, in the lee of large promontories, there are wave-smoothed boulders girding the cliffs. Where small pans have been eroded in bedrock platforms, *Hormosira banksii* is often to be found in local shelter, sometimes mingled with small soft corals.

Stable boulders with blue-green algae and *Siphonaria* on top can become cemented at their lower reaches with a mortar of *Lithophyllum*, tunnelled and hollowed by the vermetid *Dendropoma* species. At low levels very large boulders carry a stubble of a short-fronded *Sargassum spinifex*, among or above *Pterocladiella*, with small cushion stars, *Patiriella exigua*, on the pink-painted sides.

Underneath the smaller boulders lives the snail *Hinea brasiliana*, as well as *Nerita atramentosa* and the anemone *Actinia tenebrosa*. Along with the crab *Leptograpsus variegatus* a smaller, more numerous *Cyclo-*

Fig. 24.1 Zoned shore at Point Ross, Norfolk Island

a–b black lichen, *Verrucaria*; b–c *Siphonaria normalis* and *Nerita*; c–d *Lithophyllum* and algal turf; d–e *Pterocladiella* and *Sargassum*

1 *Bembicium flavescens* 2 *Hinea brasiliana* 3 *Pterocladiella capillacea* 4 *Lithophyllum* eroding *Dendropoma* 5 *Nerita atramentosa* 6 *Siphonaria normalis* 7 *Sargassum spinifex* 8 detail of Norfolk pine, *Araucaria heterophylla*.

grapsus occurs under boulders. A *Tetraclitella* barnacle lives in the boulder shade, while among the impacted vermetids, there are giant barnacles, *Megabalanus tintinnabulum linzei*.

Stony corals may be found scattered thinly among the lowest stable boulders, but the strongest coral communities live in the lagoon in front of **Kingston**. The 39 reef coral species include common *Acropora hyacinthus*, *A. valida*, *Pocillopora damicornis*, *Montipora turgescens*, *Porites lichen*, *Plesiastrea versipora*, *Goniastrea australiensis* and *Montastrea curta*.

Lord Howe Island

Lying 650 km from east Australia, at latitude 31 degrees 3 minutes south, **Lord Howe** possesses perhaps the world's most southerly coral reef. Elizabeth Pope called the island — like Hawaii — a 'Bermuda of the Pacific', and some have hailed it as the world's most beautiful island. A large accession of tropical and subtropical species has been brought in by currents from Great Barrier Reef and New Caledonia, while a significant temperate element comes from further south. Surface sea temperatures range around 25°C in summer (rising

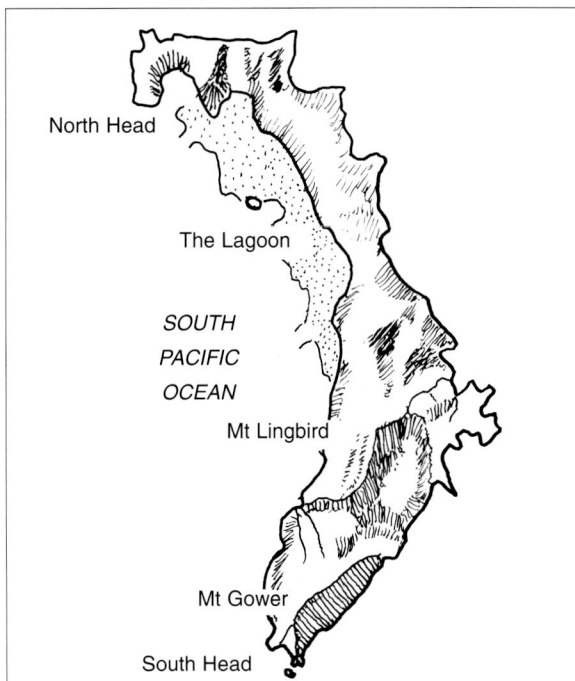

Fig. 24.2
Lord Howe Island
(**Above**) map.
(**Right**) *Lagunaria patersonii*, the hibiscus of Norfolk Island and Lord Howe Island.

to 27.5°C in the lagoon) and 17°C in winter.

Lord Howe is a boomerang-shaped island 12 km long and running almost north-south. There are two southern peaks, Mt Lingbird (765 m) and Mt Gower (866 m), that dominate its southern end and form the backdrop to a wide lagoon lying to the west, and closed off by a coral reef like the string to a bow. The lagoon is mainly sand-bottomed, 6 km long and 1.5 km across, and has rich expanses of living coral. The sandy beaches of Lord Howe are chiefly found around the lagoon. Most of the outer volcanic shoreline is of steep cliffs, continuing down to a 20 m subtidal drop, and providing a wide variety of caves, ledges, archways and fissures.

There is an extensive fringing reef along the west side of Lord Howe Island, which encloses a shallow sandy lagoon. Rich coral growth is present here locally, particularly in deeper parts of the lagoon and in erosional channels crossing the reef. Coral species diversity is much lower than on the Great Barrier Reef, but 83 reef coral species are recorded,[1] about double the number known from Norfolk Island. Most reef corals at Lord Howe Island have explanate, columnar or encrusting growth form, but arborescent *Acropora* and *Pocillopora* are present in some sheltered sites within the lagoon.

Our basic knowledge of Lord Howe fish went back

to F.H. Talbot's Australian Museum Expedition in 1973. Nearly 300 species were collected, representing more than 100 families, and with 186 new records and eight species fresh to science. The total Lord Howe list of coastal fishes has been raised by Malcolm Francis and colleagues to at least 430.[2] The main families are the wrasses (**Labridae**) with 51 species, the damsels (**Pomacentridae**) with 30, the **Gobiidae** with 24, and the butterfly fishes (**Chaetodontidae**) with 24. Almost 50% are relatively widespread in the tropical Pacific, with only 4% endemic to Lord Howe. The absence at certain times of particular species has been attributed to the vagaries of southern currents, while annual fluctuations of sea temperature are periodically reflected in faunal composition. Many have been recorded only rarely, as the result of eggs and larvae carried to Lord Howe from the Great Barrier Reef.

The shore zonation of Lord Howe, first surveyed by Elizabeth Pope in 1960, shows a biota richer and more tropical than at Norfolk Island. The **littoral fringe** is of black cyanobacteria and *Verrucaria*. Its periwinkles are the east Australian *Nodilittorina pyramidalis* and *N. unifasciata*, with the addition of the tropical *Littorina coccinea*. The **eulittoral**, with the long distance from the mainland, is — like Norfolk Island — almost destitute of barnacles, save for a sparse scatter of the Australian *Tesseropora rosea*.

Lord Howe has on its sheltered face a strong **mid-eulittoral** rock oyster belt of *Saccostrea cucullata*, and below this a vermetid band of a *Dendropoma* species. *Bembicium flavescens*, *Nerita atramentosa* and *N. plicata* are all present, with the true limpet *Cellana howensis*, as well as several *Siphonaria* species. Common lower eulittoral molluscs are *Dicathais orbita*, several cowries, and the Pacific-wide black and white-checked *Conus ebraeus*. The red anemone *Actinia tenebrosa* lives in pools, with the urchin *Echinometra mathaei* in depressions, and colonial zoanthids including *Palythoa* in sheets on the rock-face.

The **sublittoral fringe**, with its encrusting lithophylla, has a tropical assemblage of green *Caulerpa*, and turtle-weed, *Chlorodesmis fastigiata*, along with several *Sargassum* species. As well as scleractinian corals, there are alcyonaceans: the organ-pipe *Tubipora musica*, the soft corals *Lobophytum* and *Sarcophyton*, as well as a *Xenia* species. The large anemones *Stoichactis* and *Radianthus* have their attendant fishes. The echinoids include the black, needle-spined *Diadema*, red-tipped *Heliocidaris tuberculata* and *Tripneustes gratilla*. There are black *Holothuria* and yellow *Stichopus* species, with an abundance of asteroids, ophiuroids and crinoids.

Notable tropical molluscs are the giant clam *Tridacna maxima*, *Turbo cepoides* and the large cowry *Cypraea tigris*.

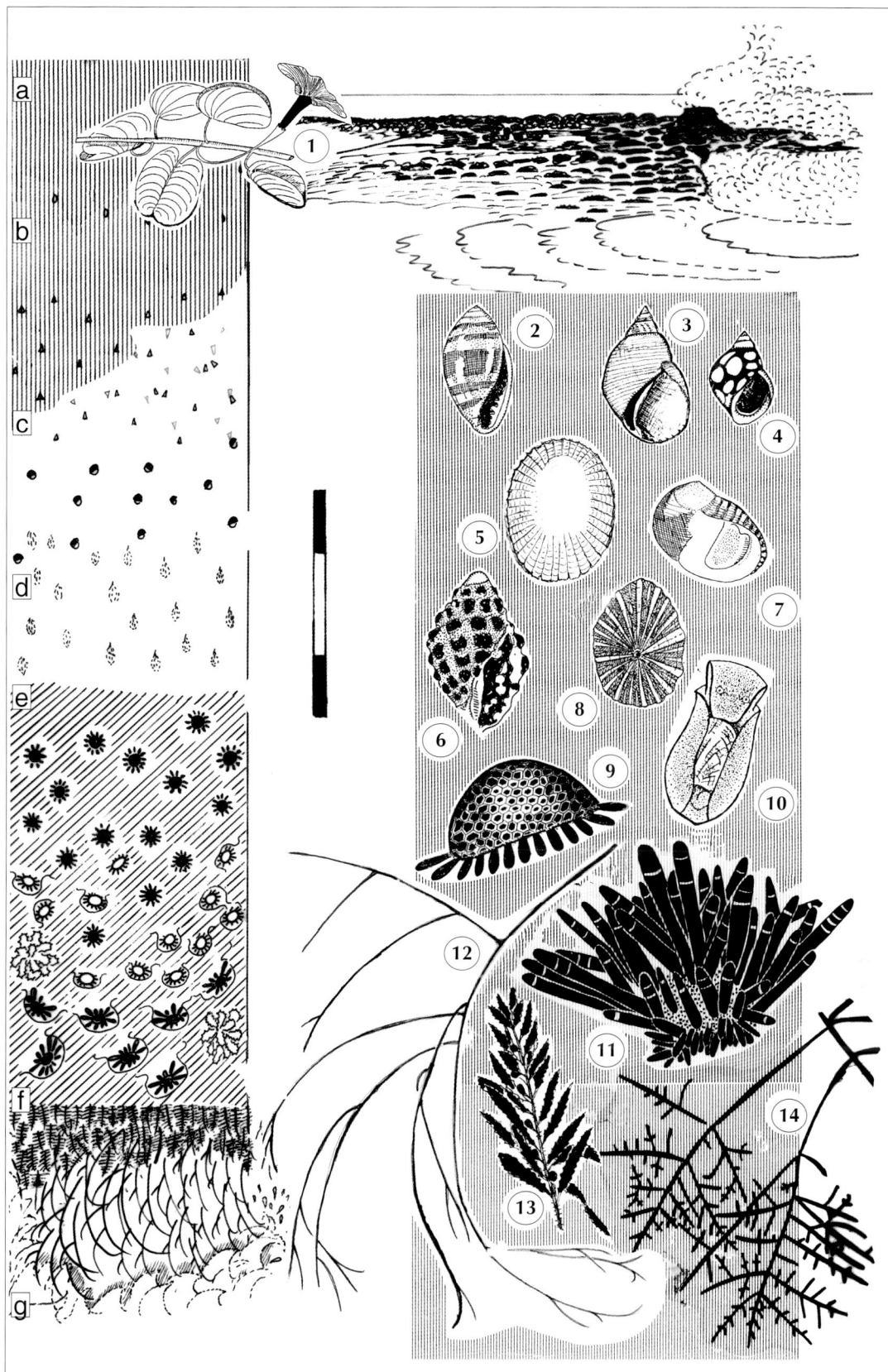

Fig. 24.3 Zoned shore at Garden Island, Kauai, Hawaii

The column (**left**) represents the wave-exposed drop of the basalt lava flow. Vertical bar = actual tidal range, (white) neap and (black) spring.

a–b *Ipomaea*; b–c littorines, *Isognomon*, ellobiids; c–d *Nerita*, *Siphonaria* and *Cellana*; d–e *Ectocarpus* and *Smaragdinella* *Sargassum* and *Acanthophora*; e–f *Podophora*, *Echinometra*, *Heterocentrotus*; f–g *Chnoospora* and *Pterocladiella*

1 *Ipomea pescaprae* 2 *Melampus fasciatus* 3 *Littorina pintado* 4 *Nodilittorina picta* 5 *Cellana sandwicensis* 6 *Morula granulata* 7 *Nerita picea* 8 *Siphonaria normalis* 9 *Podophora atratus* 10 *Smaragdinella viridis* 11 *Heterocentrotus mammillatus* 12 *Chnoospora minima* 13 *Sargassum* sp. 14 *Pterocladiella capillacea*.

Hawaiian Province

The 20 islands of this small Province form a Pliocene volcanic archipelago, extending between 18.5–23 degrees north and 154–62 degrees west, and coming under the influence of the east-west *North Equatorial Current*.

The marine fauna of Hawaii is very rich by comparison with the temperate, and much of it has been well described, with Hawaii today better documented than any other near-tropical Province. Its relatively high endemism (shore-fishes 34%, asteroids and ophiuroids 30%, crangonid crustaceans 40% and molluscs 20%) is attributed by Briggs (1974) to a long and stable climatic history, with strong geographical isolation. Except for Johnston Island 720 km south, the Hawaiian Chain is sundered from Polynesia by a deep water gap of some 1400 km.

The shore zonation (Fig. 24.3) is represented from a basaltic lava flow at Garden Island, near **Koloa**, on the island of Kauai. The intertidal shore is lacking in corals though patches of reef-forming Scleractinia are plentiful in the close-adjoining sublittoral. The basalt platform some 50 m wide falls steeply at the seaward edge, coming here under the full assault of breaking waves, with mounting surge and — at high levels — the drenching effects of splash and spray. The height of a storm-wave from trough to crest at most times exceeds the whole amplitude of the day's tide.

The back of the platform forms a **supralittoral zone** with the familiar Pacific *Ipomaea pescaprae*, oval-leaved and mauve-flowered, along with succulent aizoaceans and chenopods. The middle of the bench — with sharp-edged basalt — forms a wide, highly insolated **littoral fringe**. Small, fast-running *Cyclograpsus* crabs live here, as well as the two periwinkles, *Littorina pintado* and *Nodilittorina picta*. At this same level, shallow concavities reveal the tiny trochiform littorine, *Peasiella tantilla*. Loose scoria boulders carry on top the small bivalve *Isognomon perna* inserted into crevices. Underneath there will be found a typical gastropod faunule with an *Assiminea* and hosts of small ellobiids: *Laemodonta octanfracta*, *Pedipes sandwicensis*, and — ranging out beyond the boulders — *Melampus castanea* and *M. semiplicata*.

The **upper eulittoral** begins on bare, smooth basalt, with *Nerita picea* common, as well as *Siphonaria normalis* on dry rock and *Cellana sandwicensis* in small splash pools. Zoning barnacles are nowhere to be seen. *Nerita* and the limpets continue abundant through the **mid-eulittoral**, where the surface is constantly wave-drenched, with its prevailing colour golden yellow from the tufts of the brown filamentous alga, *Hincksia breviarticulatus*. The thaid *Morula granulata* is a predator upon other gastropods.

The small bottle-green cephalaspid, *Smaragdinella viridis*, is a notable species in this zone. As an algal browser with a semi-internal bubble shell, it is found in moderate wave energy through much of the tropical Pacific; its life under strong surge — unusual for an opisthobranch — is made possible by the fast-attaching sole and the contour of the enveloping parapodia that streamline the body against strong surge. Of comparable habit among succulent algae at low water is a much larger opisthobranch, the aplysiomorph *Dolabrifera dolabrifera*.

In the **lower eulittoral** zone — regularly visible at the drawback of waves — the surface is transformed to coralline pink, with a small black echinoid *Podophora atratus* studding the rock everywhere under full wave attack. This urchin owes its secure hold to its smooth hemispherical test with the spines replaced by a mosaic of close-fitting plates. Around this dome a fringe of spathulate marginal spines presses against the rock. The flat oral surface has strong tube-feet as fast-holding as in any surge seastar. Three species of echinoid scour deep into the *Lithophyllum* crust: *Echinometra mathaei* and *E. oblonga*, with the heavy-spined slate pencil urchin, *Heterocentrotus mammillatus*, the last generally confined to the sublittoral fringe under maximum surge.

The crustal pink is relieved by the golden brown of two species of *Sargassum*, a short one with stiff, holly-spined leaflets, and the other with smooth leaves in long tresses up to 25 cm. Tufts of the olive-brown rhodophyte *Acanthophora*, with stiff, spinose branch-lets, the pale green rhizomes of *Caulerpa* and bright green *Ulva* add their variant hues. A wide mouthed *Drupa* species attaches to the pink surface, well-streamlined against wave energy.

With the **sublittoral fringe** the ground colour again changes, with the rich wine-red of bushy *Pterocladiella capillacea*, cosmopolitan through most warm and subtropical seas. Where the wave-line briefly recedes, glimpses can be obtained of the yellow *Chnoospora minima*, a tropical brown alga of very simplified structure, with linear, dichotomous branching.

American tropical: Galapagos Islands

All too little is yet known about any of the mainland shores of the *East Pacific Tropical Region*. Two Provinces have been recognised, primarily on the basis of their in-shore fishes. The *Mexican Province* extends south from Magdalena Bay at the opening of the Gulf of California to the Gulf of Tuhuantepec (latitude 16 degrees north). The *Panamanian Province* then follows, with its southern boundary at the Gulf of Guayaquil, at

Fig. 24.4 A zoned shore at the Galapagos Islands
A composite picture has been derived from descriptions by several authors. The marine iguana *Amblyrhynchus* is shown sunning itself in the supralittoral zone from which it descends to the intertidal.
a–b *Nodilittorina, Nerita, Thais* and *Grapsus;* b–c *Tetraclita, Cerithium* and *Thais;* c–d *Lithothamnion* and large *Megabalanus tintinnabulum;* d–e *Sargassum, Blossevillea* and urchins
1 *Amblyrhynchus cristatus* 2 *Nodilittorina galapagensis* 3 *Nerita scabricostata* 4 *Grapsus grapsus* 5 *Thais patula* 6 *Thais melones* 7 *Cerithium adustum* 8 *Tetraclita squamosa milleporosa* 9 *Megabalanus tintinnabulum* 10 *Sabia trigona*, shell interior and a side profile (with scar of a smaller *Sabia*) 11 *Sargassum galapagensis* 12 jointed *Amphiroa* species 13 *Blossevillea galapagensis*.

a latitude of only 3 degrees south, where the Peru Current from the south draws away from the mainland coast.

The **Galapagos Islands,** lying across the equator (0 degrees north to 1.3 degrees south) at 900 km from Ecuador, have sufficient endemism to justify their status as a separate *Province*. A majority of their marine species, however, have probably come from the Panamanian coast (with fewer across the Pacific from the west) or from the warm temperate American mainland, like the small Galapagos penguin, *Spheniscus mendiculus.*

There are incipient coral reefs, as also on the American mainland, and mangals of American provenance, with the red mangrove *Rhizophora mangle*, the black *Avicennia germinans*, the white *Laguncularia racemosa*, and the button mangrove *Conocarpus erectus*.

The Galapagos group is of Pliocene volcanic origin, constituting a cluster of 13 large and numerous small islands. All retain the sharp youthful contours that Herman Melville captured with his 'clinker-bound tumbled masses of blackish or greenish stuff, like the dross of an iron furnace forming dark clefts and caves here and there'. Since Charles Darwin first made them known to science, the Galapagos have been acclaimed as a classical laboratory of evolution. Yet it is only recently that their intertidal shores have been well assessed. Wellington has given the best concise account of the major marine habitats including the shores and their subtidal corals.[3] Joel Hedgpeth's informal and informative ramble round the shores of a number of the islands has also contributed to the composite strip presented in Fig. 24.4.[4]

The Galapagos intertidal is at a first sighting bizarre rather than rich or diverse. The dearth of smooth surfaces, and the infrequency of tide pools (even these becoming tepid on rock rapidly rising to lethal equatorial temperatures) have resulted in an impoverished upper shore. For richer communities one must look beneath boulder cover, on shaded vertical walls or in crevices. The Galapagos tides are almost symmetrically semi-diurnal, with an extreme range of 2 m, and a mean of about 1.6 m.

Roughness of terrain or over-remoteness for larval transport could account for the absence of any chthamalid barnacles in the **upper eulittoral**. There is likewise a complete lack of limpets, either patellid or siphonariid, though a small periwinkle, *Nodilittorina galapagensis*, is to be found in pools or at the bottom of fissures. This largely barren stretch is chiefly the preserve of the grazing snail *Nerita scabricosta*, reaching to 2 m above low water. Active only in shade, this species is thick-shelled and protected with a tight-sealed operculum. At this same rather high level, there is a predatory thaid, *Thais patula*.

Any macroscopic algae at eulittoral level are regularly grazed clean or reduced to a short stubble. *Ulva lobata* is a favourite food of the marine iguana, *Amblyrhynchus cristatus*, and this large lizard — classically the best known animal of the Galapagos intertidal — descends to the shore whenever the tide is out, with obvious effects on its zoning pattern. Equally conspicuous across the shore, and almost as fully amphibious, is the reddish 'Sally Light-foot' crab, *Grapsus grapsus*, running about freely on its grazing forays.

The first well-marked zoning begins at the **middle eulittoral** with a band of *Tetraclita squamosa mille-*

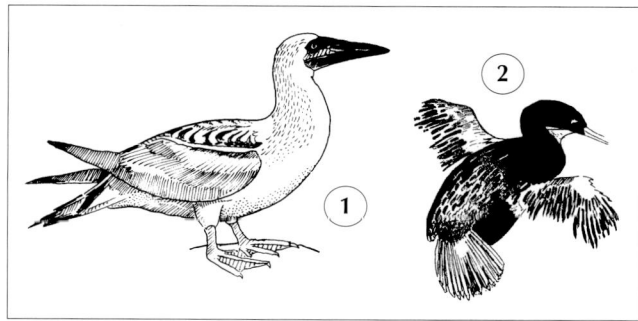

Fig. 24.5 Galapagos birds
1 blue-faced booby, *Sula dactylatra* 2 flightless cormorant, *Phalacrocorax harrisi*.

porosa. This is the principal barnacle of the Galapagian shore, abutting below with a cream to pink *Lithothamnion* crust, conspicuous round most of the islands. The common coralline is a large, coarse-jointed *Amphiroa*. Clusters of the most abundant gastropod, *Cerithium adustum*, live at the barnacle level, along with a second-carnivore, *Thais melones*.

At the bottom of the *Tetraclita* zone and invading the *Lithothamnion* there is sometimes a cover of greenish zoanthids. These may be interspersed with a brighter green anemone, while in crevices among *Tetraclita* is found a small red *Actinia*. Sand tubes of the sabellariid worm *Idanthyrsus* may form patches in the *Lithothamnion* zone. In the lowest intertidal, growing singly or clustered, lives the largest barnacle of this shore, the wide-ranging *Megabalanus tintinnabulum*. Beyond low water, the most abundant invertebrate is generally the sessile cup-shaped gastropod *Sabia trigona*.

In the **sublittoral fringe**, the algae are seen to be greatly enriched. In early days, W.R. Taylor reported a tropical assemblage with species of *Ectocarpus*, *Padina*, *Gelidium*, *Lithophyllum*, *Spatoglossum*, *Codium* and *Caulerpa*. The largest and most generally dominant seaweed is the brown *Sargassum galapagensis*, well distinguished by the chains of bladders incorporated in its pointed filiform segments. On rocky outposts exposed to increasing surf, *Blossevillea galapagensis* seems generally to take over the dominance.

Echinoderms are an important element of the pools and open surfaces. The pencil urchin *Eucidaris thouarsii* is common in pools up to half-tide level, together with the green *Lytechinus semituberculatus*, bedecked with fragments of shell and pebbles. At lower levels and less common is the almost black *Echinometra vanbrunti*. The sunstar *Heliaster cumingi* can be found scattered on rocks up to mid-tide level.

Extending subtidally is the anemone *Anthopleura dovii*, along with sea-fans *Pacifigorgia*. Grazing here becomes intense, with the work of urchins, together with the surgeon fish *Prionurus laticlavius* and the parrotfish *Scabrus ghobban*.

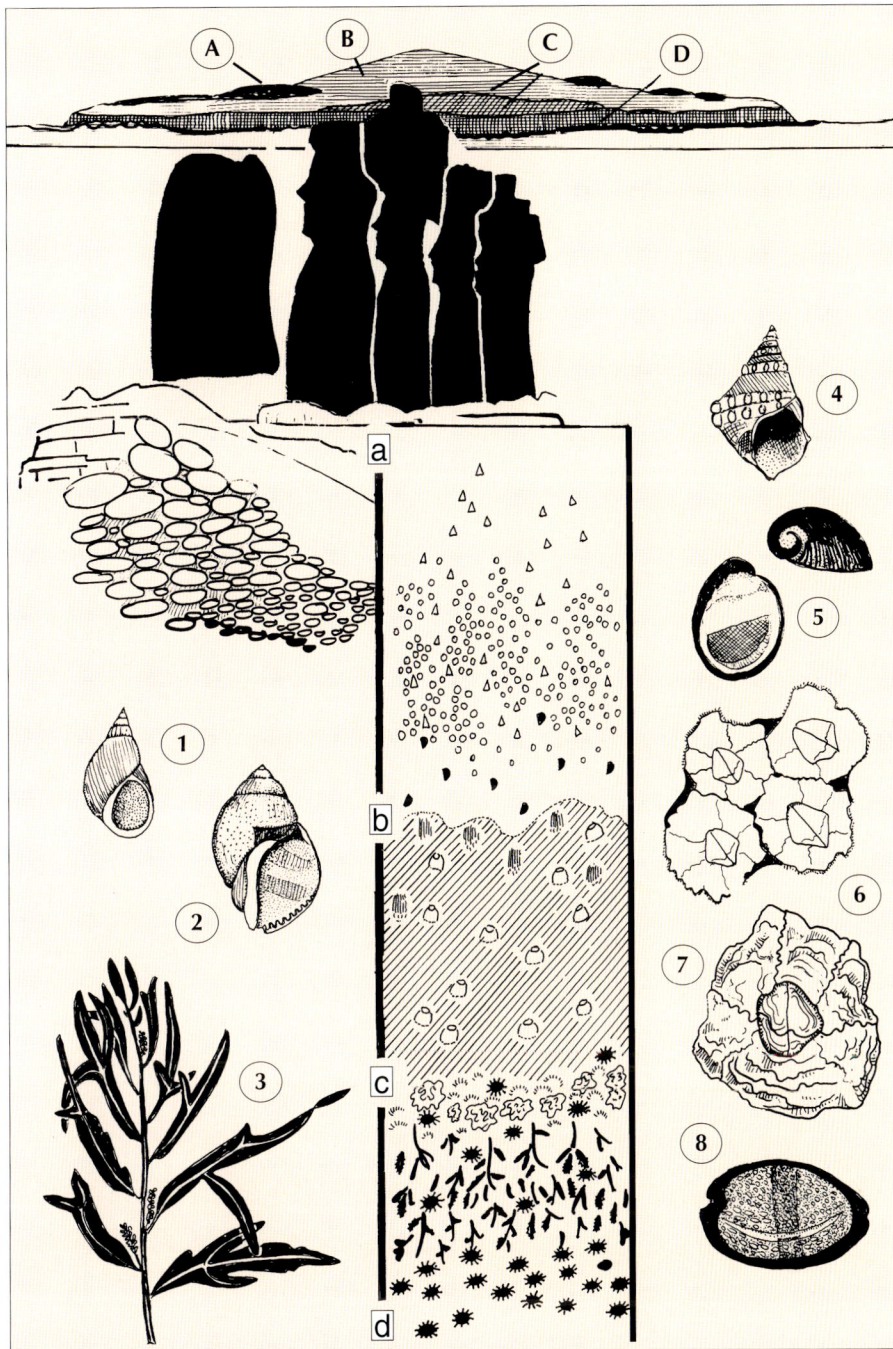

Fig. 24.6 Zonation of exposed shore at Easter Island
The profile (**top**) shows A. remnants of original forest B. grassed pasture C. basaltic cliff-line D. the zoned shore.
a–b (upper shore) *Cthamalus* and *Nodilittorina*; b–c (mid-shore) *Euraphia*, coralline crust and algal turf; c–d *Sargassum*, *Ulva* and urchins
1 *Hinea akuana* 2 *Neothais nesiotes* 3 *Sargassum scottsbergi* 4 *Nodilittorina pyramidalis pascua* in its local subspecies
5 *Nerita morio* 6 *Chthamalus belyaevi* 7 *Euraphia* sp. 8 *Cypraea caputserpentis*.
(From observations by the late Brian Foster.)

Wellington's study includes a detailed account of the incipient fringing reefs of the Galapagos. The reef-building corals are far less diverse than in the West Pacific, and include chiefly the branching *Pocillopora capitata*, and the massive *Pavona clavus*, *Pavona gigantea* and *Porites lobata*.

The biological interactions controlling the rate and extent of coral growth have been well investigated by Glynn and fellow workers. The thick-spined urchin *Eucidaris* (unlike the same species in mainland Panama) here lives directly on the tissues of corals, whose density is reduced where urchin numbers are high. In turn, the predators of the urchins are puffer fish and trigger fish, much lower in density than on the mainland.

In shallow waters, the damsel fishes, *Stegastes leucorus*, *S. arcifrons* and *Nexilosus latifrons*, maintain algal gardens for feeding, nesting and breeding on non-coral surfaces. These are defended against the entry of herbivores (parrotfish, *Scarus ghobban*; surgeon fish, *Prionurus laticlavius*) or egg predators, especially *Arothron*. Damsel fish may also at times eject urchins that enter their territories. Such garden shelters are important refuges for damsels — and sometimes urchins — against predatory fish. The damsel fishes' ejection behaviour has also the indirect effect of reducing the grazing of coral near their territories, so facilitating reef development.

Shallow water planktivorous fish — *Paranthias colonus*, *Chromis atrilobata* and *Holacanthus passer* — form the food of the blue-footed booby (*Sula dactylatra*). Numerous small benthic fish are taken by the flightless cormorant (*Phalacrocorax harrisi*), while the Galapagos penguin (*Spheniscus mendiculus*) thrives on schools of fish-fry.

Easter Island

This single island — perhaps the world's most isolated piece of land — constitutes by itself the furthest flung Province of the *Indo-West Pacific Region*. Its chief claim to speculation has been in archaeology, with its renowned shoreline statues. Geologically young at 2.5 million years, Easter Island was formed by eruptions on or near the East Pacific Rise (now moving eastwards at 10 cm per annum towards subduction beneath South America). As the island continues to sink lower down the mid-oceanic slope, and aerial erosion proceeds, it can be predicted to become eventually submarine.

Today's shoreline is heavily dissected, with pitted basalt eroded into platforms of varying width, backed with loose boulders at the base of crumbling lava cliffs. Strong waves are generated by winds over thousands of kilometres, and refracted so as maximally to pound the shore from any direction. Only two small white sand beaches exist, and there are no offshore coral reefs.

In 1976 my late colleague Brian Foster visited Easter Island and his generous help has made this account possible.[5] The intertidal rocks are generally black, relieved only by a zone of pale barnacles, *Chthamalus belyaevi*, or lower down by two species of eulittoral algae, the golden brown *Sargassum scottsbergi* and a bright green *Ulva*, bounded above by the low tidal splash-line. The chthamalid zone is within constant reach of surge and backwash. Above this, under the influence of splash and spray, extends the periwinkle *Nodilittorina pyramidalis* in its local subspecies *pascua*, most numerous among the barnacles, where it reaches 600 sq m.

The middle shore carries a veneer of encrusting corallines, with black *Nerita morio* (at 150 sq m), a scatter of a larger pink barnacle, *Euraphia devaneyi*, and a short algal turf.

On the lower shore *Sargassum skottsbergi* dominates the sublittoral fringe, growing upon rock pitted with the abraded holes of the urchin *Echinometra insularis*, which is the most important contributor to the bio-erosion of the shore.

Among the boulders at the head of small embayments are dense aggregations of the snail *Hinea akuana*, yellowish with a blue apex. Large crabs, *Pachygrapsus transversus*, live where shell sand accumulates around low tidal boulders. There are two species of black holothurians, a pencil-thick *Polycheira* (9 cm), and a *Holothuria* (17 cm), with adherent white sand grains, and also the forciculate seastar *Astrostole paschae*. The gastropods include *Mitra flavocingulata*, *Neothais nesiotes* and *Cypraea caputserpentis*.[6]

Intertidal pools at Easter Island present us with hermatypic coral, *Porites lobata*, juvenile *Pachygrapsus*, hermit crabs, carpets of *Zoanthus rapanuiensis* and the brown alga *Lobophora variegata*. Vermetid gastropods are also found here, along with the small limpet-like *Hipponyx conicus*. In deep, shady retreats live the tropic-wide chiton *Acanthopleura gemmata* and the red shade anemone *Actinoides rapanuiensis*. Under loose stones there is a hypofauna of ascidians, sponges, pink sessile foraminifera, *Spirorbis*, and further vermetid gastropods. Mobile species include two brittle stars, the xanthid crab *Trapezia areolata* and brightly coloured *Gnathophyllum* shrimps.

In contrast with the reeds in the crater lake (*Schoenoplectus californicus*) evidently from South America (where the shore statues have their likely affinity) the marine biota of Easter Island is predominantly Polynesian. The high proportion of endemic species, especially fish and molluscs, is, however, an indication of extreme isolation.

The diversity of the marine biota of the isolated South Pacific islands (c. 30 degrees south) decreases progressively from west to east (e.g. reef corals: Lord Howe [83 species], Norfolk [39], Kermadecs [17], Pitcairn [17], Easter [10]; coastal fishes: Lord Howe [433], Norfolk [254], Kermadecs [145], Easter [123]) reflecting their current dispersal from west to east.[7]

The Easter Island intertidal community was found by Foster to be neither abundant in biomass, dense in surface coverage nor of high species diversity within any one group. This is in itself a situation of biogeographic interest.

The fate of the island, in one of the least productive spots of the great South-east Pacific gyre, must eventually be erosion and submersion. The limited tidal rise and fall creates little variety of habitat, and the species that have colonised it or arisen there during the last 2.5 million years will have to disperse to new volcanic islands if they are to survive with their endemism.

Classification of Seashore Organisms

Kingdom Bacteria
Phylum Cyanobacteria (blue-green 'algae')

Kingdom Protozoa
Phylum Dinozoa (dinoflagellates)

Phylum Ciliophora (ciliates)

Phylum Foraminifera

Phylum Amoebozoa (amoeba)

Phylum Radiozoa (radiolaria)

Kingdom Fungi
Includes lichenised fungi (lichens)

Kingdom Chromista
Phylum Ochrophyta
Subphylum Phaeista (brown algae = seaweeds)
Subphylum Diatomeae (diatoms)

Phylum Haptophyta (coccoliths)

Kingdom Plantae
Phylum Rhodophyta (red algae = seaweeds)
Class Bangiophyceae
Class Florideophyceae

Phylum Chlorophyta (green algae = seaweeds)

Phylum Bryophyta
Subphylum Hepaticae (liverworts)
Subphylum Anthocerotae (hornworts)
Subphylum Musci (mosses)

Phylum Tracheophyta
Subphylum Pteridophyta (ferns)

Subphylum Spermatophyta
Infraphylum Gymnospermae (conifers)
Infraphylum Angiospermae (angiosperms)
Class Dictyledones (dicots)
Class Monocotyledones (monocots)

Kingdom Animalia
Phylum Porifera (sponges)
Class Demospongiae (demosponges)
Class Calcarea (calcareous sponges)
Class Hexactinellida (glass sponges)

Phylum Cnidaria (Coelenterata)
Class Anthozoa
Subclass Octocorallia (octocorals)
Order Alcyonacea (soft corals, gorgonians)
Order Pennatulacca (sea-pens)
Subclass Hexacorallia
Order Ceriantharia (tube anemones)
Order Actiniaria (true anemones)
Order Corallimorpharia (jewel anemones)
Order Zoantharia (zooanthid anemones)
Order Scleractinia (stony corals)
Order Antipatharia (black corals)
Class Hydrozoa
Order Leptothecata (thecate hydroids)
Order Anthoathecata (athecate hydroids and stylasterine corals)
Class Scyphozoa (jellyfish)

Phylum Ctenophora (comb jellies)

Phylum Bryozoa (moss animals)
Order Ctenostomata
Order Cheilostomata
Order Cyclostomata

Phylum Entoprocta (nodding animals)

Phylum Mollusca
Class Polyplacophora (chitons)
Class Gastropoda (snails, slugs)

Class Bivalvia (bivalves)
Class Scaphopoda (tusk shells)
Class Cephalopoda (squid, cuttlefish, octopus)

Phylum Brachiopoda (lamp shells)
Class Articulata
Class Inarticulata

Phylum Phoronida (phoronid worms)

Phylum Sipuncula (peanut worms)

Phylum Annelida (segmented worms)
Subphylum Polychaeta (marine worms)
Subphylum Clitellata (marine leeches)
Subphylum Echiura (spoon worms)
Subphylum Pogonophora (beard worms)

Phylum Nemertea (ribbon worms)

Phylum Arthropoda
Subphyllum Cheliceromorpha
Class Pycnogonida (sea spiders)
Class Arachnida (spiders, mites, ticks, scorpions)
Subphylum Uniramia
Class Insecta (insects)
Class Chilopoda (centipedes)
Subphylum Crustacea
Class Malacostraca
Order Decapoda (crabs, shrimps, crayfish)
Order Isopoda (sea lice)

Order Amphipoda (sand hoppers)
Class Cirripedia (barnacles)
Class Copepoda (copepods)
Class Ostracoda (ostracods)

Phylum Nematoda (fish parasites — flukes)

Phylum Platyhelminthes
Order Polycladida (marine flat worms)

Phylum Hemichordata (acorn worms)

Phylum Echinodermata
Class Crinoidea (sea lilies)
Class Asteroidea (seastars)
Class Ophiuroidea (brittle stars, basket stars)
Class Echinoidea (sea eggs, sand dollars)
Class Holothuroidea (sea cucumbers)

Phylum Chordata
Subphylum Urochordata
Class Ascidiacea (tunicates or sea-squirts)
Subphylum Cephalochordata (e.g. Amphioxus)
Subphylum Vertebrata
Class Chondrichthyes (sharks, rays)
Class Osteichthyes (bony fish)
Class Reptilia (lizards, skinks, geckos)
Class Aves (birds)
Class Mammalia (seals, dolphins, etc)

References

Chapter 1

1 Morton, J.E. & Miller, M.C., *The New Zealand Sea Shore*, Collins, London & Auckland, 1968.

2 Dakin, W.J., *Australian Seashores*, Angus & Robertson, Sydney, 1952.

3 Cranwell, Lucy M. & Moore, Lucy B., Intertidal communities of the Poor Knights Islands, New Zealand. *Transactions of the Royal Society of New Zealand* 67: 375–407, 1938.

4 Stephenson, T.A., The constitution of the intertidal fauna and flora of South Africa. I. *Journal of the Linnaean Society. London* 40: 487–536, 1939; II *Ann. Natal Museum* 10: 26l–358, 1944; III *Ann. Natal Museum* 11: 207–324, 1945.

5 Stephenson, T.A. & Stephenson, Anne, The universal features of zonation between tidemarks on rocky coasts. *Journal of Ecology* 37: 289–305, 1949.

6 Stephenson, T.A. & Stephenson, Anne, *Life Between Tidemarks on Rocky Shores*, Freeman, San Francisco, 1972.

Lewis, J.R., *Ecology of Rocky Shores*, London, England, University Press, 1964.

7 Oliver, W.R.B., Marine littoral plant and animal communities in New Zealand. *Transactions of the New Zealand Institute* 54: 495–545, 1923.

8 Cranwell, Lucy M. & Moore, Lucy B. Intertidal communities of the Poor Knights Islands, New Zealand. *Transactions of the Royal Society of New Zealand* 67: 375–407, 1938.

9 Dellow, Vivienne, Intertidal Ecology at Narrow Neck Reef, Auckland, New Zealand. *Pacific Science* 4: 355–374, 1950.

Dellow, Vivienne, Marine Algal Ecology of the Hauraki Gulf, New Zealand. *Transactions of the Royal Society of New Zealand* 83: 1–91, 1955.

10 Knox, G.A., The intertidal ecology of Taylor's Mistake, Banks Peninsula. *Transactions of the Royal Society of New Zealand* 81: 189–220, 1953.

11 Batham, Elizabeth J., Ecology of a southern New Zealand sheltered rocky shore. *Transactions of the Royal Society of New Zealand* 84: 447–465, 1956.

12 Batham, Elizabeth J., Ecology of a southern New Zealand rocky exposed shore at Little Papanui, Otago Peninsula. *Transactions of the Royal Society of New Zealand* 85: 647–658, 1958.

13 Powell, A.W.B., *New Zealand Mollusca*, Collins, Auckland, 1979.

14 Bergquist, P.R., *Sponges*, Hutchinson, London, 1978.

15 Adams, Nancy, *Seaweeds of New Zealand: An illustrated guide*. Canterbury University Press, 1994.

Chapter 2

1 Foster, Brian, Marine Fauna of New Zealand: Barnacles (Cirripedea: Thoracica). *New Zealand Oceanographic Institute Memoir* 69, 1976, 175 pp.

2 Luckens, P.A., Settlement and succession on rocky shores at Auckland, North Island, New Zealand. *New Zealand Oceanographic Institute Memoir* 70, 1976, 64 pp.

Chapter 3

1 Dellow, Vivienne, Marine algal ecology of the Hauraki Gulf, New Zealand. *Transactions of the Royal Society of New Zealand* 83: 1–91, 1955.

2 Walsby, J.R., Population variations in the grazing turbinid *Lunella smaragdus*. *New Zealand Journal of Marine and Freshwater Research* 11, 2: 211–238, 1977.

3 Stewart, C., de Mora, S.J., Jones, M.R.L. & Miller, M.C., Imposex in New Zealand neogastropods. *Marine Pollution Bulletin* 24: 204–209, 1992.

4 Hay, C.H. & Luckens, P.A., The Asian kelp *Undaria pinnatifida* found in a New Zealand harbour. *New Zealand Journal of Botany* 25: 329–332, 1987.

5 Knox, G.A., The intertidal ecology of Taylor's Mistake, Banks Peninsula, *Transactions of the Royal Society of New Zealand* 81: 189–220, 1953.

6, 7 Batham, Elizabeth J., Ecology of a southern New Zealand sheltered rocky shore. *Transactions of the Royal Society of New Zealand* 84: 447–465, 1956.

8 Brewin, Beryl I., Ascidians in the vicinity of the Portobello Marine Biological Station. *Transactions of the Royal Society of New Zealand* 76: 87–131, 1946.

Chapter 4

1 Crisp, D.J. & Knight-Jones, E.W., The mechanism of aggregation in barnacle populations. A note on a recent contribution by Dr H. Barnes. *Journal of Animal Ecology* 22: 360–362, 1953.

2 Ballantine, W.J., A biologically defined exposure scale for the comparative description of rocky shores. *Field Studies* I (3), 1961.

3 Ballantine, W.J. in *Marine report, Mimiwhangata Wave Exposure and Community Patterns* 57–71. Auckland University, 1973.

Chapter 6

1   Cranfield, H.J. et al., Adventive marine species in New Zealand. *NIWA Technical Report 34*, 1998, 48 pp.
2   Hayward, B.W., Introduced marine organisms in New Zealand and their impact on the Waitemata Harbour, Auckland. *Tane 36:* 197–223, 1997.
3   Nelson, W.A., Marine invaders of New Zealand coasts. *Journal of the Auckland Botanical Society* 49: 4–14, 1994.
4   Hayward, B.W., Stephenson, A.B., Morley, M.S., Riley, J. & Grenfell, H.R., Faunal changes in Waitemata Harbour sediments, 1930s–1990s. *Journal of the Royal Society of New Zealand* 27: 1–20, 1997.
5   Powell, A.W.B., Animal communities of the sea-bottom in Auckland and Manukau Harbours. *Transactions of the Royal Society of New Zealand* 66: 354–401, 1937.

Chapter 7

1   Adams, N.M., The marine algae of the Wellington area. *Records of the Dominion Museum* 8: 43–98, 1972.
2   Nelson, Wendy, Adams, Nancy M. & Fox, J.M., Marine algae of the northern South Island. *National Museum of New Zealand Miscellaneous Series* 26: 1–80, 1992.
3   Cometti, R. & Morton, J.E., *Margins of the Sea*, Auckland, Hodder & Stoughton, 1985.
4   Adams, N.M., Conway, E. & Norris, R.E., The Marine Algae of Stewart Island. *Records of the Dominion Museum* 8: 185–245, 1974.

Chapter 10

1   Bradford-Grieve, J.M., Lewis, K.B. & Stanton, J.B., Advances in New Zealand oceanography, 1967–1991. *New Zealand Journal of Marine and Freshwater Research* 25: 429–441, 1991.
2   Powell, A.W.B., New Zealand marine biotic provinces. *Tuatara* 9: 1–8, 1961.
3   Ayling, Tony & Cox, Geoff, *Guide to the Sea Fishes of New Zealand*, Collins, Auckland, 1982, 343 pp.
4   Adams, Nancy & Nelson, Wendy, The marine algae of the Three Kings Islands. *National Museum of New Zealand Miscellaneous Series* 13: 1–29, 1985.
5   Brook, F.J., The coastal molluscan fauna of the northern Kermadec Islands, Southwest Pacific Ocean. *Journal of the Royal Society of New Zealand* 28: 185–233, 1998.
6   Brook, F.J., The coastal scleractinian coral fauna of the Kermadec Islands, south western Pacific Ocean. *Journal of the Royal Society of New Zealand* 29: 435–460, 1999.
7   Francis, M.P., Grace, R.V. & Paulin, C.D., Coastal fishes of the Kermadec Islands. *New Zealand Journal of Marine and Freshwater Research* 21: 1–13, 1987.
8   Nelson, Wendy & Adams, Nancy, Marine algae of the Kermadec Islands. *National Museum of New Zealand Miscellaneous Series* 10: 1–29, 1984.
9   Knox, G.A., General account of the Chatham Islands, 1954 Expedition. *DSIR Research Bulletin* 122, 1957.
10   Kenny and Haysom, Ecology of rocky shore organisms at Macquarie Island. *Pacific Science* 16: 245–263, 1962.
11   Wild South TV documentary, *Under the Ice*, 1980s.

Chapter 13

1, 2 Vine, Peter J., The Marine Fauna of New Zealand: Spirorbinae (Polychaeta: Serpulidae). *New Zealand Oceanographic Institute Memoir 68*, 1977.

Chapter 14

1   Schuchert, Peter, The Marine Fauna of New Zealand: Athecate Hydroids and their Medusae (Cnidaria: Hydrozoa). *New Zealand Oceanographic Institute Memoir 106*, 1996.

Chapter 15

1   Whittaker, Robert H., *Communities and Ecosystems*, MacMillan, New York, 1975, 385 pp.

Chapter 16

1   Ayling, A.M., *The Role of Biological Disturbance in determining the Organisation of Subtidal Encrusting Communities in Temperate Waters.* PhD thesis, University of Auckland, 1976.
2   Ayling, A.M., Kelp Coastal Forests. *New Zealand Nature Heritage*, 579, 1974.
3   Cairns, Stephen D., The Marine Fauna of New Zealand: Scleractinia (Cnidaria: Anthozoa). *New Zealand Oceanographic Institute Memoir 103*, 1995.
4, 5 Doak, W. *The Cliff Dwellers: An Undersea Community.* Hodder and Stoughton, Auckland, 1979, 80 pp.
6   Ayling, A.M., Offshore Caves and Overhangs. *New Zealand Nature Heritage* 2: 776, 841, 1974.
7   Ayling, Tony & Cox, Geoff, *Guide to the Sea Fishes of New Zealand*, Collins, Auckland, 1982, 343 pp.
8   Cairns, Stephen D., Marine fauna of New Zealand: Stylasteridae (Cnidaria: Hydrozoa). *New Zealand Oceanographic Institute Memoir 98*, 1991.

Chapter 17

1   Womersley, H.B.S., *Marine Benthic Flora of Southern Australia.* Four parts, South Australia Government Printing, Adelaide, 1984–1996.

Chapter 18

1   Soh, C.L., Sesarmine crabs from Hong Kong. *Memoir of the Hong Kong Natural History Society* 13: 9–22, 1978.
2   Briggs, John C., *Marine Zoogeography*, McGraw-Hill, New York, 1974, 475 pp.
3, 4 Morton, B.S., The rocky shore ecology of Quingdao, Shandong Province, People's Republic of China. *Asian Marine Biology* 7: 167–187, 1990.
5   Briggs, John C., *Marine Zoogeography*, McGraw-Hill, New York, 1974, 475 pp.
6   Chihara, Mitsuo, Geographic distribution of marine algae in Japan, in *Advances of Phycology in Japan*, ed. Tokida, Jun & Hirose, Hiroyuki, 1971.
7   Golikov, A.N. & Scarlato, O.A., Soviet Union coastal ecology. In Schwartz, M.L., *Encyclopedia of Beaches and Coastal Environments*, 15: 770–780, 1982.
8   Morton, J.E., *Shore Life Between Fundy Tides.* Toronto, Canadian Scholars' Press, 1991.

Chapter 19

1 Morton, J.E., *Shore Life Between Fundy Tides*. Toronto, Canadian Scholars' Press, 1991.
2 Ricketts, E.F. & Calvin, J., *Between Pacific Tides*. Stanford University Press, 1985.
3 Abbott, I.A. & Hollenberg, G.J., *Marine Algae of California*. Stanford University Press, 1976.
4 Guiler, E.R. Intertidal belt-forming species on the rocky coasts of northern Chile. *Proceedings of the Royal Society of Tasmania* 93: 33–58, 1959.
Guiler, E.R., The intertidal ecology of the Montemar area. *Proceedings of the Royal Society of Tasmania* 93: 164–183, 1959.
5 Castilla, Juan Carlos, Guia para la observacion del litoral expedicion a Chile. *Nacional Gabriela Mistral*, 1970.
6 Luning, K., *Seaweeds, their Environment, Biogeography and Ecophysiology*. John Wiley and Sons, New York, 1990.

Chapter 20

1 Wells, J.W., Coral Reefs, in *Treatise on Marine Ecology and Paleoecology* vol. 1 Ecology. *Geological Society of America Memoir* 67, 1957.
2 Veron, J.E.N., Pichon, M. & Wijsman-Best, M., *The Scleractinia of Eastern Australia*. Parts 1 to 5.
3 Darwin, Charles. *The Structure and Distribution of Coral Reefs*. AMS Press, 1972.
4 Stoddart, D.R., Geomorphology of the Solomon Islands coral reefs *Philosophical Transactions of the Royal Society B* 235: 355–382, 1969.
5 Darwin, Charles. *The Structure and Distribution of Coral Reefs*. AMS Press, 1972.
6 Hopley, D., Coral reef ecology: structures and communities. In: Workshop on coastal processes in the South Pacific Island Nations. *SOPAC Technical Bulletin* 7: 15–26, 1991.

Chapter 21

1 Gibbs, P.E., Stoddart, D.R. & Vevers, H.G., Coral reefs and associated communities in the Cook Islands. *Bulletin of the Royal Society of New Zealand* 8: 91–105, 1971.
2 Sargent, M.T. & Austin, T.S., Biologic economy of coral reefs. US Geological Survey Professional Paper 260, 1954.
3 Odum, E.P. & Odum, H.T., Trophic structure of a windward reef community in Eniwetok Atoll. *Ecological Monographs* 25: 291–320, 1955.

Chapter 22

1 Bergquist, P.R., Morton, J.E. & Tizard, C.A., Some Demospongiae from the Solomon Islands with descriptive notes on the major sponge habitats. *Micronesica* 7: 99–121, 1971.
2 Morton, B.S., *Partnerships in the Sea: Hong Kong's marine symbioses*, illustr. Juliana Depledge, Hong Kong University Press, 1988.
3 Miller, M.C., Habits and habitat of opisthobranch molluscs of the British Solomon Islands. *Philosophical Transactions of the Royal Society B* 255: 541–548, 1969.
4 Hiatt, R.W. & Strasburg, D.W., Ecological relationships of the fish fauna on coral reefs of the Marshall Islands. *Ecological Monographs* 30: 67–127, 1960.
5 Marshall, N.B., *The Life of Fishes*, Wiedenfeld & Nicholson, London, 1965.

Chapter 23

1 McNae, W., Fauna and flora of mangrove swamps and forests in the Indo-west Pacific region. *Advances in Marine Biology* 6: 74–270, 1968.
2 Paijmans, K. (ed.), *New Guinea Vegetation*, CSIRO/Australian National University Press, Canberra, 1976.
Percival, Margaret & Womersley, J.S., Floristics and ecology of the mangrove vegetation of New Guinea. *Botany Bulletin 8*, Department of Forests, Lae, University of Papua New Guinea, 1975.

Chapter 24

1 Harriott, V.J., Harrison, P.L. & Banks, S.A., The coral communities of Lord Howe Island. *Australian Journal of Marine and Freshwater Research* 46: 457–465, 1995.
2 Francis, M.P., Check-list of the coastal fishes of Lord Howe, Norfolk and Kermadec Islands, south-west Pacific Ocean. *Pacific Science* 47: 136–170, 1993.
3 Wellington, G.M., chapter 17 in Perry, R. (ed.), *Galapagos Key Environments*, Pergamon Press, Oxford, 1984, pp. 247–268.
4 Hedgpeth, J.W., An intertidal reconnaissance of rocky shores of the Galapagos. *Wasmann Journal of Biology* 27, 1969.
5 Foster, B.A. & Newman, W.A., Chthamalid barnacles of Easter Island. *Bulletin of Marine Science* 41: 322–336, 1987.
6 Kohn, A.J. Gastropods as predators and prey at Easter Island. *Pacific Science* 32: 35–57, 1978.
7 Di Salvo, L.H., Randall, J.E. & Cea, A., Ecological reconnaissance of the Easter Island sublittoral environment. *National Geographic Research*: 451–473, 1988.

Further Reading

NEW ZEALAND
Geology and landforms
Field Guide to New Zealand Geology by Jocelyn Thornton, Reed Methuen, 1985.

Landforms of New Zealand by Jane Soons & Michael Selby (eds), Longman Paul, 2nd ed., 1992.

New Zealand Adrift: the theory of continental drift in a New Zealand setting by Graeme Stevens, Reed, 1980.

Precious Land — Protecting New Zealand's landforms and geological features by Bruce Hayward, Geological Society of New Zealand, 1996.

Reading the Rocks: a guide to the geological features of the Wairarapa Coast by Lloyd Homer & Phil Moore, Landscape Publications, 1989.

The New Zealand Coast Te Tai o Aotearoa by James Goff, Scott Nichol & Helen Rouse (eds), Dunmore Press, 2003.

The Restless Country. Volcanoes and earthquakes of New Zealand by Geoffrey Cox & Bruce Hayward, Harper Collins, 1999.

Trilobites, Dinosaurs and Moa Bones: The story of New Zealand fossils by Bruce Hayward, Bush Press, 1990.

Vanishing Volcanoes: a guide to the landforms and rock formations of Coromandel Peninsula by Lloyd Homer & Phil Moore, Landscape Publications, 1992.

General marine ecology
The New Zealand Sea Shore by John Morton & Michael Miller, Collins, 1978.

A Natural History of Auckland, by John Morton (ed.), David Bateman Ltd, 1993.

Between the Tides. New Zealand seashore and estuary life by Michael Bradstock, Reed Methuen, 1985.

Margins of the Sea by Ronald Cometti & John Morton, Hodder & Stoughton, 1985.

Nature Watching at the Beach by John Walsby, Wilson & Horton, 1990.

Reef and Beach Life of New Zealand by Michael Miller & Gary Batt, Collins, 1973.

Studying Temperate Marine Environments: A handbook for ecologists by Michael Kingsford & Chris Battershill, Canterbury University Press, 1998.

The Living Reef. The Ecology of New Zealand's Rocky Reefs by Neil Andrew & Francis Malcolm (eds), Craig Potton Publishing, Nelson, 2003.

Fiordland Underwater by Paddy Ryan & Chris Paulin, Exisle, 1998.

A Guide to the New Zealand Seashore by Dave Gunson, Viking Pacific, 1983.

Marine conservation
Seacoast in the Seventies by John Morton, David Thom & R. Locker, Hodder & Stoughton, 1972.

Marine Reserves for New Zealand by Bill Ballantine, University of Auckland Leigh Laboratory Bulletin No.25, 1991.

Marine organisms
An Encyclopaedia of New Zealand Coastal Marine Species by Steve Cook (ed.), Canterbury University Press, in press.

Powell's Native Animals of New Zealand by Baden Powell & Brian Gill (editor), 4th edition, David Bateman Ltd, 1998.

The New Zealand Inventory of Biodiversity: A Species 2000 Symposium Review by Dennis Gordon (ed.), Canterbury University Press, in press.

What's on the Beach by Glenys Stace, Viking, 1997.

What's Around the Rocks by Glenys Stace, Penguin, 1998.

Just Under the Water by Glenys Stace, Penguin, 1999.

Protista
Recent New Zealand shallow-water benthic foraminifera: taxonomy, ecological distribution, biogeography and use in paleoenvironmental assessment, by Bruce Hayward, Hugh Grenfell, Catherine Reid & Kathryn Hayward. *Institute of Geological and Nuclear Sciences Monograph* 21, 1999.

Polychaeta
The Marine Fauna of New Zealand: Spirorbinae (Polychaeta: Serpulidae), by Peter Vine, *New Zealand Oceanographic Institute Memoir* 68, 1977.

The polychaetous annelids of New Zealand. Part 1. Glyceridae, by George Knox, *Records of Canterbury Museum* 7: 219–232, 1960.

Nemertine worms
The Invertebrate Fauna of New Zealand: Nemertea (ribbon worms) by R. Gibson, *NIWA Biodiversity Memoir* 118, 2002.

Arthropods

Crabs of New Zealand by Colin McLay, *University of Auckland Leigh Laboratory Bulletin* no. 22, 1988.

Native Crabs, Nature in New Zealand by Dick Dell, A.H. & A.W. Reed, 1963.

The Marine Fauna of New Zealand: Family Sphaeromatidae (Crustacea: Isopoda: Flabellifera) by Des Hurley & K.P. Jansen, *New Zealand Oceanographic Institute Memoir* 63, 1977.

The marine fauna of New Zealand: Crustaceans of the order Cumacea, by N.S. Jones, *New Zealand Oceanographic Institute Memoir* 23, 1963.

The Marine Fauna of New Zealand: Family Hymenosomatidae (Crustacea, Decapoda, Brachyura), by M.J. Melrose, *New Zealand Oceanographic Institute Memoir* 34, 1975.

The Marine Fauna of New Zealand: Spider crabs, family Majidae (Crustacea, Brachyura), by D.J.G. Griffin, *New Zealand Oceanographic Institute Memoir* 35, 1966.

The Marine Fauna of New Zealand: Barnacles (Cirripedia: Thoracica), by Brian Foster, *New Zealand Oceanographic Institute Memoir* 69, 1978.

The Marine Fauna of New Zealand: Benthic Ostracoda (Suborder Myodocopina), by L.S. Kornicker, *New Zealand Oceanographic Institute Memoir* 82, 1979.

The Marine Fauna of New Zealand: Algal-living Littoral Gammaridea (Crustacea Amphipoda), by J. Laurens Barnard, *New Zealand Oceanographic Institute Memoir* 62, 1972.

The Marine Fauna of New Zealand: Paguridea (Decapoda: Anomura) exclusive of the Lithodidae, by J. Forest, M. de S. Laurent, P.A. McLaughlin & R. Lemaitre, *New Zealand Oceanographic Institute Memoir* 114, 2000.

The Marine Fauna of New Zealand: Pycnogonida (Sea Spiders), by C.A. Child, *New Zealand Oceanographic Institute Memoir* 109, 1998.

Sponges

The Marine Fauna of New Zealand: Porifera, Demospongiae, Part 1 (Tetractinomorpha and Lithistida), by Patricia Bergquist, *New Zealand Oceanographic Institute Memoir* 37, 1968.

The Marine Fauna of New Zealand: Porifera, Demospongiae, Part 2 (Axinellida and Halichondrida), by Patricia Bergquist, *New Zealand Oceanographic Institute Memoir* 51, 1970.

The Marine Fauna of New Zealand: Porifera, Demospongiae, Part 3. (Haplosclerida and Nepheliospongida), by Patricia Bergquist & K.P. Warne, *New Zealand Oceanographic Institute Memoir* 87, 1980.

The Marine Fauna of New Zealand: Porifera, Demospongiae, Part 4 (Poecilosclerida), by Patricia Bergquist & P.J. Fromont, *New Zealand Oceanographic Institute Memoir* 96, 1988.

The Marine Fauna of New Zealand: Porifera, Demospongiae, Part 5 (Dendroceratida and Halisarcida) by Patricia Bergquist, *New Zealand Oceanographic Institute Memoir* 107, 1996.

Cnidaria

The Marine Fauna of New Zealand: Scleractinia (Cnidaria: Anthozoa), by Stephen Cairns, *New Zealand Oceanographic Institute Memoir* 103, 1995.

The Marine Fauna of New Zealand: Stylasteridae (Cnidaria: Hydroida), by Stephen Cairns, *New Zealand Oceanographic Institute Memoir* 98, 1991.

The Marine Fauna of New Zealand: Isididae (Octocorallia: Gorgonacea) from New Zealand and the Antarctic, by Ralph Grant, *New Zealand Oceanographic Institute Memoir* 66, 1976.

The Marine Fauna of New Zealand: Athecate Hydroids and their Medusae (Cnidaria: Hydrozoa), by Peter Schuchert, *New Zealand Oceanographic Institute Memoir* 106, 1996.

The Marine Fauna of New Zealand: Hydromedusae (Cnidaria: Hydrozoa), by J. Bouillon & T.J. Barnett, *New Zealand Oceanographic Institute Memoir* 113, 1999.

The Marine Fauna of New Zealand: Leptothecata (Cnidaria: Hydrozoa) (Thecate Hydroids) by W. Vervoot & J.E. Watson, *NIWA Biodiversity Memoir* 119, 2003.

Bryozoa

Atlas of marine fouling Bryozoa of New Zealand ports and harbours, by Dennis Gordon & S.F. Mawatari. *New Zealand Oceanographic Institute Miscellaneous Publication* 107, 1992.

The Marine Fauna of New Zealand: Bryozoa: Gymnolaemata (Ctenostomata and Cheilostomata Anasca), from the western South Island continental shelf and slope, by Dennis Gordon, *New Zealand Oceanographic Institute Memoir* 95, 1986.

The Marine Fauna of New Zealand: Bryozoa: Gymnolaemata (Cheilostomata Ascophorina), from the western South Island continental shelf and slope, by Dennis Gordon, *New Zealand Oceanographic Institute Memoir* 97, 1989.

Mollusca

Marine Molluscs Part 1, by John Walsby & John Morton, Leigh Marine Laboratory, University of Auckland, 1982.

Marine Molluscs Part 2, Opisthobranchia, by Richard Willan & John Morton, Leigh Marine Laboratory, University of Auckland 1984.

New Zealand Mollusca, by A.W.B. Powell, Collins, 1979.

The Marine Fauna of New Zealand: Index to the Fauna 3. Mollusca, by Hamish Spencer & Richard Willan, *New Zealand Oceanographic Institute Memoir* 105, 1995.

The Marine Fauna of New Zealand; Octopoda (Mollusca: Cephalopoda), by Steve O'Shea, *New Zealand Oceanographic Institute Memoir* 112, 1999.

Which Seashell? by Andrew Crowe, Penguin, 1999.

483

Brachiopods

The systematics and biogeography of the living Brachiopoda of New Zealand, by Elliot Dawson, pp. 431–437 in *Brachiopods through Time* (D. Lee & D.J. Campbell eds), A.A. Balkema, Rotterdam, 1990.

Echinoderms

The Marine Fauna of New Zealand: Sea cucumbers (Echinodermata: Holothuroidea), by Dave Pawson, *New Zealand Oceanographic Institute Memoir 52*, 1970.

The Marine Fauna of New Zealand: Basket-stars and snake-stars (Echinodermata: Ophiuroidea: Euryalinida), by Don McKnight, *New Zealand Oceanographic Institute Memoir 115*, 2000.

Birds

The Reed Field Guide to Common New Zealand Shorebirds, by David Medway, Reed, 2000.

The Field Guide to the Birds of New Zealand, by Barrie Heather & Hugh Robertson, Viking, 2000.

Ascidians

The Marine Fauna of New Zealand: Ascideacea, by R.H. Millar, *New Zealand Oceanographic Institute Memoir 85*, 1982.

Fish

Coastal Fishes of New Zealand, a diver's identification guide, by Malcolm Francis, Heinemann Reed, 1988.

Guide to the Sea Fishes of New Zealand, by Tony Ayling & Geoff Cox, Collins, 1982.

The Rock Pool Fishes of New Zealand, by Chris Paulin & Clive Roberts, Museum of New Zealand, 1992.

New Zealand Fishes, by Larry Paul, Reed, 2000.

Plants

New Zealand lichens. Checklist, key, and glossary, by William M. Malcolm & David J. Galloway, Museum of New Zealand Te Papa Tongarewa, 1994.

Seaweeds of New Zealand, an illustrated guide, by Nancy M. Adams, Canterbury University Press, 1994.

Which Coastal Plant?, by Andrew Crowe, Viking, 1995.

The Native Trees of New Zealand, by John Salmon, Reed, 1986.

AUSTRALIA

Australian Seashores, by W.J. Dakin, revised by I. Bennett, Angus & Robertson, Sydney, 1987.

Australian Marine Life: The plants and animals of temperate waters, by Graham Edgar, Reed, Kew, Victoria, 1997.

Australian Marine Shells. Parts 1 & 2. Prosobranch Gastropods, by Barry Wilson, Odyssey Publishing, Western Australia, 1993, 1994.

Bivalves of Australia. Vols 1 & 2, by Kevin Lamprell et al., Crawford House Press, Bathurst, NSW, 1992.

Coral Reef Handbook, a guide to the geology, flora and fauna of the Great Barrier Reef, by Pat Mather & I. Bennett, Surrey Beatty & Sons, Norton, NSW, 1993.

EAST ASIA

The Sea-shore Ecology of Hong Kong, by Brian & John Morton, Hong Kong University Press, 1983.

WEST AMERICA

A Handbook to the Common Intertidal Invertebrates of the Gulf of California, by Richard Brusca, University of Arizona Press, 1973.

Between Pacific Tides, by Ricketts, Calvin, Hedgpeth & Phillips, Stanford University Press, 5th ed, 1985.

Intertidal Invertebrates of California, by Robert Morris, Donald Abbott & Eugene Haderlie, Stanford University Press, 1980.

Marine Algae of California, by Isabella Abbott & G.J. Hollenberg, Stanford University Press, 1976.

TROPICAL PACIFIC

Corals of Australia and the Indo-Pacific, by John Veron, Angus & Robertson, North Ryde, Australia, 1986.

Corals of the World, 3 volumes, by John Veron, Australian Institute of Marine Science, Townsville, 2000.

Fiji's Natural Heritage, by Paddy Ryan, Exisle Publishing, Auckland, 2000.

Hawaiian Reef Animals, by Edmund Hobson & E.H. Chave, University of Hawaii Press, 1990.

Indo-Pacific Coral Reef Field Guide, by Gerald Allen & Roger Steene, Tropical Reef Research, Singapore, 1994.

Marine Shells of the Pacific, Vol. 1 & 2, by Walter Cernohorsky, Pacific Publications, Sydney, 1971, 1972.

Micronesian Reef Fishes, by R.F. Myers, Coral Graphics, Guam, 1989.

Snorkeller's Guide to the Coral Reef, by Paddy Ryan, University of Hawaii Press, 1994.

The Shore Ecology of Suva and South Viti Levu, by John Morton & Uday Raj, University of the South Pacific, 2 volumes, 1985.

The Shore Ecology of the Tropical Pacific, by John Morton, UNESCO South-East Asia, Jakarta, 1990.

Tropical Pacific Invertebrates, by Patrick Colin & Charles Arneson, Coral Reef Press, Beverley Hills, USA, 1995.

Underwater Guide to New Caledonia, by P. Laboute & Y. Magnier, Les editions du Pacifique, Papeete-Tahiti, 1979.

GLOBAL BIOGEOGRAPHY

Marine Zoogeography, by John C. Briggs, McGraw-Hill, New York, 1974.

Indices

New Zealand Place, Subject and Common Name Index

New Zealand Taxonomic Index

Bold entries refer to figures

Pacific Place, Subject and Common Name Index

Bold entries refer to figures

Pacific Taxonomic Index

503